NASA SP-4014

I0495744

ASTRONAUTICS AND AERONAUTICS, 1969

Chronology on Science, Technology, and Policy

Text by
Science and Technology Division
Library of Congress

Sponsored by
NASA Historical Division
Office of Policy

Scientific and Technical Information Division
OFFICE OF TECHNOLOGY UTILIZATION   1970
NATIONAL AERONAUTICS AND SPACE ADMINISTRATION
Washington, D.C.

Foreword

History is a word with varied meanings. They range from one conveying idealistic images, to fat books of ultimate truth, to the professional's prejudices concerning history as an intellectual discipline. To those of us who have been privileged to be wholly immersed in science and technology of aeronautics and space over a number of years, history perhaps is the sense of accomplishment.

While this chronology volume is not a history, it does attempt to provide a first-cut reference to events and commentary during a most crowded year and the year that man first set foot upon an extraterrestrial body. When the *Apollo 11* astronauts Neil Armstrong and Buzz Aldrin made their lunar walk on July 20, 1969, it became one of the most vicarious events to date in world history. Over a half billion people around the world witnessed this momentous occasion live by television relayed via communications satellites. Many who did not witness it appear reluctant to admit it today. The full consequences of the seven-year Apollo endeavor are as yet in the domain of prophets and posterity despite the worldwide enthusiasm. But we have already seen evidence of the second thoughts provided by man's perspective from the moon of his own planet—a heightened awareness that spaceship Earth is perhaps unique and certainly is precious, even with the manifold problems of mankind. And we have learned about ourselves as a people. We have learned that the United States can set itself a large, difficult, long-term objective and mobilize itself and sustain its effort to the successful conclusion.

Aside from being the year that man landed on the moon, 1969 had many other significances to students and participants in aeronautics and astronautics. It was NASA's first year under the Nixon Administration and a new Administrator, Dr. Thomas O. Paine. It was the year in which the Space Task Group's report to the President reaffirmed the Nation's continued commitment to space exploration and painted in the broad outline of post-Apollo goals in space. It was the year the concept of the space shuttle emerged in detail, exciting in its potential as a practical, reusable, economical space transportation system. In space science it was the year when *Mariner VI* and *VII* flew within 2,000 miles of Mars and sent back photographs of the Martian surface and 200 times more data on Mars than had *Mariner IV* in 1964. In addition to these more spectacular events, there was solid progress in space science, exploration, and applications. All of these events and many more find their milestones recorded in this chronology.

There are both value in and special reservations about this chronology of science, technology, and public policy as related to aeronautics and space. It provides the historian or any analyst with time-oriented steppingstones toward the human and institutional stories. General items are included to help create the social environment in which the selected items took place. There seems some merit, despite inevitable bias in viewpoints, in validating

FOREWORD

entries to sources generally available. This facilitates additional research. With its detailed index, the chronology is cross-referenced to dimensions other than time and becomes a useful reference available to lay and professional inquiry.

But beyond this, history-maker, historian, observer, and student alike may become more aware of the documentation and reflection yet to be performed in comprehending more fully what has transpired.

George M. Low
Acting Administrator
National Aeronautics and Space Administration

September 15, 1970

Contents

	PAGE
Foreword	III
NASA Acting Administrator George M. Low	
Preface	VII
January	1
February	37
March	61
April	99
May	127
June	167
July	195
August	259
September	297
October	323
November	359
December	399
Appendix A: Satellites, Space Probes, and Manned Space Flights, a Chronicle for 1969	427
Appendix B: Chronology of Major NASA Launches, 1969	459
Appendix C: Chronology of Manned Space Flight, 1969	465
Appendix D: Abbreviations of References	471
Index and List of Abbreviations and Acronyms	475

Preface

The brief, chronological record of 1969 events in aerospace science, technology, and policy has been prepared as events occurred and were reported in the immediately available, open sources—the news media, press releases, speech texts, transcripts, testimony before Congress, and test and study reports. A first collection of clues to significant occurrences and background climate for future historians, the volume is also intended to serve for immediate reference uses. It does not attempt to analyze but to cite the who, what, when, and where in sequence and as near real time as possible.

Within these limitations, we make a considerable effort to ensure accuracy and comprehensiveness. Our NASA Archives, under Lee D. Saegesser, collects the current documentation. Under an exchange of funds agreement, the Science and Technology Division of the Library of Congress drafts the monthly segments in comment edition form. These are edited and augmented by the NASA Historical Division, published, and circulated for comment and use. At the end of the year the entire manuscript is reworked and augmented by the comments that have come in and by documentation that has become available since the comment edition was prepared. The Library also prepares the extensive index.

The 1969 annual volume is the work of a number of hands. The entire NASA Historical Division participated in source collection, review, and publication. The general editor was Dr. Frank W. Anderson, Jr., Deputy NASA Historian. Technical editor was Mrs. Carrie Karegeannes. At the Library Mrs. Patricia Davis, Mrs. Carmen Brock-Smith, and Mrs. Shirley Singleton prepared the monthly texts, which were circulated throughout NASA for comments as to completeness and accuracy of NASA items and then revised for annual publication. Arthur G. Renstrom prepared the index.

Appendix A, "Satellites, Space Probes, and Manned Space Flights, a Chronicle for 1969," and Appendix C, "Chronology of Manned Space Flight, 1969," were prepared by Leonard C. Bruno of the Library. Appendix B, "Chronology of Major NASA Launches, 1969," was prepared by William A. Lockyer, Jr., of the Historical and Library Services Branch, Kennedy Space Center. Appendix D, "Abbreviations of References," was prepared by Mrs. Brock-Smith. Creston Whiting of NASA's Information Services Branch, Scientific and Technical Information Division, kept the process abreast of Russian releases. At the NASA Centers the historians and historical monitors submitted local material for the chronology. Validation was the work of many busy persons throughout NASA and in other relevant branches of the Federal structure.

A chronology is but the first step toward history and even it is never completed. Comments, additions, and criticisms are always welcomed.

Eugene M. Emme
NASA Historian

January 1969

January 1: Washington *Evening Star* editorial said of success of Dec. 21–27, 1968, *Apollo 8* mission: "Modern science undercut man's bland belief that he was the center of the universe, and modern philosophy reduced him to a trivial atom of matter in the larger cosmos. To be able to sail around at will in that vast cosmos may give man back some of the confidence he once had, not the arrogance of thinking that he understands the whole pattern, but the quiet sense that he will not flinch from what he may yet learn." (W *Star*, 1/1/69, A15)

- U.S.S.R. disclosed that converted MiG fighter was prototype used for testing design features and performance of Tu-144, Soviet supersonic aircraft. Soviet aviation writer for *Pravda* K. Raspevin said four-man crew aboard Tu-144 maiden flight Dec. 31, 1968, was one of most experienced in U.S.S.R. Pilot was Eduard V. Yelyan. Copilot Mikhail V. Kozlov had won title Hero of the Soviet Union for testing Tu-22 supersonic strategic bomber. Tu-144 was constructed of light alloys with titanium on leading edges and other areas subjected to high temperatures. At cruising speed, outside skin temperature was 150° C. Air conditioning system cooled cabin. Tail unit was minus horizontal guiding surface. Crew members had catapult seats as safety precaution during test flights. (*NYT*, 1/2/69, 7)
- World Data Center A for Rockets and Satellites, established at National Academy of Sciences in June 1958, moved to location adjacent to National Space Science Data Center at GSFC. (NAS–NRC–NAE *News Rpt*, 2/69, 11)

January 2: In Washington *Evening Star*, Judith Randal said world's first successful heart transplant and *Apollo 8* mission made 1968 year "of spectacular scientific achievement" but that critics of both events had charged that technology "was being exploited at the expense of basic research and social worth." It did no harm, she said, to celebrate heart transplants and voyages to moon, "but, with the advent of a new administration, it also is worth reflecting what the price may be—when so much else needs doing—of deciding to climb Mount Everest just because Mount Everest is there and we have learned how to climb it." (W *Star*, 1/2/69, A14)

- NASA awarded Boeing Co. $32,815,000 cost-plus-fixed-fee supplemental agreement extending for additional 12 mos Apollo program technical integration and evaluation support initiated by Boeing June 15, 1967. (NASA Release 69-1)
- President Johnson announced 12 recipients of 1968 National Medal of Science, Government's highest award for distinguished achievement in science, mathematics, and engineering. Detlev W. Bronk, President Emeritus of Rockefeller Univ., past president of NAS (1950–1962) and Johns Hopkins Univ. (1948–1953), received award for "highly original research in the field of physiology and for his manifold con-

tributions to the advance of science and its institution in the service of society." Herbert Friedman, Superintendent, Atmosphere and Astrophysics Div., Naval Research Laboratory, won award "for pioneering work in rocket and satellite astronomy and in particular for his contributions to the field of gamma ray astronomy." (*PD*, 1/6/69, 11; NASA biog, 9/8/68)

January 3: H.R. 16, 17, and 204, bills to authorize award of Congressional Medal of Honor to *Apollo 8* Astronauts Frank Borman, James A. Lovell, Jr., and William A. Anders, were introduced during first session of 91st Congress. (*CR*, 1/3/69, H33–42)

- *Time* named *Apollo 8* astronauts its Men of the Year for 1968. "For all its upheavals and frustrations, the year would be remembered to the end of time for the dazzling skills and Promethean daring that sent mortals around the moon. It would be celebrated as the year in which men saw at first hand their little earth entire, a remote, blue-brown sphere hovering like a migrant bird in the hostile night of space." (*Time*, 1/3/69, 9)

- New York State Supreme Court Justice Frederick M. Marshall issued temporary injunction to block sale of Cornell Aeronautical Laboratory by Cornell Univ. to EDP Technology, Inc., of Washington, D.C., for $25 million. He directed case be given preference on trial calendar. (*NYT*, 1/4/69, 23)

January 4: At Explorers Club symposium in New York scientists, educators, community leaders, and students discussed significance of *Apollo 8* mission. William Booth, Chairman of Commission on Human Rights in New York, said, "I still am quite disturbed by the fact that we're dying at home, people are about to starve. There's overpopulation and underproduction of food in the world. We haven't been able to solve these problems and here we are going off to the moon." Dr. Robert Jastrow, Director of NASA's Goddard Institute for Space Studies, said flights were "a means of concentrating our energies toward building a technological capability." Moon flight had played "same role as Lindbergh's flight to Paris" in that it demonstrated new capability. Space program was paying economic dividends in communications, mineral exploration, and new materials. Dr. Maynard M. Miller, Chairman of Explorer Club's World Center for Exploration Foundation, said, "The word 'impossible' has a different meaning after Apollo 8. Perhaps its real contribution will be as a symbol of man's willingness to dare to do something great." (Wilford, *NYT*, 1/5/69, 26)

January 5: Venus V unmanned probe was successfully launched by U.S.S.R. on four-month journey to Venus. Tass announced that 2,491-lb spacecraft had been launched into parking orbit and then injected on trajectory toward Venus to softland, conduct extensive scientific research, and continue studies begun by *Venus IV*, which landed on Venusian surface Oct. 18, 1967. All equipment was functioning normally. Spacecraft carried pennants with bas-relief of Lenin and Soviet coat of arms and "greater range of scientific and measuring equipment, making it possible to improve the accuracy of measurements and to obtain additional data on planet's atmosphere," *Moscow News* said. (Winters, *B Sun*, 1/6/69, 1; AP, *W Star*, 1/6/69, A3; Reuters, *W Post*, 1/6/69, A3; *Moscow News*, 1/18–25/69, 3)

- Washington *Sunday Star* editorial commented on proposals made at

AAAS meeting in Dallas, Tex., Dec. 26–31, 1968, for agency to establish priorities for Federal spending in research, education, technology, weapon development, and science. Dr. James V. Shannon, former NIH director, had proposed creation of top-level council equal in stature to National Security Council and Council of Economic Advisers. Dr. Donald F. Hornig, President Johnson's Science Adviser, had recommended adding Secretary of Science to Presidential Cabinet. *Star* said, "There will be strong opposition to both of these proposals. The cry of centralization will be raised." However: "What is proposed . . . is a body of informed advisers, whose duty it is to suggest . . . to instruct the President, the Congress and public on the potentialities for good and evil that could result if a given path were followed. The President would still have to make the executive decisions. Congress would still control the purse. The public would still have the final verdict of the ballot. But all of them could use some responsible, expert guidance through the awesome and fantastic new world that lies just ahead." (W *Star*, 1/5/69, G1)

January 6: *Oao II* (launched Dec. 7, 1968) completed 30 days of flight operation and was adjudged successful by NASA. Experiments had obtained 65 hrs of scientific data over range of eight magnitudes and 4,200 Å–1,100 Å wavelengths. Smithsonian Astrophysical Observatory celescope experiment made 18 mappings and Univ. of Wisconsin experiment made detailed observations of 100 stars. Some 40,000 separate commands had been sent to *Oao II* and more than 4 million pieces of information had been collected, with 20 times more UV information from stars in 30 days than in 15 yrs of sounding rocket launchings. Satellite's 11 telescopes were studying extremely young, hot stars which emitted most of their energy in UV portion of spectrum, not visible to ground observatories because of earth's atmosphere. (NASA Proj Off; NASA Release 69–7)

- Budget squeeze had forced NASA to drop 16 institutions from its sustaining university program, *Scientific Research* said. Four others were dropped in 1968; 30 remained. Contracts of the 16 would not be renewed as they expired unless institutions devised exceptional research proposals. Then money would probably have to be taken from other universities. Grants had ranged from $75,000 to $300,000, with total saving of $4 million over FYs 1969, 1970, and 1971 expected from cancellation. Program funding had declined from $45 million in FYs 1965 and 1966 to $30 million in FY 1967, $10 million in FY 1968, and $9 million in 1969. FY 1970 budget level was expected to be $9 million. (*Scientific Research*, 1/6/69, 15–17; NASA FY 1970 Budget Briefing)

- FAA announced it had moved to reverse "escalation of aircraft noise" around airports by proposing maximum noise standards and noise objectives for new subsonic transport aircraft, including those under development. Proposal was first regulatory action taken under P.L. 90–411, which granted FAA broad authority in noise control. Noise limits on approach would be 102 to 108 effective perceived noise decibels (epndb), depending on aircraft weight; sideline noise limits would be same; and takeoff limits would be in 93- to 108-epndb range. (FAA Release 69–1; Bisen, *W Post*, 1/7/69, A3; *WSJ*, 1/9/69, 4)

- Defense Secretary-designate Melvin R. Laird announced at press conference that Stanley R. Resor would remain as Secretary of the Army, Rhode Island Gov. John H. Chafee would be nominated Secretary of the Navy, and former NASA Deputy Administrator, Dr. Robert C. Seamans, Jr., would be nominated Secretary of the Air Force in Nixon Administration. (Wilson, *W Post*, 1/7/69, 1)

January 6–8: Three-article series on "The Cost of Preparedness" by Orr Kelly in Washington *Evening Star* quoted interviews with outgoing Johnson Administration defense experts who agreed U.S. might be entering peaceful era in international relations though defense cost would remain high—at least $50 billion annually—in foreseeable future. Secretary of Defense Clark M. Clifford thought, "cautious and forceful steps" could be taken "to improve the lot of the people of the world."

USN saw post-Vietnam war need for modernization, with emphasis on nuclear-powered escort ships; USAF wanted new fighter, manned bomber, deployment of interim bomber, and revamping of U.S. air defense force.

Dr. John S. Foster, Director of Defense Research and Engineering, said: "The Soviets have four characteristics of special concern to the R&D community. They are technically advanced, they are strong economically, they have an aggressive military posture—and they work behind a veil of secrecy. The one that makes competition difficult is secrecy. To counter it, the United States must have technological superiority. We must have been there, technically, before them." (W *Star*, 1/6–8/69)

January 7: U.S. patent No. 3,420,471 was granted to John D. Bird, Howell D. Garner, Ernest D. Lounsberry, and David E. Thomas, Jr., LaRC engineers who assigned rights to NASA for jet shoes to enable astronauts to move in space. Wearer could rotate body by natural ankle and leg motions and control direction by turning body and aiming head in swimming motion. Previous devices for similar purpose required use of one or both hands or operation of complex arrangement of control jets and gyroscope sensors. Toe pressure in new shoes would release nitrogen through thruster in sole of each shoe. With electrical control, pressure would be applied to switch. Alternate method would be fluidic control, with toes pressing syringe. (Pat Off PIO; *NYT*, 1/11/69, 39)

- Once Vietnam war was over, Michael Harrington said in Washington *Evening Star*, there would be "money enough for both slums and space if the nation has the political will to appropriate it." Space exploration could provide "economic alternative to war." It was "simply not true that the United States must choose between the heavens and earth. By the mid-Seventies, this country will have achieved a $1 trillion gross national product and because of this . . . there will be almost $30 billion in 'extra' federal funds by 1972. That projection does not require any increase in taxes and it assumes that there will be a fairly high level of military spending." (W *Star*, 1/7/69, A7)

- *Christian Science Monitor* editorial asked, "Would it not be possible, as America's eventual space aim, to see the moon treated much as Antarctica today is treated?" That is "as a 'continent' where nationality does not play a significant role. There could be a research station on the moon, manned the year around—not merely by Americans but by invitation to the scientists and technicians of other interested nations." (*CSM*, 1/7/69)

January 8: Apollo 9, carrying Astronauts James M. McDivitt (commander), David R. Scott (CM pilot), and Russell L. Schweickart (LM pilot), would be launched from KSC Launch Complex 39A at 11:00 am EST Feb. 28, NASA announced. The 10-day earth orbital mission would include simulated translunar insertion; CSM separation, transposition, and docking with LM; onboard LM systems evaluation; extravehicular maneuvers between LM and CM; manned LM active rendezvous; and six SPS burns. (NASA Release 69–3)

- U.S.S.R.'s supersonic transport, Tu-144, made second test flight. Of 50-min duration, flight did not achieve supersonic speed, according to *Pravda*. (Reuters, *NYT*, 1/11/69, 65)

- NASA Acting Administrator, Dr. Thomas O. Paine, asked review board created May 17, 1968, to restudy its findings in May 6, 1968, crash of Lunar Landing Training Vehicle because of second LLTV crash, Dec. 8, 1968. Both craft were destroyed in accidents at Ellington AFB, Tex.; pilots escaped uninjured. Accident board at MSC was investigating latest crash. (NASA Release 69–5)

January 9: USAF released three-volume *Scientific Study of Unidentified Flying Objects*, report of Univ. of Colorado scientists directed by Dr. Edward U. Condon. It concluded that "nothing has come from the study of UFOs in the past 21 years that has added to scientific knowledge. . . . Further extensive study of UFOs probably cannot be justified in the expectation that science will be advanced thereby." Scientists felt "the reason that there has been very little scientific study of the subject is that those scientists who are most directly concerned, astronomers, atmospheric physicists, chemists, and psychologists, having had ample opportunity to look into the matter, have individually decided that UFO phenomena do not offer a fruitful field in which to look for major scientific discoveries."

In review of study, NAS special review panel had concluded, "On the basis of present knowledge the least likely explanation of UFOs is the hypothesis of extraterrestrial visitations by intelligent beings."

Condon report recommended DOD handle UFO sighting reports in normal surveillance operations and found no basis for contention that UFO data were "shrouded in official secrecy." Report stated, "The history of the past 21 years has repeatedly led Air Force officers to the conclusion that none of the things seen, or thought to have been seen . . . constituted any hazard or threat to national security." Report of two-year study, commissioned by USAF for $500,000, had been approved by NAS panel. (Text; Sullivan, *NYT*, 1/8/69, 1; 1/10/69, 32; Boffer, *Science*, 1/17/69, 260–2)

- NASA named Astronauts Neil A. Armstrong (commander), Michael Collins (CM pilot), and Edwin E. Aldrin, Jr. (LM pilot), as prime crew of Apollo 11 lunar landing mission scheduled for summer 1969. Backup crew would be Astronauts James A. Lovell, Jr. (commander), William A. Anders (CM pilot), and Fred W. Haise, Jr. (LM pilot). (NASA Release 69–9)

- NASA submitted to BOB proposed FY 1970 NASA authorization bill in which FY 1969 "Administrative Operations" category had been redesignated "Research and Program Management." Bill requested

$3.051 billion for R&D, $58.2 million for construction of facilities, and $650.9 million for research and program management—for total budget of $3.761 billion. (Text; NASA *LAR*, VIII/8)

- In farewell speech to JPL as President of Cal Tech, Dr. Lee A. DuBridge, Science Adviser-designate to President-elect Richard M. Nixon, said: "I am sure that under the new administration a change in the general structure of the space program may occur principally because the Apollo landings for 1969 will be carried out. . . . [And] that ends an era, so to speak, in the space program which President Kennedy started when he proposed attainment of a landing on the moon by the end of this decade." Question would be raised "in Congress and in the administration and by the people of the country, 'OK we're all through now, let's save that four billion or five billion dollars a year and settle back and do something less expensive.' I don't believe that this is a very widespread view in the top levels of the new administration. I think the Apollo 8 program came at a very critical moment" to make everyone see that "by lifting the eyes of the people to something beyond this little planet on which we live that the spiritual effect . . . the elevation of morale which has occurred, the pride which the country has taken in this sort of achievement, following on the many other achievements, is going to be a stimulus to redirecting the program, yes, but certainly not abandoning it, and deciding what are the great things that can be done in the future of the space program." (Transcript)
- NASA Aerobee 150 MI sounding rocket with VAM–20 booster, successfully launched from WSMR, carried Johns Hopkins Univ. payload to 103.2-mi (166-km) altitude. Primary objective was to measure vacuum UV spectral emission lines from Venusian atmosphere. Rocket and instruments worked satisfactorily, but experiment failed to receive STRAP acquisition and ACS failed to receive tracker lock-on. Except for terrestrial airglow, no useful spectral information on Venus or Procyon was received. (NASA Rpt SRL)
- *Apollo 8* astronauts were honored in Nation's Capital.

Astronauts Frank Borman, James A. Lovell, and William A. Anders received NASA Distinguished Service Medal from President Johnson at White House. Dr. Thomas O. Paine, NASA Acting Administrator, read citation, identical for each astronaut except for designation as commander, command module pilot, or lunar module pilot. Borman received award "for outstanding contributions to space flight, engineering, technology and exploration as Commander of Apollo 8, mankind's first venture beyond Earth into orbit around the Moon. During this flawless mission from December 21 to December 27, 1968, he made critical decisions and carried out complex maneuvers to fly into precise translunar injection, lunar orbit, and transearth injection flight paths to a successful reentry and splashdown within 5,000 yards of the recovery vessel. His scientific observations during the journey to and from the Moon and during 10 orbits of the Moon have added significantly to man's knowledge. He displayed outstanding leadership, courage, professional skill and devotion to duty in accomplishing all planned mission objectives, significantly advancing the nation's capabilities in space. As one of history's boldest explorers, he has blazed a new trail for mankind out into the vastness of extraterrestrial space."

President Johnson said: "Our space program, and this, its most spectacular achievement, have taught us some very invaluable lessons. We have learned how men and nations may make common cause in the most magnificent and hopeful enterprises of mankind. We in the United States are already engaged in cooperative space activities with more than 70 nations of the world. We have proposed a variety of adventures to expand international partnership in space exploration. This morning I renew America's commitment to that principle and to its enormous promise. The flight of Apollo 8 gives all nations a new and a most exciting reason to join in man's greatest adventure." President Johnson presented retired NASA Administrator James E. Webb as "the single man most responsible for successfully administering this program and, I think, the best Administrator in the Federal Government."

Astronauts gave President Johnson miniature copies of recent international space treaties which they had carried aboard Apollo 8 spacecraft and "picture of the ranch" (photo of earth taken from space). (*PD*, 1/13/69, 35–6)

After awards ceremony astronauts were driven in motorcade to address joint assembly of Congress attended by nine Supreme Court Justices. Borman told Congress: "The one overwhelming emotion that we carried with us is the fact that we really do all exist on the small globe. And when you get to 240,000 miles, it really isn't a very large earth." He said voyage was not just an American achievement; "we stood on the shoulders of giants," from Newton and Galileo to present day scientists and space explorers. "If Apollo 8 was a triumph at all," Borman said, "it was a triumph of all mankind."

At State Dept. Auditorium press conference, NASA Assistant Administrator for Public Affairs Julian Scheer announced Astronaut Frank Borman had been appointed Deputy Director of Flight Crew Operations at MSC.

During conference astronauts disclosed they had not seen moon during approach to lunar orbit; presented slides indicating presence of volcanics on back of moon; and announced their conclusion that conditions on pseudolanding site, B–1, indicated "lighting conditions are . . . adequate for a lunar landing, which was one of our objectives of the flight." Borman said, "I came away with the idea that the moon may be more homogeneous than I had realized . . . that you could get a spoon one place and find it just about the same as the samples somewhere else." In answer to question, he said, "I don't believe we found anything that would be of concern to future flights. We flight-tested the ground system. We flight-tested the airborne system. The command module and service module have effectively performed their designed task, their designed mission, with the exception of docking. We have got magnificent machinery. We have superb ground support."

Following press conference, astronauts and their families were guests at Smithsonian Institution dinner in Washington, D.C. (Transcript; NASA Release 69–8; Wilford, *NYT*, 1/10/69, 1; Maynard, *W Post*, 1/10/69, 1; Sehlstedt, B *Sun*, 1/10/69, 1)

January 10: U.S.S.R. launched unmanned *Venus VI* probe—second in five days [see Jan. 5]—into parking orbit and then on trajectory toward Venus. Tass said 2,491-lb spacecraft would attempt slow descent through Venusian atmosphere and softlanding on part of surface not illuminated by sun. Probe was expected to reach Venus in mid-May.

January 10

Tass said *Venus V* had completed 863,700 mi of 155-million-mile journey. *Venus VI* was last reported 40,762 mi above earth. Information radioed from both spacecraft indicated equipment was working normally. (UPI, W *Star*, 1/10/69, A8; AP, B *Sun*, 1/11/69, 1; *Moscow News*, 2/1–8/69, 11)

- New York City held ticker-tape parade for *Apollo 8* Astronauts Frank Borman, James A. Lovell, Jr., and William A. Anders, followed by presentation at City Hall Plaza of Medals of City of New York, luncheon at Lincoln Center, appearance at U.N., and Waldorf-Astoria dinner attended by 2,500 political leaders and guests. (*NYT*, 1/10/69, 30; Aarons, W *Post*, 1/11/69, A1; AP, B *Sun*, 1/11/69, A4)
- Special six-cent postal stamp honoring Dec. 21–27, 1968, *Apollo 8* mission had been approved by Post Office Dept., Postmaster General W. Marvin Watson announced. Stamp would be issued May 5, seventh anniversary of *Freedom 7*, first U.S. manned suborbital space flight, by Alan B. Shepard, Jr. (PO Dept Release 14)
- NASA released "Debrief: Apollo 8," 28-min, 16-mm color film showing first manned lunar orbit. (NASA PAO; Nelson, *Science*, 1/24/69, 371)
- National Investigations Committee on Aerial Phenomena challenged USAF's Condon Report on UFOs [Jan. 8]. At Washington, D.C., press conference Maj. Donald E. Keyhoe (USMC, Ret.), head of private committee, said investigation examined only "about 1%" of "reliable, unexplained" UFO sightings supplied to it. He said his files contained 11,000 reports of sightings, of which 3,000 were unexplained. (UPI, W *Star*, 1/11/69, 1)
- Dr. Frederick Seitz, NAS President, appointed 12-member Universities Organizing Committee for Space Sciences, chaired by Frederick T. Wall, Vice Chancellor of Graduate Studies and Research at Univ. of California at San Diego. It would serve NAS as national Board of Governors of Lunar Science Institute, Houston, Tex., establishing policy, reviewing operations and budgets, and advising Institute's director on program development. Committee also would draft objectives and procedures for consortium of universities operating the facilities for research, development, and education associated with space science and technology. (NAS–NRC–NAE *News Rpt*, 2/1969, 2)
- *New York Times* editorial commented on USAF UFO report: "Evidently many committed to the belief that reported UFO sightings prove this planet is being reconnoitered and even visited by beings from elsewhere in space will remain unpersuaded that earth has a current monopoly on space voyagers. . . . But outside the ranks of true believers, we suspect this document and its conclusions will find wide acceptance. Professor Condon and his colleagues did make a careful and extensive investigation. They enlisted specialists in the relevant branches of science, interviewed alleged witnesses, examined photographs purporting to show UFO sightings and studied cases of claimed radar detection. . . . Those believers will keep on trying, but the rest of society can dedicate themselves to worrying about more serious matters—unless and until there is new and more persuasive evidence than any now available." (*NYT*, 1/10/69, 46)
- Washington *Evening Star* commented on USAF UFO report: "Man needs his myths and his irrational beliefs—his goblins and witches and monsters. He needs to be reminded that the universe is still a wondrous,

awesome and unknown place. He needs to cling to the hope that there is, somewhere, some product of creation more frightening, more powerful and more wise than he. The UFO was the space-age thing that goes bump in the night. It should have been left alone." (W Star, 1/10/69, A10)

- FAA announced proposed rule which would require issuance to airline passengers and crew of fireproof, lightweight, plastic smoke hoods to protect against fire and smoke during evacuation following crash landing. (FAA Release 69-4)
- Japanese Cabinet approved National Defense Council's decision to produce 104 Phantom F-4E jet fighter aircraft by FY 1977 under licensing agreement with McDonnell Douglas Corp. (NYT, 1/11/69, 15)

January 11: In Philadelphia *Evening Bulletin,* J. F. Ter Horst commented on designation of Dr. Robert C. Seamans, Jr., former NASA Deputy Administrator, as Secretary of the Air Force in Nixon Administration Cabinet: "It's highly unlikely that the Nixon Administration will merge civilian and military space programs—they were back in 1958 when he was Vice President. But if it becomes easier to sell Congress a space budget with a military label than with a civilian one, Mr. Nixon undoubtedly will move in that direction. If he does, he has an uncommonly qualified administrator in Seamans." (P Bull, 1/11/69, 14)

- NASA successfully launched two Nike-Cajun sounding rockets from Point Barrow, Alaska, carrying GSFC payloads. First launch was to develop experimental techniques for determining atmospheric composition profiles in mesospheric region and to measure distribution of ozone by chemiluminescent technique and of water vapors by aluminum-oxide hygrometer in 40.4- to 12.4-mi (65- to 20-km) region. Rocket and instruments performed satisfactorily.

  Second rocket was launched to obtain data on wind, temperature, pressure, and density in support of first launch by detonating grenades and recording sound arrivals. Rocket and instruments performed satisfactorily. (NASA Rpts SRL)

- Commission on Marine Science, Engineering and Resources recommended in report *Our Nation and the Sea* creation of new agency, National Oceanic and Atmospheric Agency (NOAA), to coordinate and accelerate oceanology research and development. It proposed agency be composed initially of U.S. Coast Guard and Bureau of Fisheries, plus some functions of Bureau of Sport Fisheries and Wildlife, National Sea Grant Program, U.S. Lake Survey, National Oceanographic Data Center, and ESSA. (Pasadena Star-News, 1/12/69; Nelson, Science, 1/17/69, 263-5)

January 11-12: After being feted at Newark, N.J., airport by 1,500 persons in 15° temperature, *Apollo 8* Astronauts Frank Borman, James A. Lovell, Jr., and William A. Anders flew to Miami, Fla., for Jan. 12 Super Bowl game. Dade County, Fla., Mayor Chuck Hall presented astronauts and their families keys to county and tickets to game. (UPI, NYT, 1/12/69, 35)

January 12: U.S.S.R. launched Cosmos *CCLXIII* into orbit with 362-km (224.9-mi) apogee, 207-km (128.6-mi) perigee, 89.6-min period, and 65.4° inclination. Satellite reentered Jan. 20. (UPI, W Star, 1/13/69, A1; GSFC SSR, 1/15/69; 1/31/69)

January 12

- USAF launched an Advanced Ballistic Reentry System (ABRES) vehicle from Vandenberg AFB by Atlas-F Booster. (AP, *C Trib* 1/17/69)
- In Washington *Evening Star,* Orr Kelly said U.S.S.R. appeared to be developing new multiple warhead missile to deliver "string" of as many as 10 one-megaton nuclear bombs. Missile was similar to, but less sophisticated than, U.S. MIRV system. Both countries were reportedly at same development stage, with deployment scheduled for early 1970s. (W *Star,* 1/12/69, 13)

January 13: Apollo 8 Astronauts Frank Borman, James A. Lovell, Jr., and William A. Anders returned to Houston, Tex., for biggest parade in city's history, with quarter million spectators filling sky with ticker tape and balloons. Astronauts received city's highest honors, bronze medals for heroism with motto "per aspera ad astra." (AP, B *Sun,* 1/14/69, A8; UPI, *W Post,* 1/14/69, A4)

- In ceremony at MSC, NASA presented awards including 12 Distinguished Service Medals, recognizing contributions to *Apollo 8* space mission by groups and individuals in NASA, DOD, and industry.

  DSM, NASA's highest award, was presented to Dr. Kurt H. Debus, Director, KSC; Dr. Robert R. Gilruth, Director, MSC; Christopher C. Kraft, Jr., Director of Flight Operations, MSC; George M. Low, Manager, Apollo Program Office, MSC; Dr. George E. Mueller, NASA Associate Administrator for Manned Space Flight; Rocco A. Petrone, Director of Launch Operations, KSC; L/G Samuel C. Phillips, NASA Apollo Program Director; Dr. Eberhard F. M. Rees, Deputy Director (Technical), MSFC; Arthur Rudolph, Manager, Saturn V Program Office, MSFC; William C. Schneider, Manager, Apollo Applications Program; Gerald M. Truszynski, NASA Associate Administrator for Tracking and Data Acquisition; and Dr. Wernher von Braun, Director, MSFC.

  Exceptional Service Medals were awarded to 62 persons and Public Service Awards to 22. Group Achievement Awards went to U.S.S. *Yorktown* (CVS-10) and Embarked Air Group, Manned Space Flight Network, and NASA Office of Public Affairs. Public Service Group Achievement Award was presented to *Apollo 8* Communication Network and Certificate of Appreciation was awarded to University-NASA Scientific and Technology Advisory Committee (STAC). (NASA Special Release 1/13/69; NASA PAO)

- At nonpartisan farewell dinner given to President Lyndon B. Johnson in New York, former NASA Administrator James E. Webb read statement on accomplishments in space effort under Johnson Presidency: "Lyndon Baines Johnson has done more than lead the United States forward in space. He has stamped on our program its significant characteristics: that it be conducted in the open for all the world to see; that it be carried out so as to strengthen and not to undermine the basic institutions and values of our society; that it be dedicated to the cause of peace and the benefit of all mankind.

  "Of all the debts the American people owe President Johnson, none is likely to loom larger over time than that he started them on the road to mastery of this new, unlimited environment by means of the new rocket technology. . . . It is the lasting tribute to Lyndon Baines Johnson that he has seen from the beginning that accomplishments in space and the capability which can sustain and increase these accomplishments constitute a new barometer of the stature of our Nation." (CR, 1/17/69, S496)

- NASA announced termination of joint NASA–DOD XB–70 flight research program, for which it had assumed management responsibility in March 1967. Aircraft had been productive for studying sonic boom, flight dynamics, and handling problems peculiar to advanced supersonic aircraft. Of two XB–70s constructed by North American Rockwell Corp., one had been destroyed in June 8, 1966, midair collision. Remaining aircraft would be flown from FRC to Wright-Patterson AFB, Ohio, where it would be delivered to USAF Museum. During 2,000-mi flight NASA planned to obtain data on its handling qualities and structural response to air turbulence.

  First flight of XB–70 was made Sept. 21, 1964. Top speed of mach 3 (2,000 mph) and peak altitude of 74,000 had been attained in four-year flight program. (NASA Release 69–10)

- NSF released *Technology in Retrospect and Critical Events in Science* (TRACES), report by Illinois Institute of Technology Research Institute which traced key scientific events leading to five major technological innovations: magnetic ferrites, video tape recorder, oral contraceptive pill, electron microscope, and matrix isolation. In all five, nonmission, or basic, research "provided the origins from which science and technology could advance toward the innovation which lay ahead." Approximately 70% of key events documented were nonmission research, 20% mission-oriented, and 10% development and application. Ten years before innovation—i.e., shortly before conception of that innovation—90% of nonmission research had been accomplished. (Text)

- Defense Secretary-designate Melvin R. Laird announced retention of Dr. John S. Foster, Jr., as Director of Defense Research and Engineering. (Kelly, W Star, 1/13/69, A5; UPI, W Star, 1/14/69, A6)

- MSFC announced award of $1,311,702 contract to LTV Aerospace Corp. for construction of temperature control devices for Apollo Telescope Mount, or manned solar observatory.

  MSFC also had issued bid requests on 5 control relay packages, 4 horizon sensor scanner systems, 11 solar sensors (4 for attitude control system and 7 for solar panel control system), and 5 computer component control packages for guidance, control, and power systems for Saturn I Workshop, scheduled to be flown in 1971–72 in Apollo Applications Program. (MSFC Releases 69–6, 69–8)

January 14–18: U.S.S.R.'s *Soyuz IV*, carrying Cosmonaut Vladimir Shatalov, was successfully launched from Baikonur Cosmodrome into orbit with 224-km (139.2-mi) apogee, 213-km (132.4-mi) perigee, 88.8-min period, and 51.7° inclination. Soviet news media reported launch quickly and in detail and within one hour video recording of launch was shown on Moscow TV. Soon afterward viewers received live TV coverage from spacecraft and description of flight by Cosmonaut Shatalov. Western speculation, later confirmed, was that *Soyuz IV* would rendezvous with another spacecraft.

Soyuz V, carrying Cosmonauts Yevegeny Khrunov, Boris Volynov, and Aleksey Yeliseyev, was launched Jan. 15 into orbit with 212-km (131.7-mi) apogee, 196-km (121.8-mi) perigee, 88.6-min period, and 51.7° inclination. Tass said spacecraft would conduct joint experiments with *Soyuz IV*. Spacecraft established radio contact, coordinated

January 14–18: *First manned transfer in space. Launched on successive days*, Soyuz IV *(carrying one cosmonaut) and* Soyuz V *(carrying three) rendezvoused and docked in space. Yevegeny Khrunov and Alexey Yeliseyev performed experiments outside their* Soyuz V *and then joined Vladimir Shatalov in* Soyuz IV *for return to earth, leaving Boris Volynov to return alone in* Soyuz V. *Full-scale configuration of the docked spacecraft was photographed at Expo '70 in Japan in May 1970.* (Aviation Week *Photo*)

scientific programs, transmitted TV pictures to earth, photographed earth's surface, and conducted midcourse maneuvers.

On Jan. 16 the two spacecraft automatically approached to within 110 yds of each other and *Soyuz IV* was then steered manually until it docked with *Soyuz V*. Tass announcement said: "After the docking there was a mutual mechanical coupling of the ships, they were rigidly tightened up and their electrical circuits were connected. Thus, the world's first experimental cosmic station with four compartments for the crew was assembled and began functioning as an artificial earth satellite."

Moscow TV viewers watched as *Soyuz V* crew members Khrunov and Yeliseyev put on special spacesuits with new regenerative life-support systems and went out into space through service compartment hatch. Cosmonauts remained in space for one hour, conducting observations and experiments, and then entered service compartment of other spacecraft, *Soyuz IV*, to join Shatalov. After 4 hrs 35 min of docked flight in low, nearly circular orbit, spacecraft were uncoupled and continued their flights separately. *Soyuz IV*, with three-man crew, landed Jan. 17 and *Soyuz V*, Jan. 18. (UPI, W *Star*, 1/14–19/69; Shub, W *Post*, 1/15–19/69; Winters, B *Sun*, 1/15–18/69; *Moscow News*, 1/25–2/1/69, 3; 2/1–8/69, Supplement; GSFC *SSR*, 1/15/69; 1/31/69)

January 14: In his last State of the Union message, President Johnson told Joint Session of Congress: ". . . if the Nation's problems are continuing, so are this Nation's assets. Our economy, the democratic system, our sense of exploration, symbolized most recently by the wonderful flight of the Apollo 8, in which all Americans took great pride, the good common sense and sound judgment of the American people, and their essential love of justice." Quest for durable peace "has absorbed every Administration since the end of World War II. It has required us to seek a limitation of arms races not only among the superpowers, but among the smaller nations as well. We have joined in the test ban treaty of 1963, the outer space treaty of 1967, and the treaty against the spread of nuclear weapons in 1968." (*PD*, 1/20/69, 60–8)

- In fourth big city welcome within one week, *Apollo 8* Astronauts Frank Borman, James A. Lovell, Jr., and William A. Anders received tribute from estimated 1.5 million persons in Chicago reception at which they were made honorary citizens of city. (UPI, *NYT*, 1/15/69, 1)
- MSFC announced Dr. Arthur Rudolph, special assistant to MSFC Director, Dr. Wernher von Braun, and formerly manager of Saturn V rocket program, would retire Jan. 31. Dr. Rudolph had been awarded NASA Distinguished Service Medal Jan. 13 and on Nov. 15, 1968, had received NASA Exceptional Scientific Achievement Medal "for distinguishing himself by meritorious achievement" as manager of Saturn V program from August 1963 to May 1968. Starting career in rocketry in Germany in 1930, he later received patents for liquid-fuel rocket engines and demonstrated operation of liquid-fuel rocket. He came to U.S. with more than 100 other rocket experts in "Operation Paperclip" in December 1945. (MSFC Release 69–10)
- Secretary of the Air Force-designate, Dr. Robert C. Seamans, Jr., thought space activity should be major part of USAF, *Christian Science Monitor* said. He had said, "My prime objective will be to develop equipment for national defense, and my emphasis will be on whatever kind of equipment will be most suitable for the mission at hand." (*CSM*, 1/14/69, 5)
- Senate adopted S.R. 13 establishing numerical size of Senate standing committees for 91st Congress and adopted S.R. 14 and S.R. 15 electing majority and minority standing committee membership. Senate Committee on Aeronautical and Space Sciences was reduced from 16 to 15 members, with Republican Sens. Len B. Jordan (R-Idaho) and Charles E. Goodell (R-N.Y.) dropping off. Senate also approved appointment of Sens. Barry Goldwater (R-Ariz.), Charles McC. Mathias, Jr. (R-Md.), and William B. Saxbe (R-Ohio) to committee. Democratic assignments on committee remained unchanged. (*CR*, 1/14/69, S152–87)
- USN announced it had selected Grumman Aircraft Engineering Corp. as prime contractor for new F–14A supersonic carrier-based fighter. F–14A was expected to make maiden flight in early 1971 and to be operational with fleet in 1973. (DOD Release 33–69)
- Anglo-French Concorde supersonic airliner was undergoing final ground trials at Toulouse-Blagnac airfield in southwest France in preparation for inaugural flight expected toward end of January or early February, Reuters reported. Aircraft was expected to enter commercial service

in 1972 and to halve London-New York flight time, to 3 hrs 32 min. It had five-year lead over U.S. SST, which was still in blueprint stage. (*NYT*, 1/15/69, 77)

- NASA announced it had signed $2,919,000 supplemental agreement with Div. of Sponsored Research of MIT for fabrication and delivery of 40 inertial reference integrating gyros (IRIGs) for Apollo guidance and navigation system, bringing total contract to $81,000,000. (NASA Release 69-11)
- McDonnell Douglas Corp. received $1,000,000 initial increment to $3,900,000 fixed-price USAF contract for development, fabrication, and testing of Titan IIIC payload fairing subsystem. Contract was managed by USAF Space and Missile Systems Organization. (DOD Release 34-69)

January 15: President Johnson in message transmitting FY 1970 budget to Congress said major recommended decreases in budget authority from 1969 to 1970 included $235 million for the National Aeronautics and Space Administration, which "will provide for a program level equal to 1969 when combined with prior year funds." Major increases included "522 million for airway modernization, highways, and other activities" in DOT. Of estimated $11.6 billion increase in total budget outlays, $0.5 billion was for national defense, "largely for improvements in our strategic forces, modernization of our tactical air forces, and other increased research and development efforts needed to assure sufficient deterrent power in the future. These increases will be substantially offset by reduced outlays for Vietnam resulting from changing combat patterns and revised supply requirements. . . . In keeping with national priorities, major social programs account for largest portion" of increase.

President said: "The record of achievements of the past 5 years is an impressive one. We have witnessed a period of unprecedented economic growth, with expanded production, rising standards of living, and the lowest rates of unemployment in a decade and a half. Our military forces today are the strongest in the world. . . . Last month saw man's first successful flight to the moon. In domestic matters, the legislative and executive branches, cooperatively, have forged new tools to open wider the doors of opportunity for a better life for all Americans.

"This Nation remains firmly committed to a world of peace and human dignity. In seeking these goals, we have achieved great military strength with the sole aim of deterring and resisting aggression. We have continued to assist other nations struggling to provide a better life for their people. We are successfully pushing forward the frontiers of knowledge to outer space and promoting scientific and technological advances of enormous potential for benefit to mankind." (*PD*, 1/20/69, 70-90; *CR*, 1/15/69, S195-208)

- President Johnson sent $195.3-billion FY 1970 budget request to Congress, including total space budget of $5.946 billion. Of this sum, NASA would receive $3.599 billion (plus $7.89 million for aircraft technology and $117 million in unobligated funds from prior years to be applied to 1970 program); DOD would receive $2.219 billion; AEC, $105 million; ESSA, $10 million; NSF, $4 million; Dept. of Interior, $6 million; and Dept. of Agriculture, $4 million.

Total NASA FY 1970 budget request of $3.878 billion was below $4 billion for first time since FY 1963 request. Expenditures were budgeted to decline nearly $300 million from FY 1969 level of $4.250 billion, to $3.950 billion. Of budget request, $3.168 billion would go for R&D, $58.2 million for construction of facilities, and $650.9 million for program management. Slightly more than 50% of total FY 1970 authority—$2.008 billion—would be in manned space flight, including Saturn IB Workshop and Apollo Telescope Mount. Space science and applications were allotted $558.8 million; advanced research and technology, $290.4 million.

Larger proportion of NASA funds would go to research than to development in FY 1970 as Apollo costs declined. Increases would go to Apollo Applications (total $308.8 million in FY 1970), 1971 Mariner-Mars flight ($45.4 million), Viking project ($40 million), 1973 Mariner-Mercury ($3 million), Planetary Explorers ($8 million), Applications Technology Satellites ($44.2 million), and Earth Resources Survey program ($25.1 million). Aeronautical vehicle technology program, up from $74.9 million to $78.9 million, included $21.78 million for advanced research, $500,000 for general aviation, $11.25 million for V/STOL, $16.19 million for subsonic aircraft, $20.9 million for supersonic aircraft, and $8.28 million for hypersonic aircraft.

DOD space funding would include satellite development, certain portions of missile development and operating costs, MOL (increased to $576 million), Titan III booster, and supporting R&D. In addition, aircraft R&D funds of $1.4 billion would include $500 million for new USN F–14A fighter to replace F–111B, $1 billion for series of advanced jet aircraft, $75 million for long-range bomber to succeed B–52, and funds for USAF F–15 fighter and USN VSX antisubmarine aircraft.

AEC space funding included amounts for nuclear rocket propulsion technology and nuclear power sources for space applications, including production of isotopic fuels and aerospace safety. ESSA funds would support Earth Resources Technology Satellite (ERTS) program research. NSF's total request was up from $400 million in FY 1969 to $497 million. Of this amount, its $4-million space funds were for research in astronomy using rockets and satellite-borne observation instruments. Dept. of Interior would conduct experiments with data from ERTS spacecraft.

FAA's R&D budget would increase from $49 million in FY 1969 to $59 million in FY 1970. Principal increases were for research on air traffic control and noise abatement. President Johnson requested no additional funds for SST program. (NASA Release, 1/15/69; BOB Special Analysis Q; Dale, *NYT*, 1/16/69, 1; Schmeck, *NYT*, 1/16/69, 24; Lindsey, *NYT*, 1/16/69, 81; *W Post*, 1/16/69, A12)

- NASA released transcript of Jan. 14 briefing on NASA FY 1970 budget at which Dr. Thomas O. Paine, NASA Acting Administrator, said $3.878-billion budget approved by President Johnson had been developed "to maintain an austere but balanced NASA aeronautics and space program aimed at major program goals of high national priority."

Goals were continuing advances in space applications, including initiation of experimental earth resources technology satellites and slightly decelerated research on space environment; achieving manned lunar landing and additional Apollo moon missions, with limited provision for studies—not for development—of equipment to achieve longer lunar stay-times and mobility for future exploration; proceeding with Saturn I Workshop and Apollo Telescope Mount; proceeding with Viking Project to land instruments on Mars in 1973; initiating in FY 1970 Mariner flyby in 1973 of Venus and Mercury; producing family of small planetary Explorers for orbiting Mars and Venus; making preliminary observations of Jupiter with previously approved Pioneer spacecraft; continuing work on unmanned orbiting astronomical observatories; continuing advanced aerospace technology work at about current levels; emphasizing noise reduction, with construction of special noise research laboratory; developing full potential of civil and military aeronautics; and undertaking NERVA project for development of flight-weight nuclear engine.

Dr. Paine noted FY 1970 budget, as approved by President Johnson, "would halt a four-year downward trend in NASA budget." It was "austere and does not make full use of the aerospace capabilities that the nation has developed. . . ." But it permitted "a balanced program of useful work in critical areas." Budget left "the major new program decisions, especially in the manned flight area, for the next Administration," and was "'holding budget' that provides for progress, but defers critical program and funding decisions to the new Administration." (Transcript)

- Apollo Program Director, L/G Samuel C. Phillips (USAF), addressed National Space Club luncheon in Washington, D.C., on impact of *Apollo 8:* "Many, if not most, of the world's newspapers heralded the flight as evidence of the greatness of the United States. I'm told journalists in Germany, England and France speculated on the improved position that the U.S. would enjoy in the diplomatic arenas in which it is engaged in very important discussions. I'm told that a French paper went so far as to say that the lunar flight had vindicated Capitalism as the best system of government, and vindicated our free enterprise system as the most effective way to make progress. I'm aware from personal correspondence and discussions as well as reports in the press that scientists throughout the world have been equally impressed and that they've applauded the progress that this flight indicates for us."

Gen. Phillips described Apollo 9 mission as "far less spectacular than Apollo 8, but . . . more complex." It would be "certainly one of the most vital missions that we've had in our mission sequence over the years that leads us to a lunar landing." Risks would be different but "I personally think they're a little greater than the risks which we knowingly accepted in committing the Apollo 8 mission."

Apollo 10 would be fifth Saturn V—505—with 106 command and service module and LM-4. Crew would be Astronauts Thomas P. Stafford, John W. Young, and Eugene A. Cernan. Scheduled for May, its objectives were "to demonstrate the performance of the crew, the space vehicle, and the mission support facilities during a manned lunar mission with the Command and Service Module and Lunar Module, and to evaluate the performance of the Lunar Module in a cislunar and lunar environment." (Text)

- Astronomers at Univ. of Arizona detected for first time existence of rapidly flashing star in Crab Nebula with rhythm coinciding with that of pulsar observed by radio telescope at same position. Flashing was confirmed during week by McDonald Observatory of Univ. of Texas and by Kitt Peak National Observatory in Arizona. Presumably star and pulsar were identical. Discovery was first unequivocal observation of pulsar in visible light. (Sullivan, *NYT*, 1/21/69, 29)

- Boeing Co. announced it had made Jan. 15 deadline in submitting to FAA specifications for fixed-wing, 280-ft, titanium SST, weighing 635,000 lbs and having 141⅔-ft wingspan. General Electric Co. engines would propel aircraft to 1,800 mph. Boeing said 299-passenger aircraft's first flight was scheduled for 1972, with commercial operation possible in 1976. (*WSJ*, 1/15/69, 7; FAA Release 69–6)
- Underwater test program begun at MSFC's Neutral Buoyancy Simulator several years earlier was providing information essential for design of first U.S. space station, NASA reported. Technicians, design engineers, and professional divers in spacesuits and scuba gear were conducting tasks similar to those necessary to activate space orbiting workshop, in 1.4-million-gal water tank containing mockups of AAP cluster elements (Saturn I Workshop, lunar module ascent stage, Apollo Telescope Mount solar observatory, and airlock and multiple docking adapter), simulating weightlessness of space. Weightlessness was impossible to duplicate on earth for longer than fraction of minute. Conclusions from tests would be reflected in workshop's final design, with decision expected in May 1969. (NASA Release 69–4)
- Penn Central Railroad began electric-powered Metroliner service that would cut traveling time of 226-mi New York-Washington trip to 2 hrs 59 min—36 min faster than swiftest previous trains and, according to Penn Central Chairman Stuart T. Saunders, comparable to airplane journey which took 45 min in sky but added airport-access and airway delays. (Aug, *W Star*, 1/15/69, A1; Eisen, *W Post*, 1/16/69, B1)

January 15–17: Space Science Education Conference, to inform educational TV directors and teachers of ways NASA could assist in explaining space program to students and educational TV audiences, was held at MSFC and attended by educators and TV representatives from six states. (MSFC Release 69–11)

January 16: Secretary of Defense would exercise option to buy 57 additional C–5A aircraft from Lockheed Aircraft Corp. and General Electric Co., DOD announced. Expenditures and commitments would be limited to first 23 aircraft; decision on whether to authorize expenditures for remaining aircraft would be made later. Predicted cost for total 120 C–5As (six squadrons) was $4.343 billion. (DOD Release 43–69)

January 17: President Johnson submitted to Congress report on *U.S. Aeronautics and Space Activities for 1968*. In transmittal message he wrote: "Our astronauts have now flown 18 manned space missions, during which they experienced 3,215 man hours in space flight. Together with the activities of the Soviet Union, this makes a total to date of 28 manned flights and 3,846 man hours in space. Through this investment we have obtained new products, services, and knowledge; we have enhanced our national security; we have improved our international relations; and we have stimulated our educational system. Our Nation is richer and stronger because of our space effort. I recommend that America continue to pursue the challenge of space exploration." (*Pres Rpt 68*; *CR*, 1/17/69, H405, S524)

- *Christian Science Monitor* reported interview with Dr. Lee A. DuBridge,

President-elect Richard M. Nixon's Science Adviser-designate: "The balance between manned and unmanned [space] exploration has to be studied. But there comes a time, place, and activity where the judgment and quick reaction of men are needed to do the job or make emergency repairs. There may even be situations in which it is cheaper to have men do this than to have automated instruments. That time has not yet arrived. Manned flight is still very expensive. But I foresee the time when you will run beyond the ability of automatic instruments to do a job either adequately or economically in space exploration." (Cowen, CSM, 1/17/69)

- NASA terminated *Nimbus II* flight operations. Spacecraft, launched May 15, 1966, to flight-test instrumentation and observe region of electromagnetic spectrum not previously studied, had accomplished all primary and secondary objectives and had operated on three-axis stabilization 32 mos, greatly exceeding design lifetime. Automatic picture transmission had operated 7,900 hrs over nearly 23 mos. (NASA Proj Off)

- NASA Nike-Cajun sounding rocket was successfully launched from Kiruna, Sweden, to study sudden upper-atmosphere warming conditions by detonating grenades between 24.9- and 55.9-mi (40- and 90-km) altitudes. Launch was first in series of four scheduled under agreement between Swedish Space Research Committee (SSRC), British Science Research Council (SRC), and NASA. SSRC provided ground equipment, instrumentation, and grenade payloads and was responsible for launch operations. NASA supplied Nike-Cajun rockets and DOVAP transponders. Second launch would be conducted Jan. 19. (NASA Release 69-16; NASA Rpt SRL)

- Nike-Cajun sounding rocket launched by NASA from Wallops Station carried GSFC payload to 69.6-mi (112-km) altitude to collect data on wind, temperature, pressure, and density in 21.7- to 59.0-mi (35- to 95-km) range during atmospheric warming by exploding grenades. All 19 grenades exploded as programmed and sound arrivals were recorded. Launch was first in series of four to obtain upper-atmosphere data. (NASA Rpt SRL)

- Rep. George P. Miller (D-Calif.), Chairman of House Committee on Science and Astronautics, introduced H.R. 4046, FY 1970 NASA authorization bill, totaling $3.760 billion, in House. (CR, 1/17/69, H403)

- Arcas sounding rocket was launched by NASA from Andoya, Norway, carrying Swedish payload to study ionosphere. Rocket and instruments functioned satisfactorily. (NASA Proj Off)

- ComSatCorp announced TV coverage of Presidential inauguration of Richard M. Nixon would be transmitted via comsats across Atlantic and Pacific to viewers in Europe, Latin America, Caribbean, and Pacific areas during more than 13 hrs of overseas transmissions. (ComSatCorp Release 69-4)

- NAS–NRC Space Science Board released *Physiology in the Space Environment*, Vol. 1, *Circulation*, prepared at NASA request. Report found systematic program of ground-based and inflight biomedical experimentation was essential for planning of long-duration manned space missions, such as to Mars. Knowledge of circulatory system and effects of space flight must be greatly expanded. It recommended experimentation with animals, man, simulations, laboratory investigations, comprehensive literature studies, and physiological measurement before, after, and during flight to obtain all possible data from ground-based work. Flight experiments should be limited to those requiring weightlessness and other conditions not reproducible on ground. (Text; NAS Release)

- *New York Times* commented on Soviet linking of Soyuz spacecraft: "It is probably not too soon to begin planning for standardization of space vehicles—a move that would permit joining these vehicles regardless of their national origin. It would be tragic, for example, if a group of Soviet spacemen needed rescue and could not be saved because the only available vehicle was an American space ship impossible to link with the Soviet ship. Here is an area in which international cooperation could not only save lives but help pave the way for the joint operation of stations in space for the benefit of all mankind." (*NYT*, 1/17/69, 46)
- National Transportation Safety Board recommended to FAA new flight regulations to reduce aircraft landing and approach accidents responsible for 56% of fatal crashes since jet-age inception in 1957. During the 60 days before release, 10 airliners had crashed in U.S., Latin America, and Europe on landing or approach. Board called for review of policies, practices, and training to increase crew efficiency. It urged development of audible and visual warning devices to alert pilot when flying below safe altitude. (NTSB Release SB69-5)
- FAA announced it was considering amendment of flight recorder rules to require increase in instrument's capability so as to provide 14 additional kinds of information in accident investigation, including data on altitude, response to aerodynamic forces, flight-control surface positions, and engine performance. Underwater locator device also was proposed, to go into operation upon submersion. Proposal would require installation of new equipment on newly manufactured aircraft within three years of effective date of final rule and within five years on aircraft already in service. (FAA Release 69-9)
- NASA announced appointment of Charles G. Haynes, Director of Inspections since 1961, as Director of Hq. Administration effective Jan 19. He would succeed Alfred S. Hodgson, who retired after 35 yrs of Government service. Ralph F. Winte would serve as Acting Director of Inspections until permanent appointment was made. (NASA Ann)
- National Academy of Sciences announced Dr. Philip Handler, Chairman of Dept. of Biochemistry at Duke Univ., had been declared President-elect of NAS after tally of mail ballots. There had been no other nominee. Dr. Handler would begin six-year term July 1, succeeding Dr. Frederick Seitz, President of Rockefeller Univ. Dr. Handler, with career in enzyme research, had been National Science Board member since 1962 and its Chairman since 1966. From 1964 to 1967 he had been member of President's Science Advisory Committee. (NAS Release; *W Post*, 1/19/69, A5)

January 18: Nike-Cajun sounding rocket was launched by NASA from Point Barrow, Alaska, carrying GSFC payload to obtain data on wind, temperature, pressure, and density during period of atmospheric warming by detonating grenades and recording sound arrivals on ground. All grenades were ejected and exploded as programmed. Launch was first in series of four to be launched from Point Barrow. (NASA Rpt SRL)

- Arcas sounding rocket was launched by NASA from Andoya, Norway, carrying Swedish payload to study ionosphere. Rocket and instruments functioned satisfactorily. (NASA Proj Off)
- *Washington Daily News* editorial commented on Condon Report on UFOs: "Dedicated disciples of the little green men from Mars school no doubt will find the Condon report represents another diabolical plot to suppress truth. But most Americans will find the report something less than a surprise. Apart from wasting time, continuing study would waste taxpayer money." (*W News*, 1/18/69)

January 19: NASA Nike-Cajun sounding rocket, second in series of four [see Jan. 17] in NASA–SSRC–SRC cooperative program, was successfully launched from Kiruna, Sweden, to study sudden upper-atmosphere warming conditions by detonating grenades between 24.9- and 55.9-mi (40- and 90-km) altitudes. Vehicle underperformed but satisfactory scientific data were expected. (NASA Release, 69–16; NASA Rpt SRL)

- NASA announced appointment of Robert W. Kamm, Assistant to Director of Space Institute of Univ. of Tennessee, as consultant to Harold B. Finger, NASA Associate Administrator for Organization and Management. Kamm had been director of NASA's Western Support Office, Santa Monica, Calif., for nearly nine years. (NASA Special Release)

January 20: Administration of President Lyndon B. Johnson ended as President Richard M. Nixon was sworn in as President of U.S. Johnson had served U.S. space program continuously since *Sputnik I* in October 1957, first on Capitol Hill as Chairman of Select and then permanent Senate Committee on Aeronautical and Space Sciences. As Vice President under late President Kennedy, he had served as Chairman of National Aeronautics and Space Council, post held by Vice President Hubert H. Humphrey during Johnson Administration. (EH)

- In inaugural address following his taking oath of office as President of U.S., Richard M. Nixon said: "Those who would be our adversaries, we invite to a peaceful competition—not in conquering territory or extending dominion, but in enriching the life of man. As we explore the reaches of space, let us go to the new worlds together—not as new worlds to be conquered but as a new adventure to be shared. . . . Only a few short weeks ago, we shared the glory of man's first sight of the world as God sees it, as a single sphere reflecting light in the darkness. As the Apollo astronauts flew over the moon's gray surface on Christmas Eve, they spoke to us of the beauty of Earth—and in that voice so clear across the lunar distance, we heard them invoke God's blessing on its goodness. In that moment, their view from the moon moved poet Archibald MacLeish to write:

  " 'To see the Earth as it truly is, small and blue and beautiful in that eternal silence where it floats, is to see ourselves as riders on the Earth together, brothers on that bright loveliness in the eternal cold—brothers who know now they are truly brothers.'

  "In that moment of surpassing technological triumph, men turned their thoughts toward home and humanity—seeing in that far perspective that man's destiny on earth is not divisible; telling us that however far we reach into the cosmos, our destiny lies not in the stars but on Earth itself, in our hands, in our own hearts." (*PD*, 1/21/69, 150–154; *CR*, 1/20/69, S561)

- Inaugural parade following President Nixon's address included NASA float carrying mockup of lunar module to be used for moon landing and *Apollo 7* capsule. *Apollo 7* Astronauts Walter M. Schirra, Jr., Donn F. Eisele, and R. Walter Cunningham rode in convertible automobile in front of NASA float. (NASA PIO)

- Nike-Cajun sounding rocket was launched by NASA from Churchill Research Range carrying GSFC grenade payload to obtain data on atmospheric parameters. Rocket and instruments functioned satisfactorily. (NASA Proj Off)
- "The ability to rescue a stranded astronaut is something the U.S. does not have," John Lannan said in Washington *Evening Star*. "And—despite claims by both the Soviet Union and the foreign press—neither does the Soviet Union." NASA Deputy Director of Manned Space Flight Safety Philip H. Bolger had said NASA was not likely to have real space rescue capability before "second generation" of manned space stations. Agency was now funding rescue studies at cost of $600,000 yearly. Amount would probably rise to $1 million within two years. Immediate goal was to examine method that would fit into existing systems and bring it to hardware stage. Bolger had said "bail-out" mechanism seemed likelier than earth-based rescue system. (W *Star*, 1/20/69, A8)
- At Moscow news conference, U.S.S.R. Foreign Ministry spokesmen Leonid Zamayatin and Kirill Novikov released statement reaffirming U.S.S.R.'s readiness to discuss missile control proposals contained in July memorandum to other governments. (Shub, *W Post*, 1/21/69; *NYT*, 1/21/69)

January 21: At annual AIAA dinner, in New York, Dr. Robert C. Seamans, Jr., Secretary-designate of the Air Force and former NASA Deputy Administrator, was installed as President of AIAA for 1969, succeeding Dr. Floyd L. Thompson, who continued as a director. In interview following dinner, Dr. Seamans said NASA should maintain its open space program and its freedom for international exchange of information. USAF, with current MOL program, "has special problems which can be resolved by the Department of Defense." Value of dual space program, he said, was that one element could learn from another. Of U.S.S.R. space program, Dr. Seamans said, "They're awfully good at doubling in brass . . . getting the most out of their program," but "our program has more breadth and depth and if we're imaginative about what we're doing, we're not going to take second place."

AIAA presented its Goddard Award to Dr. Stanley Hooker, Technical Director of Bristol-Siddeley Engine Div. of Rolls-Royce, Ltd., and Perry W. Pratt, Vice President and Chief Scientist of United Aircraft Corp., for work on turbine engines.

Dr. Charles P. Sonnett, Chief of Space Sciences Div. at ARC, received annual Space Science Award of $500 for "his personal contribution as a planner, leader, and individual experimenter in major space science vehicle programs which have contributed to the field of space physics."

LaRC Director Edgar M. Cortright and Charles W. Harper, NASA Deputy Associate Administrator, Aeronautics, OART, were elected AIAA directors. (Lannan, W *Star*, 1/22/69, A13; *NYT*, 1/22/69, 33; AIA Releases)

- *Look* published interview in which former astronaut M. Scott Carpenter announced that infarcts in lower thigh bones, indicating calcification, would terminate his career as active deep-sea diver. He would remain senior aquanaut on Sealab III project as deputy on-scene commander of Sealab III command ship *Elk River.* (*Look,* 1/21/69, 68–74)

- Sen. Claiborne Pell (D-R.I.) introduced on Senate floor S.R. 33 calling on U.S. representatives to U.N. to place before U.N. Committee on the Peaceful Uses of the Seabed and Ocean Floor set of detailed principles to govern activities in ocean space of all nations of world. Resolution was referred to Committee on Foreign Relations. (*CR,* 1/21/69, S597)

January 21–28: Harris survey of 1,544 U.S. households showed widespread disenchantment with Federal commitments on space and Vietnam. Greatest number, 39%, selected space as program they would cut first, while 2% voted to keep or increase space program. Second favorite for funding cut was Vietnam war financing, with 18% of votes. Anticrime and law enforcement programs received greatest number of "keep or increase" votes—22%. (Harris, *Federal Times,* 3/5/69, 9)

January 22–29: NASA's *Oso V* (OSO–F) Orbiting Solar Observatory was successfully launched from ETR by three-stage Thor-Delta (DSV-3C) booster to study the sun and its influence on earth's atmosphere. Orbital parameters: apogee, 353.1 mi (568.2 km); perigee, 337.8 mi (543.6 km); period, 95.8 min; and inclination, 32.96°. Primary mission objective was to obtain high-resolution spectral data from pointed experiments in 1 Å–1,250 Å range during one solar rotation and conduct raster scans of solar disc in selected wavelengths. Secondary objective was to obtain useful data from nonpointed and pointed experiments for more than one solar rotation with extended observations of single lines and solar flares.

Fifth of eight spacecraft launched in NASA's OSO program to provide direct observation of sun during most of 11-yr solar cycle, *Oso V* weighed 636 lbs, carried eight experiments, was designed with six-month lifetime, and had two main sections—wheel (lower) section, which provided stability by gyroscope spinning and housed telemetry-command equipment, batteries, gas-spin control arms, and five experiment packages; and sail (upper) section, which contained solar cells and solar pointing experiments and was oriented toward sun. Experiments, designed to continue and extend work of preceding OSO spacecraft, were provided by University College (London) and Univ. of Leicester, Univ. of Paris, Univ. of Colorado, Univ. of Minnesota, Naval Research Laboratory, and GSFC.

Both tape recorders were turned on and were operating satisfactorily and all spacecraft subsystems were operating nominally. NRL wheel x-ray experiment was turned on during 11th orbit and was obtaining good data. By Jan. 29 *Oso V* had received 707 commands and had completed 102 orbits. All eight experiments had been turned on and obtained good scientific data. All spacecraft systems—including raster scan and both tape recorders—had operated satisfactorily. Data from GSFC x-ray experiment were being used to plot spectrum of sun. Data from NRL UV pointed experiment had been used to obtain *Oso V*'s first Lyman-alpha spectroheliograph.

Oso I (launched March 7, 1962) and *Oso II* (launched Feb. 3, 1965) had surpassed their six-month design lifetimes and, together, provided 6,000 hrs of scientific information. *Oso III* (launched March 8, 1967) and *Oso IV* (launched Oct. 18, 1967) continued operating satisfactorily, each providing 7½ hrs of real-time data daily. OSO program was managed by GSFC under OSSA direction. (NASA Proj Off; NASA Release 69–13)

January 22: USAF launched unidentified satellite on Titan IIIB-Agena D booster from Vandenberg AFB into orbit with 672.5-mi (1,082-km) apogee, 92.0-mi (148-km) perigee, 96.9-min period, and 106.1° inclination. Satellite reentered Feb. 3. (GSFC *SSR*, 1/31/69; 2/15/69; *Pres Rpt 70* [69])

- NASA announced it would conduct 26 major launches from ETR and WTR during 1969. First launch was *Oso V* Jan. 22. Launches from ETR would include five manned missions: Apollo 9, scheduled for Feb. 28, would place three-man crew in earth orbit for 11 days to flight-test lunar module; Apollo 10 would place three astronauts in lunar orbit and two would fly LM to within 50,000 ft of lunar surface; and Apollo 11 would land two members of three-man crew on lunar surface. Two additional lunar landings would be conducted if first landing was successful. Unmanned launches from ETR would include two Intelsat III comsats, two Mariner-Mars missions, Tiros weather satellite, Biosatellite carrying monkey, Pioneer E interplanetary spacecraft, Applications Technology Satellite (ATS), Orbiting Solar Observatory (OSO), Orbiting Astronomical Observatory (OAO), and two U.K. comsats. WTR launches would include three Tiros weather satellites, Explorer (IMP–G), Canadian International Satellite for Ionospheric Studies (ISIS), Nimbus weather satellite, and Thor-Agena (OGO–F) mission to test experimental ion-thruster. (KSC Release 19–69; NASA OSSA)

- Nike-Cajun sounding rocket was launched by NASA from Churchill Research Range carrying GSFC grenade experiment to obtain data on atmospheric parameters. Rocket and instruments functioned satisfactorily. (NASA Proj Off)

- Communist Party General Secretary Leonid I. Brezhnev told Soviet gathering in honor of four *Soyuz IV* and *Soyuz V* cosmonauts U.S.S.R. was "fully justified in saying that the successful flight . . . [see Jan. 14–18] is a great achievement of Soviet science and engineering, and a new triumph of the courage, boldness, intellect and labour of the Soviet people. The recent outstanding flight made by the American astronauts round the Moon, the confident start made by the Soviet automatic interplanetary stations 'Venus-5' and 'Venus-6' towards their distant target, and the successful flight made by the . . . [Soyuz] spaceships—all this constitutes man's new, major steps along the road to conquering the mysterious world of outer space. . . ." (*Moscow News*, 2/8–15/69, Supplement, 3–5)

- During day climaxed by shots from what U.S.S.R. Foreign Ministry called "schizophrenic" gunman, *Soyuz IV* and *V* Cosmonauts Vladimir Shatalov, Boris Volynov, Yevgeny Khrunov, and Aleksey Yeliseyev flew from Baikonur Space Center, Kazakhstan, to Moscow for Kremlin ceremony honoring success of Soyuz missions. Attack occurred as motorcade escorting cosmonauts approached Kremlin's Borovitsky Gate. Reports said driver of cosmonauts' limousine and security guard had been injured by bullets and that Cosmonaut Beregovoy had been slightly injured by flying glass. Communist Party General Secretary Leonid I. Brezhnev and Soviet President Nikolai V. Podgorny, riding several cars behind cosmonauts, were not injured. Western newsmen had already been admitted to Congress Hall for ceremony at which cosmonauts received Medal of the Order of Lenin. Reports said gunman had been apprehended. (UPI, W *Star*, 1/22/69, A13; AP, W *Star*, 1/23/69, A1; *NYT*, 1/23/69, 10; Shabad, *NYT*, 1/24/69, 1; Shub, *W Post*, 1/24/69, A1)

- State Dept. announced U.S.S.R. had accepted U.S. invitation to participate in international conference on communications satellites scheduled Feb. 24 in Washington, D.C. U.S. had notified all U.N. members it would extend "observer" invitations to any nation having "serious interest" in possibility of becoming INTELSAT member. Bulgaria and Yugoslavia also would attend. At least 80 nations were expected to participate. (Finney, *NYT*, 1/23/69, 1; AP, *W Post*, 1/24/69, A5)
- Sen. Clinton P. Anderson (D-N. Mex.), Chairman of Senate Aeronautical and Space Sciences Committee, introduced S. 539, FY 1970 NASA authorization bill totaling $3.760 billion. (*CR*, 1/22/69, S659–60)
- Saturn V 2nd stage (S–II–7) was successfully captive fired for full flight duration, 369 secs, by North American Rockwell Corp. personnel at Mississippi Test Facility. Stage developed thrust equivalent to 1 million lbs at operating altitude. (MSFC Release 69–25)
- AEC announced it had completed and successfully tested world's largest superconducting magnet at Argonne National Laboratory near Chicago. Consisting of 110-ton circular-coil assembly in 1,600-ton steel yoke, magnet formed part of world's largest bubble chamber facility for high-energy physics research. Chamber, holding 6,400 gals of liquid hydrogen, would be placed inside magnet, which was expected to operate at approximately 1/10 cost of equivalent conventional magnet. (AEC Release M–19)
- *Washington Post* editorial said: "The fact that the Russians may be able to complete a floating [space] station substantially before the United States is ready to attempt it should be of no great concern. Although the psychological impact of knowing that men are up there looking down on us constantly is bound to be great, this should be more than offset by the successes of the Apollo program. The important things are for the American space effort to proceed in a logical fashion designed to reap the largest possible scientific benefits and to remain largely under civilian control." (*W Post*, 1/22/69, A26)

January 23: U.S.S.R. launched *Cosmos CCLXIV* into orbit with 295-km (183.3-mi) apogee, 208-km (129.3-mi) perigee, 89.5-min period, and 69.9° inclination. Satellite reentered Feb. 5. (GSFC *SSR*, 1/31/69; 2/15/69)

- Nike-Cajun sounding rocket launched by NASA from Kiruna, Sweden, carried Swedish Space Research Committee (SSRC) and British Science Research Council (SRC) payload to 72.1-mi (116-km) altitude. Launch, third in series of four [see Jan. 19], was made to obtain data on atmospheric parameters of wind, temperature, pressure, and density during atmospheric warming by detonating grenades and recording their sound arrivals on ground. Rocket and instruments performed satisfactorily; 17 of 25 grenades detonated and were recorded. (NASA Rpt SRL)

- MSFC announced it had signed $2,022,500 supplemental agreement to basic contract with International Business Machines Corp. for assurance and reliability testing on Saturn IB and Saturn V instrument units. MSFC also announced modifications totaling $2,093,760 to contract with Bendix Corp. for 26 ST-124 "stable platforms," related equipment, and support in Saturn programs. (MSFC Releases 69-23, 69-24)
- FAA announced it had awarded $665,241 contract to Pratt & Whitney Div. of United Aircraft Corp. for two-year study to develop compressor/fan noise-prediction methods for design of quieter jet aircraft engines. Contract represented Government's 55% share of total $1,209,530 cost-sharing contract. Pratt & Whitney would fund remaining 45%. (FAA Release 69-11)
- Australia announced it had asked U.S. to use nuclear explosives to blast out harbor on Australia's northwest coast at Cape Keraudren, on Indian Ocean. U.S. State Department officials confirmed AEC had been authorized to begin talks with Australian officials on feasibility of using nuclear explosion to develop port. (Unna, *W Post*, 1/24/69, A1; Reuters, *NYT*, 1/24/69, 10)

January 24: NASA released *Annual Procurement Report FY 1968*. NASA procurements during FY 1968 totaled $4,133 million—11% less than in FY 1967. Approximately 83% of net dollar value was placed directly with business firms, 4% with educational and other nonprofit institutions, 5% with Cal Tech for JPL operation, and 7% with or through other Government agencies. Of latter, 90% resulted in contracts with industry. About 72% of NASA funds placed under JPL contracts resulted in subcontracts or purchases with business firms. Thus about 93% of NASA procurement dollars went to private industry. During FY 1968, 49 states and D.C. participated in NASA prime contract awards of $25,000 and over. They went to 1,299 business firms, 165 universities, and 68 other nonprofit organizations. (Text)
- NASA launched Nike-Cajun sounding rocket from Wallops Station carrying GSFC grenade experiment to collect data on atmospheric parameters. Rocket and instruments functioned satisfactorily. (NASA Proj Off)
- Mstislav V. Keldysh, President of Soviet Academy of Sciences, said during Moscow interview with Soyuz cosmonauts there were "some advantages" to joint space experiments with U.S. "We have no objection in principle," he said, "and the setting of this type of goal has some merit. Even now the two countries participate in a number of international programs." He added, "One would have to think, and choose this kind of joint program carefully. Maybe one of the flights to a planet in the future, or maybe around the earth also could be interesting. It is difficult to say exactly what I would like to see." Keldysh said Soviet scientists had not received special technical data on *Apollo 8* flight from U.S. However, "Certainly the success of such an outstanding flight, even if it does not produce any concrete new data, still gives something to all mankind." (Winters, *B Sun*, 1/25/69, A2; Shabad, *NYT*, 1/25/69, 6)
- MSFC announced it had awarded $173,000, 11-mo contract to Boeing Co. for study of cost-reduction methods in future space vehicle logistics systems, including expendable and reusable systems. Major emphasis of study would be on space station logistics missions in 100- to 300-mi

orbits, with 5,000- to 50,000-lb payloads, and capable of holding 12 passengers and 3,000- to 12,500-lb cargo. (MSFC Release 69-26)

- Sen. James B. Pearson (R-Kan.) introduced S. 608, bill to create National Aviation Planning Commission responsible for planning development of national air travel system and establishment of air transportation policy. Commission would consist of Assistant Secretary of Transportation for Policy Development, FAA Administrator, CAB Chairman, NASA Deputy Associate Administrator for Aeronautics, Assistant Secretary of Housing and Urban Development for Metropolitan Development, and not more than 10 others to be appointed by Secretary of Transportation. (CR, 1/24/69, S869-70)
- In *Science*, Karl D. Kryter, Director of Sensory Sciences Research Center of Stanford Research Institute, Calif., concluded that sonic booms from SST and Anglo-French Concorde—operating during daytime after 1975 at frequencies projected for long-distance supersonic transport of passengers over U.S.—would result in extensive social, political, and legal reactions against such flights at start, during, and after years of exposure to sonic boom from flights. (*Science*, 1/24/69, 359-67)
- *New York Times* commented on Soviet decision to participate in February INTELSAT conference [see Jan. 22] and on Soviet coverage of *Apollo 8* which was "treated more generously in the Soviet press than any earlier American space accomplishment." Editorial said: "These indications of a positive shift in Kremlin thinking seem to enhance hopes that President Nixon will have an opportunity for creative diplomacy and action of the kind he envisaged in his Inaugural Address. Thought might be given, for example, to inviting Moscow to designate a Soviet astronaut to participate in an Apollo flight late this year or early next year. Or Washington might suggest that the United States and the Soviet Union coordinate their programs of planetary exploration with one nation, say, having primary responsibility for studying Venus and the other Mars. With the landing of men on the moon now probably only months away it is certainly not too early for the two nations that have pioneered most actively in space to discuss concrete means for involving the United Nations directly in the future exploration and exploitation of the moon, as well as of the planets when men reach them." (*NYT*, 1/24/69, 46)

January 25: Apollo 9 prime crew—Astronauts James A. McDivitt (commander), David R. Scott (CM pilot), and Russell L. Schweickart (LM pilot)—held press briefing at Grumman Aircraft Engineering Corp.'s Bethpage, N.Y., plant. Describing 10-day mission scheduled to begin Feb. 28 as primarily engineering evaluation of lunar module, McDivitt said: "... we will be giving the ... LM hardware a very close scrutiny. We don't expect to find anything, but our job is to go up there and look for it. Now after we have discovered that the LM is a good vehicle, we have ... to prove the joint operations techniques that we've tried to develop on the ground over the last 3 years. It's one thing to fly one spacecraft in orbit, and have it controlled by the ground, but when you get 2 of them up there, they are trying to look at 2 vehicles simultaneously so that ... you find the ground talking to 2 spacecraft and 2 spacecraft talking back to each other and also to the ground, and it becomes a rather unwieldly communications effort." He added reminder that "only one of these vehicles has the capability to land—safely, I guess I should add. They both have the capability of landing. We only have one set of parachutes and one heat shield." He described LM as "a tissue paper spacecraft," explaining it did not have to reenter earth's atmosphere and there was no atmosphere on moon.

Scott said they now had "a new vehicle . . . a command module LM combination . . . a particularly unique situation, in that . . . we have to do the lunar orbit insertion [in lunar mission] with the two vehicles joined together with a very large mass on the end of the command module, so it's a completely new guidance task" to be checked out.

McDivitt explained separation of LM from CM on rendezvous day, with two vehicles pulling away from each other and performing maneuvers, moving up to 100 mi apart: "The object . . . is to evaluate our systems from a propulsion standpoint, electrical standpoint, the staging sequence, all of the components that we can and still get back safely to the command module."

In response to question on relation of Apollo program to life on earth, McDivitt replied: ". . . if you're not moving forward . . . the rest of the world is and they're going to pass you by. We're gaining something and we're gaining knowledge. . . . We're going to move forward on all fronts, we're not moving forward on just the space front. . . . Any organized system of intelligence moves forward in all directions, and . . . that is what we are doing." (Transcript)

- Nike-Cajun sounding rocket launched by NASA from Kiruna, Sweden, carried Swedish Space Research Committee (SSRC) and British Science Research Council payload to 73.3-mi (118-km) altitude to obtain atmospheric data by detonating grenades and recording their sound arrivals on ground. Flight, last in series of four [see Jan. 23], was successful; 24 of 25 grenades detonated and were recorded. (NASA Rpt SRL)
- Dedication ceremonies were held at site of new earth station for comsats near Cayey, Puerto Rico. (ComSatCorp Release 69–5; ComSatCorp PRO)
- NR–1, world's first nuclear-powered deep submergence research and ocean engineering vehicle, was launched at Groton, Conn. Developed jointly by USN and AEC, 140-ft-long submarine would carry five crew members and two scientists over ocean bottom to study and map ocean floor, temperature, currents, and other oceanographic parameters for military, commercial, and scientific uses. (DOD Release 64–69; UPI, P Inq, 1/26/69)

January 26: NASA launched two sounding rockets from Wallops Station. Aerobee 350 carried MSC experiment to 168.4-mi (271-km) altitude to produce artificial aurora with electron accelerator. An 85-ft-dia aluminum mylar foil, deployed as planned at 60-mi (96.5-km) altitude, acted as current selector for ionospheric electrons and electrically neutralized experiment. Series of 100 beam pulses aimed downward toward Wallops ground station by accelerator were recorded on film by very sensitive TV camera and observations of artificial aurora were observed visually by scientists on ground. Analyses were under way to determine if auroral intensity, location, and shapes were as predicted.

Nike-Tomahawk, launched 148 secs later to study acceleration beam from Aerobee 350, carried DeHavilland antenna and Langmuir probe to 11.8-mi (19-km) altitude. Second stage failed to ignite; no useful data were obtained. (WS Release 69-2; NASA Rpts SRL)

- NASA launched Nike-Cajun sounding rocket from Point Barrow, Alaska, carrying GSFC grenade experiment to collect data on atmospheric parameters. Rocket and instruments functioned satisfactorily. (NASA Proj Off)
- In *New York Times*, Walter Sullivan said data returned by *Oso V* and other OSO satellites had begun to deepen understanding of sun. "It has become sufficient so that, from information gathered in space, as well as by a globe-encircling network of stations, those in charge of the Apollo 8 flight to the moon and back last month were able to ignore a variety of ominous manifestations on the sun during the flight." He said enough radiation measurements had been made during previous "space storms" to indicate Apollo astronauts were reasonably safe as long as they remained inside their spacecraft. (*NYT*, 1/26/69, E6)
- Observers in Wisconsin, Iowa, Michigan, and Illinois reported sighting large meteorite or space debris flashing across sky and burning itself out as it entered earth's atmosphere. Several airline pilots had seen it at O'Hare International Airport, Chicago. Northwestern Univ. astronomer James Wray said it probably was large meteor breaking up in atmosphere. UFO expert Dr. J. Allen Hynek, Chairman of Northwestern's Astronomy Dept., said flash could also have been reentering debris of U.S. or Soviet space rocket. (UPI, *W News*, 1/27/69, 30; AP, *W Star*, 1/27/69, B4)
- In Washington *Evening Star*, William Hines said, "If a successful farmer suddenly started economizing on seed, his neighbors and family would begin to doubt his judgment, if not his sanity. Yet this is precisely what Uncle Sam is doing in cutting back the financing of scientific research. Despite denials from budget officials in the past weeks, federal support of science in fiscal 1970 continues on an alarming down-trend that has been apparent for several years. It has been said that in an advanced economy like ours, research expenditures are 'seed money' and the analogy is apt. Like individual seeds in a field, not every research dollar germinates, and not all those that do mature. But total return is vastly greater than outlay—and that is the story of science as well as of agriculture." (W *Star*, 1/26/69, C4)

January 27: Boosted Arcas II sounding rocket launched by NASA from Wallops Station carried GSFC experiment to 63.5-mi (102-km) altitude to evaluate rocket performance for possible use at Resolute Bay, Canada. Vehicle underperformed according to predicted trajectory; peak altitude was below predicted and tone ranging appeared too weak for Resolute Bay. (NASA Rpt SRL)

- AEC announced that S8DR nuclear reactor system developed in its SNAP (Systems for Nuclear Auxiliary Power) Program was producing 600 thermal kw at 1,300°F during tests in underground vacuum chamber at Santa Susana, Calif. Electrical power ranging from 20 to 75 kw could be generated by such a reactor to provide power for manned orbiting laboratories and bases on moon's surface. It was being considered for these uses because of its potentially high reliability, small size, and long life (two to five years) without refueling or maintenance. (AEC Release M-22)

January 28: In New York news conference, NASA Associate Administrator for Manned Space Flight, Dr. George E. Mueller, said U.S. was in danger of going "out of the manned spaceflight business" unless more funds were provided in Federal budget for projects beyond Apollo. He said U.S.S.R. probably would surpass U.S. in space exploration in 1970s. Current Soviet space expenditure was "about 50 percent greater than ours." (Wilford, *NYT*, 1/29/69, 11; *NY News*, 1/30/69)

- In speech before New York Society of Security Analysts, NASA Associate Administrator for Manned Space Flight, Dr. George E. Mueller, said: "I believe that if we wanted to we could have our space shuttle in operation by 1976.... To achieve the desired economy, it will be necessary to operate this transportation system in the successful jet transport mode. Our space shuttle will probably take off from major airports with little or no noise. It will not create a sonic boom along the route. It will go into orbit, deposit and take on crew and cargo, and return for a horizontal airport landing." He foresaw an international demand for reusable space vehicles. (Text)

- Nike-Tomahawk sounding rocket launched by NASA from Wallops Station carried Univ. of Wisconsin experiment to 139.2-mi (224-km) altitude to examine isotropic component of cosmic x-rays in wavelength region of $\lambda > 5$ Å, using collimated, thin-window gas proportional counters. Peak altitude was 3.5% over predicted but x-ray counters failed to reach design pressure. (NASA Rpt SRL)

- M2–F2 lifting-body vehicle damaged in accident May 10, 1967, would be repaired, modified, and returned to service as M2–F3, NASA announced. Modifications would include center-stabilizing fin, special equipment for use as test bed for lateral control systems research, jet reaction roll control system, and improved internal components for precise maneuvering by pilot. M2–F3 would rejoin HL–10 and X–24 in NASA–USAF flight research program to evaluate wingless vehicles for manned horizontal landings at airfields after return from space. (NASA Release 69–15)

- Eastern Airlines, Inc., Vice President A. Scott Crossfield told Aero Club in Washington, D.C., that seven-week experiment with STOL aircraft at Boston, Washington, and New York airports begun Sept. 1968 had been "unqualified success." Airline's engineers were drafting specifications for STOL aircraft to carry 125 passengers at 250 mph, capable of maneuvering at speeds of 70 mph. It could double landing capacity of airports by using taxiways and ends of unused runways. McDonnell Douglas version of French-designed Breguet 941 aircraft used by Eastern in experiments had used onboard computer-controlled system, "heart" of which was manufactured by Decca in England. It used existing navaids and was accurate within 25-ft altitude and 100-ft latitude. (Koprowski, *W Post*, 1/29/69, A8)

- S. 705 was introduced in Senate by Smithsonian Institution regent Sen. Clinton P. Anderson (D-N. Mex.), for himself and regents Sens. J. William Fulbright (D-Ark.) and Hugh D. Scott (R-Pa.) to authorize $2 million for planning and land acquisition for world's largest radio-radar astronomical telescope. Proposed 440-ft "big dish" antenna would be enclosed in 550-ft geodesic dome and cost about $37 million. It would be made available to appropriate scientists everywhere. (*CR*, 1/28/69, S967–8; *W Star*, 4/1/69, B1)

- Following conference with President Nixon, Dr. Glenn T. Seaborg, AEC Chairman since 1961, announced he would continue in that position in Nixon Administration. Current term would expire June 30, 1970. (W Star, 1/29/69, A2)

January 29: ComSatCorp announced it had applied to FCC for permission to reduce rates for TV transmissions through Atlantic satellites by about 40% and to eliminate extra charge for color TV through these satellites. Reductions were possible because of availability of *Intelsat III F-2* (launched Dec. 19, 1968) over Atlantic. (ComSatCorp Release)

- NASA selected Electro Mechanical Research Aerospace Sciences Div. of Weston Instruments, Inc. for contract negotiations on $1.37-million cost-plus-award-fee contract for spacecraft integration and ground support services for Interplanetary Monitoring Platform (IMP) missions H, I, and J. (NASA Release 69-19)
- President Nixon announced selection of Gerard C. Smith, former Assistant Secretary of State for Policy Planning, as Director of Arms Control and Disarmament Agency. Adrian S. Fisher had been acting director since resignation of William C. Foster. (PD, 2/3/69, 188)

January 30: Canadian *Isis I* (ISIS-A) International Satellite for Ionospheric Studies was successfully launched by NASA from WTR by three-stage Thrust-Augmented Improved Thor-Delta (DSV-3E) booster. Satellite entered orbit with 2,188.5-mi (3,522-km) apogee, 356.7-mi (574-km) perigee, 128.3-min period, and 88.4° inclination. Primary NASA objectives were to place *Isis I* into elliptical earth orbit that would permit study of topside of ionosphere above electron peak of F region and to extend cooperative Canadian-U.S. program of ionospheric studies initiated by *Alouette I* (launched Sept. 28, 1962) by combining sounder data with correlative direct measurements for time sufficient to cover latitudinal and diurnal variations during high solar activity.

Third in series of five satellites to improve understanding of ionospheric physics, *Isis I* weighed 520 lbs and carried six Canadian and four American experiments. First launch in series (ISIS X project, Nov. 28, 1965) orbited Canada's *Alouette II* and U.S. *Explorer XXXI.* ISIS program was joint undertaking of NASA and Canadian Defence Research Board (DRB) under December 1963 Memorandum of Understanding. DRB was responsible for spacecraft design, fabrication, electrical testing, experiment integration, and satellite control. NASA provided launch vehicles, launch facilities. (NASA Proj Off; NASA Releases 69-14, 69-22)

- NASA launched two Nike-Cajun sounding rockets from Point Barrow, Alaska, carrying GSFC experiments. First rocket was launched to obtain data on variation of temperature, pressure, and wind profile by detonating grenades at prescribed times and recording sound arrivals on ground. All 19 grenades ejected and detonated and sound arrivals were recorded. Launch was third in series of four launches from Point Barrow during stratospheric warming [see Jan. 26].

Second rocket was launched in conjunction with Jan. 11 launch to develop experimental techniques for determining atmospheric composition profiles in mesosphere and to measure ozone and water vapor distribution in 12.4- to 40.4-mi (20- to 65-km) region by separating

January 30: *Canadian Isis I International Satellite for Ionospheric Studies, launched by* NASA *from* WTR, *carried six Canadian and four U.S. experiments into orbit to study the topside of the ionosphere during a period of high solar activity. Isis I, photographed before mating to its Thor-Delta launch vehicle, combined the capability for direct and indirect ionospheric measurements in one spacecraft for the first time.*

payload from 2nd stage and deploying parachute near apogee. Ozone was measured by chemiluminescent technique and water vapor by aluminum-oxide hydrometer. All major events occurred as planned and good data were obtained. (NASA Rpts SRL)

- Nike-Cajun sounding rocket launched by NASA from Churchill Research Range carried GSFC payload to 77.1-mi (124-km) altitude to obtain data on variation of temperature, pressure, and wind profile by detonating grenades at prescribed times and recording sound arrivals on ground. All 19 grenades were detonated and sound arrivals were recorded. Launch was third in series of four rockets to be launched from Churchill during stratospheric warming period [see Jan. 22]. (NASA Rpt SRL)

- Aerobee 150 MI sounding rocket was launched by NASA from WSMR with VAM-20 booster to 111.2-mi (179-km) altitude. Objectives were to obtain stellar spectra with 1 Å resolution in 1,000–1,600 Å far UV wavelength range and to obtain photometric data on stellar fluxes in 1,050–1,180 Å, 1,230–1,350 Å, and 1,350–1,470 Å wavelength ranges. All experimental objectives were achieved and payload was recovered promptly. (NASA Rpt SRL)

- FAA Government-industry conference in Washington, D.C., discussed FAA role in STOL development, STOL noise sources, STOL operational considerations related to noise abatement, noise source reduction techniques, and review of existing aircraft certification concepts and considerations for STOL noise certification. (FAA Release 69–5)
- At White House press briefing President Nixon introduced *Apollo 8* astronauts and announced that Astronaut Frank Borman would make eight-nation goodwill trip to Western Europe. Tour would point out "what is the fact: that we in America do not consider that this is a monopoly, these great new discoveries that we are making; that we recognize the great contributions that others have made and will make in the future; and that we do want to work together with all peoples on this earth in the high adventure of exploring the new areas of space." Later, Press Secretary Ronald Ziegler announced Borman family would visit England, France, Belgium, the Netherlands, West Germany, Italy, Spain, and Portugal. (*PD*, 2/3/69, 189–90; AP, *W Post*, 1/31/69, A2)
- President and Mrs. Nixon watched *Apollo 8* films at White House showing attended by Astronauts Frank Borman, James A. Lovell, Jr., and William A. Anders. Borman narrated. (Shelton, W *Star*, 1/31/69, B1; *PD*, 2/3/69, 194)
- Dr. Thomas O. Paine, NASA Acting Administrator, presented Public Service Group Achievement Award Certificates to 12 representatives of communications organizations which had supported *Apollo 8* mission. Award to *Apollo 8* Communications Network cited "the dedication and skill of the leaders and all personnel in these organizations in maintaining reliable communications which insured the success of the first manned lunar orbit mission." British External Telecommunication Executive and Hawaiian Telephone Co. had received same award. (NASA Release 69–20)
- President Nixon issued directives to cabinet officers and agency heads. He directed DOT to establish committee to investigate all aspects of SST program and some aspects of airport development, air traffic control, and FAA regulations.

  Bureau of Budget was informed that President was disturbed by reports that Government was not fulfilling obligations to colleges and universities whose grants had been abrogated by NSF because of FY 1969 expenditure ceiling. BOB was directed to check facts, estimate cost required to eliminate inequities, and advise President on need for contingency reserve for FY 1969 and desirability of thorough budget revision for FY 1970. (*PD*, 2/3/69, 192–3)
- NASA awarded United Aircraft Corp.'s Pratt & Whitney Div. and General Electric Co. separate fixed-price contracts to design, fabricate, and test experimental quiet jet engines. Each contract would have two phases. Six-month first phase would include detailed engine design and procurement of selected engine components. NASA would have option of authorizing construction of two engines and test program of at least

250 hrs of engine operation. In second phase, expected to take 30 mos, each contractor would refurbish and deliver one engine to LeRC for additional testing. Program was expected to cost $50 million over three-year period. (NASA Release 69–21)

- Dr. George E. Mueller, NASA Associate Administrator for Manned Space Flight, told meeting of National Security Industrial Assn. at KSC, "It is very clear that just as we have had substantial payoffs from communications, navigation, geodetic, and weather satellites in the first decade of the space age, earth resources satellites will represent extremely promising investment opportunities in the second decade." They would require "close cooperation among many agencies in Washington, and in the long run with new commercial and international institutions that can bring the benefits of the space age to many people around the world." (Text)

- At his first news conference, Secretary of Defense Melvin R. Laird said he would prefer to deal from strength—including Sentinel ABM system—in future missile talks with U.S.S.R. "I think it's most important, as we go into these talks, to have defensive as well as offensive missile systems up for discussion and debate and negotiation." (Maffre, *W Post*, 1/31/69, A1)

- Moscow sources reported "Lt. Ilyin" of U.S.S.R. Army Engineers had been identified as attempted assassin who fired on Moscow motorcade carrying cosmonauts and high Soviet officials into Kremlin Jan. 22, Anatole Shub said in *Washington Post*. He reportedly had died almost immediately after taking poison and being slugged by Kremlin guards at scene [see Feb. 4]. (*W Post*, 1/31/69, A1)

January 31: Eleventh anniversary of *Explorer I*, first U.S. satellite. Since its launch Jan. 31, 1958, the 30.8-lb stovepipe-shaped satellite had completed nearly 60,000 revolutions around earth and on Dec. 31, 1968, was in orbit with 632-mi (10,170.8-km) apogee, 199.9-mi (321.7-km) perigee, and 98.1-min period. (MSFC Release 69–26)

- Apollo 9 press briefing was held at NASA Hq. Countdown would begin Feb. 22, for launch from ETR at 11 am EST Feb. 28. Ten-day earth-orbital mission would demonstrate LM manned crew performance for first time and carry out intervehicular activities between spacecraft, through-docking-tunnel activities, and EVA. Number of small aluminum-alloy brackets and fittings had been replaced or reinforced in LM–3 and LM–4 because they were sensitive to stress or corrosion. Both vehicles were ready for flight.

    Apollo Program Director George H. Hage said number of activities would be performed on Apollo 9 that had not been done before "in the sense of wringing out the spacecraft." NASA was "working the hardware launch readiness of Apollo 10 to a late April date" so that "if we have difficulty on Apollo 9 and need to repeat some element or all of the D mission, we can get that mission off as early as possible." If Apollo 9 was successful Apollo 10 could be launched as early as May 17 on lunar landing mission. (Transcript)

- NASA successfully launched two Nike-Cajun, one Arcas, and three Nike-Apache sounding rockets from Wallops Station, carrying experiments to measure meteorological, ionospheric, and composition characteristics of upper atmosphere during "winter anomaly"—unusual absorption of radio waves—which occurred during January or early February.

Launches would contribute to specimen day program, coordinated investigation of winter variability of D region of ionosphere above Wallops.

Nike-Cajuns carried GSFC payloads to 75.2-mi (121-km) and 12.4-mi (20-km) altitudes. All 19 grenades on each rocket detonated as planned and sound arrivals were recorded, but poor vehicle performance of second rocket prevented acquisition of useful data.

Arcas carried Naval Weapons Center payload to 33.2-mi (53.4-km) altitude to measure ozone concentration in 18.6- to 37.3-mi (30- to 60-km) region during parachute descent, but parachute did not deploy satisfactorily and payload descended too rapidly for recovery.

First Nike-Apache carried Univ. of Illinois-GCA Corp. payload to 141.1-mi (227-km) altitude to investigate winter variability of D region of ionosphere and measure differential absorption, Faraday rotation, and probe current to determine electron density, collision frequency, and temperature. Second Nike-Apache carried Univ. of Colorado payload to 71.5-mi (115-km) altitude to obtain vertical profile of nitric oxide density, using scanning monochromoter. Third Nike-Apache carried GSFC payload to 72.1-mi (116-km) altitude to measure degree of polarization and intensity of nitric oxide emission at 2,147 Å to determine whether resonance scattering of sunlight was responsible for emission in D and E regions. Secondary objective was to determine altitude profile in 46.6- to 65.2-mi (75- to 105-km) region. Desired spectrum was not observed, apparently because of mechanical failure in payload. (NASA Rpts SRL; WS Release 69–3)

- President Nixon, accompanied by Secretary of Defense Melvin R. Laird, visited DOD employees at Pentagon. President recalled reference made by Astronaut Frank Borman at White House *Apollo 8* briefing Jan. 30 to "400,000 men and women in the Nation who at one time or another had played a part in making this great, spectacular feat possible." President said, "I was glad to see Colonel Borman bring it home that way. Four hundred thousand made it possible for this magnificent achievement to occur. I trust that all of you can convey that kind of spirit to those who work in the Defense Department." (*PD*, 2/3/69, 194)

- Sen. Kenneth McC. Anderson, Australian Minister for Supply, accepted NASA Group Achievement Award for Dept. of Supply at NASA Hq. luncheon for "outstanding contributions in the establishment and operation of the stations and associated facilities in Australia which assured the success of the Apollo 8 mission. . . ." Sen. Anderson also received awards for *Apollo 8* support by MSFC stations at Canberra and Carnarvon and DSN station at Canberra. (NASA Release 69–23)

- FAA said preliminary figures showed its 27 air route traffic control centers handled 19.5 million aircraft in 1968, an increase of 17% over 1967. Chicago, New York, and Cleveland each logged 1.5 million operations—first time any center had reached this mark. (FAA Release 69–15)

During January: JPL Senior Staff Scientist Albert R. Hibbs summarized results of NASA's Surveyor program in *Astronautics and Aeronautics*. Experiments on five Surveyor spacecraft which successfully landed on moon between May 30, 1966, and Jan. 7, 1968, indicated surface material was granular and very fine with 10^3 dynes/cm^2 cohesion. Slightly deeper material had lower normal albedo than undisturbed surface. Chemical composition of surface material was similar to basaltic rocks on earth; mare material contained more iron elements than highland material; and highland material had higher albedo than mare material. Chemical analyses indicated material did not resemble chondritic meteorites. From observed data scientists concluded that moon had undergone significant chemical differentiation during its history and had been subjected to basaltic lava flows; surface was continually being "churned and pulverized" by meteoroid impacts; some undefined process lightened optical surface and darkened buried material; and mare areas were "surprisingly similar and offer numerous safe-landing zones for future lunar missions." (*A&A*, 1/69, 50–63)

- U.S.S.R. was testing 150-ton, 250-passenger "compound" helicopter, American Broadcasting Co. reported. Largest helicopter in Western world was 19-ton Sikorsky CH–54H Flying Crane. Soviet 47-ton Mi-10 was world's largest. New compound helicopter had wings that assumed lifting function from rotors at cruising speed; it obtained most of its thrust from conventional propulsion when it converted from vertical to cruising flight. Sikorsky had proposed 32.5-ton compound helicopter to DOD and U.S. civilian transportation authorities. (*NYT*, 1/12/69, S23)

- MIT scientist Dr. Jerome B. Wiesner in *Technology Review* said reorganization and strengthening of Federal mechanisms for planning and supporting R&D was only solution to "present antagonisms and . . . skepticism" about the value of a continued high level of R&D support. He proposed new agency with NSF at core for planning R&D and to "indicate resource allocation for all public endeavors, including foreign aid and national security." (*Technology Review*, 1/69, 15–17)

- Systems approach was needed in applying "human and technological resources to domestic problems," *Space/Aeronautics* said. Growth areas for aerospace industry spinoff included urban, environmental, surface-transportation, medical, and ocean systems. Lessons to be learned in dealing with these systems were: (1) massive problems required efforts on massive scale; (2) R&D cycle for civil system was always longer than political cycle being counted on to support it; and (3) even when system was built jurisdictional prerogatives could "make a mess of the implementation." Aerospace companies should employ their capabilities "to assess their experience in high technology and their managerial skills" and apply experience "to new systems challenges." (*S/A*, 1/69, 106–7)

February 1969

February 2: Development of laser tracking techniques permitting accurate 24-hour tracking of orbiting spacecraft was announced by NASA. New technique—particularly important in geodetic studies, which required precise angle and distance measurements between satellite and ground stations—offered greater measuring accuracy than RF methods, required only lightweight reflectors on satellite, and was less affected by transmission-impeding environmental disturbances. First operational daylight tracking with laser had been accomplished by GSFC team Oct. 21, 1968. (NASA Release 69–18; *A&A 68*)

- *Apollo 8* Astronaut Frank Borman and family departed on USAF jet for 18-day Presidential goodwill mission to Europe. Itinerary: London, Feb. 2–5; Paris, Feb. 5–7; Brussels, Feb. 7–10; The Hague, Feb. 10–11; Bonn, Feb. 11–12; West Berlin, Feb. 12–13; Rome, Feb. 13–17; Madrid, Feb. 17–19; and Lisbon, Feb. 19–21. (NASA Int Aff; W *Star*, 2/2/69, A3)
- NASA's *Apollo 8* mission and USN navigation satellite system developed by Johns Hopkins Univ. Applied Physics Laboratory had been named two of top four engineering achievements of 1968 by National Society of Professional Engineers. (W *Star*, 2/2/69, B2)
- In *New York Times Magazine*, Dr. Ralph E. Lapp, physicist, wrote: ". . . I would urge that we alter the U.S. space program as follows: First, make a firm decision to terminate the manned space program soon after the initial lunar landing. . . . I would reserve the remaining Apollo craft for future unmanned missions to the planets and I would mothball the single-purpose manned space flight facilities. At the same time, I would continue a N.A.S.A. program of long-range space development aimed at advanced modes of propulsion, compact energy sources and improved long-distance communication. High priority would be assigned to the development of nuclear energy both for propulsion and for on-board power.

    "Second, greatly expand N.A.S.A.'s present program for exploiting applications of space science and technology. The potential of satellites for communications . . . needs to be enhanced by the development of new techniques. It should not be too difficult to develop orbital systems for the control of intercontinental air traffic. Perhaps the greatest benefits from satellites are to be expected in the survey and evaluation of earth resources, such as underground water, mineral deposits and plant-forest cover. . . .

    "Third, establish a high priority within N.A.S.A. for fundamental research using unmanned space vehicles . . . 10 to 20 years in duration and . . . aimed at finding out more about our planet, the sun and the rest of the solar system. The most expensive—and probably the most dramatic—of these projects would be the planetary probes designed to fly by, orbit or land on the nearby planets." (*NYT*, 2/2/69, 32–40)

February 3: NASA announced it had extended $69,692,000 contract with Bellcomm, Inc., to provide systems analysis, study, planning, and technical support of manned space flight. Value of one-year cost-plus-fixed-fee contract extension was $11,483,000. (NASA Release 69-25)

February 3-5: London accorded *Apollo 8* Astronaut Frank Borman and family full celebrity status, including frontpage newspaper coverage, taped TV interviews, and cheers from schoolchildren. Borman lectured before Royal Society of Scientists Feb. 3 and on Feb. 4 was presented to Queen Elizabeth II at Buckingham Palace and visited Prime Minister Harold Wilson and House of Commons. Borman at U.S. Embassy presented NASA's Manned Spaceflight Group Achievement Award to Station Manager James McDowell of NASA Communications Switching Station in London and Public Service Group Achievement Award to C. James Gill, director of U.K.'s postoffice telecommunications system. (Lee, *NYT*, 2/4/69, 4; *W Post*, 2/5/69, A18; *C Trib*, 2/5/69; NASA Int Aff)

February 4: President Nixon sent directive to Dr. Lee A. DuBridge, Science Adviser to the President, asking assessment of proposal to appoint interagency committee to advise President on post-Apollo space program. Directive also asked report on "possibility of significant cost reductions in the launching and boosting operations of the space program," with judgment on "how best to assess future developments in this area." White House announcement from Key Biscayne, Fla., Feb. 8 said directive had asked assessment of recommendations that DOD and NASA be directed to coordinate activities in this area. (*PD*, 2/17/69, 249; 3/10/69, 349-51)

- Aerobee 150 MI sounding rocket launched by NASA from Churchill Research Range carried Univ. of Minnesota Institute of Technology payload to 115.6-mi (186-km) altitude to study neutral composition of polar atmosphere with neutral mass spectrometers. Rocket and instruments performed satisfactorily and experimental data showed "some extremely interesting results." (NASA Rpt SRL)

- Nike-Cajun sounding rocket was launched by NASA from Point Barrow, Alaska, carrying GSFC experiment to obtain data on variation of temperature, pressure, and wind profile by detonating grenades at prescribed times and recording sound arrivals on ground. Rocket, last in series of four launched during period of atmospheric warming [see Jan. 30], performed satisfactorily. All 19 grenades ejected and detonated as planned and sound arrivals were recorded. (NASA Rpt SRL)

- President Nixon accepted pro forma resignation of Dr. Edward C. Welsh, Executive Secretary of National Aeronautics and Space Council. Appointed by President Kennedy in 1961, Dr. Welsh had been Council's first and only appointed executive secretary. (*W Post*, 2/5/69, A7; AP, *W Star*, 2/5/69, 1)

- XB-70 supersonic research aircraft was flown from Edwards AFB, Calif., to Wright-Patterson AFB, Ohio, to be placed on exhibit at Air Force Museum. Flight had been delayed until turbulent air conditions prevailed so testing could continue until end of aircraft's service. During final flight, crew collected data on aircraft handling and structural response to air turbulence at subsonic flight. NASA had announced end of XB-70 flight research program Jan. 13. (NASA Proj Off; UPI, *NYT*, 2/5/69, 73; AP, *W Star*, 2/5/69, A19)

- In *Look*, science writer Arthur C. Clarke, Nobel Prize nuclear physicist Dr. Isidor I. Rabi, novelist C. P. Snow, and Catholic theologian Prof. Leslie Dewart wrote personal reactions to *Apollo 8* mission. Clarke said: "The Apollo 8 mission marks one of those rare turning points in human history after which nothing will ever be the same again. The immense technical achievement is already obvious to every one and has been universally praised; yet the psychological impact may be even more important and will take some time to make itself fully felt. We no longer live in the world which existed before Christmas 1968. It has passed away as irrevocably as the earth-centered universe of the Middle Ages."

  Dr. Rabi said: "It would be misleading to talk of the events that led to the journey of Apollo 8 in terms of the vast sums of money that are involved, even though it cost several times as much as the development of the first atomic bomb. What is more important and more impressive is that Apollo 8 represents the cooperation of hundreds of thousands of people over a period of years in a gigantic effort with no clearly set practical goals, except perhaps the profound desire of mankind to prove to itself that it had the knowledge and the ability to overcome its earthbound limitations."

  Prof. Dewart said: "Man has taken his first, halting steps into the cosmos beyond that earthly world in which he was born and within which he had always lived. The impact of Apollo 8 in other areas of human experience is obvious; in religion, it is much less immediately evident. And yet, in the end, it may be more significant for the development of man's religious consciousness." (*Look*, 2/4/69, 72-8)

- In letter to Astronaut Frank Borman, Board of Education of Glendale Union High School District No. 205, Glendale, Ariz., said it had named planned high school "Apollo" in "honor and appreciation of the accomplishments of the participating astronauts." It invited *Apollo 8* crew to participate in 1970 dedication ceremony. (*CR*, 2/21/69, E1216)

- USN announced award of $40,000,000 contract to Grumman Aircraft Engineering Corp. for engineering development phase of F-14A supersonic fighter (formerly VFX), replacement for F-111B. Funding during four-year development was expected to total $388,000,000. (DOD Release 92-69)

- In letter from Chairman L. Mendel Rivers (D-S.C.) to Secretary of Defense Melvin R. Laird, House Armed Services Committee informed DOD that, because of uncertainty over ABM, Committee would take no action to approve Sentinel antiballistic missile sites until Nixon Administration positively expressed interest in project. (Sell, *W Post*, 2/6/69, A1)

- In *New York Times*, Theodere Shabad said Moscow sources indicated Soviet investigators had ruled out possibility of political conspiracy in Jan. 22 shooting during Kremlin ceremonies for Soyuz cosmonauts because of amateurish behavior of gunman identified as "Lt. Ilyin" of Soviet Army. Sources denied earlier reports that gunman had taken poison after shooting and was dead. They said he was undergoing medical and psychological testing to determine his sanity and motives. (*NYT*, 2/5/69, 2)

February 5-16: *Intelsat-III F-3* was successfully launched by NASA for

ComSatCorp on behalf of International Telecommunications Satellite Consortium. The 632-lb cylindrical satellite, launched from ETR by Long-Tank Thrust-Augmented Thor (LTTAT)-Delta (DSV-3M) booster, entered elliptical transfer orbit with 23,496.9-mi (37,-814.6-km) apogee, 157.3-mi (253.1-km) perigee, 671.9-min period, and 29.8° inclination. All systems were functioning normally. On Feb. 7 apogee motor was fired to kick satellite into planned near-synchronous orbit over Pacific at 173.8° east longitude with 22,000-mi (35,719.8-km) apogee, 22,190-mi (35,703.7-km) perigee, 23-hr 56-min period, and 1.3° inclination.

Intelsat-III F-3 was second successful launch in Intelsat III series. *Intelsat-III F-2* had been launched Dec. 18, 1968, as backup to Intelsat-III F-1, which had been destroyed minutes after launch Sept. 18, 1968. Satellite began commercial service Feb. 16, handling up to 1,200 voice circuits or 4 TV channels. (NASA Proj Off; ComSatCorp Releases 69-7, 69-27; AP, B *Sun*, 2/6/69, A3; *Pres Rpt. 70* [69]; ComSatCorp PIO)

- *February 5:* DOD launched two unidentified satellites from Vandenberg AFB by Thor-Agena booster. One entered orbit with 171.0-mi (275-km) apogee, 91.7-mi (147.6-km) perigee, 88.7-min period, and 81.6° inclination and reentered Feb. 24. Second satellite entered orbit with 894.9-mi (1,439.9-km) apogee, 866.4-mi (1,394.0-km) perigee, 114.1-min period, and 80.4° inclination. (*Pres Rpt 70* [69])
- President Nixon authorized immediate $10-million increase in expenditure ceiling placed on National Science Foundation by Johnson Administration in 1968. He said: "The colleges and universities of this Nation provide a critical resource which needs to be fostered and strengthened. Our higher educational system provides the advanced training needed for tomorrow's leaders in science and technology, industry and government, and also conducts the basic research which uncovers the new knowledge so essential to the future welfare of the country. It is essential that these programs of education and research be sustained at a level of high excellence." (*PD*, 2/10/69, 224-5)
- New tempest was brewing in national scientific community over whether defense establishment absorbed exorbitant portion of U.S. scientific and technological energies, John Lannan said in Washington *Evening Star*. In New York, younger physicists had called for political activism at annual meeting of American Physical Society Feb. 3. MIT group, Union of Concerned Scientists, had scheduled day-long "research stoppage" March 4 and initiated letter campaign to spread its views to other institutions. Union's proposals included "a critical and continuing examination of government policy in areas where science and technology are of actual or potential significance"; redirection of research from defense-oriented to environment-oriented projects; opposition to antiballistic missile system; and organization of scientists into effective and vocal political action group. (W *Star*, 2/5/69, A7; Sullivan, *NYT*, 2/9/69, E7)
- Report on aviation safety for 1968 was submitted to House Committee on Interstate and Foreign Commerce by Joseph J. O'Connell, Jr., Chairman of National Transportation Safety Board. For all scheduled air carrier services there had been one fatal accident for about every 500,000 hrs, or for every 100,000 transcontinental flights. One pas-

senger had been lost for every 370 million passenger miles flown. Number of fatalities in scheduled domestic and international passenger service had been second worst of decade; but accident rates, fatal and nonfatal, continued downward for total scheduled air carriers. In general aviation, rate for fatal accidents per hours flown had increased but remained below rates of 1965 and before and was third best in decade. Total number of fatal accidents—692, killing 1,374 persons—was highest in history. (Text; NYT, 2/9/69, 94)

- In message to Senate, President Nixon urged prompt ratification of nuclear nonproliferation treaty: "I believe that ratification of the Treaty at this time would advance this Administration's policy of negotiation rather than confrontation with the USSR. I believe the Treaty can be an important step in our endeavor to curb the spread of nuclear weapons and that it advances the purposes of our Atoms for Peace program." (PD, 2/10/69, 219)

February 5–7: During two-day Paris visit *Apollo 8* Astronaut Frank Borman met with President Charles de Gaulle. At dinner given by Ambassador R. Sargent Shriver, Jr., on Eiffel Tower, Borman received offer of racing car from French manufacturer who had presented similar gift to Cosmonaut Yuri Gagarin in 1965. During Paris news conferences and on TV interview, Borman stressed international character of space exploration. He said, "I don't know why we aren't going to Russia. I would like to visit Russia. . . . I think we have some fair means of cooperation in space and I would hope to see more." (Garrison, NYT, 2/6/69, 2; 2/7/69, 3)

February 6: NASA launched four sounding rockets from Wallops Station to obtain upper-atmosphere data on normal winter day. Arcas carried Naval Weapons Center payload to 34.9-mi (56.1-km) altitude to measure ozone concentration at altitudes between 18.6 and 37.3 mi (30 and 60 km), using photometer and optical filter wheel. Failure of recovery parachute to open satisfactorily caused fast descent and prevented payload recovery. Obtaining useful information was expected to be difficult, but good data were expected.

Nike-Cajun carried GSFC payload to 79.4-mi (127.7-km) altitude to obtain temperature, pressure, density, and wind data in upper atmosphere by detonating grenades and recording sound arrivals on ground. All 19 grenades were ejected and detonated as planned and sound arrivals were recorded.

Nike-Apache carrying Univ. of Colorado payload reached 72.7-mi (117-km) altitude on flight to obtain vertical profile of nitric oxide density in 15.5- to 65.3-mi (25- to 105-km) region. Rocket and instruments functioned satisfactorily.

Second Nike-Apache carried Univ. of Illinois payload to collect data on ionosphere. Rocket and instruments functioned satisfactorily. (NASA Rpts SRL; NASA Proj Off)

- Nike-Cajun sounding rocket launched by NASA from Churchill Research Range carried GSFC payload to 80.2-mi (129-km) altitude to obtain data on variation of temperature, pressure, and wind profile by detonating grenades at prescribed times and recording their sound arrivals on ground. Rocket, last in series of four launched from Churchill during stratospheric warming [see Jan. 30], performed satisfactorily. All

19 grenades ejected and detonated as planned. Sound arrivals were recorded for 14–15 grenades because of power failure at receiving station. (NASA Rpt SRL)

- Aerobee 150 MI sounding rocket launched by NASA from Churchill Research Range carried Univ. of Minnesota Institute of Technology payload to 83.2-mi (133.8-km) altitude to study neutral composition of polar atmosphere with neutral mass spectrometers. Rocket underperformed; burnout occurred at 42 secs. All instruments worked perfectly. Useful data were obtained in 68.4- to 87.0-mi (110- to 140-km) region. (NASA Rpt SRL)
- At confirmation hearing on his appointment as Director of Office of Science and Technology before Senate Labor and Public Welfare Committee, Dr. Lee A. DuBridge said he would place his energies on analysis of weapon systems, environment and effect of technology and pollution on environment, and utilization of science and technology by Government departments. He planned to concern himself with social problems and hoped to increase social scientists on President's Science Advisory Committee from one to two. He hoped for increased funding for HUD and DOT, and regretted allocations for basic research were declining in DOD, NASA, and AEC because such agencies "will profit by good relations with universities." (Nelson, *Science*, 2/14/69, 657)
- U.S.S.R.'s *Venus V* (launched Jan. 5) and *Venus VI* (launched Jan. 10) were on course and functioning normally, Tass announced. Spacecraft were expected to reach Venus in late May. *Venus V* was 4,785,000 mi from earth; *Venus VI*, 4,050,000 mi. (Reuters, *NYT*, 2/7/69, 14)
- NASA sponsored one-day meeting to review progress in its five-year research program on fog-shrouded airports. In one test during NASA-sponsored work by Cornell Aeronautical Laboratory in Project Fog Drops, small aircraft carrying 700 lbs of salt had opened wide path in dense warm fog in five minutes. (NASA Release 69–17; Transcript)
- Sperry Rand Corp. announced election of former NASA Administrator James E. Webb to Board of Directors. He had been vice president of company's Sperry Gyroscope Div. in 1943. (Sperry Rand Release 2/6/69; SBD, 2/11/69, 140)
- Univ. of California astronomers Dr. E. Joseph Wampler and Dr. Joseph S. Miller reported they had photographed winking of pulsar in Crab Nebula—first of pulsars to be unequivocally associated with observable star—by spinning disc before star's image projected by 120-in telescope at Lick Observatory, Calif. Hole in disc, spun slower than flash rate of pulsar, permitted light from star to penetrate once each revolution. For first time star was shown photographically to be flashing on and off. Rate of light pulses was identical to that of previously observed radio pulses. (*NYT*, 2/7/69, 22; UPI, *W Post*, 2/7/69, A6)
- Cambridge Univ. announced radioastronomy team under Sir Martin Ryle, professor and astronomer, would build world's largest, most sensitive radiotelescope, to cost $4.8 million. It would be operational in two years and capable of picking up signals which started to earth 8,000 million yrs ago. Cambridge team had discovered pulsars. (UPI, *W Post*, 2/7/69, A20)
- NASA awarded Grumann Aircraft Engineering Corp. $3,438,400 supplemental agreement for changes in Apollo lunar module contract. Modifications—to documentation and reporting procedures for LM test and

checkout, to flight and ground test hardware, to test and effect analyses, and to crew safety hardware—brought total value of contract to $1.6 billion since January 1963. (MSC Release 69-14)

- Washington *Evening Star* said: "As man's horizon of space expands, the costs of maintaining an effective program expand in direct proportion. Already, the first limited steps have resulted in an economic burden that the richest nation in the world finds almost intolerable. If the adventure is to continue much longer, it will have to be as an international effort. Nixon's inaugural statement raises the possibility that some international body, a sort of United Nations for space exploration, could be established to pool the talents and the resources of all nations. It is an idea well worth pursuing. (W *Star*, 2/6/69, A10)

- DOD announced month delay in site acquisition and construction work on Sentinel ABM system. Action had been taken previous week to permit review of program. At White House news conference President Nixon said, "I do not buy the assumption that the ABM system, the thin Sentinel system, as it has been described, was simply for the purpose of protecting ourselves against attack from Communist China." System, like those U.S.S.R. already deployed, "adds to our overall defense capability." (*PD*, 2/10/69, 228; *WSJ*, 2/7/69, 6)

- State Dept. announced AEC would join Australia in exploring economic, technical, and safety aspects of producing deep-water harbor at Cape Keraudren in northwestern Australia using atomic explosives. (*W Post*, 2/7/69, A5)

February 7: U.S.S.R. launched *Cosmos CCLXV* into orbit with 457-km (284-mi) apogee, 272-km (169-mi) perigee, 91.8-min period, and 71° inclination. Satellite reentered May 1. (GSFC *SSR*, 2/15/69; 5/15/69)

- Aerobee 150 MI launched by NASA from WSMR carried Johns Hopkins Univ. payload to 101.7-mi (163.6-km) altitude to measure vacuum UV spectral emission lines from Venusian atmosphere. Experiment worked satisfactorily except for one second near end. No fine-mode acquisition was received and Vernier star-tracker could not track. No data on Venus were obtained. Terrestrial airglow data were obtained. (NASA Rpt SRL)

- Nike-Apache sounding rocket launched by NASA from Churchill Research Range carried Southwest Center for Advanced Studies payload to 826-mi (133-km) altitude to investigate auroral disturbances. Rocket and instruments performed satisfactorily and payload was recovered successfully. (NASA Rpt SRL)

- Senate confirmed appointment of Dr. Lee A. DuBridge as Director of Office of Science and Technology. (*CR*, S1536-7)

- Secretary of Transportation John A. Volpe said in Washington, D.C., that committee of academicians, committee within DOT, and committee representing other agency executives had begun extensive review for Nixon Administration to determine whether Government should continue subsidizing SST development. (Herbers, *NYT*, 2/8/69, 1; Reuters, *W Post*, 2/8/69, A2)

- Royal Crown Cola International announced former Astronaut John H. Glenn, Jr., had become its president. He had been chairman since January 1967. (*NYT*, 2/8/69)

- Committee of air traffic controllers said it had evolved program which would enable FAA to postpone restrictions scheduled to go into effect April 27 at five major airports. Professional Air Traffic Controllers Organization would petition Secretary of Transportation John A. Volpe to substitute "revamped procedures which would make operations safer and more efficient," said F. Lee Bailey, counsel. Restrictions would curtail services into New York, Chicago, and Washington, D.C. (*NYT*, 2/8/69)

- In *Science*, Walter Orr Roberts, President of University Corp. for Atmospheric Research, wrote: "Manned exploration of the moon will provide answers to age-long speculation about its nature. Perhaps even more important than what we find will be the fact that we have done it. The event will mark the successful attainment of a goal that demanded technological attainments of unprecedented complexity and difficulty. Our sights were set upon this goal nearly a decade ago by President Kennedy. I was, I confess, one who feared he had asked the impossible."

    Weather forecasting—one example of earth-oriented use of space science—would require space satellites of new and sophisticated character. "We will not solve this problem unless we can somehow inspire atmospheric scientists of all the world to commit themselves to the goal. . . . Space technology is perhaps the most important single component of the technology development needed for success. What better use could be found for our incredible talents in space? After the moon, the earth!" (*Science*, 2/7/69)

February 9: DOD's *Tacsat I* Tactical Communications Satellite was successfully launched from ETR at 4:09 pm EST by Titan IIIC booster into synchronous equatorial orbit over Pacific. Orbital parameters: apogee, 22,387 mi (36,020.7 km); perigee, 22,332 mi (35,932.2 km); period, 1,446.6 min; and inclination, 0.6°. The $30-million, 1,600-lb, cylindrical satellite would test feasibility of using satellite system to communicate over great distances with small military units such as aircraft, ships, and small ground stations. *Tacsat I* was powerful enough for ground forces to use portable receiving antennas as small as one foot in diameter. It also would test new gyrostat stabilization system. (W *Star*, 2/9–10/70; AP, *W Post*, 2/10/69, A1; GSFC *SSR*, 2/15/69; DOD Release 64–68; *Pres Rpt 70* [69])

- Supercritical wing would be flight-tested on USN F–8 fighter at FRC, NASA announced. Airfoil shape had been developed in four-year wind-tunnel studies at LaRC by Dr. Richard T. Whitcomb. If wind-tunnel performance was achieved in flight, wing could improve performance and efficiency of future aircraft, particularly jet transports. It would allow efficient cruise flight near speed of sound at 45,000-ft altitude and reduce operational cost of subsonic flights by increasing operational range or permitting less fuel and more payload on faster schedules.

    Supercritical wing shape was developed to delay rise of drag force and onset of buffeting at high speeds. Flattened top was designed to reduce intensity of airflow disturbances; downward curve at rear of wing supplied lift lost by flattening. Flight program would evaluate behavior of wing in actual flight with both high-lift maneuvering and off-design performance, and determine sensitivity of supercritical wing to wing-contour variations associated with manufacturing processes and deformations due to flight loads. (NASA Release 69–27)

February 9: NASA *announced the supercritical wing, a new airfoil shape developed in four years of wind-tunnel studies at Langley Research Center, would be flight-tested on a* USN *F-8 fighter at the Flight Research Center. Dr. Richard T. Whitcomb, inventor of the design expected to improve performance of subsonic jet transports, stood with his model in the test section of the wind tunnel at* LaRC.

- Meteorite broke into fragments in air and fell near Pueblito de Allende, Chihuahua, Mexico. Scientists at MSC Lunar Receiving Laboratory and Oak Ridge National Laboratory later reported from tests of fragments that meteorite was chondrite (C3 and C4) with opaque and microcrystalline matrices. Gamma rays from short-lived isotopes were observed in specimens brought to low-background gamma counter less than 4½ days after fall. (*Science,* 2/28/69, 928–9)
- Boeing Co. test pilot Jack Waddell flew 355-ton, $20-million prototype of 490-passenger Boeing 747 jet transport from Paine Field, near Seattle, Wash., for 1 hr 15 min of scheduled 2½-hr maiden flight. Waddell returned aircraft to field after encountering "minor malfunction" of wing surface control while lowering wing flaps to 30° angle. Later he said aircraft was "a pilot's dream" which could be "flown with two fingers" and indicated flap misalignment would not delay

further testing. The 210-ft-long 747 used only 4,500 ft of runway to become airborne at 170 mph. Spectators were impressed with quietness of its engines. (*W Post*, 2/10/69, 1; AP, W *Star*, 2/10/69, A5)

- Lunar module was "first manned spacecraft ever built that's not tough enough to survive a return to earth," said Thomas O'Toole in *Washington Post*. Vehicle from which two astronauts would descend to moon's surface in summer 1969 was 23 ft high, weighed 8,000 lbs, and carried 12 tons of propellant. It contained 25 mi of electrical wiring and more than a million parts, most of which had been designed "from scratch," held together by 216,000 "pins." Pin bent more than five degrees out of shape would have to be replaced. NASA had contracted for 15 LMs at total cost of $1.9 billion from Grumman Aircraft Engineering Corp., which had taken six years to get it from drawing board to launch pad. (*W Post*, 2/9/69, B2)
- In Brussels, Astronaut Frank Borman and family attended dinner given in his honor at palace by King Baudouin and Queen Fabiola. Borman showed *Apollo 8* film. (NASA Int Aff; AP, B *Sun*, 2/10/69)
- Hungary and Romania had issued souvenir stamps commemorating *Apollo 8* mission and astronauts, U.S. newspaper philatelic columns announced. Photograph taken from *Gemini IV* of Arabian coast provided design for new stamp in sultanate of Muscat and Oman. (Faries, W *Star*, 2/9/69, D10; AP, *W Post*, 2/9/69, K8)
- Johns Hopkins Univ. associate professor of mechanics, Dr. Robert L. Green, had designed and perfected "visualization apparatus for X-ray crystallography," device which permitted continuous observations of changes in structure of atoms in metal under stress. Device could lead to discovery of hitherto unknown properties of metals, nonmetallic crystals, and living molecules; enable scientists to study changes in internal structure of metals during deformation caused by air and water pressure; enable scientists to project image of atomic structure on closed-circuit TV screen; and result in development of stronger submarine hulls, aircraft wings, and spacecraft. (Reuters, *NYT*, 2/9/69, 92)
- FAA had awarded United Aircraft Corp. Pratt & Whitney Div. $665,241 contract for two-year study to develop design for quieter jet aircraft engines. (*NYT*, 2/9/69, 94)
- Astronaut Walter M. Schirra, Jr., and his first-grade school teacher, Mrs. Peggy Crowley, would receive 1969 Golden Key Awards from six national school organizations at annual convention of American Assn. of School Administrators, Atlantic City, N.J., Feb. 15, *Parade* reported. Awards had been founded to dramatize teacher's role in U.S. life. (*Parade*, 2/9/69, 4)
- In *Washington Post*, Thomas O'Toole said NASA Administrator was "the last big Federal post President Nixon has left unfilled." He asked, "Is it because he can't find the man he wants? Is it because no man he wants wants the job? Or is Mr. Nixon playing with the possibility of appointing [Acting Administrator Thomas O.] Paine to the post of Administrator?" Washington "space watchers" felt job could not be kept vacant much longer, "if only because the program to land American astronauts on the moon is rapidly nearing its goal." (*W Post*, 2/9/69, A11)
- *New York Times* editorial: "The Congressional pressure that spurred the

Nixon Administration to halt deployment of the Sentinel antiballistic missile system signals a healthy new disposition on Capitol Hill to challenge the military-industrial complex, against which President Eisenhower warned eight years ago." (*NYT*, 2/9/69, 12)

February 10: NAS published NRC Div. of Engineering's *Useful Applications of Earth-Oriented Satellites,* Report of the Central Review Committee of NRC Summer Study on Space Applications, prepared for NASA. Study concluded that space applications program was "too small by a factor of two or three." Benefits from program were expected to be large, "certainly larger than the costs of achieving them." However, "an extensive, coherent, and selective program" would be required to achieve benefits.

Committee recommended that NASA give greater emphasis to earth-satellite programs with promise of beneficial applications, commit additional funds to expanded R&D and prototype operations for certain applications, and commit $200 million to $300 million yearly to space applications program. Manned space programs should be justified in their own right, not in terms of space applications; near-term benefits for mankind would be achieved "more effectively and economically with automated devices and vehicles."

Noting that in meteorology and communications "satellites have already entered solidly into the area of economic usefulness," report recommended that NASA grant high priority to development of multichannel distribution system for public and private network TV; multichannel system for educational broadcasts in developing countries and for special interest groups such as physicians, lawyers, and educators; and North Atlantic satellite navigation system for traffic control of transoceanic aircraft and ships. Satellite earth-sensing was dependent on R&D in sensor signatures—form of information provided by instruments. Report recommended immediate pilot program for providing information in familiar and immediately usable form, exploration into use of side-looking radar, and start of 10- to 12-yr development plan for more sophisticated sensors. (Text; NRC Release)

- *Apollo 8* mission (Dec. 21–27, 1968) was adjudged successful by NASA. All objectives of manned circumlunar mission were attained, as well as four detailed test objectives not originally planned. (NASA Proj Off)
- MSFC announced it would manage two recently awarded $300,000 six-month contracts, one to Lockheed Missiles & Space Co. and one to General Dynamics Corp., for conceptual study of low-cost, manned logistics (space shuttle) system. Similar study contracts awarded to North American Rockwell Corp. and to McDonnell Douglas Astronautics Co. would be managed by MSC and LaRC. Integral Launch and Reentry Vehicle (ILRV) studies would investigate aspects of reusable transportation system for post-1974 use in support of proposed space stations.

  MSFC also announced $3,288,914 modification to contract with Boeing Co. for continued configuration management support on Saturn V launch vehicle program, including processing of vehicle and ground support equipment configuration changes, configuration accounting, and change integration and tracking. (MSFC Releases 69–34, 69–35)
- General Accounting Office released report to Congress, *Need for Improved Guidelines in Contracting for Research with Government-Spon-*

February 10

sored Nonprofit Contractors. It called for Government-wide guidelines on amounts and use of fees or management allowances given by DOD, NASA, and AEC Federal contract research centers. GAO found allowances paid to nonprofit organizations varied significantly, were not much used for research, and had been spent by some centers to acquire extensive capital facilities. (Text)

- USAF contract awards: $4,305,295 fixed-price contract to Computer Sciences Corp. for services and supplies to develop, install, operate, test, and maintain hardware to improve capabilities of space tracking equipment; and $1,600,000 initial increment to $4,200,000 fixed-price contract to United Technology Center for KSC launch and support services. (DOD Release 102–69; *WSJ*, 2/11/69, 17)

February 11: Initial thermal and vacuum testing of flight model of SERT II (Space Electric Rocket Test) in preparation for fall 1969 launch had been completed, LeRC announced. SERT II, second flight test in development of ion propulsion for space use and first LeRC orbital spacecraft, would be launched from WTR by Thorad-Agena booster into 621-mi (999.4-km) circular orbit to evaluate inflight performance of electron-bombardment engines for six months or more. SERT I had carried first ion thruster to operate in space on suborbital mission July 20, 1964. (LeRC Release 69–2)

- Nike-Apache sounding rocket launched by NASA from Churchill Research Range carried Southwest Center for Advanced Studies payload to 85.1-mi (137-km) altitude for comprehensive investigation of auroral disturbances during active auroral event. Rocket and instruments functioned satisfactorily and payload was recovered as planned. (NASA Rpt SRL)

- In Bonn during European tour, *Apollo 8* Astronaut Frank Borman addressed enthusiastic crowd of 1,500 students and government officials after film showing on lunar mission in Beethoven Hall: "I believe this research will teach us that we are first and foremost not Germans or Russians or Americans but earthmen." Borman met West German Chancellor Kurt Georg Kiesinger at lunch and later discussed space research with Scientific Affairs Minister Gerhard Stoltenberg. He attended evening reception given by West German Air and Space Research Institute. (Falbe, B *Sun*, 2/12/69)

- U.S.S.R. had ordered 100 space pens developed for U.S. astronauts and 1,000 special pressurized ink cartridges which enabled pen to write in weightlessness according to pen's inventor, Paul C. Fisher. When he presented models of pen to Soviet Cosmonaut Alexey Leonov at German trade fair in 1968, Leonov said Soviet cosmonauts were writing with grease pencils during space flights and incurring difficulty with their flaking. (UPI, *W Post*, 2/13/69, D24)

February 12: Pentagon sources estimated U.S.S.R. was spending equivalent of $60 billion in 1969 on national defense and space efforts, while U.S. was spending $85.2 billion, of which $29 billion was for Vietnam war. Figures left U.S.S.R. $4 billion ahead of U.S. in spending on weapon and space technology. Between 1965 and 1969, Soviet spending on offensive and defensive strategic forces increased by 40% but amount spent on intercontinental missiles and surface-to-air missile defense systems rose by 75%. (Kelly, *W Star*, 2/12/69, D4)

- Aerobee 150 sounding rocket launched by NASA from WSMR carried Naval

Research Lab. payload to 116.8-mi (187.9-km) altitude to record photographically 18 EUV spectra of solar photosphere, chromosphere, and corona, using SPARCS and flight-design verification unit of high-resolution spectrograph planned for ATM–A and ATM–B. Rocket and instruments performed satisfactorily. (NASA Rpt SRL)

- NASA launched Aerobee 150 sounding rocket from Churchill Research Range carrying Johns Hopkins Univ. payload to collect data on airglow. Mission did not meet minimum scientific requirements. (NASA Proj Off)
- During visit to West Berlin, *Apollo 8* Astronaut Frank Borman drove past U.S.S.R.'s war memorial near Berlin wall and looked across wall into East Berlin. At Tempelhof airport Borman told press, "I was here before [during 1949 Berlin airlift] amid many bags of coal. There have been many space advances in the last two decades, yet we have so many troubles here on earth." (*C Trib*, 2/13/69)
- MSFC announced it had issued $1,182,155 contract modification to Chrysler Corp. Space Div. for continued systems engineering and integration on Saturn IB launch vehicles. (MSFC Release 69–37)
- USAF F–111A piloted by Capt. Robert Earl Jobe (USAF) and instructor pilot Capt. William D. Fuchlow (USAF) failed to return to Nellis AFB, Nev., after 750-mi training mission. USAF and Civil Air Patrol were searching area between Las Vegas, Nev., and Great Salt Lake. (UPI, *W Star*, 2/13/69, 1; AP, *W Post*, 2/14/69, A4)

February 13–14: NASA successfully launched one Nike-Tomahawk and six Nike-Apache sounding rockets carrying chemical cloud experiments from NASA Wallops Station between 6:11 pm and 6:13 am EST. Rockets ejected vapor trails between 50- and 186-mi (80.5- and 299.3-km) altitudes to measure wind velocities and directions. Nike-Tomahawk launched at dusk and Nike-Apache launched at dawn carried sodium experiments which created reddish-orange trails. Other five payloads consisted of trimethylaluminum (TMA) experiments which formed pale white clouds. Data were obtained by photographing continuously motions of trails from five ground-based camera sites. Launches were conducted for GCA Corp. under GSFC contract.

In conjunction with vapor series USA Ballistics Laboratory at Aberdeen, Md., fired six projectiles containing cesium experiments to 330,000-ft altitude between 8:07 pm and 6:23 am EST for comparative study of winds. Three experiments failed to eject chemical; dispersion of cesium from remaining three projectiles was recorded by ground-based radar and ionospheric sounding stations. (WS Release 69–5; NASA Release 69–28; *NYT*, 2/14/69, 41)

February 13: President Nixon's Science Adviser, Dr. Lee A. DuBridge, announced at his first Washington press conference that overall plan for next decade of U.S. space program would be drafted at President's request by his office, NASA, NASC, and DOD for submission to President about Sept. 1. Charting "new directions, new goals and new programs for the entire United States Space program" was necessary. "Bringing to the benefit of people the marvelous space technologies that have been developed in the last decade and certainly orbiting satellites for the purpose of learning more about the earth must be an important element in our future space program," Dr. DuBridge said. "Whole problem" was balance between that enterprise and planetary and lunar

February 13

exploration and "this is the problem which our group will seek . . . to bring into perspective as we project ahead and consider the budget problems that also lie ahead."

In answer to question on White House appointments, Dr. DuBridge said, "We have not yet located the right man" for either Administrator of NASA or Executive Secretary for Space Council. (Transcript; White House Memo)

- Arthur S. Flemming Awards for 1969 were presented to 10 outstanding young men in Federal Government in Mayflower Hotel ceremony in Washington, D.C. Winners included James J. Kramer, Chief of LeRC Propulsions Systems Acoustics Branch, who kept solid rocket program "on schedule and within budgeted costs," and Dr. Norman F. Ness, head of Extraterrestrial Physics Branch, GSFC, who made "significant contributions" to understanding space through Explorer satellite program. Dr. Richard E. Hallgren, Director of Commerce Dept.'s world weather systems, was named for "imaginative leadership" in recognizing and integrating requirements of oceanographers and meteorologists. (W Star, 2/13/69, B6; LeRC Release 69–3)

- *Washington Post* reported Washington Airlines President Robert Richardson had said first scheduled STOL air shuttle in U.S. had lost more than $100,000, cut back operations 44%, and operated at less than half break-even load factor during first four months of service. He attributed most difficulties to start-up problems, including minor equipment shortcomings which had been corrected. Airline was lowering fares and could, said Richardson, break even in 12–18 mos. (Koprowski, *W Post*, 2/13/69, C9)

- At GSFC, satellite mapping authority Dr. John A. O'Keefe was preparing first precise maps of Tibet using photographs taken from 100-mi altitude by U.S. astronauts and data obtained between 1890s and 1935 by Swedish explorer Sven Hedin during only extensive survey of area by outsider. Expedition's survey sightings on mountain peaks were being applied to numerous photographs from space. Revised maps would be published in Sweden. (Sullivan, *NYT*, 2/13/69, 14)

- Intelligence briefings to high DOD officials had indicated U.S.S.R. missile defense was three-quarters complete and had been slowed in recent months to improve its radar system, said William Beecher in *New York Times*. Briefings also indicated that antimissile system around Moscow, even when finished, would not alter balance of power between U.S.S.R. and U.S. or undermine U.S. retaliatory power. (*NYT*, 2/13/69, 1)

February 14: ComSatCorp announced broadcasters had booked 40 hrs of satellite time for TV coverage of President Nixon's European trip Feb. 23–March 3. More than 17 hrs had been requested from abroad to date for coverage of Apollo 9 Feb. 28–March 3. (ComSatCorp Release 69–8; W Star, 2/16/69, C6)

- In *Science* Hudson Hoagland, President Emeritus of Worcester Foundation for Experimental Biology, commented on Condon Report on UFOs released Jan. 9, 1968: "The basic difficulty inherent in any investigation of phenomena such as those of psychic research or of UFO's is that it is impossible for science ever to prove a universal negative. There will be cases which remain unexplained because of lack of data, lack of repeatability, false reporting, wishful thinking, deluded observers, rumors, lies, and fraud. A residue of unexplained cases is not a justification for continuing an investigation after overwhelming evidence has disposed of hypotheses of supernormality, such as beings from outer space or communications from the dead. . . . Science deals with probabilities, and the Condon investigation adds massive additional weight to the already overwhelming improbability of visits by UFO's guided by intelligent beings." (*Science*, 2/14/69, 625)

- Leonard Mandelbaum in *Science* briefly examined history of U.S. decision to adopt Apollo program. "Cautious approach" to manned space flight gave way after impact of April 12, 1961, "Russian spectacular" —flight of Cosmonaut Yuri A. Gagarin—and U.S. Cuban foreign policy fiasco, Bay of Pigs. "Congress acted without hearing testimony of compelling military need. The Apollo decision was made without reference to any comprehensive and integrated national policy designed to maximize the use of scientific and technological resources for social objectives. . . . It was a typical Cold War reaction." (*Science*, 2/14/69, 649)

February 15: Project Tektite, multiagency-industry program to determine ability of men to perform scientific research mission while living isolated on ocean floor under saturated diving conditions for long period, began at St. John, Virgin Islands. Four U.S. aquanauts, Richard A. Waller, H. Edward Clifton, John G. Van Derwalker, and Conrad V. W. Mahnken jumped into sea at Beehive Cove and swam to "habitat," underwater capsule moored 42 ft below sea level for 60-day experiments. Tektite program was jointly sponsored by USN, NASA, and Dept. of the Interior, with participation by U.S. Coast Guard. Prime contractor, General Electric Co., furnished undersea habitat. NASA and USN behavioral and biomedical teams would observe aquanauts continuously to identify psychological and physiological reactions to long-term mission performed in hostile and isolated environment common to undersea and space missions. (NASA OMSF PAO; Lannan, W *Star*, 2/16/69, A3; 2/17/69, A6)

- Pope Paul VI received *Apollo 8* Astronaut Frank Borman and family for 17-min audience in Papal library. Pope said in English, "Man's reaching out to unravel the mysteries of the universe reveals more and more the wonders of God's work and shows forth His glory." Pope Paul sent personal greetings to Astronaut James W. McDivitt who had audience in 1967. (UPI, W *Star*, 2/16/69, C5)

February 16: USN's Sealab III project, in which five aquanaut teams were to spend 12 days each in 60-day test of man's ability to work under water for long periods, started early when four of first team of nine men were dispatched to repair helium leak in 57 × 12-ft habitat, 600 ft beneath Pacific Ocean off San Clemente Island, Calif. Remaining five aquanauts were scheduled to descend in pressurized personnel-transfer capsules 12 hrs later to join colleagues in performing experiments in marine biology, geology, acoustics, and ecology. (B *Sun*, 2/17/69, A7)

February 17: Tenth anniversary of *Vanguard II*, fifth U.S.–IGY satellite, launched by NASA to produce cloud-cover images using two photocells. Wobbling had prevented interpretation of data. Satellite was still in orbit. (*A&A 1915–60*; GSFC *SSR*, 2/28/69)

- Nike-Apache sounding rocket was launched by NASA from Churchill Research Range carrying Southwest Center for Advanced Studies payload to investigate auroral disturbances. Rocket and instruments functioned satisfactorily. (NASA Proj Off)

- USN suspended Sealab III project when veteran Aquanaut Berry L. Cannon was stricken while he and Aquanaut Robert A. Barth, Jr., were attempting to open habitat's hatch after Cannon's second dive to check gas leaks. He was pronounced dead of "cardiac arrest" in decompression chamber of mother ship U.S.S. *Elk River* and body was flown to San Diego for autopsy. First finding of autopsy was that Cannon did not die of heart attack. USN on Feb. 18 canceled project. (Stevens, *NYT*, 2/18/69, 1; 2/20/69, 93; O'Toole, *W Post*, 2/18–20/69; AP, W *Star*, 2/18/69, A1)
- President Nixon submitted to Senate nomination of former NASA Associate Administrator for Advanced Research and Technology James M. Beggs as Under Secretary of Transportation. (*PD*, 2/24/69, 293)
- In Madrid during European goodwill tour, *Apollo 8* Astronaut Frank Borman placed wreath at statue of Columbus and met Cristobal Colon de Carvajal y Maroto, 17th duke of Veragua and hereditary "admiral of the ocean sea," title created in 1537 for explorer's son, Diego Columbus. (AP, *C Trib*, 2/18/69)
- USAF said ground test of F–111A had revealed large crack in test version of aircraft belly section to which movable wings were attached. No F–111As would be grounded, as test did not indicate safety hazard to aircraft in service. Crack was not related to one detected Aug. 25, 1968. (UPI, *W Post*, 2/18/69, A4; AP, W *Star*, 2/18/69, A6)

February 18: Secretary of State William P. Rogers told Senate Foreign Relations Committee during hearings on nonproliferation treaty that he hoped U.S.–U.S.S.R. missile talks would be under way before it became necessary for U.S. to start deployment of proposed Sentinel ABM system. He said U.S. would have obligation under treaty to enter into strategic arms talks with U.S.S.R. and expressed hope such talks could begin within six months. (Transcript, 377–8)

- Rep. Charles H. Wilson (D-Calif.) introduced H.R. 7030, bill to encourage worldwide interest in U.S. developments and accomplishments in military and related aviation and equipment by authorizing Federal sponsorship of International Aeronautical Exposition in U.S., to be held not later than 1970. (Text)
- In *Washington Post* review of *Contact! The Story of the Early Birds* by Henry Serrano Villard, John Osgood said: "Despite the technical complexities of the recent translunar injection, the mystique of flight remains undiminished 65 years after Orville Wright managed his mere 120 feet of powered flight. Mystique or no, it is still difficult to comprehend what drove the early aeronauts to attempt feats which most often won them the contempt and ridicule of their countrymen." (*W Post*, 2/18/69, B4)

February 19: Dr. Lee A. DuBridge, Presidential Science Adviser, told Subcommittee on Science, Research, and Development of House Committee on Science and Astronautics, "Our intellectual resources—not our material resources—are the limits to what we can now achieve." During hearings on H.R. 35, bill to promote advancement of science and education of scientists through institutional grants to U.S. colleges and universities, he said: "We hear it said that if we only spent as much money on urban programs as we did, say, on the atomic bomb project or on our space program, we could quickly solve the crisis in our cities. But let us not forget that we launched the Manhattan project and the space program only after, and not before ... efforts in basic research over the previous 30 or 40 years had uncovered the knowledge which showed us how we could build atomic bombs or launch payloads into space. Neither the Manhattan project nor the space program could have been dreamed of 10 years before they started, because we did not even know enough to even formulate a development program. Now, in many ... of our present crises we are in the same position as far as technology is concerned. We do not know enough about certain technologies and ... many social phenomena to justify mounting a concentrated, technically based attack on these problems now. We must ... greatly enhance ... measures to relieve immediate suffering and injustice. But at the same time we must encourage and support new efforts to learn more, to extend our base of fundamental knowledge in science, technology, social science, so that we can move sure footedly toward long-range solutions." (Transcript)

- Rep. Louis Frey, Jr. (R-Fla.), introduced H.R. 465 "providing for the establishment of the Astronauts Memorial Commission to construct and erect with funds a memorial in the John F. Kennedy Space Center, Florida, or the immediate vicinity, to honor and commemorate the men who serve as astronauts in the U.S. space program." (*CR*, 2/19/69, H1087)

- House passed and returned to Senate S. 17, bill to amend Communications Satellite Act of 1962 to provide for apportionment of ComSatCorp directors according to percentages of stock held by public and communications corporations. (*CR*, 2/19/69, H1037–40)

February 20: NASA announced appointment of Dr. Hans M. Mark, Chairman of Dept. of Nuclear Engineering, Univ. of California at Berkeley, as Director of Ames Research Center. He would succeed H. Julian Allen, who had announced retirement Oct. 25, 1968, but had remained as Acting Director. Dr. Mark, expert in nuclear and atomic physics, was also Reactor Administrator of Univ.'s Berkeley Research Reactor, research physicist at Univ.'s Lawrence Radiation Laboratory, and consultant to USA and NSF. Clarence A. Syvertson, Director of Astronautics at ARC, was appointed to newly created position of ARC Deputy Director. Both appointments were effective Feb. 28. Because of prior commitments, Dr. Mark would spend one-fifth of his time at ARC until July 1969. (NASA Release 69–32; ARC *Astrogram*, 2/24/69, 1)

- Secretary of Defense Melvin R. Laird told Senate Foreign Relations Committee during hearings on nuclear nonproliferation treaty that U.S. should go forward with Sentinel system if DOD review found it "practical" and "effective," since U.S.S.R. was working on "sophisticated new ABM system." Curtailment in Soviet missile construction during past few months, Laird said, was due to R&D testing on more sophisticated system. U.S.S.R. had been outspending U.S. three to one in missile defense and "substantial" network around Moscow was halfway complete. (Transcript, 419–20)

- *Apollo 8* Astronaut Frank Borman and family ended official goodwill tour of Western Europe with lecture and luncheon in Lisbon. During

February 20

final European news conference previous day, he had predicted U.S. would put man on moon in summer 1969 "if everything goes well." (UPI, *W Star*, 2/20/69, A8)

- NASA launched Nike-Apache sounding rocket from Churchill Research Range carrying Southwest Center for Advanced Studies payload to investigate auroral disturbances. Mission was unsuccessful. (NASA Proj Off)
- At annual dinner of Washington Academy of Sciences, GSFC engineer Charles R. Gunn received Academy's award for "noteworthy discovery, accomplishment, or publication" in engineering field for his work as technical director of Thor-Delta launch vehicle. (GSFC Delta Proj Mgr; AP, *W Star*, 2/21/69, C10)
- First International Aviation Service Award, financed by contributions from FAA employees and established in June 1968 by retiring FAA executive Alfred Hand, was presented in Washington to Theodore C. Uebel, International Liaison Officer for FAA, for "outstanding accomplishments in furthering the interests of the United States in international aviation." (FAA Release 69-17)
- Eugene Luther Vidal, who as Director of Air Commerce of Dept. of Commerce (1933-1937) promoted growth of U.S. civil aviation, died at age 73 in Palos Verdes, Calif. He had furthered construction of airports and beacons, encouraged private flying and manufacture of small aircraft, advanced commercial aviation, and reorganized Government control of commercial flights. After leaving Commerce Dept. he had established research laboratory near Camden, N.J., where he developed process for making airframe parts from molded plywood. (*NYT*, 2/21/69, 43)

February 21: Apollo 8 Astronaut Frank Borman and family returned from European goodwill tour made on behalf of President Nixon. At Andrews AFB, Md., Borman told press on arrival that Europeans found it hard to believe U.S. "could spend all that money on its space program and still make public everything we learned." He said reception had been uniformly friendly, "but they would hesitate to ask us questions, because they assumed . . . information about the Apollo 8 flight must be classified." Borman and family reported at Capitol to Vice President Spiro T. Agnew, Chairman of NASC. Borman told press conference he had found "extreme identification of people in all walks of life in Europe with our flight. They were very well informed about it and looked on us as representatives of Earth. I hope that feeling of comradeship can continue." (AP, *W Post*, 2/22/69, A2)

- ComSatCorp reported $6.841 million 1968 net income (68 cents per share), up from 1967 net income of $4.638 million (46 cents per share). Improvement had resulted primarily from net operating income of $988,000, which contrasted with 1967 net operating loss of $642,000. (ComSatCorp Release 69-10)
- President Nixon approved "Policy on Expanded Use of Federal Research Facilities by University Investigators" which directed Federal agencies to make equipment in Federal laboratories more readily available to qualified university scientists. He directed Dr. Lee A. DuBridge to monitor execution of policy with help of Federal Council for Science and Technology, which had recommended adoption. (*PD*, 3/3/69, 304)

- FAA announced award of $35,426,283 contract to UNIVAC Federal Systems Div. of Sperry Rand Corp. for automated radar tracking systems (ARTS III) to be installed at more than 60 major U.S. airports. (FAA Release 69-22)

February 23: President Nixon arrived in Belgium at start of eight-day goodwill visit to heads of state in Brussels, London, Bonn, West Berlin, Rome, Paris, and the Vatican. In welcoming speech at Brussels National Airport, King Baudouin said: "During this year, which will perhaps be that of man's first landing on the moon, we are more than ever conscious of the gulf between the wonderful possibilities open to us and the obligations which burden the world because of war, want, injustice, and inequality. May your journey and your interviews provide an opportunity for friendly nations better to combine their efforts to solve their problems on which the very future of mankind depends." (PD, 3/3/69, 310)

February 24–28: NASA's *Mariner VI* (Mariner F) spacecraft was successfully launched from ETR by Atlas-Centaur (SLV-3C) booster on five-month, 226-million-mi, direct-ascent trajectory toward Mars—NASA's first of two attempts to conduct Mars flyby missions during 1969 launch window. Launch vehicle performance and spacecraft injection were nominal. Spacecraft separated from Centaur, deployed its four solar panels, locked its sensors on sun and star Canopus, and entered cruise mode, where it remained with all subsystems performing satisfactorily while trajectory was refined. Midcourse maneuver was successfully conducted Feb. 28 to ensure that spacecraft would fly within 2,200 mi (3,540.5 km) of Mars July 31.

Primary mission objective was equatorial flyby mission for exploratory investigations of Mars to set basis for future experiments, particularly those relevant to search for extraterrestrial life. As secondary objective spacecraft would develop technology needed for succeeding Mars missions. The 840-lb spacecraft carried six complementary experiments to provide information about Martian surface and atmosphere. Mission offered first opportunity to make scientific measurements on night side of Mars. Two onboard TV cameras would take pictures of Mars disc during approach with 15-mi optimum resolution and of surface during flyby with 900-ft optimum resolution. Infrared spectrometer and UV spectrometer would probe Mars atmosphere, and occultation experiment would obtain data on atmospheric pressures and densities. Infrared radiometer would measure surface temperatures on both light and dark sides of Mars; celestial mechanics experiment would use tracking information to refine astronomical data. Sharp increase in data returns would be achieved over 1964 Mariner missions. *Mariner VI* TV pictures would contain 3.9-million bits of information; *Mariner IV* contained 240,000 bits in 1965. *Mariner VI* would transmit science data at basic rate of 270 bps and high rate of 16,200 bps before flyby; *Mariner IV* transmitted at 8 ⅓ bps.

Mariner VI was follow-on to 1964 Mariner/Mars missions and precursor to 1971 orbital and 1973 landing missions. First Mars probe, *Mariner III* (launched Nov. 4, 1964), had failed to achieve desired orbit when shroud remained attached to spacecraft. *Mariner IV* (launched Nov. 28, 1964) had transmitted first close-up photos of Mars in July 1965. Mariner VII (Mariner G) would be launched

March 24. Mariner program was directed by OSSA Lunar and Planetary Programs Div. Project management and responsibility for spacecraft, mission operations, and tracking and data acquisition were assigned to JPL. Atlas-Centaur launch vehicle was managed by LeRC. (NASA Proj Off; NASA Release 69–26)

February 24: At State Dept. meeting of more than 60 INTELSAT member nations, U.S. delegation chairman Leonard H. Marks said, "I can think of no more important step we can take towards reducing world tensions than that of broadening communications links between power nations representing different political systems"—as U.S.S.R. and 13 other observer nations listened. In written memorandum, France had questioned whether strong centralized system desired by U.S. could or should be established and urged that any new agreement leave participating countries free to join other satellite systems. (Samuelson, *W Post*, 2/5/69, D5)

- Federal Council for Science and Technology transmitted to NASA "Policy on Expanded Use of Federal Research Facilities by University Investigators" approved by President Nixon Feb. 21. (NASA Off of Policy Memo, 3/14/69)

  Vice President Spiro T. Agnew told American Management Assn. briefing on oceanography in Washington, D.C., that Nixon Administration was not yet ready to endorse concept of "a wet NASA"—marine-oriented Government agency. As Chairman of National Council on Marine Resources and Engineering Development he was studying opinions of advocates of such an agency, as well as [Jan. 11] report by Commission on Marine Science, Engineering and Resources. (Smith, *NYT*, 2/25/69, 53)

February 24–March 3: First documented pulsar acceleration was discovered in Pulsar PSR 0833–45 in Vela constellation in southern sky by JPL radio-astronomers Paul Reichley and Dr. George S. Downs, using 85-foot dish antenna at Goldstone, Calif. While pulsars normally showed moderate but steady slowing in pulse rate, Vela's rate accelerated, then slowed at slightly faster rate than before, during week's observation. Findings in NASA-sponsored research were confirmed by Parkes Observatory astronomers in Australia. (JPL Release BB–513, 4/16/69)

February 25: Cosmos CCLXVI was launched by U.S.S.R. into orbit with 336-km (208.8-mi) apogee, 202-km (125.5-mi) perigee, 89.8-min period, and 72° inclination. Satellite reentered March 5. (GSFC *SSR*, 2/28/69; 3/15/69; AP, *W Post*, 2/26/69)

- NASA's *Oso V* (launched Jan. 22) had successfully completed more than 496 earth orbits and had satisfactorily operated all spacecraft systems, including raster scan and both tape records. Torque coil had been turned on Jan. 25 to help minimize spacecraft pitch motions and reduce gas consumption. Primary objectives had been achieved and *Oso V* had acquired scientific data from eight onboard experiments. (NASA Proj Off)

- President Nixon addressed U.S. Embassy staff in London during eight-day goodwill visit to European heads of state: "You have had a very distinguished visitor to this country, Frank Borman, a few days ago.... I recall when I was at the White House I was congratulating him in a toast for what he and his fellow astronauts had done.... He said, 'We appreciate the remarks you have made about us.' But, he said, 'I want to point out that there are 400,000 Americans who, in one way or another, contributed to the building of the Apollo spacecraft and to this program.' He said, 'I want to point out that there are 2 million parts in an Apollo spacecraft. So, if something went wrong with one of those parts, which had been created by these 400,000 Americans, that tremendous, exciting journey around the moon could not have been possible.' That, of course, is what government is about." (*PD*, 3/10/69, 341–2)

- NASA announced selection of 38 scientists organized into eight teams to assist in design and development of Martian softlander for 1973 Viking missions. Teams would participate in early instrument development, designing softlander, and planning missions. Final selection of investigations and participating scientists for both landers and orbiters making up 1973 Viking missions would be made December 1969, when initial results of Mariner flybys of Mars in summer 1969 would be available. Planetary Programs Directorate would have management responsibility for Viking Mars 1973 mission; LaRC had been assigned overall project management and direct responsibility for managing planetary lander portion; JPL had management responsibility for orbiter spacecraft. (NASA Release 69–31)
- County Coroner Robert L. Creason in San Diego, Calif., gave official cause of Feb. 17 death of Aquanaut Berry L. Cannon in Sealab III as "acute hemorrhagic pulmonary edema and congestion due to acute cardiorespiratory failure due to carbon dioxide poisoning." Earlier USN spokesman had acknowledged that one of rigs used by Cannon and colleagues on fatal dive contained canister empty of chemical used to absorb carbon dioxide from aquanauts' air supply. USN opened formal inquiry Feb. 26. (UPI, *NYT*, 2/25/69, 28; AP, *W Star*, 2/25/69, A7)
- USAF and Lockheed Georgia Co. jointly announced six-month delay in C–5A production schedule attributed to labor strikes and material shortages caused by Vietnam war. First aircraft would be delivered to USAF in December rather than June. Announcement followed successful test flight during which 250-ton aircraft reached complete stop on 1,500 ft of runway—¼ distance required by conventional 85-ton airliners. (Lindsay, *NYT*, 2/26/69; AP, *W Post*, 2/27/69, A18)
- FAA announced it had amended its Dec. 3 rule intended to ease congestion at five of Nation's busiest airports. Amendments provided for extra sections of scheduled air carrier flights without regard to established quotas at all airports except John F. Kennedy, increase in flight quotas at Kennedy between 5:00 pm and 8:00 pm, effective date June 1 instead of April 27, and termination date of Dec. 31. (FAA Release 69–23)
- Senate Foreign Relations Committee recommended U.S. ratification of nuclear nonproliferation treaty and said it would send treaty to Senate floor for action by March 6. (*W Post*, 2/26/69, A5)

February 26: NASA successfully launched *Essa IX* (TOS-G) ninth meteorological satellite in ESSA's Tiros Operational Satellite (TOS) system from ETR by three-stage Thrust-Augmented Thor-Delta (DSV-3E) booster. Primary NASA mission objective was to place and operate spacecraft in sun-synchronous orbit with local equator crossing time between 2:15

pm and 2:35 pm so that daily advanced-vidicon-camera-system (AVCS) pictures of entire globe could be obtained regularly and dependably. Satellite achieved nearly polar, sun-synchronous, circular orbit with 934.6-mi (1,503.8-km) apogee, 884.4-mi (1,423.9-km) perigee, 115.2-min period, and 101.8° inclination.

An advanced version of cartwheel configuration, 320-lb cylindrical *Essa IX* carried flat plate radiometer to measure atmosphere's heat balance and two AVCS cameras for daily global weather coverage. Photos would be stored on board satellite on magnetic tape until readout by ESSA's Command and Data Acquisition (CDA) stations at Fairbanks, Alaska, and Wallops Island, Va. Satellite was backup to ensure full coverage after failure of one AVCS camera on *Essa VII* (launched Aug. 16, 1968) and would be primary stored-data satellite in TOS system.

Spacecraft was placed in wheel mode and spin rate was adjusted. Only anomaly was 20 rpm spin rate (rather than expected 10 rpm) after spacecraft spin-down.

ESSA financed and managed TOS system and would operate spacecraft after NASA completed checkout in month. GSFC was responsible for procurement, launch, and initial checkout of spacecraft in orbit. (NASA Proj Off; ESSA Release ES-69-9)

- U.S.S.R. launched *Cosmos CCLXVII* from Baikonur Cosmodrome. Orbital parameters: apogee, 331 km (205.7 mi); perigee, 202 km (125.5 mi); period, 89.8 min; and inclination, 65°. Satellite reentered March 6. (GSFC *SSR*, 2/28/69; 3/15/69; *SBD*, 2/27/69, 212; *C Trib*, 2/27/69)
- LeRC announced it had completed assembly of Brayton Cycle space power generating system, which appeared promising as source of electrical power for space flights up to five years long. Self-supporting, closed-loop system operated when mixture of helium and xenon was heated to 1,600°F and circulated to drive turbine. Turbine operated alternator providing electric power and also compressor that helped circulate gas through system. Cycle would undergo tests in simulated space environment in summer. (LeRC Release 68-9)
- Secretary of Transportation John A. Volpe announced President Nixon had nominated John H. Shaffer, Vice President of TRW Inc., as Federal Aviation Administrator. Shaffer would replace Acting FAA Administrator, David D. Thomas, who would remain as Deputy Administrator. (DOT Release 2469)
- MSFC announced it had extended contract with Mason-Rust for continued support services at Michoud Assembly Facility for six months. Contract modification amounted to $3,786,203. (MSFC Release 69-46)

February 27: NASA postponed Apollo 9 earth-orbital mission, scheduled for launch Feb. 28, after intensive medical examinations of prime crew revealed viral infections. (W *Star*, 2/27/69, A1)
- Nike-Apache sounding rocket was launched by NASA from Churchill Research Range carrying Rice Univ. payload to conduct auroral studies. Rocket and instruments functioned satisfactorily. (NASA Proj Off)
- White House announced President Nixon had established interdepartmental ad hoc committee to review SST program's technology, commercial potential, schedule and costs, and environmental side-effects, par-

ticularly sonic boom phenomenon. Under Secretary of Transportation James M. Beggs was designated chairman of 11-member committee, which also included Presidential Science Adviser, Dr. Lee A. DuBridge; Secretary of the Air Force, Dr. Robert C. Seamans, Jr.; and NASA Deputy Associate Administrator Charles W. Harper. (*PD,* 3/3/69, 329-30)

- Commemorative stamp to be issued May 5 in honor of Dec. 21-27, 1968, *Apollo 8* mission would include phrase "In the beginning God . . ." on photo of earth as seen from moon, taken by *Apollo 8* crew. Postmaster General Winton M. Blount said phrase, read from Genesis by Astronaut William A. Anders during lunar orbit Christmas Eve 1968, would be included in response to many requests. Stamp would be first U.S. stamp with religious wording since 1961. (UPI, *W Post,* 2/28/69)
- FRC announced award to Serv-Air Inc. of one-year, cost-plus-award-fee contract for administrative technical support services. Contract, estimated at $750,000 per year, included provision for two one-year extensions. (FRC Release 7-69)

February 28: NASA and British Science Research Council (SRC) had agreed to conduct cooperative project to launch fourth Ariel satellite, NASA announced. Ariel IV would be launched by Scout booster from WTR in late 1971 or early 1972 carrying one U.S. and four U.K. experiments to explore interactions among plasma-charged particle streams and electromagnetic waves in upper atmosphere. SRC would be responsible for spacecraft design, fabrication, and testing; NASA would provide Scout launch vehicle. Both agencies would participate in tracking, data acquisition, and data reduction. (NASA Release 69-35)

- Tenth anniversary of DOD's 1,450-lb *Discoverer I* satellite successfully launched into polar orbit by Thor-Agena booster. Tracking acquisition was hampered by stabilization difficulties and satellite reentered in early March 1959.

  Agena launch vehicle—most widely used booster in U.S.—had completed more than 250 successful flights in DOD and NASA operations since its first mission Feb. 28, 1958, and had carried first spacecraft to achieve circular orbit, first to be controlled in orbit by ground command, and first propelled from one orbit to another. It had been continually updated and used as versatile, multipurpose vehicle. (*A&A 1915-60; Space Propulsion,* 2/28/69, 199)
- Canadian Black Brant IIIB sounding rocket launched by NASA from Wallops Station reached 134-mi (215.6-km) altitude on first of two flights to evaluate improved Black Brant IIIB single-stage rocket and to provide data for payload environmental test specification [see May 1]. (NASA Rpt SRL)
- LaRC issued RFPs for design and financial proposals for planetary lander and project integration portions of NASA's Viking project. Viking spacecraft—consisting of lander and orbiter—were to be procured for two planned flights to Mars to search for scientific data in 1973. (NASA Release 69-36)
- NASA announced it would negotiate with North American Rockwell Corp. for modifications to four Apollo spacecraft for Apollo Applications program. Combined value of spacecraft and modifications was estimated at $340 million. (NASA Release 69-34)
- In *Science,* Bryce Nelson reviewed *Science Policy in the USSR,* study

February 28

sponsored by Directorate for Scientific Affairs of Organization for Economic Cooperation and Development (OECD). It indicated, he said, that Soviet scientists and political leaders "need to spend considerable time thinking about how to correct imbalances in their R&D system." U.S.S.R. had succeeded outstandingly in aviation rocketry, space exploration, atomic energy, machine tools, and iron and steel technology but its R&D system seemed sluggish. Main bottleneck was relative unavailability of testing facilities. Central planning system in U.S.S.R. reinforced separation between R&D establishments and industry and contributed to reluctance of factories to innovate. Increasing use of contract system, with industries placing growing number of R&D contracts with institutions of higher learning, was helping bridge gap between research centers and industry. (*Science*, 2/28/69, 917–8)

During February: In *Astronautics & Aeronautics* editorial written just before his appointment as Secretary of Air Force, incoming AIAA President, Dr. Robert C. Seamans, Jr., said: "I believe that to understand adequately the challenges that confront those of us in aeronautical and aerospace activities, we must take as our perspective the commitments that challenge the nation as a whole. President Eisenhower, President Kennedy, and President Johnson, each in his own way, had a major impact on aeronautics and astronautics. And for each, his support of aerospace was a function of his belief that such efforts were instrumental in the accomplishment of national goals. In the future, as in the past, governmental support of aerospace will be based largely on its demonstrated relevance to the needs of the nation." (*A&A*, 2/69, 26–7)

March 1969

March 1: Terminal countdown for Apollo 9 mission, scheduled for launch March 3, began at 10:00 pm EST. (NASA Proj Off)

- ComSatCorp submitted *Report to the President and the Congress for the Calendar Year 1968.* Highlights included completion of three new ground stations in U.S. and seven in foreign countries, successful launch and operation of *Intelsat-III F-2* (Dec. 18, 1968), increase in INTELSAT membership to 63 nations, award of $72-million contract to Hughes Aircraft Co. for four Intelsat IV satellites, and phasing out of regular service of *Intelsat I* (Early Bird) after 42 mos of commercial service with 100% reliability.

  In 1968 ComSatCorp realized net income of $6,841,000 (68 cents per share), 47% increase over $4,638,000 (46 cents per share) earned in 1967. Revenues for 1968 totaled $30,495,000; they were $18,464,000 in 1967. Utilization of comsat system continued to increase, with ComSatCorp leasing 941 full-time circuits at end of 1968, up from 717 at end of 1967 and 73 at end of 1966. Demand for TV coverage of world news events increased, with 666 hrs of TV transmitted via satellite during 1968—nearly three times as many hours as in 1967. (Text; *Annual Rpt to Shareholders*)

March 2: Sud-Aviation chief test pilot André Turcat flew Anglo-French supersonic Concorde 001 prototype airliner in successful 27-min maiden flight from Toulouse-Blagnac Airport, France. Inclement weather, which had delayed event originally scheduled for Feb. 28, forced holding 193-ft-long, 200,000-lb aircraft to altitudes below 3,000 ft and maximum speed of 350 mph. Concorde was designed to fly at 1,400 mph at 12,000-ft altitude. Turcat pronounced flight "very satisfactory" and said aircraft "behaved perfectly" in 90° sweep around area. U.K. prototype would fly in six weeks and air worthiness certificates were hoped for by manufacturers Sud-Aviation and British Aircraft Corp. by end of 1972, so aircraft could enter service in 1973. (BAC PIO; AP, W *Star*, 3/3/69, A7; Wentworth, W *Post*, 3/3/69, A3)

- Dr. Charles A. Berry, Director of Medical Research and Operations at MSC, told preflight press conference Apollo 9 astronauts were "in a real fine state of health" for March 3 launch. Although two astronauts still had some minor throat infection, it would not interfere with planned launch time. Three-day postponement of launch from original Feb. 28 date had made possibility of inflight illness "exceedingly slim." Only addition to spacecraft's standard medical kit—which already included nasal emolient—might be throat lozenges. (Transcript)

- President Nixon addressed U.S. Embassy staff in Paris before departure for visit with Pope Paul VI at Vatican and return to U.S. after goodwill tour: ". . . the success of a policy depends upon thousands of people around, in an embassy like this and an establishment like this, and millions around this world," in same way that success of *Apollo 8*

March 2: *Concorde 001*, Anglo-French supersonic prototype airliner, lifted off on its maiden flight, from Toulouse-Blagnac Airport, France. (British Aircraft Corp. photo)

had depended on 400,000 Americans working on project. (*PD*, 3/10/69, 355)

- U.S. authorities reported U.S.S.R. had conducted mid-February test-firing of defense rocket that could intercept attacking missiles at 100 to 450 mi from its launch site. Rocket appeared comparable to U.S. Spartan interceptor planned for U.S. ABM system. U.S.S.R. also was reported making progress on phased-array radar judged essential for swift detection and handling of several attacking missiles at once. (Corddry, B *Sun*, 3/3/69, A1)
- Thomas O'Toole in *Washington Post* observed similarities among astronauts. Of 23 who already had flown in space, 21 were either only sons or eldest sons. Pattern tied in, he said, with psychologists' beliefs that only and eldest children tended to achieve more in life because they were disciplined more and trained and treated better by parents. Astronauts also were athletic, showed academic excellence, and had intense love of flying, O'Toole said. (*W Post*, 3/2/69)

March 3–13: NASA's *Apollo 9* (AS–504), first manned flight of Apollo lunar module, was successfully launched from KSC Launch Complex 39, Pad A, at 11:00 am EST by Saturn V booster—for extensive LM tests, extravehicular activity, and CSM–LM separation, rendezvous, and docking to simulate activities after lunar landing. Flight carried LM–3 and CSM–104. Launch had been postponed three days because crew had virus respiratory infections.

Primary objectives were to demonstrate crew, space vehicle, and mission support facilities performance during manned Saturn V mission with CSM and LM; demonstrate LM and crew performance, demonstrate performance of nominal and selected backup lunar orbit rendezvous (LOR) mission activities; and assess CSM/LM consumables. Multispectral photography experiment was carried for first time

March 3–13: Apollo 9, *first manned flight of the Apollo lunar module, successfully tested in space CSM–LM separation, rendezvous and docking, and extravehicular activities to simulate actions after a manned lunar landing. David R. Scott, pilot of the CM, stood in the open hatch of the docked CM with the earth as a backdrop, photographed by LM pilot Russell L. Schweickart from the porch of the LM.*

to provide photos of earth resources using several different film-filter combinations.

Launch events occurred as planned and spacecraft, carrying Astronauts James A. McDivitt (commander), David R. Scott (CM pilot), and Russell L. Schweickart (LM pilot), entered initial orbit with 119.5-mi (192.3-km) apogee and 117.6-mi (189.3-km) perigee. After post-insertion checkout CSM, code-named "Gumdrop," separated from Saturn V 3rd stage (S-IVB) and LM, code-named "Spider." Crew successfully transposed CSM and docked with LM, and docked spacecraft was separated from 3rd stage with RCS burn. Two S-IVB burns placed stage on earth-escape trajectory. Crew conducted first docked SPS burn.

On second day crew tracked landmarks, conducted pitch and yaw roll maneuvers, and increased apogee by firing SPS engine three times. On third day, McDivitt and Schweickart entered LM through docking tunnel, evaluated LM systems, transmitted first telecast, and conducted first manned firing of LM descent propulsion system (DPS). They then returned to CSM and conducted fifth SPS burn to circularize orbit. McDivitt and Schweickart reentered LM on fourth day and transmitted second telecast. Schweickart, recovered from earlier nausea, spent 37 min outside spacecraft, walking between LM and CSM hatches, maneuvering on handrails, and standing in "golden slipper" foot restraints. He commented on sun's brightness, photographed spacecraft and earth, and described rain squalls over KSC before he and McDivitt returned to CSM. Scott opened CM hatch and retrieved thermal samples from CSM exterior.

McDivitt and Schweickart reentered LM on fifth day to perform CSM-LM rendezvous. Scott separated CSM from LM and fired CSM reaction control system thrusters to place spacecraft about 3.4 mi (5.5 km) apart. LM DPS was ignited twice to set up rendezvous. LM descent stage was jettisoned and LM ascent propulsion system (APS) was fired to set up conditions for circularization. Although problems were encountered with crewman optical alignment sight (COAS) because of extremely bright reflections, radar and optical sightings backed up by earth tracking enabled spacecraft to dock successfully after being up to 114 mi (183.5 km) apart during 6½-hr separation. After McDivitt and Schweickart returned to CSM, crew jettisoned LM ascent stage and maneuvered to safe distance while stage burned to propellant depletion and entered orbit with 4,309-mi (6,934.5-km) apogee and 142.2-mi (228.8-km) perigee. By end of fifth day 97% of *Apollo 9* objectives had been successfully accomplished.

On sixth through ninth days crew conducted sixth and seventh SPS burns to alter apogee, tracked NASA's *Pegasus III* meteoroid detection satellite (launched July 30, 1965), took multispectral photos of earth, tracked landmarks, exercised spacecraft systems, and prepared for reentry.

Final SPS burn for deorbit on 10th day was delayed one revolution because of unfavorable weather in planned landing area. CM-SM separation, parachute deployment, and other reentry events were nominal; spacecraft reentered during 152nd revolution and splashed down in Atlantic 180 mi east of Bahamas at 12:53 pm EST March 13, 241 hrs 53 secs after launch, within sight of recovery ship U.S.S. *Guadalcanal*. Crew was picked up by helicopter and flown to recovery ship within one hour after splashdown.

Astronauts were welcomed by *Guadalcanal* crew and received congratulatory telegram message from President Nixon which said: "The epic flight of Apollo Nine will be recorded in history as ten days that thrilled the world. You have by your courage and your skill helped to shape the future of man in space. The three of you and the great team which enabled you to complete your successful mission have shown the world that the spirit of man and his technological genius are eager to begin an age of adventure, an age which will benefit all the people on this good earth."

All primary *Apollo 9* objectives were achieved and anomalies were not serious enough to alter mission operations or flight plan significantly. First manned flight of LM qualified last major component for lunar landing mission.

Apollo 9 was sixth Apollo mission to date and third manned Apollo mission. Earlier unmanned Apollo flights had yielded all spacecraft information possible without crew on board. *Apollo 4* (launched Nov. 9, 1967) and *Apollo 5* (launched Jan. 22, 1968) had both been highly successful, completing inflight tests of all major pieces of Apollo hardware. *Apollo 6* (launched April 4, 1968), despite launch vehicle problems, had attained four of five primary objectives with spacecraft recovered in excellent condition. First manned Apollo mission, *Apollo 7* (Oct. 11–22, 1968), had achieved all primary objectives and verified operation of spacecraft for lunar-mission duration. Second manned mission, *Apollo 8* (Dec. 21–27, 1968), proved capability of Apollo hardware and systems to operate out to lunar distances and return through earth's atmosphere.

Apollo program was directed by NASA Office of Manned Space Flight; MSC was responsible for Apollo spacecraft development, MSFC for Saturn V launch vehicle, and KSC for launch operations. Tracking and data acquisition was managed by GSFC under overall direction of NASA Office of Tracking and Data Acquisition. (NASA Proj Off; NASA Releases 69–29, 69–33; *PD*, 3/17/69, 400)

March 3: Following successful *Apollo 9* launching, President Nixon issued statement: "The successful launching of the Apollo 9 spacecraft marks another milestone in the journey of man into space. The hopes and prayers of mankind go with Col. James A. McDivitt, Col. David R. Scott, and Mr. Russell Schweickart on their courageous mission. The genius of the American scientist and technological community, which created and designed the Saturn V, the command ship, and the lunar module, once again stirs the imagination and gratitude of the world. We are proud of this American adventure; but this is more than an American adventure. It is an adventure of man, bringing the accumulated wisdom of his past to the task of shaping his future. The 10-day flight of Apollo 9 will, we hope, do something more than bring America close to the moon; it can serve to bring humanity closer by dramatically showing what men can do when they bring to any task the best of man's mind and heart." (*PD*, 3/10/69, 356)

- After watching *Apollo 9* launch at KSC, Vice President Spiro T. Agnew —NASC Chairman—told press he would be special advocate for space program. "I will lend whatever thrust I can to nudge the President into an awareness of what I consider of overriding importance." His interest in space was heightened by "the wonderful experience of visiting with astronauts, preparing for future missions," and seeing dedication of workers in all jobs connected with program. (B *Sun*, 3/4/69, A6)

- Aerobee 150 MI sounding rocket launched by NASA from WSMR with

VAM-20 booster carried GSFC experiment to 101.5-mi (163-km) altitude to search for cosmic x-ray radiation near Scorpius and north pole of galaxy. Rocket and experiments performed satisfactorily. (NASA Rpt SRL)

- USAF announced it had selected TRW Inc. for initial increment of $14 million to estimated $37,653,090 contract to design new, synchronous comsats for Phase II of Defense Satellite Communications System (DSCS). New satellites would be used with small surface terminals. Steerable, narrow-beam antennas would focus portion of satellite energy to areas of 1,000- or 2,000-mi dia and could be steered to different locations on earth's surface in minutes. Satellites could be moved to new position in days as needed. (DOD Release 148-69)

March 4: USAF launched unidentified satellite from Vandenberg AFB by Titan IIIB-Agena booster into orbit with 279.6-mi (449.9-km) apogee, 96.3-mi (155.0-km) perigee, 90.2-min period, and 92.0° inclination. Satellite reentered March 18. (GSFC *SSR*, 3/15/69; 3/31/69; *SBD*, 3/7/69, 30; *Pres Rpt 70* [69])

- NASA Acting Administrator, Dr. Thomas O. Paine, testifying before House Committee on Science and Astronautics on NASA FY 1970 authorization request, outlined goals in space and aeronautics toward which U.S. should move in next decade:

  "*First*—We should do all we can to understand and put into early use the promise of space for people here on earth. We should increase our scientific knowledge of the vital earth-sun relationship and study the earth itself from space. We should develop and experiment with new and improved practical applications of satellites, particularly in earth resources. We should continue to foster prompt introduction into the economy of space applications and technology.

  "*Second*—We should follow up the first Apollo landing with a sound program of manned lunar exploration.

  "*Third*—We should proceed with the development and experimental operation of a permanent U.S. space station in earth orbit.

  "*Fourth*—We should move out steadily in the exploration of deeper space, exploring the planets with unmanned probes and the sun, stars, and galaxies from orbital observatories outside the atmosphere.

  "*Fifth*—We should provide the technology for developing the full potential of U.S. civil and military aeronautics.

  "*Sixth*—We should maintain a strong momentum of broad technological advance in all aerospace disciplines."

  Although NASA's 1970 budget was "'holding' budget," Dr. Paine said request did include funds for starting three principal new programs: Earth Resources Technology Satellite Program, with start of ERTS A and B development; NERVA flight-weight engine development, postponed from 1969; and series of planetary explorers for future flights to Venus and Mars.

  In period of "retrenchment and declining resources," FY 1969 operating budget was $762 million below FY 1968 budget and over $1 billion below FY 1967. Nationwide employment on NASA work had decreased from earlier peak of 420,000 to 270,000 at end of FY 1968, to about 215,000 at end of FY 1969, and under FY 1970 budget to about 190,000. (Testimony)

- Apollo program after lunar landing was discussed by NASA Associate

Administrator for Manned Space Fight, Dr. George E. Mueller, before House Committee on Science and Astronautics hearing on NASA FY 1970 appropriations: "A thorough exploration plan has been evolved by the scientific community which will be initiated with the remainder of the fifteen Saturn V launch vehicles and Apollo spacecraft available under the Apollo program. Three initial phases of lunar exploration have been defined. The first phase will consist of landings that sample and observe the major classes of regions on the moon. To establish these norms, it will be necessary to land, carry out geological prospecting, and obtain rock and soil samples for return to earth from four separate sites. . . .

"The second phase would include the investigation of the major classes of lunar anomalies . . . volcanic types, sinuous riverlike channelways, fracture zones and impact craters. Six additional sites have been identified as the minimum . . . to provide answers to basic questions about the moon and to evaluate locations of potential resources, building materials or underground shelter openings. The third phase would be to tie together this information from 10 or more sites by making a remote sensing survey of the moon from lunar orbit."

Apollo Applications FY 1970 budget provided for "continuation of flight hardware development and for integration of modified subsystems into hardware for a set of five earth-orbital flights." Their completion in 1972 "terminates the manned flight activity until other manned flight programs are established." (Testimony)

- Philadelphia *Evening Bulletin* editorial on *Apollo 9* mission: "In the first Apollo launchings, it was the taming of sheer, brute power that awed the on-looker.

"At the moment of ignition, it was not what lay beyond the astronauts that gripped the millions watching TV. It was the question whether the huge Saturn V booster would respond to command, whether it would hurl the astronauts into orbit or collapse, toppling slowly into a furnace of its own making.

"But with yesterday's flawless Apollo 9 launch, the Saturn V booster seemed to emerge as a proven piece of space hardware. The preoccupation now is with the complexity, sophistication, the intricate workings of the most complicated of the several machines the United States has put together for the conquest of the moon. . . . In its sophistication and vulnerability, [the LM] is . . . an extension of man himself." (P *Bull*, 3/4/69)

- A number of MIT scientists, in day-long work stoppage, gathered to discuss uses and misuses of scientific knowledge, including military research, university-Government relations, disarmament, and responsibilities of intellectuals. Similar programs were held on 30 campuses across country, and Univ. of Pennsylvania in Philadelphia canceled all undergraduate classes for day. (Reinhold, *NYT*, 3/5/69, 1)

March 5: U.S.S.R. successfully launched two Cosmos satellites. *Cosmos CCLXVIII,* launched from Kapustin Yar, entered orbit with 2,161-km (1,342.8-mi) apogee, 211-km (131.1-mi) perigee, 109.1-min period, and 48.4° inclination and reentered May 9, 1970. *Cosmos CCLXIX,* launched from Plesetsk, entered orbit with 542-km (336.8-mi) apogee, 525-km (326.2-mi) perigee, 95.2-min period, and 74° inclination. (GSFC *SSR*, 3/15/69; 5/31/70; *SBD*, 3/7/69, 30)

March 5

- President Nixon, at White House ceremony, presented National Space Club's Robert H. Goddard Memorial Trophy to *Apollo 8* astronauts and announced nomination of Acting Administrator, Dr. Thomas O. Paine, to be NASA Administrator.

    Astronaut James A. Lovell, Jr., accepted award for himself and Astronauts Frank Borman and William A. Anders. Citation: "In an epic journey man for the first time in December 1968 soared out of the earth gravitational field, flew unerringly into a close orbit of the moon, then back to a precise and safe landing. This historic voyage performed at times before the largest television audience in history, and open for coverage by the world's press, reflects the utmost credit on the United States Space Program, Congress, NASA, and thousands of companies and employees in industry representing all these, the courageous, competent crew of Apollo 8."

    Following award presentation, President announced Dr. Paine's appointment: "There has been a great deal of interest as to who would be the new head of NASA. . . . we have searched the country to find a man who could take this program and give it the leadership that it needs, as we move from one phase to another. This is an exciting period, and it requires the new leadership that a new man can provide. But after searching the whole country for somebody, perhaps outside of the program, we found . . . that the best man in the country was in the program." Dr. Paine in his response said: "I believe in the space program. I believe in this country, and I think that this country should indeed be the preeminent nation in spacefaring, and . . . I am sure that we can go ahead to . . . see that the NASA program in the second decade of space will even out-perform the accomplishments in the first." (*PD*, 3/10/69, 369–71)

- At 12th annual Goddard Memorial Dinner in Washington, D.C., sponsored by National Space Club, Presidential Science Adviser, Dr. Lee A. DuBridge, said: ". . . we are witnessing another spectacular example of the utilization of scientific knowledge, accumulated by many generations of scientists, some famous and some obscure, who worked away in their laboratories trying to probe the secrets of nature. We have seen how once these secrets . . . have been revealed, engineering skills could be put to work. We see in our space program also an example of the reverse process. . . . New technological developments lead to new techniques and new instruments which . . . speed up our basic work in science and lead to new ways of uncovering new secrets of nature. Thus, during these past ten years the advance of science has been enormously aided by the advance of space technology." (*CR*, 3/12/69, S2755)

    Astronautics Engineer Award was presented to NASA's L/G Samuel C. Phillips (USAF) in absentia since *Apollo 9* mission was still under way. Citation read: "For his personal direction of the Apollo program throughout development and into the final phases of flight to the Moon dramatically illustrated by the two successful all-up flights of the Saturn V and the error-free flights of Apollo 7 and the extraordinary flight of Apollo 8 to the Moon and back. His engineering skill and leadership throughout the development and execution of this complex program have made possible an outstanding American success."

    Richard J. Allen of NASA Apollo Program Office Test Div. was awarded Hugh L. Dryden Memorial Fellowship presented to NASA employee adjudged "a deserving individual in disciplines applicable to science, astronautics and space administration."

Mitchell R. Sharpe of Systems Safety and Manned Flight Awareness Office, MSFC, received certificate, trophy, and $500 award for winning entry in 1969 Robert H. Goddard Historical Essay competition. (His *Development of the Lifesaving Rocket: A Study in 19th Century Technological Fallout* was released by NASA June 10 as MSFC Historical Note 4.)

National Space Club Press Award "for penetrating, consistently informed and lucid writing on all phases of the national space program" was awarded Evert Clark, Washington science correspondent, *Newsweek*. North American Rockwell Corp. received Nelson P. Jackson Award "for its major contribution to the success of Apollo during 1968." (Program; MSFC Historical Note 4)

- Dr. John E. Naugle, NASA Associate Administrator for Space Science and Applications, told House Committee on Science and Astronautics in testimony on NASA FY 1970 budget request: "We are moving toward a number of important milestones in the Space Applications Program, the most important of which is the Earth resources survey area, where we are requesting funds for the design and construction of a research satellite, Earth Resources Technology Satellite (ERTS). We think the ERTS Program will be a very valuable addition to mankind's tools for handling the natural and cultural resources of the world."

  FY 1970 OSSA program required $559 million in new obligational authority. Although increased over FY 1969, program was well below FY 1963 through 1967. Increases were in planetary and space applications program, with planetary increases primarily due to costs in FY 1970 of program authorized in FY 1969. In addition, NASA was proposing Planetary Explorer program and dual planet mission to Venus and Mercury for 1973. Space applications increase was for experimental ERTS satellite and prototype of operational Synchronous Meteorological Satellite (SMS). (Testimony)

- Harold B. Finger, NASA Associate Administrator for Organization and Management, testifying on budget request before House Committee on Science and Astronautics, described employment restrictions under FY 1970 operating plan: In May 1968, NASA had restricted employment to avoid "large and disruptive" personnel reduction in FY 1969. "As a result . . . the on-board manpower complement is being reduced by 1,285 in Fiscal Year 1969 and 559 new positions are being established . . . [at GSFC] for support service operations. From our highest civil service employment level of 34,126 in July 1967, the number of NASA's permanent employees will have decreased by 2,526 to 31,600 at the end of Fiscal Year 1970. During this same period, 965 positions will have been established specifically to convert certain support service functions carried out under contract with industry to civil service operations. This results in an effective decrease in manpower of 3,491, with a reduction from our planned manpower of 4,374. We are planning all of these reductions in personnel by attrition. . . . This is becoming increasingly difficult since the rate of separation has slowed appreciably. . . . The overall separation rate during the current year is only about two-thirds of the rate for Fiscal Year 1968." (Testimony)

- NASA Deputy Associate Administrator for Advanced Research and Technology Bruce T. Lundin, testifying on FY 1970 budget request, told House Committee on Science and Astronautics there was "a large opportunity to transfer the electronics technology developed for the space program to the problems of aeronautics, such as for collision avoidance and the development of all-weather capabilities." ERC was studying use of proximity-warning devices to avoid midair collisions and "possibilities of an aircraft-satellite link as a means of communication and precise navigation over water areas where present capabilities are limited. In the future, this method of navigation could provide precise position fixing in crowded domestic operations."

  Increased FY 1970 request for aeronautics research, $187 million, would fund program strengthening base of aeronautical advanced research, increasing technology base of short-haul transports, and increasing research and technology for aircraft noise abatement and safety. (Testimony)

- FAA released forecast of aviation growth through 1980. Airline passenger traffic would more than triple by 1980, with 470 million passengers compared with 152.6 million in 1968. Average annual growth would be 10%, less than in recent years. Revenue passenger-miles flown would reach 379 billion, from 106.5 billion in 1968. More than 90% of 3,600 airline aircraft in use would be jets; in 1968, 50% of 2,452 airline aircraft were jets.

  General-aviation fleet would total 214,000 by 1980, up from 114,186 in FY 1968. Civil aircraft production would more than double FY 1968 total of 15,044, reaching 33,950. Air carrier transport aircraft production would decrease gradually from FY 1968 record of 625 to 250. (FAA Release 69-29)

- Dr. James G. Harlow, President of West Virginia Univ., was sworn in as consultant to NASA Administrator-designate, Dr. Thomas O. Paine. He would serve on Management Advisory Council and had held similar position in 1961 and 1962. (NASA Release 69-39)

- MSFC announced it had signed agreement with North American Rockwell Corp.'s Rocketdyne Div. for extension of J-2 engine production through April 30, 1970, at reduced rate of one engine per month instead of three. Engines would not be used as rapidly as originally planned because of overall extension of launch vehicle production schedule. Modifications amounted to $8,423,454. (MSFC Release 69-70)

- North American Rockwell Corp.'s Space Div. had earned incentive award fees of $1,100,000 and $270,000 under NASA contracts for Apollo command and service modules and 2nd stage (S-II) for Saturn V launch vehicle, NASA announced. Awards were determined by Performance Evaluation Board on basis of NAR's achievement of management objectives specified in contracts from Sept. 1, 1967, through Dec. 31, 1968, which encompassed S-II contract activities from Aug. 4, 1968, through Dec. 28, 1968—first of three time increments extending through completion of Apollo program. (NASA Release 69-38)

March 6: *Cosmos CCLXX* was launched from Plesetsk by U.S.S.R. into orbit with 330-km (205.01-mi) apogee, 200-km (124.3-mi) perigee, 89.8-min period, and 65.4° inclination. Satellite reentered March 14. (GSFC *SSR*, 3/15/69; *SBD*, 3/7/69, 30; UPI, W *Star*, 3/6/69)

- Aerobee 150 MI sounding rocket launched by NASA from WSMR carried Columbia Univ. payload to 99.6-mi (160-km) altitude to study polarization of x-rays from Crab Nebula in 10- to 25-kev energy region, using x-ray polarimeter with lithium scattering blocks and gas proportional counters. All systems were perfect; good data were obtained. (NASA Rpt SRL)
- Members of Subcommittee on Manned Space Flight of House Committee on Science and Astronautics visited MSFC for inspection tour and hearings. Chairman Olin E. Teague (D-Tex.) was ranking Representative. (MSFC Release 69–72)
- MSFC announced it had awarded $1,954,999 cost-plus-incentive-fee contract to IBM Federal Systems Div. for designing, developing, and building five general-purpose digital computers for Apollo Telescope Mount project. (MSFC Release 69–71)

March 7: Aerobee 150 sounding rocket was launched by NASA from WSMR carrying Columbia Radiation Laboratories payload to conduct stellar spectra studies. Rocket and instruments functioned satisfactorily. (NASA Proj Off)

- *Time* magazine said Astronaut Frank Borman had amused audiences during February 1969 European tour by claiming *Apollo 8* astronauts deserved overtime pay because they had aged about 300 microseconds more than people on earth during moon mission. At NASA's request, Univ. of Maryland physicist Carroll Alley had calculated effects on astronauts of phenomena described in Einstein's relativity equations—time ran slower for object as its speed increased, and time accelerated for object as it moved away from body exerting gravitational force. Alley found *Apollo 8* spacecraft speed was predominant factor when it was within 4,000 mi of earth; time slowed and astronauts actually aged more slowly. Beyond 4,000 mi, effects of earth's gravity lessened as Apollo's time passed 300 microseconds faster than earth's.

  Despite Alley's calculations, said *Time*, Borman's claim was valid only for Astronaut William A. Anders, who made his first space flight on *Apollo 8*. Astronauts Borman and James A. Lovell had been crewmates on Dec. 4–18, 1965, *Gemini VII*, when time dilation effect was dominant for entire two weeks. They had aged less than those on earth by 400 microseconds. Lovell's time also had been slowed during four-day *Gemini XII* mission Nov. 11–15, 1966, by about 100 microseconds. "Thus," said *Time*, "during all their missions in space, Lovell and Borman respectively spent 200 and 100 microseconds less time than was recorded on earth—which means they were paid for more time than they actually worked." (*Time*, 3/7/69, 42)

March 7–8: NASA launched series of three Nike-Tomahawk sounding rockets from Dew Line station Pin Main, Cape Parry, Canada, carrying GSFC payloads to analyze electric fields from observed motions of neutral and ionized barium clouds during disturbed magnetic conditions. Four barium clouds on each rocket were released during disturbed magnetic conditions as planned. Good photographic coverage was obtained from all sites. (NASA Rpts SRL)

March 8: Anglo-French Concorde supersonic airliner made successful second flight at altitudes to 15,840 ft and speeds to 345 mph with droop nose lifted in flight. Sud-Aviation test pilot André Turcat termed flight "very satisfactory." (UPI, *NYT*, 3/10/69, 91)

- *Christian Science Monitor* congratulated "Gumdrop, Spider, and their gallant crew" for successful *Apollo 9* mission to date: "In a series of smoothly executed maneuvers, they have opened the way for landing on the moon. They have removed any doubt that the machinery and tactics designed to set men down and recover them safely can indeed do the job." Spectacular view from space symbolized "men's ability to surmount human and natural obstacles to reach the most difficult goals when they want to. It bespeaks a spirit running counter to the rivalries, hatreds, and selfishness that often seem to prevent a similar attack on tough human problems on earth. The spirit now has brought the moon within mankind's grasp. It could bring a better life on earth within their grasp, too." (CSM, 3/8–10/69)

March 9: While spaceborne, *Apollo 9* Astronaut James A. McDivitt was selected to receive Ancient Order of Hibernians' John F. Kennedy Medal for National Civic Service. Award would be presented by Irish-descent society in Newark, N.J., May 10. (Sehlstedt, B *Sun*, 3/10/69, A1; *W Post*, 3/10/69, A5)

March 10: Nike-Tomahawk sounding rocket launched by NASA from Churchill Research Range carried Univ. of Colorado payload to measure Vegard-Kaplan (1,500–1,700 Å and 2,300–3,000 Å) and Lyman-Birge-Hopfield (1,300–1,500 Å) bands of nitrogen in UV auroral emissions. Rocket and instruments performed satisfactorily. Payload penetrated aurora, and data were excellent. (NASA Rpt SRL)

- "Man has three sets of capabilities that make him extremely hard to replace by any machine," NASA Associate Administrator for Manned Space Flight, Dr. George E. Mueller, said in address before NRC Engineering Div.'s annual meeting in Washington, D.C.: "1) he has a very wide-band set of sensors for acquiring information; 2) he has in his head a built-in memory and computer that cannot yet be matched by our largest and fastest machines; and finally he has a remarkably versatile capability for action, and physical operations with his body, hands, and tools. These three capabilities make man such a valuable element in space science and applications that we need to take advantage of him at the site of operations whenever this is feasible. . . . With the prominence that manned flight has gained, it is surprising to realize that while 589 unmanned satellites have been launched, there have been only 29 manned flights, eleven by the USSR and eighteen by the USA. These few experiments have shown that we have the capability to extend man's genius into the new dimension of space, as far out as the moon. In the years ahead I expect that man will exploit his bridgehead into space and use this new territory for his own good on earth." (Text)

- NASA released *Space Resources for Teachers: Biology, Including Suggestions for Classroom Activities and Laboratory Experiments*. Curriculum project was prepared by Univ. of California at Berkeley to introduce high school teachers and students to scientific advances in space biology. (Text)

March 10–12: During AIAA 3rd Flight Test, Simulation and Support Conference in Houston, Tex., *Apollo 7* Astronauts Walter M. Schirra, Jr., Donn F. Eisele, and R. Walter Cunningham received Haley Astronautics Award for "exceptionally meritorious service and outstanding contributions to the advancement of manned space flight during the

11-day flight of Apollo 7." Award—$500 honorarium, medal, and certificate—was presented annually for "an outstanding contribution by test personnel who undergo personal risk in the advancement of space flight."

Lockheed Aircraft Corp. test pilot William C. Park received Octave Chanute Award for "flight test development of Mach 3+ aircraft." He was first test pilot to fly SR–71 and YF–12A at design speed. (AIAA Release 3/5/69)

March 11: Apollo 10 spacecraft, atop Saturn V launch vehicle, rolled out of Vehicle Assembly Building at KSC to Launch Complex 39, Pad B, for May liftoff to lunar orbit. (AP, *NYT*, 3/12/69, 14; UPI, *Huntsville News*, 3/12/69)

- Dr. George E. Mueller, NASA Associate Administrator for Manned Space Flight, testified on Apollo program, space flight operations, and advanced manned missions before House Committee on Science and Astronautics' Subcommittee on Manned Space Flight. He explained importance of lunar exploration and outlined plans for new space station.

  President's Science Advisory Committee had said answers to questions about moon "may profoundly affect our views of the evolution of the solar system and its place, as well as man's in the larger scheme of things." Fact that earth's moon was largest in relation to its planet, Dr. Mueller said, "implies that the two bodies may have been formed in the same manner at the same time. If true, the moon may be a book containing the secret of the earth's first billion years of life. This record is lost on the earth which is subjected to . . . erosion by atmosphere and water. . . . By comparing similarities and contrasting differences, man may be able to arrive at a greater understanding of the fundamental processes that affect the earth; for example, the mechanisms that cause earthquakes and volcanic eruptions, and the processes responsible for concentrating ore deposits. The orbits of Apollo 8 and the Lunar Orbiters were disturbed by mass concentrations beneath the circular lunar seas. These may be huge meteors that struck the moon with such force that they melted and sank into the interior, or they may be iron deposits.

  "Another objective of lunar exploration is to learn about man as a space explorer—his capabilities and limitations. Some day man will move on to other planets; the moon is a training ground. . . . We don't have the basic information which early lunar landings will furnish and we can only speculate today about the feasibility of the moon as a base for an observatory or a permanent science station—about exploiting its environment of low gravity and high vacuum—about its potential for natural resources. . . . A long-range goal like the lunar base would direct technological advances, stimulate public interest, and attain subsidiary objectives with earth application such as food synthesis, environmental control, and recovery of useful elements from rock."

  Within new program category, Space Flight Operations, NASA was bypassing intermediate space station, launch vehicles, and logistic craft and proposing to move directly to new, semipermanent space station and low-cost earth-to-orbit transportation system. Space station "should be in being by the mid-1970s." With FY 1969 funds, contractor definition efforts were being initiated. FY 1970 funds would continue definition, preliminary design, and supporting work.

Dr. Mueller described space complex 200 to 300 mi above earth planned for 10 yrs' continuous operation and adaptable for crew size, additional laboratory facilities, or other special-purpose equipment through selection, design, and arrangement of component modules. Crew would rotate at three- to six-month intervals, ferried between station and earth by reusable shuttles. Station's electric power would come from solar panels or small nuclear generator. (Testimony; NASA Release 69-49)

- Lee R. Scherer, Director of NASA's Apollo Lunar Exploration Office, testifying on FY 1970 budget request before Subcommittee on Space Science and Applications of House Committee on Science and Astronautics, described extravehicular activity (EVA) planned for first lunar landing.

Emerging on lunar surface, astronaut would acquire and seal bagful of surface material and place it in spacesuit pocket, inspect LM, and with geological tools, fill large sample container. Location would be photographed before and after sampling. Passive seismometer would be emplaced to measure "moonquakes" and permit estimates of moon's internal structure for comparison with that of earth. Instrument, operated on solar power, would record and transmit lunar data to earth for about one year. Laser retroreflector would be emplaced to permit measurement of earth-to-moon distance and monitoring of distance variation. Measurements would be repeated several times daily for year or more and precise times determined by atomic clock. From data on relative motion of moon and earth and of each about own center of gravity, scientists could refine "knowledge of size and shape of the Moon, detect small movements that may occur between the Earth's continents, and perhaps even test gravitational theories." If time permitted, second sample return container would be filled more selectively, with greater effort to document and to pack samples individually. (Testimony)

- Dr. Thomas O. Paine, NASA Administrator-designate, discussed possibility of integrated European space program alongside that of U.S. and U.S.S.R. in Thomas A. Edison Memorial Lecture at Naval Research Laboratory, Washington, D.C. "The space-age challenge to Europe in my view is to find new ways of organizing and managing the great wealth of that continent to overtake, if you like, major American and Soviet space and other programs. This is the space-age challenge to Europe: not the 'technology gap' but the 'management gap.'"

Dr. Paine also said, if 1969, 1971, and 1973 space probes should find "very exciting things about Mars, it is possible that we or the Russians might want to move a manned interplanetary expedition to a higher order of national priority. At the present time, the best guess is that this would not take place until the mid 1980's but new priorities might lengthen or shorten this period."

In reply to question on man-in-space justification, he said: "For almost any simple specific experiment . . . an automated probe can be developed to do it at lower cost. The basic question here is the difference between short-range research projects to achieve simple specific objectives, and complex research aimed at developing general exploratory capability and broad technological positions which will let you do things in the future which you can't yet fully foresee. For the latter you will need men."

Dr. Paine expected U.S.S.R. "very likely" would put space station into orbit before U.S. In lunar exploration, he expected U.S. to stay ahead a while. In planetary probes, he expected U.S.S.R. to stay ahead. In earth applications, he expected U.S. to stay ahead. (Text)

- Dr. Robert C. Seamans, Jr., Secretary of the Air Force and former NASA Associate and Deputy Administrator, delivered 1969 Minta Martin Lecture at MIT as 1968–69 Jerome Hunsaker Professor of Aeronautical Engineering. Describing origins and development of U.S. space program as well as decision-making and implementation in commitment to manned lunar landing, paper would be presented also at Univ. of Maryland March 20 and at LeRC April 24.

Citing need for set of national goals, Dr. Seamans defined national agenda for allocation of R&D that might be acceptable to man in street: understanding, forecasting, and controlling environment; supplying basic resources of food, fuel, minerals, and water; improving quality of life; improving transportation; improving communications; encouraging economic growth; and assisting international development and providing national security.

Space program contributed to all categories. Studies of sun's transmissions, of Mars and Venus atmospheres, of moon's origin, and of earth itself contributed to understanding of environment for practical use of man. Space R&D might aid search for alternative basic resources; it could establish facilities for detecting available resources and speed communications to meet problems. Biological and medical investigations in space program held greatest promise in study of reactions of biological specimens, animals, and man. Impact on education had been cited by some as greatest value of space exploration. Program had required in-depth investigation of waste management, fire prevention, materials development, and microminiaturization. Space technology influenced new vehicles being tested for transportation—hydrofoil ships, surface-effect and air-cushion vehicles, high-speed trains, electric cars, VTOL and V/STOL aircraft, jumbo subsonic aircraft, and supersonic aircraft.

Communications satellites, already in commercial use, ultimately could broadcast directly to home receivers. In international cooperation, NASA was working with 64 nations in space activity. Returns from NASA's scientific satellites and communications satellites, development of manned space flight capability, and tracking facilities contributed to national defense. (Text)

- In Houston, Tex., press interview Astronaut Walter M. Schirra, Jr. (Capt., USN), said he had turned down offers for Naval promotion to admiral and NASA Hq. executive position, to become president of Regency Corp. in Denver, Colo., financial complex leasing industrial equipment worldwide. One of original astronauts and only veteran of Mercury, Gemini, and Apollo missions, Schirra said he did not want to "stick around as a half astronaut." He had rejected aerospace offers because "I might be limiting the contribution of which I feel I'm capable." (AP, *W Post*, 3/12/69, A8)

March 12: Lunar Science Institute and MSC's Lunar Receiving Laboratory were preparing for "lunar knowledge explosion" expected to follow first manned lunar landing, Washington *Evening Star* reported. Laboratory would quarantine astronauts and lunar samples brought back from landing, storing samples for study and distribution to scientists. Institute, funded by NASA and administered through NAS by Rice Univ., would facilitate lunar research by nonspace scientific community, Administrator Robert C. Wimberley said in interview. Directed by William C. Rubey, Institute was soon to be turned over to consortium of universities.

U.S. Public Health Service, NAS, Dept. of Interior, Dept. of Agriculture, and NASA had planned 30-day quarantine and study of moon material and 21-day quarantine of astronauts. Samples would be sealed on lunar surface and would be flown from recovery carrier to Laboratory after spacecraft return. Astronauts were to be transferred from spacecraft into sealed van through plastic tunnel aboard carrier [see May 16]. Van would be delivered to nearest port and then flown to Houston to sealed-off laboratory. Objective was to prevent importing viruses, some of which had been known to survive thousands of years under unlikely conditions.

During three-week debriefing, crew would live in glassed-in quarters including medical area. Samples would be maintained under vacuum seals until opened in high-vacuum chambers containing remotely controlled equipment. Once photographed, catalogued, and processed, content and structure would be studied to determine composition and origins. (W *Star*, 3/12/69, A13)

- NASA advocated "balanced, broad-based Planetary Exploration Program" as "feasible and practical" part of U.S. space program, Donald P. Hearth, Director of Planetary Programs, OSSA, testified before House Committee on Science and Astronautics' Subcommittee on Space Science and Applications. Basic goals were to increase understanding of origin and evolution of solar system and life in solar system and understanding of dynamic processes shaping man's terrestrial environment—"increasing our understanding of the planet Earth, how it has evolved, and how it may evolve in the future through a comparative study of the other planets in the solar system."

  Approach would be broad-based exploration of several planets—Jupiter, Mercury, Venus—over period of time, combined with direct measurements of Mars surface in orbital and lander missions. Pioneer F and G spacecraft (planned for launch in 1972 and 1973) and new cooperative project with West Germany, Helios, would open "new era" in solar system exploration, studying effects of radial distance on interplanetary medium close to sun. (Testimony)

- Milton Klein, Manager, NASA Space Nuclear Propulsion Office, testified on nuclear rocket capabilities before Subcommittee on Advanced Research and Technology of House Committee on Science and Astronautics: Benefits to be realized included "significantly extended performance of launch vehicles which now exist or may be operational by the late 1970's, feasibility of certain near-Earth missions otherwise impractical," and "increased potential for mission success." One basic propulsion system could be developed to yield all these benefits. (Testimony)

- NASA announced appointment of Gerhard B. Heller as Director of MSFC

Space Sciences Laboratory, replacing Dr. Ernst Stuhlinger, who had been promoted to new position of MSFC Associate Director for Science. Heller had worked in rocket R&D since 1940, when he joined Wernher von Braun group in Germany. For more than 10 yrs he had directed fluid and thermodynamics research at Laboratory and its predecessors. He was responsible for thermal design aspects of several satellites, including *Explorer I*, first U.S. satellite. (MSFC Release 69–75)

- MSFC announced $48,142,823 modification to contract with North American Rockwell Corp. Space Div. to provide for 14-mo extension of S–II program as part of overall Apollo program stretchout. S–II was 2nd stage for three-stage Saturn V launch vehicle.

  MSFC also announced $1,896,916 modification to Apollo Applications program contract for further Harvard College R&D of modified UV scanning spectrometer to fly in space on Apollo Telescope Mount (ATM). (MSFC Releases 69–76, 69–77)

- U.S. Equal Employment Opportunity Commission opened three-day hearings in Los Angeles on extent of racial discrimination in aerospace industry. Commission's figures showed only 177, or 0.9%, of 20,000 officials and managers in Los Angeles aerospace industry in 1968 were black. (*Pasadena Star-News*, 3/13/69)

- USN ended 17-day investigation into Feb. 17 death of Sealab III aquanaut Berry L. Cannon. USN had disclosed that one of rigs worn by four divers did not contain chemical to filter carbon dioxide from recycled helium-oxygen breathing mixture, but photographic experts had been unable to determine which rig was worn by Cannon. Findings were not expected to be released for some time. (UPI, W *Star*, 3/13/69, A3)

March 13: Shortly after *Apollo 9* splashdown [see March 3–13] Dr. George E. Mueller, NASA Associate Administrator for Manned Space Flight, told press at MSC, mission was "as successful a flight as . . . any of us have ever seen." It had "fully achieved all of its primary objectives and in numerical count, we accomplished more than the planned number of detailed test objectives." Apollo program would move toward greater scientific content each mission as NASA developed its capabilities. ". . . we have been remarkably successful thus far in the Apollo program. The hardware has worked better than anyone should have expected, and better than any of us did expect." Mission profile for Apollo 10, scheduled for launch May 17, would be released March 24 after careful evaluation of *Apollo 9* results.

"We are doing the planning for . . . 10 vehicles beyond Apollo 11." After first landing NASA would stretch out flight schedule to allow time to make modifications and to understand results, with about two or three landings per year. NASA had equipment for scientific payloads for first four Apollo vehicles—of which two would land in lunar maria and two in highland areas—which would provide capability for later pinpoint landings at points of unique scientific interest outside Apollo landing zone. NASA would emplace on lunar surface network of 10 seismographs, series of heat-measuring probes, cameras, and other geophysical instruments to determine environment of lunar surface. Astronauts would wear new, "constant volume" spacesuits which would enable them to move around lunar surface very freely to conduct experiments. (Transcript)

March 13

- Dr. Lee A. DuBridge, President's Science Adviser, told AIAA board luncheon in Washington, D.C., that President's Space Task Group had forged ahead with plans for new space program by Apollo program's end but he did not see specific mechanism in Nixon Administration for handling "tremendous" aviation problems. He had raised question at first Task Group meeting, but budget squeeze had cut DOT request for civil aviation research by two-thirds. Noise would be dealt with by Cabinet-level committee on environmental quality. President Nixon was "anxious to extend our technical and scientific collaboration with other nations." Recent visit to U.S.S.R. by U.S. scientists had led to plans for more extensive collaboration in high-energy physics. (*A&A*, 4/69, 5; AIAA PIO)
- NASA Aerobee 150 sounding rocket launched from WSMR with Naval Research Laboratory experiment collected data that led to discovery of x-ray pulsar in Crab Nebula. Sounding rocket carried several UV detectors and gas proportional counters to 103-mi (153.5-km) altitude to obtain x-ray spectra in Crab Nebula, region in galactic plane, region near cluster of galaxies, and region toward earth. All detectors gave good data and NRL scientists later reported pulsations were observed at frequency closely matching radio and optical pulsations with 5% of total x-ray power of nebula appearing in pulsed component. Pulsations included main pulse and interpulse separated by about 12 milliseconds. (NASA Rpt SRL; Fritz, Henry, Meekins, Chubb, Friedman, *Science*, 5/9/69, 709–12)
- Orr E. Reynolds, NASA Director of Bioscience Programs, OSSA, testified before Subcommittee on Space Science and Applications of House Committee on Science and Astronautics that species used in NASA *Biosatellite II* radiation experiments—drosophila, lysogenic bacteria, flour beetles, neurospora, and spiderwort—had been used as biological experiments on *Zond V*. "The fact that both nations are using many of the same species of organisms for space radiation studies," he said, "offers a considerable advantage to both for comparison and confirmation of experimental results."

    Reynolds also said that space flight offered bioscientists only known keys to number of life phenomena, "some fundamentals of life that must be sought here in . . . the space program, because scientists know of no other experimental environment that will serve." Role of gravity in life processes, cyclical behavior of living organisms, and origin and character of life in universe were areas for which NASA had unique research capabilities. (Testimony)
- Senate approved nuclear nonproliferation treaty without change by vote of 83 to 15, in what Joseph Sterne in Baltimore *Sun* called most "lopsided" margin accorded to major pact involving U.S.S.R. "since the cold war began." It gave President Nixon "clear signal to pursue his policy of 'negotiation rather than confrontation' with the Kremlin." Pact had been signed by 87 nations and ratified by 9. It would go into effect when ratified by U.S., U.S.S.R., U.K., and 40 other nations. (*CR*, 3/13/69, D180; B *Sun*, 3/14/69, A1)
- President Nixon transmitted to Congress first annual plan for U.S. participation in World Weather Program: "This project, and our role in it . . . have great political significance." Program had "developed into a most impressive example of international cooperation. On

a scale never attempted until this decade, scientists and governments in many countries are joining hands across national boundaries to serve the entire human community. Their example should be instructive for all of us as we pursue lasting peace and order for our world." (PD, 3/17/69, 399–400)

- MSFC announced it had completed negotiations with North American Rockwell Corp. Rocketdyne Div. on $4,075,490 contract modification extending F-1 engine deliveries through June 1970 to align engine effort to stretchout in production rate of Saturn V boosters. F-1 engines for initial order of 15 Saturn V boosters had been slated for April 1969 delivery.

  MSFC also announced award of $1,239,045 contract modification to IBM for work in connection with 31 launch-vehicle data adapters and 31 launch-vehicle digital computers. Adjustments, necessitated by changes in Apollo launch schedule, provided for streamlining contractor's plan for product identification control and accounting system. (MSFC Releases 69–78, 69–81)

- Senate confirmed nomination of former NASA Associate Administrator for Advanced Research and Technology James M. Beggs as Under Secretary of Transportation. (CR, 3/13/69, S2833)

- Dr. Robert C. Seamans, Jr., Secretary of the Air Force, resigned as president of AIAA. Letter of resignation said, "It is not in the best interest of either the Air Force or the AIAA for me to continue with AIAA responsibility." AIAA board of directors elected Dr. Ronald Smelt, Vice President and Chief Scientist, Lockheed Aircraft Corp., to replace Dr. Seamans. (AIAA Release)

- *Krasnaya Zvezda* (*Red Star*) claimed world heavy-lift helicopter record for U.S.S.R., citing 68,266-lb payload carried to 9,675-ft altitude by V–12 helicopter at 595-ft-per-min rate of climb. Four world records in 15-, 20-, 25-, and 30-ton-weight categories at 2,950-m altitude also were claimed. (*AFJ*, 3/22/69, 6)

March 14: NASA Wallops Station recovered parachute and payload launched on two-stage Sidewinder-Arcas meteorological sounding rocket. Recovery—by fixed-wing Cessna 206 aircraft—was third success to date. Flight investigated performance of Sidewinder-Arcas rocket system and demonstrated deployment of 16-ft-dia, disc-gap-band parachute at high altitude. Wallops scored first aerial recovery Nov. 9, 1966, when payload launched on Argentina Orion II sounding rocket was snatched by helicopter. (WS Release 69–7)

- Aerobee 150 sounding rocket launched by NASA from WSMR carried GSFC-Univ. of Wisconsin-Smithsonian Astrophysical Observatory payload to 109.8-mi (176.7-km) altitude. Objective was to provide precisely calibrated stellar observations of stars Regulus, Spica, Denebola, and Benetnasch, which had been viewed by NASA's *Oao II*—using 12.9-in-dia Dall-Kirham telescope, plane-grating spectrograph, and STRAP III stellar pointing attitude control system. Rocket and instruments performed satisfactorily, but attitude control system malfunctioned. First target was acquired and data were obtained; second and third targets were not acquired; and fourth target was acquired but not identified. (NASA Rpt SRL)

- Senate Committee on Aeronautical and Space Sciences ordered favorably reported the nomination of Dr. Thomas O. Paine as NASA Administrator. Nomination awaited approval by Senate. (NASA *LAR*, VIII/42)

- President Nixon proposed deployment of $6- to $7-billion modified, "Safeguard" ABM system using components developed for Sentinel, but altering deployment to provide local defense of selected Minuteman sites, area defense to protect bomber bases and command and control authorities, defense of continental U.S. against accidental attack, and "substantial protection against the kind of attack which the Chinese Communists may be capable of launching throughout the 1970's." Deployment would not place missile and radar sites near major cities (except Washington, D.C.). President said safety of country "requires that we should proceed now with the development and construction of the new system in a carefully phased program. . . . The program is not provocative. The Soviet retaliatory capability is not affected by our decision." (*PD*, 3/17/69, 400–9)
- FAA announced signing of two-year agreement with Air Transport Assn. of America to participate in ATA's airborne collision avoidance program. FAA would take part in testing and evaluation of airborne CAS systems supplied by industry in ATA-funded program. (FAA Release 69–34; ATA Release 22)
- Soviet youth newspaper *Komsomolskaya Pravda* praised NASA's *Apollo 9* mission: "A thorough check of the entire equipment necessary for a lunar landing has been attained, thanks to the courage and gallantry of the three American astronauts. At the same time the Apollo 9 flight showed that it is impossible in terrestrial conditions to envisage all difficulties astronauts are encountering in real flight."

    Soviet Academician Prof. Anatoly Blagonravov complimented NASA's planning for lunar landing: "The fact that the Americans earlier made a flight around the moon and now conducted the Apollo 9 experiment, in my opinion, is evidence of a rather complete solution of the problem of dependability," though some risk always remained. (UPI, W *Star*, 3/14/69, A6)
- Lockheed Aircraft Corp.'s Hummingbird experimental VTOL aircraft, officially designated XV–4B, crashed 22 mi from Dobbins AFB, Ga., during research flight. Civilian test pilot Hal J. Quamme parachuted to safety. Aircraft was being tested and developed for USAF. (UPI, P *Inq*, 3/15/69, 1; AP, W *Star*, 3/16/69, A8)
- *Los Angeles Times* editorial observed space industry employment drop from 400,000 in mid-1960s to current 200,000—at which rate it would sink to 50,000 by 1972. ". . . the pool of scientific and technical expertise which has been brought together in the manned space program is disintegrating. If the President wants to save what is really an invaluable national resource, he and Congress cannot wait much longer to assign new projects to the space agency. . . .

    "A sound space program deserves a high place on the scale of priorities. And, both economic and military considerations dictate that the emphasis should be heavily on activities in the space near earth rather than on esoteric exploration of such deep planets as Mars and Venus. The proposal for construction in earth orbit of a large, permanent scientific laboratory . . . seems to deserve serious consideration as the major space project of the 1970s." (*LA Times*, 3/14/69)

March 15: U.S.S.R. launched *Cosmos CCLXXI* from Plesetsk. Orbital parameters: apogee, 310 km (192.6 mi); perigee, 187 km (116.2 mi); period, 89.7 min; and inclination, 65.4°. Satellite reentered March 23. (GSFC *SSR*, 3/15/69; 3/31/69; *SBD*, 3/18/69, 78)

- In press interview Transportation Secretary John A. Volpe said he favored building SST. "I certainly don't want to see our country play second fiddle . . . and if this is feasible and economical to build and operate and the economy of the nation budget-wise will permit it, I probably would like to see it go ahead." (UPI, W *Star*, 3/16/69, A11)

- In letter to Ambassador Gerard C. Smith, Head of U.S. Delegation to Geneva Disarmament Committee Meeting and Director of Arms Control and Disarmament Agency, President Nixon issued instructions: Delegation was to seek discussion of international agreement prohibiting placement of weapons of mass destruction on seabed, support conclusion of comprehensive test ban adequately verified, press for agreement to end production of fissionable materials for weapons and transfer materials to peaceful uses, explore proposals for control of chemical and biological weapons, understand that actual reduction of armaments remained U.S. goal, and remember U.S. hoped international situation would permit U.S.–U.S.S.R. talks on strategic arms limitation soon. (*PD*, 3/24/69, 434–5)

- First Soviet press and radio reports of President's ABM announcement spotlighted—without comment—statement that proposed program should not complicate U.S.–U.S.S.R. nuclear arms control talks. Moscow Radio described decision as "compromise" after "exceptionally fierce" congressional struggle over any "antimissile defense system" with "monopolistic corporations and Congressmen faithful to them" favoring powerful system. (Marder, *W Post*, 3/16/69, A1)

March 15–17: NASA released *Apollo 9* pictures—including detailed shots of CM, LM, S–IVB, and Astronaut Schweikart during EVA; 70-mm color still pictures; and 16-mm color movies. Pictures of earth showed clear views of major cities across southern U.S. and U.S. East Coast from North Carolina to Long Island. (UPI, *C Trib*, 3/18/69; W *Star*, 3/16/69, A1; *W Post*, 3/18/69, A1)

March 16: Apollo 9 mission had gathered new evidence of clearer visibility in space and "marvelous" acuity of human eye, Associated Press reported. Astronauts had told officials onboard recovery ship after splashdown they could see much farther in space than in earth's atmosphere—both with telescopes and with naked eye. They had tracked several orbiting space objects up to 1,600 mi away with telescope and had tracked jettisoned 2½-ft-long, 14-ft-dia LM ascent stage to 1,000 mi. (*W Star*, 3/17/69, A4; *W Post*, 3/17/69, A4)

- More than 2,000 Americans had made reservations with Pan American World Airways and Trans World Airlines on first commercial trips to moon, said Joe McCarthy in *This Week*. Downpayments were not being accepted, but lunar reservations were being confirmed, acknowledged by letter, and placed in order on waiting list. PAA spokesman had said, "It will undoubtedly be an expensive trip. When we finally start asking the passengers for money, a lot of them will probably drop off the list." (*This Week*, 3/16/69, 9–10)

- In *Washington Post*, Victor Cohn said fight for ideas "which once would have been labeled 'wild' " had begun when NASA began congressional committee hearings on its $3.7-billion FY 1970 budget [see March

4]. It had continued with announcement of "dramatic plan for ten more manned expeditions to the moon after the first men land there in July." Americans had responded to "exciting" Apollo program "with heightened fervor for a man in space." But new fight for funding would probably be "NASA's toughest." Americans were asking, "How about our needs here on earth?" Many scientists felt unmanned instruments could do cheaper and nearly as effective job of space exploration. Congress had been calling NASA's present spending level "enough." Still, NASA had succeeded in marshaling "powerful scientific support."

As yet unreleased report of Dr. Charles H. Townes' space task force appointed by President Nixon urged both vigorous manned space program and development of reusable space shuttle. Presidential Science Adviser, Dr. Lee A. DuBridge, had declared himself for "a really solid, many-faceted program." NASA officials were optimistic but funds actually appropriated for FY 1970 budget would finance only five more Apollo flights plus first hardware for Apollo Applications program. Public thinking was "just about 50–50 today, and new U.S. consensus has not yet crystallized." (*W Post*, 3/16/69, A1)

- First royalty-bearing license under NASA's foreign patent program had been granted to Nippon Electric Co., Ltd. in Tokyo, NASA announced. Invention bearing NASA-owned patent No. 484,436 and made by GSFC's Joseph G. Haynos, was concerned with connections between solar cells that permitted flexibility and low weight. Company had made initial payment to NASA for exclusive manufacturing rights in Japan and agreed to continue royalty payments for duration of license. (NASA Release 69-40)

- Bitter battle was raging in Nixon Administration over construction of SST, said Robert H. Phelps in *New York Times*. Opposition to 1,800-mph aircraft had been rising since Feb. 7 appointment by President Nixon of 11-member interdepartmental study committee. Indications were that majority would recommend shelving project until technical, economical, and environmental problems, particularly aircraft noise, were closer to solution. President Nixon had inherited controversy from Johnson Administration, which had not earmarked funds for SST. He would have to decide whether to ask Congress to appropriate $212 million to $247 million to keep project on schedule. (*NYT*, 3/16/69, 1)

March 17: U.S.S.R. launched *Cosmos CCLXXII* from Plesetsk into orbit with 1,210-km (751.8-mi) apogee, 1,180-km (733.1-mi) perigee, 109.3-min period, and 73.9° inclination. (GSFC *SSR*, 3/31/69; *SBD*, 3/19/69, 86; AP, *W Star*, 3/18/69)

- Canister containing S–16 barium cloud experiment was successfully ejected from ESRO's *Heos I* satellite (launched by NASA Dec. 5, 1968) at 43,495.9-mi (70,000-km) altitude. Cloud, 1,864.1 mi (3,000 km) long and visible to the naked eye, had lasted 20 min and yielded information on magnetic fields in space. (NASA Proj Off)

- Apollo 10 mission would be launched May 18, one day later than originally planned, NASA announced. Change from first to second day of launch window would permit observation and collection of data on Apollo landing site 2, as area of primary interest, and observation of site 3 after sunrise on moon. Final decision on specific nature of mission would be made after review of *Apollo 9* mission. (NASA Release 69–41)

- Nike-Tomahawk sounding rocket launched by NASA from Fairbanks, Alaska, carried Univ. of Alaska payload to 134.2-mi (216-km) altitude. Objectives were to examine spatial distribution of atmospheric currents near visual, stable, homogeneous and pulsating auroral forms and to examine relationship between electron and proton precipitation and visual auroral distribution produced. Forward experiment portion of payload functioned satisfactorily, but data reduction was difficult. Some data loss occurred on rear portion of payload and additional data reduction was required. One flashing light failed at apogee. (NASA Rpt SRL)
- Eleventh anniversary of launch of *Vanguard I*, 3¼-lb, 6½-in-dia U.S. IGY satellite which proved earth was slightly pear-shaped and examined composition of upper atmosphere. Satellite had stopped transmitting in May 1964, but was expected to remain in orbit at least 200 yrs longer. (EH; KSC Release 63–68)
- Vice President Spiro T. Agnew announced Astronaut James A. Lovell, Jr., would head $4.5-million summer day camp program for youth to be administered by National Collegiate Athletic Assn. under HEW contract. (Text)
- Dr. Leo S. Packer, former Assistant Postmaster for Bureau of Research and Engineering, became Special Assistant to NASA Associate Administrator for Advanced Research and Technology. (NASA Ann, 3/21/69)
- White House released letter to President Nixon from Dr. Lee A. DuBridge, Presidential Science Adviser, on proposed Safeguard ABM system: It "eliminates the serious defects of the old Sentinel plan, focuses on the reasonable, feasible and necessary defense of our deterrent force, provides time for more thorough testing of an operating system and phases future deployment to progress of arms control negotiations and the changing information on the nature and imminence of potential threats to our security. I shall endeavor to make clear to my scientific colleagues that the Safeguard plan represents a sound and a reasonable approach to our strategic defense problem." (PD, 3/24/69, 430)

March 17–19: At Sixth Space Congress sponsored by Canaveral Council of Technical Societies, James R. Williams of MSFC Engineering Laboratory presented "Space Manufacturing Modules," paper describing NASA's work on manufacturing-in-space experiments. Paper outlined future possibilities, including development of manufacturing module to be attached to future space station. Phase 3 module would contain work space for at least two astronauts, equipment, raw materials, and manufacturing process chambers. It would be designed to dock with earth-orbiting space station proposed for mid-1970s launch, weigh about 23,000 lbs, and provide for continuous investigation of manufacturing processes and for production of small quantities of special items best produced in zero-gravity environment. (MSFC Release 69–80)

March 18: *OV 1–17, OV 1–17A, OV 1–18,* and *OV 1–19* research satellites, carrying 41 experiments, were successfully launched by USAF pickaback on Atlas-F booster from Vandenberg AFB. *OV 1–17* weighed 312 lbs and carried experiments to measure incoming solar electromagnetic radiation and its reaction with earth's upper atmosphere, make electric-

March 18

field and VLF energy propagation studies, evaluate ionized meteor trails in calibration of ground-based radars, and make engineering tests on spacecraft materials and solar-cell power system. Orbital parameters: apogee, 287.7 mi (462.9 km); perigee, 246.7 mi (396.9 km); period, 93.1 min; and inclination, 99.1°. *OV–17* reentered March 5, 1970.

OV 1–17A, 487-lb payload consisting of *OV–17* propulsion module and Naval Research Laboratory two-beacon ORBIS–CAL II experiment to study unusual transmission of radio waves through ionosphere, entered orbit with 233.0-mi (347.9-km) apogee, 106.9-mi (172.0-km) perigee, 89.8-min period, and 99.0° inclination and reentered March 24.

OV 1–18 weighed 275 lbs and carried experiments to study ionosphere and measure radio-wave interference, electric-field intensity, and hazardous radiation. Orbital parameters: apogee, 362.3 mi (582.9 km); perigee, 288.3 mi (463.9 km); period, 95.0 min; and inclination, 98.8°.

OV 1–19 weighed 273 lbs and entered orbit with 3,592.9-mi (5,780.9-km) apogee, 288.0-mi (463.4-km) perigee, 153.5-min period, and 104.7° inclination, where it would study events resulting in and sustaining trapped radiation in Van Allen belts and hazards to man. (GSFC *SSR*, 3/31/69; 3/31/70; OAR *Research Review*, 6–7/69, 23–4; *Pres Rpt 70* [69]; *W Post*, 3/19/69)

- Though "highly successful subsonic jet transports designed and manufactured in the United States are symbols of United States leadership on the airlines around the world," Albert J. Evans, Director of Aeronautical Vehicles, NASA OART, told House Committee on Science and Astronautics' Subcommittee on Advanced Research and Technology, "both the British/French SST and the Russian SST are in flight test. Many short-haul aircraft used by United States commuter and feeder airlines are foreign aircraft and in the VTOL and STOL area the United States severely lags its foreign competition." STOL aircraft were "within our grasp." Two concepts showed promise, one for propeller-driven aircraft and one for jet-powered. Move to flight test in FY 1970 was planned. First was rotating-cylinder-flap STOL research aircraft. Second, jet-augmented wing, would be applicable to jet-powered STOL aircraft. (Testimony)

- Dr. John E. Naugle, NASA Associate Administrator for Space Science and Applications, testified before House Committee on Science and Astronautics' Subcommittee on Space Science and Applications: "We have been able to maintain this country in the forefront of planetary exploration by the quality of both our engineering talent and our Nation's scientists, and by the way in which they are involved in planning and executing our program. By making the most effective use of our best people we have been able to make better use of the limited resources we have had available for planetary exploration than the Russians. If the Russians continue to commit a major share of their space effort to planetary exploration—and the two recent launches to Venus indicate they are—and if they begin to involve their very best scientists in the planning and execution of their mission so that they fly first rate reliable experiments—and there is evidence from our scientific colleagues that they are—then it is clear that this Nation cannot continue in the forefront of space exploration in the next decade." (Text)

- NASA announced MSFC award of $98,200,000 definitive contract for payload integration in support of Apollo Applications program to Martin Marietta Corp. Work—which started under Jan. 16, 1968, letter contract—would extend through Nov. 30, 1972, and cover flights 1, 2, 3, 3A, and 4.

  MSFC also awarded North American Rockwell Corp.'s Rocketdyne Div. $1,142,294, 26-mo contract for additional work on aerospike (plug-nozzle) engine evaluated by Rocketdyne under previous MSFC contract. Ring-like engine had tiny throats along circumference that discharged engine exhaust down sides of plug, making aerodynamic extension for plug, providing nozzle during launch ascent, and simulating performance characteristics of full-length nozzle without increased weight and length. (MSFC Releases 69–43, 69–87)

- If JPL were taxed as business it would bring in estimated $840,000 and reduce property tax rate in La Canada-Flintridge, Calif., district from $4.20 to $2.56, according to La Canada School Superintendent, Dr. Donald Ziehl. Instead, he told School Board, district would receive $85,000 from Government under fixed-amount system used in lieu of property tax on Federal facilities. (Michals, *Glendale News-Press*, 3/19/69)

- USAF announced award to North American Rockwell Corp. of $700,000 initial increment to $2,473,000 cost-plus-fixed-fee contract for research on feasibility of using advanced composite materials for production of aircraft wings. (DOD Release 200–69)

March 18–19: In testimony before House Committee on Science and Astronautics' Subcommittee on Space Science and Applications, Leonard Jaffe, NASA Director of Space Applications Programs, urged commencement of Earth Resources Technology Satellite program: "We have reached a stage in our supporting investigations of Earth resources phenomenology that can be measured from space, where we have confidence that an experimental Earth Resources Technology Satellite can show us the way to the future operational satellite system that will make major contributions to the management of our resources. The state of technology in sensors and spacecraft systems has also reached a sufficiently advanced stage, so that we can with confidence embark this year on the development of our first experimental Earth resources satellite system."

Proposed design and selection of candidate sensors for ERTS had been closely coordinated with other Government agencies. . . . NASA was requesting FY 1970 funds to initiate development of flight hardware for two experimental satellites, ERTS–A and B, with ERTS–A scheduled for launch in late 1971 or early 1972.

NASA also would proceed with techniques for infrared sounding of atmosphere with Nimbus B–2 and D launches in 1969 and 1970 and had begun construction of Nimbus E and F for development of microwave sounding techniques. Prototype second-generation operational meteorological satellite Tiros-M was being prepared for launch in few months. Study to define third-generation Tiros-N would start in

FY 1970. NASA planned to accelerate basic studies of remote sensor signature relationships, instrument development, data processing, user decision models, and aircraft use.

Synchronous satellite ATS-E would be launched in 1969 with gravity gradient experiment. ATS-F and G would not be ready for first launch before late 1971 or early 1972. Synchronous Meteorological Satellite (SMS), new FY 1970 effort for 1971 and 1972 launches, would use existing technology and be prototypes for National Operational Meteorological Satellite System (NOMSS).

National Geodetic Satellite program would complete objectives with GEOS-C launch in 1970. Cooperative Applications Satellite (CAS), French satellite, would be launched on NASA Scout booster in 1970 to track some 500 meteorological balloons to determine wind direction and speed. (Testimony)

March 19: USAF launched two unidentified satellites from Vandenberg AFB by Thor-Agena booster. One entered orbit with 156.6-mi (252.0-km) apogee, 102.5-mi (164.9-km) perigee, 86.6-min period, and 82.9° inclination and reentered March 24. Second entered orbit with 318.8-mi (513.0-km) apogee, 312.8-mi (503.3-km) perigee, 94.7-min period, and 83.0° inclination. (GSFC *SSR*, 3/31/69; *Pres Rpt 70* [69])

- Secretary of Defense Melvin R. Laird presented to Senate Armed Services Committee FY 1970 defense budget reduced from Johnson Administration proposal of $80.6 billion to $78.5 billion. He said FB-111 bomber program would be cut off with aircraft already on order and work would be speeded up on advanced manned strategic aircraft expected to replace B-52s. (Testimony)
- Project Tektite civilian aquanauts Richard A. Waller, Conrad V. M. Mahnken, John G. Van Derwalker, and Dr. H. Edward Clifton set 32-day record for remaining under water in "habitat" off St. John Island in Caribbean. Previous record was held by former Astronaut M. Scott Carpenter, who had remained submerged 30 days in 1965. In Project Tektite, aquanauts were trying for 60-day submersion. (DOD Release 204-69; UPI, *W Post*, 3/20/69, A2)
- At Geneva luncheon during meetings of Disarmament Conference, U.S. and U.S.S.R. opened exploratory talks on proposed treaty banning use of seabed for stationing nuclear weapons. (UPI, *W News*, 3/20/69, 2)

March 19-21: Air Force Assn. held 23rd Annual Convention in Houston, Tex. Dr. Thomas O. Paine, NASA Administrator-designate, said national defense aspects of permanent space station in earth orbit was "a question which should not be considered only from the narrow standpoint of special operational systems requirements that can be clearly established as necessary today. The lead times are too great. The uncertainties as to the future world situation and the precise nature of future defense needs are also too great. I believe that the approach should be to consider carefully what our long-term national security posture would be ten to fifteen years from now if by then the United States has developed a space station capability and the logistics or 'space shuttle' system necessary to utilize it, and compare this posture to what it might be if we do not then have the capability, and the Soviet Union has developed and are fully utilizing such capability." NASA planned to work closely with Defense establishment for "national program for the development of a permanent space station and a logistic support system in a way that will not only serve the needs and goals of the civilian space program, but will also be capable of supporting future defense research and developments as needs become clearly defined." (Text)

Dr. Robert C. Seamans, Jr., Secretary of the Air Force and former NASA Deputy Administrator, said: "Although similar space technology is used in Apollo Applications and the Manned Orbiting Laboratory, there is no unnecessary duplication in the experiments planned. These activities require different equipment, different orbits and different timing. I believe that any attempt to combine the two programs would jeopardize the returns to each agency and would ultimately increase the cost. There is a distinct need to continue with manned space operations under both military and nonmilitary auspices. Cooperation between NASA and DOD has been close, and will continue." (Text)

AFA's highest award, Henry H. Arnold Trophy, was presented to *Apollo 8* Astronauts Frank Borman, James A. Lovell, Jr., and William A. Anders. NASC Executive Secretary, Dr. Edward C. Welsh, received Gill Robb Wilson Trophy for Arts and Letters for his part in interpreting aerospace role in modern society. AFA's Citation of Honor went to L/G Samuel C. Phillips (USAF) for his management of NASA Apollo program, and to Maj. William J. Knight (USAF) for "outstanding contributions to the Air Force and the nation for being the first to achieve hypersonic winged flight while piloting the X-15 aircraft to a speed of 4,534 miles per hour." (AFSC *Newsreview*, 3/69, 1; *AFJ*, 3/8/69, 30)

March 20: Senate confirmed nomination of Dr. Thomas O. Paine as NASA Administrator. (*CR*, 3/20/69, S2949)

• In first test of NERVA inflight configuration at Jackass Flats, Nev., engine was held to 1/10 maximum power development of 1,100 mw or 55,000 lbs thrust, during three 25-sec warmups. Full power tests would be held in April by Aerojet-General Corp. and Westinghouse Astronuclear Laboratory, when NERVA was expected to produce electrical energy equal to one fourth that developed by Hoover Dam. Test—to 110 mw and 5,500 lbs thrust to duplicate engine startup procedures in space and verify system performance with liquid hydrogen fuel—marked first test of rocket with nozzle down and thrust blasting into water-cooled firing pit. Engine had produced sufficient energy to boil ton of water every two seconds. (AEC/NASA Release M-54; *LA Times*, 3/23/69)

• Appointment of NASA Associate Administrator for Organization and Management Harold B. Finger as first Assistant Secretary for Urban Research and Technology was announced by Housing and Urban Development Secretary George W. Romney. From 1960 until 1967, Finger had been manager for AEC-NASA Space Nuclear Propulsion Office and, since 1961, director of nuclear system for OART. (UPI, *NYT*, 3/16/68, 39; AP, *W Post*, 3/21/69, A17)

• MSFC announced it had signed $4,095,000 contract with Radio Corp. of America for engineering and logistics support for ground computer systems and other equipment in Saturn V program. (MSFC Release 69-84)

• NASA announced completion of negotiations for one-year $513,293,000 extension, through 1969, of cost-plus-fixed-fee contract with Philco-Ford Corp. Education and Technical Services Div. for engineering support and related services at Wallops Station. (NASA Release 69-8)

March 21: Dept. of Commerce, DOD, Dept. of Interior, DOT, AEC, NASA, and NSF released *News from BOMEX*, report of Barbados Oceanographic and Meteorological Experiment to be conducted in May, June, and July in cooperation with government of Barbados. Scientific study of joint behavior and interactions of atmosphere-ocean system in subtropical and tropical waters was part of Federal Air-Sea Interaction Research Program and major U.S. contribution to Global Atmospheric Research Program (GARP). BOMEX would study continuous exchange of energy, momentum, gases, particulates, and electrical charges at air-sea interface and study ways in which energy and other properties were transported from area by atmosphere and ocean.

NASA would test concepts of satellite sensors for weather and oceanographic observations. Its experimental weather and communications satellites (*Ats III, Nimbus III*, and ESSA satellites) and devices being developed under Earth Resources Survey Satellite program would be directly engaged in BOMEX. Photos taken by *Apollo 9* astronauts with hand-held Hasselblad camera of 900 sq mi of equatorial Atlantic Ocean off Barbados would be used in project. (Text; NASA News; *Science*, 3/28/69, 1435-6)

- First decade in space corresponded with early years of aviation following Wright brothers flight at Kitty Hawk, NASA Associate Administrator for Manned Space Flight, Dr. George E. Mueller, told Annual Kiwanis Dinner in Milledgeville, Ga. "I recently calculated the costs per seat mile of their first flight for comparison with our Apollo flight to the moon. They probably had ten thousand dollars invested in first Kitty Hawk flight that went one fifth of a mile, giving a cost per seat mile for Orville of fifty thousand dollars. We will have about twenty-four billion dollars invested in our first five-hundred-thousand-mile trip for three astronauts to the moon, giving a cost of only about sixteen thousand dollars per seat mile. We are ahead of the Wright Brothers—but we have a long way to go to catch up with the DC-8 or 707. If subsequent flights to the moon cost two hundred million dollars each, that's less than one hundred and fifty dollars a seat mile, so we are going in the right direction." (Text)

- INTELSAT conference of 67 countries and observers from Communist bloc and underdeveloped nations ended at State Dept. without agreement on method of sharing control over international satellite communications. Committee of INTELSAT members was appointed to work through summer on alternate drafts of final agreement to be presented to conference in November. Leonard H. Marks, American chairman of conference, had announced his resignation and would return to private law practice in Washington, D.C. (Lydon, *NYT*, 3/23/69, 31; Rpt of US Del)

- Aviation Progress Committee announced appointment of former Sen. A. S. Mike Monroney as consultant and adviser to committee and stated its purpose was "to work with interested individuals and organizations to . . . seek and support ways to improve and expand the National Aviation System of airports and airways." As Chairman of Aviation Subcommittee of Senate Commerce Committee, Monroney had

been largely responsible for legislation which created FAA in 1958. (Committee Release; W *Star*, 3/23/69, C5)
- Secretary of Transportation John A. Volpe announced award of $544,302 FAA contract to Laboratory for Electronics, Inc., to develop two low-cost, solid-state, microwave-instrument landing systems (ILS) for STOL aircraft operations at FAA's National Aviation Facilities Experimental Center in Atlantic City, N.J., and at suitable operational STOLport such as Dulles International Airport near Washington, D.C. (FAA Release 69-38)
- Harvard Univ. biologist Dr. George Wald and MIT physicist Dr. Bruno Rossi disclosed that some 251 NAS members out of 806 had signed letter to President Nixon expressing continued opposition to deployment by U.S. of ABM system. Letter said, "Russia is as well prepared to build such devices as we are. This can only introduce a new and perhaps disastrous spiral in the arms race. Our science and technology, rather than being used to add further to the present 'balance of terror,' needs to be redirected to solving pressing problems of poverty, malnutrition, control of population, and improvement of the human environment for our own people and people everywhere." (AP, *W Post*, 3/22/69, A5)

March 22: *Cosmos CCLXXIII* was launched by U.S.S.R. into orbit with 335-km (208.2-mi) apogee, 199-km (123.6-mi) perigee, 89.8-min period, and 65.4° inclination. Satellite reentered March 30. (GSFC *SSR*, 3/31/69)
- Five segments of *Apollo 9*'s 6,500,000-lb "stack" remained in space, making total 1,613 objects floating in earth orbit after *Apollo 9*, said James J. Haggerty, Jr., in *Armed Forces Journal*. Descent stage of LM, in eccentric low orbit, would soon be captured by earth gravity and burn on reentry; ascent stage might remain in high orbit 20 yrs, along with LM adapter [GSFC reported later that LM descent stage had reentered March 22; ascent stage was still in orbit]. S-IVB 3rd stage mated to instrument unit was directed into solar orbit. Of 1,613 orbiting objects listed by NORAD, 356 were payloads, both active and expired. Remaining 1,257 pieces were debris.

Items in solar orbit included 3rd-stage and instrument-unit combinations from *Apollo 8* and *9*, four U.S. Pioneers and *Mariner VI* still sending useful data, and U.S. and U.S.S.R. planetary explorers launched before 1969. About 20 solar orbiting objects would remain in space a long time, along with 28 spacecraft which had crashed on other celestial bodies, including U.S.S.R.'s *Venus III* and *IV* on Venus, 9 Soviet Lunas, 5 U.S. Rangers, 7 Surveyors, and 5 Lunar Orbiters. (*AFJ*, 3/22/69, 15; GSFC *SSR*, 3/31/69)
- DOD announced name of planned U.S. antiballistic missile defense had officially been changed from Sentinel to Safeguard. (AP, *W Star*, 3/23/69, A4)

March 23: NASA announced release of *The Book of Mars*, one-volume digest of facts and theories about Mars by Dr. Samuel Glasstone, illustrated with photos taken during *Mariner IV* mission 1964–65, when spacecraft flew within 6,000 mi of Mars surface. Book traced history of man's acquisition of Mars data and compared Mars with other planets. (NASA Special Release)

- William Hines in Washington *Sunday Star* commented on "deadly dullness" of last five days of *Apollo 9* mission: "The flight plan of Apollo 9—which most people do not see and therefore do not understand—was carefully divided into six 'activity periods,' of which five were each about 24 hours long and the sixth five days in duration.

  "It is no criticism of the . . . crew that after five days of brilliant spacemanship they took it easy for the rest of the flight. Nor is it criticism of the space program to say that the last half of Apollo 9 was dull. It was planned that way, and if it had turned out otherwise the United States very likely would not be attempting a moon landing in July." (*W Star*, 3/23/69, C4)

March 24: U.S.S.R. launched *Cosmos CCLXXIV* into orbit with 300-km (186.4-mi) apogee, 208-km (129.3-mi) perigee, 89.5-min period, and 64.9° inclination. Satellite reentered April 1. (GSFC *SSR*, 3/31/69; 4/15/69)

- JPL astronomer Dr. Ronald Schorn announced "definite and conclusive" evidence of water on Mars had been found in five-year study with Stephen Little of Univ. of Texas and JPL scientist Dr. C. B. Farmer at McDonald Observatory in Ft. Davis, Tex. Noting Mars generally should be compared with driest deserts of earth, he said: "If you took all the water we found and laid it out over the whole planet, it would be only 1,000th of an inch thick. Mars may still not be a great place to live, but there's a chance of life there." Photos to be taken by *Mariner VI* and Mariner VII spacecraft, Dr. Schorn said, would help solve question of whether enough water existed to sustain life on Mars.

  New spectra, superior to any previously available, were made possible by NASA-supported improvements in McDonald 82-in Struve reflecting telescope and its large spectrograph. Observations of Mars were beginning with more powerful, 107-in reflecting telescope at McDonald, constructed with funds from NASA, NSF, and Univ. of Texas. (NASA Release 69–48; AP, *W Post*, 3/25/69, A5)

- Apollo 10 mission would be launched from ETR May 18 carrying Astronauts Thomas P. Stafford (commander), John W. Young (CM pilot), and Eugene A. Cernan (LM pilot) on eight-day lunar orbital mission. Final decision to fly mission as previously planned followed review of technical and operational data from *Apollo 9* (March 3–13). Mission would provide additional experience in combined system operation and, with exception of actual landing on lunar surface, was same as for lunar landing mission. While spacecraft circled moon at 69-mi (111.0-km) altitude, Stafford and Cernan would separate LM from CSM, pilot LM twice to within 10 mi (16.1 km) of preselected landing site, and return to CSM. Crew would then make landmark sightings, take photos, and transmit live TV views of moon, earth, and spacecraft interior before returning to earth. (NASA Release 69–46)

- Gas from solar flare wiped out much of earth's outer radiation belt and caused auroral displays—normally seen only over subpolar regions—to move to lower latitudes. Residents of metropolitan New York area flooded weather bureau and newspapers with inquiries. Radiation belts' discoverer, Dr. James A. Van Allen, said display might have been caused by class 2–B flare on sun. ESSA's space disturbance warning center, Boulder, Colo., said several smaller flares had occurred March 21 and 22. Auroras were to be expected, Walter Sullivan said in *New*

York Times, because sun was near maximum of its 11-yr cycle of sunspot and eruptive activity. (*NYT,* 3/25/69, 30)
- MSFC contract awards: $1,521,500 supplemental agreement with IBM for design improvement of control signal processors for Apollo/Saturn V guidance instrument units; and $1,334,931 contract modification to McDonnell Douglas Corp. to develop, design, manufacture, and test restarting oxygen/hydrogen burner. (MSFC Releases 69–88, 69–89)
- USN announced award of $28,161,681 modification to contract with LTV Aerospace Corp. for improvement changes on F–8B and F–8C aircraft. (DOD Release 212–69)

March 25: Apollo 9 commander James A. McDivitt told press at Washington, D.C., news conference that *Apollo 9* spacecraft performance had been outstanding and procedures had been "as near perfect as anything I could possibly imagine," without deviating "from a single step in rendezvous . . . and I don't recall a single procedure that we recommended be changed."

Astronaut Russell L. Schweickart said new spacesuit had increased mobility significantly. Activities like manipulating camera were "probably the most challenging . . . and I found that after taking a whole series of pictures and all the various motions you go through . . . my hands were far less tired than I would have anticipated." (Transcript)
- House Committee on Science and Astronautics' Subcommittee on Manned Space Flight voted to add $230.5 million to $2.212-billion authorization requested by NASA for manned space flight in FY 1970. Of raise, $168 million was earmarked for development of scientific payloads for six manned lunar landings between 1970 and 1973 and $66 million for space flight operations to cover flying men in earth orbit for month at a time, in 1970 onward. Flights were to be forerunners of orbiting space stations. (*CR,* 3/26/69, D233; Committee member)
- White House submitted to Senate nomination of former NASA Associate Administrator for Organization and Management Harold B. Finger to be Assistant Secretary of Housing and Urban Development. (*PD,* 3/31/69, 489)
- MSFC announced award of $3,657,000 contract to Sanders Associates, Inc., for continuation of engineering and logistics support for operational display systems for Saturn V program. Contract covered Oct. 1, 1968, through June 1970. (MSFC Release 69–91)

March 26: U.S.S.R. launched *Meteor I* satellite from Plesetsk into orbit with 686-km (426.3-mi) apogee, 632-km (392.7-mi) perigee, 97.9-min period, and 81.1° inclination. Tass said satellite carried solar batteries and would obtain information about weather prospects. (GSFC *SSR,* 3/31/69; AP, *NYT,* 3/28/69, 5; *Interavia,* 11/69, 1751)
- In NASA Hq. *Apollo 9* awards ceremony Vice President Spiro T. Agnew, as NASC Chairman, presented NASA Distinguished Service Medals to Astronauts James A. McDivitt, David R. Scott, and Russell L. Schweickart for individual contributions to "the Nation's manned space flight program and the advancement of space technology" as participants in "this historic mission, the first manned flight involving the Command Module, Lunar Module, and extravehicular mobility unit." NASA Exceptional Service Medal went to Carroll H. Bolender, Manager for Lunar Module, Apollo Spacecraft Program Office, MSC, and Eugene F. Kranz, *Apollo 9* Flight Director, Flight Control Div., MSC.

March 26

NASA Public Service Award was presented to Llewellyn J. Evans, President of Grumman Aircraft Engineering Corp., LM manufacturer. (NASA Awards summary)

- House Committee on Science and Astronautics' Subcommittee on Advanced Research and Technology voted increase of $11.8 million over $818.8-million NASA request for research. (CR, 3/26/69, D233)
- NASA reported experiments with yaw-damper system developed at FRC, consisting of small gyroscope, magnetic-clutch actuator, and electronics control assembly in package occupying 1/10 cu ft and weighing less than two pounds. It could cost as little as $200 and be installed on light general-aviation aircraft to control "dutch roll" or yawing. Commercial jet aircraft were equipped with advanced systems to eliminate yawing. (NASA Release 69-44)
- Boeing Co.'s Vertol Div. was conducting wind-tunnel tests of $250,000 model of V/STOL aircraft with tilting wing and large cyclic-pitch propellers, *New York Times* said. Tests would provide data for advanced aircraft, including Light Intratheater Transport (LIT) for which USAF was funding studies. Tilt wing rotated from horizontal to vertical flight in takeoff or descent. Cyclic pitch would control pitch (nose-up-or-down movement) by acting on all four of aircraft's propellers simultaneously. According to Boeing engineers, cyclic pitch would eliminate need for horizontal tail rotor for pitch control in hovering and slow flight. Large propellers—26 ft in dia on full-sized aircraft—would reduce engine horsepower requirements. (*NYT*, 3/2/69, 93)
- In *Washington Post* Thomas O'Toole said: ". . . man's flight to the moon and beyond could be the only means he has left of renewing his dwindling spirits. . . . The voyage of Apollo 8 around the moon last Christmas did more than any other single event last year to restore man's faith in himself—and that flight will pale beside an actual moon landing when it comes. . . . It could just be that when man walks the moon for the first time it will be felt round the world as such a triumph of the human heart that its beat shall go on for a million years." (*W Post*, 3/26/69, A27)

March 27–April 8: NASA's *Mariner VII* (Mariner G) spacecraft was successfully launched from ETR by Atlas-Centaur (AC–19) booster on four-month, 193-million-mi, direct-ascent trajectory toward Mars—NASA's second mission to fly past Mars during 1969 launch window. Launch vehicle performance was nominal. Spacecraft separated from Centaur, deployed its four solar panels, and locked its sensors on sun and star Vega. Because of several minor spacecraft anomalies during launch, spacecraft was kept in sun-Vega cruise while performance was evaluated. Star-lock override command was transmitted to spacecraft April 1; Canopus was acquired as planned initially and *Mariner VII* entered stable cruise mode. Midcourse maneuver was conducted April 8 to ensure that spacecraft would fly within 1,900 mi (3,057.7 km) of Mars Aug. 5.

Primary mission objective was to fly by southern hemisphere and polar regions of Mars to set basis for future experiments, particularly those relevant to search for extraterrestrial life. As secondary mission spacecraft would develop technology needed for succeeding Mars missions. The 900-lb spacecraft carried six complementary experiments to provide information about Martian surface and atmosphere. *Mariner*

March 27–April 8: NASA *launched* Mariner VII *on an Atlas-Centaur booster as the second of two spacecraft in the Mariner Mars 1969 mission to take photographs and collect data about the surface and atmosphere of Mars.* Mariner IV *was launched Feb. 24. An April 8 maneuver ensured that* Mariner VII *would pass Mars Aug. 5.*

VII mission, except for flyby area, was almost identical to mission of *Mariner VI* launched Feb. 24 for investigation of equatorial region and scheduled to arrive at Mars July 31. (NASA Proj Off; NASA Releases 69-26, 69-42)

March 27–29: NASA launched four Nike-Cajun sounding rockets carrying GSFC payloads to obtain data on atmospheric parameters. Two rockets launched from Arenosillo, Spain, March 27–28 reached 75.2- and 75.8-mi (121- and 122-km) altitudes, each exploding 19 grenades with sound arrivals recorded on ground. Rockets launched from Wallops Station March 28–29 reached 75.4- and 69.1-mi (121.3- and 111.2-km) altitudes and exploded 19 grenades each, with sound arrivals recorded on ground. Data would be analyzed and compared. (NASA Rpts SRL)

March 27: LaRC had awarded 10-mo, $155,000 feasibility study contract to North American Rockwell Corp. to design two-man, lunar emergency, escape-to-orbit vehicle which could be carried aboard LM on Apollo missions. (*SBD*, 3/27/69, 125; NAR *Skywriter*, 4/4/69, 1)

- AEC–NASA Space Nuclear Propulsion Office awarded Aerojet-General Corp. $47,447,601 contract extension for completion of preliminary engine and component design for NERVA I and initiation of procurement or fabrication of component development hardware. Extension, from Oct. 1, 1968, through Sept. 30, 1969, brought total value of cost-plus-fixed-fee contract to $500,015,527. (NASA Release 69–47)

- MSFC announced $7,384,543 modification to contract with Chrysler Corp. Space Div. for assembly of two boosters for Saturn IB rockets 213 and 214 for use in Apollo Applications program. (MSFC Release 69–93)

March 28: U.S.S.R. launched *Cosmos CCLXXV* into orbit with 761-km (472.9-mi) apogee, 275-km (170.9-mi) perigee, 95.1-min period, and 70.9° inclination. Satellite reentered Feb. 7, 1970. (GSFC *SSR*, 3/31/69; 2/7/70)

- NASA Science and Technology Advisory Committee for Manned Space Flight, chaired by Dr. Charles H. Townes, published *Proceedings of the Winter Study on Uses of Manned Space Flight, 1975–1985*, Vol. I—*Proceedings:* NASA program for next two decades must project state of technology for that period. Program must be balanced in use of manned and automated operations. "The benefits to the nation, both internal and international, dictate that the United States remain in the forefront of all major categories of space activities," space sciences, exploration of solar system, manned space flight capability, and economic applications of space flight.

    Study said it was reasonable to use $\frac{1}{2}\%$ to 1% of GNP to support civilian space flight program of which major elements were: (1) aggressive automated planetary exploration program as recommended by NAS–NRC Space Science Board, with options for manned phase to follow early automated phase; (2) economic applications program as recommended by 1968 Summer Study on Space Applications by NAS; (3) continuation of lunar exploration after Apollo landing as recommended by Lunar and Planetary Missions Board of NASA; (4) vigorous program of astronomical observations in earth orbit as recommended by NASA Astronomy Missions Board; and (5) extension of manned space flight capability in earth orbit to longer duration for scientific and technological purposes. Achievement of manned low-cost transportation system deserved high priority. Study recommended use of long-duration manned space station designed to support men in weightless condition "unless unexpected biomedical problems are encountered or overwhelming engineering advantages for artificial gravity are discovered." It agreed on advisability of placing observatories and laboratories in earth orbit but noted, "Relative emphasis among these activities and the extent of manned attendance desirable in each, must be decided by appropriate studies and experiments."

"To provide a focal point for structuring a manned planetary program a target date of 1982 and the Mars landing mode . . . are assumed here for discussion. Achievement of the operational capability in that year would require initiation of system design in about 1975." (Text)

- President Nixon received report of 10-member ad hoc committee appointed Feb. 7 to review all aspects of SST program. Under Secretary of Transportation James M. Beggs, chairman, said no votes were taken by committee on recommendations to President; each member had submitted his suggestions to Transportation Secretary John A. Volpe and they were included in report. (UPI, *W Post*, 3/29/69, A2)

March 29: While Nixon Administration warned of $2- to $3-billion cut in FY 1970 budget, NASA was asking $100-million increase with good chance of approval, James J. Haggerty, Jr., said in *Armed Forces Journal*. "NASA officials report a generally warmer reception on Capitol Hill as a result of near-flawless performance on all of the manned Apollo missions." Apollo 1970 funding represented penultimate installment on basic program and "Congress can at last see light at the end of the tunnel." Most important, "extra $100-million is a real bargain price for the potential benefit . . . one of those deals that the buyer can't afford to turn down. If you invest a hundred thousand in a magnificent home and the contractor demands another thousand to put a roof on it, you have little option." Since U.S. had committed $25 billion to program, "it would be questionable economy to settle for less than maximum benefit by withholding what amounts to two-fifths of one percent of the total." (*AFJ*, 3/29/69, 21)

March 30: Thomas O'Toole in *Washington Post* said plans for handling lunar samples [see March 12] had stirred bitter scientific controversy. Tests scheduled at NASA's Lunar Receiving Laboratory would last at least two months while scientists waited for "the chance to study what amounts to a Rosetta stone that could hold the clue to the origin of life itself." At close of examination period, NASA would parcel out moon rocks to 110 scientists from group of 600 and even "chosen few" had no guarantee of receiving piece of moon. LRL official had explained that if NASA found, for example, that there was absolutely no evidence of radioactive argon, "It would be quite foolish to waste a sample on a man . . . whose primary goal was to look for radioactive argon." Scientists also were disturbed over involvement of other Federal agencies in lunar sample handling. (*W Post*, 3/30/69, A1)

March 31: Library of Congress Legislative Reference Service published *United States and Soviet Rivalry in Space: Who Is Ahead, and How Do the Contenders Compare?* by Dr. Charles S. Sheldon II, Senior Specialist in Space Transportation and Technology, Science Policy Research Div. By June 30, 1969, U.S. expected to have spent $50.6 billion on space. Because of greater weight of Soviet hardware, "one could assume that their program is at least of the same magnitude . . . and may be larger." While "there is no reason to believe their total aerospace industry is as fully equipped as our own," missiles in Moscow parades indicated existence of multiple design and development team for space work. Each country seemed adequately equipped with launch pads and ground support. While U.S. had worldwide tracking system supplemented by ships and electronics-carrying aircraft, U.S.S.R. accepted "certain constraints," relying on ships to fill gaps in tracking system. U.S. launch vehicles ranged from those lifting 20 lbs to Saturn V capable of 285,000-lb lift to low earth orbit. U.S.S.R. still used basic ICBM vehicle introduced in 1957, with upper stages added to improve performance.

While NASA program was run on open basis, there was less "openness" in DOD. U.S.S.R. "holds to a minimum advance notice of flights, limits information . . . but at least makes a prompt announcement, assigns a name, and gives orbital parameters" of successful launches. There seemed little difference between space programs "as to general purpose and direction." Both placed emphasis on military, though little was known of Soviet organization pattern. All "reasonable analyses are fairly convincing that up to this time the Soviet Union has not placed nuclear bombs in orbit."

In space applications U.S. "has held a clear lead from the earlier days." U.S.S.R. "has made a greater relative effort in . . . flights committed to lunar and planetary work than the United States" but "has not gained as good results."

In future programs, both nations "undoubtedly have to face hard budget choices before actual hardware can begin." Soviet goal was "comprehensive exploitation of space technology including the exploration and settlement (where practical) of the planets, along the way exploring the Moon in great detail, and using Earth orbital stations for a host of practical purposes." While large orbital station would be within NASA's technical capabilities in late 1970s, beginning of such capability probably already existed in U.S.S.R., but "it would be risky to predict whether such a station will appear soon or only after some years."

Cooperation of U.S. and U.S.S.R. in space already existed in exchange of information, treaties, plans for trading space-collected weather pictures, plans for joint textbook on space biology, and joint efforts in geomagnetism, but no assessment could be made of future prospects. (Text)

- At American Cancer Society Seminar in New Orleans, La., Clarence D. Cone, Jr., head of LaRC's Molecular Biophysics Laboratory, discussed his discovery of intercellular linkages and its application to understanding behavior of certain cancer types. His basic research into effects of space radiation on body cells had enabled him to observe phenomenon in which dividing cancer cell appeared able to induce connected cells to divide by transmitting chain-reaction stimulation through thin linkage of cytoplasm. Networks of these stimulus-transmission bridges permitted continued division of cells and, in human body, might constitute basic mechanism for cancer spread. (NASA Release 69–45)

During March: NSF published *Scientific Activities of Nonprofit Institutions, 1966* (NSF 69–16). Full-time equivalent number of R&D scientists and engineers employed by independent nonprofit institutions in January 1967 totaled 24,300, annual increase of 12.4% from 1954 to 1967. In 1966 expenditure for R&D in independent nonprofit institutions was $800 million—about seven times the $100 million in 1953. Federal

R&D expenditures contracted to nonprofit institutions reached $540 million in 1966, nine times 1953's $60 million. (Text)
- NSF published *R&D Activities of Local Governments, Fiscal Years 1966 and 1967* (NSF 69–14). Local governments spent $20.3 million in 1966 and $28.8 million in 1967 for R&D and $1 million in 1966 and nearly $3 million in 1967 on R&D plant. Federal Government provided 56% of funds in 1967 and local governments, 35%. Health and hospitals accounted for approximately 50% of total local R&D expenditures in both years, with education second at 10%. (Text)
- *Engineer*, journal of Engineers Joint Council, published "The Engineering Profession: A New Profile." Management, not design or development, was most common job function of engineers. Only about 10% of engineers worked directly for Federal Government and 12% in total aircraft, missile, and rocket area. Median age of engineers was 43 and profession was more than 99% male. (Text)

April 1969

April 1: Reporting findings from four months of *Oao II* data to 129th meeting of American Astronomical Society in Honolulu, Univ. of Wisconsin astronomer Dr. Arthur D. Code said: "It puzzles me to see so much ultraviolet light from the Andromeda galaxy (M31) and so little from M81 because they are rather old 'garden variety' galaxies. . . . There is a possibility that we might have discovered an old quasar a few million light years away."

Full impact of *Oao II* data would not be felt for couple of years. "Some theories on cosmology will have to be modified and others discarded. Practically all phases of optical astronomy will be affected." Temperature figures assigned to young, hot stars with masses more than 15 times sun's would require alteration. "These stars are considerably hotter than 20,000 degrees absolute. They are aging about twice as fast as we thought and are burning hydrogen at a very rapid rate." OAO data thus far provided argument against steady-state theory of universe which maintained universe always looked same, from any point at any time, Dr. Code said.

At same meeting, Joseph Purcell, OAO Project Manager at GSFC, said *Oao II's* observatory control system had exceeded its pointing accuracy requirement of one minute of arc by factor of two. "A subsequent OAO will be 100 times more stable." (NASA Release 69-51; Lannan, W *Star*, 4/4/69, A5)

- NASA's *Mariner VII* Mars probe (launched March 27) obeyed radio command to lock its sensors on star Canopus. Spacecraft would fly past Mars night of Aug. 4. Command was radioed from NASA tracking station in Woomera, Australia. (AP, *W Post*, 4/2/69; UPI, *C Trib*, 4/3/69)
- Pakistan had successfully launched her first rocket, a two-stage vehicle to investigate upper atmosphere, Pakistan Space and Upper Atmosphere Research Committee announced. (AP, *W Post*, 4/2/69, A7)
- Secretary of Defense Melvin R. Laird told House Armed Services Committee he planned to cut FY 1970 defense budget by at least $1.1 billion because of Nation's "extremely difficult and dangerous economic and fiscal situation." At press interview following testimony, he said DOD budget submitted to Congress included cutback in B-52 raids over South Vietnam from 1,800 to 1,600. (Transcript of Press Conference; Corddry, B *Sun*, 4/2/69, A1)

April 2: National Academy of Engineering announced election of 44 U.S. engineers to membership. Included were Dr. John S. Foster, Jr., Director of Defense Research and Engineering, honored for "technological leadership in defense research and engineering," and Edward Wenk, Jr., Executive Secretary of National Council on Marine Resources and Engineering Development, for "major contributions to the design of military submarines and other underwater equipment." (NAE Release)

April 3: Dr. Thomas O. Paine was sworn in by Vice President Spiro T. Agnew as NASA Administrator in ceremony in Vice President's office. Dr. Paine said: "I am particularly anxious to see that in the second

April 3: *Dr. Thomas O. Paine (right) was sworn in as Administrator of the National Aeronautics and Space Administration by Vice President Spiro T. Agnew.*

decade of space we bring down to the people on earth more of the benefits that can be obtained from this wonderful new technology. . . . Such areas as navigation, communications and particularly earth resources are things that are very much on our minds as we look out to the moon and beyond." (NASA Hq *WB*; Sehlstedt, B *Sun,* 4/4/69; AP, *W Post,* 4/4/69)

- National Geographic Society's Hubbard Medal for research, exploration, and discovery was presented by Vice President Spiro T. Agnew to *Apollo 8* Astronauts Frank Borman, James A. Lovell, Jr., and William A. Anders in Constitution Hall, Washington, D.C. Vice President Agnew said: "The possibilities of space exploration are so infinite they overwhelm the mind. The speed with which we have reached this new threshold of hope itself is awesome. Less than half a century has passed since the Hubbard Award was presented to Colonel Charles A. Lindbergh for his solo flight from New York to Paris. Less than a decade has passed since Colonel John Glenn was awarded the first Hubbard Medal for exploration in space."

Award, one of Nation's highest honors, had been given only 24 times in 63 years, for outstanding achievement in geography and exploration. First medal had gone to Robert E. Peary in 1906 for Arctic explorations.

Earlier in day at National Geographic headquarters, *Apollo 8* astronauts had received General Thomas D. White Space Trophy for 1968, awarded to military or civilian USAF member making most outstanding contribution to U.S. progress in aerospace. (NGS Release)

- MSFC announced two contract awards. Definitive $38,340,000 contract to North American Rockwell Corp. Rocketdyne Div. for J-2 engine operational and flight support Jan. 1, 1969, to June 30, 1970, replaced letter contract which earlier authorized $9 million. Contract modification of $15,253,945 was given to Chrysler Corp. Space Div. for extending delivery schedule of Saturn IB boosters and stage storage. (MSFC Releases 69-97, 69-96)

April 4: U.S.S.R. launched two Cosmos satellites from Plesetsk. *Cosmos CCLXXVI* entered orbit with 371-km (230.5-mi) apogee, 200-km (124.3-mi) perigee, 90.1-min period, and 81.3° inclination and reentered April 11. *Cosmos CCLXXVII* entered orbit with 466-km (289.6-mi) apogee, 267-km (165.9-mi) perigee, 91.8-min period, and 70.9° inclination and reentered July 6. (GSFC *SSR,* 4/15/69; 7/15/69; AP, *C Trib,* 4/5/69; SBD Space Log Supplement, 4/15/69)

- USAF X-24A lifting-body vehicle, piloted by Maj. Jerauld R. Gentry (USAF), was carried to 45,000-ft altitude by B-52 aircraft during first captive flight. All systems functioned satisfactorily and vehicle was deemed flight worthy in NASA-USAF program. (NASA Proj Off)

- NAA announced *Apollo 8* Astronauts Frank Borman, James A. Lovell, Jr., and William A. Anders would receive 1968 Robert J. Collier Trophy for significant achievement in aeronautics and astronautics, as representatives of "entire United States space flight team for the successful and flawless execution of the first manned lunar orbit mission in history." Trophy would be presented by President Nixon at Washington, D.C., luncheon May 7. (NAA Release)

- Astronaut James A. Lovell, Jr., received Distinguished Service Medal at DOD ceremony for service as *Apollo 8* command module pilot. (*W Post,* 4/5/69, A7)

- *Apollo 8* Astronauts James A. Lovell, Jr., and William A. Anders, both U.S. Naval Academy graduates, presented Academy flag carried aboard *Apollo 8* spacecraft to brigade representing 4,000 midshipmen at Annapolis. (AP, B *Sun,* 4/3/69, A11)

- INTELSAT conference ended March 21 had made it clear "that Intelsat, in its brief 5-year history, has been an extraordinary success," wrote Robert J. Samuelson in *Science*. Its transoceanic satellites, transmitting telephone signals primarily, had tended to depress cost of communications by multiplying available channels and pressuring carriers to lower rates. Satellites had given "third world" nations in Asia, Africa, and Latin America, previously dependent upon "confused mixture of radio and cable channels," chance to join advanced world's communications system. Attendance of U.S.S.R. as conference observer was "sure sign" of INTELSAT success.

Soviet decision to join INTELSAT might hinge on organization's decision about its future. Issue was INTELSAT's formal structure and

April 4

problem was "determining how large a role the United States should play." U.S. domination stemmed from its economic and technological power. U.S. firm, ComSatCorp, was made manager of INTELSAT under 1964 agreement, to oversee satellite design, contract with NASA for launchings, and supervise operation in space.

Month-long meeting had not resulted in accord on even draft agreement. Europeans wanted to replace ComSatCorp management with international secretariat to subcontract technical tasks to ComSatCorp and other organizations as they demonstrated genuine competence. U.S. approach was, Why "tamper with a successful formula?" Fundamental issue lay deeper. "Technological superiority . . . creates its own foreign policy problems. America's Intelsat partners are pushing for a Space Age which—if not truly international—is at least more multinational." (*Science*, 4/4/69, 56–7)

- FAA released *Air Traffic Activity Report* for 1968: Los Angeles International Airport had climbed to position of second busiest U.S. airport, from sixth place in 1967. Los Angeles had logged 594,486 takeoffs and landings. O'Hare International in Chicago, which had ranked first every year since 1962, still led with 690,810. Van Nuys, Calif., retained third position with 567,973 total and led in general aviation with 317,816 operations. Instrument-flight-rule aircraft handled by FAA air route traffic control centers had more than doubled in decade, to 19.4 million in 1968. (FAA Release 69–43)

April 5: NASA adoption of field sequential color TV system to relay pictures from moon on future Apollo flights was "solace" to inventor Dr. Peter C. Goldmark, president of CBS Laboratories, *New York Times* said. System, employing small revolving filter disc to inject primary colors in front of camera, had been rejected by industry and FCC some 15 yrs earlier in favor of all-electronic compatible system now in worldwide use. Advantage of Goldmark system on moon flights was ability to pick up usable color images under exceptionally low levels of light intensity—important where stars might be major illumination source. Images received from space would then be converted through compatible system for home reception. Dr. Goldmark said system might be "first technological breakthrough that is 28 years old." (*NYT*, 4/5/69, 30)

April 6: In interview published by *This Week*, Apollo 8 Astronaut Frank Borman said many things could be learned from moon: "One . . . I hope will be international cooperation, such as now exists in Antarctica. Even though the moon struck me as a very desolate, forbidding expanse, it will be very beneficial for men to work together to unlock its many mysteries and secrets. I hope that both the moon and the large permanent space stations from earth will be citadels for international cooperation and that the people who visit them will really be internationalists in the truest sense." (*This Week*, 4/6/69)

- Baltimore *Sun* said Indiana Univ. had announced plans for experiments with NASA to determine problems in making future lunar colonies self-supporting. Studies to start in August would probe feasibility of growing earth plants in lunar soil to provide food for manned space stations and possibly fodder for animals transported to moon colonies in 20 to 30 yrs. "Moon grow" experiment would use lunar soil retrieved by lunar missions and would expand as more material became available.

Experiments would be conducted by astrobotanist Dr. Paul Mahlberg and team at MSC laboratory and later at university. (Perkinson, B *Sun*, 4/6/69)

April 7: NASA's Lunar Landing Training Vehicle (LLTV), piloted by NASA test pilot Harold E. Ream, successfully completed six-minute flight at Ellington AFB. LLTV flights had been suspended since Dec. 8, 1968, crash. (AP, B *Sun*, 4/8/69, A3; 4/7/69, A9)

- MSFC announced engineer Chester B. May would be member of oceanologist Dr. Jacques Piccard's six-man crew on Gulf Stream Drift Mission, scientific undersea journey aboard submersible vessel *Ben Franklin* (PX-15). Vessel would drift with Gulf Stream current from Miami, Fla., to Halifax, Nova Scotia, from four to six weeks beginning in June. May would study vessel's operation and evaluate analogies between it and future NASA space station. Mission, covering 1,450 nm, would be conducted at 1,000-ft average depth, with periodic excursions to 300 ft and 2,000 ft. PX-15, designed by Dr. Piccard, would remain submerged throughout journey. Crew would experience space station characteristics: isolation, confinement, and stressful environment. (MSFC Release 69-100; *Marshall Star*, 4/9/69, 1)

- "World's only jet-powered personal jet propulsion system"—jet belt developed by Bell Aerosystems Co. under DOD sponsorship—made its first free flight near Niagara Falls International Airport. Device would provide quick-response, individual aerial mobility. (DOD Newsfilm Release 185-69; *AFJ*, 6/14/69, 20)

- In U.S.'s *Annual Review of National and Co-Operative International Space Activities*, Soviet Government report said Soviet scientists had paid "great attention" to developing methods of detecting signs of life on other planets. "With the development of space research, the problem of detecting life on the celestial bodies closest to the earth by means of space craft is becoming a priority matter. The considerable difference between conditions on the surface of the moon, Venus and Mars and those in which terrestrial life exists makes it necessary for us to extend our knowledge of the limits within which terrestrial life and life in general can exist. In this connexion Soviet scientists are investigating the possible limits of the existence of life. The absence of systematic processes for the movement of matter on the moon obviously makes active life on its surface impossible. On Mars, where free liquid water is probably absent, life is possible using matter transfer by frost, ice in the soil, water vapour and the wind. Examination of the temperature limits for the existence of life gives rise to a number of considerations which allow us not to exclude the possibility of the existence of life, in for example, the polar regions of Venus." (Text)

- Space Publications, Inc., reported its poll of Senate showed 48 Senators opposed to or "leaning against" deployment of proposed Safeguard ABM system; 46 for or leaning toward deployment; and 6 uncommitted, of which 4 had record of voting for former Sentinel system and 2 against. (*SBD*, 4/7/69, 167)

April 8: NASA's *Mariner VII*, launched from ETR March 27, successfully executed midcourse maneuver in response to radio commands from Woomera, Australia, tracking station. Spacecraft, more than 2.5-million mi from earth, would fly within 2,000 mi of Mars Aug. 5. (Sehlstedt, B *Sun*, 4/9/69, A5; Reuters, *W Post*, 4/9/69, A7)

April 8

- Saturn V 2nd stage (S–II–8) was successfully captive-fired at Mississippi Test Facility for 385 secs, with only four outboard engines functioning for final 86 secs. Center engine was intentionally cut off early to evaluate early cutoff as suppressor of longitudinal oscillations (pogo effect) which had occurred on *Apollo 8* and *9* flights. If test data were satisfactory, center engine might be cut off early during Apollo 10 mission in May. (MSFC Release 69–107)
- Aerobee 150 MI sounding rocket launched by NASA from WSMR carried American Science & Engineering, Inc., payload to 109.4-mi (176-km) altitude. Objective was to obtain high-resolution pictures of general x-ray emission from solar corona in quiescent state, using 9-in-dia x-ray mirror and 12-frame rotating camera with variety of filters. Rocket and instruments performed satisfactorily; x-ray exposures of considerable scientific value were anticipated. (NASA Rpt SRL)
- President Nixon announced nomination of NASA Assistant Administrator for Industry Affairs Philip N. Whittaker to be Assistant Secretary of the Air Force for Installations and Logistics. He also announced appointment of former Gov. William W. Scranton of Pennsylvania as U.S. Representative to INTELSAT Conference with rank of Ambassador, replacing Ambassador Leonard H. Marks, who resigned March 21. (*PD*, 4/14/69, 533–4; W *Star*, 4/9/69, A8)
- MSFC announced $8,391,052 modification to contract with North American Rockwell Corp. Rocketdyne Div. for support services to J–2 rocket-engine program from June 1, 1969, through April 30, 1970. (MSFC Release 69–104)
- Merger of military and civilian man-in-space programs was major decision facing Nixon Administration, said Howard Benedict of Associated Press. Many observers felt it necessary because, under separate courses being taken by NASA and DOD, costs of competing hardware systems "might explode out of proportion." Congressmen had charged there was costly duplication in NASA's orbiting workshop and DOD's MOL, both planned for 1971 launch, but with work on both so far advanced there was no turning back. Air Force Secretary, Dr. Robert Seamans, Jr., former NASA Deputy Administrator, had said any attempt to combine two programs "would jeopardize the returns to each agency and would ultimately increase the cost. . . . These activities require different equipment, different orbits and different timing." NASA Associate Administrator for Manned Space Flight, Dr. George E. Mueller, had said classified study to determine MOL's capabilities to accomplish NASA's long-duration earth orbit objectives had shown extended MOL "too limited to provide a significant, cost-effective step toward achieving NASA's long duration objective."

    Many observers believed technology and information for building national space station housing military and civilian personnel might emerge from the two programs. Another possibility was use by both agencies of similar rocket and spacecraft on separate scientific and military missions with equipment standardization providing "considerable saving." (*Huntsville Times*, 4/8/69)
- "Where the Legend Starts," film depicting life of late Cosmonaut Yuri Gagarin, was being prepared in U.S.S.R., Tass announced. Gagarin, first man in space on *Vostok I* April 12, 1961, died in aircraft crash March 27, 1968. (AP, *C Trib*, 4/9/69)

- Federally sponsored TurboTrain passenger demonstration began with one return trip daily from Boston to New York on Penn Central's New Haven Region. Within year service would permit elimination of 45 to 50 min from schedules. Developed by United Aircraft Corp., Turbo-Trains were leased by DOT for two-year Government experiment. (DOT Release 4769)

April 9: *Cosmos CCLXXVIII* was launched from Baikonur by U.S.S.R. into orbit with 318-km (197.6-mi) apogee, 203-km (126.1-mi) perigee, 89.6-min period, and 65.4° inclination. Satellite reentered April 17. (GSFC *SSR,* 4/15/69; 4/30/69; *SBD,* 4/10/69, 190)

- British Aircraft Corp. chief test pilot Brian Trubshaw flew Anglo-French Concorde 002 supersonic airliner on successful 24-min maiden flight from Filton Airfield, near Bristol. Aircraft, built to reach 1,400-mph speed, flew at 300 mph during flight, which copilot John Cochran termed "marvelous." French version Concorde 001 made maiden flight March 2. (W *Star,* 4/10/69, A13; AP, *W Post,* 4/10/69, A12)

- MSFC announced issuance of RFP's for 10-mo study to establish design concepts and development requirements for nuclear rocket stage to replace Saturn V 3rd stage for advanced missions in late 1970s and 1980s and payload design concepts and development requirements for flight test and early operational applications of stage using NASA-AEC NERVA. Study, for which proposals were due April 17, also would investigate payloads for nuclear-stage test flights including interplanetary meteoroid experiment (IME) and barium cloud experiment (BCE). IME would gather information on meteoroid environment in interplanetary space, particularly in asteroid belt between Mars and Jupiter. BCE would create artificial plasma cloud in space to simulate comet's tail and to study motions of ionized particles in earth's magnetosphere. Work would be done at MSFC. (MSFC Release 69–105)

- Astronaut R. Walter Cunningham was named a director of American Systems Inc., Los Angeles electronics firm. Cunningham, LM pilot on Oct. 11–22 *Apollo 7* mission, would continue duties as astronaut. (Reuters, *W Post,* 4/10/69)

- MSFC shipped 20,000-lb, 20-ft-tall F–1 and 225,000-lb-thrust J–2 Saturn V rocket engines from New Orleans to France as part of NASA exhibit at Paris Air Show, May 29–June 8. Other items in display would include *Apollo 8* spacecraft and an Apollo lunar module. (MSFC Release 69–106; MSFC PIO)

April 9–11: NASA and National Science Teachers Assn. sponsored Youth Science Congress at LaRC to encourage original scientific research by outstanding high school students. (*Langley Researcher,* 4/18/69, 1)

April 10: Prime crew for Apollo 12 mission was announced by NASA: Astronauts Charles Conrad, Jr. (commander), Richard F. Gordon, Jr. (CM pilot), and Alan L. Bean (LM pilot). Backup crew would be Astronauts David R. Scott, Alfred M. Worden, and James B. Irwin. Apollo 12 would land on moon four to six months after July 1969 Apollo 11 mission. (NASA Release 69–53)

- NASA reported research project to develop TV tube with completely black face to give pilots sharp-contrast picture of aircraft instrument readings, providing accurate information on rapid scanning of instrument panel. Filters would absorb prevailing cockpit light and prevent back reflection. NASA believed high-contrast cathode-ray tube could be used

April 10

in commercial TV receivers as well if production cost could be reduced. (NASA Release 69–52)

- NSF released *Federal Support to Universities and Colleges, Fiscal Year 1967*, prepared for Office of Science and Technology. Federal support to universities and colleges totaled $3.3 billion in 1967, up 9% over 1966 but below increases of 32% and 42% in previous two years. Federal support of academic science activities reached $2.3 billion, up 6%, with two-thirds for science education and institutional development and one-third for operating and plant costs for R&D projects. Nonscience activities reached $987 million, or 30% of total. While HEW, Dept. of Agriculture, AEC, and NSF increased aid to higher education, NASA decreased spending by $11 million (8%) and DOD by $37 million (12%). (Text)
- USA issued "cure notice" giving Lockheed Aircraft Corp. 15 days to prove it could solve technical problems of new high-speed AH56 Cheyenne helicopter production or face cancellation of 375 on order. Cost estimate had soared from $1.5 million each to $2.25 million with final contract terms not yet negotiated. (*WSJ*, 4/14/69, 15)

April 11: U.S.S.R. launched *Molniya 1–11* comsat to relay telephone and telegraph communications and TV broadcasts to "the far north," Siberia, Central Asia, and Far East. Orbital parameters: apogee, 39,595 km (24,603.2 mi); perigee, 483 km (300.1 mi); period, 712.1 min; and inclination, 64.9°. Equipment was functioning normally. (GSFC *SSR*, 4/15/69; *SBD*, 1/14/69, 204; AP, W *Star*, 4/11/69, A1)

- Terms of Reference for joint NASA/DOD study of space transportation systems were approved by NASA Administrator, Dr. Thomas O. Paine, and Secretary of the Air Force, Dr. Robert C. Seamans, Jr. Two-part study to assess practicality of common system would correlate needs of both agencies, assess technical feasibility of various systems, compare costs, assess economic sensibility of systems, and recommend concepts of space transportation system with rationale for each concept. Group would report to President's Space Task Committee June 15. (Terms of Reference)
- U.K.'s Institute for Strategic Studies predicted U.S.S.R. would overtake U.S. in ICBM production by mid-1969 but U.S. would retain overall lead in nuclear weapons because of greater submarine and air forces. U.S. international role in 1970s could become smallest since pre-World War II. (UPI, W *Star*, 4/11/69, A12)

April 12: NASA's *Oao II* orbiting astronomical observatory (launched Dec. 7, 1968) refused to accept commands from NASA's Santiago, Chile, tracking station. Satellite began tumbling out of control and its solar cells were unable to receive energy from sun to charge its batteries. While project officials tried to determine exact nature of anomaly, satellite recovered, accepting command from Australian station within few hours of battery depletion. *Oao II* was placed in sunbathing mode while batteries recharged. (Memo, NASA Asst Director for Projects; NASA Release 69–55)

- USAF launched unidentified satellite from ETR by Atlas-Agena booster into near polar orbit with 24,391-mi (39,245.1-km) apogee, 20,302-mi (32,665.9-km) perigee, 1,436.0-min period, and 10.2° inclination. (GSFC, *SSR*, 4/15/69; *Pres Rpt 70* [69]; *W Post*, 1/13/69, A14)

- President Nixon announced he would submit to Congress April 15 proposed anti-inflation revisions in FY 1970 budget including $4-billion reduction in Federal spending, to $192.9 billion; $5.5-billion reduction in appropriations requests and other budget authority; and $5.8-billion budget surplus—largest since 1951. (*PD*, 4/21/69, 553–4)
- Unpublished DOD estimate put Federal funding for C–5A transport at $5,202,400,000, Bernard D. Nossiter wrote in *Washington Post*. Figure was $77.2-million increase over quotation by USAF in January and total $2.1 billion (66%) increase since original 1964 estimate. (*W Post*, 4/12/69, A2)
- *April 12–14:* NASA successfully launched series of three Nike-Apache sounding rockets from Churchill Research Range carrying GSFC payloads to study energy spectra and relative abundances of various charge species of solar cosmic radiation during period of solar maximum. Each rocket carried three nuclear emulsion stocks and solid-state detector sensitive to protons above 30 mev. Rockets reached 98.4-mi (158.4-km), 96.4-mi (155.2-km), and 100.0-mi (161.0-km) altitudes and instruments performed satisfactorily. Payloads were recovered in good condition. (NASA Rpts SRL)

April 14: NASA's *Nimbus III* (Nimbus-B2) meteorological satellite was successfully launched from WTR by Long-Tank, Thrust-Augmented Thor (Thorad)-Agena D booster after three-day postponement because of fuel leak. Satellite entered orbit with 703-mi (1,131.1-km) apogee, 662.2-mi (1,065.5-km) perigee, 107.3-min period, and 80.1° inclination.

Nimbus III carried USA's *Egrs XIII* (also called *Secor XIII*) Sequential Collation of Range satellite as secondary payload on Agena 2nd stage and injected it into orbit with 704-mi (1,132.7-km) apogee, 667-mi (1,073.2-km) perigee, 107.3-min period, and 99.9° inclination.

Primary objectives were to inject *Nimbus III* into orbit and demonstrate satisfactory operation of active, three-axis, earth-oriented spacecraft for at least three months and to acquire representative global samples of infrared spectra for vertical temperature profiles of atmosphere. As secondary objectives spacecraft would make global maps of radiative energy balance of earth atmosphere and cloud cover over at least one seasonal cycle; demonstrate feasibility of surface pressure and tropospheric wind measurements by infrared interferometer spectrometer system and temperature profile determination by infrared spectrometry; make global maps of earth and day-and-night cloud cover for three months from image-dissector camera system and high-resolution infrared radiometer; and demonstrate SNAP–19 system as auxiliary power system for three months. Spacecraft carried seven meteorological experiments—most ever carried on U.S. meteorological satellite—and was first capable of measuring emitted infrared energy that would permit inference of atmospheric profile on global basis. Interrogation, recording, and location system (IRLS) would pinpoint position of special electronic platforms on fixed land sites and moving objects such as buoys, balloons, aircraft, and elk in Yellowstone National Park.

Butterfly-shaped 1,269-lb *Nimbus III* was fourth in series of seven spacecraft designed to develop significantly improved meteorological satellite, prove applicability of instrumentation, and fulfill special data

April 14

requirements of atmospheric sciences research community. It was replacement for Nimbus B, which failed to enter orbit May 18, 1968, when launch vehicle malfunctioned. *Nimbus I* (launched Aug. 28, 1964) had operated for one month before solar-array drive system malfunctioned. *Nimbus II* (launched May 15, 1966) had exceeded six-month lifetime, operating successfully until Jan. 18, 1969. Nimbus program was managed by GSFC under OSSA direction. (NASA Proj Off; NASA Release 69–50)

- NASA's Apollo 10 lunar orbital mission was proceeding well toward launch readiness for 11:49 am EST May 18, Deputy Apollo Program Director George H. Hage told NASA Hq. briefing. Flight readiness test had been successfully completed April 9. Countdown demonstration test (CDDT) would begin April 27, completing cryogenic propellant flow May 2; crew participation in CDDT would begin May 3. Astronauts Thomas P. Stafford (commander), John W. Young (CM pilot), and Eugene A. Cernan (LM pilot) would be launched from KSC Launch Complex 39, Pad B—being used for first time—on eight-day mission in which crew would enter lunar orbit, separate LM from CSM, and pilot LM to within 50,000 ft of lunar surface.

    Mission would include 11 different crew operations on TV and Westinghouse Electric Corp.-developed "experiment involving color TV if we can develop it and get it prepared to fly in time to support this mission. If we are able to . . . we would do most of the pictures with color TV rather than black and white. . . ." (Transcript)

- Astronaut Neil A. Armstrong would be first man to step on moon, according to NASA plans for July 16 Apollo 11 mission, Apollo Program Manager George M. Low told MSC press conference. Armstrong, mission commander, followed by Edwin E. Aldrin, Jr., LM pilot, would leave LM and spend 2 hrs 40 min walking on lunar surface, gathering rock samples, setting up experiments, and taking pictures within 100 ft of LM. Astronauts would not go farther, Low said, because "all that we need to carry out in deploying the experiment in doing all of the activities that we have to do on the surface the first time . . . can be done within the first 50 or 100 ft from the LM and we see no reason to go any further and use up a lot of energy walking as opposed to doing those things that we would like to do. . . ." (Transcript)

- Apollo 11 CSM and LM were mated with Saturn V (SA–506) launch vehicle at KSC in preparation for July 16 lunar landing mission. (*SBD*, 4/15/69, 209)

- Aerobee 150 MI sounding rocket with VAM–20 booster launched by NASA from WSMR carried GSFC payload to 116.5-mi (187.5-km) altitude. Primary objective was to obtain solar x-ray spectra from 2 to 400 Å. Secondary objective was to obtain integrated solar flux from 8 to 20 Å. Data would be used to check calibration of spectrometer on board *Oso V* (launched Jan. 22). Rocket and instruments functioned satisfactorily. (NASA Rpt SRL)

- Nixon Administration was responding to NASA requests for $200 million supplemental R&D funds by "emphatically ordering future reductions" in FY 1970 budget, said William J. Normyle in *Aviation Week & Space Technology*. BOB had told NASA Administrator, Dr. Thomas O. Paine, to plan on $140-million loss despite conclusions of House subcommittees that NASA's R&D should be increased $234.4 million. Until

Nixon decisions on U.S. space future, cut would "wipe out" work on space stations, shuttle/logistics vehicles, and manned lunar exploration. Feasible future programs were being studied by President's ad hoc committee which would report in September. (*Av Wk*, 4/14/69, 27–8)
- General aviation's impact on U.S. economy in 1980 would be $7.1 billion, 222.7% above $2.2-billion industry contribution to GNP in 1967, according to *The Magnitude and Economic Impact of General Aviation*, study by R. Dixon Speas Associates for AIA. Study said 1980 airports and airways system must be prepared to accept 260,000 general-aviation aircraft making 241 million takeoffs and landings and carrying 317 million passengers. (AIA *Fact Book*; Bramley, *Amer Av*, 4/14/69, 17–9)
- In Duke Univ. lecture, Dr. Peter van de Kamp, Director of Sproul Observatory at Swarthmore College, Pa., described discovery of fourth planet-like body found outside solar system. Existence had been deduced from 30 yrs telescopic observation of irregularities in Barnard's Star, six light years from earth. (*NYT*, 4/15/69, 16; AP, *W Post*, 4/17/69, F13)

April 15: USAF launched unidentified satellite from Vandenberg AFB by Titan IIIB-Agena D booster. Satellite entered orbit with 292.7-mi (471-km) apogee, 78.9-mi (127-km) perigee, 89.9-min period, and 108.7° inclination and reentered April 30. (GSFC *SSR*, 4/15/69; 4/30/69; *SBD*, 4/25/69, 264)
- U.S.S.R. launched *Cosmos CCLXXIX* into orbit with 350-km (217.5-mi) apogee, 205-km (127.4-mi) perigee, 89.8-min period, and 51.8° inclination. West German Institute for Space Research later reported spacecraft was a Soyuz space capsule and that it reentered and soft-landed in U.S.S.R. April 24. (GSFC *SSR*, 4/15/69; AP, *C Trib*, 4/25/69)
- President Nixon submitted to Congress FY 1970 budget amendments [see April 12]. White House, DOD, and NASA released details. NASA funds were reduced $45 million from $3.878-billion Johnson proposal to $3.833 billion (recommended $3.716 in new obligational authority plus $117 million carried over from prior years).

Apollo Applications program would be cut by $57 million but new obligational authority of $46 million for resumption of Saturn V rocket production and $40 million for lunar exploration would be added, for net increase of $29 million in manned space flight funds.

Space science and applications would be cut by $41 million, to $517.8-million new total. Advanced research and technology would be cut $13 million, to total $277.4 million; and tracking and data acquisition, $20 million, to total $278 million. University affairs funding would remain at $9 million, technology utilization at $5 million, and NERVA funding under nuclear rocket program at $36.5 million.

Funds proposed for construction of facilities and for research and program management remained unchanged.

At NASA budget briefing, NASA Administrator Dr. Thomas O. Paine said: "The reductions we have been required to make will make necessary difficult program adjustments and will result in reduced accomplishments in many areas. However . . . in a context clearly requiring that Government spending be held to a minimum, the Administration has recognized the importance to the United States of a strong and continuing program in space and aeronautics. . . . Last January, I characterized President Johnson's FY 1970 Budget as a 'holding' Budget . . . deferring to President Nixon's Administration the decisions the nation faces on the future of manned flight programs." President Nixon's recommendations, "if approved by the Congress, will ensure that the nation can continue a scientifically effective program of manned lunar exploration and avoid foreclosing our ability to continue large-scale space operations in the future by allowing the capability to produce Saturn V launch vehicles . . . to lapse beyond the point where it can economically be resumed."

DOD spending was reduced $1.1 billion and requested new obligational authority, $3.1 billion, including $51 million from MOL. Other science budget cuts: AEC funding, $78.6 million; agricultural and natural resources conservation, $345 million; HEW university facilities, $107 million; and NIH, $47.4 million. NSF budget remained at $495 million. Nixon budget made available $92.7-million carry-over for SST R&D but no funds for prototype construction. Overall reduction in space and atomic energy funding was $140 million. (PD, 3/21/69, 561–3; NASA Budget Briefing Transcript; DOD Transcript; W Post, 4/15–16/69; NYT, 4/16/69; Science, 4/25/69)

- Project Tektite Aquanauts Richard A. Waller, Conrad V. W. Mahnken, John G. Van Derwalker, and H. Edward Clifton were brought to surface and placed in decompression chamber for 19 hrs, after record-breaking 59 days on ocean floor off St. John, Virgin Islands. They had submerged Feb. 15 in successful experiment to determine how men functioned for extended periods underwater. At news conference later, aquanauts revealed they had spent 25–40 hrs outside underwater habitat during first two weeks, moving no farther than 300 ft; later they averaged 70 hrs per week, swimming up to 3,000 ft from habitat to study marine life, ocean currents, and geology. (Lyons, NYT, 4/14/69, 17; AP, W Star, 4/15/69, B11; W Post, 4/16/69, A9; 4/17/69; 4/19/69, A6)

- At American Chemical Society meeting in Minneapolis, Univ. of California at Berkeley nuclear scientist Albert Ghiorso reported discovery of element 104 isotopes 104–257 and 104–259 and possibly 104–258. Discoveries, made by bombarding target in heavy ion linear accelerator (HILAC), were announced by Lawrence Radiation Laboratory, operated at Univ. of California for AEC. (AEC Release M–87)

April 15–17: Conference on technology of food management for aerospace vehicles was sponsored by NASA, NAS, and Univ. of South Florida at Tampa, Fla. Discussions included Mercury, Gemini, and Apollo mission experience by Dr. Charles A. Berry, MSC Director of Medical Research and Operations, and feeding system requirements for Manned Orbiting Laboratory and Apollo Applications program. Food specialists had found they needed to improve methods on all types of aerospace flights with emphasis on more palatable food and less food preparation time in flight. (NASA Special Release)

April 16: NASA would require $5- to $5.6-billion annual budget "within three years" for simultaneous development of orbiting space station and lunar exploration in 1970 and onward, NASA Associate Administrator for Space Science and Applications, Dr. Homer E. Newell, said in *Washington Post* interview. (Cohn, W Post, 4/17/69, A3)

- Aerobee 150 MI sounding rocket launched by NASA from WSMR with VAM-20 booster carried Naval Research Laboratory payload to 117.0-mi (188.3-km) altitude to record white-light corona of sun from three to nine solar radii with package containing two externally occulted coronagraphs and three related experiments. Rocket and instruments functioned satisfactorily. (NASA Rpt SRL)
- NASA noted new level of maturity in conduct of *Apollo 8* mission, NASA Associate Administrator for Space Science and Applications, Dr. Homer E. Newell, told National Space Club in Washington, D.C. "We got the feeling that the people . . . handling the operations had now a new tool that fit the hand." Maturing of total space capability was seen in use of Pioneers and earth orbiting satellites during *Apollo 8* "to keep track of what the sun was doing so that the directors of the Apollo operation could know what radiation conditions in space were relative to . . . crew safety." Mission also used weather satellites and communication satellites, "not only to transmit pictures . . . but also as an integral part of the operation, to transmit data. . . ."

  Asked NASA's priority on space station versus lunar exploration in view of increase in funding request for lunar landing program, Dr. Newell said: "Now you have touched upon what I think is going to be the most difficult question for the country to resolve in the months ahead." There was general agreement that manned space flight should continue. Debate would be over whether it would be done by continued lunar exploration, earth orbital operations, or both. "Our own feeling is that the country ought to do both. Certainly after having put all the investment into landing a man on the moon and developing the capability to do so we must continue to explore the moon. And we cannot foresee any reasonable or rational national program in which we do not continue that exploration. At the same time . . . we haven't finished developing the manned space flight capability. We have got to . . . get that permanent foothold in space and that is where the space station comes in." (Transcript)
- LaRC and JPL announced formation of management teams to direct Viking Mars 1973 program to send two instrumented lander-and-orbiter spacecraft to Mars in 1973. LaRC had responsibility for overall project management and for lander portion of spacecraft. JPL would manage orbiter portion and tracking and data acquisition. James S. Martin, Jr., was LaRC project manager, with Henry W. Norris named Viking orbiter manager at JPL. (NASA Release 69-54; JPL Release 512)
- MSFC announced award of eight-month identical $400,000 contracts to Grumman Aircraft Engineering Corp. and Bendix Corp. for preliminary design and definition studies of dual-mode lunar roving vehicles. DLRV would provide mobility for one astronaut on lunar surface and could be operated by remote control from earth while making automated, long-range traverses of large lunar areas. DRLV would be delivered to moon aboard Apollo LM. After astronauts left, it would be placed in remote control for geological and geophysical trips of 600 mi or more for one year, during which it would collect up to 200 lbs of lunar samples and measure lunar terrain. It would then rendezvous with manned spacecraft and transfer samples for return to earth for analysis. (MSFC Release 69-110)
- At closed session of Senate Armed Services Committee Dr. Robert C.

Seamans, Jr., Secretary of the Air Force, and Gen. John P. McConnell, Air Force Chief of Staff, said development of new intercontinental supersonic bomber was "perhaps our most urgent requirement." Aircraft, with proposed ABM system, would "provide insurance against unexpected Soviet developments." (Homan, *W Post*, 4/17/69, A9)

- U.S.S.R. announced it would conduct series of rocket tests in Pacific from April 17 to June 15. Aircraft and ships were asked to avoid 55-nm-dia circular area north of Midway Island with center at 35°23′ north latitude and 172°24′ west longitude. (UPI, *W News*, 4/17/69, 9; *SBD* Space Log supplement, 4/18/69)

- At Vienna peace conference, following three-day secret talks, AEC member Gerald F. Tape and Dr. Yevgeny K. Fedorov, head of Soviet Weather Bureau, issued joint communique expressing concern about amount of harmful radioactivity that would be released by nuclear explosions when used for such projects as canal digging. Tape predicted widespread use of nuclear explosion for benefit of nonnuclear countries was still five years away; Fedorov said Soviet technology would permit general use of some types before 1974, but applications would come later. (Hamilton, *NYT*, 4/17/69, 8)

- MIT and USN unveiled in Boston computerized guidance and control system to enable pilot of deep-diving rescue vessel to rendezvous and dock with disabled submarines at depths to 5,000 ft. It would be used in Deep Submergence Rescue Vehicles—new class of USN submarines scheduled for first test dive in June 1969. (Wilford, *NYT*, 4/17/69, 94; *W Post*, 4/17/69, A4)

April 17: *Maj. Jerauld R. Gentry (USAF) piloted USAF X-24A lifting-body vehicle on its first glide flight. The wingless craft, which depends on shape and speed for aerodynamic lift, was air-launched from a B-52 aircraft at 45,000-foot altitude and mach 0.66. Designed for maximum speed of mach 2 and altitudes to 100,000 feet, the X-24A was one of three wingless experimental vehicles in the joint USAF–NASA research program studying concepts for reusable and maneuverable reentry spacecraft.*

April 17: X–24A lifting-body research vehicle, piloted by Maj. Jerauld R. Gentry (USAF), successfully completed first glide flight from Rogers Dry Lake, Calif. Vehicle was air-launched from B-52 aircraft at 45,000-ft altitude and mach 0.66 for pilot checkout and data on longitudinal trim, lift, and drag. (NASA Proj Off)

- NASA's HL–10 lifting-body vehicle, piloted by NASA test pilot John A. Manke, successfully completed 15th flight. Primary objectives were to expand flight envelope to mach 0.9 and to determine control characteristics at mach 0.9. Vehicle was air-launched from B–52 aircraft north of Four Corners, Calif., at 45,000-ft altitude and mach 0.7. Manke ignited three chambers of XLR–11 engine, rotated vehicle, climbed to 55,000-ft altitude, and sustained flight at mach 0.9 for 100 secs. (NASA Proj Off)
- Rep. George P. Miller (D-Calif.) introduced in House H.R. 10251, new NASA FY 1970 authorization bill totaling $3.716 billion in line with President Nixon's April 15 amended budget request. Bill was referred to House Committee on Science and Astronautics. (Text; *CR*, 4/17/69, H2806)
- Nike-Tomahawk sounding rocket launched by NASA from Churchill Research Range carried TRW Systems Inc. payload to 161.6-mi (260-km) altitude. Objectives were to measure total flux and energy, including spectrum of precipitated energetic (1–20 kev) H atoms and precipitated energetic protons and electrons; fluctuating DC electron fields; H_B light intensity altitude profile; and location and intensity of ionosphere current systems. Rockets and instruments functioned satisfactorily; good data were obtained. (NASA Rpt SRL)
- Aerobee 150 MI sounding rocket launched by NASA from WSMR with VAM–20 booster carried Naval Research Laboratory payload to 113.7-mi (182-km) altitude to record white-light corona of sun from three to nine solar radii with package containing two externally occulted coronagraphs and three related experiments. Rocket and instruments functioned satisfactorily. (NASA Rpt SRL)
- Nike-Apache sounding rocket launched by NASA from NASA Wallops Station carried Univ. of Illinois and GCA Corp. payload to 130.5-mi (210-km) altitude to measure electron density, collision frequency, and temperature in lower ionosphere at vernal equinox during sunspot maximum. Rocket and dual-frequency propagation experiment for airglow photometer performed satisfactorily. Langmuir-Smith probe and UV experiment produced no data and payload did not come out of calibration until near impact. (NASA Rpt SRL)
- MSFC announced modification of $3,057,503 to contract with Chrysler Corp. Space Div. for work on mechanical ground support equipment for Saturn IB and Saturn V launch vehicles from Dec. 1, 1968, through March 31, 1970. MSFC also announced one-year $5,704,116 extension to contract with RCA Service Co. Div. of Radio Corp. of America for technical services in support of MSFC Management Services. (MSFC Releases 69–115, 69–114)
- DOT released *Study of Air Cargo and Air Passenger Terminal Facilitation* by Simat, Helliesen, & Eichner, Inc., and TransPlan, Inc.—source document for Transportation Facilities Committee's industry-Government task forces. It forecast worldwide passenger traffic increase of nearly 10% annually and air cargo increase of nearly 20% annually during

next decade. Documentation, processing, and handling delays were contributing to nearly six-day delivery time of air cargo despite six-hour Atlantic crossings, and Government clearance procedures were critical problem for international cargo operations. Report recommended: further consideration of high-speed rail, STOL, and VTOL services; computerized processing of cargo documentation and high-speed communication to facilitate advanced clearance; off-airport cargo terminals; automated passenger ticketing and baggage handling; and streamlined, mechanized border formalities. (DOT Release 5869)

- Rep. Emilio Q. Daddario (D-Conn.) told House Dr. Franklin A. Long of Cornell Univ. had been asked to withdraw from nomination as NSF Director after refusing to support Administration's ABM system. "It is unfortunate that the Nixon administration is sacrificing the National Science Foundation on the altar of the ABM, and, by so doing, seriously affecting its unique capability to be of service to our country." (CR, 4/17/69, H2759)

April 18: Apollo 11 Astronauts Neil A. Armstrong and Edwin E. Aldrin, Jr., dressed in full landing attire, rehearsed simulated lunar landing at MSC for the 2 hrs 40 min they would spend walking and working on lunar surface in July. Astronauts practiced scooping rock and soil samples, unfurled umbrella-like antenna, and deployed seismometer to detect quakes and array of mirrors to serve as laser target for astronomers on earth. (Wilford, *NYT*, 4/19/69; UPI, *P Bull*, 4/19/69)

- MSFC announced it had issued RFPs for assistance in producing 320 completed solar "arrays" to convert solar energy into electrical power to operate Saturn I Workshop. Two wings covered with 120 modules each, for total 1,200-sq-ft area, would produce 12,000 w for Workshop. Pre-proposal conference was scheduled at MSFC May 1. (MSFC Release 69–116)

- INTELSAT Consortium had selected Atlas-Centaur launch vehicle for Intelsat IV program, ComSatCorp announced. First Intelsat IV—18-ft-high, 8-ft-dia comsat with 5,000 two-way voice grade circuits—would be launched in early 1971. ComSatCorp would negotiate with NASA for purchase and launch of two Atlas-Centaur boosters with option for two more. Atlas-Centaur was manufactured by General Dynamics Corp. Convair Div. and managed by LeRC. (ComSatCorp Release 69–19; *Lewis News*, 4/25/69, 1)

- ComSatCorp announced earnings of $1,525,000, or 15 cents per share, for first quarter of 1969. It had earned $1,798,000 (18 cents per share) for similar period in 1968. Expected decline followed increased operating expenses from expansion of satellite and earth station system. (ComSatCorp Release 69–18)

- Having started from "near zero" in 1961, West Germany was spending about $90 million annually on space activities, said D. S. Greenberg in *Science*, with rise to $150 million expected within few years. About 40% was in international programs like ELDO and ESRO but emphasis was on building domestic facilities. All-German satellite was being built for launch by NASA and Germans were working with French on comsat for 1972 Munich Olympic Games coverage. "With no manned space effort or aspirations to divert their resources, and with the military barred from space, whatever the Germans muster in space activity goes directly into scientific research or commercial applications," Greenberg said. (*Science*, 4/18/69, 281–3)

April 20: AH–56A Cheyenne helicopter—integrated, aerial, vehicle-armament-avionics-fire control gunship capable of firing machine guns, grenades, rockets, and missiles—could become latest DOD-purchased aircraft to incur Congressional and public criticism because of massive cost overruns, major technical difficulties, and hints of improper procurement practices, Robert Walters said in Washington *Sunday Star*. Lockheed Aircraft Corp.'s California Co. was prime contractor on Cheyenne project which "has been in trouble almost since its inception in August 1964." Total cost for Cheyenne program, including $195.3 million estimated for R&D, was set as $1.06 billion and Lockheed contract allowed further increases. (W *Star*, 4/20/69, A8)

April 21: NASA's *Explorer XXXVIII* (launched July 4, 1968), orbiting at 3,640-mi altitude, had discovered earth, like Jupiter, sporadically emitted low-frequency radio signals, Radio Astronomy Explorer project scientist Dr. Robert G. Stone told 1969 spring meeting of U.S. National Committee of International Union of Radio Science in Washington, D.C. "In the radio frequency range below 10 megahertz, the Earth looks very much like Jupiter. This suggests that the same processes that account for the Jovian radio noise may also be active in the Earth's magnetosphere." Signals from both planets appeared sharply beamed or directed in narrow cone, were quite intense and increased in intensity rapidly when observed toward lower frequencies, and were impulsive, occurring in rapid but sporadic bursts.

Explorer XXXVIII also had revealed sun was more active source of radio outbursts in lower frequencies than expected, providing unique means of studying sun's outer atmosphere to 36-million-mi distance of Mercury orbit. "Such information," Stone said, "could shed further light on mystery of Sun's part in cause of low frequency radio storms on the Earth." *Explorer XXXVIII*'s four 750-ft-long antennas had provided most comprehensive and detailed measurements of cosmic radio noise at low frequencies yet available. Information was providing first low-frequency maps of radio emissions in Milky Way galaxy, showing that most radio emissions originated in plane or disc of Milky Way. Satellite had remained stable since orbit despite repeated movement and had continued to operate successfully. (NASA Release 69–57; AP, B *Sun*, 4/22/69, A6; O'Toole, W *Post*, 4/22/69, A1)

- President Nixon's amendments to proposed FY 1970 space budget would support post-Apollo manned flight at expense of unmanned planetary space flight, William J. Normyle wrote in *Aviation Week & Space Technology*. In effect, NASA had received "almost all it wanted for planning post-Apollo manned lunar exploration." Argument was expected in House Committee on Science and Astronautics over relative apportionment. (*Av Wk*, 4/21/69, 16–7)

- Senate Committee on Banking and Currency approved nomination of NASA Associate Administrator for Organization and Management Harold B. Finger as Assistant Secretary of Housing and Urban Development. (*CR*, 4/21/69, D289; W *Star*, 4/22/69, A3)

April 22: NASA named Brian M. Duff, Vice President for Communications of National Urban Coalition, as Public Affairs Officer for MSC. He would succeed Paul Haney who was named Special Assistant to Julian Scheer, Assistant Administrator for Public Affairs, NASA Hq. Duff had been Director of Special Events in NASA Office of Public Affairs before going to National Urban Coalition. In new position, Haney would coordinate development of news media materials, with concentration on Apollo manned lunar landing. Haney later told press he had been ordered to new position because of personal differences with Scheer. "I'm definitely uncertain whether to go and have consulted a lawyer," he said. Scheer said on April 23 transfer was not personal matter but "case of using two very good people where they can best serve our needs." (NASA Release 69–59; *H Chron*, 4/22–23/69; *H Post*, 4/23/69, 1; *W Post*, 4/23/69, A3)

- Rep. Charles W. Whalen, Jr. (R-Ohio), inserted in *Congressional Record* results of annual opinion poll of his constituents which showed 51% preferred continuation of space funding at present level, 21% preferred cutback with reallocation of space funds to social welfare programs, 16% wanted acceleration of space program with increased funding if necessary, and 12% wanted none of these. (*CR*, 4/22/69, E3230)

- DOD reported Soviet SS–9 missile had been test-fired with multiple reentry vehicles for first time over U.S.S.R.'s Pacific testing area at 5,000-mi range. Missile, of which U.S.S.R. had deployed 200, was capable of carrying warhead of up to 25 mt or three warheads of 5 mt and was only Soviet missile credited with hard-site destruction capability. Portions of its booster had been used in testing FOBS. (DOD Release 310–69; Homan, *W Post*, 4/23/69, A26)

April 22–25: Discovery of six new mascons (mass concentrations of dense material) beneath moon's surface was reported by JPL scientists at 50th Annual Meeting of American Geophysical Union in Washington, D.C. William L. Sjogren, Paul M. Muller, and Dr. Peter Gottlieb reported discovery that brought to 12 total mascons mapped on moon's near face and leading and trailing edges and that were expected to refine gravity model used in Apollo navigation.

Dr. Gottlieb said latest gravity model produced significant agreement with analysis of tracking information from *Apollo 8* mission Dec. 21–27, 1968. JPL researchers were working with MSFC to predict accurately landing sites several orbits before spacecraft landing. Most data available had been taken from spacecraft in lunar orbit of about 60 mi. JPL team expected new and possibly higher resolution data from Apollo 10 when lunar module orbited at lower altitude. (JPL Release 514; NASA Release 69–61)

April 23: *Cosmos CCLXXX* was launched by U.S.S.R. into orbit with 251-km (156-mi) apogee, 198-km (123-mi) perigee, 88.1-min period, and 51.0° inclination. Satellite reentered May 6. (GSFC *SSR*, 4/30/69; 5/15/69; *SBD*, 4/25/69, 261)

- Briefing on *Apollo 9*'s earth resources survey experiment [see March 3–13] was held at NASA Hq. Dr. Leonard Jaffe, Director of Space Applications Programs, stressed importance of satellite photos for earth resources program: ". . . we have taken advantage of the Apollo and Gemini, as we did with Mercury, opportunities to get pictures of the earth and its environment with cameras, largely held by hand, by the astronauts. These pictures have been a very large source of experimental data for our Earth Resources Survey Program."

Advantage of using space for earth resources survey was twofold. One was "large view that it gives you of the earth." Second was "ability to see the same phenomenon, or the same area of the earth time and time again, to be able to arrive over this particular spot on the earth periodically and observe temporal changes." *Apollo 9* experiment had obtained some 120 or 130 useful frames of multispectral data, all of which had been distributed to principal investigators for analysis. (Transcript)

- House Committee on Science and Astronautics accepted subcommittee reports on H.R. 10251, FY 1970 NASA authorization [see April 17] and added $258 million to amounts requested by President Nixon. Committee would meet again April 29. (*CR*, 4/23/69, D302; Lannan, W *Star*, 4/24/69)
- National Telemetry Conference of Institute of Electrical and Electronic Engineers, Inc., presented award for telemetry achievements to NASA team responsible for directing, planning, engineering, implementing, and operating Manned Space Flight Network in support of Apollo program. At Washington, D.C., luncheon, award was accepted by GSFC Assistant Director for Manned Flight Support, Ozro M. Covington. (GSFC Release G-1-69)
- Australian Prime Minister John G. Gorton announced in Canberra that Australian government had accepted U.S. proposals for "a joint United States-Australian defense space communications facility" at Woomera, South Australia, site of Australian Weapons Research Establishment. (Sehlstedt, B *Sun*, 4/24/69, A1)

April 24: NASA's *Mariner VI* and *Mariner VII* spacecraft were 10 million mi and 6.47 million mi from earth en route to Mars. *Mariner VI* Canopus tracker had failed to change position properly April 20 and search was underway to find substitute for Canopus. (JPL Status Bull)

- President Nixon announced appointment of five new members to 19-member President's Science Advisory Committee: Dr. John D. Baldeschwieler, professor of chemistry at Stanford Univ.; Dr. Richard L. Garwin, Director of IBM Watson Laboratory at Columbia Univ.; Dr. Murray Gell-Mann, professor of theoretical physics at Cal Tech; Dr. Patrick E. Haggerty, President of Texas Instruments, Inc.; and Dr. Gerald F. Tape, President of Associated Universities, Inc. (*PD*, 4/28/69, 602)
- Signing of $8,802,472 supplemental agreement with McDonnell Douglas Corp. defining 18 change orders affecting S-IVB quality maintenance program was announced by MSFC. Agreement included reliability and quality reviews, documentation, and expanded production acceptance tests. (MSFC Release 69-122)
- FAA announced initiation of Airport Data System (ADS) to collect and validate data on facilities and service available at nation's airports in centralized section of FAA—National Flight Data Center (NFDC). Data would be made available to U.S. Coast and Geodetic Survey and commercial chart producers. (FAA Release T 69-20)

April 25: NASA's HL-10 lifting-body vehicle, piloted by NASA test pilot William H. Dana, was air-launched from B-52 aircraft at 45,000-ft altitude and glided to successful landing. Purpose of flight, 16th in series at FRC, was pilot checkout. (NASA Proj Off)

- Electrostatic zero-gravity workbench experiment by Chrysler Corp. at

Michoud Assembly Facility might provide substitute for gravity in small space-station area, NASA reported. If successful, experimental equipment—three-foot-square workbench with coated metal top, high voltage and variable power supply, and ion source with interconnected cabling—would enable astronaut to manipulate loose objects like tools in weightlessness of orbiting space station. Ion source and high-voltage power supply, directed downward, would create force field to hold tools to bench top. Ground experiments would be completed in six months. (NASA Release 69-58)

- Sen. Clinton P. Anderson (D-N. Mex.), for himself and Sen. Margaret C. Smith (R-Me.), introduced S. 1941, NASA authorization bill, similar to H.R. 10251 [see April 17]. Bill was referred to Senate Committee on Aeronautical and Space Sciences. (Text; *CR*, 4/25/69, S4118)
- Senate confirmed nomination of Harold B. Finger, NASA Associate Administrator for Organization and Management, as Assistant Secretary of Housing and Urban Development. (*CR* 4/25/69, S4189)
- Paul Haney, described by press as "Voice of Apollo," announced resignation after 10 yrs with NASA following NASA Hq. discussion of his April 22 appointment as assistant to Associate Administrator for Public Affairs Julian Scheer. Haney said decision to resign resulted from personal differences with Scheer, who told press he had tried to persuade Haney to remain with agency. (AP, *NYT*, 4/26/69, 2; O'Toole, *W Post*, 4/29/69, A3)

April 26: Aerobee 150 MI sounding rocket launched by NASA from WSMR carried MIT payload to 97.7-mi (157.9-km) altitude to study x-ray spectrum of Crab Nebula at long wavelengths and to search for weak x-ray sources associated with supernova remnants or with galaxies outside Milky Way. Rocket and instruments performed satisfactorily. (NASA Rpt SRL)

- Nike-Apache sounding rocket was launched by NASA from Thumba Equatorial Rocket Launching Station (TERLS) carrying Indian-Japanese payload for x-ray astronomy studies. Rocket and instruments functioned satisfactorily. (NASA Proj Off)
- Testifying before Subcommittee on Space Science and Applications of House Committee on Science and Astronautics, Joseph B. Mahon, NASA's Director of Launch Vehicle and Propulsion Programs, OSSA, said NASA planned to use Titan IIIC and Titan Centaur to carry out high-velocity Viking mission in FY 1970. They would provide capability for both orbital and lander scientific experimentation and would expand launch window. After proof test flight in 1972 of integrated Improved Centaur and Titan III, configuration would be flown in support of 1973 Viking mission and other planetary missions, as well as earth synchronous missions using 4,000- to 8,000-lb spacecraft. Titan IIIC also would be used on two synchronous-altitude missions, ATS-F and ATS-G. (Transcript)

April 27: Several thousand gallons of fuel escaped from prevalves in Saturn V 1st stage (S-IC) while it was being prepared for Apollo 10 countdown demonstration test (CDDT) at KSC. Prevalves opened while leak was being repaired in nitrogen pressurization system on mobile launcher. CDDT, scheduled to begin April 28, was delayed 24 hrs while vehicle was examined for damage. (O'Toole, *W Post*, 4/29/69, A3; *Marshall Star*, 4/30/69, 1)

- FRC announced retirement of NASA's two X-15 rocket research aircraft, designed for manned hypersonic flight research at speeds up to 4,000 mph and altitudes of 50 mi. X-15 No. 1, which had made first flight June 8, 1959, would be displayed at Smithsonian Institution in Washington, D.C. No. 2 would be displayed at Air Force Museum at Wright-Patterson AFB, Ohio. No. 3 had been destroyed in accident Nov. 15, 1967. During $300-million NASA–USAF–USN program X-15 had set two unofficial world records, reaching 354,200-ft altitude and 4,520 mph (mach 6.7). Aircraft served as reusable manned platforms for wide range of experiments that helped advance development of vital spaceflight systems. Final flight, 199th, had been Oct. 24, 1968, and NASA had announced completion of program Dec. 20. (FRC Release 9–69)
- William Hines said in Washington *Evening Star:* "It now seems entirely possible that NASA may sneak into a Mars program without a specific go-ahead of the type given for the moon adventure in 1961. A case can be made that exploration of the planets is but a logical extension of exploration of the moon; that once the space frontier is crossed in 1969 everything else is evolutionary, not revolutionary. This sort of gradualism, which is analogous to a girl's becoming a little bit pregnant, is just as effective in the long run as the flamboyant setting of spectacular goals, and probably a good bit more feasible in a time when there is already much grumbling about pre-occupation with other worlds when our own is in such a sorry state." (W *Star*, 4/27/69, G4)
- Dr. Nicholas E. Golovin, technical adviser for aviation and space science and technology in Office of Science and Technology, died of heart attack in Washington, D.C., at age 57. He had been deputy Associate Administrator of NASA in 1960. After returning to private industry for a year, he had rejoined NASA as director of Large Launch Vehicle Planning Group. Before coming to NASA he had been Chief Scientist at White Sands Missile Range for DOD and then Director of Technical Operations Div. of ARPA. (*W Post*, 4/30/69, B14; W *Star*, 4/30/69, B6)

April 28: Dr. Thomas O. Paine, NASA Administrator, testified in FY 1970 authorization hearings before Senate Committee on Aeronautical and Space Sciences that he believed "the greatest significance" of President Nixon's decision to include in budget amendment amounts for continued lunar exploration after first Apollo landings and for continued Saturn V production was "that it recognizes the fundamental fact that the United States should not and does not plan to stop its further development and use of manned space flight."

Early manned lunar landing would allow additional savings of perhaps as much as $39 million in amended Apollo budget. Budget included funds for improved scientific equipment for more advanced missions to moon and for limited Apollo Applications program. Fund cuts would mean some program reorientation as well as delay in Workshop, previously scheduled for mid-1971.

Amended budget supported continuation of plans for 1971 and 1973 Mars missions and first mission to Mercury in 1973, with Venus swing-by. In space applications, "where we are placing special emphasis," Earth Resources Technology Satellite program would enable NASA, with other agencies, to test practical use of space to gather information on water levels, crops, forests, and other resources. "Despite

April 28

the sizable overall reduction . . . we have not reduced the earth resources project."

Budget included funds to proceed with NERVA development and continued to give "high priority to our work in aeronautics." New construction would include Aircraft Noise Reduction Laboratory at LaRC, "unique national facility for studying the fundamentals of noise generation and reduction."

U.S.S.R. was continuing vigorous program in both aeronautics and space. "The Soviet commitment represents . . . the application of resources at about the same rate as that we have averaged in recent years," with "significantly higher percentage of the gross national product." Soviets were "proceeding in manned space flight programs directed both at sending men to the moon and at substantially enlarged and extended manned operations in earth orbit. . . . Automatic rendezvous and docking flights with Cosmos 186–188 and 212–213 and the manned Soyuz 4–5 mission in January, with manned docking and crew transfer, demonstrated the increasing Soviet capability. . . . They appear to be pointed to a future capability for assembly in earth orbit of large space stations and of manned spacecraft to send to the moon and beyond." U.S.S.R. had "made attempts to launch one or more spacecraft to the planets at almost every opportunity—three or four times the number of attempts we have been able to make." (Testimony)

- Mascons might be excess mass deposited by water and supported by internal strength of rigid moon, JPL scientist Paul M. Muller said in address before American Physical Society in Washington, D.C. Muller was codiscoverer of mascons [see April 25]. Mascons probably were not floating on liquid lunar interior as were earth's continents and mountains, but were held there by moon with structural strength. He illustrated with photos taken by Lunar Orbiters and *Apollo 8* theory that lunar features might have been cut by primordial lunar water. (NASA Release 69–62; JPL Release 515)

- NASA issued RFPs for design and planning studies of 12-man, earth-orbital space station for 1975 with 10-yr operational life, subject to resupply of expendables and crew rotation with logistics vehicles. Work also would include conceptual design of 50-man space base of specialized modules assembled in low earth orbit in late 1970s and early 1980s and comparable to scientific and technical research, development, and operations center on earth. Modified Apollo and Gemini spacecraft would be considered as initial logistics systems if space shuttle did not become available in early phase of space station operations. Data from studies would be available for final design of future space station if program were approved for development. Proposals were due June 9. (NASA Release 69–65)

- Nike-Apache sounding rocket was launched by NASA from Thumba Equatorial Rocket Launching Station (TERLS) carrying Indian-Japanese payload for x-ray astronomy studies. Rocket and instruments functioned satisfactorily. (NASA Proj Off)

- LeRC said its engineers had built and were operating world's largest acoustical testing facility for fan portion of compressors on turbofan engines—key element in aircraft noise. It could test fans up to six feet in diameter to collect basic noise information and help determine fan configuration for NASA's quiet engine program. Facility was powered by supersonic wind tunnel's drive motor, whose 23,000 hp spun fan to 3,500 rpm. Noise created was recorded by strategically placed microphones to determine differences in level producing by fan blade configurations and effects of noise-deadening inlet lining materials. (LeRC Release 69–19)

- At international airline conference in Dublin, U.S.S.R.'s First Deputy Minister of Civil Aviation, Gen. Leonid V. Zholudev, denied Tu-144 supersonic transport had been in accident, *American Aviation* said. It was "undergoing tests according to our program," Gen. Zholudev said. He declined to speculate when Soviet aircraft would go into service and said "many complex problems must be solved." Tu-154, 170-passenger trijet, would enter service "in near future"; An-22 turboprop, with reported maximum takeoff weight of 551,160 lbs, was being used exclusively as cargo carrier. (*Amer Av*, 4/28/69, 17)

- President Nixon met with National Science Board members and NAS council and promised to nominate NSF director without regard to his ABM position. He asked for names for possible nominee and promised to consider only scientific qualifications. President disclosed he had asked Cornell Univ.'s Dr. Franklin A. Long to consider having his name resubmitted to Senate after original decision not to nominate Dr. Long [see April 17] but Dr. Long had declined. (Kilpatrick, *W Post*, 4/29/69, A1)

April 28–30: At 106th annual meeting of National Academy of Sciences in Washington, D.C., Dr. Eugene N. Parker of Univ. of Chicago and Dr. J. P. Wild of Commonwealth Scientific and Industrial Research Organization in Sydney, Australia, received Henry Arctowski Medal for studies of solar activity changes and their effects upon ionosphere and terrestrial atmosphere. Dr. Parker was honored for "contributions to theoretical understanding of interaction between magnetic fields of sun and earth and surrounding ion plasmas"; Dr. Wild, for contributions to solar radio astronomy, including development of technique of studying solar disturbances through moving pictures of sun "photographed" in radio spectrum. Dr. Jürgen K. Moser of New York Univ. received James Craig Watson Medal for mathematical contributions to dynamical astronomy. (NAS Release 4/23/69)

Cal Tech scientist Kip S. Thoren suggested pulsars were subject to quakes which accelerated their pulsation. At press interview during annual meeting, he said pulsars' insides were probably superfluids—more liquid than liquid—and were gradually cooling, with crust crystallization requiring adjustments. Adjustments, he said, would cause "pulsequakes" and could result in sudden speedups in pulsar periodicity.

Cornell Univ. astronomer Dr. Thomas Gold said observations of pulsars indicated they slowed down as they aged, with oldest finally slowing billionths of seconds per year until they reached periods up to almost four seconds. He suggested pulsar radiation might account for luminosity of whole galaxies and even produce all cosmic rays detected throughout universe, including earth.

Cornell Univ. astronomer Dr. Frank D. Drake, former director of Cornell's Arecibo Ionospheric Observatory in Puerto Rico, said space clocks were emerging as practical scientific tools, especially in testing Einstein's theories. Dr. F. Graham Smith of Nuffield Radio Astronomy

Laboratories, Manchester, England, said space clocks were being used to investigate interstellar medium—gas clouds between earth and other parts of solar system. (Lannan, *W Star*, 5/1/69, A6; O'Toole, *W Post*, 5/1/69, A1)

April 29: Dr. Lee A. DuBridge, Presidential Science Adviser, said at NAS dinner in Washington, D.C.: "The relations between science, technology, government and the various elements of our society are enormously complex. Science and technology are no longer separable from political and social problems. . . . Whether we like it or not, science is in politics and politics is in science. . . . The fact is that today science is too important to our nation to stay out of politics. Clearly we all—politicians and scientists—must find ways of adapting ourselves to a new era—an era which began . . . really on Hiroshima day in 1945." (Text)

April 29–May 1: MSFC held workshop on optical telescope technology to exchange technical information on design of future space telescopes and identify research and technology efforts needed to support future missions. (MSFC Release 69-117)

April 30: At KSC briefing on NASA FY 1970 budget for community leaders, KSC Director, Dr. Kurt H. Debus, said: "It is clear that we will continue the present pace of launch operations in the Apollo program until the first lunar landing has been accomplished. . . . The continuation of space effort, however, must somehow take into account the same constraints on Government spending which now affect the entire federal budget. . . . The total KSC budget is being revised from $455 million for all purposes to $410 million. I am taking action to accomplish this reduction without crippling the essential launch team capability or drastically impacting the local economy." Savings would be effected by: gradual cutback in efforts of contractors supporting design engineering; phasing down of Boeing technical integration effort; deceleration of Apollo/Saturn V launch rate to three manned vehicles annually, reducing employment; and greater use of five-day week and two-shift schedule, permitting stabilization of employment level at 18,500 by June 30, 1970, instead of current 23,500.

KSC strength would remain at 80% of current level, with reduction in support and stage contractors. Civil service cadre would drop from 2,920 to 2,880 by June 30, 1970. Employment already had been frozen at current level.

Saturn IB Apollo Applications missions would require increased employment at Launch Complexes 34 and 37 in 1971, with estimated increase in contract jobs of 2,600 to handle nearly simultaneous or dual launches. KSC also would participate in earth orbital space station studies. (Transcript)

• KSC announced selection of Dr. Kurt H. Debus, KSC Director, to receive 1969 Career Service Award of National Civil Service League at Washington, D.C., banquet June 13. Citation said: "The development of Complex 39, the installation from which the Apollo program launches take place, is the crowning achievement of Dr. Kurt H. Debus' career. His leadership was pivotal in both the design and construction of the famed Apollo launch complex. Further, he has been responsible for many of the technical advances in launch technology, and for the formation of the government-industry launch team which has carried out more than 150 successful launches, including several notable firsts." Award—consisting of $1,000, inscribed gold watch, and citation—was one of most prestigious for which civil service career employees were eligible.

Also among 10 selected was George S. Moore, Associate Administrator for Operations, FAA, for "an extraordinary record in the development of up to date methods of evaluation of aircraft worthiness." (KSC Release 154–69; W Star, 4/30/69, A2)

- NASA was engaged in "comprehensive continuing program" to gain "better understanding of the structural loads due to buffeting and the buildup of buffeting intensity for flight conditions above buffet boundary," Acting Associate Administrator for Advanced Research and Technology Bruce T. Lundin testified before Senate Committee on Aeronautical and Space Sciences. FRC program, in which F–111A was being used, also included verification in flight of favorable effects of flaps in alleviating buffeting. Program was supported and complemented by wind-tunnel studies at LaRC and ARC. (Testimony)

- Dr. George E. Mueller, NASA Associate Administrator for Manned Space Flight, spoke before student seminar at California Museum of Science and Technology in Los Angeles: "Besides serving as a subject of scientific exploration for its own secrets, the moon may be an important base for outward looking space science programs of the future." It might eventually support large optical telescopes. There was strong evidence "that the most ideal location for large radio telescopes will eventually be the far side of the moon. This may be the only place within our convenient reach where the earth, which will become increasingly noisy as a radio source, may be completely screened out. In addition, the lunar surface presents us with a very large stable base, with only 1/6 gravity, no wind disturbance and no atmosphere absorption at any wave length. . . . Another attractive possibility is to use stations on both the moon and earth as a radio interferometer baseline for highly precise directional radio astronomy." (Text)

- NASA announced it would sponsor, in 1970, 90-day test of four-man life-support system with continuous regeneration of water and oxygen without resupply, under $586,885 contract with McDonnell Douglas Corp. Advanced waste management subsystem would be used. Experiment, scheduled to begin in March 1970, would demonstrate crew's ability to function during long period of uninterrupted confinement. (NASA Release 69–60)

- Univ. of Miami and Chrysler Corp. Space Div. had conceived $100-million undersea laboratory "Atlantis" to be constructed on ocean floor near Miami, AP said. Laboratory would be testing ground for future USN centers directing submarine warfare and test bed for industrial equipment to mine ores and drill for undersea oil. It would consist of 80-ft-long cylinder-shaped habitat housing 10 to 12 scientists at initial depth of 1,000 ft. Later it could be moved to 6,000- or 12,000-ft depths. (B Sun, 4/30/69, A3)

- President Nixon had apparently yielded to "top level economizers" and returned SST and airways and airports programs to DOT for further study, said Associated Press. (AP, W Star, 4/30/69, A11)

- Secretary of Defense Melvin R. Laird issued statement expressing concern over C–5A contract and procurement actions and announcing that

April 30

new Assistant Secretary of the Air Force (Installations and Logistics) had been nominated [Philip N. Whittaker had been nominated to replace Robert H. Charles] and new Assistant Secretary of the Air Force (Financial Management) would be nominated [to replace Thomas H. Neilson]. "I am determined to insure that full and accurate information on C-5A procurement, and on all other procurement matters, is given to the Congress and to the public promptly. I am also determined to insure that past mistakes in the procurement of this transport aircraft will not be repeated." (DOD Release 340-69; AP, W Star, 5/1/69, A1; Nossiter, W Post, 5/1/69, A1)

- Rep. William F. Ryan (D-N.Y.) presented to House petition by 729 scientists, engineers, and students in these fields, submitted by Scientists for Social and Political Action, care of Dr. Martin L. Perl of Stanford Univ. It called for open Congressional hearings to review substance and direction of entire military R&D program including ABM system. (CR, 4/30/69, H3220)

During April: NAS-NRC Space Science Board published *Sounding Rockets: Their Role in Space Research,* report by Committee on Rocket Research recommending that NASA increase annual expenditures on rocket research to $27 million by 1971 and thereafter restore its earlier policy of increasing support 12% per year; that NSF, Naval Research Laboratory, and USAF immediately increase their support by 36%, and then maintain an annual 12% increase until 1975; and, additionally, that support for rocket research be increased by appropriate ratio to compensate for any reduction in satellite support.

Report noted sounding rockets were responsible for dozens of major scientific discoveries, including discovery of solar x-rays emanating from millionth-degree corona of sun; for mapping of solar spectrum and structure of earth's atmosphere; and for existence of three new branches of astronomy—UV, x-ray, and gamma-ray. Greatest single advantage of sounding rockets for studying upper atmosphere was unique ability to obtain direct, vertical profiles at altitudes of 24.8 to 124.3 mi (40 to 200 km). (NAS-NRC-NAE *News Rpt,* 4/69, 2; NAS PIO)

- Aerospace Industries Assn. released *Annual Report 1968:* Aerospace industry had gained "in virtually every major category": sales reached record $30.1 billion, up 10.5% over 1967, with turbine-powered commercial aircraft sales accounting for $3.73 billion, up 211% over 1965. General-aviation aircraft sales reached 13,698 units valued at $426 million. Civil helicopter production increased to 528 units valued at $59 million. DOD sales reached $16.9 billion, up nearly 7% over 1967; nonmilitary sales decreased 3.7%, to $4.047 billion. Backlog at end of third quarter was due primarily to commercial transport orders. (Text)

- NASA issued *Relay Program Final Report* (NASA SP-151), prepared by GSFC. It was sequel to *Relay I Final Report* (NASA SP-76) and summarized operations of *Relay II* satellite (launched Jan. 21, 1964), including communications and other experiments. It compared *Relay I* (launched Dec. 13, 1962) and *II* performances and documented aspects of *Relay I* operations and experiment results not covered in *Relay I* report. *Relay I* transmitted last useful data Feb. 10, 1965; *Relay II* operated normally for last time Nov. 20, 1966. (Text)

- NAS announced appointment of Dr. T. Keith Glennan, first NASA Administrator (1958–1961) and Assistant to Chairman of the Urban Coalition, as Chairman of NRC's Committee on Radio Frequency Requirements for Scientific Research. Dr. Glennan had been President of Case Institute of Technology, AEC Commissioner, and President of Associated Universities, Inc. (NAS PIO; NAS–NRC–NAE *News Rpt*, 4/69, 10)

May 1969

May 1: Canadian Black Brant IIIB sounding rocket was launched by NASA from NASA Wallops Station to 133.6-mi (215-km) altitude. Primary objective was to evaluate improved Black Brant IIIB single-stage rocket, using flight-performance instrumentation. Flight, second of two (first was Feb. 28), was successful and sounding rocket was recommended for consideration as operational vehicle in NASA sounding rocket program. (NASA Rpt SRL; NASA Release 69–9)

- NASA was developing novel technique for photographing faint x-ray sources, NASA Associate Administrator for Space Science and Applications, Dr. John E. Naugle, told Senate Committee on Aeronautical and Space Sciences. Testifying on supporting research and technology program, he said technique would reduce required exposure time 1/10 to 1/1000 and was "likely to revolutionize x-ray observations where telescopes are not usable." Based on Princeton Univ. development and laser image-enhancement technique, it was "major technological advance as an outgrowth of the space program and should find applications in all fields of x-ray science." GSFC was developing advanced gamma-ray telescope to observe center of galaxy and other celestial objects. Univ. of Colorado was developing advanced coronagraph to observe solar corona from earth orbit six months instead of the few minutes per year possible during solar total eclipse by ground-based observation. (Testimony)

- National Aviation Club's highest honor, Award for Achievement, was presented at Washington, D.C., ceremony to M/G Jewell C. Maxwell (USAF), Director of Supersonic Transport Development for FAA, for "laboring so magnificently and conscientiously in the public and national service." (FAA Release T 69–25)

- U.S.S.R. celebrated May Day without military participation. In traditional address from atop Lenin's tomb in Moscow, Communist Party General Secretary Leonid I. Brezhnev said: "The Soviet Union will further consistently uphold the cause of peace and security of people, the Leninistic principles of peaceful coexistence of states with different social systems. It will come out for the solution of unsettled international problems by way of talks." (Kamm, *NYT*, 5/2/69, 1)

May 2: USAF launched two unidentified satellites from Vandenberg AFB by Thor-Agena booster. One entered orbit with 202.1-mi (325.2-km) apogee, 104.6-mi (168.3-km) perigee, 89.5-min period, and 64.9° inclination and reentered May 23.

Second satellite entered orbit with 283.2-mi (455.7-km) apogee, 255.9-mi (411.7-km) perigee, 93.2-min period, and 65.7° inclination. Satellites reentered Feb. 16, 1970. (GSFC *SSR*, 5/15/69; 5/31/69; 2/28/70; *Pres Rpt 70* [69])

- Data from *Mariner V* flyby Oct. 19, 1967, indicated Venus was layered with deadly compounds of mercury, GSFC scientist Dr. S. Ichtiaque

May 2

Rasool said in interview. Although bulk of its atmosphere was primarily carbon dioxide, mercury and halides—reactive chemicals including iodine, bromine, and chlorine—had boiled off planet's surface and combined to form clouds of poisonous gas and dust, often covered by water vapor. Findings, Dr. Rasool said, upset scientific notions of origin of Venus' heat as "greenhouse" effect—which might be occurring on earth with addition of carbon dioxide to atmosphere through burning of coal and oil. Clouds of carbon dioxide were assumed to retain heat, but planet covered with four layers of heavily colored atmosphere would never feel sun's heat. (Lannan, W Star, 5/3/69)

- Basic research program conducted by Dr. Wilhelm Rindner had led to development of cardiovascular pressure transducer, ERC reported. Tiny device had been placed in arteries and hearts of laboratory animals to measure blood pressure while using 100 times less power than conventional devices. Medical team headed by Harvard Univ. cardiologist Dr. Bernard Lown, in collaboration with NASA, said device was significant advance in monitoring important blood flow changes. Eventually it should be possible to implant device in human beings to observe blood flow and pressure in persons suffering from hypertension and other blood pressure disorders, including heart attacks.

    Discovery of properties of device was made during ERC study of effects of pressure on semiconductors. Properties would also be important in space applications; sensing of acceleration already had been demonstrated. (ERC Release 69-9)

- NASA unloaded eight-ton airlock at MSFC for ground testing to qualify it as part of orbiting space station. Part of Apollo Applications program cluster to be launched in mid-1970s, 65-in-dia, 17-ft cylindrical unit was flown from McDonnell Douglas Corp.'s St. Louis plant to be joined to multiple-docking adapter. It would provide interconnecting passageway between S–IVB rocket stage and multiple-docking adapter in flight, condition environmental gases, and provide instrumentation, data management, intercommunication, and other services. (MSFC Release 69–124)

- MSFC announced it had issued RFPs for experiment modules to be used with proposed manned space station to orbit earth in 1970s. Study, under eight-month contract, would analyze scientific and engineering need for experiment modules and develop concepts for least number of modules needed. Study tasks included further defining candidate experiment groupings, developing preliminary module concepts, defining minimum number of concepts, developing preliminary design and cost analysis for each module concept, and making proposed plan schedule. Proposals were due May 22. (MSFC Release 69–125)

- At DOD news conference, Secretary of the Air Force, Dr. Robert C. Seamans, Jr., said current estimate of cost to Government of C–5A aircraft was $4.348 billion, increase of $882 million (25%) over original estimate of $3.466 billion on which Lockheed Aircraft Corp. contract was based.

    Assistant Secretary of the Air Force Robert H. Charles said USAF had not disclosed Feb. 1, 1967, "cure order" to Lockheed on aircraft procurement contract because of concern over its effects on financial community. He maintained data on cost overruns had been developed too early in program to be solid enough for publication and that subsequent firmer cost estimates had been made public. (Transcript; Corddry, B *Sun,* 5/3/69, A4)

- In editorial directed to "Americans who think Soviet Russia is 'mellowing'" and who opposed President Nixon's Safeguard ABM plan, *New York News* noted U.S.S.R. had paraded no military gear in Moscow May Day parade and limited oratory "to a peace-it's-wonderful speech" by Brezhnev. Paper then quoted April 30 issue of *Partinaya Zhizn* in which Marshal Matvey V. Zakharov, U.S.S.R. Chief of Staff, described Soviet stockpile of nuclear missiles: "'These rockets are always prepared for immediate firing. Global rockets have unlimited range, and are able not only to carry colossal payloads but to overcome the antimissile defense of the adversary.'" (*NY News,* 5/2/69)

May 3: Press conference on results of NASA's *Oao II* orbiting astronomical observatory (launched Dec. 7, 1968) was held at NASA Hq. OAO Project Manager Joseph Purcell of GSFC said spacecraft's five months of orbital operations had been "a fabulous success" and all spacecraft systems were operating normally. As of last week, he said, "we had 137 mission days. [Univ. of] Wisconsin [experiment] has been pointed to 846 different locations in the sky [and] 344 of those were unique objects that they were studying. SAO, the Smithsonian package, has been pointed at 483 separate locations and taken some 1,172 pictures." (Transcript)

- NASA's *Explorer XXXIV* (IMP–F) Interplanetary Monitoring Platform (launched May 24, 1967) reentered earth's atmosphere. More complex than previous IMP spacecraft, *Explorer XXXIV* had carried 11 experiments and obtained more than 170,000 hrs of data on solar activity, near-earth environment, and magnetosphere. (GSFC *SSR,* 5/15/69; NASA Release 69–63)

May 4–11: London *Daily Mail* sponsored eight-day "Great Transatlantic Air Race of 1969" between top of New York's Empire State Building and top of London's General Post Office to commemorate 50th anniversary of first nonstop transatlantic flight by John W. Alcock and Arthur W. Brown in 1919. Nearly 400 entrants devised combinations of land, sea, and air transportation to compete for 21 prizes worth total $150,000 in separate categories. Fastest west-to-east time—5 hrs 11 min 22 secs—was made by L/C P. M. Goddard (RN) in U.S.-built F–4 Phantom jet, motorcycle, and helicopter. Air time was 4 hrs 46 min. Royal Navy later claimed this was new world record for New York-London flight. Previous unofficial record had been set during race by Phantom which made flight in 4 hrs 53 min. Best east-to-west time in race was made by S/L Tom Lecky-Thompson (RAF) in RAF Harrier VTOL jet, which flew from London to New York in 6 hrs 11 min 57 secs. (*NYT,* 5/6/69, 93; 5/12/69, 93)

May 5: At dedication ceremony in Houston for *Apollo 8* stamp, Postmaster General Winton M. Blount said: "The fact that this is the fifth stamp the Post Office Department has issued commemorating accomplishments of the space program signifies the extent to which space exploration has captured the imagination of the American people. . . . The nation's concerted effort to reach the moon and outer space reflects two traditional aspects of our national character. One is scientific: the search for truth, knowledge, discovery. The other is psychological: the strong urge for adventure—an urge kindled by the unknown. We are all explorers at heart. The space program has made an entire nation explorers in fact. The flight of *Apollo 8* around the moon and back was perhaps the greatest technological achievement of man to date. Certainly it was the most dramatic." (Text)

- NASA announced Bernard Moritz, Assistant Administrator for Special Contracts Negotiations and Review, would serve as Acting Associate Administrator for Organization and Management since Harold B. Finger had assumed new duties with HUD. (NASA Ann)
- Senate confirmed nomination of Philip N. Whittaker, former NASA Assistant Administrator for Industry Affairs, to be an Assistant Secretary of the Air Force. (*CR*, 5/5/69, S4668)
- U.S. returned 14-in metal sphere from Soviet spacecraft to U.S.S.R. under terms of space rescue treaty. Later, State Dept. spokesman said object—gas storage tank—was washed up on Alaskan coast in late 1968. Delay in return was due partly to efforts in identifying it. (Reuters, *W Post*, 5/8/69, A25)
- In letter to Attorney General, Sen. William F. Proxmire (D-Wis.) asked that Justice Dept. investigate possible violations of Federal law in handling of Government's contract with Lockheed Aircraft Corp. for production of C-5A aircraft and urged that DOD immediately freeze funding for C-5A. (Proxmire Off; Nossiter, *W Post*, 5/6/69, A3)
- Jerald R. Kubat, former Director of NASA Manned Space Flight Program Control Office, died in Seattle, Washington. He had joined Apollo Program Office in 1964. (NASA Hq *WB*, 5/12/69, 5)

May 6: Milton Klein, Manager of AEC–NASA Space Nuclear Propulsion Office and Director of Space Nuclear Systems, AEC, testified in NASA authorization hearings before Senate Committee on Aeronautical and Space Sciences that for second decade "what is clear is that regardless of the specific directions the space program may take, if it is to be a progressive one, nuclear energy will play an increasingly important role. As payloads become larger and energy requirements become greater to move these payloads farther from the earth, the nuclear rocket will become a workhorse propulsion system. As we move farther from the sun or as the power requirements increase for the more sophisticated payloads, electric power generated from radioisotopes or nuclear reactors will similarly become more and more important. (Transcript)

- *Apollo 9* mission to prove capability of LM to operate in space (March 3–13) was adjudged successful by NASA. Overall performance of launch vehicle, spacecraft, flight crew, ground support and control facilities, and personnel was satisfactory and all primary mission objectives were accomplished. (NASA Proj Off)
- At annual meeting of Aerospace Medical Assn. and interview which followed in San Francisco, MSC Director of Medical Research and Operations, Dr. Charles A. Berry, said it was almost certain that at least one Apollo 11 astronaut would develop illness during 21-day quarantine period following return from moon. "We have to face the fact that we've had post-flight illness on every Apollo mission so far. Our problem will be to determine whether any illnesses that show up . . . are due to the stresses of space flight or to some micro-organism picked up on the moon." Among difficulties noted during Apollo program were loss of exercise capacity for period after return to earth, motion sick-

ness in flight, and damage to red blood cells from atmosphere of 100% oxygen. Blood cell damage had been alleviated by addition of nitrogen to spacecraft atmosphere. (*Huntsville Times*, 5/7/69)

- Dr. John S. Foster, Jr., Director of Defense Research and Engineering, discussed aeronautical portion of DOD FY 1970 program before Senate Committee on Aeronautical and Space Sciences hearing on NASA appropriations. Upsurge in DOD aircraft weapon systems development was "reflected in the DOD fiscal year 1970 budget of $1,372 million for aircraft and related equipment R.D.T. & E., an increase of about $387 million" over FY 1969. High-priority programs were USN's F-14A air defense fighter and S-3A carrier-based antisubmarine warfare aircraft; USAF's F-15A air superiority fighter, advanced manned strategic aircraft (AMSA), and AX close-support aircraft; and USA's heavy helicopter (HLH). Comsat program included Defense Satellite Communications System, Tactical Satellite Communications Program, and Very Low Frequency Propagation Satellite. (Transcript)

- American Security Council published *The ABM and the Changed Strategic Military Balance, U.S.A. vs. U.S.S.R.* Although U.S. GNP ran almost twice that of U.S.S.R., latter was investing two to three times more in strategic military forces annually. Report concluded: "Antiballistic missile defense is not a cure-all for the security of the United States. It is not the ultimate defense system, for technology knows no limits and each decade produces fresh challenges and fresh need for response on the part of free nations. But anti-missile defense is an essential component in the network of military systems designed to give the American people a seamless garment of security in an age of acute danger. . . . We firmly believe that an American ABM system is the soundest insurance for peace and against war that the United States can buy in 1969, for the 1970s. . . . It may well be . . . the single most important step the United States can take toward a real and lasting peace at this moment in history." (Text)

- Sen. Edward M. Kennedy (D-Mass.) released report on Safeguard ABM system prepared at his request by Harvard law professor Abram Chayes and MIT provost Dr. Jerome B. Wiesner and scientists George W. Rathjens and Steven Weinberg. It said that "the Sentinel/Safeguard anti-ballistic missile system should not be deployed at this time." Recommendation was based on conclusions system was unlikely to perform according to specifications under nuclear attack, was susceptible to penetration, and was not well adapted to perform missions assigned to it. Deployment would probably start "new round in the arms race" and "seriously impede the conclusion of an arms control agreement." (Kelly, *W Star*, 5/7/69, A8; Chapman, *W Post*, 5/7/69, A1)

May 7: NASA announced establishment of task group on manned space station under Dr. George E. Mueller, NASA Associate Administrator for Manned Space Flight, and of task group on space shuttle under Charles W. Mathews, Deputy Associate Administrator for Manned Space Flight. Reporting to Dr. Mueller would be LeRoy E. Day, former Director of Apollo Test. His group would develop NASA material for report on space shuttles to President's Space Task Group by June 15. Frank Borman, former Deputy Director of Flight Crew Operations at MSC and *Apollo 8* commander, would report to Mathews as Field Director for Space Station effort. (NASA Release 69-70)

May 7

- Robert J. Collier Trophy for 1968 was presented to *Apollo 8* Astronauts Frank Borman, James A. Lovell, Jr., and William A. Anders by Dr. Thomas O. Paine, NASA Administrator, at Statler Hilton Hotel ceremony in Washington, D.C. National Aeronautic Assn. award was made annually for greatest achievement in aeronautics or astronautics in U.S. toward improving performance, efficiency, and safety of air or space vehicles. (NAA *News*; W *Star*, 5/8/69, A3)
- Nike-Cajun sounding rocket launched by NASA from Wallops Station carried GSFC payload to 80.4-mi (129.4-km) altitude to provide data on wind, temperature, pressure, and density in 21.8- to 59.0-mi (35- to 95-km) range during atmospheric warming. All 19 grenades ejected and exploded as programmed and sound arrivals were recorded on ground. Mission was launched in conjunction with Nike-Cajun launch from Arenosillo, Spain. (NASA Rpt SRL)
- MSC announced Astronaut Alan B. Shepard, Jr., had been cleared medically for return to space flight status following correction of inner ear disorder which had grounded him in 1963. (*NYT*, 5/8/69, 2; *W Post*, 5/8/69, A7)
- George J. Vecchietti, NASA Director of Procurement, would serve as Acting Assistant Administrator for Industry Affairs, replacing Philip N. Whittaker, who had assumed new duties as Assistant Secretary of the Air Force, NASA announced. (NASA Ann)
- Associated Press said NATO nations U.K., West Germany, Canada, Italy, Belgium, Holland, and Norway had agreed to participate in tactical satellite communications experiments with U.S. France was interested but would not participate; system was for use of closely integrated forces and she had withdrawn her troops from integrated NATO commands. Satellite for exclusive use of alliance was scheduled for launch by end of year. (W *Star*, 5/7/69, A9)
- Four contract modifications totaling nearly $10 million, for change orders issued in connection with Saturn V 2nd stage, had been awarded to North American Rockwell Corp., MSFC announced. (MSFC Release 69–127)
- V/A Hyman G. Rickover (USN), addressing Convocation on Ecology and the Human Environment at St. Alban's School in Washington, D.C., said that "keeping our small crowded planet inhabitable" was "of utmost importance and great urgency.... We have been brought to this critical situation by the scientific-technological revolution, and can extricate ourselves only by a change of direction in thought and action so drastic it would rate the term counterrevolutionary." Science, "pure thought," harmed no one. "But technology is *action* ... often potentially dangerous action. Unless it is made to adapt itself to human interests, needs, values, and principles, more harm will be done than good. Never before ... has man possessed such enormous power to injure himself, his human fellows, and his society.... That is why it is important to ... recognize clearly that ... technology can have no *legitimate* purpose but to serve man—man in general, not merely some men...." (Text; W *Star*, 5/11/69, E2)
- Securities and Exchange Commission said it had begun inquiry into all phases of Government's contractual dealing with Lockheed Aircraft Corp. on C-5A procurement. (Nossiter, *W Post*, 5/8/69, A1; AP, W *Star*, 5/8/69, A3)

May 8: X–24A lifting-body research vehicle, piloted by Maj. Jerauld R. Gentry (USAF), successfully completed second glide flight over South Rogers Lake Bed, Calif., to obtain additional data on handling qualities. (NASA Proj Off)

- New direct value for sun's thermal radiation of 125.7 w per sq ft, obtained from *Mariner VI* and *VII* en route to Mars, was nearly twice as accurate as old value of 129.5 w per sq ft, NASA announced. JPL Mariner project engineer Joseph A. Plamondon said change in probable error in new measurement was one-half that of old figure. Results of inflight measurements obtained with temperature-control flux monitors (TCFM) monitoring solar radiation since Mariners' launch Feb. 24 and March 27 would be compared with preflight predictions of solar radiation and spacecraft temperature variations in flight, to establish new standard for temperature-control design and testing. Data from TCFM agreed with data obtained by NASA from high-altitude experiments on Convair 990 research aircraft and X–15. (NASA Release 69–69; JPL Release 518)
- Nike-Cajun sounding rocket was launched by NASA from Arenosillo, Spain, carrying grenade payload for Spain to conduct meteorological studies. Rocket and instruments functioned satisfactorily. (NASA Proj Off)
- Smithsonian Institution celebrated 50th anniversary of May 8, 1919, takeoff of first aircraft to cross Atlantic Ocean with display on Washington, D.C., mall of original NC–4 (restored by Smithsonian Air Museum Curator Paul Garber) and with presentation to Institution of Plaque and aircraft's log. Glenn Curtiss-built, long-range seaplane ordered by USN as bomber during World War I but completed too late to see service had been flown by USN crew from Rockaway Beach, N.Y., to Plymouth, England, with two sister aircraft which crashed in Azores. NC–4, known as "Nancy," made journey in 53 hrs 58 min flying time and 23 days elapsed time. (Lydon, *NYT*, 5/1/69, 22; Durbin, *W Post*, 5/8/69, C1)
- Sen. Warren G. Magnuson (D-Wash.) submitted to Senate S.C.R. 23, expressing sense of Congress that U.S. participate in international decade of ocean exploration. Measure was referred to Senate Commerce Committee. (*CR*, 5/8/69, S4688)

May 9: NASA's HL–10 lifting-body vehicle, piloted by NASA test pilot John A. Manke, reached 54,000-ft altitude and 724 mph (mach 1.1)—first supersonic flight by HL–10—after 45,000-ft-altitude air-launch from B–52 aircraft. Primary purpose of 17th flight, made north of Four Corners, Calif., was to obtain stability and control data. (NASA Proj Off)

- Nike-Cajun sounding rocket launched by NASA from Wallops Station carried GSFC payload to 79.4-mi (127.7-km) altitude to obtain data on wind, temperature, pressure, and density in 21.8- to 59.0-mi (35- to 95-km) range during atmospheric warming. Seventeen of 19 grenades ejected and exploded as programmed and sound arrivals were recorded. Mission was launched in conjunction with Nike-Cajun launch from Arenosillo, Spain. (NASA Rpt SRL)
- Astronomers at 60-in optical telescope at Cerro Tololo, Chile, began two-week alert in attempt to photograph Scorpio X–1, brightest x-ray star, best seen in southern skies. When flare-up occurred, they would

radio message to astronomical teams in Hawaii, which would launch two Nike-Tomahawk rockets above atmosphere to photograph x-rays from giant star. At 200-in Palomar, Calif., telescope, astronomers would try to photograph star's visible and infrared light during flare period, while astronomical team in Wisconsin would order *Oao II* to observe UV light from star. Astronomers hoped to match all photos of flare-up to determine element in star which excited x-rays. (*W Post*, 5/4/69, A3; Hines, C *Sun-Times*, 5/5/69)

- Dr. Lee A. DuBridge, Presidential Science Adviser, testified before Senate Committee on Aeronautical and Space Sciences during NASA FY 1970 authorization hearings: "Nothing can do more harm to support for the space program than to have a series of missions for which there are no clear objectives—such as a series of manned revisits to the moon without providing the capability to perform new scientific experiments and to explore interesting new lunar features." When Space Task Group considered urgent items in manned space flight area for FY 1970, it "gave high priority to the provision of additional science payloads for lunar flights and increased capability for man on the lunar surface, to support Apollo missions after the first four landings. Funding for this purpose is included as part of the budget amendment to the NASA request for fiscal year 1970.

  "An additional item . . . is funding for maintaining the production of the Saturn V launch vehicle. Although specific commitment to a particular mission or missions has not been made for the initial vehicles to be produced under this budget amendment, it was the judgment of the Space Task Group that this vehicle represented a unique and valuable resource that we would undoubtedly wish to continue to use, at least through the mid-1970s. Because of the long lead times involved in a vehicle of this size, action is necessary now if we are to have follow-on vehicles produced and available by 1973 and after." (Transcript)

- Harold R. Kaufman, Assistant Chief of Electromagnetic Propulsion Div., LeRC, would receive James H. Wyld Propulsion Award for "outstanding leadership in the field of electric propulsion, including the conception design, and development of the world's most successful ion rocket" at AIAA 5th Propulsion Joint Specialist Conference in Colorado Springs, Colo., June 9–13, AIAA announced. (AIAA *News*; *Lewis News*, 5/9/69, 1)

- Associated Press said Astronaut-Aquanaut M. Scott Carpenter (Cdr., USN) would retire from USN July 1 to enter private business in oceanography field. He was second U.S. astronaut to orbit earth, during May 24, 1962, Mercury mission in *Aurora 7*. (*W Post*, 5/10/69, A3)

- Tom Barker, owner of bingo hall in Cardiff, Wales, had written to American and Soviet embassies in London for permission to open first amusement and bingo hall on moon, Reuters said. U.S. Embassy spokesman had replied: "There are no proposals to colonize the moon and many factors inhibit large-scale development." (*NYT*, 5/9/69, 16)

- *New York Times* editorial: "Now it appears that the solution to the cosmic ray mystery may be intimately related to the explanation for the strangest astronomical phenomena discovered in recent years, if not all history, the pulsars." Present favored explanation "views pulsars as neutron stars composed of matter packed so tightly that a mass the weight of the earth would be a sphere with a diameter of a few hundred feet. The extremely swift rotation of a neutron star, it is now theorized, produces both the periodic radio emissions of the pulsars and the super-energetic cosmic rays." (*NYT*, 5/9/69, 46)

May 10: Loose wire in 1st stage of Delta launch vehicle's guidance system had been identified by final failure review committee as most probable cause of Intelsat-III F–1 mission failure Sept. 18, 1968, NASA announced. Report did not eliminate possibility that electrical failure in unrecovered pitch gyro or interconnecting wiring had caused failure. (NASA Release 69–71)

- Nike-Cajun sounding rocket was launched by NASA from Arenosillo, Spain, carrying grenade payload for Spain to conduct meteorological studies. Rocket and instruments functioned satisfactorily. (NASA Proj Off)
- Comsat station for communications with Europe was opened at Yamaguchi, Japan, to replace telephone and telegraph relay through U.S. (Reuters, *W Post*, 5/11/69, A3)

May 11: Jet Propulsion Laboratory announced appointment of Dr. Robert J. Mackin, Jr., as Manager of Space Sciences Div., succeeding Dr. Donald P. Burcham. Dr. Burcham had been named Research and Advanced Development Manager for Space Science, JPL Office of Research and Advanced Development. (JPL Release 517)

- Vice President Spiro T. Agnew announced President's Council on Youth Opportunity and NASA would explain mechanics of scheduled July Apollo 11 lunar landing to high school and junior high school pupils in 50 cities under summer program to be held in city streets, playgrounds, and classrooms. (*W Star*, 5/12/69, A3)
- Rise of costs of Mark II electronic equipment for F–111 aircraft of more than 100% above original contract with Autonetics Div. of North American Rockwell Corp. were described by Bernard D. Nossiter in *Washington Post* article based largely on memo from DOD official. Memo had warned, "If it fails to enforce the contract, the Air Force and the entire Department of Defense can count on many more years of misleading promises from contractors and failures to meet contractual requirements." (*W Post*, 5/11/69, A1)
- Fire at AEC plutonium-handling facility at Rocky Flats, Colo., might halt U.S. nuclear missile production for remainder of 1969, Associated Press reported testimony released by Senate Appropriations subcommittee had disclosed. Most nuclear weapons needed plutonium to trigger atomic warheads. (*W Post*, 6/24/69, A3)

May 12: NASA launched two Nike-Apache sounding rockets from Wallops Station: first carried GSFC payload to study ionospheres; second carried Univ. of Michigan payload to conduct studies on atmospheric structure. Rockets and instruments functioned satisfactorily. (NASA Proj Off)

- Melvin S. Day became NASA Acting Assistant Administrator for Technology Utilization, succeeding Dr. Richard L. Lesher, who had resigned to accept position in industry. (NASA Ann)
- Author Norman Mailer's total publishing rights on book on lunar landing would exceed $1 million when book was published by Little, Brown & Co. in January or February 1970, according to his agent, Scott Meredith. If film rights were sold, total could approach the $1.5

million paid for Lyndon B. Johnson memoirs. Mailer planned to visit KSC during Apollo 11 launch to interview astronauts and describe space center operations. He also planned chapter on philosophical and technological implications of lunar landing. Meredith said he was surprised at "phenomenal competition among foreign publishers for book and magazine rights." (Raymont, *NYT*, 5/13/69, 44)

- Science students, younger scientists, and many older professors of physics and physiology were engaging in what Harvard Univ. political scientist Prof. Don K. Price called "a new kind of rebellion," linked only in part with radical activists on campuses, said Victor Cohn in *Washington Post*. It was rebellion against ABM "and other costly military-technological systems, against 'weaponeering' at secret laboratories on or near campuses and, in many cases, against doing any research, secret or non-secret, to help the military." It had helped cause Stanford Univ. to decide to phase out 50% of secret projects at Applied Electronics Laboratory, made Stanford's trustees place moratorium on new chemical and biological warfare contracts at Stanford Research Institute, caused MIT moratorium on new secret contracts, and forced American Univ. to cancel partly secret USA research contract. Movement and student protests had, in past year, forced DOD to cut from 400 to 200 its classified R&D contracts on U.S. campuses. (*W Post*, 5/12/69, A1)

- In *American Aviation*, Eric Bramley called 1969 year of "cautious optimism" for air transport industry. Deliveries of new aircraft to U.S. carriers would drop from 478 in 1968 to 309. Trunk traffic was expected to grow at same 14% rate as 1968, with available seat-mile increase of 17%. CAB-approved fare increases would add $194 million to revenues with profit level and rate of return expected to improve. (*Amer Av*, 5/12/69, 40–1)

May 12–24: At 10th annual meeting of Committee on Space Research (COSPAR) in Prague, *Apollo 8* Astronaut Frank Borman received medal from Czechoslovak Academy of Sciences.

NAS–NRC submitted *United States Space Science Program*, comprehensive summary of scientific research in space science in U.S. during 1968. "Although the principal concern of the space science program in the United States continues to be with the Earth, its environment, the Sun, interactions of solar and terrestrial phenomena, the Moon and planets, and the biological effects of weightlessness and radiation, there is a trend toward increasing emphasis on the use of space vehicles for stellar and galactic astronomy, especially in areas of the electromagnetic spectrum for which the atmosphere is essentially opaque. The successful operation of the Orbiting Astronomical Observatory satellite and the rapid development of improved instruments and techniques for ultraviolet and x-ray astronomy . . . are examples of this trend."

In interview with press, NAS–NRC Space Science Board member Dr. Richard W. Porter said U.S. would probably have to review its expensive prophylactic measures in planned Mars landings if U.S.S.R. landed there first with same techniques and precautions used in Venus shots. Outgoing contamination of planets might well be bigger problem than contamination of incoming spacecraft. There was little likelihood of spacecraft landing on Mars and Venus and returning for at least 10 yrs, Dr. Porter said, and risk from lunar bacteriological contamination was infinitesimal. But contamination of planets was serious problem because it could spoil man's first chances to make a pure investigation of biological evolution elsewhere in the solar system. (Text; *W Post,* 5/23/69, A15)

May 13: Cosmos CCLXXXI was launched by U.S.S.R. from Plesetsk into orbit with 301-km (187.0-mi) apogee, 188-km (116.8-mi) perigee, 89.3-min period, and 65.4° inclination. Satellite reentered May 21. (GSFC *SSR,* 5/15/69; 5/31/69; *SBD,* 5/14/69, 61)

- Countdown for NASA's Apollo 10 mission, scheduled for launch May 18, began at KSC. Astronauts completed three-hour physical examinations and were reported to be in good health and good spirits. (*W Post,* 5/13/69, A7; Sehlstedt, B *Sun,* 5/14/69, A3)

- NASA and Australian Dept. of Supply and Dept. of Education and Science announced that 210-ft-dia radiotelescope at National Radio Astronomy Observatory in Parkes, Australia, might be used to relay TV signals from moon during Apollo 11 lunar landing mission in July. Signals from Apollo 11 antennas were scheduled to be received by Goldstone Tracking Station. If mission were delayed and moon were not visible from Goldstone while TV was scheduled, signals would be received at Parkes, transmitted to Sydney, and transmitted to NASA's Mission Control Center at Houston via *Intelsat-III F-3* over Pacific. (NASA Release 69-72)

- ComSatCorp President Joseph V. Charyk told Annual Meeting of Shareholders in Washington, D.C., key goal of establishing global comsat system was "within immediate view." When *Early Bird [Intelsat I]* "was launched just four years ago, there were only a handful of experimental stations in Europe and the United States. Today, there are 25 earth stations operating in 15 different countries of the world, with many more nearing completion. It is expected that a total of 43 stations will be in service by the end of this year and that 26 different countries will have direct access to all forms of high quality communications that the global system of satellites makes economically available to them." By 1972 "there will be more than 70 stations operating in nearly 40 countries of the world, thereby making this high quality means of communications available directly to practically every nation on earth."

  At end of first quarter of 1969 1,209 full-time circuits were being leased, up from 777 at end of 1967. Leased voice and record tariffs published by international carriers represented 40% reduction in Atlantic and Pacific areas since advent of comsats. Transmission of TV via satellite increased from 225 hrs in 1967 to 666 hrs in 1968, with 40% reduction in TV rates. (Text)

- *New York Times* editorial urged President Nixon to "undo a mistake and strike a blow for the more rational ordering of Federal spending priorities" by supplanting Government subsidy of SST program with SST Authority. "The Government's S.S.T. contribution should be converted to a preferred equity interest in a new public corporation with variable proportions of the total common stock being reserved for the airlines and the investing public. To the extent necessary, the S.S.T. authority would be authorized to raise development funds by selling bonds . . . guaranteed by the Federal Government." (*NYT,* 5/13/69)

May 14: Rep. George P. Miller (D-Calif.) introduced in House H.R. 11271, substitute NASA FY 1970 authorization bill increasing total from $3.716 billion of April 15 amended budget request (as reflected in H.R. 10251 introduced April 17) to $3.966 billion.

Bill added $258 million to R&D funds for new total of $3.264 billion, including increase of $75.7 million for Apollo program, to total $1.767 billion. Of this increase, $32.1 million was for Saturn V improvements and $4.6 million for lunar exploration. Within new total of $354.8-million NOA for manned space flight operations, bill restored $57 million cut from Apollo Applications by budget amendment and added $66 million for space station and shuttle and $6.2 million for Saturn V production.

In space science and applications, bill reduced funds for supporting research and technology by $12 million and deferred funding for four proposed Explorer satellites and for Mariner-Mercury 1973 mission, but restored biosatellite program to originally requested $18 million plus $1.6 million restored for Delta launch vehicle for Biosatellite-F. Earth Resources Technology Satellite funds were increased by $10 million, with transfer of funds to other projects prohibited.

Advanced research and technology funds were increased by $31.5 million, including $13.5-million increase for nuclear rocket program, $5.25 million for chemical propulsion, and restoration of $1.20 million for aeronautics.

Research and program management total was cut by $7.15 million, to $643.75 million. Construction of facilities total remained unchanged. (Text; *CR,* 5/15/69; House Rpt 91–255)

- In press statement, Secretary of Defense Melvin R. Laird said GAO team investigating Lockheed Aircraft Corp. books had estimated cost of 120-aircraft C–5A program exceeded advance estimates by about $550 million. DOD analysts had computed $450-million overrun and USAF had used $350-million figure. Laird denied there was $2-billion overrun. Figures had been turned over to Congressional committee. (AP, B *Sun,* 5/15/69, A4)

- USAF's Arnold Engineering Development Center at Tullahoma, Tenn., had "brought much of the universe down to pocket-size" for scientists working on Apollo 10, Henry J. Taylor wrote in *Washington Daily News.* Tests to establish spacecraft's ability to withstand lunar environment were under way in Center's 216,000-hp wind tunnel, which produced 8,000-mph winds and was "largest hypersonic wind tunnel in the free world." (*W News,* 5/14/69, 31)

- Eugene S. Burcher, NASA Tektite Program Manager, OMSF, received Navy Distinguished Public Service Award for "distinguished and outstanding service to the United States and to the Department of the Navy as a participant in Project TEKTITE I mission, as well as its planning and implementation." (NASA Hq *WB,* 6/2/69, 6)

- Charles L. Lawrence Award of Aviation/Space Writers Assn. was presented to Volta Torrey, Publications Officer, NASA Scientific and Technical Information Div., for "efforts to inform the public of NASA's activities." Citation and silver tray were presented at Dayton, Ohio, banquet. (ASWA letter of notification, 4/15/69; NASA Sci and Tech Info Div)

- U.S. District Court Judge Gerhard A. Gesell had turned down appeal of

Aircraft Owners and Pilots Assn. to prevent enforcement of FAA regulation limiting nonscheduled flights at five major airports, Washington *Evening Star* said. AOPA had asked for preliminary injunction to stop regulations from becoming effective June 1. (W *Star*, 5/14/69, G8)

May 15: Aerobee 150 A sounding rocket launched by NASA from NASA Wallops Station carried payload containing two white rats to 97.9-mi 157.6-km altitude in fourth of four experiments to study rats' behavior in artificial gravity field and determine minimum level of gravity needed by biological organisms during space flight. During free fall rats selected artificial gravity levels created through centrifugal action by walking along tunnel runway in extended arms of payload. Data on their position and movement were telemetered to ground stations. Last flight in series had been Nov. 21, 1968. (NASA Rpt SRL; WS Release 69-10)

- Dr. John E. Naugle, NASA Associate Administrator for Space Science and Applications, announced reorganization of Hq. Space Applications Programs Office in recognition of increasing importance of applications satellite programs. Leonard Jaffe, former Director of Space Applications Programs, had been named Deputy Associate Administrator for Applications, responsible for near-term and long-range planning and interagency policy coordination. As Acting Director of Earth Observations Programs Office—one of two new program offices into which Space Applications Programs Office was being divided—he would supervise R&D efforts in meteorology and earth resources survey including TIROS, Nimbus, Synchronous Meteorological Satellite, ERTS, and sounding rocket programs in support of meteorology. Dr. Morris Tepper would continue to direct scientific activities.

  Communications Programs Office—concerned with R&D for communications, navigation, traffic control, and geodetic satellites; ATS program; and COMSAT support—would have as its Director Dr. Richard Marsten, Manager of Advanced Programs Technology, RCA Astro-Electronics Div. Dr. Marsten's NASA appointment would become effective June 23. (NASA Release 69-76)

- ERC Director James C. Elms discussed NASA's role in computer R&D at Spring Joint Computer Conference in Boston. While Apollo guidance computer was best known among major computer activities undertaken by NASA in its first decade, other onboard computer developments "of considerable technical challenge" had been pursued. Versatile test-bed multiprocessor EXAM was oriented toward use of "hierarchy of memories" to increase capability to evaluate advanced technology for very large bulk-storage systems, particularly for spaceborne computers. TRIM—for Transformation of Imagery—was experimental tool for advancing state of art in imagery processing; computer-driven flying-spot scanner with color capability had auxiliary display terminal permitting man-machine-interactive operation. SOFIX—for Software Fix—was coordinated university research program on problems in developing computer software at same pace as hardware. (Text)

- Soviet astronomer Dr. Nikolay A. Kozyrev had revealed detection by Pulkovo Observatory of volcanic activity on moon, *Space Business Daily* reported. Two spectrograms of Aristarchus crater's western side taken April 1 had shown "an unusual red spot of approximately 102 kilometers [63.4 mi]" which was "result of the emission of gases—

molecular nitrogen and cyanogen." Emissions, which had occurred one day after earthquakes in U.A.R. and Japan, affirmed link between tectonic phenomena on earth and moon and showed "that the moon had 'responded' to this terrestrial phenomenon." (*SBD*, 5/15/69, 67)

- AFSC announced new UHF communications terminal developed by Electronic Systems Div. might solve problems in relaying messages between spacecraft and ground stations. Designed for USAF aircraft as part of joint-service Tactical Satellite Communications Program, terminals would be installed and tested on USAF jet tracking aircraft used for communications support in Apollo program. They would allow aircraft to relay spacecraft information and recovery operations reliably from parts of globe where communications had posed problem. (USAF Release 51.69)
- Strategic threat to U.S. security was rapidly increasing, Secretary of the Air Force, Dr. Robert C. Seamans, Jr., told Chamber of Commerce Armed Forces Day luncheon in New Orleans. U.S.S.R. had built intercontinental missile force from 250 in 1966 to 1,150 operational or under construction. "With their large SS-9 missiles, the Soviets will soon have about twice as much missile payload . . . as our missile force, even including our advantage in submarine launched missiles. In addition, the Soviets have already deployed an antiballistic missile system that includes some 60 long-range ABM missiles." (Text)

May 16: U.S.S.R.'s *Venus V* planetary probe (also designated *Venera V*) entered atmosphere of planet Venus at 2:01 am EDT and ejected instrumented capsule [see June 4]. Probe decelerated from 6.9 mps to 688.8 fps, deployed parachute, and transmitted data during 53-min descent through dense clouds to night side of Venusian surface. Launched Jan. 5, *Venus V* had traveled 217-million-mi trajectory. Tass said: "The instrument capsule was automatically jettisoned from the station before entry into Venusian atmosphere. The aerodynamic deceleration of the apparatus in the atmosphere began . . . and was accompanied by a sharp decrease in overloads and a growth of temperature on the craft's outer surface. . . . During the 53-min parachute descent, measurements of the temperature, pressure and chemical composition of Venusian atmosphere were made. This information was uninterruptedly transmitted to earth." *Venus IV* (launched June 12, 1967) had reached Venus Oct. 18, 1967; *Venus VI* (launched Jan. 10) was scheduled to reach planet May 17. (Winters, B *Sun*, 5/17/69, A1; Bausman, *W Post*, 5/17/69, A3; *SBD*, 5/19/69, 77; Gwertzman, *NYT*, 5/17/69)

- *Intelsat-III F-3* comsat, launched Feb. 5 and in orbit over Pacific, had lost some of potential capacity and would be moved to 62.5° east longitude over Indian Ocean, ComSatCorp announced. In new position satellite would link directly all countries with appropriate earth stations in Western Europe, Near East, Africa, Asia, and Australia. Move would be made soon after launch of Intelsat-III F-4 May 21. Future satellites would be modified for additional redundancy. (ComSatCorp Release 69-27)
- Aerobee 150 MI sounding rocket launched by NASA from WSMR carried GSFC payload to 125.5-mi (202-km) altitude to obtain solar EUV spectra from 40 to 390 Å and from 10 to 390 Å using BBRC-SPC 330D solar-

pointing control and recovery system. Rocket and instruments performed satisfactorily. (NASA Rpt SRL)

- President Nixon submitted to Senate nomination of *Apollo 8* Astronaut William A. Anders as Executive Secretary of NASC to succeed Dr. Edward C. Welsh. It was highest Government post ever offered to an astronaut. (PD, 5/19/69, 705; Kirkman, *W Post*, 5/15/69, A25)
- MSC announced revised quarantine procedures for Apollo 11 after landing. To prevent back contamination from moon, astronauts would dispose on moon under containment conditions equipment exposed there; brush, vacuum-clean, and bag other equipment and clothing for return; prevent dust from being transferred from LM to CM; and continuously filter CM atmosphere during return trip to remove dust particles.

  Under original plans Apollo 11 crew would have remained in CM after splashdown while it was hoisted onto recovery ship. After reviewing loads to be lifted in transferring CM to deck, reliability of shipboard cranes, and capacity of available load-limiting elastic tackle, NASA decided to retain helicopter lift used on previous Apollo missions. Crew would emerge from CM to raft, where they would put on biological isolation garments that would cover them completely and provide high-efficiency air outlet filter. Interagency Committee on Back Contamination—NASA, Dept. of Agriculture, HEW, Dept. of Interior, and NAS—had agreed that helicopter lift, combined with other prelanding procedures, would provide maximum achievable precautions against back contamination. (MSC Release 69–47)

- In *Science*, Leonard D. Jaffe, Surveyor data analysis manager at JPL, cited important findings of five Surveyor spacecraft which softlanded on lunar surface: surface of both maria and highlands was covered with layer of particulate material of 10-micrometer particles scattered with rocks and clods; layer was few meters deep in maria and varied from few centimeters to tens of meters in highlands, with density and other properties varying with depth; particulate material had cohesion; fine material moved gradually downhill; freshly exposed fine material from below surface was darker than previously exposed surface material; density of surface rock was 2.8 ± 0.4 gr per cc; composition of surface material was approximately that of basalt (mare material had elemental composition like high-iron basalt; highland material, like low-iron basalt; not more than one-quarter volume percent of metallic iron was present); lunar surface material had experienced extensive melting and chemical differentiation. (*Science*, 5/16/69, 775–8)
- FAA and USAF announced that M/G Jewell C. Maxwell (USAF), Director of Supersonic Transport Development for FAA, would become Commander of Armament Development and Test Center at Eglin AFB, Fla. His successor at FAA had not yet been selected. (FAA Release 69–56)

May 17: U.S.S.R.'s *Venus VI* planetary probe (also designated *Venera VI*) launched Jan. 10 landed on night side of planet Venus at 2:03 am EDT, 186.4 mi (300 km) from *Venus V* after 51-min descent [see June 4].

Tass said landing of both probes "was accomplished with tremendous precision." Throughout flight "the necessary temperature in the compartments and the permanent orientation of solar batteries on the Sun were ensured. During radio communication sessions the directional parabolic antennae were oriented on Earth. As envisaged by plan, the detachable capsules of both stations descended through the planet's atmosphere on its night side. In the course of descent, scientific equipment measured the chemical composition, pressure, density and temperature of the planet's atmosphere. Experts analysed the graph of temperature and pressure changes while the apparatus was approaching . . . and found in it a resemblance with the . . . picture of the landing of Venera 5. The apparatus performed with the same precision all major operations of the landing." (*SBD*, 5/20/69, 84; AP, *W Post*, 5/18/69; UPI, W *Star*, 5/18/69, A9)

- Apollo 10 astronauts would each carry out about two hours of head exercises early in mission in attempt to prevent motion sickness which plagued previous Apollo astronauts, MSC Deputy Director of Medical Operations, Dr. A. Duane Catterson, said. Exercises—which included nodding, rocking, and twisting head—would be done to point just below threshold of illness until normal adaptation occurred. Since pilots who regularly performed drastic maneuvers in aircraft seldom suffered motion sickness, astronauts had flown aerobatic jet flights in barrel rolls and high-gravity maneuvers during week before launch. (Cohn, *W Post*, 5/18/69, A8)

May 18-26: NASA's *Apollo 10* (AS–505), first lunar orbital mission with complete Apollo spacecraft, was successfully launched from KSC Launch Complex 39, Pad B, at 12:49 pm EDT by Saturn V booster. Flight carried three-man crew and CSM–106 and LM–4. Primary objectives were to demonstrate crew, space vehicle, and mission support facilities during manned lunar mission with CSM and LM and to evaluate LM performance in cislunar and lunar environment.

Launch events occurred as planned and spacecraft—carrying Astronauts Thomas P. Stafford (commander), John W. Young (CM pilot), and Eugene A. Cernan (LM pilot)—entered initial parking orbit with 118.1-mi (189.9-km) apogee and 114.6-mi (184.4-km) perigee. Checkout followed lunar trajectory insertion; then CSM, code-named *Charlie Brown*, separated from Saturn V 3rd stage (S–IVB) and LM, code-named *Snoopy*. Crew successfully transposed CSM and docked with LM. Excellent quality color TV coverage of docking sequences was transmitted to Goldstone tracking station and seen on worldwide commercial TV. Crew extracted LM from S–IVB and conducted 1st SPS burn. All launch vehicle safing activities were performed as scheduled and successful propellant dump provided impulse to S–IVB for slingshot maneuver to earth-escape velocity.

On second day, first midcourse maneuver was not required. Crew conducted midcourse maneuver number two, which was so accurate that third and fourth maneuvers were canceled. Five color TV transmissions totaling 72 min and showing excellent views of receding earth and spacecraft were made during translunar coast. Spacecraft entered moon's sphere of influence on fourth day, May 21, at 61:50 GET. Crew conducted first lunar orbit insertion maneuver with 356-sec SPS burn to reduce speed to 5,474 fps and place spacecraft in initial lunar orbit with 196.1-mi (315.5-km) apolune and 68.6-mi (110.4-km) perilune. Second LOI maneuver, 13.9-sec SPS burn, circularized orbit with 70.8-mi (113.9-km) apolune and 67.8-mi (109.1-km) perilune. Crew tracked lunar landmarks and transmitted 29-min color TV of lunar surface. Cernan transferred to LM at 81:55 GET for two hours of housekeeping activities and communications test.

On fifth day Stafford and Cernan entered LM and checked out all systems before firing SM reaction control system thrusters to separate CSM and LM about 30 ft and again for 2.3-mi separation. LM descent propulsion system burn propelled LM to within 9.6 mi of lunar surface over landing site No. 2. Crew had no difficulty identifying landmarks and Stafford said, "It looks like all you have to do is put your tail wheel down and we're there. . . . The craters [around No. 2 landing site] look flat and smooth at the bottom. It should be real easy" for Apollo 11 landing. LM crew took numerous photos of lunar surface and provided continuous commentary on their observations after cameras malfunctioned. Astronauts described volcanoes and light-colored craters that glowed as if lit by radioactive substance.

Crew conducted LM radar test during low-altitude pass which indicated 47,000-ft pericynthian (lowest point in orbit). DPS phasing burn raised apocynthian (orbital high point) to 218.6 mi (351.7 km). LM descent stage was jettisoned and RCS separation maneuver and staging were accomplished. Anomaly in automatic abort guidance system caused LM ascent stage to undergo extreme gyrations. By taking over manual control, Stafford reestablished proper attitude. APS insertion maneuver burn at pericynthian established equivalent of standard LM insertion orbit of lunar landing mission (51.8 by 12.9 mi), where LM coasted for one hour. Concentric sequence initiation at apocynthian, constant-delta-height maneuver, and terminal maneuver were conducted. LM successfully docked with CSM at 106:33 GET, after eight-hour separation; and LM crew returned to CSM.

On sixth day LM ascent stage was jettisoned; its batteries burned to depletion and it entered solar orbit. Crew made 18 landmark sightings and took extensive stereo and oblique photos of moon. Two scheduled TV periods were deleted because of crew fatigue and crew rested and prepared for return to earth. SPS burn at 137:36 GET injected CSM into transearth trajectory after 61.5 hrs (31 orbits) in lunar orbit. Maneuver was so accurate that two other scheduled midcourse maneuvers were not necessary. During return to earth astronauts made star-lunar landmark sightings, live color TV transmissions, star-earth horizon navigation sightings, and CSM S-band high-gain antenna reflectivity test. Pictures of moon from receding spacecraft were spectacular. Scheduled 10- and 29-min color TV broadcasts of earth, moon, and spacecraft interior were later followed by unscheduled TV transmission, which provided beautiful pictures of earth and brought total color TV broadcasts to 19 transmissions totaling almost six hours.

On eighth day crew prepared for reentry and SM separated from CM on schedule. Parachute deployment and other reentry events occurred as planned. *Apollo 10* splashed down in Pacific at 12:52 pm EDT May 26, 3.4 mi from recovery ship U.S.S. *Princeton* 192 hrs 3 min after launch and precisely on time. Crew was picked up and reached recovery ship at 1:31 pm EDT.

All primary *Apollo 10* mission objectives and detailed test objectives were achieved. All launch vehicle and spacecraft systems performed according to plan, with only minor discrepancies, which were corrected. Flight crew performance was outstanding; all three crew members

May 18–26: Apollo 10, *first lunar orbital mission with complete Apollo spacecraft, carried Astronauts Eugene A. Cernan, John W. Young, and Thomas P. Stafford (left to right above) around the moon for the first demonstration of lunar orbit rendezvous. Apollo 10 CM (at right), carrying Young, was photographed by the LM with Stafford and Cernan aboard, 60 miles above the moon's far side. Craters below were unnamed.*

remained in excellent health and their prevailing good spirits were continually evident. Accomplishments included evaluation of LM steerable antenna at lunar distances; demonstration of lunar landing mission profile; low-level evaluation of lunar visibility; inflight demonstration of Westinghouse color TV camera; testing of landing radar in near-lunar environment; and manned navigational, visual, and photographic evaluation of lunar landing sites 2 and 3, and in addition other possible landing sites in highland areas.

Apollo 10 was seventh Apollo mission to date, fourth manned Apollo mission, largest payload ever placed in earth and lunar orbits, and first demonstration of lunar orbit rendezvous. Mission acquired major quantities of photographic training materials for Apollo 11 and subsequent missions and numerous visual observations and photos of scientific significance. First manned Apollo mission, *Apollo 7* (Oct. 11–22, 1968), had achieved all primary objectives and had verified operation of spacecraft for lunar-mission duration. First manned lunar orbital mission, *Apollo 8* (Dec. 21–27, 1968), had proved capability of Apollo spacecraft and hardware to operate out to lunar distances and

return through earth's atmosphere at lunar velocity. *Apollo 9* (March 3–13, 1969) had proved capability of manned LM to operate in space. Apollo program was directed by NASA Office of Manned Space Flight; MSC was responsible for Apollo spacecraft development, MSFC for Saturn V launch vehicle, and KSC for launch operations. Tracking and data acquisition was managed by GSFC under overall direction of NASA Office of Tracking and Data Acquisition. (NASA Proj Off; NASA Release 69–68; *W Post*, 5/19–27/69, A1; Sehlstedt, B *Sun*, 5/19–27/69, A1; W *Star*, 5/19–27/69)

May 18: On NBC TV program "Meet the Press" NASA Administrator, Dr. Thomas O. Paine, said there was possibility U.S.S.R. would land men and instruments on all planets before U.S. "The Russians have publicly taken the position that they are extremely interested in landing on the planets, and . . . one of the popular songs in the Soviet Union right now is a song about little apple trees growing on Mars." NASA felt with Apollo lunar landing "we are really taking a lead in the ability to operate on all other bodies. . . . the technology we are developing . . . is a kind . . . that would be required." If July lunar landing succeeded, "we will have enough hardware for nine additional flights to begin the exploration of the lunar surface. In parallel with that, we see activities in earth orbit which will be the precursors to the eventual emplacement of a large permanent space station, a laboratory in the sky."

NASA expected to find "surprising amount" to study on moon. "In fact, we have already radically changed our views of the moon just with the activities we have carried out in preparation for the Apollo landing. For example . . . men are beginning to seriously question whether there may have been water at one time on the surface of the moon. Indeed, whether the large mare areas, the smooth areas . . . may even have been the beds of ancient seas. As we get more and more familiar with the moon we realize how little we know about it. It will take those ten flights and many other trips to the moon before man really begins to understand his twin planet."

Dr. Paine said USAF's MOL and NASA orbiting workshop were "two very different projects." NASA's was "longer range program aimed at a very substantial facility which would be really a university campus type of research station in orbit." MOL was "program that is well advanced, and is designed to find out the military applications of space." (Transcript)

- On ABC radio-TV program "Issues and Answers," Secretary of Defense Melvin R. Laird said he did not favor increased military participation in U.S. space program. "We've had very good cooperation between the military and our civilian programs and I think that's the way it should continue." (B *Sun*, 5/19/69, A1)
- Norwegian explorer Thor Heyerdahl had rejected NASA request to install communications equipment aboard his papyrus boat for July hookup between Apollo 11 and Heyerdahl's voyage into the past, his navigator, Norman Baker, said. Also rejected, Baker said, was request to install satellite-controlled navigation system on replica of 4,700-yr-old vessel in which Heyerdahl hoped to reach Mexico from Morocco in four months, to reinforce theory that Egyptian adventurers reached Americas more than 2,500 yrs before Christ. Heyerdahl felt craft could not safely carry 400-lb communications equipment load. He had refused satellite navigation system because Egyptians had managed without navigational aids. (AP, *W Post*, 5/19/69, A15)
- Tass quoted unidentified Soviet space scientist, described as chief designer of automatic interplanetary stations, as saying U.S.S.R. favored manned space flight, but only in earth orbit "so far." It would continue probing planets with automatic apparatus. (Reuters, B *Sun*, 5/19/69, A1)
- Soviet Embassy Second Secretary Oleg M. Sokolov said in Washington, D.C., that U.S.S.R. definitely would display supersonic Tu-144 airliner at 28th Paris Air Show, May 29–June 8. He said aircraft would beat Anglo-French Concorde into operation and plans were under way to set up worldwide logistics support for Tu-144. DOT officials said West

would be watching closely to see whether U.S.S.R. could fulfill pledge and, if so, would take good look at Tu-144, which could hurt commercial market for SST if it had worldwide logistics backup. (Bentley, B *Sun*, 5/19/69, A5)

May 19: At dedication of Robert Hutchings Goddard Library of Clark Univ., Worcester, Mass., Sen. Edward M. Kennedy (D-Mass.) said: "I am for the space program. But I want to see it in its right priority: One which will let it continue into the future and not have to be cut back or abandoned because the nation that supports it is hobbled by internal disorder. And so, once the lunar landing and exploration are completed, a substantial portion of the space budget can be diverted to the pressing problems here at home. We should develop a plan for an orderly programmed exploration of outer space. But we no longer need an accelerated program. . . . We should continue an orderly and rational space program for the advancement of man's knowledge of the universe and for the considerable benefits it will bring us here at home." NASA program "has been the first time, outside wartime, in which the nation has organized its scientific and industrial disciplines and techniques.

"Our challenge today is to use the same techniques and the same discipline: To lower the cost of production of home building . . . ; to develop command and control systems in the fight against crime; to apply the versatility of computers to education and worker training and the vital work of neighborhood health centers; to organize government and scientific resources to find new and cheaper ways to end the destruction of our environment. The American team of government, industry and labor has been able to achieve Dr. Goddard's impossible dream: Certainly the same industries, the same employees, the same techniques and support can be applied to the urgent business here at home." (Kennedy Off Release)

Honorary Doctor of Laws degrees were awarded to Sen. Edward M. Kennedy (D-Mass.); J. Leland Atwood, President of North American Rockwell Corp. and general chairman of Goddard Library program; and Jack S. Parker, General Electric Co. Vice Chairman; and Doctor of Science degree to Apollo 11 Astronaut Edwin E. Aldrin, Jr. Mrs. Robert H. Goddard, widow of U.S. rocket pioneer, spoke during ribbon-cutting ceremony at which Dr. Charles G. Abbot, 97-yr-old Smithsonian Institution Secretary Emeritus and Dr. Goddard's close friend, received ovation from audience of 4,000. MSFC Director, Dr. Wernher von Braun, and North American Rockwell Corp. Vice President Francis D. Tappaan received Clark Univ. chairs during luncheon after ceremonies. (Program)

- House Committee on Science and Astronautics favorably reported, without amendment, H.R. 11271, NASA FY 1970 authorization bill introduced May 14. (House Rpt 91-255)
- William R. Frye described magnificence of *Apollo 10* liftoff in Philadelphia *Evening Bulletin:* "TV cameras do not do it justice. It is like 100 claps of thunder, each following the other with machine-gun speed.

"The flame that leaps from behind the rocket could have come straight from Dante's inferno. It is too bright to be seen with comfort by the naked eye. The earth trembles beneath the feet, two miles away. Then the towering rocket, nearly twice as high as Niagara Falls, two-thirds the height of the Washington Monument, creeps with agonizing slowness the first few feet off the ground, enveloped by a white cloud. "Then it is gone—and man is left to wonder and to pray." (P *Bull*, 5/19/69)

- NASA announced it would close transportable tracking station near Toowoomba, Queensland, Australia, following August launch of ATS–E and relocate it later for use in advance versions of ATS series scheduled for late 1972 launch. Emplaced in 1966, $6-million station included 40-ft parabolic antenna and trailers containing electronic gear to maintain communications and receive telemetry from ATS satellites. (NASA Release 69–77)

- Secretary of the Army Stanley R. Resor announced termination of production phase of AH–56A Cheyenne armed helicopter program for default of contractor, Lockheed Aircraft Corp., and said USA might issue "cure notice" on R&D contract with Lockheed for same aircraft. USA had concluded that any aircraft delivered in accordance with contractual schedule would fail to meet performance specifications, particularly those for safe speed and maneuverability. (DOD Releases 416–69, 417–69)

May 19–20: Officials of Eurocontrol, seven-nation organization established under Convention of Cooperation for the Security of Air Navigation, met with DOT and FAA in Washington, D.C., to exchange information on air traffic control and other aviation developments. (FAA Release T 69–30; FAA PIO)

May 20: U.S.S.R. launched *Cosmos CCLXXXII* from Plesetsk into orbit with 321-km (199.5-mi) apogee, 201-km (124.9-mi) perigee, 89.7-min period, and 65.4° inclination. Satellite reentered May 28. (GSFC *SSR*, 5/31/69; SBD, 5/22/69, 98)

- NASA's HL–10 lifting-body vehicle, piloted by NASA test pilot William H. Dana, reached 50,000-ft altitude and mach 0.9 after air-launch from B–52 aircraft at 45,000-ft altitude north of Four Corners, Calif. Flight objectives were to complete pilot checkout and to obtain stability and control data. (NASA Proj Off)

- Dr. Thomas O. Paine, NASA Administrator, said to press representatives in Houston he was "surprised and disappointed" by May 19 speech of Sen. Edward M. Kennedy (D-Mass.) suggesting slowdown in U.S. space program after lunar landing and exploration. "He is wrong," Dr. Paine said. "The United States should not weakly yield technological supremacy in space to the Soviets. We should not ground our astronauts after Apollo." He told press he did not want *Apollo 10* astronauts in flight to moon to hear news of speech and would not include item in news reports sent up to spacecraft. (*W Post*, 5/21/69, A12)

- Dr. Thomas O. Paine, NASA Administrator, presented paper "Space-Age Management and City Administration" at 1969 National Conference on Public Administration in Miami: "Mobilizing modern science, technology and management to accomplish bold ventures in space is clearly far simpler than better organizing the extraordinarily complex human interactions that comprise a modern metropolis. NASA's spectacular advances in space are undoubtedly exacerbating public frustration with urban failures, but . . . they are encouraging the nation to tackle its more complex human problems with greater confidence on a bolder scale. If America can go to the moon, it can indeed do much better here on spaceship earth.

"NASA's range of management approaches is nearly as broad as the range within an urban complex." Urban manager, like NASA manager, "can and should directly manage only a limited part of the complex interacting human enterprise for which he has responsibility. For the important remainder he must structure a 'Darwinian Discipline' system to encourage essential contributions from industry, from universities, and from the entrepreneur, the free wheeler, the scientist, the brilliant innovator, the gifted teacher, and other committed individuals. . . . The greatest single achievement of the space age may have been the formation of NASA; the rest followed as the energies and talents of America were released and given direction." (Text)

- Stacked spacecraft and Saturn V launch vehicle for Apollo 11, first lunar landing mission, rolled out to Launch Complex 39, Pad A, at KSC. (KSC Hist Off; McGehan, B *Sun*, 5/21/69)
- Philadelphia *Evening Bulletin* editorial on *Apollo 10* color TV pictures: "Of all the visions man sees from his new and precarious vantages in space, the most compelling is still the planet from which he comes. In the eye of Apollo 10's color TV camera, Earth is indeed the fairest object, the 'oasis' the Apollo 8 astronauts saw last Christmas on Earth's first manned mission to the moon.

   "For a stranger entering the solar system from the outer reaches of the Universe, surely Earth's mist-shrouded blues, browns and reds would be a goal to satisfy the utmost yearning. Set against the cold blackness of space, it would be a prize to draw bold and venturesome inhabitants of other planets across incredible distances. It would be a goal courageous strangers would endure incredible hardships to win. . . .

   "The awe expressed by the intrepid Apollo 10 astronauts . . . is further reminder that the greatest space prize presently within man's comprehension is already in his keeping. And it is one to leave man wondering whether beings on other planets strive and dream as he does. For the moment certainly, it would not seem so. For what people on another planet could resist the vision in the eye of Apollo 10's TV camera?" (P *Bull*, 5/20/69)

- Baltimore *Sun* noted *Apollo 10*'s first day in space enabled men to be "as near as they will ever come to being in two places simultaneously—there in their own living rooms with their television sets before them and, at the same moment more than 22,000 miles away, observing the planet on which they live.

   "What we saw with such marvelous cold clarity was, of course, a round and mottled swirl of blue, brown and white, a small fragment of the cosmos which until less than a decade ago had fixed absolute limits upon all of mankind's history. Now suddenly we saw it as a unity, a whole, as the habitation common to all of us, just as it would be seen by a non-human visitor approaching it as the astronauts are approaching the moon, silent, mysterious and seemingly lifeless and motionless. The observer had to remind himself that this was indeed the earth he knew. . . ." (B *Sun*, 5/20/69)

- *Philadelphia Inquirer* cited possible danger to Apollo spacecraft from "drifting junk" in space. "One notable sidelight to the Apollo 10 flight came in a report that the spaceship brings the number of man-made objects in space to 1691. Although the chance of Apollo 10 colliding with any of the objects is infinitesimal, it is significant that there is so much hardware floating around in space 12 years after Sputnik I. . . .

"U.S. and Soviet scientists should find ways of bringing back or destroying rockets and satellites and their separate components after they have become inoperable. If some sort of solution isn't found, it may not be long before a tragic collision will occur." (*P Inq*, 5/20/69)

- NASA announced it had invited commercial and educational broadcast organizations interested in experimental use of Applications Technology Satellites (ATS) to send representatives to June 13 briefing at NASA Hq. to learn possibilities for working with ATS. *ATS I* (launched Dec. 6, 1966) and *ATS III* (launched Nov. 5, 1967) were in orbit but had largely filled basic technical assignments. Their facilities could be made available for additional experiments. Third ATS, scheduled for August launch, might be available for additional experimental use on completion of technical missions assigned. (NASA Release 69-74)
- AIA released results of survey which showed expected decline of 4.5% in aerospace industry employment between September 1968 and September 1969, from 1,416,000 to 1,353,000, because of continuing decline in civilian space program and decreasing sales of civilian aircraft. Employment in aircraft production and R&D plants was expected to decline 4.6%, transport aircraft employment, 7%; general-aviation aircraft employment, 0.3%; and missile and space employment, 5.7%. Helicopter and nonaerospace employment, including oceanographic research, was expected to increase slightly. Scientists and engineers would continue to account for 16% of total aerospace employment. (Text)
- USAF announced issue of $1,616,000 initial increment to $5,370,750 cost-plus-incentive-fee contract with Lockheed Aircraft Corp. for prototype development and testing of system to improve navigation and guidance of space vehicles. (DOD Release 415-69)
- Lockheed Aircraft Corp. laid off 700 workers and reassigned 1,800 others as result of May 19 USA cancellation of contract for AH-56A Cheyenne helicopter. Company's stock fell $2.50 per share to $32.37 on New York Exchange. (AP, B *Sun*, 5/21/69, A6)
- At Wings Club luncheon in New York world aviation leaders honored 80th birthday of aeronautical pioneer Igor I. Sikorsky and presented him with silver goblets and tray. (A&A, 7/69, 110)
- Sen. George Murphy (R-Calif.) introduced S. 2204, bill to establish National Oceanic Agency. It was referred to Senate Commerce Committee. (*CR*, 5/20/69, S5403)

May 21–23: Intelsat-III F-4 was successfully launched by NASA for ComSatCorp on behalf of International Telecommunications Satellite Consortium. The 632-lb cylindrical satellite, launched from ETR by Long-Tank, Thrust-Augmented Thor (LTTAT)-Delta (DSV-3E) booster, entered elliptical transfer orbit with 22,802.7-mi (36,689.5-km) apogee, 183-mi (294.4-km) perigee, and 29.1° inclination. All systems were functioning normally. On May 23 apogee motor was fired to kick satellite into planned near-synchronous orbit over Pacific with 22,164.3-

mi (35,644.2-km) apogee, 21,887.2-mi (35,216.5-km) perigee, and 5° inclination.

Intelsat-III F-4 was third successful launch in Intelsat III series. *Intelsat-III F-3* had been launched Feb. 5, and *Intelsat-III F-2*, Dec. 18, 1968. Intelsat-III F-1 had been destroyed minutes after launch Sept. 18, 1968. New satellite was scheduled to begin commercial service June 1, handling up to 1,200 voice circuits or four TV channels. (NASA Proj Off; ComSatCorp Release 69-27)

May 21: USAF's C-5A Galaxy jet became heaviest aircraft flown, in test flight from Dobbins AFB, Ga., with 728,100-lb takeoff weight. Manufacturer, Lockheed-Georgia Co., said weight exceeded design gross takeoff load by 100 lbs. Its previous record was 703,826 lbs. (AP, W *Star*, 5/22/69, A5)

- NASA announced it had issued 12 RFPs for definition and design of Earth Resources Technology Satellite system including study of ground data-processing system. Responses were due June 18. First of two planned spacecraft, ERTS–A was scheduled for late 1971 or early 1972 launch as R&D satellite to test new technology to verify effectiveness of earth resources survey from space. ERTS–A sensors would obtain image data in regions of near and infrared spectrum. Satellite, weighing 1,000 lbs, would also carry experimental data-collection system for measurements of remote, unattended sites. It would be placed in sun-synchronous, near-polar orbit at 500-mi altitude to view entire earth in 100-mi-wide increments in less than three weeks for at least one year. (NASA Release 69-73)

- AIAA submitted to President Nixon's Science Advisory Committee *The Post-Apollo Space Program: An AIAA View*. While "remarkable progress of the Apollo-Saturn lunar program has erased almost all doubt about man's ability to travel in space and return safely," program's magnitude had overshadowed "very solid accomplishments" of unmanned satellites. It was based largely on technology available at its inception; neglected "growing accumulation of feasible, but undeveloped technology" in space vehicle design that could affect space transportation costs; and failed to specify goals beyond manned lunar landing.

  Report rejected single national space objective for next decade. It recommended programs to determine man's usefulness in space over prolonged periods and to reduce cost of manned operations and urged Government to give high priority to multifaceted applications satellite program. It urged planning and funding for communications data relay, meteorology data, earth resources data, and navigational aids satellite programs and "well-integrated inter-agency plan to develop data-management subsystems."

  AIAA considered Apollo Applications program and MOL of "substantially greater importance" than last four or five lunar landing missions and encouraged "their timely continuance." It urged designs proceed for extensions of capability in AA and MOL orbital hardware to permit continuation of manned orbital program after 1973, supported retention of at least one crew-carrying vehicle with increased capability for 1973–1975; encouraged early steps to commit to flight demonstration partially reusable low-cost space transportation system for 1974–1976; encouraged accelerated study of space station hardware to succeed AA program and MOL; and encouraged early steps toward commonality of NASA and DOD subsystems. It considered "commitment to an entirely new station" was "less urgent than commitment to a new logistics system."

AIAA recommended continued Apollo lunar program through at least two or three missions and then evaluation, as well as immediate beginning of "vigorous study and controlled funding" of advanced subsystems for 1973–1975 to permit continuation if early success provided support for extension. It recommended manned planetary exploration commitments await evaluation of current programs.

Search for extraterrestrial life was "perhaps the most exciting and spectacular of all space-science program objectives" and might well serve as one of "central themes for set of balanced space goals for the 1970s." (A&A, 6/69, 39–46)

- U.S.S.R. publicly demonstrated Tu-144 supersonic airliner in 90-min test flight from Moscow's Sheremetyevo Airport. Test pilot Eduard V. Yelyan said aircraft had not yet exceeded speed of sound although it was designed for speeds to 1,600 mph (mach 2). At airport press conference Boris Savchenko, head of U.S.S.R. aircraft export agency, said production had started on 120-seat, 130-ton airliner. In *New York Times*, Bernard Gwertzman said observers believed purpose of demonstration was to dispel Western reports of accident to aircraft. (*NYT*, 5/22/69, 94; UPI, *W Post*, 5/21/69, C11)

- Vice President Spiro T. Agnew, Chairman of NASC, issued statement on Administration's attitude toward space program: "In response to those who would denigrate our space effort, I think it is clear that this Administration has already demonstrated its belief in the strength and potential of America's space program." Administration was taking steps to "evaluate the costs and alternatives available to us in extending the program once man has been placed on the moon and returns." (AP, *W Post*, 5/22/69, A5)

- Pope Paul VI hailed *Apollo 10* flight and said man's presence in cosmos was sign of God's presence "in our world and our life." Pope told 30,000 persons at St. Peter's Basilica in Rome, "Even more than the face of the moon, the face of man shines before us; no other being known to us, no animal, even the strongest and most perfect in its vital instincts, can be compared to the prodigious beings we men are." (*NYT*, 5/22/69)

- Press commented on *Apollo 10* mission:

    Washington Post editorial: "The mission of Messrs. Cernan, Stafford and Young, as dramatic and daring as it is, is only an interim step between the first trip to the moon and the first landing on the moon. It is, however, a crucial step since any major problems in this mission or any major unanticipated discoveries about the moon's gravity might well delay the ultimate landing. Because it is both so crucial and so risky, the Nation will wait with special concern when they disappear behind the moon for the first time this afternoon and when the lunar landing craft breaks away from the mother ship tomorrow afternoon for its descent toward the moon's surface. The hope hardly needs to be expressed that these maneuvers, like those in the other Apollo flights,

May 21: *The supersonic Soviet airliner Tu-144 following its landing at Sheremetyevo Airport, Moscow, after a public-demonstration test flight. (AP wirephoto)*

will be executed with the same precision that marked the early stages of this trip." (*W Post*, 5/22/69, A24)

John Lannan in Washington *Evening Star:* "Where manned space flight once was fraught with fright and peril, it seems to have moved into an era of fun and games. . . . The astronauts have learned to fly their spacecraft, the ground crews to launch them and the Defense Department to recover them. What remains is to use this accumulated knowledge, and that's what NASA is attempting to do." It all pointed to fact that "space flight has come of age." (W *Star*, 5/21/69, A4)

Marquis Childs in *Washington Post:* "When the moon landing . . . is completed the cost to the government will be just under $24 billion. The achievement represents an unprecedented orchestration of the resources of science and technology. The contrast with the failures here on earth to begin to try to cure rudimentary ills could hardly at this moment of grave uncertainty be sharper." (*W Post,* 5/21/69, A25)

Christian Science Monitor editorial: *Apollo 10* mission "has again posed the old, old question: If mankind can achieve so spectacularly in space, why cannot we improve matters faster here on earth? The answer, surely, is that it is simpler to mobilize the complex electronics and space-science gadgetry to rocket a spacecraft to the moon than it is to coordinate the manifold and conflicting human emotions, ambitions, and processes necessary for cleaning up the cities, thrusting the black revolution forward with a minimum of friction, and banishing earth's pollutants. The earth problem is more complex than the moon problem. Yet success in the moon venture will offer assurance that the earth challenge can be met.

"So let no one call the moon venture a waste of ambition, treasure, achievement. The whole brilliant enterprise is immensely horizon-widening, thought-expanding." (CSM, 5/21/69)

May 22: Dr. Thomas O. Paine, NASA Administrator, presented NASA Distinguished Service Medal to Jesse L. Mitchell, Director of Physics and Astronomy in OSSA, and Joseph Purcell, OAO Project Manager at GSFC, for their contributions to *Oao II,* at GSFC ceremonies. Dr. Frederick Seitz, NAS president, received Distinguished Public Service Medal, special award, for leadership in solid-state physics. Without solid-state circuitry, *Oao II* would not have been possible. Additional awards were presented to 13 persons from Government, industry, and universities for *Oao II* efforts. GSFC OAO project team, Atlas/Centaur personnel from LeRC, and launch operations personnel from KSC received Group Achievement awards. (NASA Release 69–78)

- MSFC announced award of $4,620,310 contract modification to Chrysler Corp. Space Div. for vehicle systems engineering and integration on Saturn IB vehicles scheduled for NASA AA program flights. Work begun Jan. 1, 1969, would extend through March 31, 1970. (MSFC Release 69–133)

- NASA Wallops Station announced selection of Aerojet-General Corp. to fabricate and support launch of two Orbiting Frog Otolith spacecraft for basic research on frog's balance mechanism under $1,676,000 cost-plus-fixed-fee contract. Project was part of NASA's human factor systems program to investigate functioning of primary balance mechanism within inner ear under zero g conditions. (WS Release 69–11)

- In Washington *Evening Star,* Crosby S. Noyes said: ". . . it would be a mistake to consider the space program itself as nothing more than a kind of inspirational stunt to show us what we can do if we really put our minds to it. The idea that once the demonstration is over, we should divert all of its resources to domestic problems is excessively simple-minded. Of all the resources that have gone into the space program, the money that is so much on everyone's mind is undoubtedly the least critical. The major resources . . . are people and organization and a continuing process of research and application. These resources cannot be 'diverted' into new areas; nor can a program such as this

- LeRC announced it had acquired F–8 single-seater supersonic jet from USN for use as chase aircraft during flights by its F–106 research jet in program to evaluate advanced inlets and exhaust nozzles. Although F–8 could achieve mach 1.5 plus speeds, it would be flown at mach 1.2—top speed in F–106 flight plan. (LeRC Release 69–24)
- U.S. submitted draft treaty to Geneva Disarmament Conference to prohibit emplacement of nuclear weapons, other weapons of mass destruction, and fixed launching platforms "on, within or beneath the seabed and ocean floor." Proposal completed U.S. rejection of Soviet draft treaty that would prohibit not only such weapons but also "all objects of a military nature." U.S., Canada, Italy, and other coastal states opposed ban on submarine detection devices included in Soviet treaty. (Hamilton, *NYT*, 5/23/69, 12)
- F–111A fighter-bomber crashed in northern Arizona on training mission from Nellis AFB, Nev. USAF said both pilots had ejected safely. (AP, *B Sun*, 5/23/69, A6; *W News*, 5/23/69, 3)

May 23: USAF Titan IIIC booster launched from ETR successfully orbited five unmanned satellites—two Vela nuclear detection satellites and three orbiting vehicle research satellites. Launch was 17th for Titan IIIC and last in development program.

Vela IX entered orbit with 69,387-mi (111,643.7-km) apogee, 68.653-mi (110,462.7-km) perigee, 6,718.5-min period, and 32.7° inclination. *Vela X* entered orbit with 69,614-mi (112,008.9-km) apogee, 68.774-mi (110,657.4-km) perigee, 6,707.6-min period, and 32.8° inclination. Velas would monitor nuclear weapon detonations and natural radiation sources.

OV V–5 (ERS–29), *OV V–6*, and *OV V–9* entered orbits with 69,427-mi (111,708.0-km) apogee, 10,480-mi (16,862.3-km) perigee, 3,119-min period, and 32.9° inclination to study particles and fields and solar processes. (*Pres Rpt 70* [69]; GSFC *SSR*, 5/31/69; UPI, *NYT*, 5/24/69, 6)

- *Apollo 8* Astronaut Frank Borman ended four-day visit to Czechoslovakia —first visit by U.S. astronaut to any Communist country. (UPI, *W Post*, 5/24/69, A3)
- Press commented on personal qualities of *Apollo 10* astronauts during crises and technological marvels.

    New York Times: "Their courage and high technical skill were evident. There was total absence of posturing or pomposity. Notable, too, was the absence of false patriotism or of any attempt to use the space feat as the basis for claims of national or ideological superiority. The astronauts' personal behavior added a warm human luster to the superb scientific and technological feats they were and are performing." (*NYT*, 5/23/69)

    Washington *Evening Star:* ". . . the flight of Apollo 10 has shown something . . . about the durability of human nature. . . . However rigorous the training, however unworldly and unreal the surroundings, man is still capable of awe, error, fright, outrage and—when the occasion calls for it—profanity. It's good to know." (*W Star*, 5/23/69, A16)

- Project research method of supporting principal investigators' research within universities accounted for about 70% of NASA funds obligated to universities and was serving NASA and schools well, Dr. George E. Mueller, NASA Associate Administrator for Manned Space Flight, said in speech at Ohio State Univ. More than 10% of project research support funds had been invested in equipment in university laboratories for continuing education and research. "More than 90% of balloon borne experiments, more than 40% of sounding rocket experiments, and more than 50% of satellite experiments flown on NASA vehicles had principal investigators or co-investigators in our universities. A large share of the significant discoveries in space science were made in university originated experiments." NASA "supports about 13,000 project oriented research grants and contracts in universities . . . [and] 32 universities in 21 states are now working with NASA on various aspects of the earth resources satellite program." (Text)
- At Second Advanced Marine Vehicles Meeting in Seattle, Wash., Frank E. Rom, Chief of LeRC Advanced Nuclear Concepts Branch, discussed technical developments in reactor design which could make nuclear propulsion feasible for use on hovercraft. Water-moderated nuclear reactor would heat helium, which then would pass through heat exchanger where water would be boiled. Resultant steam would drive 6,500-hp steam turbines to power lift and thrust fans. Nuclear propulsion would increase hovercraft range, reduce cargo hauling costs, and make vehicle competitive with freighters. (LeRC Release 69-26)
- FAA Administrator John H. Shaffer announced allocations of $34,144,479 to construct and improve 177 U.S. airports under FY 1970 Federal-aid Airport Program. Appropriations, based on $30 million authorized by Congress and carryover funds from previous years, represented last year of funding authorized under current Federal Airport Act. Program stressed preservation and expansion of facilities at existing airports to accommodate high-performance, sophisticated aircraft; increasing airport capacity; relieving congestion; and continuing construction of airports initiated under earlier programs. (FAA Release 69-59)
- Use of new "alphanumeric" system—computer-originated display of letters and numbers on radarscopes indicating aircraft identification, direction, altitude, speed, and flight attitude at FAA's Atlanta, Ga., control center—was described by Robert Lindsey in *New York Times*. System, in which each airliner constantly radioed flight data to ground where it was processed through computer and then displayed on radar screen, would eventually be used by FAA throughout its traffic control network. (*NYT*, 5/23/69, 92)
- U.S.S.R. announced completion of rocket tests begun in Pacific April 17. Tests had been scheduled to end June 15. (*SBD*, 5/26/69, 109; *W Post*, 5/24/69, A12)

May 23–24: NASA Astronomy Missions Board, chaired by Dr. Leo Goldberg of Harvard Univ., met at MSFC to evaluate potential astronomy missions for NASA. Board would submit formal recommendations for space astronomy to NASA later in year. (MSFC Release 69–135)

May 24: Apollo 11 Astronauts Neil A. Armstrong, Michael Collins, and Edwin E. Aldrin, Jr., practiced splashdown and anticontamination procedures they would use after return from moon in July. Astronauts donned plastic-coated biological isolation garments and sprayed each other with Betadine disinfectant before leaving dummy spacecraft in Gulf of Mexico. (UPI, W *Star*, 5/25/69, A6)

May 25: Excited U.S. residents called air control towers, police departments, and newspapers to report citing NASA's *Apollo 10* spacecraft circling moon. Weather Bureau explained bright object actually was planet Jupiter, which was approaching its nearest distance to earth. (W *Star*, 5/26/69, A7)

- *New York Times*—while urging precautions against contamination on Apollo 11 lunar landing mission—praised *Apollo 10* mission as "triumphant scientific climax" of Apollo effort: "The breathtaking virtuosity of Apollo 10's equipment and crew leaves little doubt that similar apparatus can deposit properly trained men on the moon and bring them back safely to earth—always barring the possibility of unexpected mechanical or other malfunction. At the cost of more than $20 billion the United States has acquired the capability of manned travel to the moon. Whatever the wisdom of concentrating such vast resources on the space race, the accomplishment is brilliant and merits awed congratulations for all those whose work and talent made it possible." (*NYT*, 5/25/69, E16)

May 26: President Nixon telephoned congratulations to *Apollo 10* crew following successful splashdown after lunar mission for its "magnificent achievement" and invited astronauts and their wives to dinner at White House. "This is a proud moment for the country," President said. (*NYT*, 5/27/69, 29; *PD*, 6/2/69, 775)

- Soviet Embassy praised *Apollo 10* mission as event that "inspires into us pride for man." Message, written by academician Boniface Kedrov, called astronauts "20th Century Columbuses" and said mission was an "immeasurably more complex, dangerous and almost unrealizable aim compared with that Columbus set before himself at the dawn of the new era." Moscow TV showed splashdown. (*W Post*, 5/27/69, A9)

- Scientific and Technical Subcommittee of U.N. Committee on the Peaceful Uses of Outer Space had recommended new U.N. initiatives to promote access to applications of space technology by small and nonspace powers, Richard S. Kahn said in *Newsweek*. Recommendations included appointment of U.N. official to serve as contact point for member states; panel meetings for promoting collaboration; U.N. assistance for survey missions, panel meetings, and fellowships; investigation of use of earth resource satellites; and dissemination of information on opportunities for education and training in space-related fields.

    Points of debate between large and small powers were on whether new technical assistance was required or whether existing machinery was sufficient. NASA Assistant Administrator for International Affairs Arnold W. Frutkin, as U.S. Representative on subcommittee, had observed that the "only two applications of space technology substantially available today are in . . . meteorology and communications, in both of which U.N. agencies are active." (*Newsweek*, 5/26/69, 57–61)

- Lockheed Aircraft Corp. founder Allan H. Lockheed died in Tucson, Ariz., at age 80. He had begun career at 16 as auto mechanic, taught himself to fly, and in 1915 established aircraft manufacturing firm with brother Malcolm. In 1926 he formed partnership with John K. Northrop which developed into Lockheed Aircraft Corp. They pooled resources to produce Lockheed Vega aircraft, which set 27 records from 1928 to 1932. Lockheed resigned in 1929 but served as adviser to several aviation companies. (UPI, W *Star*, 5/28/69, B7; *W Post*, 5/28/69, C10)

May 27: U.S.S.R. launched *Cosmos CCLXXXIII* into orbit with 1,501-km (932.7-mi) apogee, 196-km (121.8-mi) perigee, 102.0-min period, and 81.9° inclination. Satellite reentered Dec. 10. (GSFC *SSR*, 5/31/69; 12/15/69; *NYT*, 5/28/69, 16)

- Special message to *Apollo 10* crew from five Soviet cosmonauts was released by Soviet Embassy: "We Soviet cosmonauts followed your difficult work very closely. We sincerely admire the high accuracy with which you carried out all the maneuvers planned, your excellent preparedness and courage." Message was signed by Cosmonauts Gherman S. Titov, Andrian G. Nikolayev, Aleksey Leonov, Georgy Beregovoy, and Vladimir Shatalov. (UPI, *W Post*, 5/28/69, A13)

- International comment on NASA's *Apollo 10* mission:

    U.N. Secretary General U Thant said flight was "a thrilling compound of great skill, boundless courage and fabulous technology, as impressive for its perfection as for its informality and its great humor. . . . It is refreshing to have been able to turn for a moment from all our troubles on earth to this magnificent spectacle of man's extraordinary capacity for achievement and peaceful quest."

    Sir Bernard Lovell, Director of U.K.'s Jodrell Bank Experimental Station, said mission represented almost miraculous achievement attainable only by finest technology and engineering in world: "Every part of the Apollo 10 concept now appears to have been performed perfectly. . . . We are nearly about to enter an epoch when men and materials can be transferred to other planets in the solar system."

    U.K. Prime Minister Harold Wilson termed flight "a great triumph in both human and technical terms."

    Soviet space scientist, Dr. Vassily V. Parin, called mission "big event in the history of cosmonautics," impressive because of "the accuracy of all its maneuvers."

    Mrs. Indira Gandhi, Prime Minister of India, said of *Apollo 10* crew, "These men who could have come back with moon dust on their feet are leaving that for others in the true spirit of detachment of great pioneers."

    Heinrich Luebke, President of West Germany, said mission "brings the United States to the brink of an historical high point—the landing of the first man on the moon." (*NYT*, 5/27/69, 29; AP, B *Sun*, 5/27/69, A1)

- *Apollo 8* Astronaut Frank Borman, as Field Director of Space Station Studies for NASA, told Pasadena, Calif., press conference before addressing annual meeting of Chamber of Commerce there were five valid reasons for continuing space program despite high cost: (1) need for program that challenged U.S. in only way it could in time of peace, (2) educational impact of space technology in lower grades as well as among graduates, (3) scientific findings in space, (4) sheer quest and exploration, and (5) tendency of all countries to cooperate and perhaps realize earth's fragility. Borman said, "I hope we can isolate successes and failures from funding because I think space exploration is an important phase of American life." (*LA Times*, 5/28/69; *Pasadena Star-News*, 5/28/69)

- NASA's *Pioneer IX*, launched into heliocentric orbit Nov. 8, 1968, to collect data on electromagnetic and plasma properties of interplanetary medium, was adjudged successful by NASA. Spacecraft had transmitted more than 6 billion bits of data and was continuing to transmit useful data from all scientific experiments. *Pioneer IX* had passed through inferior conjunction Jan. 30 and had reached perihelion of 0.75 au April 7. It would pass through superior conjunction in November 1970, when special experiments utilizing spacecraft-earth radio communication frequencies would be conducted. (NASA Proj Off)
- Discovery of microscopic evidence in lava on Deception Island, Antarctica, indicated algae, fungi, and minute bacteria had begun to thrive in previously sterile lava within 13 mos after volcanic blasts, NASA reported. JPL scientist Dr. Roy E. Cameron and Virginia Polytechnic Institute biologist Dr. Robert Benoit brought back samples in February of lava rubble from Dec. 4, 1967, volcano-earthquake. Discovery indicated sterile material could withstand invasion of growing things for only limited time. (NASA Release 69–80)
- U.S. patent No. 3,446,999 was granted to Dr. Athelstan F. Spilhaus, AAAS President-elect, for rolling device—toy that could move around circular track. Same propulsion method—attraction of electromagnets in car to circuitry in rails, providing continuous revolving movement—could be adapted to larger equipment. Patent had been assigned to Experimentoy Corp. (Pat Off PIO; Jones, *NYT*, 5/31/69, 29)

May 28: NASA's HL–10 lifting-body vehicle, flown by NASA test pilot John A. Manke, successfully completed 19th flight after air-launch from B–52 aircraft at 45,000-ft altitude north of Four Corners, Calif. Manke fired engine at full power for 66 secs and at half power for 40 secs, reaching 64,500-ft altitude. Objectives—expansion of flight envelope to mach 1.2 and collection of stability, control, and performance data —were met. (NASA Proj Off; UPI, *W Star*, 5/29/69, A16; *SBD*, 6/4/69, 152)

- Nike-Apache sounding rocket was launched by NASA from Wallops Station carrying Dudley Observatory payload to collect micrometeoroids. Rocket and instruments functioned satisfactorily. (NASA Proj Off)
- NASA released first of hundreds of photos and moving pictures taken by *Apollo 10* crew. Pictures showed moon from variety of positions and CSM as seen from LM. Photo of Triesneck Crater showed network of broad rilles which resembled tracks left by large snowballs rolled over snow-covered plain. Photo of Sea of Tranquility, prime landing site, had only few rugged features, including medium-sized Moltke Crater and Hypatia Rille. Pictures showed striking resemblance to aerial photos of Antarctica. Films included scenes of moon taken from LM at pericynthian and of Astronaut John W. Young shaving in CSM. (AP, *B Sun*, 5/29/69, A1; *W Post*, 5/29/69, A4)
- Soviet Deputy Minister of Aviation Vasily Kazakov told press in Paris on eve of 28th Paris Air Show that U.S.S.R. would not bring Tu-144 supersonic airliner to show. U.S.S.R. would exhibit its An-22 700-passenger turboprop and would make major effort to promote sales of 10 competitive Soviet aircraft. Cosmonauts Vladimir A. Shatalov and Aleksey S. Yeliseyev said two Zond moon-orbiting capsules launched

Sept. 15, 1968, and Nov. 10, 1968, were large enough to have carried men. Shatalov said he hoped in future U.S. and Soviet space crews would be able to work together. (*NYT*, 5/29/69, 78)

- In letter to stockholders, Lockheed Aircraft Corp. said it had initiated appeal to Armed Services Board of Contract Appeals against cancellation of its Government contract to produce AH-56A Cheyenne helicopter for USA and asked USA to defer demand for $50 million in repayment of progress payments until appeals board ruling. (UPI, *W Post*, 5/29/69, A7; *WSJ*, 5/29/69, 7)

May 29: Cosmos *CCLXXXIV* was launched by U.S.S.R. into orbit with 297-km (184.6-mi) apogee, 204-km (126.8-mi) perigee, 89.5-min period, and 51.7° inclination. Satellite reentered June 6. (GSFC *SSR*, 5/31/69; 6/15/69)

- Specific objectives of lunar exploration were discussed in testimony by NASA Associate Administrator for Manned Space Flight, Dr. George E. Mueller, before Senate Committee on Aeronautical and Space Sciences. Evaluation of natural resources on 14.6-million-sq-mi lunar surface would include minerals of yet undetermined nature and unique combination of high vacuum and gravitational field one-sixth as strong as earth's. "On the results of our evaluation will depend our decision some years from now as to whether there is sufficient potential to justify establishing a lunar base." NASA also would investigate use of moon as "island near our shores to which we can voyage . . . to develop man's potential to function as an explorer throughout the solar system." (Testimony)

- NASA selected Martin Marietta Corp. for $280,000,000 cost-plus-incentive-fee/award-fee contract for Viking lander system and technical integration of project to send two instrumented spacecraft to Mars during summer of 1973. (NASA Release 69–82)

- At USN Symposium on Military Oceanography in Seattle, Wash., Naval Oceanographic Office scientists Paul E. La Violette and Sandra E. Seim said pictures taken by astronauts during Mercury, Gemini, and Apollo missions had been of greater value than expected. "The high resolution of these color photographs has shown a wealth of detail impossible to duplicate by television pictures." Surface and near-surface conditions "appear as sea scars, rips, sea state, bathymetric features. . . ." Many of these features had been shown to exist over large areas on a scale previously unimagined. (AP, *NYT*, 6/1/69, 82)

- NASA published *Significant Achievements in Space Science 1967* (SP–167). Among achievements described were discovery of strong x-ray-emitting objects in stellar astronomy; controlled, quantitative testing of biological hypotheses provided by *Biosatellite II* data; increased use of remote-sensing radio techniques in ionosphere and radio physics; acquisition of data on surface temperature, total pressure, atmospheric composition, exospheric temperature and composition, and strength of magnetic field of Venusian atmosphere by Soviet *Venus IV* and simultaneous flybys of U.S. *Mariner V* in planetology. Developments in solar physics had led to revision of existing set of numbers and had repercussions on interpretation of measurements from sun. (Text)

- If NASA's Apollo 11 successfully landed on moon July 20, "it will be a proud moment for Americans and a costly one for British bookmakers, who will pay out at least $172,000 in bets they wish they had never made," Karl E. Meyer reported in *Washington Post*. Big winner would be David Threlfall of Preston, Lancashire, who in April 1964 bet £10 ($24) that man would land on moon before January 1971. Since odds were 1,000 to 1, Threlfall would collect $24,000 from William Hill Organization betting firm. As one broker explained: "When you think about it, it's a bit ridiculous. This is one of the few times we've made a mistake—the man in the street knew more about space than we did." (*W Post*, 5/29/69, A4)

- Terre Haute [Ind.] *Star* said: "In backing Columbus, according to Samuel Eliot Morison, historian, Queen Isabella had two motives: To make a buck in the spice trade, and to open new territories for her Catholic missionaries. It did not occur to her that she was about to change the history of mankind. That is the way of most turning points in human history. The Manhattan Project which resulted in the atomic bomb was basically a defensive move against the danger of Germany's doing it first. It launched the atomic age. Johann Gutenberg found a way to print with movable type for the simple reason that he wanted more people to read the Bible. He had no idea that he was introducing mass literacy. These great human adventures had two things in common: They were done for practical reasons, and most contemporaries said, 'who needs it?' The space program began for practical, everyday reasons. Russia's Sputnik scared the pants off most Americans. The initial goal of catching up with the Russians has been achieved, and this is part of the reason why there is now a slowing interest. However . . . this country now has the potential of changing the world. . . . The U.S. should continue in the forefront of space exploration, with a well-funded and stable program. Space should continue to have a permanent, though not extravagant, position in the priority of national goals." (Terre Haute *Star*, 5/29/69)
- Subcommittee on Science, Research, and Development transmitted to House Committee on Science and Astronautics *Centralization of Federal Science Activities*. Report, prepared by Library of Congress Science Policy Research Div., described centralization and potential organization of Federal science activities, summarized arguments for and against centralization, examined major functions of Federal Government in dealing with science and technology and present organization, and presented historical summary of evolution of Federal organization for science and of proposals for reorganization and consolidation. (Text)
- Wendell F. Moore, assistant chief engineer at Bell Aerosystems Co. and developer of rocket belt which could lift man and carry him length of football field, died in Niagara Falls, N.Y., at age 51. He had won John Price Wetherill Medal of Franklin Institute of Philadelphia in 1964 for his invention, first flight-tested in 1961. (UPI, *NYT*, 5/30/69)

May 29–June 8: 28th Salon Internationale de l'Aéronautique et de l'Espace —Paris Air Show—featured nearly 550 exhibitors representing 14 nations. U.S., with largest pavilion, emphasized space achievements, taking "Countdown Apollo" as theme. On opening day biggest display attraction, said United Press International, was *Apollo 8* spacecraft, which *Apollo 9* Astronauts James A. McDivitt, David R. Scott, and Russell L. Schweickart unveiled in ceremony attended by U.S. Ambassador to France, R. Sargent Shriver. U.S. exhibit also included F-1 and J-2 engines of 1st and 2nd stages of Saturn V rocket and full-scale model of Apollo 11 LM.

French prototype of Concorde 001 supersonic transport was flown over Paris for first time and later took its place on apron at Le Bourget Airport among 150 aircraft, helicopters, and gliders.

Opening day also marked signing of agreement for joint development of short-haul airbus by French Transport Minister Jean Chamant and West German Economics Minister Karl Schiller. Spokesman for U.K. manufacturer Hawker Siddeley said firm was still negotiating on building wings for 250-seat airbus. U.S. exhibited 490-passenger Boeing 747 but did not show Lockheed C-5A, world's largest aircraft. U.K. exhibits at Air Show included Concorde 002, British prototype; Hawker Siddeley Harrier VTOL jet fighter; and Hawker Siddeley Nimrod, maritime reconnaissance version of Comet. French exhibits included Dassault Mirage fighters and Dassault Hirondelle turboprop. U.S.S.R.'s chief entry was 500-passenger An-22 turboprop airliner. (*Amer Av*, 5/26/69, 33–6; UPI, W *Star*, 5/29/69, A10; Reuters, *NYT*, 5/30/69, 40)

May 30: NASA's Biosatellite III, scheduled to carry monkey on 30-day earth orbital mission June 18, was damaged when unexplained pressure blew top off spacecraft at KSC. Accident, which severed electrical wire harnesses and caused minor structural damage, might delay launch. (AP, W *Star*, 5/31/69, A1; 6/1/69, A8)

• *Science* magazine published letter from Rep. Joseph E. Karth (D-Minn.), Chairman of Subcommittee on Space Science and Applications of House Committee on Science and Astronautics, clarifying his views on cost effectiveness evaluation of Earth Resources Satellite system. He reiterated opinion he stated in Dec. 1968 *Earth Resources Satellite System* report: "Precise determination of cost effectiveness at this early stage . . . is not possible. . . . The magnitude of the economic benefits simply cannot be calculated in the absence of the type of data which the system is designed to produce." Conclusions of studies already completed had constituted "strong evidence that precision was not possible. Yet, I am reassured by the fact that all such studies . . . have concluded that the potential economic benefits will exceed the costs of such a system by a substantial margin and some predict that benefits will someday be measured in billions of dollars annually. It is my personal conviction that an operational ERS system will ultimately prove highly cost-effective."

Karth said he considered it "NASA's responsibility to experiment with new space systems that appear to have potential, and to conduct the necessary research and development which will lead to a firm foundation for a subsequent determination as to whether operational systems should be built. In this context, I believe cost effectiveness is not an appropriate standard to apply in advance to NASA's experimental work, though it is certainly applicable when the time comes to decide whether to go forward with an operational system." (*Science*, 5/30/69, 1009)

• In *Science* editorial, Kenneth V. Thimann said: ". . . there is no doubt that some of our most thoughtful young people see science as a destructive force. Some of this disillusionment stems from a preoccupation with the failings of science, and especially the failings of technology." But people complaining had forgotten past history of far worse air and water pollution and malnutrition. "On the contrary, the record of steady progress can give us confidence that the residual blemishes and pockets will indeed be wiped out as the power of science and technology is increasingly brought to bear on them." In some fields "scientist wields almost unlimited power for good." International Rice Research Institute, with staff of 16 Ph.D.s, had apparently changed "whole nutritional future of Asia in a scant 5 years" by doubling or even tripling rice yields. Discovery of penicillin and streptomycin had saved countless lives. (*Science*, 5/30/69, 1013)

- There was "growing conviction that Soviet authorities have taken administrative measures to punish the noted physicist Andrey R. Sakharov," said Bryce Nelson in *Science*. Washington, D.C., sources had said Sakharov was summoned for verbal criticism after July 11, 1968, publication in *New York Times* of his essay "Progress, Coexistence, and Intellectual Freedom." Discipline was thought to have been performed in early 1969. Fragmentary accounts in Western publications indicated he had been deprived of work as consultant to ministry, removed from position as chief consultant at State Committee for Atomic Energy and from work in restricted physics institute at Chernogolovka, barred from research institute at Dubna, and possibly expelled from Soviet Academy of Sciences. (*Science*, 5/30/69, 1043–4)

May 31: *New York Times* said DOT had received detailed proposals for construction of 150- to 300-mph, air-cushioned vehicle guided by track or guideway. Agency hoped to design vehicle that could avoid problems of steel train wheels and rails, which lost traction and spun at speeds of 150–200 mph. DOT expected to award contract for Tracked Air Cushion Vehicle (TACV) within three months, with completion of prototype and several miles of test track by mid-1971. (*NYT*, 5/31/69, 46)

During May: *Space/Aeronautics* said: "Reshaping of the budget, in combination with a flurry of new activity among NASA, the Air Force and the President's in-house and specially commissioned science advisors, has left no doubt that the Administration favors continuation of a strong manned space flight program and a total space effort much more national in character than the current one. The latter point involves greater pressure on the Air Force and NASA to bring their future programs together, particularly in the space station and support areas.

"Although manned space flight was the clear victor in the Republican amendments to the NASA budget, the surgery on the unmanned sectors was artfully performed." Although $41 million was cut from OSSA, none of it came from Earth Resources Survey satellite program. Although $14 million was cut from OART, none came at expense of NERVA program. Only notable individual reductions in areas other than manned space flight came in deferral of Sunblazer program, cancellation of Biosatellite-F, and deferral for one year of new Planetary Explorer project. In net increase in OMSF, NASA essentially traded off slippage in AA program for resumption of Saturn V production and insurance that lunar exploration would continue into early 1970s. (*S/A*, 5/69, 31–6)

During May

- NAS–NRC Space Science Board had formed 13-member standing Committee on Space Medicine to respond to requests from NASA on problems in manned aspects of national space effort, NAS–NRC–NAE *News Report* said. Chairman was Dr. Shields Warren of Cancer Research Institute of New England Deaconess Hospital, Boston, who was noted for his work on effects of radiation. (NAS–NRC–NAE *News Rpt*, 5/69, 1)
- OAR *Research Review* summarized 1968 research activities of Air Force Cambridge Research Laboratories: During 1968 AFCRL scientists had "placed more instruments on board more research vehicles than any other research group" in U.S. Experiments included 46 research rockets, 75 small meteorological rockets, 110 research balloons, and 374 research flights by 6 flying laboratories. Eight of nine satellite experiments were successfully orbited; most significant was 600-lb *OV 1–16* ("Cannonball") low-altitude-density satellite launched July 11 to measure atmospheric density. (OAR *Research Review*, 5–6/69, 9)
- NSF published *R&D in the Aircraft and Missiles Industry, 1957–68* (NSF 69–15). In 1967 aircraft and missiles industry spent record $5.6 billion for R&D—34% of all industrial R&D spending and 116% increase from 1957 level of $2.6 billion. Federal Government had continued to finance more than 80% of industry R&D. In 1967 this was $4.5 billion, of which estimated $2.7 billion was supplied by DOD, $1.6 billion by NASA, and $0.2 billion by all other Federal agencies combined. However, Federal spending in 1967 was $100 million less than in 1964. Companies' own R&D funds rose from $445 million in 1964 to $1.1 billion in 1967, with growing emphasis on nonmilitary and nonspace areas, particularly commercial aircraft and general-aviation fields. (Text)
- *Flying* magazine issued special report on F–111, including "The People vs. the F–111" by John Fricker and "The F–111—a Pilot's Verdict" by Richard B. Weeghman. Fricker called F–111 "not guilty" of charges that concept of commonality was invalid, that selection of General Dynamics Corp. instead of Boeing Co. as prime contractor was result of "political consideration," that F–111 suffered from excessive flight restrictions, that it was unsafe, and that it was "operational flop." To charge that cost escalation of F–111 program had been excessive, Fricker delivered verdict "Guilty, with mitigating circumstances"—factory, engineering, and research costs had risen twice as much as originally estimated. (*Flying*, 5/69)
- Copy of original tape recording of excited voices of astronomers as they discovered first optical pulsar on night of Jan. 15–16 had been deposited in Niels Bohr Library, American Institute of Physics *Newsletter* noted. Tape, made accidentally during moment when optical pulses from Crab Nebula were discovered, had been preserved by discovery team, W. J. Cooke, M. J. Disney, and D. J. Taylor at Steward Observatory, Univ. of Arizona. (AIP *Newsletter*, 5/69)
- In Communist Party cultural weekly *Kultura*, Warsaw, Janusz Wilhelm said: "Once more the world is experiencing a sense of exultation over the universe. Man's latest cosmic achievements have caused talk, writing and speculation everywhere. Moreover they are almost personally experienced by all." Exultation over man's ability to cope with universe "surpasses all national and political boundaries." It was rare for

people to react "just as human beings without any special differences or distinctions." Moon flight was not going to solve "dramatic problems and conflicts besetting earth," but exultation represented "the essence of rationalism and pragmatism to a much greater extent than most of our emotions." What we felt was "the unity (or oneness) of humanity. . . . So it carried with it a hope." (*Atlas*, 5/69, 23)

June 1969

June 1: Special Task Force report submitted to President Nixon Jan. 8 but not released by White House called for NASA revamping and shift in space priorities, John Lannan said in Washington *Sunday Star*. Panel, chaired by Univ. of California at Berkeley physicist Dr. Charles H. Townes, included Dr. Robert C. Seamans, Jr., Secretary of the Air Force, then NASA Deputy Administrator; Dr. James A. Van Allen of Univ. of Iowa; Dr. Harry H. Hess, chairman of NAS-NRC Space Science Board; and Dr. Walter Orr Roberts of National Center for Atmospheric Research.

Report recommended continuation of $6-billion space effort, with $2 billion for DOD and rest for NASA; disapproved of any commitment to large orbiting space station; and urged commitment to unmanned planetary probes. It considered NASA's present structure inappropriate for post-Apollo program and urged bringing "an outstanding scientist into its top administrative ranks." It was desirable to avoid manned versus unmanned operations argument and to focus on search for most appropriate role for human being in entire system. NASA organization was not "adapted to this approach."

NASC should be chaired by President rather than Vice President. Panel advocated lunar exploration and gave high priority to use of space for commercial and civil benefits. It urged space spending at $1/2$% to 1% of GNP and proposed U.S. intensify efforts toward international cooperation in space, Lannan reported. (W *Star*, 6/1/69, A1)

- NASA's *Mariner VI* spacecraft (launched Feb. 24) was 21,731,091 mi from earth and would fly past Mars July 30. *Mariner VII* (launched March 27) was 19,526,893 mi from earth and would fly past Mars Aug. 4. Both spacecraft were operating normally. (JPL Release 521)
- Atomic scientist Dr. Edward Teller thought nuclear explosion on moon would be scientifically useful, Associated Press reported after New York interview. Vibrations would be source of seismographic measurement for study of moon's interior. "The best information on earth" came from nuclear explosions, "because the energy-generating event is confined very sharply both in space and in time." Factors making moon extremely inhospitable to life—absence of air and water—were highly desirable to researchers, since "change that has taken place billions and billions of years ago is still visible today." Dr. Teller also favored development on moon of research station powered by nuclear reactor that heated lunar rocks to high temperature and liberated oxygen for breathing purposes. There was probability rocks also contained water, which reactor could reduce to hydrogen and oxygen for making rocket fuel to power short-range rocket trips on moon and soft-landing interplanetary spacecraft. Moon's environment might lead to advances in low-temperature physics and surface chemistry in electronics, which could result in development on earth of smaller, more

efficient, and more sophisticated electronic equipment, particularly for information-storing and information-reordering. (Nicholson, AP, W Star, 6/1/69, A1)
- Retiring Chief Justice of Supreme Court Earl Warren said at Lincoln Univ. commencement in Oxford, Pa., "We're going to be on the moon —perhaps by July, they tell us. But it would be better if our universities taught us how to live in our great cities." (AP, W Post, 6/3/69, A9)

June 2: NASA announced preliminary flight plan for Apollo 11 lunar landing mission. Spacecraft, carrying Astronauts Neil A. Armstrong (commander), Michael Collins (CM pilot), and Edwin E. Aldrin, Jr. (LM pilot), would be launched from KSC Launch Complex 39, Pad A, by Saturn V booster at 9:32 am EDT July 16, with touchdown on moon's Sea of Tranquility at 2:22 pm EDT July 20. At 12:12 am EDT July 21 Armstrong would step onto lunar surface, followed hour later by Aldrin. Astronauts would collect up to 50 lbs of lunar surface samples for return to earth, take photos, and deploy experiments package before leaving moon at 12:00 pm EDT July 21 and returning to CSM piloted by Collins. They would complete eight-day mission with splashdown in Pacific at 12:52 pm EDT July 24, 195 hrs 20 min 42.2 secs after launch. (NASA Release 69-83)
- *Apollo 8* Astronaut Frank Borman acknowledged that he had discussed possibility of running for Governor of Arizona or U.S. Senate with Rep. Morris K. Udall (D-Ariz.) but said he did not "foresee right now" that he would do so, Associated Press reported. (W Post, 6/4/69)
- Prearranged meeting at Paris Air Show between *Apollo 9* Astronauts James A. McDivitt, David R. Scott, and Russell L. Schweickart and Cosmonauts Aleksey S. Yeliseyev and Vladimir Shatalov and wives developed from brief technical exchange into what U.S. officials called an epic of all space meetings. After inspecting interior of *Apollo 8* spacecraft and joining astronauts for drinks in VIP lounge at U.S. pavilion, cosmonauts escorted astronauts through U.S.S.R. pavilion, provided technical explanation of 1968 Soyuz missions, and entertained with vodka and caviar in Soviet trijet Yak-40 on display field and later in 500-passenger An-22. (NYT, 6/3/69, 78; AP, B Sun, 6/3/69, A1)
- Dr. Thomas O. Paine, NASA Administrator, received honorary Doctor of Science degree from Brown Univ., his alma mater. (NASA Off of Administrator)
- X-ray, one of science's foremost photographic tools, was being supplemented by revolutionary process of neutron radiography called "neutrography," said *New York Times*. It had been used to check safety of components in *Apollo 10* spacecraft and was subject of Government-supported research in U.K., France, West Germany, and Japan. In U.S., commercially oriented studies were being pursued by General Electric Co., Aerojet-General Corp., and North American Rockwell Corp. Process—in which object to be radiographed was placed in large, high-density beam of neutrons that passed through object and registered data concerning its internal structure on film—had applications in inspection of pyrotechnic devices and nuclear reactor fuel and detection of excessive moisture or minute cracks. Critical welds, guidance components, and "honeycomb" bonding used in NASA program

could also be inspected—as well as home TV sets, telephones, radios, missiles, and SST. (*NYT*, 6/2/69, 39)
- North American Rockwell Corp. announced it had reduced its activity on USAF's Advanced Manned Strategic Aircraft (AMSA) but was maintaining team effort in connection with program. Reduction was made to permit maximum attention to F–15 fighter weapon system competition. (NAR Release NN–28; Wilson, *W Post*, 5/31/69, A1)

June 3: USAF launched unidentified satellite from Vandenberg AFB by Titan IIIB-Agena D booster. Satellite entered orbit with 265.3-mi (426.9-km) apogee, 86.4-mi (139.0-km) perigee, 89.8-min period, and 110.0° inclination and reentered June 14. (GSFC *SSR*, 6/15/69; UPI, *W Post*, 6/4/69, A18; *Pres Rpt 70* [69])
- U.S.S.R. launched *Cosmos CCLXXXV* into orbit with 493-km (306.3-mi) apogee, 266-km (165.3-mi) perigee, 92.1-min period, and 71.0° inclination. Satellite reentered Oct. 7. (GSFC *SSR*, 6/15/69; 10/15/69; AP, *NYT*, 6/4/69, 5)
- International team of scientists might man first U.S. permanent space laboratory, Dr. Thomas O. Paine, NASA Administrator, said at London news conference. Work on project would begin about 1975 and additional sections would be added each year for decade, eventually producing laboratory for about 50 scientists. Dr. Paine, on his way to Paris Air Show, said it was too early to say if there would be permanent U.S. lunar space station or to predict if man would land on Mars. He had no information to support rumors of imminent Soviet moon landing. (Reuters, *W Post*, 6/5/69, E5)
- Boeing Co. pilot Don Knutson flew 362-passenger version of Boeing 747—world's largest passenger aircraft—on 9-hr 8-min maiden Atlantic crossing from Seattle-Tacoma Airport, Wash., to Le Bourget Airport, Paris, for 28th Paris Air Show. Aircraft was fourth 747 off assembly line and acquired one-third of its 27 hrs flying time during transatlantic flight at average 570-mph and maximum 656-mph speeds. (*NYT*, 6/4/69, 74; *Amer Av*, 6/9/69, 16–7)
- House adopted resolution electing Rep. Barry M. Goldwater, Jr. (R-Calif.), to Committee on Science and Astronautics. (*CR*, 6/3/69, H4401)
- London *Times* published four-page space supplement *On the Edge of the Moon*. Man had been traveling to moon for centuries in transport which was "romantic, ingenious, foolish and brilliant: chariots of swans, giant guns, artificial clouds and enormous metal springs; even rockets." Journeys, "dreams that ranged between ludicrous fantasy and prophetic imagination," had not been recorded much before second century A.D. "But later, as writers discovered science fiction and the appetite men had for it, the stories proliferated." At times, either by luck, reasoning, knowledge of science, or uncanny inspiration, they foresaw details of voyages like Apollo 10's and that planned for next month." Article traced 1,800 yrs of space travel "from dream to reality." Supplement also described stage sequences planned for NASA's lunar landing, specifications of lunar module and its achievements, data which scientists hoped to extract from lunar explorations, and possible construction of lunar observatory. Costs of observatory would be justified "only as part of a space programme much larger than what is envisaged for the immediate future, and too large, perhaps,

for the resources of any one nation." (London *Times*, 6/3/69, I–IV)

June 3–5: NASA held Spacemobile Conference in Washington, D.C., to familiarize all personnel with Vice President's Summer Space Education Program for the Cities conducted by NASA in cooperation with President's Council on Youth Opportunity. (Program)

June 4: President Nixon addressed Air Force Academy commencement in Colorado Springs, Colo.: "A nation needs many qualities, but it needs faith and confidence above all. Skeptics do not build societies; the idealists are the builders. Only societies that believe in themselves can rise to their challenges. Let us not, then, pose a false choice between meeting our responsibilities abroad and meeting the needs of our people at home. We shall meet both or we shall meet neither.

"This is why my disagreement with the skeptics and the isolationists is fundamental. They have lost the vision indispensable to great leadership. They observe the problems that confront us; they measure our resources and then they despair. When the first vessels set out from Europe for the New World these men would have weighed the risks and they would have stayed behind. . . .

"Our current exploration of space makes the point vividly, here is testimony to man's vision and man's courage. The journey of the astronauts is more than a technical achievement: it is a reaching-out of the human spirit. It lifts our sights; it demonstrates that magnificent conceptions can be real.

"They inspire us and at the same time they teach us true humility. What could bring home to us more the limitation of the human scale than the hauntingly beautiful picture of our earth seen from the moon?

"When the first man stands on the moon next month every American will stand taller because of what he has done, and we should be proud of this magnificent achievement.

"We will know then that every man achieves his own greatness by reaching out beyond himself, and so it is with nations. When a nation believes in itself—as Athenians did in their Golden Age, as Italians did in the Renaissance—that nation can perform miracles. Only when a nation means something to itself can it mean something to others.

"That is why I believe a resurgence of American idealism can bring about a modern miracle—and that modern miracle is a world order of peace and justice." (*PD*, 6/9/69, 797–802)

- In Huntsville interview, Dr. Wernher von Braun, MSFC Director, said accomplishment of first lunar landing would not necessarily mean U.S. was first in space race. ". . . whether the Russians have this particular objective in their program, I just don't know." It was no longer possible to decide in simple terms who was ahead. ". . . today the space program has so many facets that it may be impossible for all eternity from now on to be ahead of them in all fields. And . . . impossible for them to be ahead of us in all fields." It was still possible for U.S.S.R. to reach moon first if July launching date for Apollo 11 was delayed. Russians now had rocket more powerful than Saturn V, which would allow direct lunar flight. "If this rocket is flown in the very near future—which it might—they may still have a chance of landing a man on the moon in the latter part of 1969." As for Soviet unmanned lunar landing, Dr. von Braun said, "I think the Russians very definitely have the capability as far as their equipment is concerned to soft land enough payload on the moon to take a sample of lunar soil and fly it back to earth." (UPI, *NYT*, 6/6/69, 18)

- Tass released reports which revealed that neither *Venus V* (launched Jan. 5) nor *Venus VI* (launched Jan. 10) had reached Venusian surface intact in May. Data radioed from two spacecraft suggested terrain was very uneven, with height differences of more than 50,000 ft. Reports also conceded that *Venus IV*, which purportedly had landed on Venus Oct. 18, 1967, had not relayed data from Venusian surface, but had apparently been crushed during descent by extreme atmospheric pressure. "The pressure might have pushed in the upper lid of the instrument department and affected the instruments of the radio complex," Tass said. Data suggested that *Venus V* had descended over deep basin where temperature was almost 1,000°F and pressure was 140 times that on earth. *Venus VI* had descended over plateau where temperature was 750°F and pressure was 60 times that on earth. Since spacecraft had not been designed to withstand pressures greater than 25–27 times that on earth, they had not sent data from below 12 mi. (Sullivan, *NYT*, 6/5/69; *W Post*, 6/5/69, A25)

- ComSatCorp announced selection of General Telephone & Electronics International as contractor for construction of earth station for satellite communications near Talkeetna, Alaska. Contract price was $3,558,000. (ComSatCorp Release 69–32)

June 5–8: NASA's 1,393-lb *Ogo VI* (OGO–F) Orbiting Geophysical Observatory, carrying 25 experiments to study sun's influence on earth's near-space environment during period of maximum solar activity, was successfully launched from WTR at 7:42 am PDT by Thorad-Agena D (SLV–2G) booster. Spacecraft entered orbit with 682.4-mi (1,098.2-km) apogee, 246.4-mi (396.6-km) perigee, 99.8-min period, and 82.0° inclination.

Primary mission objective was to conduct correlative studies of latitude-dependent atmospheric phenomena during period of maximum solar activity. Secondary objectives were to search for celestial hydrogen Lyman-alpha radiation, conduct neutron and cosmic-ray observations, measure solar UV and x-ray radiation, make detailed observations of VLF radio emissions, and exceed one year of active, three-axis stabilization. By June 8, all experiments except one—Naval Research Laboratory's x-ray spectrometer, which was expected to have humidity problems—had been turned on and were operating satisfactorily. Two 30-ft antennas had been deployed.

Ogo VI was sixth and last spacecraft in NASA's OGO series. Scientific instrumentation for *Ogo VI*'s 25 experiments had been provided by 10 U.S. universities, 1 foreign university, 4 Government centers, and 5 private companies. *Ogo VI* joined four other operational OGO's— *Ogo I* (launched Sept. 4, 1964), *Ogo III* (launched June 6, 1966), *Ogo IV* (launched July 28, 1967), and *Ogo V* (launched March 4, 1968)—which had provided more than 1.2-million hrs of scientific data on earth-sun relationships and on near-earth environment. *Ogo II* (launched Oct. 14, 1965) had been turned off in November 1967. Results from OGO program included: first observation of protons responsible for ring of current surrounding earth at distance of several earth radii during magnetic storms; first satellite global survey of earth's magnetic field, resulting in proposed new magnetic field model

June 5–8: Ogo VI—*sixth and last of* NASA's *Orbiting Geophysical Observatory series— carried 25 experiments into orbit to study the sun's influence on the earth's near-space environment during a period of maximum solar activity. The satellite was launched from* WTR *by a Thorad-Agena D booster into near-polar orbit. By June 8, all but one experiment had been turned on and were operating satisfactorily.*

for International Geomagnetic Reference Field; clear identification of controlling influence of earth's magnetic field on ion population; verification of existence of inward boundary (plasma-pause) surrounding region of stable trapped radiation; first evidence that region of low-energy electrons completely enveloped trapped radiation regions; first observation of daylight auroras; and first worldwide map of airglow distribution.

OGO program was managed by GSFC under OSSA direction. LeRC was responsible for Thorad-Agena D launch vehicle, and KSC for launch operations. (NASA Proj Off; NASA Releases 69–81, 69–92)

June 5: Electronic disorder in NASA's *Oao II* orbiting astronomical observatory (launched Dec. 7, 1968), first noted June 2, had been successfully adjusted from ground, NASA announced. Spacecraft's inability to receive ground commands or maintain correct orientation was similar to malfunction April 12 which had almost caused spacecraft's death. GSFC controllers corrected problem by adjusting ground command transmitters and computer programs and switching attitude control system from gas-operated to gyro stabilization. (NASA Release 69–88)

- Investigating board headed by Astronaut Walter M. Schirra, Jr., reported primary cause of Dec. 8, 1968, crash of NASA's lunar landing training vehicle (LLTV) No. 1 was "that the vehicle entered a region of flight where aerodynamic moments overpowered the control system . . . such that attitude control was lost. The source of the control problem was not identified . . . in time to add (use) a second control system which could have restored control capability." Crash did not involve any malfunctions of systems. Adverse region of flight was entered because the aerodynamic limitations of LLTV were not completely understood, wind conditions were insufficiently accounted for, and displays in LLTV and support van were inadequate for conditions. Board made 11 recommendations—including wind-tunnel tests to assess LLTV aerodynamic characteristics—for improved safety. (NASA Release 69–87; AP, *H Chron*, 6/6/69)

- U.S.S.R.'s Tu-144 supersonic aircraft exceeded mach 1 for first time during flight test, according to Tass. No further details were released. (*InteraviaAirLetter*, 6/9/69, 5)

- At Paris Air Show Sud Aviation test pilot André Turcat demonstrated Concorde 001. During engine checks before takeoff, noise level at 300 m to side and behind was not excessive even with afterburners switched in. Fly-pasts at various speeds and configurations showed good handling characteristics and low noise levels. Aircraft landed smoothly and stopped in relatively short distance on wet runway with brake chute. (*InteraviaAirLetter*, 6/6/69, 4)

- Dr. Eugene G. Fubini, former Assistant Secretary of Defense, was sworn in as consultant to NASA Administrator. He would advise NASA senior officials on scientific and engineering aspects of agency programs and review and advise on work of President's Space Task Group and NASA–DOD cooperation on space shuttle. Before joining NASA Dr. Fubini had been Vice President and Group Executive at IBM. (NASA Release 69–85)

- NASA announced selection of United Aircraft Corp. for $4-million, three-year contract to design and develop life support and environmental control systems to sustain astronauts for long-duration earth orbital space flights. (*WSJ*, 6/5/69)

- Secretary of State William P. Rogers told Washington, D.C., news conference Nixon Administration was resolved to continue test firings of advanced strategic missiles. He said tests would not affect chances for U.S. success in disarmament talks with U.S.S.R., scheduled for summer, and that they should be continued even after negotiations had begun. (Grose, *NYT*, 6/6/69, 1)

June 6: NASA's HL–10 lifting-body vehicle, piloted by Maj. Peter Hoag (USAF), successfully completed 20th flight over Buckhorn, Calif. Vehicle was air-launched from B–52 aircraft at 45,000-ft altitude and glided to landing. (NASA Proj Off)

- At New York meeting sponsored by Goddard Institute for Space Studies, MSC, and Columbia Univ.'s Lamont-Doherty Geological Observatory, MSC scientist Wilbur R. Wollenhaupt said *Apollo 8* computer had erred in prediction of spacecraft's position by 15,000 ft in range and 1,500 ft in elevation. *Apollo 10* computer erred only 2,000 ft in range and 500 ft in elevation with programming of more accurate model of moon's asymmetrical gravity field. *Apollo 10* results made it likely Apollo 11 could be guided to extremely accurate landing in July.

    Dr. Richard Lingenfelter of UCLA described study of meandering tracks across moon's surface which showed evidence that at least 130 river-like rilles around large circular lunar mare had been formed by flowing water. Evidence was presented from gravity studies and magnetic observations in nearby space that moon's interior was homogeneous rather than subdivided into heavy core surrounded by lighter mantle. (Sullivan, *NYT*, 6/7/69, 16)

- In *Life*, Hugh Sidey said: "It was just exactly eight years ago that John Kennedy set the moon goal and called the nation into 'the exciting adventure of space.' There have been great space moments in these years, but they have faded rather rapidly as the earth problems pressed in. Now there is a lasting excitement which will build to the big launch [Apollo 11] this summer and probably will linger for months or years. . . . History suggests that man, despite his obvious and obsessive miseries, craves something to lift him beyond himself. War too often has been one outlet. Americans in particular have needed a quest, across the mountains or the continent, into the sky and the sea, to the poles or inside the atom." This pointed up "classic dilemma in presidential leadership." Did national pride in space achievement and its technological and military benefits mean more to nation than plans for aid to education, welfare programs, or feeding the hungry? "It could be that the world's ills are not too great to allow such dreams. It could also be that Americans cannot live without them." (*Life*, 6/6/69, 4)

- President Nixon announced intention to appoint Stanford Univ. physicist Hubert B. Heffner as Deputy Director of Office of Science and Technology, succeeding Ivan L. Bennett, Jr., who had resigned. Nomination was submitted to Senate June 9. (*PD*, 6/9/69, 806; 6/16/69, 845)

- NAS and NAE issued *Scientific and Technical Communication: A Pressing National Problem and Recommendations for Its Solution.* Committee on Scientific and Technical Communication (SATCOM) emphasized need for maintaining pluralistic, diverse communication activities in science and engineering as opposed to monolithic, centralized system and recommended 55 methods for meeting accelerating growth of technical data—product of $27-billion R&D enterprise in U.S. Recommendations included creation of Joint Commission on Scientific and Technical Communications responsible to NAS–NAE councils to stimulate greater coordination among private organizations and facilitate interaction with government. (Text; NAS Release; NAS–NRC–NAE *News Rpt*, 5–7/69, 1)

- Florida Legislature passed concurrent resolution asking President Nixon and Congress to restore name "Cape Canaveral" to Cape Kennedy, subject to agreement by Sen. Edward M. Kennedy (D-Mass.), brother of late President John F. Kennedy. Original Spanish name meant "plantation of cane." (AP, W *Star*, 6/7/69, A1)

June 7: World's largest passenger airliner, 629,000-lb Boeing 747, landed at Dulles Airport, Washington, D.C., en route from Paris Air Show. Pan American World Airways would take delivery of its first 747 in September and inaugurate passenger service shortly after first of year. (W *Star*, 6/8/69, A23)

- Since "any contamination of the earth from the moon would affect all men and all nations," said *New York Times*, protective arrangements "should be approved by an international group, preferably by a formal United Nations committee. In the future men will go to Mars and other parts of the solar system where the prospects of finding living organisms are much greater than they are on the moon. If Americans now monopolize the key decisions regarding protection of the earth's environment, they will have no grounds for objecting later on if Russians, Chinese, Germans, Japanese, Brazilians or others monopolize similar decisions affecting human beings returning from more distant celestial bodies." (*NYT*, 6/7/69, 32)

- In *Nature*, Stanford Univ. astronomer Dr. Edward K. Conklin reported recording earth's motion using background radiation believed to have been produced at early stage in universe's expansion. If theory was correct, radiation defined extremely distant reference frame for measurement of earth's motion. Recording showed 100-mi-per-sec movement in direction midway between direction of Big Dipper and star Arcturus. (*Nature*, 6/7/69, 971–2)

June 8: NASA's *Echo II* comsat, launched Jan. 25, 1964, reentered atmosphere at 60.3° north latitude and 148.1° east longitude, north of Siberian Sea of Okhotsk after orbiting earth more than 28,000 times. Launched as passive comsat and air-density research satellite, 532-lb, 135-ft-dia, laminated mylar plastic and aluminum balloon had been used as reflector for bouncing radio transmissions between ground points and for geodetic studies. (NASA Release 69–90)

- *New York Times Magazine* profile quoted NASA Administrator, Dr. Thomas O. Paine: Late President John F. Kennedy's decision to try to put man on moon by end of 1960s was "bold act that is standing the test of time damned well," Dr. Paine had said. "Our ability to function now in a new environment a decade after Jack Kennedy is going to be a very challenging test for us. Do we understand that environment? Can we achieve a new consensus without the 'Pearl Harbor' of a Russian lead in space? We're not really talking about the space program anymore. The space program is finished. You wouldn't speak about

Columbus's voyage as the sail-powered water craft program. What Columbus's journey was all about had nothing to do with water. It was the extension of man's dominion, new life styles, new forms of government, new societies." Dr. Paine believed U.S. had no choice but to push ahead. "A nation that turns down a challenge like this is a nation that's on its way out." (Buckley, *NYT Magazine*, 6/8/69, 34–63)

- William Hines in Washington *Sunday Star* criticized NASA's Lunar Receiving Laboratory and plans to protect U.S. from lunar contamination: ". . . there is ample doubt that (1) the quarantine will really be air-tight and (2) it will make very much difference if the contamination shield leaks a little." He also commented that, earlier, "great lip service was paid to the necessity for avoiding contamination of the moon and other celestial bodies by lifeforms from Earth. The rationale was pragmatic, not moral, and purely anthropocentric: If we contaminate the surface we won't be able to say with certainty whether the lifeforms we eventually find are native or imported. This line is still being hewed to—after a fashion—in the case of Mars, but for the moon the game has proved too costly to be played with strict attention to rules." (W *Star*, 6/8/69, C4)

June 9: Nike-Apache sounding rocket was launched by NASA from Wallops Station carrying Dudley Observatory payload to collect micrometeoroids. Rocket and instruments functioned satisfactorily. (NASA Proj Off)

- FAA announced there were 10,470 airports, heliports, and seaplane bases on its records on Dec. 31, 1968: net increase of 344 over 1967. They included 555 heliports, 411 seaplane bases, and 28 landing facilities outside U.S. Of total, 3,986 were publicly owned and 6,848, privately. Airlines served 183 airports with turbojets and were expected to extend this service to additional 215 airports by 1973. (FAA Release 69–68)

June 10: House passed by vote of 328 to 52 H.R. 11271, FY 1970 NASA authorization of $3.966 billion, allocating $3.26 billion for R&D, $58.2 million for construction of facilities, and $643.8 million for research and program management. House had adopted amendment canceling $327 million authorized for FYs 1967, 1968, and 1969 for which appropriations had not been made. It also adopted amendment requiring emplacement of U.S. flag, exclusively, on moon or any other planet by U.S. astronauts during visits financed entirely by Government funds. Act would be symbolic gesture of national pride in achievement, not declaration of national appropriation by claim of sovereignty. (Text; *CR*, 6/10/69, H4615–56)

- Deputy Secretary of Defense David Packard announced cancellation by DOD of Manned Orbiting Laboratory (MOL) program because of "continuing urgency of reducing Federal defense spending" and "advances in automated techniques for unmanned satellite systems." Cancellation would save "several hundred million" of $525 million proposed for MOL in FY 1970 budget authorization. Remainder would be needed for termination costs and USAF unmanned space programs. Cancellation also would save $1.5 billion in FY 1970 through 1974.

Since 1965 initiation of MOL program, DOD had accumulated much experience in unmanned satellite systems and "profited from both manned and unmanned space exploration of NASA" for "the many, advanced technologies in the MOL effort." Some MOL technology and hardware would be used in other DOD unmanned space programs and DOD was exploring with NASA "the usefulness of some MOL developments to NASA programs." (DOD Release 491-69)

- In Bonn Dr. Thomas O. Paine, NASA Administrator, and West German Science Minister Gerhard Stoltenberg signed Memorandum of Understanding calling for NASA and West German Ministry for Scientific Research cooperation on Helios. Most advanced international scientific space program, Helios would consist of two solar probes carrying 10 scientific experiments 28 million mi—closer to sun than any other yet scheduled—in 1974-75 to provide new understanding of fundamental solar processes and sun-earth relations by studying solar wind, magnetic and electric fields, cosmic rays and cosmic dust. NASA would launch two German-built spacecraft on Atlas-Centaur vehicles one year apart. Seven experiments would be provided by German scientists and three by GSFC in cooperation with U.S., Australian, and Italian experimenters. (NASA Releases 69-86, 69-91)
- At Smithsonian Institution ceremony, X-15 No. 566670, one of three rocket-engine aircraft built to test flight environment in upper atmosphere, took its place near Wright brothers' *Kitty Hawk Flyer* and Charles A. Lindbergh's *Spirit of St. Louis.* Aircraft had been officially retired with completion of joint NASA-USAF X-15 program in December and flown from Edwards AFB, Calif., as cargo to be refurbished by Smithsonian. First X-15 built, it made first captive flight March 10, 1959, and flew first glide and power flights June 8, 1959, and Jan. 23, 1960. It completed last flight in test program Oct. 24, 1968, to total 81 free flights and 142 flights with B-52 mothership.

  Of three X-15s built by North American Rockwell Corp. and Thiokol Chemical Corp., No. 3 had been destroyed in Nov. 15, 1967, crash which killed pilot, Maj. Michael J. Adams (USAF); No. 2 was being displayed at Air Force Museum, Wright-Patterson AFB, Ohio. X-15 program had cost $300 million and established records for 354,200-ft altitude and for 4,520 mph (mach 6.7) speed.

  At ceremony Dr. Robert C. Seamans, Jr., Secretary of the Air Force, said it was difficult to believe designer's dream of 15 yrs ago had already found its way into museum. (Program; NASA Release 69-56; DOD Release 327-69; *NYT*, 6/15/69, 70)
- MSFC announced award of $1,712,000 contract change to Bendix Corp. for construction of three additional control computers for Apollo Telescope Mount project, to be delivered from April 1970 through July 1970. (MSFC Release 69-141)
- *Space Business Daily* said poll of 1,400 U.S. adults conducted after *Apollo 10* splashdown by A. Singlinger & Co. had found 51.3% in favor of lunar exploration program. Of those polled, 39% disapproved of program and 9.7% had no opinion. (*SBD*, 6/10/69, 174)
- Rep. Lester L. Wolff (D-N.Y.) proposed in House that Apollo 11 spacecraft be commissioned "The John F. Kennedy." (*CR*, 6/10/69, H4639)
- In Washington *Evening Star* Crosby S. Noyes noted: "There are . . . a number of questions about interplanetary travel that remain to be answered, the most obvious being why take the trouble. It is, no doubt, a magnificent conception. But whether it can or should be made real is still open to some doubt." (*W Star*, 6/10/69, A11)

June 11: NERVA nuclear experimental engine (XE) was successfully ground-tested by NASA and AEC in Jackass Flats, Nev., reaching full power for first time under simulated altitude conditions. Reactor operated at 50,000 lbs thrust for 3½ min during 13-min test. (AEC/NASA Release M-144)

- NASA Associate Administrator, Dr. Homer E. Newell, and Assistant Associate Administrator for Advanced Research and Technology John L. Sloop presented paper "Planning Space Technology for the 1970's" to National Security Industrial Assn. in Washington, D.C. Technology readiness for manned missions beyond the moon depended "very heavily upon (1) carrying the technology through proof-of-concept or prototype phase in order to assure long-life, reliable flight equipment and operations, and (2) precursor missions, particularly an Earth orbiting laboratory and lunar exploration that provide an opportunity to obtain the needed technology." Technology needed for manned Mars exploration "represents capabilities that are very useful in many space missions and for some non-space applications as well, particularly with regard to long-life equipment and man's performance under stress." (Text)

- Dr. Raymond L. Bisplinghoff, MIT Dean of Engineering, received FAA's highest honor—Extraordinary Service Award—in Washington, D.C., ceremony for service as technical adviser on SST program. Gold medal, lapel ribbon, and citation for exceptional contribution were presented by Under Secretary of Transportation James M. Beggs. He noted Dr. Bisplinghoff's extraordinary competence and knowledge of aeronautics had played significant role in analyzing complex technical aspects of SST development program. Dr. Bisplinghoff had served as technical adviser to FAA Administrator on SST program since April 1966, had held key scientific posts with NASA since 1962, and was member of NASA Historical Advisory Committee. (FAA Release 69-69)

- Congressional sources quoted by John Finney in *New York Times* said White House had ordered cancellation of DOD's MOL program [see June 10] over DOD and USAF objections and in response to mounting congressional pressure to hold down military spending. (*NYT*, 6/12/69, 1)

- *Washington Post* Federal Diary column noted Astronaut Neil A. Armstrong would receive $2-per-day travel allowance as civil servant during Apollo 11 mission. As GS–16 at NASA, Armstrong collected maximum per diem of $16 when traveling on duty. But for Apollo 11 mission, it had been ruled that he would be enjoying Government billeting and subsistence. (Clopton, Causey, *W Post*, 6/11/69, B11)

June 12: L/G Samuel C. Phillips (USAF), Apollo Program Director, announced NASA would proceed with plans for July 16 Apollo 11 launch. He stressed, however, that NASA would not hesitate to postpone launch if officials did not feel "ready in every way. Nor, once the voyage has begun, would we hesitate to bring the crew home immediately if we encounter problems." (NASA Special Release)

- Bullpup Cajun sounding rocket launched by NASA from Wallops Station carried GSFC payload to 45.1-mi (72.5-km) altitude to study capability of Bullpup Cajun as sounding rocket system and to test prototype ozone payload. Rocket performed satisfactorily but loss of signal at payload separation prevented analysis of payload performance. Parachute deployed as planned but payload was not recovered. (NASA Rpt SRL)

- NSF released *Scientists, Engineers, and Physicians from Abroad, Fiscal*

Years 1966 and 1967 (NSF 69-10). More than 12,500 scientists and engineers had been granted immigrant status in U.S. in FY 1967, increase of 74% from 1966 and 134% from 1956. Immigrant physicians and surgeons increased 30% from 1966 and 65% from 1965, to 3,300. Number of immigrant scientists and engineers was estimated roughly at one-tenth of gross addition to domestic science and engineering manpower. (Text)

- New York *Daily News* editorial said: "For some years, the U.S. Air Force has had an entirely feasible project for orbiting by 1972 a 30,000-lb. space laboratory carrying two men, at a cost of around $3 billion. About $1.3 billion having been spent on the program, Deputy Defense Secretary David M. Packard announced Tuesday that it has been junked, scrapped, scrubbed, in order to save the taxpayers some money. We're hot for government economy. But this looks to us like a most dubious move in that direction. You can bet that Soviet Russia, poverty-stricken though it is, is not skimping in its drive to make space serve the Kremlin militarily." (NY *News*, 6/12/69)

June 13: In letter to NASA Associate Administrator for Space Science and Applications, Dr. John E. Naugle, ComSatCorp President Joseph V. Charyk offered plan for NASA-ComSatCorp cooperation in demonstrations of TV and other satellite services between U.S. areas, including Alaska, using existing earth stations at Brewster, Wash., and Paumalu, Hawaii, plus two new small stations, and NASA ATS satellite or possibly in-orbit commercial satellite. (ComSatCorp Release 69-33)

- Aerobee 150 sounding rocket launched by NASA from Natal, Brazil, carried Brazil-Univ. of California payload to conduct stellar x-ray studies. Rocket and instruments functioned satisfactorily. (NASA Proj Off)

- Aerobee 150 MI sounding rocket was launched by NASA from WSMR with VAM-20 booster to 106.6-mi (171.5-km) altitude carrying Univ. of Colorado payload to measure height profile of nitric oxide, nitrogen, and ionized nitrogen and to test Mariner-Mars UV spectrometer. Rocket and instruments functioned satisfactorily. (NASA Rpt SRL)

- NASA announced addition of two lunar orbits to Apollo 11 flight plan, which would increase revolutions to 30 and total time in lunar orbit to 59 hrs 30 min. Addition of orbit before LM/CSM undocking would improve communications during critical maneuver by bringing LM within radio sight of 210-ft dish antenna at Goldstone Tracking Station during its descent to lunar surface. Addition of orbit after redocking and before LM jettison would allow astronauts two more hours for decontamination of equipment exposed to lunar environment. (NASA Release 69-83A)

- Catastrophic contamination of earth by returning Apollo 11 astronauts and lunar samples was "extremely unlikely," Philip H. Abelson explained in *Science*. "One argument is that a form of life adapted to the absence of H_2O, O_2, and organic compounds could scarcely be expected to survive on earth, much less infect earth's creatures. The most compelling argument, however, is that the lunar-return experiment has been conducted many times in the past. It has been estimated that millions of tons of unsterilized lunar material have reached the earth as a consequence of meteor impact. . . . Sterile containment of lunar specimens during the journey to Houston is assured. . . .

"Procedures involving the astronauts are more controversial. Careful effort to keep to a minimum the amount of adventitious material returned to earth is a substantial factor in the procedures that have been adopted. The astronauts face a difficult and dangerous mission. Were their procedures to be made even more complex because of panicky, last-minute objections, their chances of a safe return could be needlessly jeopardized." (*Science*, 6/13/69, 1227)

- Discovery of approximately 621.4-mi-dia (1,000-km-dia) mascon—mass concentration of gravitational pull—on moon's far side and of fact that Mare Marginis at eastern edge was flooded fraction of mascon basin was reported in *Science* by Cornell Univ. radiophysicists Dr. Malcolm J. Campbell, Dr. Brian T. O'Leary, and Dr. Carl Sagan. Discoveries were made from study of Lunar Orbiter and *Apollo 8* photos and gravitational data. New mascon was 2.8 times heavier than mascons associated with Mares Imbrium and Serenitatis on moon's near side. If mare material was confirmed in basin, discoverers proposed calling it Mare Occulum (Hidden Sea). Dr. Campbell said they believed mascons explained moon's entire "out-of-balance" appearance as seen from spacecraft. While some Apollo mission planners believed moon to be pear-shaped because of its effects on spacecraft, Cornell team believed moon was nearly perfect sphere. Mascons explained unexpected variations in lunar gravity which, according to MSC officials, had dragged *Apollo 10* off course. With mascons, scientists were close to answering question of origin of lunar seas, Dr. Campbell said, "But we haven't quite gotten the whole story." (*Science*, 6/13/69, 1273–5; Cohn, *W Post*, 6/14/69, A1)

- President Nixon submitted following nominations to Senate: Gen. John D. Ryan (USAF) to be USAF Chief of Staff, Adm. Thomas H. Moorer (USN) to be Chief of Naval Operations for additional two-year period, and Gen. Earle G. Wheeler (USA) for reappointment as Chairman, Joint Chiefs of Staff, for additional one-year term. (*PD*, 6/16/69, 845)

- Disruption of circadian rhythms—cycles in life processes dependent on biological mechanism operating like internal clock—placed stress on long-distance air traveler, said FAA medical officials Dr. Peter V. Siegel, Dr. Siegfried J. Gerathewohl, and Dr. Stanley R. Mohler in *Science*. In modern aviation environment man was exposed abruptly to disruptions, particularly during long east-to-west and west-to-east flights. (*Science*, 6/13/69, 1249–55)

June 14: Madrid ceremony marked takeover by Spanish crew of operations at U.S. Deep Space Network tracking facility. Attending were NASA Administrator, Dr. Thomas O. Paine; Spanish Space Research Council President, Gen. Luis Azcarrago; U.S. Ambassador Robert C. Hill; and Gen. Antonio Perez-Marin, President of Spanish Instituto Nacional de Técnica Aeroespacial. Dr. Paine sent final U.S. signal to *Pioneer VIII* (launched Dec. 13, 1967) orbiting sun. Short time later Gen. Azcarrago sent first signal under Spanish control. Dr. Paine said, "Spanish determination to participate in this exciting 20th Century form of exploration reminds us that five centuries ago Columbus' great voyage of exploration was carried out under the flag of Spain." (NASA Release 69–93)

- *Washington Post* published letter from John M. Raymond, Jr., of Washington, D.C., which praised decision of Florida Legislature to ask for return of original name "Canaveral" to Cape Kennedy [see June 6].

"It is for us today a thrilling thought that men will leap to the moon from a cape discovered by Spanish adventurers early in the 16th century—a cape with one of the very earliest American place names to be retained to the present day. Or almost to the present day. Let the NASA center bear the name Kennedy. It is an appropriate tribute to the man who set us on the course to the moon. But let us restore to the cape the proud name it carried for 400 years." (*W Post*, 6/14/69, A24)

- *Washington Post* commented on DOD's cancellation of MOL program: "While few tears . . . will be shed for MOL, the process by which it fell from grace deserves scrutiny. Many of the suggestions for taming the defense budget have posited some kind of non-Pentagon review mechanism, either inside or outside government. Yet MOL lost its place not through such a review but through intense general pressure, which became focused inside the Pentagon upon this particular item. The choice of which project to save, which to sacrifice, was made by the military on the basis of an obscure bureaucratic struggle with high Darwinian overtones. Like the brontosaurus, MOL came upon the tougher conditions of a new environment and was found unfit to survive. This may be an effective way to exercise a measure of occasional control over a swollen defense budget, but it is a crude way and one not at all guaranteed to leave the right elements intact. Both the spenders and the critics ought to keep looking for a more refined and selective approach." (*W Post*, 6/14/69, A24)

June 15: Cosmos CCLXXXVI was successfully launched by U.S.S.R. from Plesetsk. Orbital parameters: apogee, 322 km (200.1 mi); perigee, 197 km (122.4 mi); period, 89.7 min; and inclination, 65.4°. Satellite reentered June 23. (GSFC *SSR*, 6/15/69; 6/30/69; *SBD*, 6/17/69, 205; Reuters, *W Post*, 6/16/69, A16)

- U.S.S.R. was watching U.S. missile debate "with keen interest but without any sign of serious concern," said Bernard Gwertzman in *New York Times*. Moscow diplomats were pessimistic about chances of U.S.–U.S.S.R. disarmament agreement being reached within three years, which meant U.S.S.R. "must go ahead with new systems." Even possibility of eventual agreement on slowdown in arms spending was clouded by emergence of Communist China as possible missile threat to U.S. and U.S.S.R. (*NYT*, 6/15/69, E3)
- London *Sunday Times* article by Francis James said Communist China would soon test nuclear missile with 6,000-mi range that could afford second-strike nuclear capability in 1970s. (*W Post*, 6/16/69, A14)
- Current Book-of-the-Month Club selection, *The Andromeda Strain* by Michael Crichton, would "hardly gladden the hearts" of those at NASA, said Walter Sullivan in *New York Times*. It dramatized dangers of back contamination that had "suddenly become a subject of sharp debate on the eve of the Apollo 11 mission to the moon." (*NYT*, 6/15/69, E8)
- *New York Times* said Dr. Carlos Varsavsky, Director of Argentine Radio Astronomy Institute at La Plata, and 23 team members engaged in international research in radioastronomy faced dismissal for joining general strike May 10 in defiance of Argentine government. They had received notifications of termination of appointment. (*NYT*, 6/15/69, 24)

June 15–16: Astronaut Neil A. Armstrong successfully completed four simulated lunar landings in third lunar landing training vehicle (LLTV) at Ellington AFB, Tex. Armstrong piloted LLTV to 100- and 300-ft altitude and practiced touchdown maneuvers in preparation for lunar landing during Apollo 11 mission, to be launched July 16. Armstrong told newsmen LLTV did "excellent job in simulating the landing characteristics of the lunar module." Flights were first for Armstrong since crash of second LLTV Dec. 8, 1968. First LLTV, piloted by Armstrong, had crashed May 6, 1968. (UPI, *NYT*, 6/16/69, 1; MSC Release 69–49; W *Star*, 6/17/69, A5)

June 16: Apollo 11 preparations were proceeding well toward launch to moon at 9:32 am EDT July 16, Apollo Mission Director George H. Hage told Washington, D.C., press conference. Hypergolic propellant loading would begin June 18, wet phase of countdown demonstration test (CDDT) would begin July 1, and crew participation in CDDT without propellants would begin July 2.

After landing on moon astronauts would descend ladder to lunar surface. When Astronaut Neil A. Armstrong reached second rung of ladder, he would pull D-ring to activate camera for TV coverage of descent to lunar surface. Astronaut Edwin E. Aldrin, Jr., would descend about 15 min after Armstrong. Entire $2\frac{1}{2}$-hr period during which astronauts explored lunar surface, set up experiments, and collected lunar samples within 100 ft of landing site would be telecast live to TV viewers on earth. (Transcript)

- President Nixon sent message to Congress urging approval of 5-million, 10-yr program for expanding planning effort and construction and improvement of airports. He called for increased taxes on users to fund major part. Levies would include increase from 5% to 8% in tax on domestic flight tickets; new $3 tax on tickets for international flights originating in U.S.; new 5% tax on air freight waybills; and increase from two to nine cents per gallon on fuels used by general aviation. (*PD*, 6/23/69, 861–5)
- First stage of Saturn V (SA–508) launch vehicle, to be used on Apollo 13, reached KSC. Second stage, scheduled to leave MTF June 25, would reach KSC June 30. Third stage had arrived at KSC June 13. Instrument unit would be flown from MTF to KSC July 7. (MSFC Release 69–148)
- In *Physical Review Letters*, Univ. of Maryland physicist Dr. Joseph Weber described detection of gravity waves from unknown source but in two places simultaneously. Coincidences were observed on gravitational-radiation detectors over 1,000-km base line at Argonne National Laboratory and at Univ. of Maryland. Probability that coincidences were accidental was "incredibly small." NSF-supported study provided first real evidence of existence of gravity waves postulated by Prof. Albert Einstein more than 50 yrs ago. (*Physical Review Letters*, 6/16/69, 1320–24; Lannan, W *Star*, 6/15/69, A25)

June 17: Apollo Program Director, L/G Samuel C. Phillips (USAF), gave go-ahead to Apollo 11 lunar landing mission for launch July 16. Nine-hour flight readiness review had revealed only one major problem—in guidance system. "Although this problem is not completely resolved at this time," Gen. Phillips said, "I am confident this will not become a constraint to the July launch." (AP, W *Star*, 6/18/69, A1)

- Dr. Arthur Rudolph, former Saturn V Program Manager at MSFC, was honored by W. Randolph Lovelace II Award at American Astronautical

Society banquet in Denver, Colo. Award was presented for his sustained contribution to space travel in directing Saturn V launch vehicle program 1963 through 1968. It was accepted for Dr. Rudolph by Dr. Helmut G. Krause of MSFC. (Release 69-144)

- Dr. Robert C. Seamans, Jr., Secretary of the Air Force, discussed need for improved manned bomber at joint national meeting of American Astronautical Society and Operations Research Society of America, in Denver, Colo.: "Those who criticize the bomber as an obsolete system in the missile age are often the same people who refer to our alleged 4-to-1 superiority over the Soviets in individually targeted warheads." Ratio would be "nearly 1-to-1, with total payload running heavily against us, if it were not for our bomber force with its multiple weapons on each aircraft. If our bombers are to continue to provide deterrence, they must be able to survive an attack and then penetrate the ever-improving Soviet defenses. The B-52 is still a good aircraft, but the prototype was flying in 1952 and the latest models were produced in 1962. An advanced bomber will take advantage of the many improvements that have been made in airframe and engine design. It would have short take-off and landing capability needed for dispersal and payload, structure, and speed necessary for penetration."

    In security terms, space age presented dangers, but also afforded opportunities for increasing strategic stability. "Each generation of space vehicles will provide additional improvements in our ability to monitor enemy activities. We are now working on a satellite early warning system that would detect missiles as they are launched from land or sea. With the aid of such a warning system a dispersed bomber force would be able to take off from its bases before the impact of enemy weapons, even if the time of flight of the latter were greatly reduced." (Text)

- Australian marathon runner Bill Emmerton left Houston, Tex., on 1,034-mi jog to Cape Kennedy, Fla., where he would watch Apollo 11 launch. He would travel on foot approximately 40 mi per day, arriving morning of July 16. Purpose of run was to publicize benefits of physical conditioning, pay tribute to fitness of lunar crew, and commemorate Apollo 11 flight. (*Spaceland News,* 6/69, 11; PMR Release 916-69)

- Sen. Edward W. Brooke (R-Mass.), for 37 cosponsors, introduced on Senate floor S.R. 211, "sense of the Senate" resolution urging President Nixon to propose to U.S.S.R. immediate suspension by U.S. and U.S.S.R. of flight test of multiple missile warheads and strongly supporting prompt negotiations with U.S.S.R. on weapon issues. (*CR,* 6/17/69, S6538)

June 18: NASA successfully deployed and recovered 40-ft-dia parachute in final Project SHAPE (Supersonic High Altitude Parachute Experiment) test to evaluate possible use of parachute for aerodynamic deceleration for soft landings on planets with thin atmospheres. Parachute was ejected from five-foot-long canister carried to 33-mi altitude and 1,800 mph (mach 2.7) by three-stage rocket. Previous tests had been conducted Oct. 23 and Dec. 11, 1968. (NASA Release 69-95)

- USAF announced that data from preliminary investigation of effect of SST wing shape in creating sonic boom indicated that varying shape to reduce pressure field beneath wing might decrease wing's flight efficiency. Tests had been completed at Arnold Engineering Development

Center in Tullahoma, Tenn. Tests also had been made to determine lift-to-drag characteristics of wing shapes upon which aircraft range and efficiency were dependent. (AFSC Release 100.69)

June 19: NASA's HL-10 lifting-body vehicle, piloted by NASA test pilot John A. Manke, successfully completed 21st flight after air-launch from B-52 aircraft at 45,000-ft altitude west of Rosamond, Calif. Objectives were to obtain stability, control, and performance data at speeds up to mach 1.35. (NASA Proj Off)

- House Committee on Appropriations favorably reported H.R. 12307, Independent Offices and HUD appropriations bill, which provided $3.696 billion for NASA—$63.544 million below original budget, $18.544 million below revised budget, $298.290 million below 1969 appropriation, and $269.394 million below total authorizations approved by House. Bill provided $3 billion for NASA R&D, $53.233 million for construction of facilities, and $643.750 million for research and program management. With unobligated carryover from 1969 of $117.473 million, total $3.117 billion would be available for obligation in 1970.

  H.R. 12307 also provided $418 million for NSF—$79 million below amount originally budgeted but $18 million above FY 1969 appropriation. It provided $500,000 for NASC, $24,000 below budget estimate. (Text; Committee Rpt 91-316)

- Senate confirmed nomination of *Apollo 8* Astronaut William A. Anders as Executive Secretary of National Aeronautics and Space Council. (CR, 6/19/69, S6736)

- President Nixon announced intention to nominate Dr. William D. McElroy, Chairman of Johns Hopkins Univ. Dept. of Biology, as Director of National Science Foundation. Dr. McElroy would replace Leland J. Haworth, whose six-year term had expired. He was director of McCollum-Pratt Institute at Johns Hopkins, former member of President's Science Advisory Committee, and member of NAS, AAAS, American Chemical Society, and American Society of Biological Chemists. (PD, 6/23/69, 877)

- Library of Congress announced acquisition of more than 500 kinescope films of "Meet the Press" TV series covering 1949-65, presented by producer and panel moderator Lawrence E. Spivak on behalf of NBC News. Collection included interviews with aviation expert Maj. Alexander de Seversky; MSFC Director, Dr. Wernher von Braun; atomic energy expert Dr. Vannevar Bush; NASA pioneers Dr. T. Keith Glennan and Dr. Hugh L. Dryden; former NASA Administrator James E. Webb; AEC Chairman, Dr. Glenn T. Seaborg; Sen. Thomas J. Dodd (D-Conn.); and members of Senate Committee on Aeronautical and Space Sciences. (LC *Info Bull*, 6/19/69, 1; LC Motion Picture Dept)

June 20: Aerobee 150 MI sounding rocket launched by NASA from WSMR with VAM-20 booster carried Princeton Univ. observatory payload to 112.5-mi (181-km) altitude to study UV radiation of hot stars in constellation Scorpius with 1 Å and 0.3 Å resolution. Rocket and instruments functioned satisfactorily, but camera and stabilization were completely demolished because of fall after parachute failed. (NASA Rpt SRL)

- "There is a good chance there is some form of life on Mars," Los Angeles *Herald-Examiner* quoted Dr. William H. Pickering, JPL Director, as saying in interview. "The planet has a thin atmosphere, the climate is tough but not completely unreasonable." Changes of colors on Mars could be vegetation. Mars' seasons were like earth's but twice as long. Its days were same length as earth's. (Smith, LA *Her-Exam*, 6/19/69)

- S–II stage for ninth Saturn V launch vehicle was successfully test-fired at MTF for almost six minutes. All test objectives were met. (*Marshall Star*, 6/25/69, 2)

- MSFC announced appointment of Saverio F. Morea, former manager of F–1 and J–2 engine projects, as manager of new lunar roving vehicle project. Small manned vehicle would weigh 400 lbs and would be carried on board LM in 1971 to provide lunar surface transportation for two astronauts, hand tools, lunar samples, and other equipment. (MSFC Release 69–150)

June 21–26: NASA successfully launched 174-lb *Explorer XLI* (IMP–G) Interplanetary Monitoring Platform from WTR by Thrust-Augmented Improved Thor-Delta (DSV–3E) booster at 1:48 am PDT during major electric power failure. Spacecraft entered nearly polar orbit with 110,722.5-mi (178,191-km) apogee, 213.1-mi (343-km) perigee, 4,906-min period, and 87° inclination. Primary objectives were to place spacecraft into orbit with apogee of at least 92,000 mi (148,028 km) and to obtain for 90 days adequate measurements from plasma and energetic particle experiments for continuation and extension of studies of environment within and beyond earth's magnetosphere during period of high solar activity. Solar proton data would be transmitted to MSC as needed to support Apollo missions. *Explorer XLI* carried 12 experiments—greatest number ever carried by IMP spacecraft—provided by universities, NASA centers, and industry to measure cosmic rays, solar plasmas, and magnetic fields in interplanetary space.

Explorer XLI was seventh in series of 10 IMP spacecraft planned by NASA. Two of six previously orbited satellites—*Explorer XXXIII* (IMP–D), launched July 1, 1966, and *Explorer XXXV* (IMP–E), launched July 19, 1967—were still providing scientific data. IMP program was managed by GSFC under OSSA direction. (NASA Proj Off; NASA Release 69–89)

June 21: Cancellation of MOL program had left USAF with $1 billion in space hardware and 14 highly trained astronauts, Ralph Dighton of Associated Press wrote. Most of $1 billion already spent on program had been for undisclosed number of Titan IIIM boosters and satellites built or on order. USAF had said launch facilities would be finished on schedule in September because they would have to be paid for anyway. They could be used for unmanned spacecraft. The $1.3-million MOL administration building at Vandenberg AFB, Calif., and $1.6-million medical and training structure could be adapted for office space. MOL astronauts had been offered to NASA but no decision had yet been reached. Aerospace workers in seven firms had been hit by MOL cancellation, with McDonnell Douglas Corp. most affected. It had 7,200 employees on project. (AP, W *Star*, 6/21/69, A5)

- Soviet academician Anatoly A. Blagonravov said in Moscow interview: "Exploration of moon and planets is a most noble task and our generation can rightly be proud it has opened the space era. Any scientific achievement accomplished in any country in the long run becomes an achievement of world science. Space efforts of the United States and

the USSR sometimes complement one another." (UPI, P *Bull*, 6/22/69)
- NASA was combining Apollo 11 spectacular with "bit of spectacular lobbying," Washington *Evening Star* said. It had invited all 533 members of House and Senate to fly at NASA's expense to witness Apollo 11 launch from Cape Kennedy. Total bill could come to nearly $28,000. (*W Star*, 6/21/69, A3)

June 22: NASA had raised limit on amount of samples Apollo 11 crew could bring back from moon, *Washington Post* writer Thomas O'Toole reported in *Los Angeles Times*. Instead of 50 lbs of samples originally set as maximum, astronauts would be permitted to collect as much as they could carry—probably up to about 100 lbs. (*LA Times*, 6/22/69)
- Aerobee 150 MI sounding rocket launched by NASA from Natal, Brazil, with VAM-20 booster carried Univ. of California at Berkeley payload to 102.3-mi (164.6-km) altitude. Objective was to search sky for diffuse and point x-ray sources in $\frac{1}{8}$- to 10-kev range and for possible existence of soft x-ray galactic corona and x-ray emission from Magellenic clouds. Rocket—first live Aerobee 150 launched from rail launcher—functioned satisfactorily. One source counter failed after one-third of flight; other counter operated perfectly, but door failed to open. Objective was not achieved, but some data on x-ray sources were obtained. (NASA Rpt SRL)
- William Hines in Washington *Sunday Star* said June 10 cancellation of USAF MOL program would eliminate "wasteful, ill-conceived and costly project set in motion four years ago principally to still the clamoring of aerospace lobbyists for 'military presence in space,' whatever that might mean." Longer term effect would be "inevitable blurring of a meaningless and arbitrary—but still restrictive—line separating U.S. civilian and military space activities." (*W Star*, 6/22/69, C4)

June 23: *New York Times* interview of M/G Jewell C. Maxwell (USAF), Director of Supersonic Transport Development, reflected optimism about SST's future despite fact program had slipped "two quarters" while Administration deliberated allocation of funds for prototype construction. Gen. Maxwell said program currently was "marking time" at cost of $11 million monthly and had $90 million to $100 million left, but he had no doubts U.S. would push ahead with construction. Gen. Maxwell was leaving project to become Commander of Armament Development and Test Center at Eglin AFB, Fla. (Phelps, *NYT*, 6/23/69, 62)

June 24: U.S.S.R. launched *Cosmos CCLXXXVII* from Baikonur into orbit with 254-km (157.8-mi) apogee, 186-km (115.6-mi) perigee, 88.9-min period, and 51.7° inclination. Satellite reentered July 2. (GSFC *SSR*, 6/30/69; 7/15/69; *SBD*, 6/25/69, 244; *W Star*, 6/24/69, A1)
- House passed by vote of 388 to 6 H.R. 12307, FY 1970 Independent Offices and HUD appropriations bill which provided NASA $3 billion for R&D, $53.233 million for new facilities, and $643.750 million for research and program management—to total $3.696 billion. Bill cut NSF budget request by $80 million to total $418 million. (Text; *CR*, 6/24/69, H5154-5; *W Post*, 6/27/69, A4)
- Senate Committee on Aeronautical and Space Sciences ordered favorably reported with amendment in form of substitute bill H.R. 11271, FY 1970 NASA authorization of $3.716 billion. Committee recommended $250.851 million reduction from amount authorized by House [see

June 10], adjusting total to President's revised budget—$45 million less than authorization requested in his original budget. Recommended authorization would provide $3.020 billion for R&D, $58.2 million for construction of facilities, and $637.4 million for research and program management. Recommendation was lowest made by Committee since 1962—$435 million less than its FY 1969 recommendation. (*CR*, 6/24/69, D543; Committee Rpt 91–282)

- With Apollo 11 launch "minus-23 days" MSFC Director, Dr. Wernher von Braun, visited 3,000-yr-old temple of Apollo at Delphi, Greece. After consultation with oracle, Dr. von Braun said, "I am convinced that we will succeed because no other space operation was ever so well prepared in advance." Oracle, he said, "was ambiguous, as usual." (*Time*, 7/4/69, 35; AP, *W Post*, 6/25/69)

- MSFC issued RFPs on eight-month study of integration of Centaur and Saturn S–IVB stages for possible use for future unmanned high-velocity missions. Proposals for study, which would include six launch vehicle configurations, were due July 10. (MSFC Release 69–153)

- Willy Ley, space author and rocket expert, died at age 62, apparently from a heart attack. Ley—also planning consultant to NASA—had been one of founders of German Rocket Society, fellow of British Interplanetary Society, and member of American Institute of Aeronautical Science. Among his last published books were *Rockets, Missiles and Men in Space* and *Watchers of the Skies*. (AP, W *Star*, 6/25/69, B5; *NYT*, 6/25/69, 41)

June 25: NASA published *Lunar Orbiter I Preliminary Results* (SP–197), including assessment of lunar terrain and results of secondary experiments in selenodesy, micrometeoroids, and radiation. Launched Aug. 10, 1966, spacecraft had been first U.S. vehicle to orbit moon, first to obtain detailed photographic coverage of near and far sides of moon, and first to photograph earth from moon's vicinity. Photos showed fractured and faulted lunar crust with mass-wasting where large boulders had tumbled into craters. Moon appeared to have been highly dynamic and affected by volcanic activity, but despite overall roughness, some photos showed regions of relative smoothness. Surface of far side appeared much rougher than near side with higher terra-to-mare ratio. Meteoroid sensors registered no impacts during mission, indicating meteor activity near moon was no greater than that near earth. Radiation dose rate during transit to moon corresponded to that produced by galactic cosmic rays, but dose rates as high as 70 mrad per hr and 7 rad per hr were experienced during solar flares Aug. 26 and Sept. 2, 1966. (Text)

- U.S. and Spain exchanged notes in Madrid confirming 10-yr extension—to Jan. 29, 1984—of 1964 agreement establishing NASA space tracking and acquisition facility near Madrid. Since June 1965, facility had supported all Surveyor and Lunar Orbiter flights to moon, four Mariner flights to Mars and Venus, four Pioneer interplanetary probes, and all manned Apollo flights. It would support Apollo 11 and passage of *Mariners VI* and *VII* near Mars during summer; 210-ft-dia parabolic antenna for tracking and communication in interplanetary space would be built during next three years. (NASA Release 69–97)

- NASA announced appointment of Astronaut James A. McDivitt as Manager for Lunar Landing Operations in MSC's Apollo Spacecraft Program

Office. McDivitt, who would remain in USAF, would be responsible for planning lunar landing missions subsequent to first landing and would no longer be candidate for space flight crew assignments. (NASA Release 69-96)

- At Salzburg news conference preceding celebration of his 75th birthday, German rocketry pioneer Prof. Hermann Oberth proposed that man extract usable raw materials from moon and store them suspended in gravity-free zone between moon and earth. Materials could then be retrieved from area and brought to earth. He suggested erection of giant concave "space mirrors" to gather celestial light and reflect it on earth as heat to melt polar caps and improve earth's climate. He predicted development of electrically propelled spacecraft and electromagnetic catapults to launch spacecraft without consuming fuel. During celebration, sponsored by Hermann Oberth Society of Nuremberg, Dr. Wernher von Braun, MSFC Director, said Oberth's ideas on rocketry published in 1923 remained valid to date. (*NYT*, 6/29/69, 3)
- FAA announced it had proposed rule establishing "area navigation routes" to relieve air congestion. Multiple flight paths had been made possible by increasing availability of computerized airborne navigation equipment. (FAA Release 69-70)

June 26: Javelin sounding rocket launched by NASA from Natal, Brazil, carried Southwest Center for Advanced Studies experiment to 481.6-mi (775-km) altitude to study ionosphere-protonosphere transition region by measuring vertical profiles of ionospheric parameters. Rocket and instruments functioned satisfactorily. Excellent data on electron temperature, ion temperature, and ion composition were obtained. (NASA Rpt SRL)

- M/G Edmund F. O'Connor, Director of Program Management at MSFC, would return to duty with USAF after Apollo 11 mission, MSFC Deputy Director, Technical, Dr. Eberhard F. M. Rees, announced. Maj. O'Connor would be replaced by Lee B. James, Manager of Saturn Program Office. (MSFC Release 69-155)
- JPL Director, Dr. William H. Pickering, announced appointment of Dr. Clarence R. Gates as manager of JPL's newly established Mission Analysis Div. New division would incorporate Systems Analysis section, Systems Analysis Research section, and JPL Navigation Program which Dr. Gates had headed since 1968. (JPL Release 524)
- Saturn V 1st stage (S-IC-11) caught on fire in test stand at Mississippi Test Facility during acceptance test, scheduled to last 125 secs. Test was terminated automatically after 96 secs when temperature on No. 3 engine turbopump exceeded limit. Fire was extinguished by fire-control system built into test stand after burning for over half hour. (MSFC Release 69-156)
- NAS and NAE formed joint committee chaired by Gen. Bernard A. Schriever (USAF, Ret.) to advise HUD on scientific and technical aspects of "Operation Breakthrough"—HUD program to develop low-cost, mass-produced housing—and to encourage broad industrial and professional participation in program. (NAS-NRC-NAE *News Rpt*, 8–9/69, 1; NAS PIO)
- Sealab III medical officer Cdr. Paul G. Linweaver said extreme cold—result of breathing helium gas under pressure—was major contributor

to Feb. 17 death of Aquanaut Berry L. Cannon in USN's Man-in-the-Sea project. Autopsy reports had indicated Cannon had been asphyxiated by carbon dioxide from faulty breathing apparatus. Linweaver said Cannon was so cold he did not know anything was wrong with apparatus. (AP, *NYT*, 6/27/69, 17)

June 27: U.S.S.R. launched *Cosmos CCLXXXVIII* into orbit with 270-km (167.7-mi) apogee, 199-km (123.7-mi) perigee, 89.2-min period, and 51.7° inclination. Satellite reentered July 5. (GSFC *SSR*, 6/30/69; 7/15/69)

- Nike-Apache sounding rocket launched by NASA from NASA Wallops Station carried Univ. of Colorado experiment to 65.9-mi (106-km) altitude to measure density of hydroxyl radical between 43.5- and 62.1-mi (70- and 100-km) altitudes, using scanning UV monochromater. Rocket and instruments performed satisfactorily and all experimental objectives were achieved. (NASA Rpt SRL)

- In *Science*, JPL scientist Dr. Leonard D. Jaffe said despite successful landing of seven unmanned spacecraft on moon controversy over density of lunar surface material continued. Further analysis of data showed relation of density of lunar surface layer to depth was best determined from spacecraft measurements of bearing capacity as function of depth. Comparison of these values with laboratory measurements of bearing capacity of low-cohesion particulate materials as a function of percentage of solid indicated bulk density at lunar surface was approximately 1.1 gr per cc at depth of 5 cm. (*Science*, 6/27/69, 1514–6)

- NASA announced selection of Collins Radio Co. to provide two 210-ft space communications antennas and supporting concrete pedestals for NASA Deep Space Network stations near Canberra, Australia, and Madrid, Spain. (NASA Release 69–98)

- U.S., U.S.S.R., and 26 other nations agreed during Geneva meeting of subcommittee of Committee on Peaceful Uses of Outer Space that international law, rather than national law, should determine liability for damage caused by spacecraft. Question of whether to fix ceiling on damage claims remained unanswered. (Reuters, *NYT*, 6/29/69, 2)

- Philip M. Boffey in *Science* said nomination by President Nixon of William D. McElroy to head NSF [see June 19] "was particularly interesting because it seemed to carry out a pledge made by Nixon on 28 April that politics would play no part in selection of a new NSF director." Choice had been "greeted with enthusiastic praise and a sign of relief by leaders of the scientific community," since it had been difficult to find scientist willing to take the $42,500-a-year post. (*Science*, 6/27/69, 1504–6)

- AEC announced it had implanted two compact, 10-w, nuclear-power generators in Pacific Ocean off San Clemente Island, Calif., in depths of 60 and 130 ft to subject devices to marine growth in one- to two-year test to determine their long-term behavior in ocean environment. Deepwater testing would follow in AEC program to develop second generation of highly reliable, long-endurance, economic, radioisotope-power source for terrestrial and marine applications. (AEC Release M–152)

June 28: NASA's *Biosatellite III* (Biosatellite-D) primate experiment was successfully launched from ETR at 11:16 pm EDT by two-stage, Long-

June 28

Tank, Thrust-Augmented Thor-Delta (DSV–3N) booster into orbit with 245.1-mi (394.4-km) apogee, 224.4-mi (361.1-km) perigee, 92.1-min period, and 33.6° inclination.

The 1,535-lb spacecraft carried 15-lb male pigtail monkey named Bonny in two-gas atmosphere similar to earth's (20% oxygen and 80% nitrogen)—being used for first time—with sea-level pressure (14.7 psi) and 75°F temperature. Primary objective was to provide suitable physiological environment for instrumented monkey and measure functioning of central nervous system or cardiovascular and metabolic systems. Secondary objective was to evaluate monkey's performance in orbit. Monkey was carefully instrumented so that scientists could monitor wave patterns from 10 brain areas for first detailed studies of brain activity in orbit ever made. Scientists would also record heart action and respiration, monitor circulatory and urinary systems, and observe performance on two behavioral tasks—short-term memory and eye-hand coordination. Some 80% of the experimental data would be radioed to earth by high-speed telemetry at rate of 22,400 bps 18–26 times per day. Spacecraft would be retrieved in midair or off ocean surface after reentry. Monkey would be flown to Hawaii laboratories, where scientists would examine him for changes in bone density, muscle tone, blood cell mass, fluid balance, and reproductive system. [See July 7–8.]

Biosatellite III was third and last spacecraft in Biosatellite series. *Biosatellite I* (launched Dec. 14, 1966) had failed to deorbit on command after three days in orbit because of retrorocket system failure and had not been recovered. *Biosatellite II*, successfully launched Sept. 7, 1967, and recovered Sept. 9, 1967, had demonstrated that plants required gravity to maintain orientation and showed effect of radiation on living organisms. Biosatellite project was managed by ARC under OSSA direction. (NASA Proj Off; NASA Release 69–79)

- U.K.'s three-stage Black Arrow booster exploded 50 secs after launch from Woomera Rocket Range during first full-scale test. (UPI, W *Star*, 6/28/69, A3)

- White House announced President Nixon would observe Apollo 11 splashdown and recovery aboard U.S.S. *Hornet* and proceed to Philippines, Indonesia, Thailand, India, and Pakistan. He would meet with Asian leaders "to reemphasize his longstanding concern with peace and progress in Asia." Announcement was released in New York City. (*PD*, 7/7/69, 926)

June 29: AEC's SNAP–3A nuclear generator, launched on board USN's *Transit IV–A* navigational satellite June 29, 1961, completed eight years in orbit—three years longer than five-year design lifetime—after circling earth 40,530 times and traveling more than 1 billion mi. First nuclear generator to operate in space, 5-in-dia, 5½-in-high SNAP–3A converted heat given off by plutonium directly into 2.7 w of electricity. Five SNAP (Systems for Nuclear Auxiliary Power) generators had been launched to date, including two on NASA's *Nimbus III* (launched April 14). Atomic-fueled generator would be placed on lunar surface by Apollo 12 astronauts to power instrument package. (AEC Release M–150)

- *New York Times* editorial said: "The Administration's action in ordering production started on MIRV multiple warhead missiles before opening negotiations for a moratorium with Moscow touches the most sensitive point in the projected missile-curb talks. Equally grave is the manner in which the Air Force quietly awarded the $88-million contract to General Electric, on the day President Nixon spoke favorably of proposals for a Soviet-American moratorium on MIRV flight testing to head off production and deployment of the weapon by either side. Given this situation, the country is entitled to an explanation from the President of his intentions in the Soviet-American missile-control talks, which he has personally held up for more than seven months." (*NYT*, 6/29/69, 10)

June 30: At small White House dinner, *Apollo 10* Astronauts Thomas P. Stafford, John W. Young, and Eugene A. Cernan presented President Nixon and Vice President Agnew with four wrinkled flags which had been carried aboard spacecraft during mission. Stafford told President, ". . . these flags have been to the moon and 31 times around it, so we thought you'd like to have them just the way we brought them back. That's why we didn't press out the wrinkles when we had them framed." (Dean, W *Star*, 7/1/69, D1)

- *Intelsat I* (*Early Bird*) comsat, which had been retired in orbit during December 1968, had been reactivated and was working with *Intelsat-II F-3* to provide communications between North America, Europe, and Latin America, ComSatCorp announced. Reactivation would compensate for failure of *Intelsat-III F-2*, which malfunctioned when mechanically despun antenna locked. (ComSatCorp Release 69-37)

- Univ. of Chicago scientist Anthony Turkevich had found by analysis of data from three Surveyor spacecraft that rocks on lunar surface contained sufficient oxygen to maintain life without supplemental sources, UPI reported. In interview Turkevich had said that, with nuclear or solar power sources, oxygen extraction from moon might cost less than shipping oxygen supplies to moon from earth. Also, there was little danger that moon rocks and dust carried into LM by returning astronauts would create explosion hazard because of oxygen lack. He had been unable to determine whether moon had sufficient hydrogen to allow chemical creation of water by future colonists or valuable minerals in commercially exploitable quantities and had found no evidence of fossil fuel supplies. (W *Star*, 6/30/69, A5)

- USAF announced award of $718,009 increment to $1,177,125 cost-plus-incentive contract with Avco Corp. for design, fabrication, test, and support through orbital infancy of satellite for investigation of fundamental processes of magnetic storms. (DOD Release 554-69)

- L/G Ira C. Eaker (USAF, Ret.) criticized MOL cancellation in *Detroit News*: Although 80% of U.S. space budget had been devoted to peaceful purposes in space, "it has been recognized generally that prudence dictated that we should ultimately possess the capability of intercepting, inspecting and, if need be, destroying hostile weapons in space. . . . Cancellation of the MOL project concedes to the Russians control of space. After about 1972 the Russians will have the capability of overhauling and destroying our reconnaissance satellites, and they will also be capable of placing weapons in space which we can neither intercept, identify nor disarm." (*Detroit News*, 6/30/69)

- In *Aviation Week* Robert Hotz wrote: "There need be no tears shed over the passing of the U.S. Air Force manned orbiting laboratory (MOL)."

Program "has been so stretched out by funding cuts and low keyed management that its technology has become obsolete and its costs astronomical. It is a classic example of what happens to a major technical development program that is not permitted to pursue its goals at the maximum pace possible. In contrast, the ICBM development and Apollo lunar landing programs have proved what can be achieved in a relatively short period at relatively economical funding." (*Av Wk*, 6/30/69)

During June: NASA's plans for two three-planet Grand Tours—8- to 11-yr missions to outer planets—were described by JPL scientist James E. Long in *Astronautics and Aeronautics*. Envisioning 1,200-lb spacecraft launched by Titan-Centaur, Long described missions that had been identified and analyzed: four-planet missions, including Jupiter, Saturn, Uranus, and Neptune, from 1976 to 1979 (earth-launch dates), with 1977 and 1978 giving best combination of closest-approach altitude, flight time, and launch energy; three-planet missions to Jupiter, Saturn, and Pluto from 1976 to 1979, with 1977 and 1978 preferred; and three-planet missions to Jupiter, Uranus, and Neptune from 1977 to 1980. Long said, "The fortuitous concurrence of mission technology and experience suitable for the challenges of missions to the outer planets, with a unique outer-planet alignment in the 1976–80 period, should make exploration of these planets, as a class, a high-priority candidate for program support." (*A&A*, 6/69, 32–47; NASA Release 69–84)

- Laser range-finding equipment was installed at AFCRL's Lunar Laser Observatory near Tucson, Ariz. Constructed largely with NASA funding, Observatory was built specifically for lunar laser-ranging experiments. (OAR *Research Review*, 3–4/70, 31)
- President Nixon's Space Task Group had established that less expensive space operations in future depended on NASA and USAF development of lower cost boosters, J. S. Butz, Jr., said in *Air Force and Space Digest*. Joint effort on booster selection would be made within understanding signed by Dr. Thomas O. Paine, NASA Administrator, and Dr. Robert C. Seamans, Jr., Secretary of the Air Force. Two major complicating factors existed: "First NASA and the Air Force must overhaul their bureaucracies." Second was "requirement for a compromise between military and civil needs." While USAF needed vehicle which could be launched quickly in large numbers and carry sufficient fuel for extensive space maneuvers, NASA wanted larger vehicle to carry more people and large cargo volume. More difficult would be choice of reentry vehicle. Both NASA and USAF wanted winged configuration for operational flexibility and airliner-style landings, but development costs would be high. (*AF/SD*, 6/69, 79–81)
- Kurt Stehling reviewed *Space Age Management* by former NASA Administrator James E. Webb in AIAA *Astronautics & Aeronautics:* "Despite the bumpy course of NASA's history—technically, fiscally, and managerially; despite Webb's motherhood ways, as reflected in this book; and particularly despite the Johnson Administration's retrenchment of the space program and our sudden awareness of the backstage noises in our society which have moved up front so discordantly (it would have taken a superhuman individual to have foreseen these)—if we see a manned lunar landing next month we will owe it in no small

measure to the managerial role played by James Webb and his associates. And in assessing the event, the historian will be forced to try to make sense of this book." (*A&A*, 6/69, 74)

- Commentary in West Berlin's independent *Tagesspiegel*: "Some people take off for the moon while others try to learn to operate a farm tractor. Our culture will depend on mastering this schizophrenic situation." (*Atlas*, 7/69, 16)

- Aerospace Industries Assn. released *1969 Aerospace Facts and Figures*. During 1968 industry sales reached $29.5 billion, up 8.1% over 1967, with civil transport aircraft accounting for major share. These sales were expected to decline in 1969 when current models were phased out before third-generation jet transports were delivered in substantial quantities. Total industry sales were expected to decline to $28.7 billion in 1969 but 1968 backlog for 60 major companies was $31 billion—approximately 2½ times that in 1960.

    Overall space program expenditures during FY 1969 were estimated $6.3 billion—$4.1 billion for NASA, $2.1 billion for DOD, $117 million for AEC, and $34 million for other agencies. Space expenditures declined approximately $300 million from 1967. Aerospace R&D, including NASA's, reached $8 billion. (Text)

- NSF published *Scientific and Technical Personnel in the Federal Government, 1967* (NSF 69-26). Professional scientific and technical personnel in Federal Government numbered 204,200 in October 1967— 5% increase over October 1966. Engineers, numbering 81,200, were largest of three major groups—scientists, engineers, and health professionals—comprising 40% of 1967 total. DOD continued as major Government employer, with 76,900 scientific and technical employees, of which 93% were engineers and scientists. (Text)

July 1969

July 1: Apollo 8 Astronaut Frank Borman and family flew from New York for nine-day tour of U.S.S.R. Institute for Soviet-American Relations in Moscow had extended invitation through Soviet Embassy in Washington, D.C. Itinerary included Moscow, Leningrad, Novosibirsk, and Crimea. (*W Post*, 7/1/69, A15; AP, *W Star*, 7/1/69, A4)

- Preliminary investigation had revealed leak in small fuel line on Saturn V 1st stage (S–IC–II) No. 3 engine had caused June 26 fire, MSFC announced. Board had been convened to conduct further investigations and recommend preventive measures. Stage's No. 3 and No. 5 engines would be replaced; other three engines received minor damage and would be repaired in place. Accident would not affect launch preparations for Apollo 11 mission; inspection of Apollo 11 vehicle SA–506 had confirmed that its high-pressure fuel lines were in good condition. (MSFC Release 69–156)

- North American Rockwell Corp. consolidated its Rocketdyne Div. and Atomics International Div. into new Power Systems Divs. headed by Jay D. Wethe, Vice President of Aerospace and Systems Group. (NAR Release N–14)

- U.K. Defence Ministry said it had transferred its nuclear strike force from delta-wing bombers to Polaris submarines. Seven eventually would be brought into service. (Reuters, *B Sun*, 7/2/69, A2)

July 2: Preliminary countdown demonstration test (CDDT) for July 16 Apollo 11 launch was successfully completed at KSC. Except for 3-hr 18-min hold during which technicians repaired leaky fuel valve, 5½-day test had run smoothly. (AP, *B Sun*, 7/3/69, A4)

- Unofficial Communist sources said U.S.S.R. would launch Luna spacecraft July 10, which would attempt to scoop up lunar sample and return it to earth, Associated Press reported. Sources said launch would be third attempt to conduct successful mission; first had reportedly exploded on launch pad at Baikonur in early April, and second had exploded in flight June 14, when 2nd stage ignited. One source said Soviet space officials were "very disturbed over the success of the American Apollo program. Losing the moon race will be a terrible blow to them." (*B Sun*, 7/3/69, A1)

- Cosmonauts Gherman S. Titov, Konstantin P. Feoktistov, and Georgy T. Beregovy were among Soviet officials who met Astronaut Frank Borman and family on arrival at Moscow's Sheremetyevo Airport at start of nine-day U.S.S.R. visit. Asked if Soviet cosmonaut might visit U.S., Borman said, "I'm sure that will be discussed. Cooperation in space activities is an important aspect of the space program." Bormans breakfasted in Moscow and returned to airport for flight to Leningrad. Schedule called for visit to Zvezdny Gorodok—Star City—where cosmonauts lived and to space communications center in Crimea; nothing

in program indicated visit to U.S.S.R. launching center at Baikonur in Kazakhstan. (AP, W *Star*, 7/2/69, A14)
- Time-Life, Inc., would pay minimum of $400,000 for exclusive book rights of lunar landing story to the combine established by NASA astronauts in 1959, Don Kirkman said in *Washington Daily News*. Money would be split equally into 60 shares for 52 active astronauts and widows of 8 deceased. (*W News*, 7/2/69, 7)
- NASA announced award to Bendix Field Engineering Corp. of $30 million, one-year, cost-plus-award-fee contract extension for operation and maintenance of major portion of Manned Space Flight Network. Extension was third exercised under option and brought total funding to $139,215,832. (NASA Release 69-100)

July 3: European Launcher Development Organization (ELDO) attempt to place Italian ELDO F-8 spacecraft into polar orbit from Woomera Rocket Range failed when West German 3rd stage of Europa booster malfunctioned. U.K. 1st stage and French 2nd stage performed satisfactorily. (*SBD*, 7/16/69, 14; AP, W *Star*, 7/3/69, A3; NASA Int Aff)
- Apollo 11 booster, spacecraft, and Astronauts Neil A. Armstrong, Michael Collins, and Edwin E. Aldrin, Jr., completed final countdown rehearsal test. Astronauts achieved simulated liftoff at 9:32 am EDT—exact time of scheduled July 16 launch. Final countdown for manned lunar landing mission would begin July 10. (AP, B *Sun*, 7/4/69, A10)
- Apollo 11 astronauts would leave three items on lunar surface to commemorate landing, NASA announced. Silicon disc, 1½-in-dia, would carry statements by Presidents Eisenhower, Kennedy, Johnson, and Nixon; messages of goodwill from leaders of 73 countries; list of leaders of Congress and members of four congressional committees responsible for NASA legislation; and names of NASA's top management, past and present. Statements, messages, and names had been etched on disc by process used to make microminiature electronic circuits. Each message had been reduced 200 times, to barely visible dot.

Three- by five-foot nylon American flag with tubing along top edge would be erected on eight-foot aluminum staff on airless moon. Two other U.S. flags and flags from 136 nations and 50 U.S. states would be carried to moon and returned to earth. Plaque left on LM descent stage would bear images of two hemispheres of earth and inscription "Here men from the planet earth first set foot upon the moon July 1969, A.D. We came in peace for all mankind." It would bear names of Apollo 11 crew—Astronauts Neil A. Armstrong, Edwin E. Aldrin, Jr., and Michael Collins—and President Nixon. (NASA Releases 69-83E, 69-83F, 69-83H)
- At Leningrad news conference during U.S.S.R. tour, Astronaut Frank Borman said he hoped U.S. and Soviet spacemen would fly together in joint mission by mid-1970s. (Reuters, *W Post*, 7/4/69, A3)
- Editorial in *Washington Post* entitled "Our Man on the Moon" criticized White House decision to leave on moon plaque on *Apollo 11* LM descent stage with signature of President Nixon: "The proposed plaque would state that 'we came in peace for all mankind.' That message, together with the names of the three brave men who made the voyage would seem to us to be enough." Editorial erroneously cited April 1968 article by NASA Historian Eugene M. Emme, "Historical Perspectives on Apollo," saying that nowhere did Mr. Nixon's name appear. Name

did appear with reference to post-Sputnik statements in October 1957 and to promises of lunar landing by 1971 in 1960 election campaign. (*W Post*, 7/3/69, A14; *Journ of Spacecraft and Rockets*, 4/68, 369–81)

- Apollo 11 might signal end to KSC area's economic boom, *Wall Street Journal* said. NASA had announced slack in Apollo launchings and cut in KSC employment from 23,500 to 18,500 persons. Brevard County (site of KSC) housing construction had fallen some 40%, from 3,438 units in 1967 to 2,080 in 1968, and was currently down another 40%. (Prugh, *WSJ*, 7/3/69, 28)
- Message from President Nixon was read at opening of summer session of 18-nation Disarmament Committee in Geneva: ". . . draft agreements have been submitted by the United States and by the Soviet Union to prevent an arms race on the seabeds. Although differences exist, it should not prove beyond our ability to find common ground so that a realistic agreement may be achieved that enhances the security of all countries. . . . Our goal should be to present a sound seabed arms control measure to the 24th General Assembly of the United Nations." (*PD*, 7/7/69, 929–30)
- At Paris press conference Sud-Aviation President Henri Ziegler denied reports that France was dropping Concorde supersonic transport project for economic reasons. Milan aerodynamic system developed for Mirage supersonic fighter-bomber was being tested on French prototype. It consisted of two small nose wings which shortened takeoff and landing runs and retracted in flight to reduce resistance. (*NYT*, 7/5/69, 28)

July 4: NASA officials ordered technicians to repaint Saturn V 3rd stage (S–IVB) after they discovered old coating had begun to peel. Thermal paint would help protect super-cold hydrogen fuels from sun's heat. Repainting of stage, scheduled to boost manned Apollo 11 spacecraft toward moon July 16, would not affect launch date. (AP, W *Star*, 7/5/69, A13)

- At U.S. Embassy Independence Day Party in Moscow *Apollo 8* Astronaut Frank Borman signed autographs with Cosmonauts Gherman S. Titov, Georgy T. Beregovoy, and Konstantin P. Feoktistov. Among 1,000 persons attending reception given by U.S. Ambassador Jacob D. Beam were Vasily V. Kuznetsov, U.S.S.R. First Deputy Foreign Minister, and Mikhail P. Georgadze, Secretary of the Presidium of the Supreme Soviet. They were highest ranking Soviet officials to attend annual July 4 reception since 1964 attendance of Nikita S. Khrushchev as head of government and Communist Party. (Clarity, *NYT*, 7/5/69, 28; AP, B *Sun*, 7/5/69, A2)
- Dr. Lee A. DuBridge, Presidential Science Adviser, addressed Independence Day celebration at Dearborn, Mich.: "For untold millions of years the human animal was chained to the earth. Sixty years ago he found a way of soaring into its atmosphere. Ten years ago he learned to break the chains of gravity and to soar out into space. This month the first man will set foot on another world. Later this month two spacecraft will reach Mars and send back new information about that Planet. Americans will have no reason to be ashamed of their nation on those days. Is it worth while? Is it worth while to lift the spirits of millions of human beings? If not, what else is worth while?"

Developing lunar landing technology was relatively easy. "The laws of nature which made it possible have been well known for a long time. The engineering skills required . . . were available and were brilliantly organized. Hundreds of thousands of Americans worked together to make this dream come true. They had faith and they had hope.

"The problems of our cities and the other social problems which beset us are not all that easy. In this area human beings are not working together but are in conflict. We find that we do not yet know the cause of these troubles nor do we yet have the mechanisms for curing them. Hence we must study, we must experiment, we must try and we will often fail. . . . And we shall learn from our failures." (Text; *CR*, 7/29/69, E6415–7)

- Analysis of lunar surface would provide key to earth's history by indicating whether moon's origin was catastrophic or noncatastrophic, Dr. H. Alfvén and Dr. G. Arrhenius of Univ. of California at San Diego said in *Science*. Radiometric dating of igneous lunar rocks might provide information on time of their solidification. If catastrophic alternative was correct, rocks should date to less than 4.5 eons, minimum age of moon, and predominant age should be approximately 0.7 eon, with major surface and subsurface features less than 0.7 eon. If noncatastrophic alternative proved correct, predominant age of lunar rocks should exceed 4 eons, at least, since it was likely moon predated earth. (*Science*, 7/4/69, 11–7)

- Japanese freighter had been hit by wreckage of Soviet spacecraft, Japanese diplomats reported to five Western delegations on legal subcommittee of U.N. Committee on Outer Space meeting in Geneva. June 5 damage to *Dai Chi Chinei* while outside territorial waters and near Siberian coast had previously been attributed to unidentified object. It was believed to be first authenticated case of terrestrial damage caused by falling space objects. (Hamilton, *NYT*, 7/5/69, 28)

July 5: Apollo 11 Astronauts Neil A. Armstrong (commander), Edwin E. Aldrin, Jr. (LM pilot), and Michael Collins (CM pilot) held press conference at MSC. Astronauts were seated 50 ft from nearest newsmen and were partially enclosed in plastic booth as part of plan to limit crew's contacts during 21 days immediately preceding flight and prevent development of illness. Collins told press that from CM viewpoint, Apollo 11 should not be very different from previous manned Apollo missions. He did not feel "slightest bit frustrated" about going to moon without landing on it: "I'm going 99.9 . . . percent of the way there, and that suits me fine."

LM flight plan would pick up where *Apollo 10* left off with phasing maneuver, Aldrin explained. There would be number of "firsts": "the ultimate test," actual touchdown; 1/6 g environment; new thermal conditions; two-man EVA on lunar surface; sleeping in LM on moon; star sighting technique with alignment telescope on lunar surface; and powered ascent from moon with seven-minute engine burn. One of important early activities after exiting from LM on moon would be determining best pace for moving about: ". . . there have been several different techniques employed in the partial zero gravity training. And, it looks like you can walk conventionally one foot after another. It also looks as though you can do a two-footed hop—kangaroo style." In training in aircraft flown at 1/6 g, "a fairly rapid pace" appeared quite easy to perform. "It looks like we shouldn't have too much difficulty in moving at something like 6, 8, or 10 miles per hour."

Armstrong said crew would use "somewhat hybrid methods of manual and automatic" for descent to moon. "The predicted method at this point, although we have a great deal of flexibility and choice, based on the situation at the time, would be to maintain manual control of attitude and automatic control of throttle through the final descent from an altitude of somewhere between 500 and 1000 feet until such time as the automatic throttle rated descent was unsatisfactory, at which time we'll go full manual on the throttle . . . flying it in a manner like a normal VTOL machine."

Code names for CM and LM had been selected as "representative of the flight, the nation's hope," Armstrong revealed. LM would be called "Eagle" for U.S. national emblem, and CM would be called "Columbia" for U.S. symbol, statue on top of Capitol, and Jules Verne's fictional spacecraft, "Columbiad," which flew to moon 100 yrs ago. (Transcript; O'Toole, *W Post*, 7/6/69, A1)

July 5–6: In Moscow *Apollo 8* Astronaut Frank Borman placed wreaths at tombs of Vladimir I. Lenin, founder of Soviet state; rocket designer Sergey Korolev; and Cosmonauts Yuri A. Gagarin and Vladimir M. Komarov. Later he placed wreath at tomb of Soviet Unknown Soldier.

Borman and family visited Star City, home of cosmonauts outside Moscow, where he presented color film of *Apollo 8*. Cosmonauts presented Borman with model of *Vostok 1*, first manned spacecraft. Later Borman toured major space tracking station at Eupatoria near Yalta in Crimea. (AP, *W Star*, 7/5/69, A13; UPI, *W Star*, 7/7/69, A2)

July 6: NASA's *Mariner VI* (launched Feb. 24) and *Mariner VII* (launched March 27) were performing well and had traveled 41 million mi and 39 million mi from earth. *Mariner VI*, scheduled for July 31 flyby, was 9 million mi from Mars; *Mariner VII*, scheduled for Aug. 5 flyby, was 11 million mi from Mars. (NASA Release 69-102)

- Dept. of Commerce announced it had successfully tracked free-drifting buoy in deep ocean with satellite telemetry in test off east Florida coast. Navigational data were relayed via satellite to GSFC for processing. Experiment proved ocean currents could be traced accurately and atmospheric and oceanographic data could be obtained from sensors on drifting buoy and transmitted with navigational information from remote regions of world. (NASA Release 69-41)

- In Washington *Sunday Star*, David Van Praagh discussed President Nixon's planned Asian tour, to follow mid-Pacific recovery of Apollo 11 crew: Nixon would find "the problems of this planet's most populous continent present a striking contrast. They are not subject to quick, rational, scientific or dramatic solutions. Usually they can not even be tackled through modern communications. The vast majority of Indians and Pakistanis for example, can't read or write and do not own a tiny transistor radio or TV set to monitor the Pacific splashdown. They live in a rather backward age and most of them are hungry." (*W Star*, 7/6/69, D14)

July 7: *U.S. News & World Report* published interview with Dr. Thomas O. Paine, NASA Administrator. Apollo 11 lunar landing would be "culmination of America's satisfying everyone that it is indeed the leading technological nation that it thought it was before Sputnik blazed across the skies." U.S.S.R. would continue to put great stress on space and move ahead steadily. ". . . there's always the danger that we may feel we can relax now—having attained the lunar goal—and perhaps slack off. . . . if we were not to start new programs now, I think the situation might well reverse and the Soviets might once again develop superior technological capabilities in space." Space technology could affect future defense posture. In the past, "wherever man has flown farther and higher and faster, wherever he has developed new capability to observe from higher areas, to carry out operations in new media, this has had a major effect on the equations of international power. We're quite confident that this will probably be true again in space." Application of space to defense area was DOD's job. "We do not consider Apollo applications as any kind of substitute for MOL."

Journey of man to another solar system was "completely out of the realm of possibility" for next generation; "but in the more distant future, if it were ever possible . . . to control the energy of nuclear fusion and adapt it in some efficient way to the propulsion of spacecraft, it might be possible to think in terms of longer voyages to another star." Fundamental breakthrough would have to be made.

Chance that life existed in other solar systems seemed "absolutely 100 per cent." (*US News*, 7/7/69)

- Bonny, pig-tailed monkey launched on board NASA's *Biosatellite III* June 29, was showing marked decrease in interest and efficiency. Although he was still in satisfactory physical condition, Bonny was becoming much less energetic and was consuming less food and water. (AP, W *Star*, 7/5/69, A3; W *Post*, 7/5/69, A6)
- *Apollo 10* commander Thomas P. Stafford received Flying Tiger Pilot Trophy, presented every two years by Flying Tigers, group of World War II veterans. Trophy was presented during 27th reunion in Ojai, Calif. (AP, W *Star*, 7/7/69, A2)
- White House announced President Nixon had canceled plans to dine with Apollo 11 astronauts July 15, eve of launch. MSC Director of Research and Medical Operations, Dr. Charles A. Berry, had expressed concern that crew might catch earthly illness from President, which could complicate lunar landing mission. (AP, B *Sun*, 7/8/69)
- Apollo 11 lunar samples would not be first moon material to reach earth and to undergo scientific examination, *Los Angeles Times* quoted Dr. Dean R. Chapman, Chief of ARC Thermo- and Gas-Dynamics Div., as saying in interview. Tektites—molten pieces of lunar surface in form of chunks of black glass—had fallen to earth when meteorites struck moon with tremendous force. Most recent tektite shower had occurred 700,000 yrs ago. While tektites' origin was matter of scientific dispute, Dr. Chapman believed most commonly held theory—lunar origin. In working out shower's trajectory, he had determined tektites came from Tycho crater on moon's southern hemisphere. He believed Tycho to be 700,000 yrs old and that crusts of earth and of moon were intimately related. (Getze, *LA Times*, 7/7/69)
- *Newsweek*'s 42nd space age cover story since October 1957 contained comments from "opinion makers" on Apollo 11 mission.

    Dr. Robert Jastrow, Director of NASA's Goddard Institute for Space Studies, said scientific basis for mission was to discover secret of earth's past through study of lunar rocks. Resolution of "cold moon" versus "hot moon" theory controversy—whether moon was formed cold, or cooled off shortly after its birth, or whether it was like earth molten or partly molten inside with volcanic surface—"may ride with Apollo 11."

Southern Christian Leadership Conference President, the Rev. Ralph D. Abernathy, said: "A society that can resolve to conquer space; to put man in a place where in ages past it was considered only God could reach; to appropriate vast billions; to systematically set about to discover the necessary scientific knowledge; that society deserves both acclaim and our contempt . . . acclaim for achievement and contempt for bizarre social values. For though it has the capacity to meet extraordinary challenges, it has failed to use its ability to rid itself of the scourges of racism, poverty and war, all of which were brutally scarring the nation even as it mobilized for the assault on the solar system."

Anthropologist Dr. Margaret Mead said: "This can be a first step, not into space alone, but into the disciplined and courageous use of enhanced human powers for man, ennobled as he is today, as the first men step on the moon."

Philosopher Lewis Mumford said: "Space exploration . . . is strictly a military by-product; and without pressure from the Pentagon and the Kremlin it would never have found a place in any national budget." Best hope was "that this colossal perversion of energy, thought and other precious human resources may awaken a spontaneous collective reaction sufficient to bring us down to earth again. Any square mile of inhabited earth has more significance for man's future than all the planets in our solar system." (*Newsweek*, 6/7/69, 3, 60-1)

- Original equipment of field-sequential color TV system which would be used by Apollo 11 to transmit pictures from moon [see April 5] was presented to Smithsonian Institution by inventor Dr. Peter C. Goldmark, President of CBS Laboratories. Apollo 11 would carry three-pound miniaturization of system in Westinghouse camera. (Schaden, W *Star*, 7/8/69, B1)
- DOD announced award of $356,713,045 fixed-price contract to McDonnell Douglas Corp. for F-4 Phantom II high-performance jet fighter aircraft for USN and USAF use. (DOD Release 568-69)

July 7-8: NASA terminated *Biosatellite III* mission to determine long-term effects of weightlessness on living organisms when Bonny, pig-tailed monkey on board, registered extremely low metabolic state and refused to drink water after receiving 10 emergency water commands. Spacecraft had been scheduled to remain in orbit 30 days after launch June 29, but monkey's condition—as indicated by steadily lowering body temperature, reduced heart rate, shallow breathing, substantial periods of sleep during day, and general sluggishness—had declined steadily for several days.

Spacecraft separated and parachute deployed successfully, but recovery aircraft was unable to retrieve spacecraft in midair as planned because of clouds and rainstorms. Capsule was recovered from Pacific off coast of Kauai at 7:36 pm EDT, minutes after splashdown, and flown to Hickam AFB, Hawaii, laboratories, where monkey was removed from capsule immediately and given intensive care. Without prior warning from changes in physiological parameters being recorded, Bonny died suddenly at 6:04 am EDT July 8. Detailed analyses of data would be made during next six months and formal report would be issued after Jan. 1, 1970. Despite curtailed mission, experimenters expected significant information. (NASA Proj Off; ARC *Astrogram,* 7/17/69, 1)

July 8: Rep. John V. Tunney (D-Calif.) introduced H.J.R. 810, "designating the day which man lands on the moon, and the anniversary of that day each year thereafter as a national holiday to be known as 'Space Exploration Day.'" Resolution was referred to House Judiciary Committee. (*CR,* 7/8/69, H5725)

- CBS Enterprises Inc. announced first agreement for regular satellite transmission of news stories had been reached with CBS Newsfilm subscribers in Australia and Japan. It would eventually lead to daily, instantaneous, intercontinental transmission of TV news by satellite, company said. (CBS Enterprises Release, 7/8/69)

July 9: Apollo 8 Astronaut Frank Borman met in Moscow with Mstislav V. Keldysh, President of Soviet Academy of Sciences, and spent 40 min with U.S.S.R. President Nikolay V. Podgorny in talk which Borman said was "encouraging and beneficial" in efforts to achieve U.S.–U.S.S.R. cooperation in space. (AP, *W Post,* 7/10/69, A23)

- *Apollo Program Management: Staff Study* was submitted to House Committee on Science and Astronautics by Subcommittee on NASA Oversight. It identified key concepts contributing to successful evolution of NASA-industry management team and areas where additional studies would be useful in application of its expertise. Key factors included clear definition of primary objective, monitoring and auditing systems that allowed vertical and horizontal information flow, refinement in program-control techniques using incentive contracts, correlation and definition of multiple-program interfaces by use of systems-oriented staff groups, real-time and flexible management reporting system, and balance between governmental in-house capability and industrial capability. (Text)

- Die proof of 1.05- by 1.80-in 10-cent airmail stamp commemorating "First Man on the Moon," attached to envelope, would be carried by Apollo 11 and canceled on Moon by Astronauts Neil A. Armstrong and Edwin E. Aldrin, Jr., Postmaster General Winton M. Blount announced. On return to earth, die would be used to produce commemorative stamps for August issue. Hand-canceled "Moon Letter" would undergo 21-day decontamination period at LRL and be returned to Post Office Dept. for display in Washington, D.C., and later throughout U.S. and abroad. Stamp was designed by Paul Calle, modeled by Robert J. Jones, and engraved by Edward R. Felver and Albert Saavedra. It depicted astronaut stepping from spacecraft onto lunar surface. (PO Dept Releases 107, P-37; *W Post,* 7/10/69, A13)

- Some observers on earth might be able to see moon-bound Apollo 11 on two occasions, NASA announced. At 2:44:18 GET, exhaust plumes from S–IVB firing in parking orbit over Gilbert Islands in South Pacific would be visible to naked eye for several minutes over large part of sky. For several hours after translunar injection burn, CSM/LM, S–IVB, and four spacecraft-lunar-module-adapter (SLA) panels would be visible through telescope to observers in U.S., Mexico, Central and South America, and western Africa. (NASA Special Release)

- As NASA prepared for Apollo 11 lunar landing, space contractors, engineers, and scientists cited thousands of "space technology transfers," down-to-earth rewards from space program, *Wall Street Journal* said. Side benefits ranged from medical innovations and safer highways to new management techniques, commercial products, and industrial tools. They included liferaft with bucket keel to prevent capsizing in rough water and inner tube that inflated automatically to keep craft afloat if outer skin was punctured, computer system to track down fathers behind in child support payments, inertial navigation systems that were standard equipment on new 360- to 400-passenger Boeing 747, and thermal mapper developed for satellites, being used to seek oil formations, diagnose cause of sinking airport runways, and find sources of water pollution. Other space age spinoffs were plastic resin marketed as commercial laminates, adhesives, and coatings; devices to monitor internal stress in dams during earth tremors; data-processing techniques to record train traffic and to match power-generating capacities to demand; electromagnetic hammer that smoothed and shaped metal without weakening it; and luminous devices for aircraft exit signs, map reading, and gun sites. Medicine was benefiting from miniaturized electronic devices in cardiac pacemakers, remote-handling and manipulation equipment that had improved prosthetic devices like artificial limbs, space-helmet-like hoods to measure oxygen consumption while patient exercised, and computer to provide sharper x-ray photos. (Tanner, *WSJ*, 7/6/69)
- MSFC announced it would exercise option in existing contract with Computer Sciences Corp. to allow continued support services through June 30, 1970, at cost of $6,081,887. (MSFC Release 69–157)

July 10: Apollo 11 countdown began at KSC at 8:00 pm EDT in preparation for launch toward moon at 9:32 am EDT July 16. (Apollo 11 Status Rpt)

- U.S.S.R. launched *Cosmos CCLXXXIX* from Plesetsk into orbit with 324-km (201.3-mi) apogee, 208-km (129.2-mi) perigee, 89.6-min period, and 65.4° inclination. Satellite reentered July 15. (GSFC *SSR*, 7/15/69; *SBD*, 7/15/69, 5)
- Four-stage Pacemaker rocket launched from NASA Wallops Station carried 58-lb instrumented payload to 65,000-ft altitude and reentered atmosphere at 7,000 mph. Primary objective was to evaluate performance of carbon phenolic, synthetic resin, as ablative material. Secondary objective was to evaluate performance of low-density ablative materials—pyronne foam, polymer blend, and phenolic nylon—for possible use on manned lifting-body reentry vehicles. (WS Release 69–12; WS PIO)
- American Academy of Achievement presented 1969 Gold Plate "Man of Achievement" Awards to Dr. William H. Pickering, JPL Director, and to *Apollo 8* Astronauts William A. Anders and Frank Borman. (LA *Her-Exam*, 6/19/69; *AFJ*, 6/21/69, 30)
- Sen. Spessard L. Holland (D-Fla.), for himself and Sen. Edward J. Gurney (R-Fla.), introduced S.J.R. 133 "to redesignate the area in the state of Florida known as Cape Kennedy as Cape Canaveral." Measure

was referred to Senate Committee on Interior and Insular Affairs. (*CR, 7/10/69,* S7819)

- National Geographic Society cartographer and "backyard stargazer" David Moore was one of few amateur astronomers selected by NASA to help nearly 200 professionals who had volunteered to attempt sightings of Apollo 11 spacecraft, Washington *Evening Star* said. Through telescope in yard of his Wheaton, Md., home he would watch for "small brilliant flashes when rocket engines are turned on or 'burned' or . . . when waste water is ejected from the spacecraft." In ejection, water froze instantly and resultant ice crystals flared in sunlight. NASA had credited Moore with one of few sightings of earthbound *Apollo 10.* (Radcliffe, *W Star,* 7/10/69, D2)

- Apollo 11 and current nationwide water shortage were "two illustrations of man's efficiency in achieving the thrills of life and man's inefficiency in not achieving the necessities of life," Drew Pearson said in *Washington Post.* "At Cape Kennedy, the United States is about to launch the most carefully rehearsed, most expensive, most unnecessary project of this century by which man will reach a piece of drab, radioactive, lava-like real estate hitherto romantic because of distance—the moon. The launching will succeed because a vast amount of money and the best scientific brains in America over a period of seven years have been lavished on this moon shot. Meanwhile, up the Atlantic Coast, the Capital which voted the $20 billion to reach the moon is desperately short of the second essential to man's life—water—all because of lack of planning, lack of foresight, and lack of money—the same ingredients which have put the moon shot on the verge of success." (*W Post,* 7/10/69, F11)

July 11: Apollo 11 Astronauts Neil A. Armstrong, Michael Collins, and Edwin E. Aldrin, Jr., underwent last major preflight medical examination at KSC and were cleared for July 16 launch. (Apollo Status Rpt; UPI, *W Post,* 7/12/69, A4)

- At Cape Kennedy press conference, *Apollo 8* Astronaut Frank Borman termed "totally ridiculous" cancellation of President Nixon's dinner with Apollo 11 crew on eve of launch to avoid contaminating crew with presidential germs. He had delivered invitation to dinner to White House and President Nixon had accepted when MSC Medical Director, Dr. Charles A. Berry, criticized dinner [see July 7].

  Borman said his talks with U.S.S.R. officials during recent tour had left him convinced they planned lunar landing soon but, "from the people on the subways to their president, all I heard was that they are wishing success for Apollo 11." (Greider, *W Post,* 7/13/69, A4)

- NAS published *Plan for U.S. Participation in the Global Atmospheric Research Program.* It recommended five-year effort including Pacific test of global weather observing system and large-scale atmospheric study in 1973, series of small regional studies beginning in 1969 or 1970, and experiments to improve numerical models of atmosphere for computer forecasting, with continued development of computer 100 times faster than currently available. Total effort would require 10 yrs, with plans for second portion to be based on information gained during first 5 yrs.

  Report, prepared by NRC committee, said developments in computers and satellites had made it possible "to advance toward the goal of accurate two-week forecasts and, eventually, toward intelligent modification of the weather." Use of satellites such as *Nimbus III*, launched April 14, and expansion of other observing systems made it technically and economically possible to provide adequate global observations for long-range forecasts. Recommended test of global observing system would require 2 satellites with advanced instrumentation, nearly 1,000 balloons, and 135 instrumented buoys. Simultaneous cloud-cluster study would require 12 additional aircraft, several ground stations, and computer facility. Participating Government agencies would be responsible for determining program costs. (Text; NRC Release)

- NASA and USAF announced cooperative flight test program using two USAF YF-12A aircraft and spares, ground equipment, maintenance personnel, and base support at Edwards AFB, Calif. NASA would budget for and fund $10 million for program through FY 1974. About $4 million had been made available by completion of X-15 and XB-70 flight programs. USAF purpose in two-part program was to gather data on aircraft operational factors, procedures, limitations, and possible bomber penetration tactics. NASA would seek data on altitude-hold at supersonic speeds, boundary layer noise, heat transfer under high speed, airframe-propulsion system interactions, and other characteristics. (DOD Release 581-69)

- MSFC issued RFPs for design, development, test, and delivery of four flight models of manned lunar roving vehicle. Four-wheeled, 400-lb vehicle would be carried to moon on board LM in 1971, to transport astronauts, tools, lunar samples, and other equipment and experiments. (NASA Special Release)

- Emperor Haile Selassie of Ethiopia toured MSC during five-day visit to U.S. (Reuters, *W Post*, 7/10/69, A23; Apollo Status Rpt; NASA PAO)

July 12: NASA program of returning man from lunar landing was based on conclusion there was no risk, Stanford Univ. geneticist Dr. Joshua Lederberg said in *Washington Post*. "We could not mount an effective quarantine against a real peril of global infection unless we were prepared to sacrifice the astronaut, which is unthinkable." Arguments for zero risk were "quite persuasive"—lack of atmosphere on moon, "an absolutely necessary condition for life to flourish," and fact that earth had experienced lunar material samples from secondary meteorites. Main purpose of quarantine was "to protect the samples from earthly contamination—not altogether successfully, in view of the exhalations from the landing rocket and from the astronauts' space suits. It was then reasonable to add on whatever additional precautions against back-contamination were possible without impeding the mission." Project had helped show lunar arrangements would be "quite inappropriate to a real risk, for example a sample return from Mars." For Mars program, "we must learn a great deal more by instrumented observations left there, before we can begin to design the precautions needed for samples, or men, returned to earth." (*W Post*, 7/12/69, A15)

- "Poor People's Campaign" Director, the Rev. Hosea Williams of Southern Christian Leadership Conference, said "hungry" people from five southern states would demonstrate at Cape Kennedy July 15 on eve of Apollo 11 launch and would try to get "as close as possible" to launch site with mules and wagons. "We're not against things like the space shot, but there's been a miscalculation in priorities." (Reuters, *W Post*, 7/13/69, A5)

- NASA said Soviet Ambassador Anatoly F. Dobrynin had rejected U.S. invitation to watch Apollo 11 launch. U.S.S.R. originally had accepted, but Soviet Embassy in Washington said Dobrynin would be out of the country. (AP, *W Star*, 7/13/69, A9)
- USN reported eight-ship Soviet Naval fleet was heading south 25 mi east-southeast of Miami, Fla., on course that could provide view of Apollo 11 launch. U.S. carrier aircraft and destroyer escort *Gary* shadowed squadron, officially en route to Havana for July 26 commemoration of Cuban revolution. (Homan, *W Post*, 7/12/69, A1; AP, *W Post*, 7/13/69, A5)

July 13: U.S.S.R. launched *Luna XV* unmanned spacecraft from Baikonur into selenocentric orbit to conduct "further scientific studies of the moon and near lunar space," Tass announced. Spacecraft was expected to reach moon late July 16—scheduled date of launch of NASA's Apollo 11 manned lunar landing mission. There was speculation that *Luna XV* was Soviet attempt to land spacecraft on moon and return it to earth with sample of lunar soil before U.S. landed. (*W Post*, 7/14/69, A1; *SBD*, 7/18/69, 22; *B Sun*, 7/14/69, A1; GSFC *SSR*, 7/15/69)

- Washington *Sunday Star* published Associated Press interview with Dr. Charles A. Berry, MSC Director of Medical Research and Operations: While 4,514 hrs of weightlessness endured by U.S. astronauts in space had produced no serious medical problems, on moon "we will be placing men in an entirely new environment." After four days of weightlessness, they would step onto surface where gravity field was one-sixth that of earth.

  At Mission Control Center in Houston, Dr. Berry would be watching Apollo 11 astronauts' heart rate, oxygen consumption, and temperature of water that cooled spacesuits. Preflight physicals had enabled doctors to draw metabolic profile of each astronaut, including work capacity on earth at various heart rates, oxygen consumption, and body heat generated. "We know the heat production level which the portable life support system can handle without being overburdened. If it reaches that point for five minutes, we will tell the astronauts to stop and rest." Because of spacesuits' bulk astronauts would start with simple tasks and work up to tougher ones. Excitement could affect ability to sleep in four-hour rest period planned before lunar walk. "We might have to make a real-time decision on whether to give them a sleeping pill or perhaps a stimulant."

  Apollo had taught one "amazing medical fact—that the loss of red-blood-cell mass apparently is caused by a pure oxygen atmosphere." Results of using mixed nitrogen-and-oxygen atmosphere in spacecraft since January 1967 Apollo fire had indicated nitrogen apparently protected cells. (Benedict, AP, *W Star*, 7/13/69, A9)
- From summer residence, Castel Gondolfo, Pope Paul VI asked Christians worldwide to pray for Apollo 11 astronauts and said mission showed man was a "giant." (AP, *B Sun*, 7/14/69, A5)

  At White House religious service the Rev. Paul H. A. Noren of Mount Olivet Church in Minneapolis led 300 people in prayer: "We

ask Thy divine protection for our space pioneers who will soon make footprints on the moon." (AP, *NYT*, 7/14/69, 23)
- *New York Times* editorial: "This is the week of the moon. The countdown is on at Cape Kennedy and, if all goes well, a week from today a manned vehicle will for the first time alight on another celestial body. . . . all mankind will share in the exhilaration of discovery. Ever since man evolved he has been exploring, extending his domain over all parts of his planet. Now that insatiable curiosity is bursting its terrestrial bounds to provide our first personal knowledge of the nearest neighbor in the cosmos. It is an inspiring adventure, a testimony not only to man's imagination in amassing knowledge of nature, but to his courage, his perseverance and his indomitable spirit." (*NYT*, 7/13/69)
- In Washington *Evening Star* William Hines said: ". . . Space Administrator Thomas O. Paine was dead right when he acclaimed Project Apollo as 'a triumph of the squares.'" While word "square" was in disrepute, "you will find no umbrage taken by the clean-cut stars of this week's cosmic drama if you called them squares. They are, and probably proud of it. There was no fight from Neil Armstrong when Congress told him to plant an American flag on the surface of the moon. . . . The Apollo program is not only run by squares, but for squares, as well; its thrills and glories appeal to the vast majority of Americans who, at the bottom, are just as square as any Armstrong on Earth—Jack or Neil or any other." (W *Star*, 7/13/69, D2)
- Wing of Lockheed C–5A static test specimen cracked during stress tests at point below aircraft's contract specifications but above its design limit. USAF later said cause of crack was overloading of wing area where spar attached to lower rear beam cap; it would not require extensive redesign. It was first major performance failure reported for C–5A. Contractor was planning modification and retesting of static specimen. (USAF Memo 8/18/69; *W Post*, 7/15/69, A2)

July 14: Apollo 11 Astronauts Neil A. Armstrong, Michael Collins, and Edwin E. Aldrin, Jr., appeared in nationally televised press conference. Interview with four newsmen was conducted over closed-circuit TV, with astronauts at KSC and press 15 mi away. TV cameramen allowed in auditorium with crew had undergone thorough medical examinations. Armstrong, mission commander, said that after decade of planning and hard work astronauts were "willing and ready to attempt to achieve our national goal. This is possible because very many Americans across the nation have dedicated themselves to quality craftsmanship and ingenuity."

In response to question on astronauts' attitudes toward mission, Armstrong said fear was not unknown, but added: "Fear is characteristic particularly of a knowledge that there may be something that you haven't thought of and feel that you might be unable to cope with. I think our training and all the work that goes into the preparation for a flight does everything it can towards erasing those kinds of possibilities and I would say that as a crew we . . . have no fear of launching out on this expedition." (Wilford, *NYT*, 7/15/69, 1, 20)
- Chances of U.S.S.R.'s *Luna XV* successfully returning to earth with lunar sample were small because of complexity of operations required, NASA Associate Administrator for Manned Space Flight, Dr. George E.

Mueller, told KSC Center Directors' Briefing. Landing, deploying equipment, collecting and storing samples, and then lifting off "are not simple things to do . . . and doing it remotely is more difficult than doing it with men in space. I don't think by any means impossible, but . . . the chances of being able to carry it out on the first mission are relatively low compared to the kind of probability that we would associate with our own landings."

If *Luna XV* were able to successfully retrieve lunar sample, feat would be "significant technological step and one that represents a considerable degree of prestige," he said, but "each country [U.S. and U.S.S.R.] will obtain its proper share of credit. . . . The first sample returned if it were possible to do so and the first man landing on the moon are significant events, each in their own right, and are to be treated as such." (Transcript)

- IAF announced official endorsement of absolute world's records for Dec. 21–27, 1968, *Apollo 8* mission's 10 lunar orbits: altitude, 234,672.5 mi; greatest mass lifted into earth orbit, 282,197 lbs; total time in space for an astronaut, James A. Lovell, Jr., 572 hrs 10 min 16 secs. *Apollo 8* world class records: duration of lunar mission, 146 hrs 59 min 49 secs; duration in lunar orbit, 20 hrs 14 min 13.2 secs.

  To obtain IAF certification of Apollo 11, crew would be given torn halves of four $1 bills for comparison with other halves on return as proof same men returned as took off. NAA would submit claim for absolute world record for extravehicular activity (EVA) for successful Apollo 11 mission. (AP, *NYT*, 7/15/69, 20)

- *New York Post* published results of Louis Harris poll which showed American people favored manned lunar landing by 51% to 41%. In February poll public opinion had been opposed by 49% to 39%. Harris attributed change to feeling "if we have gone this far, we ought to finish the job." He said 56% of 1,607 adults polled from June 16 to 22 were opposed to annual $4-billion outlay for space program, while 37% favored it—little change from 55% for to 34% against in February. Reaction to *Apollo 10* flight had been generally favorable. (*NYT*, 7/15/69, 20)

- Expectation of one million tourists to witness Apollo 11 launch had led to extraordinary precautions at Cocoa Beach, Fla., *Washington Post* said. Tank truck would be stationed at City Hall to fuel police cars; airboats would stand by to rush casualties to hospital if ambulances could not penetrate automobile traffic; and officials were concerned about scores of aircraft circling overhead to glimpse spacecraft. (Greider, *W Post*, 7/14/69, A1)

- Washington *Evening Star* special supplement, "Voyage to the Moon," commented: "Hanging in the sky, attracting man's attention for untold generations, the moon has been the reputed home of gods and goddesses of all religions, primitive and modern. If all these deities lived there at any one time, the reasons for its battle-scarred appearance would be obvious. But assuming that none did . . . that pock-marked face still poses more questions than it answers." (*W Star*, 7/14/69)

- Aerospace Systems Laboratory had been established at Princeton Univ. to investigate U.S. space program and other broad areas of applied research, including transportation systems, *New York Times* said. Project was assisted by NASA and other Federal agency grants. (*NYT*, 7/14/69, 23)

- NASA announced availability of 16-in-dia globe of moon prepared by USAF Aeronautical Chart and Information Center from NASA photos made by Lunar Orbiter series. Lovell Observatory, Ariz., prepared art work with exaggerated color tones showing lunar landscape bathed in morning sunlight and large Ring Plains, or explosive craters, on far side. One globe had been presented to President Nixon by *Apollo 10* astronauts. (NASA Release 69–83G)
- Harold W. Adams, Deputy to Vice President-Chief Engineer of Douglas Aircraft Co., received AIAA Aircraft Design Award of citation and $500 honorarium at AIAA Aircraft Design and Operations Meeting in Los Angeles. Citation read: "In recognition of your outstanding contributions to the safety and economic practicality of commercial air transportation during the past 38 years by development of aircraft design principles for high reliability and ease of maintenance." Adams was specialist in electric and hydraulic systems. (AIAA Release, 7/9/69)
- Oceanographer Jacques Piccard cast off in 48-ft research submarine from West Palm Beach, Fla., for rehearsal of 1,500-mi Gulf Stream Drift to study ocean depths [see April 7]. If four- to five-day trial run was successful, team would remain submerged for 30 days and drift to Boston. (UPI, *W Star*, 7/14/69, A10)

July 15: President Nixon sent telegram to Apollo 11 astronauts: "On the eve of your epic mission, I want you to know that my hopes and my prayers—and those of all Americans—go with you. Years of study and planning and experiment and hard work on the part of thousands have led to this unique moment in the story of mankind; it is now your moment and from the depths of your minds and hearts and spirits will come the triumph all men will share. I look forward to greeting you on your return. Until then, know that all that is best in the spirit of mankind will be with you during your mission and when you return to earth."

President also telephoned astronauts: ". . . as you lift off to the moon, you lift the spirits of the American people as well as the world. . . . You carry with you a feeling of good will in this greatest adventure man has ever taken. . . ." (*PD*, 7/21/69, 997)

- First notables to arrive at Cape Kennedy on eve of Apollo 11 launch included former President and Mrs. Lyndon B. Johnson and Southern Christian Leadership Conference President, the Rev. Ralph D. Abernathy. Johnsons arrived in military aircraft assigned by President Nixon, to attend luncheon honoring James E. Webb, former NASA Administrator.

Abernathy led 25 poor southern families to protest Federal funding priorities. Dr. Thomas O. Paine, NASA Administrator, met group of 150 poor people outside KSC gate where Abernathy requested 40 VIP passes to launch, asked Dr. Paine to join fight against poverty, and urged that NASA technology be converted to finding new ways to feed poor. Dr. Paine agreed to admit members of group to launch and pledged to do what he could to adapt space-developed food concentrates to aid undernourished. "It will be a lot harder to solve the problems of hunger and poverty than it is to send men to the moon." But, "if it were possible for us not to push that button tomorrow and solve the problems you are talking about, we would not push the button." He said space program and science could be used to help solve poverty problems. "I want you to hitch your wagon to our rocket and tell the people the NASA program is an example of what this country can do." The poor people said they would pray for Apollo 11 astronauts.

By evening 500,000 tourists had arrived in Brevard County, site of KSC, with total one million expected by early morning. Air traffic had quadrupled, with 10 local airfields handling over 1,200 small aircraft, and 200 private jets. Aircraft were to bring Vice President Spiro T. Agnew, over 200 Congressmen, 60 ambassadors, 19 governors, 40 mayors, and other public figures July 16. More than 1,000 police struggled to control road traffic, and hordes settled to sleep on beaches from which they could see illuminated spacecraft on launch pad. (Weinraub, *NYT*, 7/16/69, 22; Greider, *W Post*, 7/16/69)

- Proximity of probable date of lunar landing to date of arrival of *Mariner VI* and *VII* cameras near Mars surface would provide U.S. TV viewers with "double space feature," NASA said. Gerald M. Truszynski, NASA Associate Administrator for Tracking and Data Acquisition, credited feat to advances in electronics through which streams of signals could be returned from moon and from Mars into tracking centers and switching points on earth, thence via comsats into TV networks throughout globe. Apollo 11 mission would include eight color telecasts from spacecraft. Lunar telecasts would be black and white since LM would lack power for color TV. Mars telecasts from *Mariner VI* would produce 50 photos; *Mariner VII* would deliver 91. Best resolution from closeup would be 900 ft; it had been 2 mi in 1965 *Mariner IV* photos and was 100 mi by best optical means from earth. (NASA Release 69-83I)

- Europeans were "as excited as many Americans" about Apollo 11 launch, *New York Times* reported. But "only the sharpest observer of the Soviet news media could guess, as he went to bed tonight, that Americans will try to send men to the moon tomorrow," according to Baltimore *Sun*. Last mention of Apollo 11 in Soviet press had been July 9 meeting of President Nikolay V. Podgorny with Astronaut Frank Borman.

  In U.K., BBC and commercial TV were planning extensive Apollo 11 coverage, some live via comsat. British newspapers were competing with special space supplements and guides. Exceptions to generally "adulatory" reportage was *The Times* of London article in which philosopher Lord Russell had said: "Men will not be content to land upon the moon and try to make it habitable. They will land simultaneously from Russia and the United States, each party, complete with H-bombs and each intent upon exterminating the other."

  American Embassy in Warsaw was packed every day with Poles viewing space films. Spain's *Evening Daily Pueblo* had sponsored contest to send 25 readers to Apollo 11 launch. In France 22-page space supplement issued by *France-Soir* had sold 1.5 million copies at $1 each. *Bild Zeitung* in Germany had noted 7 out of 57 Apollo supervisors were of German origin. Austrian press had lionized Dr. Wernher von Braun during recent visit to Salzburg.

  Volume of Western European newspaper space devoted to lunar

landing mission rivaled that in U.S., *New York Times* said, and "the whole story of the moon effort is improving the 'prestige' of the United States. . . . But . . . respect voiced by individuals is often for America's technological power, not her humanity or civilization." (Lewis, *NYT*, 7/16/69, 20; B *Sun*, 7/16/69, A8, Mills, A9)

- Across U.S. on eve of Apollo 11 launch, newspaper editorials commented on lunar landing mission:

    Los Angeles *Herald-Examiner:* "It is with an almost breathless sense of awe that we await tomorrow's blast-off from Cape Kennedy—the launching of three space explorers on the most ambitious and fearsome adventure in all human history. Mere words cannot capture the immensity of the flight of Apollo 11. Quite literally, man will be attempting a final break of the chains which have bound him to this earth." (LA *Her-Exam*, 7/15/69)

    Newport News, Va., *Times-Herald:* "Now, this triumph of human courage and knowledge stands poised on the threshold of accomplishment. For a few fleeting moments, the attention of the world will follow the Eagle as it ferries its two astronauts toward a destiny until now only dreamed of in our history. Then, most probably, our attention will filter back to the pressing problems on earth." (*Times-Herald*, 7/15/69)

    Milwaukee Journal: "Apollo 11 is providing insight into the meaning of life and the imperatives of human society. It is forcing us to face the grim paradox of exploiting human reason and the marvels of machinery to soar into the majesty of space while the world becomes fragmented into selfish national sovereignties—some armed, some arming, with the hideous capacity to end life itself." (*Milwaukee Journal*, 7/15/69)

    Denver Post: "The Soviet attempt to send an unmanned spaceship to the moon in advance of Apollo 11 is a bold bid to draw attention to Soviet space prowess. But even if it succeeds . . . in mechanically scooping up samples of the moon and returning to earth, the Soviet project will not overshadow the American mission. Instead, the Soviet flight will serve to underscore the expensive duplication of effort created by the space race. If the Russians and Americans had cooperated, rather than competed, the risks and the costs involved in landing a man on the moon would have been far less." (*Denver Post*, 7/15/69)

    Washington *Evening Star:* Soviet *Luna XV* seemed strangely timed. During their Moscow discussions on space cooperation, U.S.S.R. President Nikolay V. Podgorny had not given Astronaut Frank Borman "slightest hint that the Kremlin was planning to send an unmanned spacecraft to the moon to coincide with the history-making Apollo 11 American mission." Was it really possible "to work together in space exploration with a country that seems to be playing tricks with ours at a moment when we are engaged in a historic effort to land men on the moon?" (W *Star*, 7/15/69, A12)

- San Francisco Mayor Joseph Alioto urged San Franciscans to fly U.S. flag from Apollo 11 blastoff to splashdown and to sound every bell, siren, and whistle in the city at splashdown. (AP, *W Post*, 7/17/69, A27)

- Nike-Tomahawk sounding rocket launched by NASA from Wallops Station carried Univ. of Wisconsin payload to 129.9-mi (209-km) altitude to examine auroral directions and intensities of isotopic component of cosmic x-rays. Magnetometer and startracker functioned as planned but doors and covers shielding proportional counters failed to eject and no x-ray data were received. (NASA Rpt SRL)

- NASA awarded General Electric Co.'s Aircraft Engine Group $18.7-million, fixed-price contract with performance-award provision to construct and test two experimental quiet jet aircraft engines. To cut development costs, CF–6 and TF–39 engines developed for DC–10 and C–5A aircraft would be used as core of new engine. Engines would produce 4,900-lb thrust at cruise and 22,000-lb thrust for takeoff. Work was part of OART's Quiet Engine Research Program to develop turbofan engine with noise level 15–20 db below present engines. Contract would be managed by LeRC. (NASA Release 69–103)

July 16–24: Apollo 11 (AS–506) manned lunar landing mission flown by NASA achieved eight-year national goal set by President Kennedy May 25, 1961. On July 20, spacecraft's LM–5, *Eagle,* landed on lunar surface and first man stepped out onto moon. Two astronauts performed assigned tasks on lunar surface before reentering LM to lift off from moon, redock with CSM–107, *Columbia,* and return safely to earth.

July 16–19: Mission began at 9:32 am EDT July 16, when spacecraft was launched from KSC Launch Complex 39, Pad A, by Saturn V 506 booster. Liftoff was relayed live on TV to 33 countries on 6 continents, watched by estimated 25 million TV viewers in U.S., and heard on radio by millions of listeners. Launch events occurred as planned and spacecraft carried Astronauts Neil A. Armstrong (commander), Michael Collins (CM pilot), and Edwin E. Aldrin, Jr. (LM pilot), into circular parking orbit with 118.5-mi (190.7-km) altitude. After post-insertion checkout CSM separated from Saturn V 3rd stage (S–IVB) and LM *Eagle.* Crew successfully transposed CSM and docked with LM, ejected CSM/LM from S–IVB, and conducted first SPS burn. Successful propellant dump provided impulse to S–IVB for slingshot maneuver to earth-escape velocity. Translunar injection maneuver was so accurate that first midcourse correction was not required. Midcourse correction No. 2, at 26:45 GET, was so accurate that third and fourth maneuvers were not necessary.

Crew conducted two unscheduled color TV broadcasts—for 16 min beginning at 10:32 GET (taped for 11:26 GET transmission) and for 50 min beginning at 30:28 GET—and one scheduled 36-min transmission beginning at 33:59 GET. Broadcasts were very clear and showed earth, onboard computer keyboard, and crew. At 55:08 GET (4:40 pm EDT July 18) crew began 96-min color TV transmission with excellent picture resolution, coverage, and general quality. Viewers in North America, South America, Japan, and Western Europe saw live pictures of CSM and LM interiors, CSM exterior, and earth and watched crew removing probe and drogue, opening spacecraft tunnel hatch, preparing food, and housekeeping LM.

Apollo 11 passed into moon's sphere of influence at 61:40 GET, 214,546.8 mi (345,205.8 km) from earth, traveling at 2,990 fps relative to earth. Spacecraft entered lunar orbit with 194.3-mi (312.6-km) apolune and 70.5-mi (113.4-km) perilune at 75:56 GET (1:28 pm EDT July 19) after first SPS burn. During second lunar orbit, live color TV transmission showed spectacular views of lunar surface and ap-

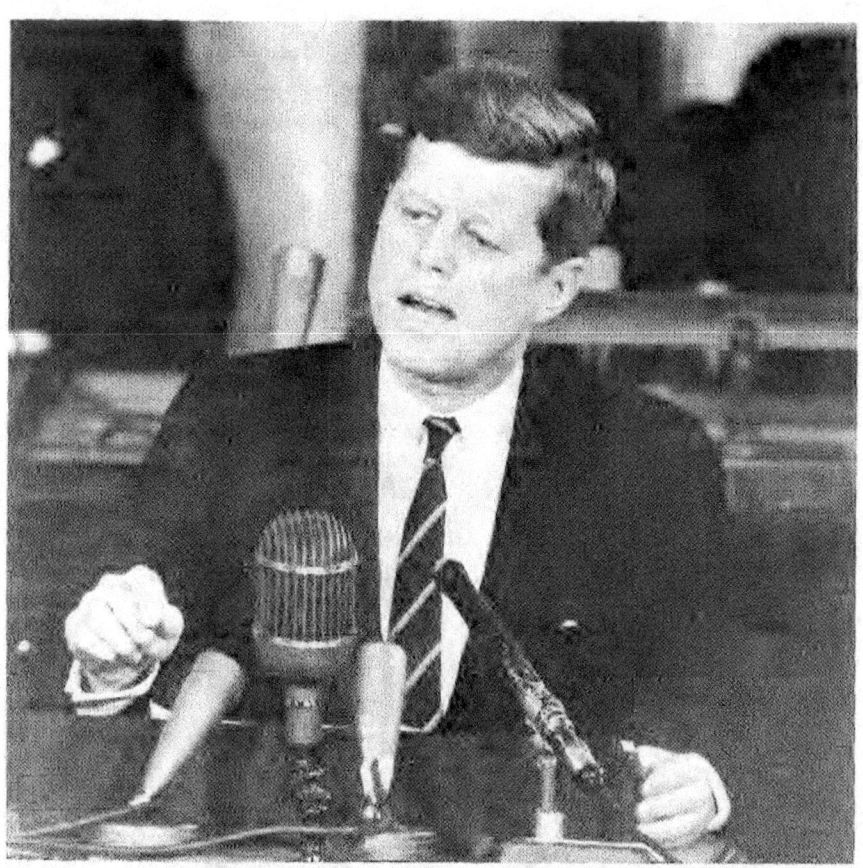

July 16–24: Apollo 11's *successful manned lunar mission achieved the national goal set eight years earlier by President John F. Kennedy, who told Congress May 25, 1961: "I believe this nation should commit itself to achieving the goal, before this decade is out, of landing a man on the Moon and returning him safely to Earth."*

proach path to Site 2. Armstrong pointed out lunar landmarks and described unexplained glow near crater Aristarchus which some scientists believed to be volcanic activity. Second SPS burn circularized orbit with 75.6-mi (121.6-km) apolune and 61.9-mi (99.6-km) perilune. Aldrin transferred to LM for two-hour housekeeping, voice and telemetry test, and oxygen-purge-system check.

July 20–21: Armstrong and Aldrin reentered LM at 95:20 GET and checked out all systems before firing SM reaction-control-system thrusters to separate CSM and LM on far side of moon. LM descent-propulsion-system engine propelled LM to within 9.8 mi (15.8 km) of lunar surface. Because LM-powered descent maneuver—initiated at perilune of descent orbit—was about 4.6 mi (7.4 km) downrange from planned position, landing point was also shifted downrange. During final approach phase, crew noted that landing point to which spacecraft

July 16: Saturn V (left) thrust Apollo 11 spacecraft toward the moon, watched by former NASA Administrator James E. Webb (below at left), former President Lyndon B. Johnson, and Vice President Spiro T. Agnew at Kennedy Space Center. In KSC Launch Control Center (above) mission officials relaxed after launch; left to right were Charles W. Matthews, NASA Deputy Associate Administrator for Manned Space Flight; Dr. Wernher von Braun, Director of MSFC; Dr. George E. Mueller, Associate Administrator for Manned Space Flight; and L/G Samuel C. Phillips, Apollo Program Director.

was heading was in center of large, rugged crater with 5- to 10-ft-dia boulders. Consequently crew flew past crater to more suitable touchdown point by controlling attitude manually and adjusting descent rate and horizontal velocity. Officials later attributed change in course to malfunction in onboard radar and subsequent critical overload of computer, which caused warning alarms and could have aborted mission.

LM landed on moon at 102:46 GET (4:18 pm EDT July 20) in Sea of Tranquility, 20,800 ft west and 4,000–5,000 ft south of center of planned landing ellipse. Landing-point coordinates were approximately 23.5°E and 0.64°N and site altitude was approximately 8,600 ft below moon's mean radius.

Armstrong reported: "Houston, Tranquility Base here—the Eagle has landed."

Mission Control replied: "Roger, Tranquility. We copy you on the

ground. You got a bunch of guys about to turn blue. We are breathing again. Thanks a lot."

Armstrong said landing area contained numerous boulders up to two feet in diameter, some apparently fractured by LM engine exhaust, and surface color varied from very light to dark gray. Crew immediately adapted to one-sixth (earth) gravity in LM and moved with ease. About two hours after landing astronauts requested and were granted permission to perform extravehicular activities (EVA) on moon's surface before sleep period—about 4½ hrs earlier than originally scheduled.

After postlanding checks, Armstrong opened LM hatch, descended LM ladder, and deployed modularized equipment stowage assembly (MESA) containing camera, which recorded his descent to lunar surface. Aldrin remained inside LM and recorded Armstrong's descent with 16-mm Maurer camera.

Armstrong took man's first step on moon at 109:24 GET (10:56 pm EDT July 20). Some 600 million viewers on earth—one-fifth of world population—watched live TV transmission and heard him describe feat as "one small step for a man—one giant leap for mankind."

Collins, orbiting moon alone in CSM *Columbia*, was unable to see

July 20–21: "One small step for a man—one giant leap for mankind." Astronaut Neil A. Armstrong (top left) descended the LM ladder to set the first foot on surface of the moon. Astronaut Edwin E. Aldrin, Jr. (bottom left), photographed by Armstrong, deployed the passive seismic experiments package, with the laser ranging retroreflector and the Eagle in the background. The lunar surface TV camera was in the far left. Below, the flag of the United States remained on the surface of the moon.

landing and subsequent EVA because CSM was not equipped to receive TV transmission. Armstrong said moon had "stark beauty all its own" like desert in southwestern U.S. Lunar surface was "fine and powdery" and could be kicked up loosely. "It adheres like powdered charcoal . . . but I only go in a small fraction of an inch. I can see my footprint in the moon like fine grainy particles." Armstrong checked LM exterior and reported penetration of LM footpads into lunar surface was three to four inches and that strut collapse was minimal. Blast of descent engine had not formed crater in surface and about one foot of clearance remained between engine bell and lunar surface. He reported only problem was seeing his footing in darkness of LM shadow. He emplaced microdot containing messages from world leaders, collected contingency sample of lunar soil near LM ladder, and reported that, although surface consisted of soft loose material, material six or eight inches under surface was very hard and cohesive. Rocks were very slippery, apparently because vesicles (pores) were filled with powdery surface material.

Armstrong photographed Aldrin's descent to lunar surface at 11:15 pm EDT and astronauts unveiled plaque mounted on strut behind ladder and read its inscription to their worldwide TV audience: "Here men from the planet earth first set foot on the moon July 1969, A.D. We came in peace for all mankind." Armstrong then removed

TV camera from MESA, obtained panorama, and placed camera on tripod 40 ft from LM to view subsequent EVA. Aldrin experimented to assess mobility on moon by walking, running, leaping, and doing two-footed kangaroo hops between LM and camera. He indicated some difficulty in maintaining balance but said that his agility was better than expected and that he was able to move with great ease. Mass of backpack affected inertia and caused "slight tendency . . . to tip backwards. If I'm about to lose my balance in one direction, recovery is quite natural and easy. You've just got to be careful landing in the direction you want to go in."

Aldrin deployed solar wind composition experiment in sunlight north of LM and joined Armstrong in erecting three- by five-foot American flag on eight-foot aluminum staff. Astronauts saluted flag and then talked by radiotelephone with President Nixon at White House in what President called "most historic telephone call ever made from the White House." President said: "Because of what you have done the heavens have become a part of man's world. As you talk to us from the Sea of Tranquility, it inspires us to redouble our efforts to bring peace and tranquility to earth. For one priceless moment in

the whole history of man all the people on this earth are truly one—one in their pride in what you have done and one in our prayers that you will return safely to earth." Astronauts saluted President and expressed honor and privilege of representing U.S. and world on moon.

Continuing EVA, Armstrong collected bulk sample of assorted surface material and selected rock chunks, inspected LM, deployed passive seismic experiment package and laser ranging retroreflector, and collected two core samples and 20 lbs of discretely selected material. Throughout EVA continuous black-and-white coverage of crew activity provided live documentation, with telemetered data and voice comments. Lunar surface photography included both still and sequence coverage using Hasselblad, Maurer data-acquisition, and Apollo lunar surface close-up cameras. Astronauts completed EVA, transferred film and samples to LM, reentered LM and jettisoned equipment according to plan, closing hatch by 111:39 GET (1:11 am EDT July 21). Armstrong and Aldrin rested inside LM seven hours and checked out systems.

At 124:22 GET (1:54 pm July 21) LM successfully lifted off moon, after 21 hrs 36 min on lunar surface. All lunar ascent and rendezvous maneuvers were nominal. *Eagle* reported to Mission Control: "*Eagle*

July 24: *Welcome back to earth:* Apollo 11 *Astronauts Neil A. Armstrong, Michael Collins, and Edwin E. Aldrin, Jr. (left to right inside the mobile quarantine facility), were greeted by President Nixon aboard the U.S.S. Hornet after their splashdown.*

is back in orbit, having left Tranquility Base, and leaving behind a replica from our *Apollo 11* patch with an olive branch." LM docked with CSM at 128:03 GET. Crew transferred with samples and film to CSM, and LM ascent stage was jettisoned into lunar orbit. SM reaction-control-system separation maneuver placed CSM into orbit with 72.0 mi (115.9-km) apolune and 63.0-mi (101.4-km) perilune, where crew rested and prepared for return to earth.

July 22–24: Crew fired SPS engine at 135:24 GET (12:55 am EDT July 22), injecting CSM into transearth trajectory after 59 hrs 28 min (30 revolutions) in lunar orbit. Midcourse correction No. 5, at 150:30 GET, was so accurate that sixth and seventh corrections were not necessary. During 18-min color TV transmission, crew demonstrated weightlessness of food and water and showed scenes of moon and earth. Aldrin showed in-space preparation of ham sandwich and Collins showed viewers how to drink water from teaspoon and from water gun. Final, 12½-min broadcast at 177:32 GET sent message of appreciation from each astronaut to all who helped make *Apollo 11* mission possible.

CM separated from SM on schedule at 194:49 GET. Because of deteriorating weather in nominal landing area, splashdown point was moved 247.4 mi downrange, where weather was excellent. Parachute deployment and other reentry events occurred as planned.

Apollo 11 splashed down in mid-Pacific at 12:51 pm EDT July 24, 15 mi from recovery ship U.S.S. *Hornet*, 195 hrs 19 min after launch. Swimmers attached flotation collar and seven-man raft to spacecraft and helicopter dropped biological isolation garments, which were donned by astronauts inside CM and by one swimmer. Two other swimmers moved upwind of CM on another large raft. Postlanding ventilation was turned off and CM powered down, and astronauts climbed out and helped swimmer close hatch. Swimmer then decontaminated all garments, hatch area, flotation collar, and area around postlanding vent valves with Betadine disinfectant. Helicopter carried astronauts to recovery ship, where they entered 32-ft-long mobile quarantine facility (MQF) with recovery physician and technician. They were congratulated by President Nixon and Dr. Thomas O. Paine, NASA Administrator, who were on board ship. Crew, physician, and technician remained inside MQF until it was delivered to Lunar Receiving Laboratory (LRL) in Houston July 27.

CM was retrieved, placed in dolly on board recovery ship, moved to MQF, and mated to transfer tunnel. From inside MQF/CM containment envelope, MQF engineer removed lunar samples and equipment through decontamination lock and CM was sealed until delivery to LRL. Sample return containers, film, and other data were flown to Johnston Island and to Houston for transport to LRL.

Primary *Apollo 11* mission objective—to perform manned landing on moon and return—and all detailed test objectives were achieved. All launch vehicle and spacecraft systems performed according to plan, with only minor, corrected discrepancies. Flight crew performance was outstanding; all three crew members remained in excellent health and their prevailing good spirits and proficiency were major factors in mission's success. Accomplishments included first manned lunar landing and return; first lunar surface EVA; first seismometer, laser reflector, and solar wind experiment deployed on moon; first lunar soil samples returned to earth; largest U.S. payload ever placed into lunar orbit (72,037.6 lbs at lunar orbit insertion); acquisition of numerous visual observations, photos and TV transmissions of scientific and engineering significance; and first operational use of MQF and LRL.

Apollo 11 was eighth Apollo mission to date, fifth manned Apollo mission, and first manned lunar landing mission. Mission acquired major quantities of data for subsequent Apollo flights. First manned Apollo mission, *Apollo 7* (Oct. 11–22, 1968), had verified operation of spacecraft for lunar-mission duration. First manned lunar orbital mission, *Apollo 8* (Dec. 21–27, 1968), had proved capability of Apollo spacecraft and hardware to operate out to lunar distance and return through earth's atmosphere. *Apollo 9* (March 3–13, 1969) had proved capability of LM to operate in space under manned conditions. *Apollo 10* (May 18–26, 1969) had successfully operated complete Apollo spacecraft on lunar orbital mission and had provided major quantities of scientific and training materials for *Apollo 11*. Apollo program was directed by NASA Office of Manned Space Flight; MSC was responsible for Apollo spacecraft development, MSFC for Saturn V launch vehicle, and KSC for launch operations. Tracking and data acquisition was managed by GSFC under overall direction of NASA Office of Tracking and Data Acquisition. (NASA Proj Off; NASA Release 69–83K; *NYT*, 7/16–25/69; *W Post*, 7/16–25/69; W *Star*, 7/16–25/69; B *Sun*, 7/16–25/69; *PD*, 7/28/69, 1016)

July 16: U.S.S.R.'s *Luna XV* (launched July 13) entered lunar orbit at 3:00 pm Baikonur time (6:00 am EDT) with all systems functioning normally, Tass announced. Sir Bernard Lovell, Director of U.K.'s Jodrell Bank Experimental Station, said spacecraft was transmitting telemetry data but no photographic signals. (*SBD*, 7/18/69, 22; AP, W *Star*, 7/17/69, A1)

- In Cape Kennedy interview before *Apollo 11* launch, Vice President Spiro T. Agnew said, "It is my individual feeling that we should articulate a simple, ambitious, optimistic goal of a manned flight to Mars by the end of this century. Whether we say it or not, someone's going to do it."

  After liftoff, Vice President told NASA launching team, ". . . all the time I was out there I couldn't help thinking of you, the people in here and all over NASA who have done such a brilliant job in putting together the combined effort behind those three gentlemen who are off on this historic mission. . . . I bit the bullet for you today as far as Mars is concerned. But on the other hand . . . I may be a voice in the wilderness."

  In Washington, D.C., Senate Majority Leader Michael J. Mansfield (D-Mont.) told press, "I think we have a lot of problems here on earth that we must face up to and when we settle those we ought to consider future space ventures." Senate Majority Whip Edward M. Kennedy (D-Mass.) said, "The Apollo program is for landing a man on the moon and exploration and should take another one to two years. I think after that the space program ought to fit into our other national priorities." (Witkin, *NYT*, 7/17/69, 1; Transcript, Agnew statement to NASA launch crew; Unna, *W Post*, 7/17/69, A1)

- At White House, President Nixon proclaimed July 21 National Day of Participation. "Apollo 11 is on its way to the moon. . . . Never before has man embarked on so epic an adventure. . . . As the astronauts go . . . we on earth will want, as one people, to be with them in spirit . . . and to support them with prayers that all will go well." All Executive departments and Government agencies would be closed and U.S. flag would be displayed on public buildings.

  With many members at Cape Kennedy, Senate and House met briefly and conducted only routine business. *Congressional Record* was filled with comments on *Apollo 11* and wishes for Godspeed to astronauts. (*PD*, 7/21/69, 997–8; *CR*, 7/16/69)

- During CBS TV interview at Cape Kennedy following *Apollo 11* launch, former President Lyndon B. Johnson said, "If our industrial people, these great managers of industry, the laboring people of the country, the government, the scientists, all with the help of Congress, can get together and do a job like this there's just nothing we can't do." To world's ills, "we must apply some of the great talent that we've applied to space." There wasn't "a single thing that our country does, that our government does, that our people do, that has greater potential for peace than the space effort." (UPI, *NYT*, 7/17/69, 20)

- Between 750,000 and 1 million persons crowded Brevard County, Fla., to witness launch of *Apollo 11*, including 5,000 dignitaries headed by Vice President Spiro T. Agnew and former President Lyndon B. Johnson. The Rev. Ralph D. Abernathy and 40 representatives of Poor People's Campaign watched launch from bleacher seats with 10,000 guests including families of Apollo program personnel, while other representatives marched outside KSC. *Paris Match* had brought 105 European businessmen. Some 3,100 press members were at special stand. As Apollo spacecraft lifted from launch pad there was some applause, but most spectators stared in silence until Saturn V rocket disappeared overhead. Afterward many were caught in monumental traffic jams. Banana River, five miles south of Launch Complex 39, was clogged with several thousand boats registered from New England to Texas. (Greider, *W Post*, 7/17/69, A1; Weinraub, *NYT*, 7/17/69, 21; Lyons, *NYT*, 7/17/69, 21)

- *Apollo 11* launch brought mood of reflection across Nation, *New York Times* said. Dawn was breaking in western U.S. when blastoff occurred. Workers in San Francisco's open air fish markets stood in silence to hear radio report. In San Diego motorists crossing U.S.-Mexican border listened to countdown on car radios.

    In mid-America, classes were postponed at Air Force Academy in Colorado Springs, Colo., while cadets watched launch on TV. Cowhands at northern Wyoming ranch, inaccessible to radio or TV, interrupted work to honor *Apollo 11*. Ranch owner Dr. Oakleigh Thorn II said, "We feel so close to the moon shot out here, because we're so close to the stars and sky."

    In Biloxi, Miss., harbor fishermen paused on wharf to hear countdown. In Tennessee, tobacco farmers listened to transistor radios in fields.

    Washington, D.C., schoolteacher said, "The astronauts didn't just go to the moon. All our minds went to the moon and intellectually man's horizons have jumped leaps and bounds beyond the historical situation they've always been confined to." (Fosburgh, *NYT*, 7/17/69, 1)

- Worldwide audience focused on *Apollo 11* launch:

    At summer residence, Castel Gondolfo, Italy, Pope Paul VI asked for prayers for U.S. astronauts a few hours before launch toward moon.

    U.S.S.R. radio and TV gave factual accounts of *Apollo 11* launch but maintained third day of silence on *Luna XV*. Major Soviet news program at 8:30 pm Moscow time showed tape of *Apollo 11* liftoff taken from live comsat coverage.

    In U.K. TV viewers saw launch via transatlantic satellite. BBC scheduled live coverage through July 24 splashdown and would relay broadcasts to continent by cable. London newspapers frontpaged *Apollo 11*. *Daily Express* headline read, "Ho Hum—Anyone for the Moon Today?" over report on relaxed astronauts.

    Polish TV viewers saw launch via 45-min transmission directly from Cape Kennedy. Hundreds of Germans and Americans crowded *Apollo 11* exhibit in Mannheim, Germany, department store.

Swedish TV viewers were advised by state broadcasting company not to turn off sets Sunday night—so they could be awakened for scheduled moon landing Monday.

Hippies in Iran held milk and honey pots in Teheran restaurant to toast astronauts. In Egypt, Moslem world's leading moon expert, Sheikh Ahmand Haredi said, "The Koran urges Moslems to look up from their earthly abode to what lies behind the moon and stars."

Japanese department stores featured models of Apollo command module. In Greece, Aspis-Pronoia insurance company issued first outerspace life insurance policy, to cover *Apollo 11* crew at $10,000 each. In Spain people called event most interesting since Columbus discovered America.

Israel's state radio broadcast in Hebrew from Cape Kennedy while Israelis stood around TV sets and portable radios in streets. U.S. Embassy in Tel Aviv and U.S. Consulate in Jerusalem opened *Apollo 11* information offices. *Apollo 11* reaction was "generally tepid" in Lagos, Nigeria. Radio Nigeria reported launch seven minutes into its morning broadcast. Later it became number one newscast item.

Most of Latin America missed launch on TV because of failure of *Intelsat-III F–2* June 29. Latin American newspapers and TV correspondents traveled to U.S. to cover launch and were reported to be outraged by absence of TV coverage in their countries. In Colombia, government asked TV manufacturers to put sets in all town squares. Bogota students would have July 21 off to watch lunar landing. (*NYT*, 7/17/69, 21, 22; Bishop, *C Trib*, 7/17/69)

- Harry F. Guggenheim said in Washington *Evening Star* article that rocket expert Dr. Robert H. Goddard "was to the moon rocket what the Wright brothers were to the airplane." Guggenheim, administrator of Daniel Guggenheim Fund for the Promotion of Aeronautics during period it helped support Dr. Goddard's research, traced career of "Father of modern rocketry" from early experiments in 1908. Among Goddard's inventions were: first liquid-fuel rocket, first smokeless powder rocket, and first practical automatic steering device for rockets. It was no wonder American Rocket Society had conceded to Goddard "the almost single-handed development of rocketry 'from a vague dream to one of the most significant branches of modern engineering.'" He had left "testimony to the power of one solitary individual to effect change and to transform the future." While Dr. Goddard had died without fame which had accrued to Wright brothers in their lifetime, "he died still believing that man would one day shatter the fetters of Earth's gravity and stride majestically into the vast reaches of space. I wish he were here now to share this moment. It belongs to him." (AP, *W Star*, 7/16/69, A15)

- As part of NASA and Washington National Gallery of Art program, Eyewitness to Space, group of artists attended *Apollo 11* launch to paint facets of mission. Program originated in 1963 when artists were invited to cover *Mercury 9* mission. Among those commissioned to record *Apollo 11* were Peter Hurd, Robert Rauschenberg, Lamar Dodd, and James B. Wyeth. Since program started, 25 artists had produced more than 500 sketches and paintings. (Casey, *W Post*, 7/13/69, G1; Hicks, *NYT*, 7/15/69, 33; *W Star*, 7/17/69, A12)

- *Apollo 11* was producing noticeable effect on business and consumer products, *Washington Post* said. Snoopy the Astronaut dolls were selling out; sales of color TV sets had risen in some stores; and sales of "moon maps and globes, as well as toy rocket ships and lunar exploration vehicles had also lifted skyward." Two Washington, D.C., department stores were offering Japanese telescopes ranging from $19.99 to $1,000.00. One toy store manager said sales of space-related toys had jumped 70% or 80% in two months. Rockets propelled by solid-fuel inserts sold for $1.50 to $5.00 complete with recovery parachute. One Washington store had sold out supply of $10 spacesuits. Demand for rental of color TV sets in Washington area had been "terrific," according to area dealer. (Cushing, *W Post*, 7/16/69, D11)
- NAS announced formation of Universities Space Research Assn. (USRA) —national consortium of 48 universities—to foster cooperation among universities, other research organizations, and Government for advancement of space research [see Jan. 10]. It would acquire, plan, construct, and operate laboratories and other facilities for R&D and education in space science and technology and had submitted proposal to NASA for management of Lunar Science Institute in Houston, Tex. Existing contract between NASA and NAS would expire in autumn. (NAS Release)
- U.S. newspaper editorials hailed *Apollo 11* launch.

    Miami News: "All America, represented by three lonely men in space, is on its way to the moon. In fact, this is a people's effort, arousing the interest and participation of all the people of this country. This is evidenced by the more than one million persons on hand at Cape Kennedy . . . for the start of the moon voyage and by the many millions who join in the adventure by television. Today's magnificent launch, and the elan stirred in our people by it, makes this one of America's most splendid hours." (*Miami News*, 7/17/69, A16)

    Washington *Evening Star:* countdown which culminated in *Apollo 11* liftoff, "regardless of NASA's official records," had begun, "when primitive man first looked up into the night sky to gaze at the moon, and to feel the first stirring of wonder." (*W Star*, 7/16/69, A22)

    Huntsville Times: Manhattan Project had climaxed with July 16, 1945, explosion of world's first successful atomic bomb. "Men, it seems, can only pray that the consequences of the quest of the planets may be better than those born in the irreversible explosion on a New Mexico desert 24 years ago." (*Huntsville Times*, 7/16/69)

    Chicago Sun-Times: "Man has always looked upward to the stars, first in fear and awe, then in need to know. Today the first great step to the firmament will be taken. If it is successful man will stand on the threshold of outer space—and standing there will reach outward." (*C Sun-Times*, 7/16/69)
- *Svenska Dagbladet*, Stockholm, Sweden, welcomed *Apollo 11* launch: "One of the greatest adventures of human history begins today. . . . Studies of the moon will to a great degree enrich our knowledge of both the earth and space. Among other things it will be possible to make comparisons which will propel science by leaps in various disciplines. . . . While we can predict much that may result from conquest of the moon, there will in all likelihood be many results

which we cannot even imagine now. All great discoveries and bold undertakings have brought advances which no one could have foreseen from the outset."

Arbetet, Malmo, Sweden: "There is an irrational element in these feats of discovery which fortunately dominates the prosaic calculation of gains. Then one can regret that man's fantasy seems incapable of being fired for such a tremendous task as eliminating starvation from our earth, or for bringing peace to Biafra or for eliminating the U.S. Negro ghettos. . . . Three men will be lifted to world acclaim today on the crest of mankind's greatest ever coordinated effort. . . ." (Am Embassy, Stockholm)

July 17: White House announced *Apollo 11* crew on way to moon was carrying Soviet commemorative medals brought back to U.S. by Astronaut Frank Borman, who had received them from widows of Cosmonauts Yuri A. Gagarin and Vladimir M. Komarov during his Moscow visit. *Apollo 11* also carried Apollo 204 crew patch and commemorative medals struck for families of Astronauts Virgil I. Grissom, Edward H. White, and Roger B. Chaffee before astronauts died in Jan. 27, 1967, fire.

President Nixon said, "The names of Gagarin and Komarov, of Grissom, White, and Chaffee, share the honors we pray will come to Armstrong, Aldrin, and Collins. In recognizing the dedication and sacrifice of brave men of different nations, we underscore an example we hope to set: that if men can reach the moon, men can reach agreement." (PD, 7/21/69, 999)

- Aerospace industry was having its greatest week in history with *Apollo 11* launch, said *New York Times*, but aerospace stocks remained in doldrums. Wall Street was "bearish about the industry and, from an investment standpoint, unenthusiastic about space." Security analysts interviewed agreed *Apollo 11* would have little effect on long-depressed stocks, which commenced decline in 1968; many were selling near lows for year. Aerospace industry was chief beneficiary of space program funds, but largest portion of $34 billion spent since 1960 had been allocated before "really spectacular shots" occurred. While Apollo program had been "tremendous boon to the aerospace industry and to the advancement of technology," it represented small part of total industry revenues and outlook was for further decline. (Hammer, *NYT*, 7/17/69, 63)

- Teletype from German ship *Vegesack* reported numerous pieces of Saturn V launch vehicle from *Apollo 11* were sighted dropping into sea around ship. *Vegesack* had been at position some 375 mi east-northeast of Cape Kennedy when *Apollo 11* lifted off toward moon July 16. (*W News*, 2/3/70)

- *Apollo 11* launch continued to draw wide editorial comment in foreign and national press.

New York Times: "One could hardly watch the magnificent spectacle of the liftoff, let alone contemplate the feats of human ingenuity that made it possible, as well as the courage and skill of the flyers, without some reflection upon the meaning of this event. . . . The temptation is strong to fall back upon lyricism. The poetry of the thing has yet to find its expression in any of the earnest, proficient Americans who have ventured away from the Earth; yet, the stunning beauty of man's most marvelous creation, as it rose in its majesty toward the unknown, toward the future, could be matched only by the profound sense of having been present at an end to something and therefore necessarily at a beginning." (Wicker, *NYT*, 7/17/69)

Chicago Tribune: "The Apollo 11 blastoff was as beautiful a one as we've seen. It displayed every bit of the precision and the careful planning which we have come to expect from NASA." One of most "satisfying" things was that, "like our earlier launchings, it took place within the sight of anybody who wanted to go to the Florida coast to watch it, and was broadcast live to countless millions more in every corner of the world. People will not fail to contrast this with the secrecy of Russia's unmanned Luna 15, which may reach the moon today on a mysterious mission of its own." (*C Trib*, 7/17/69)

Christian Science Monitor: "And although it is an American mooncraft, bearing American men . . . the venture is, in the best sense, a universal one. It is the result of American technology putting to use the knowledge, techniques and discoveries in which all nations and races have participated. . . . all nations and peoples are taking part." (*CSM*, 1/17/69)

Seattle Times: "The space program has yielded immense new resources in . . . scientific and technological advances which . . . make the program worth while even beyond the explorations and discoveries —and national pride—offered by the ventures into outer space. It strikes us, therefore, that the time is at hand for these so-called byproducts of the space program, which hold such promise for utilization in behalf of mankind, to be put to work for that purpose." (*S Times*, 7/17/69)

Bulgarian Telegraph Agency report carried in Bulgarian newspapers *Rabotnichesko Delo, Narodna Mladezh, Trud,* and *Kooperativno Celo* commented: "In the coming days all humanity will follow this flight with interest and tension. And surely there is no one on our old planet who will not ask himself this question: 'Will it succeed?'" (Am Embassy, Sofia)

- Florida Legislature had neglected to send President Nixon copy of its June 6 resolution asking him to restore original name "Cape Canaveral" to Cape Kennedy, *Orlando Sentinel* said. Fate of project seemed to rest with joint resolution introduced in Congress July 10 for same purpose. (*Orlando Sentinel*, 7/17/69, 14A)
- DOD estimates in transcript of closed session of U.S. Senate revealed that by 1974 U.S.S.R. could have 420 SS–9 missiles, or total of 1,260 warheads. Even if Phase I of U.S. Safeguard were deployed by that time, 1,000 arriving Soviet warheads would be able to knock out all but 135–150 Minuteman missiles—far below DOD estimates of what was needed for adequate U.S. second strike capability. (*AFJ*, 12/6/69)

July 18: In response to telephone inquiry by Astronaut Frank Borman, Mstislav V. Keldysh, President of Soviet Academy of Sciences, sent telegram guaranteeing that *Luna XV*, orbiting moon, would not interfere with *Apollo 11* mission and assuring Borman that he would be notified of any changes in spacecraft's course. Under 1967 U.N. Outer Space Treaty, U.S. and U.S.S.R. were bound to furnish each other this kind of information. (Wilford, *NYT*, 7/19/69, 1)

- Apollo passive seismic experiment, part of extravehicular activity to be performed by *Apollo 11* astronauts on moon, was described in *Science* as "the most exciting experiment in seismology." Dr. G. Latham and Dr. M. Ewing of Columbia Univ.'s Lamont-Doherty Geological Observatory, Dr. F. Press of MIT, and Dr. G. Sutton of Univ. of Hawaii explained objective was to detect naturally occurring seismic events on lunar surface through early Apollo scientific experiment package (EASEP) planted on moon. Package weighed 105 lbs and would transmit data to earth one year (or maximum two years), during lunar days because its solar cell panels required illumination to provide power. Complete Apollo lunar surface experiments package (ALSEP), containing at least three additional experiments for measurements of solar wind and magnetic field, would be included on Apollo 12 for day and night operation.

In *Apollo 11* experiment astronaut would remove instrument from LM to smoothest area within 6.6–9.8 ft (20–30 m) of LM, unfold solar panels, adjust package level to within 5°, orient it with azimuth for maximum illumination of solar panels, and aim antenna toward earth. MSC would issue commands to uncage and level seismometers and select proper gain. Expected sources of lunar seismic activity included several hundred monthly moonquakes, thermal stresses produced by rapid temperature variations at surface; tidal stresses exerted by earth and sun; and meteoroid impacts. By end of Apollo program, scientists hoped to have achieved crude curves of travel time for body and surface waves and beginning of seismicity map of moon.

During post-Apollo period, seismologists wanted to achieve wider distribution of detectors to map seismically active belts in greater detail; study mechanisms of energy release; lower minimum detectable ground motion of individual seismometer; and improve performance of long-period seismometer systems at ultralong-period end of spectrum for recording surface waves from moonquakes, free oscillations of moon, and lunar tides. (*Science*, 7/18/69, 241–50)

- White House confirmed President Nixon would talk with *Apollo 11* astronauts over two-way TV hookup as they first set foot on moon. Nixon and Astronauts Neil A. Armstrong and Edwin E. Aldrin, Jr., would be visible on split screen to earth TV viewers. President could watch on White House TV, but astronauts would have no view of him. President Nixon planned to spend evening of July 20 watching *Apollo 11* progress on TV with former Astronaut Frank Borman, White House liaison with NASA. (Lyons, *W Post*, 7/19/69, A9)
- *Apollo 10* mission (May 18–26), first lunar orbital mission with complete Apollo spacecraft, was adjudged successful by NASA. Mission had achieved all objectives; systems had performed according to plan with only minor anomalies and crew had acquired major quantities of photographic training materials for subsequent Apollo missions. (NASA Proj Off)
- *Izvestia* gave first U.S.S.R. report of President Nixon's July 17 announcement that medals of two dead Soviet cosmonauts would be placed on moon by *Apollo 11* astronauts. Factual account of mission carried no comment. (*W Post*, 7/19/69, A10)
- Pride Inc. Operations Director Marion Barry called on black community to work during July 21 National Day of Participation declared by President Nixon in honor of *Apollo 11* lunar landing. During Wash-

ington press conference he said, "Why should blacks rejoice when two white Americans land on the moon when white America's money and technology have not even reached" the inner city? "Why should blacks celebrate Monday . . . when President Nixon didn't feel that Dr. Martin Luther King's assassination deserved to be observed?" (Paka, *W Post*, 7/19/69, A9)

- Richmond, Va., *News-Leader* editorial approved Vice President Spiro T. Agnew's calling for flight to Mars by end of century [see July 16]: "One day, man will go beyond the planets, to other solar systems; right now . . . that is not within our technological reach. But Mars is, and so are the other planets. The moon is in earth's, and man's, own crib. Plans and commitments should be made—now—for man to take grown-up strides in the real world of space." (R *News-Leader*, 7/18/69)
- After four years of "running at top speed," MSC had failed to turn Houston, Tex., into "science city," said Thomas G. Plate in *Science*. Houston area, as largest petrochemical industry area in U.S., was "going its own booming way" while 4,600 NASA people and 9,000 employees of 125 private firms working on NASA business in area helped to shape space age community at MSC. "The injection of $140 million a year in NASA money and the impact on the life of the area of NASA workers—some 2500 of them R&D scientists and engineers—and of the 9000 employees of . . . high-technology firms serving MSC has so far had surprisingly little effect. But meanwhile the space community has developed its own special character with its own style of life and its own special goals." (*Science*, 7/18/69, 265–9)
- ComSatCorp reported second quarter earnings of $1,976,000 (20 cents per share); earnings had been $1,506,000 (15 cents per share) in similar 1968 period. Earnings for first six months of 1969 totaled $3,501,000 (35 cents per share) and $3,405,000 (33 cents per share) in 1968. (ComSatCorp Release 69–43)

July 19: Montreal, Canada, *Gazette* commented on *Apollo 11* mission: "Lyndon Johnson, more than any other man, is responsible for meeting the moon-shot deadline this week. . . ." (Am Consul, Montreal)
- Pittsfield, Mass., *Berkshire Eagle* editorial said: "It subtracts nothing from the extraordinary human and technical achievement represented by Apollo 11 to say that the projected lunar landing is an occasion not only for awe and pride but also for a thoughtful reappraisal of our whole approach to the new frontier of space." (*B Eagle*, 7/19/69)

July 20: "We have entered a new era," Dr. Thomas O. Paine, NASA Administrator, told press in Houston following *Apollo 11* lunar landing. "The significance of the trip is that mankind is going to establish places of abode outside of his planet earth."

In telephone call to White House, Dr. Paine had told President Nixon, "It is my honor on behalf of the entire NASA team to report to you that the Eagle has landed on the Sea of Tranquility and our astronauts are safe and looking forward to starting the exploration of the moon." Dr. Paine said President Nixon had spoken with "excitement and awe in his voice" and mood was that of "considerable tension relieved." NASA planned tentative six additional manned lunar missions over next few years. Dr. Paine praised U.S.S.R.'s cooperation in providing *Luna XV* information to Astronaut Frank Borman [see July

18]. He also said if Astronaut Neil A. Armstrong had not assumed manual control of LM to steer it from crater during lunar landing, "we might . . . have had considerable difficulty." (McGehan, B *Sun*, 7/21/69, A1)

- CBS presented interview with former President Lyndon B. Johnson which had been taped July 5. President Johnson credited space program with sparking "revolution of the 60s" and said, "We can't discard space. We're just beginning." U.S. had enough money "to do all the things we need to do" in space, education, and health. "What we must have is the determination to do it." He said his last act as president had been to send *Apollo 8* photos of earth to 186 leaders of foreign governments. (*W Post*, 7/21/69, A7)

- Astronaut Frank Borman repeated *Apollo 8* reading from Genesis at White House service attended by President and family, Vice President, Cabinet members, Congressmen, and members of Joint Chiefs of Staff, and of diplomatic corps. During sermon, Dr. Paul S. Smith, President of Whittier College and member of Religious Society of Friends, said: "It was a philosopher . . . who, two thousand years ago, first recounted a voyage to the moon. Lucian called it *The True History* but confessed in the preface that he wrote 'of things which are not and never could have been.' It was a political satirist's precautionary disclaimer because his real subject was the stupidity of human warfare. His lunar voyagers got caught up in internecine strife between the moonmen and the sunmen over the colonization of Venus! If there is something instructive in the thought, it may be the implication that after two millennia of philosophy men are still fighting over real estate and still dying in the name of philosophical abstractions, but that a voyage to the moon is just as feasible (though somewhat more expensive) as a trip to Timbuktu." (Wiegers, *W Post*, 7/21/69, B1; *CR*, 7/22/69, H6189–90)

- Hours before lunar landing attempt by *Apollo 11* Astronauts Neil A. Armstrong and Edwin E. Aldrin, Jr., Pope Paul VI said at Castel Gondolfo, Italy: "In the ecstasy of this prophetic day, a real triumph for means produced by man for the domination of the universe, we must not forget man's need to dominate himself. Admiration, enthusiasm and passion for instruments, for the products of man's hand, fascinate us, perhaps to the point of madness. . . . This is the danger: We must beware of this worship." (Schmick, B *Sun*, 7/21/69, A4)

- Tass announced that *Luna XV* was still functioning normally in lunar orbit with 109.4-km (68-mi) apolune, 16.1-km (10-mi) perilune, 1-hr 54-min period, and 127° inclination. Sir Bernard Lovell, Director of U.K.'s Jodrell Bank Experimental Station, said *Luna XV* had conducted two midcourse corrections and speculated that spacecraft was preparing either to land or to observe *Apollo 11* landing. (AP, B *Sun*, 7/21/69, A1)

July 20–21: White House was flooded with congratulatory cables and telephone calls on *Apollo 11* landing, from heads of state throughout world. *Washington Post* estimated half billion persons had watched lunar touchdown on worldwide TV, and NBC said 123 million in U.S. saw it, mostly in their own homes. But 35,000 baseball fans in New York had learned of landing's success when words "They're on the moon" flashed on scoreboard at Yankee Stadium. In New York's

Harlem, many of 50,000 attending soul music festival booed lunar landing announcement. At massive "Moon In" at Central Park, enthusiastic crowd of young people watched landing on huge outdoor TV screen in steady downpour and bought "lunar dogs," "Apollo rock candy," and "moon picnic" boxes.

Composer and band leader Duke Ellington made singing debut with "Moon Maiden," song he wrote to celebrate *Apollo 11* success, taped for ABC. *New York Times* sold out 950,000 copies of July 21 issue announcing lunar landing and announced it would reprint entire edition July 24 as souvenir. Special *Florida Times-Union* edition datelined "Moonday, July 21" sold out in Jacksonville within two hours. Estimated 8,000 Western Electric Co. employees left work or failed to show up in protest against being denied access to TV or radios on job during lunar landing. Des Moines, Iowa, TV stations received some complaints from viewers over absence of regular programs.

Crime rate fell in Los Angeles, while in Savannah, Ga., 17 prisoners sawed their way out of Chatham County prison branch while guards watched *Apollo 11* on TV.

At MSC, Houston Welfare Rights Organization members demonstrated around display of LM, calling on U.S. to set new goal—elimination of poverty. (AP, B *Sun*, 7/22/69; *W Post*, 7/21/69; 7/22/69; *Apollo 11* Mission Commentary, 7/21/69; *NYT*, 7/17/69, 7/27/69)

- Millions around world hailed *Apollo 11* landing:

Soviet Premier Alexsey Kosygin complimented U.S. on lunar landing and expressed interest in widening U.S.–U.S.S.R. space cooperation during July 21 Moscow discussion with former Vice President Hubert H. Humphrey, who was ending Soviet visit. Soviet TV did not carry live coverage of *Apollo 11* lunar landing July 20; Tass announcement was read by newscaster and carried in two-paragraph item on *Pravda*'s front page. Evening paper, *Izvestia*, accorded story more space and featured photo of astronauts on moon. On TV, Cosmonaut Konstantin P. Feoktistov described landing as "major landmark" and said crew had coped "brilliantly" with mission. Georgy Petrov, Director of Soviet Institute for Cosmic Research, called *Apollo 11* "outstanding achievement" but said more data per ruble could have been gathered by unmanned probes.

Statue dedicated to *Apollo 11* astronauts was unveiled July 21 in sports stadium at Cracow, Poland.

In U.K., Queen Elizabeth watched lunar landing on TV, then cabled President Nixon "warmest congratulations." Prime Minister Harold Wilson expressed "heartfelt relief." At Jodrell Bank Experimental Station astronomers applauded and director, Sir Bernard Lovell, said that "the future has been revolutionized." David Threlfall collected $24,000 on five-year-old bet that man would land on celestial body before 1971. Betting shop had given him thousand-to-one odds [see May 29].

In Wollongong, Australia, local judge heard cases while watching *Apollo 11* lunar landing on portable TV set.

Czechoslovakia issued two postage stamps July 21 commemorating lunar landing, while record crowds at U.S. Embassy exhibition tapered off after exhausting supply of Apollo giveaway materials.

Five thousand Hungarians walked through American Embassy in Budapest July 21, picking up USIA pamphlet *Man on the Moon.*

In Romania, bouquets were tossed through U.S. Embassy fence to foot of flagpole and several Romanians reported large numbers of Bulgarians were crossing border to watch live TV coverage of *Apollo 11.*

Cuban government decided not to jam Voice of America broadcast of *Apollo 11* lunar landing. In Algiers news was ignored except for announcement in government-controlled newspaper that "the man is on the moon." In Ghana, village chief listening to VOA broadcast feared astronauts might fall off moon if not careful.

In Bangkok, freedom for 622 pardoned prisoners was delayed because guards refused to leave TV sets showing *Apollo 11.*

Lunar landing stole top play in Israel and Egypt, from accounts of their fierce fighting at Suez Canal.

In Singapore, girl born half hour after lunar landing was named Luna. In Pakistan, boy baby was named Apollo.

Prime Minister, Mrs. Indira Gandhi, and Indian Parliament gave standing ovation to *Apollo 11* astronauts at opening of day's business in New Delhi July 21.

In Japan, Emperor Hirohito called off customary daily stroll and interrupted lunch to watch *Apollo 11* on TV.

Iroquois Indians in Brantford, Ontario, Canada, feared lunar landing might plunge earth into darkness and release monsters from earth's core. Their medicine man and chief, Joseph Logan, Jr., had said moon was sacred to his people and "we are not supposed to disturb her."

In Taipei, Formosa, Nationalist China Parliament member Hsieh Jen-chao invited *Apollo 11* astronauts to attend Moon Festival honoring rabbit which Chinese legend said lived on moon and could provide eternal life.

Some devout Moslems in Somalia refused to believe *Apollo 11* lunar landing was reality. Following radio, press, and word-of-mouth announcement, fist fights broke out July 21 in Mogadiscio streets between believers and disbelievers. Parents of baby boy born on lunar landing day broke with Muslim tradition and named child Armstrong Abdurahman Osman.

In Brussels workers in radio and TV studios suspended strike during transmission of *Apollo 11* mission film.

In Brazil several thousand persons cheered as they witnessed televised lunar landing at Museum of Modern Art in Rio de Janeiro while church bells rang outside. In Santiago de Chile people rushed out of restaurants to look at moon, forgetting it was midafternoon when they learned of lunar landing.

While rest of world focused on lunar landing, one quarter of world's population labored through sixth moon of Chinese lunar year unaware of event. Approximately 800 million people in Communist China had heard no news of lunar landing. Only deviation from "total blackout on space exploration" was July 17 story of Astronaut Frank Borman's visit to Moscow, reported by New China News Agency. (*C Trib,* 7/22/69; *W Post,* 7/21–22/69; *W Star,* 7/22/69; *NYT,* 7/22/69;

B *Sun*, 7/21–22/69; Am Embassy, Prague, Bucharest, Brussels, Budapest, Mogadiscio)
- Press in U.S. and around the world underscored *Apollo 11*'s landing on moon and man's first steps on another planet.

St. Louis Post-Dispatch: "There is no doubt that the United States should continue to support a substantial spacefaring program. Anything else would be a denial of the scientific spirit of the century and the qualities that have made America what it is. But its scope should be measured by findings and probabilities—and one other factor. Future spacefaring ought to be a co-operative effort of all nations able to participate, with the benefits to be shared by all." (*St. Louis P–D*, 7/20/69)

Washington *Sunday Star:* "A creature that can stand where Armstrong and Aldrin stand tonight—that can, in the future, move among the spheres and literally explore new worlds . . . is unlikely to give up on the hard task of perfecting himself and his life in his natural environment on earth. The God who brought him thus far from a blob of squirming protoplasm . . . is unlikely . . . to let man blow it all now. Here . . . must be the answer to the national debate as to whether we go ahead in space, or whether we tend to our knitting at home. We are bound to do both. . . . The progressive expansion of the physical and spiritual domain of man inevitably will intensify our determination and ability, in concert with other nations, to build a home world where hunger, fear and violence no longer have a place." (W *Star*, 7/20/69, G1)

William Hines in Washington *Sunday Star:* "One cannot question the majesty of conception or magnitude of effort that made Apollo 11 possible." But one could ask, "Is this trip really necessary?" One saw in Apollo "that fundamental failing called hubris, which got so many protagonists into hot water in the old Greek mythology. Hubris in English is usually taken to mean prideful arrogance; in ancient Greek the word meant simple insolence. The Apollo enthusiast rejects the concept of hubris; he says we go to the moon not because we are arrogant, but because we are driven, and thereby implicitly rejects the concept of free will and substitutes sappiness for sassiness. The majority asks, 'But if we didn't go, what?' and the minority responds, 'If we didn't go, so what?'" (W *Star*, 7/20/69, G2)

Humorist Art Buchwald in *Washington Post:* "Sometimes one gets the feeling that the right hand germs in the Government don't know what the left hand germs are doing. This was brought home to me . . . when I read about the millions of dollars that were being spent to see that the astronauts did not bring back a single germ from the moon. Unfortunately, across the page from that story was another that the Army was going ahead with open air testing of nerve gases and germ warfare." (*W Post*, 7/20/69, B6)

Los Angeles *Herald-Examiner:* "America's moon program has benefited all mankind. It has brought better color television, water purification at less cost, new paints and plastics, improved weather forecasting, medicine, respirators, walkers for the handicapped, laser surgery, world-wide communications, new transportation systems, earthquake prediction system and solar power. . . . The Mars goal should bring benefits to all mankind even greater than the tremendous contributions of the moon program." (LA *Her-Exam,* 7/20/69)

Baltimore *Sun:* ". . . it is still almost incredible that in the afternoon of a Sunday on earth two humans found themselves within a vehicle resting on the surface of the moon. Nothing could quite prepare one's mind for that, or for the subsequent moment of climax, the actual setting of a human foot on the substance of our barren satellite. One of the mysteries that had engaged the infinitely inquisitive mind of man is now made tangible. Others remain beyond our planet and upon it." (B *Sun,* 7/21/69, A16)

Chicago Daily News: "These have been moments to savor—moments in which uncounted millions have shared the immediacy of a turning point in history. This time there was no lapse of weeks or months, waiting for the event to be confirmed. We were all there, bound together by the miracle of communication that intertwined all the other miracles of technology that marked man's first step on a celestial body." (*C Daily News,* 7/21/69)

Milwaukee Journal: "Superlatives pale before the magnificence of the achievement. . . . but how many years before the astounding performance of Armstrong and Aldrin will seem as primitive as the pioneering work of the Wright brothers?" (*MJ,* 7/21/69, 14)

Cleveland *Plain Dealer:* "Man's store of scientific knowledge will be vastly enriched by the landing on the moon. In no other single event in history has there been greater opportunity to unlock the mysteries of the universe." (*Plain Dealer,* 7/21/69)

London *Daily Sketch:* "America's moon triumph offers this old world's bickering and jealous people a parable of hope." (B *Sun,* 7/22/69, A1)

Montreal Star: "The deepest hope for a world starved for some form of symbolism, of an attempt at harmony in place of selfishness and narrow nationalism, came from the astronauts," CFOX Radio, Montreal, broadcast. "Eliminate war? Yes! Eliminate poverty? Yes! But the exploration of space will help us, not impede us, in reaching these goals." (Am Consul, Montreal)

Arbetet, Malmo, Sweden (principal organ of Social Democratic Party): "No Soviet politician has ever before used such conciliatory tones toward the U.S.A. as did Foreign Minister Gromyko recently in his speech before the Supreme Soviet. . . . This Russian position seems generally to be based on fears of a confrontation with China. . . . One of the side effects can be increased Russian interest in broader scientific cooperation in space research. Nothing else could be better designed for global cooperation, since nothing else gives us clearer testimony that we live in one world." (Am Embassy, Stockholm)

July 21: U.S.S.R.'s *Luna XV* (launched July 13) had landed on moon at 6:45 pm Moscow time (11:45 am EDT) and had ended its work, Tass announced. Spacecraft had "reached the moon's surface in the preset area" after 52 revolutions around moon and 86 communications sessions during which "the work of the new systems of the station was checked, the parameters of the trajectory of the movement was measured, and scientific research was conducted." Tass said *Luna XV* had demonstrated capability to land on various areas of lunar surface by changing selenocentric orbit and that mission had yielded important data on spacecraft systems.

Sir Bernard Lovell, Jodrell Bank Experimental Station Director, said signals from spacecraft had ended suddenly and estimated craft might have landed in Sea of Crises, about 500 mi from Sea of Tranquility. "If we don't get any more signals, we will assume it crashlanded. But we don't make that assumption at the moment." (Gwertzman, *NYT*, 7/22/69, 1, 29)

- Univ. of Texas astronomers reported second unsuccessful attempt to bounce laser beam off reflector left on moon by *Apollo 11* astronauts. McDonald Observatory Director, Dr. Harlan Smith, said he expected eventual success. (AP, B *Sun*, 7/22/69, A8)
- Galabert International Astronautics Prize for 1969 was awarded in Paris to *Apollo 11* astronauts. Award of $4,000 was presented annually for notable contributions "to human progress for the advancement of all sciences and techniques associated with astronautics." (AP, B *Sun*, 7/22/69, A8)
- HUD Secretary George W. Romney addressed International Platform Assn. in Washington, D.C.: "I do not propose that we now abandon our efforts to extend man's reach still further beyond our planet, any more than we abandoned our domestic goals while we were reaching for the moon. But I do believe the time has come for a revision—in fact, a reversal—of our national priorities. I believe that in the decades ahead, the public interest and indeed our national survival require us to assign our housing and urban goals a high priority—at least comparable to the priority we gave our space program in the decade just ending." (*HUD News*; Hutchens, W *Star*, 7/22/69, A6)
- South Korea dedicated its first superhighway, linking Seoul with Inchon. It was named Apollo in honor of U.S. moon landing. (AP, W *Post*, 7/23/69, C5)

July 22: U.S.S.R. launched two unmanned satellites. *Cosmos CCXC*, launched from Plesetsk, entered orbit with 323-km (200.7-mi) apogee, 192-km (119.3-mi) perigee, 89.6-min period, and 65.4° inclination and reentered July 30. *Molniya I–12* comsat, launched from Baikonur, entered orbit with 39,526-km (24,560.3-mi) apogee, 496-km (308.2-mi) perigee, 711.0-min period, and 64.9° inclination. (GSFC *SSR*, 7/31/69; *SBD*, 7/28/69, 62; *NYT*, 7/23/69, 26)

- Scientists at MSC, monitoring seismometers left on lunar surface by *Apollo 11* astronauts, recorded five-minute tremor they said could have been internal activity—moonquake—or meteoroid strike on surface. Scientists expressed concern that seismometer was overheating, probably because of damage to protective cover from LM exhaust, and might not survive heat of lunar moon. (McGehan, B *Sun*, 7/23/69, A1; Cohn, W *Post*, 7/24/69, A15)
- Scientists at Lick Observatory in California unsuccessfully tried for third consecutive night to bounce ruby laser beams off reflector left on lunar surface by *Apollo 11* astronauts. They admitted difficulty in pinpointing reflector's exact location and speculated that it might have been knocked down by LM exhaust during ascent. (AP, W *Star*, 7/23/69, A7)
- NASA announced revised plans for first orbital workshop, with 1972

launch using first two stages of Saturn V to launch workshop and Apollo Telescope Mount together. Workshop would be outfitted on ground and would arrive in 253-mi circular orbit equipped for immediate occupancy by astronauts and with ATM attached. Program objectives remained same as when NASA intended to use Saturn IB 2nd stage as 1971 workshop: to provide environment in which man could live and work for extended periods in space and to study man's physiological and psychological responses and capabilities in space. ATM would permit man to operate high-resolution astronomical telescopes in space, free from earth's atmosphere.

Saturn V hardware from Apollo program was available for revised plan. (NASA Release 69–105; Simons, *W Post*, 7/22/69, A1)

- President Nixon addressed 2,000 American Field Service students from 60 countries on White House lawn: ". . . in the year 2000 we will, on this earth, have visited new worlds where there will be a form of life. I know this will happen, and I want to tell you as I look forward and dream about that future . . . this is the kind of world I would like to see and the kind of exploration of that new world that I know all Americans want. I hope that when the next great venture into space takes place that it will be one in which Americans will be joined by representatives of other countries." (*PD*, 7/28/69, 1016–7)

- U.K. radioastronomer Sir Bernard Lovell told press at U.K.'s Jodrell Bank Experimental Station that *Apollo 11* and *Luna XV* increased hopes for U.S.–U.S.S.R. space cooperation because "this is the first time the United States has been demonstratively superior in a vital part of the space program. American approaches for collaboration may be received with sympathy in the Soviet Union as they can no longer regard themselves as masters." (AP, *B Sun*, 7/23/69, A4)

- Wall Street brokerage houses were watching effect of *Apollo 11* success on stocks as market resumed trading after July 21 holiday. Some firms believed lunar landing would generate enthusiasm, although its impact would be restrained by uncertainties over surtax extension, House committee vote to cut oil depletion allowance (major tax benefit of petroleum industry), and apparent standoff at Vietnam peace conference. (UPI, *W Star*, 7/22/69, C7)

- *Washington Post* said it found intellectuals "deeply divided" on implications of lunar landing. Univ. of California physicist Dr. Owen Chamberlain had said achievement showed "mankind can be in charge of his destiny. . . . We should now come back and put our emphasis on the surface of the globe" to achieve peace, lessen poverty, control overpopulation, and preserve our environment.

Univ. of California chemist Dr. Harold C. Urey said if some of space effort reliability rubbed off on industry, "spin-off" would be enormous and space program would pay for itself. Less than ½ of 1% of GNP was spent on space and if lessened there was no guarantee it would be spent on necessary domestic programs.

Harvard Univ. biochemist Dr. George Wald had said: "What should have been a great flight of the human spirit comes to us heavy with threat. Those almost miraculous guidance systems that so uncannily find their targets, will they one day be guiding missiles to find us?" Dr. Wald wondered if *Apollo 11* had opened new horizons for his students. "I am afraid that they see in this an exercise in great wealth and power, heavy with military and political overtones. I am afraid that they feel a little more trapped; a little more disillusioned, a little more desperate."

Most overseas intellectuals tended to concur with historian Prof. Arnold J. Toynbee's judgment, "If we are going to go on behaving on earth as we have behaved here so far, then a landing on the moon will have to be written off as just one more shocking misuse of mankind's slender surplus product."

But Oxford Univ.'s Prof. A. J. Ayer had said, "I doubt if Prof. Toynbee has any evidence that men are being prevented in any large numbers from turning their minds to meaningful pursuits by the part which they play, or the interest which they take, in the exploration of space. . . . I think that these spatial explorations . . . are intellectually stimulating, especially to young people." (*W Post*, 7/22/69, A14)

- Australian Civil Aviation Minister Reginald Swartz said passengers on transpacific Qantas Airlines flight would see *Apollo 11* reentry July 24 when command module would parallel their aircraft for four minutes during descent near Gilbert and Ellice Islands. (Reuters, *W Post*, 7/23/69, A12)
- U.S. Patent Office issued patent No. 3,456,387 to Clyde A. Tolson, Associate Director of Federal Bureau of Investigation, for equipment to operate emergency windows and exits in aircraft and space vehicles. Without action of occupants, sensors would detect abnormal conditions and computer would weigh considerations before opening appropriate escape exits.

  Patent No. 3,456,445 was issued to Curtiss-Wright Corp. for improved version of astronaut maneuvering unit, Cap Pistol, intended to propel man outside space vehicle by capsules spaced along tape strip and fired by engine in pistol fashion. Inventors were Joseph F. Loprete, Max Beniele, and Richard E. Biehl. (Pat Off PIO; Jones, *NYT*, 7/26/69, 31)
- Goodyear Aerospace Corp. had invented USAF Pilot Airborne Recovery Device (PARD) to keep ejecting jet fighter pilot aloft and out of range of enemy ground fire until his midair retrieval by rescue aircraft. Ballute (balloon-parachute) attached to main parachute had burner suspended below and fueled from propane tank on pilot's back. At 250°F, hot air kept parachute above ground for 30 min. System could be operated automatically to carry pilot 6,000 ft or manually to 10,000-ft hovering altitude. (*NYT*, 7/22/69, 58)
- National and international press continued comment on *Apollo 11* lunar landing.

  Philadelphia Inquirer: "Will this magnificent accomplishment serve as inspiration, urging Americans and all mankind on to a genuine 'giant leap' forward, not merely into the infinite reaches of space but into the infinite possibilities of achievement on earth where the space age has recorded many more failures than successes? Or will the inspiration be abandoned before the veiled censure of those who seem to suggest the solution of all human dilemmas lies in turning away from space to other priorities?" Cutbacks at hour of triumph would be only waste of investment in technology which could be helped in solving earth problems. "This is no time to falter, our astronauts should come home to a world and nation determined to fulfill the prophecy in Commander Armstrong's words." (*P Inq*, 7/22/69)

Washington Post: It was foolish "to leap from this historic moment to eager expectations of the day when men will live and work in space, when colonies will be established, food raised and industrial products built on heavenly bodies other than the earth. These things will doubtless come in their own good time. But this is not the occasion on which to make a new national commitment in space that would keep NASA's program going at the frantic pace which fulfilled President Kennedy's great promise for the moon. Now is the occasion, rather, to establish a steady program of space development, one removed from the political debate over national priorities, which will ensure that we establish a firm base for future generations to build upon while creating at home . . . a kind of society which will allow them to use fully the new opportunities opened up by the three new American heroes and the tens of thousands of other people who made their flight possible." (*W Post*, 7/22/69, A24)

Handels Och Sjofartstidning, Goteborg, Sweden: "This is a small step for a man, but a great one for humanity. Neil Armstrong's commentary when he stepped down onto the surface of the moon has every prospect of becoming one of those winged expressions which generations of school children will commit to memory. . . . Now should be the time to replace the extraordinarily costly space race with cooperation between the Soviet and the U.S.A." (Am Embassy, Stockholm)

Stockholm *Expressen:* "The 'moonshot' . . . was imposing. But it also gives a horrible feeling to think that the U.S.A. can handle tremendous technical problems with such ease while it is considerably more difficult to cope with those of a complicated social, political and human nature." (Am Embassy, Stockholm)

Canadian *Montreal Star:* "The scientific information which results from Apollo 11 is an extra dividend from an enterprise which has produced its own benefits for the human spirit and, perhaps, for human solidarity." (Am Consul, Montreal)

July 23: USAF launched unidentified satellite from Vandenberg AFB by Thor-Burner booster into orbit with 531.5-mi (855.2-km) apogee, 488.4-mi (785.8-km) perigee, 101.3-min period, and 98.8° inclination. (GSFC *SSR*, 7/31/69; *Pres Rpt 70* [69])

- NASA's HL-10 lifting-body vehicle, piloting by NASA test pilot William H. Dana, reached 68,000-ft altitude and mach 1.2 during 22nd flight west of Rosamond, Calif. Purpose was to obtain performance, stability, and control data. (NASA Proj Off)

- Scientists monitoring seismometer left on lunar surface by *Apollo 11* astronauts told press at MSC five-minute event recorded July 22 was either meteoroid strike or moonquake similar to mild California earthquake recorded on East Coast. MIT geologist, Dr. Frank Press, said tremor would have magnitude of four or five according to Richter scale, on which major earthquake registered seven or eight. Seismic reading was strong indication that moon was layered with outer crust and inner mantle like earth and supported theories that moon was formed near or torn from earth. Layering, he said, "would imply that at one time there was enough heat so that the heavier rocks went to the interior and the lighter ones to the surface." (McGehan, B *Sun*, 7/24/69, A1; Lyons, *NYT*, 7/24/69, 1)

- NASA announced selection of McDonnell Douglas Corp. and North American Rockwell Corp.'s Space Div. to conduct parallel $2.9-million, 11-mo design and planning studies of 12-man earth orbital space station which could be developed by 1975 and have 10-yr lifetime. Companies would also include conceptual design of 50-man space base composed of specialized modules assembled in low earth orbit in late 1970s and early 1980s to serve as centralized scientific and technical facility in orbit.

    Aerojet-General Corp., General Electric Co., and Hughes Aircraft Co. had been selected for final competitive negotiation of contract to develop advanced optical communications experiment. Companies would compete for one $5-million contract to develop wideband laser communications system to be placed on board Applications Technology Satellite ATS–F, scheduled for launch in 1972, for communications between satellite and transportable ground station. (NASA Releases 69–108, 69–109)

- Canadian *Isis I* International Satellite for Ionospheric Studies (launched Jan. 30) was adjudged successful by NASA. Nine of ten experiments were operational; ion mass spectrometer had been turned off after one week of operation, when it developed high-voltage problems, and since had been used only for short periods to collect engineering data. Low-frequency receiver experiment had been providing indirect ion data, thus compensating partially for IMS loss. Onboard tape recorder was providing excellent topside ionograms of Antarctic area and other previously inaccessible areas. (NASA Proj Off)

- Full-color lunar photos from *Apollo 11*, including one of man first setting foot on moon, would be released by NASA to press and TV four days after splashdown, following two-day decontamination of film, NASA announced. Superintendent of Documents, GPO, was taking orders from public for photos to be filled in late August. Series of reproductions of paintings by American artists recording space program, "Eyewitness to Space," also would be released. (NASA Release 69–83J)

- Successful *Apollo 11* mission was expected to spur reservations on first lunar passenger flight, Washington *Evening Star* said. Before launch Pan American World Airways held 30,000 reservations and Trans World Airlines, 5,000. Pan Am spokesman said rush began after film "2001: a Space Odyssey" was first shown in 1968. In letters acknowledging reservations, Pan Am was saying, "Starting date of service is not yet known. Equipment and route will, probably, be subject to government approvals." TWA was saying, "We will be in contact with you again, as soon as technological advances develop to the point where we can project departure date." (W *Star*, 7/23/69, A7)

- In *Pravda* Soviet academician, Prof. Leonid I. Sedov, said space research was developing in so many different directions that realization of future projects would require huge material expenditure and concentration of creative efforts of "countless highly qualified workers and specialists." He said, "Not one individual country can afford the practical implementation of all the technically feasible and worthwhile projects." While scientists had said unmanned spacecraft could not always be substituted for manned vehicles, "flights by automatic stations have preceded and will continue to precede manned flights." Human feelings and observations, "especially when something turns up unexpectedly and unforeseen, cannot be completely replaced by automatic stations." But unmanned probes would continue as pathfinders because they were "cheaper, more simple and less dangerous vehicles for research." (Reuters, *W Post*, 7/24/69, A15)

- U.S. delegate to U.N. William B. Buffum, responding to Soviet tribute to *Apollo 11* astronauts by U.S.S.R. delegate Aleksey V. Zakharov, said before Security Council he hoped "fraternal spirit" demonstrated by astronauts and cosmonauts would lead to greater cooperation on earth also. (*NYT*, 7/25/69, 31)
- In his fourth reference to *Apollo 11* within week, Pope Paul VI said at summer palace, Castel Gondolfo, Italy: "Catholic faith, not only does not fear this powerful confrontation of its humble doctrine with the wonderful riches of modern scientific thought, but it desires it . . . because truth although diverse on various levels . . . is one and because such a confrontation is of mutual advantage to faith and to study in every field." (AP, *W Post*, 7/24/69, A15)
- Rep. Louis Frey, Jr. (R-Fla.), introduced for himself and Rep. William Chappell (D-Fla.) H.J.R. 834 "to redesignate the area in the State of Florida known as Cape Kennedy as 'Cape Canaveral.'" Measure was referred to House Committee on Science and Astronautics. (*CR*, 7/23/69, H6238)
- Czechoslovakian Communist Party Central Committee's weekly *Tribuna* said of *Apollo 11* landing: "It would be premature today to try to attempt a detailed evaluation of the historical significance of this act. Surely its influence will be no smaller than that of Columbus' travels many centuries ago." (Am Embassy, Prague)

July 24: President Nixon welcomed returned *Apollo 11* astronauts aboard U.S.S. *Hornet*: "I think I am the luckiest man in the world . . . not only because I have the honor to be President of the United States, but particularly because I have the privilege of speaking for so many in welcoming you back to earth." Washington had received messages from more than 100 foreign governments: "Emperors, Presidents, Prime Ministers, and Kings, have sent the most warm messages that we have ever received. They represent over 2 billion people on this earth, all of them who had the opportunity, through television, to see what you have done." Week of mission had been "the greatest week in the history of the world since the Creation, because as a result of what happened in this week, the world is bigger, infinitely, and also, as I am going to find on this trip around the world . . . as a result of what you have done, the world has never been closer together before." (*PD*, 8/4/69, 1032–3)

- At MSC news conference following *Apollo 11* splashdown, Dr. George E. Mueller, NASA Associate Administrator for Manned Space Flight, said: ". . . we now stand at what is undoubtedly the greatest decision point in the history of this planet." *Apollo 11* had proved "that man is no longer bound to the limits of the planet on which for so long he has lived. We will return to the moon first in November and then at regular intervals in the coming year. But these trips are only the first step. . . . Will we press forward to explore other planets or will we deny the opportunity to the future? To me, the choice is clear. We must take the next step. . . . This is the time for decision. . . . The knowledge possessed by men is sufficient, the resources are adequate for the task of carrying out this next step. . . .

"In this moment of man's greatest achievement, it is timely for us to dedicate ourselves to the unfinished work so nobly begotten by three of us. To resolve that this nation, under God, will join with all men in the pursuit of the destiny of mankind will lead to the way to the planets."

In answer to questions, Dr. Mueller said next major step should be manned landing on Mars which would be possible "sometime after 1980."

L/G Samuel C. Phillips (USAF), Apollo Program Director, told press Apollo team was "strongest team that's ever assembled in the history of man. It has the strength of technical and engineering confidence, scientific competence, and management competence that's unexcelled. It has the dedication that's necessary to be able to tackle an almost impossible job and bring it through" and an exciting future in lunar exploration.

Second manned lunar landing mission, Apollo 12, would be launched from KSC Nov. 14 toward touchdown on Site 7 in moon's Ocean of Storms. Primary objective would be to deploy Apollo lunar surface experiment package (ALSEP), explore and survey mare area, and return samples to earth. Secondary objective, if LM softlanded on target, would be to examine *Surveyor III* spacecraft (launched April 17, 1967), which was resting on moon near planned Apollo 12 touchdown point. Astronauts would have two periods for extravehicular activities (EVA), during which they would explore surface and conduct experiments for over three hours and walk farther away from spacecraft than had *Apollo 11* crew. Maximum lunar stay time would be 28–32 hrs. Schedule called for planning to fly follow-on missions through Apollo 15 at four-month intervals and missions after that at five-month intervals. (Transcript)

- USAF launched unidentified satellite from Vandenberg AFB by Thor-Agena booster. Orbital parameters: apogee, 136.1 mi. (219 km); perigee, 110.6 mi (178 km); period, 88.4 min; and inclination, 74.9°. Satellite reentered Aug. 23. (GSFC *SSR*, 7/31/69; 8/31/69; *InteraviaAir-Letter*, 7/25/69, 5; *Pres Rpt 70* [69])
- In nationwide reaction to safe return of *Apollo 11* astronauts, New York Stock Exchange went wild though stocks continued to fall. Numbers on annunciator boards flapped in unison as message "New York Stock Exchange shares the world's joy at the safe return of Apollo from the moon—Astronauts Armstrong, Aldrin, and Collins—So proudly we hail you" appeared on tape and illuminated on screen. Along Fifth Avenue church bells rang. Hayden Planetarium suspended usual program to throw "splashdown party" with champagne and live color telecast of *Apollo 11* recovery operations flashed on blackened dome.

San Franciscans exploded firecrackers and threw ticker tape from windows, and 10-story-high figure "11" was fashioned in lighted windows at MIT in Boston. Des Moines, Iowa, rang Liberty Bell reproduction for first time since its 1950 installation on State House grounds.

In Astronaut Neil A. Armstrong's home town, Wapakoneta, Ohio, high school band marched playing moon songs. Montclair, N.J., theater marquee read, "Congratulations Buzz Aldrin—Montclair's Man on the Moon."

In Huntsville, Ala., Dr. Wernher von Braun, MSFC Director, was hoisted on shoulders of four local councilmen while thousands at MSFC site cheered and waved banners saying "Huntsville is Rocket City."

City of Houston planned "Texas size" celebration for *Apollo 11* astronauts Aug. 16, including ticker-tape parade and huge program in city's Astrodome coliseum. (Sloan, Weinraub, Hicks, Borders, UPI, *NYT*, 7/25/69, 67, 29, 69, 31, 30; B *Sun*, 7/25/69, 45)

- Trans World Airlines filed first application with Civil Aeronautics Board for routes between earth and moon. Airline said it had received 1,200 reservations during final four days of *Apollo 11* mission. (TWA Release)
- Safe landing of *Apollo 11* in Pacific made "splash applauded around the world," *New York Times* said. In U.S.S.R. TV viewers had live coverage for first time during mission as Moscow TV station hooked into Eastern Europe's Intervision network for live transmission of astronauts being deposited on carrier *Hornet*. Later, station devoted first two-thirds of final newscast to *Apollo 11* and announced that Soviet President Nikolay V. Podgorny had sent telegram to President Nixon offering "our congratulations and best wishes to the space pilots." Soviet Academy of Sciences president Mstislav V. Keldysh called voyage "a big contribution to space exploration and further progress of world science." Cosmonauts sent message to *Apollo 11* crew: "We . . . closely followed your flight. We wholeheartedly congratulate you on the completion of your wonderful journey to the moon and safe return to earth."

In London Lloyd's of London's Lutine Bell tolled twice for good news of splashdown of *Apollo 11*. Sir Bernard Lovell, Jodrell Bank Experimental Station Director, said, "The successful conclusion of this immense project marks the beginning of a new phase when man must concern himself with the greatest issues of peaceful coexistence in extraterrestrial space."

Thunderstorm in Paris drove many people off streets at time of splashdown. On Riviera, bells tolled for five minutes and ancient cannon boomed.

Mayor Pascal Rossini of Ajaccio, Corsica, sent invitation to astronauts to visit Corsica during 1969 bicentennial of Napoleon's birth.

In Warsaw crowd of 300 Poles broke into applause at U.S. Embassy.

Over Pacific on Qantas airliner flying under Apollo reentry point, crew and 80 passengers saw space capsule reenter. In Canberra Prime Minister John Gorton invited astronauts to visit Australia.

Pope Paul VI sent telegram to President Nixon with prayer "that this immense achievement may foster peace and prosperity and scientific and moral progress for all mankind." Congratulatory messages were sent by President Giuseppe Saragat of Italy, President Yahya Khan of Pakistan, Prime Minister Eisaku Sato of Japan, President Chung Hee Park of South Korea, U.N. Secretary General U Thant, President Gustav Heinemann of West Germany, and Prime Minister John Gorton of Australia. (Collier, *NYT*, 7/25/69, 31; Mills, B *Sun*, 7/25/69, A6; AP, B *Sun*, 7/25/69, A6)

- More TV coverage of *Apollo 11* mission had been transmitted overseas via satellites to worldwide audience than of any previous event, ComSatCorp announced. More than 230 hrs of satellite time for 200 programs were transmitted during nine-day mission. Previous record was 225 hrs, set by Mexico Summer Olympic Games during 18 days in October 1968. Broadcasters estimated that 500 million persons were able to watch *Apollo 11*'s TV broadcasts in more than 40 countries on 5 continents. (ComSatCorp Release 69-46)

- During stop at Hickam AFB, en route to MSC from *Apollo 11* splashdown, *Apollo 8* Astronaut Frank Borman said it would be "helpful and hopeful for U.S. and U.S.S.R. to cooperate in space missions. He saw "indications" during his tour of U.S.S.R. that Russians would be interested, but "talk is cheap" and U.S.S.R. "is still supplying 85 per cent of the munitions to North Vietnam." He said U.S. had gone "95 per cent of the way" toward promoting cooperation. It was up to U.S.S.R. to do the rest. (UPI, *NYT*, 7/26/69, 12)

- USAF promoted *Apollo 11* Astronaut Michael Collins to full colonel. In congratulatory message Gen. John P. McConnell, Air Force Chief of Staff, said *Apollo 11* mission was "indeed a momentous achievement" and promotion was "token of appreciation for the part you played." (UPI, *NYT*, 7/25/69, 28)

- NASA Office of Space Science and Applications announced establishment of Earth Resources Research Data Facility at MSC, containing documentation from NASA and user agency investigators in Earth Resources Survey Program over past three years. Information was available for examination in facility by all interested persons. (NASA Ann)

- Rep. Louis Frey, Jr. (R-Fla.), introduced House Joint Resolution "providing for the establishment of the Astronauts Memorial Commission to construct and erect with funds a memorial in the John F. Kennedy Space Center . . . to honor and commemorate the men who serve as astronauts in the U.S. Space Program." Measure, cosponsored by House Committee on Science and Astronautics, was referred to Committee on House Administration. (*CR*, 7/24/69, H6293)

July 25: NASA launch from ETR of *Intelsat-III F-5* failed to reach planned synchronous orbit when 3rd stage of Long-Tank Thrust-Augmented Thor-Delta booster malfunctioned. Satellite entered low earth orbit with 3,354.8-mi (5,399-km) apogee, 167.2-mi (269-km) perigee, 146.7-min period, and 30.3° inclination instead of elliptical orbit with 23,000-mi (37,007-km) apogee and 175-mi (281.6-km) perigee. Mission, originally scheduled for launch in October 1969, had been rescheduled for July 17 to replace *Intelsat-III F-2*, which had stopped operating over Atlantic June 29. Launch had been delayed for technical reasons. (NASA Release 69-119; *SBD*, 7/29/69, 65; GSFC *SSR*, 7/31/69)

- *Apollo 11* recovery physician, Dr. William R. Carpentier, reported from inside Mobile Quarantine Facility onboard U.S.S. *Hornet* that astronauts had completed preliminary medical examination and were "fine." Astronaut Neil A. Armstrong's slight ear infection had disappeared and all three astronauts were in excellent condition. (Wooten, *NYT*, 7/26/69, 1)

- Two boxes of lunar samples from *Apollo 11* arrived at Lunar Receiving Laboratory in Houston, where they would be examined and used in experiments. (Wilford, *NYT*, 7/26/69, 1)

July 25

- Dr. Thomas O. Paine, NASA Administrator, told news conference aboard U.S.S. *Hornet* he expected U.S.S.R. to land men on moon within 18 mos. "My guess is it'll be much sooner than most people think." He thought U.S.S.R. had lost race "by keeping their program so secret." U.S. had encouraged suggestions from scientists throughout non-Communist world, while details of Soviet program were known only to "small elite." *Apollo 11* success would eventually lead to closer cooperation with U.S.S.R. in space exploration. "I don't look for any early change in the attitude . . . but a steady interest on their part. I don't see joint efforts but cooperation from time to time." (UPI, *NYT*, 7/25/69, 30)
- President Nixon arrived at Guam International Airport after flight from carrier *Hornet*. He said, "As I stand here and think of what happened today, the completion of that historic flight to the moon and the landing on the moon, I can say that I am sure all of us—all of the American citizens around the world—are proud today of what has happened. . . ." (*PD*, 8/4/69, 1033)
- Senate unanimously adopted S.R. 224, introduced by Sen. Michael J. Mansfield (D-Mont.) for himself and Sen. Everett M. Dirksen (R-Ill.), expressing gratitude on behalf of Senate and of all American people for "dedication, devotion, courage and effort of all associated with the Apollo program and with the Apollo 11 mission." (*CR*, 7/25/69, S8575)
- In telephone interview, evangelist and presidential religious adviser Billy Graham took issue with July 24 statement of President Nixon in welcoming *Apollo 11* astronauts back to earth. Graham told UPI, ". . . as a Christian, I would contend that there have been three much much greater days" than those of lunar landing and moon walk. They were first Christmas, day on which Christ died, and first Easter. While he did not wish to detract from "magnificent achievement," he felt "President was speaking extemporaneously. And I've found from years of speaking extemporaneously that in the excitement and emotion of a moment, you don't think through every statement you make." Associated Press later quoted Graham as saying, "I know that President Nixon agrees that the greatest single event in history was the coming of Christ"; he was sure President Nixon meant moon walk was probably man's greatest accomplishment. (*W Post*, 7/26/69, A10)
- Plans for proposed $1-million Neil A. Armstrong Aerospace Museum at *Apollo 11* astronaut's birthplace, Wapakoneta, Ohio, called for completion in 1970, Ohio Historical Society Director Daniel R. Porter said. (UPI, *W Post*, 7/26/69, B7)
- National and international press commented on successful completion of *Apollo 11* mission:

    Washington Post: "It has been eight days of triumph for America, eight days of triumph for mankind. Much more will undoubtedly follow as the secrets of space bow to the advances of science. But it is enough now—more than enough for an entire lifetime when you think about it—to have seen the first men walk on the moon and then, less than four days later, to welcome them back home safely." (*W Post*, 7/25/69)

    New York Times: "For the first time in history, men have gone from this earth to another celestial body, landed there and returned home, even bringing back with them extraterrestrial matter. Not since the human race evolved has there been a comparable event, nor one so capable of lifting all mankind's horizons, dreams and aspirations. What was fantasy to preceding generations is now accomplished fact. The achievement will be remembered so long as civilization survives."

Of President Nixon's round the world tour, *Times* said: "The spectacular success of Apollo 11 has vastly increased good feeling toward the United States throughout the world. The President obviously wants to capitalize on it both for foreign and domestic political purposes." (*NYT*, 7/25/69, 46)

Cleveland, Ohio, *Plain Dealer*: *Apollo 11* mission "closes out one aspect of the exploration program but opens wide the door of what can be an almost endless journey. . . . Although the Apollo program is not complete, the lure of Mars, 5 million miles away, grows in bold anticipation of the future. . . . the race has only just begun." (Cleveland *Plain Dealer*, 7/25/69)

Newport News, Va., *Times Herald*: "All of the money poured into the space program would appear justified if one of the side products was the kind of cooperation [with the U.S.S.R.] now possible." (Newport News *Times Herald*, 7/25/69)

El Rai El Amm, Khartoum, Sudan: "America achieved a victory for the human mind by sending the first man from the earth to the moon. . . . But America, the great power that achieved this astonishing big success, must stop doing things that are far below these standards." (Am Embassy, Khartoum)

Somali News, Mogadiscio, Somali: "It is true that the responsibility for the Apollo-Eleven is entirely American, but the message left behind on the moon for posterity by the astronauts . . . acknowledges the universal aspect of such a feat. We think . . . of those courageous astronauts not . . . as Americans but as worthy representatives of the human race on whose total achievement they relied in carrying out their mighty and splendid mission." (Am Embassy, Mogadiscio)

- Motion picture footage of *Apollo 11* lunar landing mission would be released for sale to commercial producers after quarantine period, NASA announced. Two 600-ft rolls would be made available initially: one would include prelaunch, launch, and recovery operations; other would include all usable onboard footage. (NASA Release 69-83L)

July 26: *Apollo 11* astronauts, enclosed in mobile quarantine facility (MQF), arrived at Pearl Harbor, Hawaii, where they were greeted by 12,000 cheering people and Mayor of Honolulu Frank F. Fasi. MQF was then transported to aircraft which would carry it to Lunar Receiving Laboratory in Houston. At LRL, scientists opened first of two boxes of lunar samples and made preliminary examinations of samples in one box. (Wooten, Wilford, *NYT*, 7/27/69, 47, 1)

- At lunar landing celebration dinner in Huntsville, Ala., Dr. Wernher von Braun, MSFC Director, said: "We worked together and together we accomplished our part of the mission. The moon is now accessible. And someday, because of the beginning that we have made here, the planets and the stars may belong to mankind. This reach toward the heavens, toward the stars, can eventually loose the human race from the confines of this earth and maybe even this solar system and give it immortality in the immense and never-ending reaches of space." For first time, "life has left its planetary cradle and the ultimate destiny of mankind is no longer confined. When the Mayflower landed on American shores the pilgrims did not envision the nation that would eventually evolve. Neither can we truly say what will eventually spring from the footprints around Tranquility Base." (Text)

- At state dinner in Manila, Philippine President Ferdinand E. Marcos exchanged toasts with President Nixon and commented on *Apollo 11*: ". . . we participate in the celebration of this achievement as man aspires for the stars, the stars outside of this world and the stars within himself and within his spirit. It is the hope of humanity, as it is the hope of the Philippines, that this vision and this genius, this courage and this ingenuity shall be utilized for the solution of man's problems." (*PD*, 8/4/69, 1036–7)

- *New York Times* interview quoted Dr. William H. Pickering, JPL Director: "Now that Apollo has been accomplished, rather than set another ambitious goal we should have a period of consolidation," during which "the balance should be increased toward unmanned effort." There was talk of exploring universe, "but the solar system is only a small part and it's going to be a long time before we venture out. We are making a very local exploration." He believed solar system exploration would pay off in understanding of history and evolution of solar system and, possibly, discovery of life on another planet and in social benefits. "The trouble with the social world is that we cannot agree on goals. We talk of weather control . . . but control for whom? The farmer or the sportsman or the businessman?" (Reinhold, *NYT*, 7/27/69, 47)

- Creation of U.N. Space Institute was urged by Columbia Univ. law professor Richard N. Gardner in *New York Times*. It would be "center for the cooperative planning of space exploration in which all U.N. members would be invited to take part." U.S. and U.S.S.R. could divide responsibilities for instrumented landings on different planets. There should be "United Nations Space Station" in outer space manned by astronauts from all U.N. nations and trained at U.N. Space Institute. It would gather information about solar system and universe and be used for practical earth applications. (*NYT*, 7/26/69, 24)

- London *Economist* editorial: "When Europe drew pride and status from its colonies, the Americans had none: the tables are turned now. While the United States rings July 21st red on its calendar, Europe faces the probability that when the planets are opened up we Europeans will have no part in doing it. The idea, at this late stage, of a European manned space programme is nonsense. The policy that would make more sense would be to approach the United States to see if the Administration will accept some foreign collaboration in the hugely expensive next years of its space programme. If the next American objective is Mars, a sensible Administration may welcome help and participation—especially if this excludes pressure to co-operate with the Russians. . . . There will be no opportunity in this generation that it would cost us more to miss." (*CSM*, 8/1/69)

- *Federal Register* published rule signed by NASA Administrator, Dr. Thomas O. Paine, which made unauthorized manufacture, sale, reproduction, or possession of official Apollo flight insignia, "or any colorable imitation thereof," misdemeanor punishable by $250 fine and six months in prison. (*Federal Register*, 7/26/69, 12332-4)

July 27: Lamont-Doherty Geological Observatory's Dr. Gary Latham said in Houston his team had detected 14 "unusual seismic events" from seismometers left on moon by *Apollo 11* astronauts. They believed walls of lunar craters had been falling in as different parts became hotter than others during highest lunar temperatures and felt they might be observing "initial stages of the process by which fresh new craters are transformed to old." (*W Post*, 7/28/69, A6)

- President Nixon toured Jakarta Fair during Indonesian visit. He offered to send Indonesian President Suharto and other world chiefs of state "a piece of the moon as a souvenir." In evening at state dinner in Jakarta, President Suharto said: "I underline Mr. Armstrong's momentous enunciation, when he, as the first human being, put his feet on the moon, declaring: 'These are small human steps which form a great leap to mankind.' This leap has occurred in the outer space, a very expansive space full of mysteries, but it has not taken place in this world of ours, which seems to be contracting and is relatively simpler. . . . It is the task of all nations in this world to realize peace and unity." (*NYT*, 7/28/69, 18; *PD*, 8/4/69, 1043-6)

- *Apollo 11* flight was public relations man's and reporter's dream, James Clayton said in *Washington Post*. NASA had kept "very little, if anything" from hundreds of U.S. and foreign press. More than 3,500 sets of press credentials had been issued at MSC and Cape Kennedy. Most went to Americans, but 55 foreign countries were represented, including Czechoslovakia, Yugoslavia, and Romania. There were 111 newsmen representing Japan among 800 foreign newsmen, several of whom had been waiting in Houston since *Apollo 10* flight May 18-26. Voice of America joined in transmitting news abroad. At peak, *Apollo 11* story was going out in 22 languages to every world area except some Communist countries. "Even those had the radio beams directed at them." (*W Post*, 7/27/69, B6)

- Psychological, technical, and political factors had combined to enable U.S. to win lunar landing race over U.S.S.R., said Harry Schwartz in *New York Times*. Moscow had shown overconfidence in underrating American capabilities, ignored lunar rendezvous technique adopted by U.S., and purged Nikita S. Khrushchev, who had been "enamored of space exploits and the propaganda they gave him." New Soviet leaders had changed priorities to concentrate on domestic problems. Since U.S. lunar landing, however, "two very different reactions are visible in the Soviet Union." Scientists, engineers, and many ordinary people were overcome with admiration. Ideologists and Soviet propaganda managers were deeply unhappy, "and their regret that it was not Soviet cosmonauts who went to the moon is scarcely hidden." (*NYT*, 7/27/69)

- Washington *Sunday Star* editorial said: "Apollo 11 has cast a harsh light on life on earth, showing man's failures in sharp contrast to his breathtaking technical achievements. It is a vision that should . . . be exploited as an incentive to get the vitally needed jobs done on earth." But, the U.S. could not withdraw from space. "The complete Apollo program . . . should be funded. Beyond that, serious consideration should be given to the establishment of permanent manned stations on the moon so that we may truly explore and perhaps exploit the new world that we have already conquered. The manned orbiting station that NASA has proposed should be provided to test the ability of men to live and work for long periods in weightlessness. And far more emphasis should be placed . . . on unmanned probes of the planets." Minimum requirement should be enough momentum in program to prevent it from falling apart through disuse. Every effort should be made "to enlist the cooperation, the technical help, and the financial support of any nation that is willing to contribute to the adventure that must, finally, be seen as the collective achievement of all mankind." (W *Star*, 7/27/69, E1)

- In Washington *Sunday Star* William Hines said: "Considering how very little he had to do with the whole enterprise, it is remarkable how much political mileage President Nixon got out of the flight of Apollo 11. The plaque, the phone call and the trip to greet the returning heroes all were benefits Nixon inherited rather than earned." Official NASA space age history *This New Ocean*, published by GPO in 1966, mentioned Nixon only once in 648 pages "and there hardly as an aggressive champion of manned space flight." Book said Nixon, as Vice President and as presidential candidate running against John F. Kennedy, had defended Eisenhower Administration's attitude toward space which ruled out manned flights to moon in foreseeable future. "The new President's belated enthusiasm blurs memories of the olden days," Hines said. "But 'This New Ocean' remains, proving perhaps that all government-sponsored history books should be armed to self-destruct whenever a change of administration occurs." (W *Star*, 7/27/69, E4)

July 28: JPL engineers sent signals to *Mariner VI* to turn on TV camera and scientific experiments that would measure Mars surface characteristics and atmosphere. Spacecraft (launched Feb. 24) began tracking Mars and would begin taking first of 33 far-encounter pictures 771,500 mi from Mars early July 29. Full-disc photos would be received at JPL July 29. (AP, B *Sun*, 7/29/69, A5)

- Geologists at Lunar Receiving Laboratory held press conference on *Apollo 11* samples and expressed surprise at discovery of tiny glass-like crystals in lunar dust. Analyses had revealed samples were crystalline, igneous, fragmented, scoriaceous, and vesicular. They confirmed theory based on *Surveyor V* data that lunar material contained titanium and indicated presence of number of minerals. Columbia Univ. scientist Dr. Paul Gast said, "The most exciting discovery to date has been that of the glass. There is something going on on the moon far different than on the earth." He said scientists speculated impact of meteoroids on moon had vaporized lunar material and caused it to rain back on surface in small drops which formed tiny yellow, brown, and clear pieces of glass few tenths of millimeter in diameter. (Lyons, *NYT*, 7/29/69, 1; Sehlstedt, B *Sun*, 7/29/69, A1)

- U.S. applied to Astronautic Committee of IAF for six world records based on *Apollo 11* achievements: duration of stay on lunar surface outside spacecraft, Astronaut Neil A. Armstrong, 2 hrs 21 min 15 secs; duration in lunar orbit, Astronaut Michael Collins, 59 hrs 27 min 55 secs; duration of stay on lunar surface, Astronauts Armstrong and Edwin E. Aldrin, Jr., 21 hrs 36 min 16 secs; duration of stay on lunar surface inside spacecraft, Aldrin, 19 hrs 45 min 52 secs; greatest mass landed on moon, Armstrong and Aldrin, 7,211 kg (15,897 lbs); greatest mass lifted into lunar orbit from lunar surface, Armstrong and Aldrin, 2,648 kg (5,837 lbs). Records would not be acknowledged officially until NASA presented confirming data and Federation officials approved. (*NYT*, 7/29/69, 16)

- At state banquet in Bangkok, Thai King Bhumibol Adulyadej toasted President Nixon: "Last week's breathtaking achievement of Apollo 11 and its brave American crew cannot be measured solely in scientific terms, for it also indicates man's ability to look beyond his earthbound problems and to set his sights on new horizons in quest of wider knowledge and deeper understanding of himself and his environment." (*PD*, 8/4/69, 1049–50)
- Gloom and embarrassment over *Apollo 11* success and crash of *Luna XV* on moon had caused controversy among Soviet leaders, including Communist Party Secretary Leonid I. Brezhnev and President Nikolay V. Podgorny, at July 21–23 meeting of Eastern European leaders in Warsaw, *New York Times* said. Reports of enthusiastic public response to Apollo feat across Eastern Europe had been interpreted as sign of lingering and latent sympathy for U.S. It was strongest in technologically advanced East Germany and Czechoslovakia, but had been noted as well in Poland, Hungary, and Romania. (Hofmann, *NYT*, 7/28/69, 7)
- U.K.'s Royal Geographical Society awarded special gold medal—its first for space exploration—to Astronaut Neil A. Armstrong for leading *Apollo 11* mission. Other gold medalists included Capt. Roald Amundsen, first to reach South Pole; Adm. Robert E. Peary, first to reach North Pole; Sir Edmund Hillary, conqueror of Mt. Everest; and Sir John Hunt, leader of Everest expedition. (AP, *W Star*, 7/28/69, A5)
- Senate Committee on Banking and Currency favorably reported S.J.R. 140 with amendments, providing for striking of medals honoring U.S. astronauts who had flown in outer space. (*CR*, 7/28/69, D681)
- MSFC announced resignation of M/G Edmund F. O'Connor (USAF), Director of Program Management, would be effective July 31. Gen. O'Connor, on loan to NASA from USAF for past five years, would become Vice Commander of Air Force Aeronautical Systems Div. He would be succeeded by Lee B. James, Saturn V Manager, MSFC. (MSFC Release 69–166)
- USAF released *Air Force Review of the C–5A Program*. Total cost of 120 Lockheed C–5A aircraft had increased from $3.369 billion at 1965 contract award to total $5.125 billion, overrun of $1.756 billion. At DOD press conference Air Force Secretary Robert C. Seamans, Jr., criticized "ambiguities and deficiencies" in original contract and hinted remaining 39 aircraft in 120-plane package might not be purchased unless revisions were made in contract. (Text; Phillips, *W Post*, 7/29/69, A1)
- *Il Mattino del Lunedi*, Asmara, Ethiopia, said of *Apollo 11* mission's completion: ". . . today we not only admire, but exult. Because this 'almost superhuman' exploit has been accomplished by a society which is free and pluralistic, by a society which has no close and oppressive traditions, by a society which has founded its political and constitutional structure not on a totalitarian ideology but on the democratic philosophy of the Declaration of Independence. It has been accom-

plished by a nation, the American nation, whose characteristic . . . is the fusion of the spirit of precision and discipline . . . with the spirit of freedom. . . . This is the reason why we today exult. Because we know that the conquest of Apollo–11 is in the service of man and not to oppress him." (Am Consul, Asmara)

- *Norrlandska Social-Demokraten,* Boden, Sweden, editorial commented that Russian press was surprisingly generous with praise of men behind *Apollo 11* and American space research in general during mission, but now press seemed to fear landing might have increased respect for U.S. around the world. "It is surely disturbing for *Pravda* and the Russian Party leaders that the American conquest of the moon . . . witnessed by the greater part of the Communist world, crushed the myth of the Communist system's superiority." (Am Embassy, Stockholm)

- Within 76 hrs after *Apollo 11* splashdown, Bantam Books and *New York Times* published *We Reach the Moon,* 416-page paperback account of U.S. space program from 1961 through *Apollo 11*'s success. Early publication was effected by nearly 2½ yrs of planning. Book went to press immediately after July 24 splashdown while aerospace reporter John Noble Wilford was completing text. Final copy was telexed to Chicago printer July 25. First printing comprised 375,000 copies. Hardcover edition would be published by W. W. Norton & Co. in September. (*NYT,* 7/29/69, 16)

July 29: First pictures of Mars taken by NASA's *Mariner VI,* launched Feb. 24 to fly by Mars equator, were received at JPL. Full-disc photos, taken between 771,500 and 450,000 mi from Mars, were flashed on screen every five minutes. They showed Mars as dull, gray, egg-shaped body with crack in surface and bright spot—southern polar cap—with ragged edge. Better pictures were expected as spacecraft traveled closer to Mars. (AP, *B Sun,* 7/30/69, A1; Lannan, *W Star,* 7/30/69, A3)

- JPL radar readings which showed 8.3-mi altitude variation in Mars' north equatorial zone and included corrected figures for Mars' ephemeris, or orbital path, were expected to ensure accuracy of TV cameras aboard *Mariners VI* and *VII,* NASA said. Experimenters hoped to obtain photos identifying objects 900 ft across at close approach and pictures were expected to be 500 times better than those taken to date by earth-based telescope cameras. Readings were obtained at NASA's Goldstone Tracking Station in California by team directed by Dr. Richard M. Goldstein during planet's closest approach to earth (Mars had been within 45 million mi of earth June 9). They would be of great interest to astronomers because they showed that areas which appeared light to telescopes might be either high or low in elevation. Optically dark areas appeared to be of medium elevation to radar-scanners. (NASA Release 69–111; JPL Release 530)

- NASA released first photos taken by *Apollo 11* astronauts on and near lunar surface, including four color stills and 16-mm film of LM descent. Film opened as LM swung low and curved slightly over area pocked with craters and rocks and showed dust being scattered by exhaust as LM touched down safely. It then showed Astronaut Neil A. Armstrong as he descended ladder to surface, took first step on moon, and deployed initial equipment.

Still photos showed closeup of brownish surface sprinkled with footprints, silhouettes of LM and U.S. flag, Armstrong inside LM, and earth with Europe, Africa, and Asia visible. (Witkin, *NYT*, 7/30/69, 1; Cohn, *W Post*, 7/30/69, A1, A3)

- LRL scientists continued examining lunar samples and preparing them for experiments on living organisms. Experiments, scheduled to begin July 29, would be delayed one day to repair cracked glove which permitted scientists outside vacuum box to handle objects inside and to allow more time for grinding samples to uniform size. (AP, *NYT*, 7/30/69, 19)
- NASA Wallops Station announced award of 40-mo, $936,311 contract to Rice Univ. to investigate relationship between field-aligned currents and auroral particle fluxes and document and summarize findings. Rice would construct and test suitable flight and ground instrumentation for three Nike-Tomahawk sounding rocket payloads; prepare and preflight-test payloads; and acquire, record, reduce, analyze, and publish resulting magnetic and auroral particle data. (WS Release 69–14)
- FCC, at White House request, decided to delay for 60 days decision on establishment of domestic comsat system to enable Nixon Administration to study issues and make recommendations. (Aug, W *Star*, 7/29/69)
- Rep. William G. Bray (R-Ind.) introduced H.J.R. 844, providing for distribution of *Apollo 11* lunar samples to Governors of 50 states. (*CR*, 7/29/69, H6486)
- New York weathermen were being deluged with calls blaming 10 days of rain and overcast weather in northeastern U.S. on *Apollo 11*, Associated Press said. WCBS radio news meteorologist Dr. Robert Harris had said, "We've had an abundance of calls from all sorts of people who are absolutely certain, through their Bible studies, that the Lord has taken the sun away from us." (AP, B *Sun*, 7/30/69, A6)
- National Assn. of Government Employees President Kenneth T. Lyons told House Interstate and Foreign Commerce Committee landing on moon would soon be safer than landing at most U.S. airports. "Do we have to have NASA take over from the FAA in order to get a little sense into our airport and aircraft traffic management jumble?" (Bentley, B *Sun*, 7/29/69, A5)

July 30–31: NASA's *Mariner VI*, launched Feb. 24 on Mars equatorial flyby mission, approached Mars and completed 17 pictures of planet taken at about 111,400-mi altitude, which showed ragged edges of polar cap, W-shaped cloud, and seas, deserts, and craters seen by *Mariner IV* in 1965.

As spacecraft neared and swung around Mars it took 24 close-up pictures from about 2,000 mi at closest point. Pictures were so sharply defined and detailed that they were shown live on TV instead of being refined and released later in photographic prints as originally planned. Pictures—enhanced by computers at JPL to clear out static, highlight images, adjust contrast and brightness, and exaggerate features—were spectacular. They showed that Mars was heavily cratered and looked very much like moon. One photo showed 11-mi-dia crater closely resembling moon's Copernicus crater and diagonal ditch resembling lunar rille. During closest approach, onboard TV cameras took 12 high-resolution and 12 medium-resolution pictures, stored some on board for later playback, and transmitted some immediately to ground stations for conversion to images at JPL. Three of four onboard experiments—TV to take pictures, UV spectrometer to identify and measure gases in upper atmosphere, and infrared radiometer to measure planet's temperature—functioned satisfactorily. Only anomaly was failure in cooling of one channel on infrared spectrometer, designed to identify gases in lower Martian atmosphere, which prevented proper acquisition of data.

JPL controllers temporarily lost contact with second Mariner, *Mariner VII*, en route to Mars, at 6:00 pm EDT July 30. Engineers speculated that spacecraft had been thrown out of alignment when struck by micrometeoroid traveling at 40 mps and had locked on planet Jupiter or another bright object. Contact with *Mariner VII* was regained seven hours later by switching from one antenna to another and proper attitude was restored by rolling spacecraft around until it locked on star Canopus. Although some of data being transmitted appeared to be abnormal, flyby mission was still expected to succeed.

Mariner VI would continue taking pictures and play back recorded near-encounter data during final phase of mission. Data would be compared with data from *Mariner VII* (launched March 27), which would fly past Mars polar region Aug. 4. (NASA Release 69-26A; Sullivan, *NYT*, 8/1/69, 1; Auerbach, *W Post*, 8/1/69, A1; Lannan, *W Star*, 7/31/69, A5; NASA News Release, 9/11/69)

July 30: On arrival in Saigon, Republic of Vietnam, President Nixon said: "I am happy that the moon landing, which in its universality signifies a symbolic drawing together of all mankind, has provided an occasion for me to meet with President Thieu in the capital of his country."

Later, after discussions with President Nixon, President Nguyen Van Thieu said, "The Vietnamese people fully concur in the message of peace which the three brave American astronauts deposited on the moon for all mankind." (*PD*, 8/4/69, 1051-4)

• During *Apollo 11* celebration, credit should be given to former NASA Administrator James E. Webb, "whose organizational skill, vision and drive played a major part in its success," MIT Provost, Dr. Jerome B. Wiesner, and MIT physicist Jerrold Zacharias said in letter to *New York Times*. "There never was any question regarding the technical feasibility of a manned lunar landing. The real question was whether or not we could organize and manage so large and complex a program on the time schedule laid down by President Kennedy." Webb had organized, defended, and managed program, "and as the world celebrates this great technical and human achievement we should also honor the man who directed its accomplishment." (*NYT*, 8/5/69, 32)

• LRL scientists began injecting pulverized lunar samples into sterile white mice in attempt to discover germs or chemicals hazardous to human beings. Mice, born by Caesarean section and raised in sterile environment so that they would be extremely sensitive to infection, would also have samples mixed in their food and air. (UPI, *W Star*, 7/31/69, A5; AP, *B Sun*, 7/30/69, A1)

• World Health Organization Director General, Dr. M. G. Candau, and Dr. Karel Raska, Director of WHO's Communicable Disease Div., said in

Houston that Soviet scientists had "initiated" plans for lunar receiving laboratory. Soviet delegates to international conferences had discussed subject but no details were available. WHO officials were in Houston to observe LRL at U.S. Government invitation. (*W Post*, 7/31/69, A3)

- After two-hour inspection of Tu-144 at Moscow's Sheremetyevo International Airport, Pan American World Airways president Najeeb E. Halaby said Soviet supersonic transport had left group of U.S. aviation experts "very, very impressed." U.S.S.R. apparently had progressed further in testing than U.K. or France with Concorde, and Tu-144 had reached 900 mph, breaking sound barrier several times. Concorde hoped to reach mach 1 in six months and U.S. SST was at least five years behind. Aeroflot planned to put Tu-144 in service by 1973. Pan Am would review all information available before deciding whether to order aircraft as hedge against competition. Halaby liked Tu-144's design and advanced instrumentation and was impressed with amount of titanium used in construction. Russians had told him aircraft's noise level was low in landings and takeoffs. (*NYT*, 7/31/69, 58)
- Senate passed S.J.R. 140, providing for striking of medals honoring American astronauts who had flown in outer space. (*CR*, 7/30/69, S8786)
- Subcommittee on Science, Research, and Development of House Committee on Science and Astronautics published *Science, Technology, and Public Policy During the Ninetieth Congress*. Report covered 1967–1968, giving details behind 94 public laws passed that authorized, funded, or otherwise affected R&D in U.S. and 45 additional bills on which Congress took legislative action. It included reviews of U.S. policy for science and technology by Organization for Economic Cooperation and Development and by NSF for United Nations Educational, Scientific, and Cultural Organization. Both reviews showed pluralistic nature of U.S. public policy for science, built up by laws, executive orders, and other expressions of policy as they occurred. (Text)

July 31: USAF launched unidentified satellite from Vandenberg AFB by Thor-Agena booster into orbit with 333.1-mi (536-km) apogee, 288.9-mi (464.8-km) perigee, 94.6-min period, and 75.0° inclination. (GSFC *SSR*, 7/31/69; *Pres Rpt 70* [69])

- At state dinner in New Delhi, India, Acting President Mohammed Hidayatullah exchanged toasts with President Nixon and congratulated him: "The epic flight to the moon and back by three of your countrymen has amazed the world and marks a new stage in science and technology. On behalf of the Government and people of India, and myself, I congratulate you, and through you, the people of your country on this historic occasion. ... We are glad to know that you are sharing the knowledge you have gained with the rest of the world." (*PD*, 8/4/69, 1056–9)
- Sequence of five color photos of *Apollo 11* Astronauts Neil A. Armstrong and Edwin E. Aldrin, Jr., performing extravehicular activities on lunar surface were released by NASA. Vivid sequence showed Aldrin descending ladder to surface, walking near LM, posing near U.S. flag, deploying seismometer, and walking with Armstrong's reflection visible in his visor. NASA also released two-part 16–mm film which showed moon

fading away as LM ascended and LM's rendezvous with CSM in lunar orbit. It also showed Astronaut Michael Collins shaving inside CSM. (*W Post*, 8/1/69, A7; Witkin, *NYT*, 8/1/69, 16)

- Hans H. Maus, Director of Executive Staff at MSFC, and Dr. George N. Constan, Director of Michoud Assembly Facility, retired after combined total of 51 yrs Government service. Maus, expert in rocket development and production engineering, had received USA's Exceptional Civilian Service Award and number of citations for development of manufacturing methods, process automation, assembly, and tooling concept development. Dr. Constan had served with USA at Milan, Joliet, and Redstone Arsenals before his appointment to Michoud in 1961. (MSFC Release 69–167)

- Soviet Academician, Dr. Anatoly A. Blagonravov, conceded that competition with U.S.S.R. might have been major factor in U.S. determination to reach moon and said that in space there was no way to declare a winner, *Space Business Daily* reported. "I don't preclude the idea that such a boosted preparation of the Apollo project was in some measure the result of competition with us. Basically a healthy competition is no obstacle to success.... Science is boundless in its development and it cannot be compared to a horse race—there is no finishing line. The interests of science are bound to win anyway...." U.S.S.R. would continue research in "several major scientific areas," make "extensive use of automatic devices for exploring outer space," and pay "due attention" to moon and to both manned and unmanned missions. Cosmos, Zond, and Proton spacecraft would continue to be used for research and Soyuz spacecraft would be converted into "modules of orbital space laboratories designed for research in lengthy flight." (*SBD*, 7/31/69, 79)

- Man's knowledge of Venus, Mars, and moon had been enormously enhanced by unmanned Mariner missions, *New York Times* editorial said. They were relatively inexpensive and did not risk human lives. "Nevertheless, American political leadership has been so obsessed with sending a man to the moon that unmanned probes of the planets became the stepchildren of the national space program. There were times when even the continued existence of the Jet Propulsion Laboratory—the center for these unmanned flights—seemed in doubt. Now, in the new phase of American space exploration begun in the wake of *Apollo 11*'s historic achievement, the major cost-benefit advantages of Mariner type unmanned flights need to be more fully appreciated by Washington policy makers, and even more intensively exploited than in the past, even as the manned exploration of the moon continues." (*NYT*, 7/31/69, 32)

During July: NASA-appointed Astronomy Missions Board recommended long-range program in space astronomy to NASA. Board of 19 leading U.S. astronomers chaired by Dr. Leo Goldberg, Harvard College Observatory Director, had been appointed in 1967 to propose programs for 1970s. NASA published Board's report in November [see Nov. 9]. (Text; NASA Release 69–149; NASA OSSA)

- Data were relayed by two Vela nuclear detection satellites launched May 23 that led to discovery of x-ray star between constellations Centaurus and Lupus [see Aug. 14]. (Sullivan, *NYT*, 8/14/69, 7)

- NAS published *Physics of the Earth in Space: The Role of Ground-Based*

Research, report of study by Committee on Solar-Terrestrial Research of Geophysics Research Board of NRC. Among recommendations were new facility to exploit incoherent-scatter techniques for investigation of ionospheric and magnetospheric dynamics, program of controlled sounding of magnetosphere based on new VLF transmitter facility near 60° invariant latitude in Antarctic, and relocation of Stanford Research Institute's incoherent-scatter facility at Palo Alto, Calif., to auroral zone at College, Alaska, to measure F-region electron density and temperature and ion temperature. Report advocated increased support for specific solar radioastronomy techniques, improvements in balloon technology, and acquisition of ground-based geophysical data via satellite links, possibly using comsat channel. (Text)

- NAS report to Congress, *Technology: Processes of Assessment and Choice,* was published by House Committee on Science and Astronautics. Panel headed by Harvey Brooks of Harvard Univ. strongly urged creation of "constellation of organizations, with components located strategically within both political branches, that can create a focus and a forum for responsible technology-assessment activities throughout government and the private sector." Such organizations "must be separated scrupulously from any responsibility for promoting or regulating technological applications." (Text)

- House Committee on Science and Astronautics published *A Study of Technology Assessment: Report of the Committee on Public Engineering Policy, National Academy of Engineering.* Report recommended establishment of technology assessment task force of members of public and private organizations with knowledge of subject under assessment, including behavioral and political scientists. (Text)

- NSF published *Research and Development in Industry, 1967: Funds, 1967; Scientists and Engineers, January 1968* (NSF 69-28): Historically, about 70% of U.S. R&D had been performed by industrial sector. In 1967, industry spent $16.4 billion for R&D, four and half times 1953 level of $3.6 billion. Federal agencies financed 51% of 1967 total, or $8.4 billion. Ratio was down from 1959 high of 59%. NASA and DOD furnished 89% of Federal funds to industry in 1967 and supported 89% of 158,000 R&D scientists and engineers working on Federal programs in January 1968. NASA's R&D cost was $55,400 per scientist or engineer, while DOD spent $51,600. Industry spent $8 billion of its own funds for R&D in 1967, 11% more than in 1966 and 265% more than in 1953.

    In January 1968, 387,900 full-time-equivalent R&D scientists and engineers were in industry sector, of which 59% worked on company-financed R&D projects. Industrial spending for basic research reached $655 million in 1967, 5% higher than previous year and more than four times 1953 level. (Text)

- MOL cancellation "should at most be a 'postponement,'" Dr. Edward C. Welsh, former NASC Executive Secretary, said in *Air Force/Space Digest.* "Contrary to assertions made by people who should know better, the MOL was not planned as a weapon system and would not have been a threat to any other nation." MOL observations would be "as peaceful as those obtained on the NASA Gemini and Apollo flight. Men on board the spacecraft can be justified by the contributions men make in matters of choice of observations, maintenance, and communication with earth." MOL would not duplicate NASA's Apollo Applications program. "To try to combine the Air Force and NASA manned programs would waste much of the investments already made, would delay both programs, would increase the total cost over the long run, and would violate the sound administrative principle of having the experts do what they have been trained to do. Failure to get a maximum return from this national-security system would seem to be woefully shortsighted and wasteful." (*AF/SD*, 7/69, 60–1)

- American Embassy science attaché in New Delhi reported completion of India-U.S. project to erect 48-in telescope at Hyderabad. Project was started in 1955 and completed just before U.S. lunar landing. (O'Neill, *W Post*, 8/31/69, D5)

August 1969

August 1: Univ. of California's Lick Observatory successfully recorded first hits on laser reflector left on moon by *Apollo 11* astronauts. Hits, which came after 2,000–3,000 unsuccessful attempts by Lick and Univ. of Texas's McDonald Observatory, were made by Lick's 120-in telescope —world's second largest. Scientists fired 500 pulses with pure red beam of ruby laser. Each pulse lasted 15–20 billionths of a second, reached moon in 1.3 secs, and bounced back in same time. Target— 18-in-square panel of 100 three-faced prisms of fused silica—was hit about three-fourths of time. (GSFC Historian)

- JPL engineers reported *Mariner VII*, en route to Mars, had suffered sudden change in velocity—possibly because of gas leak from pressure can in infrared spectrometer—that could throw off its approach to Mars. Also, 20 of 92 telemetry channels, including one that aimed TV camera platform toward Mars, had not operated properly after controllers lost contact with spacecraft July 30. When engineers turned on TV cameras, however, they began taking pictures that appeared to be of Mars. (Auerbach, *W Post*, 8/2/69, A4; NASA News Release, 9/11/69)

- Dr. Wilmot N. Hess, Director of Science and Applications at MSC, announced he would leave NASA in September to become Director of Research Laboratories for ESSA in Boulder, Colo. His successor had not yet been selected. Dr. Hess said he was taking new position because job was challenging one in growing organization with important mission. "We have passed a milestone in the manned space flight program by the recent lunar landing. We have put the Lunar Receiving Laboratory into operation and it is performing its mission well. We have placed instruments on the moon successfully and have the scientific program for the next several lunar missions well organized." Before going to MSC, Dr. Hess served as Chief of Laboratory for Theoretical Studies at GSFC, 1961–1967. (MSC Release 69–54)

- *Intelsat-III F-2* comsat (launched Dec. 18, 1968), which had stopped operating June 29 when mechanically despun antenna malfunctioned, was restored to service. *Intelsat I* (*Early Bird*), reactivated June 30, had worked with *Intelsat-II F-3* to provide service during interruption. (ComSatCorp Release 69–49)

- President Nixon arrived at Lahore, Pakistan, during round-the-world journey. Pakistan President Yahya Khan said at airport, "The City of Lahore is happy to receive you on its historic soil and to share your joy at the most recent and the most memorable triumph of human courage, determination, and scientific skill which was achieved by your astronauts when they were first to land on the moon." (*PD*, 8/4/69, 1060)

- Dr. Thomas O. Paine, NASA Administrator, outlined possible 1981–1982 manned mission to Mars in speech before Commonwealth Club in San

Francisco. With "window opening" on Nov. 12, 1981, expedition "would set forth from earth orbit [where spacecraft were assembled] in two six-man vehicles, each propelled by three nuclear rockets." At end of Mars injection burn by two outer rockets, rockets would be disengaged and return to earth orbit for later reuse. Third rocket would remain unfired as spacecraft coasted to Mars. On nine-month journey spacecraft could be joined and spun to provide artificial gravity.

"On August 9, 1982, the craft would arrive at Mars, and the unused rocket fired to brake each ship into Mars orbit. About three months would be spent orbiting Mars while two surface landers . . . took astronauts down to surface for a month-long stay. These landers would also be three-man laboratories in which men could live and work productively on the surface of Mars. At the end of their surface research the astronauts would rendezvous again with the spacecraft overhead, and then begin the return voyage by firing the nuclear engine again on October 28, 1982." Spacecraft would swing by Venus Feb. 28, 1983, using Venus' orbital motion around sun to retard it and sling it toward earth. "Returning to earth on August 14, 1983, the nuclear rocket would fire for the third time to put each space ship into earth orbit. The crews would return to earth via the shuttle. After refurbishing, the space ships would be available for the next voyage."

To hold option open for Mars voyages in 1980s, U.S. should in 1970s develop reusable shuttle for flight between earth and low earth orbit, permanent orbiting space station, and nuclear rocket propulsion. (Text)

- LRL technicians Ronald J. Buffum and George E. Williams, accidentally exposed to lunar samples when glove used to examine samples cracked, were placed in quarantine with *Apollo 11* astronauts, 2 doctors, and 12 technicians, cooks, and other employees. (*W Post*, 8/2/69, A4)
- James L. Stamy, Deputy Manager of Michoud Assembly Facility since 1962, became Acting Manager, replacing Dr. George N. Constan, who retired July 31. (MSFC Release 69–169)
- U.S. Army Collateral Investigation Board appointed to investigate March 12 crash of AH–56A Cheyenne helicopter, in which civilian pilot David A. Beil lost his life, issued report. It found accident was caused by divergent, low-frequency, main-rotor oscillation and pilot had been killed by rotor blades. Manufacturer, Lockheed-California Co., had "failed to exercise due care and judgment in the planning and execution of flight 288 and in so doing failed to adhere to an acceptable level of sound industrial practice." (Text)
- M/G James T. Stewart, former Vice Director of USAF's Manned Orbiting Laboratory (MOL) program, had been named AFSC Deputy Chief of Staff for Systems, Gen. James Ferguson, AFSC Commander, announced. Gen. Stewart would replace M/G John L. Zoeckler, who retired from USAF July 31. (AFSC Release 130.69)
- USN announced award of $461-million contract to Lockheed Aircraft Corp. for development of S–3A carrier-based antisubmarine-warfare aircraft, formerly designated VSX. Contract, to be funded over five years, was for 6 R&D aircraft with option to procure 193 production models, contingent upon successful development phase. (DOD Release 647–69)

August 2: Press conference on *Mariner VI* preliminary results was held at JPL. Scientists reported experiments had revealed that recurring white blob seen in previous Mars pictures was 300-mi wide crater with peak in center; linear features known as canals were actually large, irregular, low-contrast splotches without specific detail; temperature in equatorial area ranged from 75°F to —100°F; atmosphere was almost nonexistent; Mars had no sharply defined borders separating light and dark areas; and Martian surface was more heavily cratered than previously believed.

Dr. Charles A. Barth of Univ. of Colorado said any life on Mars would be very different from life on earth, perhaps form that used carbon dioxide. He said UV spectrometer had found atomic carbon and carbon monoxide, but no traces of nitrogen—essential to life on earth.

Dr. George C. Pimentel of Univ. of California at Berkeley said infrared spectrometer had detected presence of unknown compound related to methane, building block of life on earth. He also reported detection of super-thin layer of water ice hanging in atmosphere above Mars equator. (Auerbach, *W Post,* 8/3/69, A3; Lannan, *W Star,* 8/3/69, A5)

- NASA's *Mariner VII* televised two good test pictures before start of its first series of 34 approach shots more than 1 million mi from Mars. (AP, *W Star,* 8/2/69, A3; NASA News Release, 9/11/69)
- Initial results of tests at MSC's Lunar Receiving Laboratory in which mice were exposed to lunar samples showed no indication of life on ported. All 24 sterilized mice that had lunar dust injected into their stomachs July 31 and 240 mice inoculated Aug. 1 were "alive and kicking.... They have shown no untoward reaction to the sample and seem to be in very good health." (AP, *W Star,* 8/3/69, A5)
- Lick Observatory scientists said at news conference they had measured distance between earth and moon to be 226,970.9 mi, based on data from Aug. 1 test in which laser beam successfully hit reflector on moon, LRL preventive medicine specialist Dr. Norman D. Jones remoon. Figure was accurate to within 150 ft and eventually might be pinned down to inches. (AP, *NYT,* 8/4/69, 13)
- Romanian President Nicolae Ceausescu met President and Mrs. Nixon on arrival at Otopeni Airport, Bucharest. President Nixon replied to welcome: "... this significant moment in the history of relations between our two countries coincides with a great moment in the history of the human race. Mankind has landed on the moon. We have established a foothold in outer space. But there are goals we have not reached here on earth. We are still building a just peace in the world. This is a work that requires the same cooperation and patience and perseverance from men of good will that it took to launch that vehicle to the moon." (*PD,* 8/4/69, 1065)
- *Washington Post* editorial: "It is not often that the public has a chance to share in the day to day unraveling of scientific mysteries. The men and women who engage in basic research prefer to work quietly in laboratories and eventually announce their findings in the atmosphere of scholarly meetings or academic publications. But at Houston and Pasadena [MSC and JPL] these days, the public has become a silent observer of the plodding work that goes into basic research. Regardless of the drama that is involved, the study of the rocks brought back by

Apollo 11 from the moon and of the pictures being transmitted back by Mariners 6 and 7 as they fly past Mars is simply basic research. Although results are trickling out each day, the dimensions of each discovery are hard to measure and an understanding of their cumulative impact is likely to be long in coming." (*W Post*, 8/2/69, A12)

August 3: At Andrews AFB, on return from world tour, President Nixon said: "In Bucharest I noted that so many, particularly of the young people, held up a newspaper picture of the astronauts landing on the moon, and everywhere we went it was the same. Some way, when those two Americans stepped on the moon, the people of this world were brought closer together. . . . I really feel in my heart that it is . . . the spirit of Apollo, that America can now help to bring to all relations with other nations. The spirit of Apollo . . . can bring the people of the world together in peace." (*PD*, 8/4/69, 1071–2)

- *New York Times* published interview in which Grumman Aircraft Engineering Corp. President Llewellyn J. Evans expressed concern over possibility of failure in future space missions. "It has been one big gamble up to this point. This country must come up with rescue hardware. It would be shocking if someone got stuck in orbit someplace." He saw need for four space facilities: space station in earth or lunar orbit, shuttle for travel between earth and space laboratory, space "tug" to go between nonatmospheric orbits, and rescue vehicle. (Kampel, *NYT*, 8/3/69, F7)

- *New York Times* editorial commented on *Apollo 11* lunar landing and *Mariner VI* Mars mission: "Future generations may well regard the last two weeks of July 1969 as the most revolutionary and significant fortnight of the entire twentieth century. Not for 300 years has any comparable quantum leap in man's knowledge of the cosmos taken place in so brief a time." (*NYT*, 8/2/69, 10)

- There was no question that manned Mars mission could be "organized, equipped and flown, possibly by 1985 or 1986," William Hines said in Washington *Sunday Star*. "But the cost of such a flight would be tremendous." Apollo had cost $25 billion over eight years. Project Mars "would cost four times as much over a period twice as long." Taxpayers and legislators "should listen to the professional pitchmen of space with a dubious ear, demanding facts instead of the sort of rhetoric Dr. George E. Mueller delivered on Apollo 11 splashdown day." (W *Star*, 8/3/69, C4)

August 3–4: Photos of Mars taken from 65,000-mi altitude by NASA's *Mariner VII* were received by JPL and shown live on TV. Although pictures were clear, canals were barely visible as dark splotchy areas, indicating they were not sharply defined features as previously believed. Viewers saw 100-mi-wide, 750-mi-long dark streak identified as Agathadaemon canal, Cerberus canal in light Plateau Elysium area, and Martian south pole with craters filled with substance resembling snow or ice. Pictures showed white grid pattern around Nix Olympica, identified by *Mariner VI* photos as 300-mi-wide crater. Absence in *Mariner VII* photos of bright streak on Tempe desert near Mars north pole that had been visible in *Mariner VI* photos suggested meteorological phenomenon similar to earth's seasonal changes. South polar cap, which was 2,500 mi across in *Mariner VII* photos, shrank to 250 mi across in Martian summer and increased to 3,500 mi across in winter. (Auerbach, *W Post*, 8/5/69, A1)

August 4: Scientists at Lunar Receiving Laboratory opened last box of *Apollo 11* lunar samples containing charcoal-gray dust and assorted rocks ranging from gravel to size of orange. NASA geologist Dr. Jeffrey L. Warner described rocks as "different from anything we have on earth." Some of rocks had flat faces and appeared to have been broken off larger chunks of material. Rocks in first box of samples had been rounded. Some rocks contained unidentified crystals that sparkled; others had "an unusual smattering of what appeared to be metallics," possibly ilmenite (iron-titanium mineral oxide), important source of titanium. (UPI, *W Post*, 8/5/69, A6; AP, *B Sun*, 8/5/69, A1)

- NAS–NRC Space Science Board issued *The Outer Solar System: A Program for Exploration.* Report detailed program for unmanned exploration from 1974 to early 1980s; reaffirmed goals set by earlier study emphasizing experiments contributing to understanding of origin and evolution of solar system, of life, and of dynamic processes in terrestrial environment; and agreed exploration would concentrate on planets but time in flight would permit study of interplanetary medium. Missions recommended were 1974 Jupiter deep-entry probe and flyby, 1976 Jupiter orbit, 1977 earth-Jupiter-Saturn-Pluto probes, 1979 earth-Jupiter-Uranus-Neptune probes, and earth-Jupiter-Uranus entry probes in early 1980s. Vigorous national program could be developed for small fraction of total NASA program cost and increased portion of space budget should be devoted to planetary exploration. Report, originating from June 1968 study chaired by Dr. James A. Van Allen of Univ. of Iowa and Dr. Gordon J. F. MacDonald of Univ. of California at Santa Barbara, recommended NASA include long-term outer solar system exploration plan in 1971 congressional budgetary presentation. (Text)

- NASA's *Pegasus III* meteoroid detection satellite, launched July 30, 1965, reentered earth atmosphere at 2:04 am CDT over Indian Ocean at 3.4°N. latitude and 56.7°E. longitude. *Pegasus III* was last in series of three Pegasus satellites with 96-ft-long detector panels launched to determine frequency of meteoroids in near-earth environment. All three had been turned off in 1968 after operating for more than double design lifetime. Few hours before reentry, controllers commanded *Pegasus III* beacon to begin operating again and beacon functioned satisfactorily until satellite was destroyed by reentry heat. (MSFC Release 69-170; GSFC *SSR*, 8/15/69)

- ERC announced it had developed and successfully flight-tested "Flying Baton," simple, low-cost device to provide eye-level artificial horizon for pilots. Developed by Center's William J. O'Keefe, device could contribute to more "head-up" flying, be used for precision attitude flying, and allow pilot more time to look outside aircraft. (ERC Release 69-19)

- DOT and HUD announced $166,734 project for studies to recommend short- and long-term relief from aircraft noises at John F. Kennedy International Airport, New York; O'Hare International Airport, Chicago; Bradley International Airport, Hartford, Conn.; and Cape Kennedy Regional Airport, Fla. Studies were to define noise problems, to identify activities affecting problems, to identify approaches to land

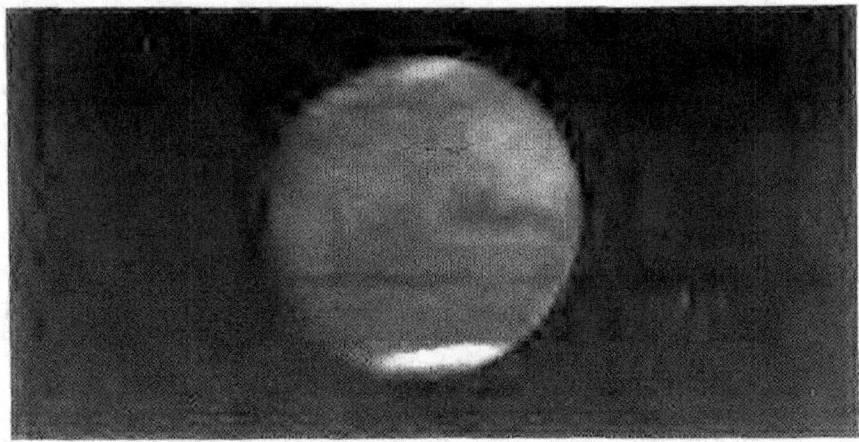

August 2: JPL *press conference reported new information on Mars from the Mariner VI flyby. The far-encounter photo above, taken 463,250 miles from Mars July 29, showed Meridiani Sinus and Sabacus Sinus as a dark feature near the equator. The bright area at the top was Cydonia. The south polar cap showed at the bottom of the planet.*

August 4–5: Mariner VII *transmitted first close-up photo of Mars' south polar cap. In the photo below, taken 3,300 miles from Mars Aug. 4, the south pole was believed to be at the lower right. Three large craters showed partly bare floors. Snowdrift-like formations and an irregular cloud-like object (upper left) were apparent.*

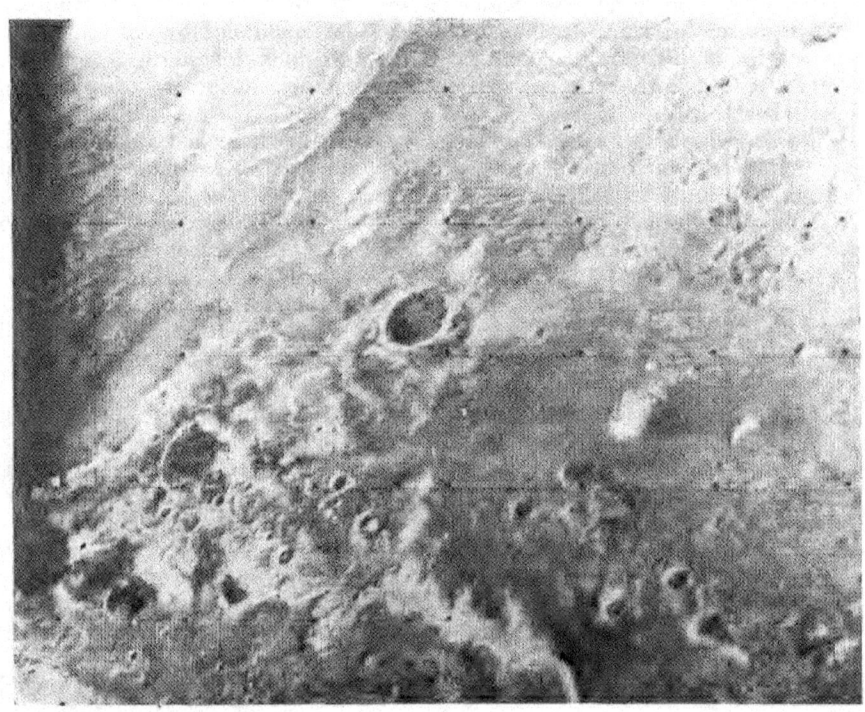

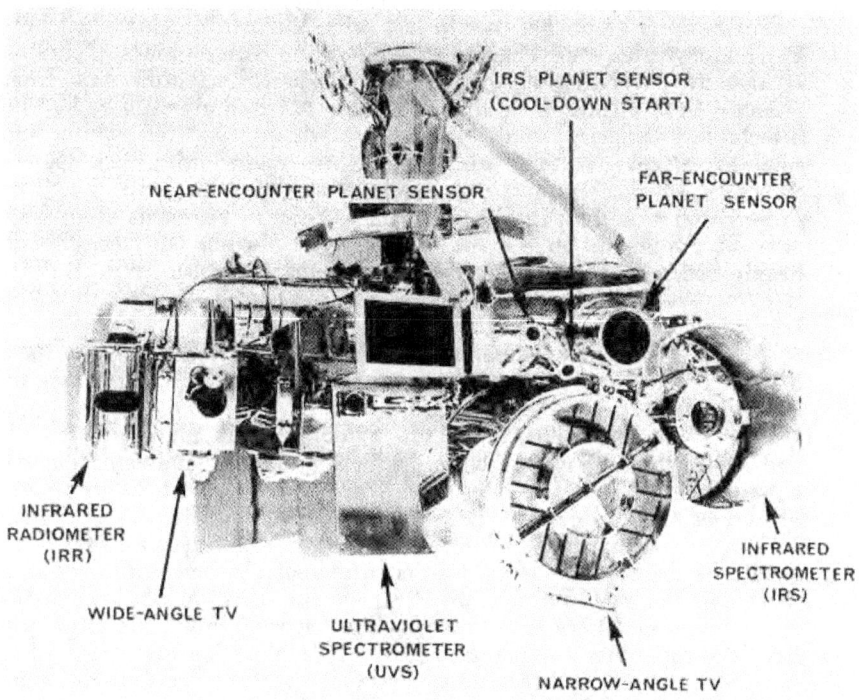

August 4–5: *Scientific instruments on board* Mariner VII *collected detailed data on the Mars surface and atmosphere as the spacecraft made its close-encounter flyby of the southern hemisphere and polar regions.* Mariner VI *used similar instruments to gather information on the equatorial regions of Mars on its July 30–31 flyby.*

use compatible with airport locations, and to analyze feasibility of compatible land development in high-noise areas. (DOT Release 18369)
- *Apollo 8* commemorative medallions containing metal carried on mission were being distributed to NASA employees "as a token of appreciation for each individual's efforts in making the United States lunar program possible," NASA Hq. *Weekly Bulletin* said. (NASA Hq *WB*, 8/4/69, 1)

August 4–5: NASA's *Mariner VII* transmitted first closeup photos of Mars south pole as it flew within 2,100 mi of planet. Dr. Robert P. Sharp, geologist at Cal Tech, said 31 photos might look like "baby pictures of Earth. This is what the Earth might have looked like some four billion years ago before it developed an atmosphere and oceans to weather its surface and nurture life." Photos taken on pass Aug. 4 were transmitted to JPL Aug. 5.

Polar cap, which had appeared gleaming white in more distant photos, looked dull gray in closeups. Pictures showed south polar cap with snow-like substance—possibly frozen carbon dioxide—piled up in vast dunes; pocked with deep, steeply walled craters; and much darker than desert to north. Floor of bright circular Hellas desert area, believed to be shallow crater or collapsed area, was strangely free of meteorite impact craters.

Mariner VII (launched March 27) with *Mariner VI* (launched Feb. 24) had provided most detailed information to date on Mars, including 198 photos covering 20% of planet and detailed scientific data from onboard experiments. Data indicated: thin Martian atmosphere had no detectable nitrogen; south polar cap, which appeared white and smoothly circular in telescope pictures, was ragged with dark splotch in center; surface temperatures ranged from 75°F to —100°F; some of narrower thin dark lines called canals might be segments of rubbled rims of craters up to 300 mi across; and Martian surface, though heavily pocked, was not as rugged as lunar surface. (AP, B *Sun*, 8/6/69, A1; AP, W *Star*, 8/6/69, A7; Auerbach, W *Post*, 8/6/69, A3)

August 5: Dr. Thomas O. Paine, NASA Administrator, and other top NASA officials testified on future space programs before Senate Committee on Aeronautical and Space Sciences.

Introducing programs, Dr. Paine said: "The decade of the 1970's and 1980's should have a program as bold in concept and as productive as we have had in the decade of the 1960's. . . . we need to have clear objectives to focus our work and a commitment, subject . . . to annual review, as to what these achievements will be. Our general goal area should be the continued exploration of the solar system while deriving the maximum scientific and practical benefits here on earth from the space program. There is no question that, at some future time, we will have the capability for manned planetary exploration and we need to face now some of the decisions that will not bear fruition for more than a decade. Although I do not believe that we will see manned exploration of the planets in the 1970's in the United States . . . I do think this could come in the 1980's. It is by no means clear that for the Soviet Union the decision may not be made to mount a crash program and bring this in before the end of the decade of the 1970's."

Dr. Wernher von Braun, MSFC Director, described possible 1981–1982, 12-man, 2-ship expedition to Mars [see Aug. 1]. Each spacecraft would weigh 1.6 million lbs at departure from earth orbit and would be 270 ft long, "smaller than what we are flying already to the moon."

NASA was using "concept of reusability" in planning, to improve and reduce cost of operating in space, Dr. George E. Mueller, NASA Associate Administrator for Manned Space Flight, told Committee. Reusability could be achieved "through the reuse of launch and space vehicles and . . . through the reuse of a mission module such as a space station" put into orbit and used over 10-yr or even 20-yr period. Space shuttles would be designed to run 100 or more flights. Modules and vehicles would be designed for multiple applications in earth, lunar, and synchronous orbits. Space tug would permit travel from space station to other spacecraft and back again—"general purpose . . . equipment." In earth orbital operations, "it permits us to fly off from the space station over to . . . an OAO, orbiting astronomical observatory, either to repair or check the OAO, or to bring it back to a space station where it can then be loaded on the space shuttle for return to earth and then brought back into orbit after repairs." (Transcript)

• Four LRL technicians—Miss Heather A. Owens, Chauncey C. Park, Roy G. Coons, and Riley Wilson—were placed in isolated area under quarantine after being exposed to lunar material when line carrying contaminated material from vacuum chambers to disposal area burst, spraying lunar material into examining room. Mishap, second in which LRL technicians were exposed to lunar material [see Aug. 1], brought total number of persons under quarantine to 23. (AP, *W Post*, 8/6/69, A3; MSC Hist Off)

- NASA announced resignation of Astronaut F. Curtis Michel, effective Aug. 18. Dr. Michel, who had been on one-year leave of absence from NASA to do scientific research at Rice Univ. in Houston, said that—although he was reluctant to leave NASA and prospect of flight in space—he wanted to devote full time to research at Rice. Resignation reduced number of NASA astronauts to 48. (MSC Release 69-55)
- NASA notified Instituto Geofisico del Peru that $2-million NASA tracking station near Lima, Peru, would be closed because of shifting program requirements and economic reasons. Station, to be phased out by November, had participated in more than 75 satellite missions since 1957. (NASA Release 69-117)
- Dept. of Interior announced grant of $100,225 for research into health, safety, and water pollution in coal mining operations. Island Creek Coal Co. would determine if miners equipped with self-contained breathing apparatus similar to astronauts' could work efficiently in mines filled with nitrogen or other inert gas. (DOI Release 17784-69)

August 6: U.S.S.R. launched *Cosmos CCXCI* from Baikonur into orbit with 527-km (327.5-mi) apogee, 143-km (88.9-mi) perigee, 91.2-min period, and 62.2° inclination. Satellite reentered Sept. 8. (GSFC *SSR*, 8/15/69; 9/15/69; SBD, 8/7/69, 109)

- NASA's HL-10 lifting-body vehicle, piloted by NASA test pilot John A. Manke, reached 78,000-ft altitude and mach 1.55 after air launch from B-52 aircraft at 45,000-ft altitude west of Rosamond, Calif. Objective of flight, 23rd in series, was to obtain data on performance, stability, and control—especially roll control. (NASA Proj Off)
- NASA named flight crews for Apollo 13 and 14 lunar landing missions. Prime crewmen for Apollo 13 were Astronauts James A. Lovell, Jr. (commander), Thomas K. Mattingly II (CM pilot), and Fred W. Haise, Jr. (LM pilot). Backup crew was composed of Astronauts John W. Young, John L. Swigert, Jr., and Charles M. Duke, Jr. Apollo 14 prime crewmen were Astronauts Alan B. Shepard, Jr. (commander), Stuart A. Roosa (CM pilot), and Edgar D. Mitchell (LM pilot). Backup crewmen were Astronauts Eugene A. Cernan, Ronald E. Evans, and Joe E. Engle.

  Both missions would include lunar exploration and deployment of Apollo lunar surface experiment packages (ALSEP). Total lunar surface stay time would include two EVA periods of three hours each and would not exceed 35 hrs. Flights would be first for Astronauts Mattingly, Haise, Roosa, and Mitchell. (NASA Release 69-115)
- *Apollo 11* Astronauts Neil A. Armstrong and Edwin E. Aldrin, Jr., quarantined in LRL, discussed lunar surface activities with about 40 scientists and geologists over closed-circuit TV. Astronauts said lunar surface was rich with interesting rocks. They described small, walnut-size fragments which appeared translucent or transparent, with reflecting surfaces like quartz crystals; spatters of glass on rocks, especially on rocks on bottom of craters; and rocks shaped like automobile dis-

tributor caps, which appeared to be weathered or eroded and sculptured at top in cylindrical shape. Astronauts said they regretted not being able to retrieve more lunar rocks and suggested that future Apollo astronauts be equipped with extra pocket or shoulder bag so they could collect interesting rocks as they saw them. (*W Post*, 8/7/69, A4; AP, B *Sun*, 8/7/69, A8)

- Atheist Madalyn Murray O'Hair—on behalf of herself, her husband Richard O'Hair, and her "Society of Separationists"—filed suit in Austin, Tex., District Court against NASA Administrator, Dr. Thomas O. Paine, to prevent U.S. astronauts on duty from practicing religion on earth, in space, or "around and about the moon." She objected specifically to *Apollo 8* Christmas Eve 1968 Bible reading by Astronaut Frank Borman and placing by *Apollo 11* Astronauts Neil A. Armstrong and Edwin E. Aldrin, Jr., of disc containing microfilmed prayers on moon July 20. (UPI, *W Post*, 8/7/69)

- Dissatisfaction with substandard performance in other parts of society was "something worthwhile that the Space Program is contributing to the United States," NASA Administrator, Dr. Thomas O. Paine, said in speech before National Press Club in Washington, D.C. "I hope we have spurred our society and our people to . . . demand higher performance, to set bolder goals, and then to have the gumption to stand up before the whole world and demonstrate whether or not the goals are achieved." Space program had also taught "need for broadly enlisting not only American capability, but the best people from around the world willing to throw their competence and a portion of their careers into challenging endeavors." It had "unleashed in the 1960's the talents and energies of a technological generation." U.S. must continue "to put together bold programs that will release the creative energies of our people in productive channels."

    Dr. Paine thought history would record as "the great contribution of our generation" astronauts' blazing of "trail for all future generations of men who want to . . . conquer new worlds. . . . Through man's brains, energy and resources life can—and life will—extend itself through the solar system. . . . The 1980's are very clearly the decade in which both we and the Soviet Union, with reasonable-sized space programs in the 1970's, will develop a technological capability for landing on Mars."

    In response to questions, Dr. Paine said: "It seems clear to me that there are increasing opportunities for all nations to work together in space exploration and application. Certainly we and the Russians can and should cooperate more closely in space science so our two programs can produce greater results than the simple sum of their outputs." (Transcript)

- Future space program was described by Dr. George E. Mueller, NASA Associate Administrator for Manned Space Flight, before National Space Club in Washington, D.C. Reusable nuclear vehicle to serve as space shuttle between space station in earth orbit and space station in lunar orbit would be "final link that would permit us to reduce the cost of operation to something like $200 for moving a pound of material from the earth's surface to the lunar surface and return as compared to something like $100 thousand a pound using today's techniques. Similar reductions in the cost of transportation to the earth orbital station will permit us for the first time to consider processing materials in space, to use space for the kind of laboratory work that we now associate with ground-based laboratories." By end of 1970s "we would find so many uses for operations in synchronous orbit both for observing the universe and for observing the earth that we would have established a space station in synchronous orbit which would be regularly supplied by a nuclear shuttle system and which would provide us with direct television broadcasting and direct radio broadcasting to the homes of all people in the world, as well as providing us with great stellar observatories and a viewing platform for air traffic control, navigation and for a permanent weather watch.

". . . this approach to using space is one that is readily extended, once the shuttle capability has been developed, to a corresponding approach for planetary exploration and . . . the same nuclear shuttle system together with the space station modules need only to be supplemented by a Mars landing module to permit us to carry out the first manned planetary expedition to Mars." (Text)

August 7: Scientists at JPL presented conflicting opinions to newsmen on preliminary data from *Mariner VI* and *Mariner VII* flybys of Mars. Dr. George C. Pimentel and Dr. Kenneth C. Herr of Univ. of California at Berkeley said data from infrared spectrometer indicated presence of gaseous ammonia and methane in Martian atmosphere. "We are confident that we have detected solid carbon dioxide that is not on the surface; that is, it is suspended as a cloud above the polar cap. Our data are consistent with and suggest that the polar cap is composed of water ice and probably not solid carbon dioxide near the polar cap edge." If life did exist on Mars, they said, it could be in region near edge of polar icecap where "polar ice provides a reservoir of water" and solid carbon dioxide cloud "provides protection from ultraviolet radiation."

Dr. Gerry Neugebauer of Cal Tech said temperature of Mars was "strong circumstantial evidence that the polar caps are in fact predominantly made of carbon dioxide." Infrared radiometer experiment, which measured temperatures on Martian surface, indicated that temperature of south polar cap was close to —253°F, temperature at which carbon dioxide would solidify in thin Martian atmosphere.

UV spectrometer experiment had found large amount of UV radiation reflected from south polar icecap, indicating that UV light from sun was penetrating thin Martian atmosphere and reaching surface. Dr. Charles W. Hord of Univ. of Colorado said strong UV radiation reaching surface "would destroy many of the important molecular bonds of organic compounds." If life did exist on Mars, he said, it "must be pretty strong stuff," or it must have some means of protection against UV rays.

Dr. Robert B. Leighton of Cal Tech said one of most striking results of Mariner photos was indication of dynamic process occurring on Martian surface. Unlike rest of Martian surface, which was heavily cratered and closely resembled moon, Hellas area appeared to be smooth and free of craters. "Hellas is the first non-lunar-like feature" discovered by *Mariner VII*, he said. Apparently there was "activity in

August 7

that region . . . obliterating craters as fast as they are being formed." (Bishop, *WSJ*, 8/8/69; Sullivan, *NYT*, 8/8/69, 1; Lannan, W *Star*, 8/8/69, A4)

- Lunar Receiving Laboratory scientists issued first comprehensive report on preliminary study of lunar samples collected by *Apollo 11* astronauts. Experiments indicated there was no life in sample and traces of organic material reported earlier were probably from astronauts' spacesuits and containers, rubber gloves, and tools used to handle material. (Traces of hydrocarbons in two samples of lunar dust had been reported Aug. 6.) Autopsies performed on 48 mice injected with lunar dust and then killed had shown no signs of germs or illness. Detailed analyses of samples would begin in late September when material would be released from quarantine and distributed to 146 principal investigators in 9 countries. (AP, W *Star*, 8/8/69, A4; Sullivan, *NYT*, 8/7/69, 1)

- Sen. Ralph W. Yarborough (D-Tex.) offered amendment to H.R. 11271, FY 1970 NASA authorization bill [see June 24], to increase funds for NASA R&D and program management from amount reported by Senate Committee on Aeronautical and Space Sciences to amount passed by House. Increases would total $256.50 million in R&D and $6.35 million in research and program management. He said: "My amendment authorizes the bare minimum that we, as a nation, should commit to space. Its adoption is vital to the proper balance in our national priorities; it is vital to the future of our exciting and promising space program; and it is vital, in my opinion, to the interests and well-being of our country." (*CR*, 8/7/69, S9383)

- *Washington Post* published results of July 26–28 Gallup survey, which found public lukewarm about Government funding of manned Mars landing. While majority of young adults favored idea, majority of those 30 or over opposed it. Generally, 39% of those polled favored attempt to land man on Mars, 53% opposed, and 8% had no opinion. Blacks opposed by 3-to-1 ratio. (*W Post*, 8/7/69, F4)

- At White House dinner for West German Chancellor Kurt G. Kiesinger, Dr. Wernher von Braun, MSFC Director, told press putting man on Mars by 1982 posed less risk than putting man on moon because most technical problems had been solved. In time space travel would become commonplace, with spacecraft carrying passengers. On Martian surface man could move from home to car or office in completely controlled environment. (Shelton, *W Post*, 8/8/69, D2)

- MSFC announced award of two contracts. Eight-month, $400,000 contract had been given to General Dynamics Corp. to study experiment modules for proposed manned space station. Study, which would complement space station investigations being conducted by McDonnell Douglas Corp. and North American Rockwell Corp., would examine variety of experiments suitable for manned space station, analyze scientific and engineering community's need for experiment modules, and develop concepts for least number of modules needed to meet these requirements.

   Martin Marietta Corp. had been awarded $1,170,000 contract to fabricate, test, and deliver 15 Saturn V workshop rate gyro processors and 1 module test set and to retrofit 22 Apollo Telescope Mount rate-gyro processors for Apollo Applications program. Work, expected to take 18 mos, would be done in Orlando, Fla. (MSFC Releases 69–172, 69–173)

- With encouragement from President Nixon and Vice President Spiro T. Agnew, NASA had "begun drumming up pressure for the huge sums required to send men to Mars in the early 1980's," *New York Times* editorial said. "But the latest Mariner information makes the probability of life on Mars much less than it seemed even a week ago, thus removing much of the original motivation for such a project. The shift of emphasis now proposed to unmanned satellites would be far cheaper; scientifically it would also be far more productive." (*NYT*, 8/7/69, 32)

August 8–14: *Zond VII* automatic space station was launched by U.S.S.R. from Baikonur with "powerful carrier rocket" and placed on free-return lunar trajectory from parking orbit. Tass said mission objectives were to study moon and near-lunar space further, photograph lunar surface, and test improved onboard systems and design of "rocket-space complex." All equipment was functioning normally.

On Aug. 11 Tass announced that spacecraft had circled moon on flight plan similar to that of *Zond V* (launched Sept. 15, 1968) and *Zond VI* (launched Nov. 10, 1968), had photographed lunar surface, and was returning to earth. *Zond VII* reentered atmosphere by skipping across outer layers of atmosphere to reduce its entry speed and then descended and softlanded in predetermined area near Kustanay in northern Kazakhstan Aug. 14. (*SBD*, 8/11/69, 120–1; 8/18/69, 152; *NYT*, 8/9/69, 26; 8/12/69, 6; 8/15/69, 14; GSFC *SSR*, 8/15/69)

August 8: NASA announced selection of Heliodyne Corp. and Wolf Research and Development Corp. for final negotiations leading to one-year $1-million, cost-plus-award-fee contract with two one-year options to operate National Space Science Data Center at GSFC. (NASA Release 69–118)

- In *Washington Daily News* column Sen. Everett M. Dirksen (R-Ill.) said: "Unknowing voices clamor to us to give up the search into the unknown. They ask us to spend the money on things here on earth. They ask for something that already has been done. Where do you think the money is spent that sent Apollo 11 to the moon? It wasn't spent on the moon. There are no creatures there to benefit from the billions spent to finally land Neil Armstrong and Buzz Aldrin in the Sea of Tranquility. The money was spent here on earth, where it enriched the laborers, the craftsmen, the technicians, the engineers, the scientists—and their neighborhoods. It enriched the millions and millions of people who always benefit from industry. . . . (*W News*, 8/8/69, 23)

- *Washington Post* editorial: "There was a certain logic in playing down the purely scientific aspects of the Apollo program in the past since the effort was to land men on the moon before the Russians did. But that day is past. The scientists of space, as contrasted with its engineers and technicians, have been forced into the back seat of the manned space program. It is time now to make them the navigators. The choice of missions—for future flights to the moon and for future operations that will lead some day to a trip to Mars and eventually other planets—should be largely in their hands. They, far better than the men who created the hardware and the knowledge necessary to make space travel possible, know the areas most appropriate for exploration in terms of gaining knowledge." (*W Post*, 8/8/69)

August 9–15: NASA's *Oso VI* (OSO–G) Orbiting Solar Observatory was successfully launched from ETR at 3:52 am EDT by two-stage Thor-Delta N booster to study sun and its influence on earth's atmosphere. Orbital parameters: apogee, 348.0 mi (560 km); perigee, 307.6 mi (495 km); period, 95.2 min; and inclination, 32.96°. Primary mission objective was to obtain high-resolution spectral data from pointed experiments in 10- to 20-kev range and 1 to 1,300 Å range during one solar rotation and make raster scans of solar disc in selected wavelengths. Spacecraft would obtain data from nonpointed and pointed experiments for more than one solar rotation for extended observations of single lines and solar flares.

Pac (Package Attitude Control) system, carried pickaback on Delta 2nd stage, was ejected into orbit with 340-mi (547.1-km) apogee, 300-mi (482.7-km) perigee, 94.2-min period, and 32.9° inclination. Primary objective was to flight-test long-life, low-power, three-axis *Pac* earth-stabilized control system for Delta 2nd stage and to demonstrate feasibility of using stage as experimental platform.

Oso VI was spin-stabilized, weighed 640 lbs, carried seven experiments, was designed with six-month lifetime, and had two main sections—wheel (lower), which carried nondirectional scanning experiments and basic support equipment, and sail (upper), which carried pointed experiments. It was similar to previous OSOs but had unique capability which enabled two sun-pointing telescopes to study in detail UV and x-ray spectra at any point on solar disc and would provide greater knowledge of solar atmosphere (chromosphere) as well as outermost layer (corona) visible only through special instruments or during total solar eclipse. Experiments, designed to continue and extend work of preceding OSO spacecraft, were provided by Harvard College Observatory, Naval Research Laboratory, Rutgers Univ., Los Alamos Scientific Laboratories, Univ. of New Mexico, Univ. of Bologna, and University College (London).

Both tape recorders were operating at liftoff and were still operating satisfactorily. Spacecraft stabilized and acquired sun as scheduled shortly after entering orbit. By Aug. 15 all experiments had been turned on and were operating satisfactorily. Two minor anomalies—higher than planned use of current by motor that provided fine elevation pointing and lower than expected spacecraft operating temperature—were not expected to affect spacecraft operation adversely.

Oso VI was seventh in series of eight OSO spacecraft designed to provide direct observation of sun during most of 11-yr solar cycle. *Oso I* (launched March 7, 1962) and *Oso II* (launched Feb. 3, 1965) had surpassed their six-month design lifetimes and together provided more than 8,600 hrs of scientific information. OSO–C (launched Aug. 25, 1965) had failed to reach orbit when booster malfunctioned. *Oso III* (launched March 8, 1967) and *Oso IV* (launched Oct. 18, 1967) continued operating satisfactorily, each providing 7½ hrs of real-time data daily. *Oso V* (launched Jan. 22, 1969) had both tape recorders and seven of eight experiments operating satisfactorily after six months in orbit. OSO program was managed by GSFC under OSSA direction. (NASA Proj Off; NASA Releases 69–112, 69–112A, 69–123; *Pres Rpt 70* [69])

August 9: "Scientists who have long felt that their role was secondary to that of engineers in the Apollo project" were complaining openly and trying to force greater emphasis on science in planning future lunar landing missions, John Noble Wilford reported in *New York Times.* "Their argument is that, with the success of Apollo 11, the project's goal should be to learn as much as possible about the moon and not merely to repeat the demonstration that moon landings are possible."

Dr. Elbert A. King, curator of Lunar Receiving Laboratory, had said in interview that NASA administration did not have "enough sympathy with, or understanding of, scientific objectives." Casting science in "piggyback role" for first manned lunar landing was understandable, he said. "No one really criticizes that, because . . . getting men to the moon and back had to be a massive engineering effort. But now that we have accomplished that goal, the justification for future lunar exploration is largely science. There has to be a shift of emphasis." Scientists were pressing for more active role in mission planning, return of larger amounts of lunar samples, selection of scientists for flight crews, and more time between missions in which to evaluate data for applications to future experiments. (*NYT*, 8/10/69, 44)

- Soviet scientist Dr. Valery A. Krasheninnikev and academician Dr. Aleksandr P. Lisitzin had returned to San Diego from 55 days with U.S. Deep Sea Drilling Project aboard drilling ship *Glomar Challenger* convinced their findings were "more important to man than the samples from the moon," Associated Press reported.

  Concentrated drilling between Honolulu and Guam had produced rocks and sedimentary cores showing microorganisms in perfect state of preservation. They might provide history of earth's creation. Project was cooperative venture of Scripps Institution of Oceanography, Woods Hole Oceanographic Institution, Lamont Geological Observatory of Columbia Univ., Univ. of Miami Institute of Marine Science, and Univ. of Washington. (*NYT*, 8/10/69, 45)

August 10: Apollo 11 Astronauts Neil A. Armstrong, Edwin E. Aldrin, Jr., and Michael Collins and 20 other persons quarantined in Lunar Receiving Laboratory were released shortly after 9:00 pm CDT—four hours earlier than planned. Dr. Charles A. Berry, Director of Medical Research and Operations, MSC, said astronauts, who had been confined since July 24, showed no signs of any possible infection from exposure to moon. He cautioned that astronauts might become ill after release—not from lunar contamination, but from earth organisms to which they were highly susceptible after long period of isolation. Report on health of persons under quarantine would be presented by NASA Aug. 11 to Interagency Committee on Back Contamination, which had approved early release. (Sehlstedt, B *Sun*, 8/11/69, A1; AP, *W Post*, 8/10/69, A7)

- NASA had assured Post Office Dept. that master die for 10-cent moon landing commemorative stamp [see July 9] had gone all the way to lunar surface as planned. It had returned to earth in CM and been rushed to Washington, D.C., July 31 after decontamination at MSC. However, moon letter envelope with die proof of moon landing stamp had not been postmarked on lunar surface. Because of tight schedule for lunar EVA, letter had been left with Astronaut Michael Collins in CM *Columbia* while Astronauts Neil A. Armstrong and Edwin E. Aldrin, Jr., worked on moon. "Moon Landing" postmark had been applied during return voyage. Letter had been decontaminated at MSC and returned to Postmaster General Winton M. Blount Aug. 5. (Fairies, W *Star*, 8/10/69, H11)

- Space program spinoffs of medical benefit to mankind were described by Howard A. Rusk, M.D., in *New York Times*. NASA's Scientific Information Div. provided stockpile of knowledge indexed in computer tapes and distributed on microfiche. Collection of 250,000 documents increased by approximately 75,000 items annually. NASA Technical Utilization Div. selected inventions, ideas, and new techniques for use in nonaerospace activities, including medicine, and distributed them through eight regional centers. Under interagency agreement, NASA and HEW Social and Rehabilitation Services Administration reported results of aerospace research to solve problems of 4 million physically or mentally disabled Americans of working age. Research had developed technique for sharpening x-rays, micrometeorite sensor to record Parkinsonian tremors, and technique for applying electrodes with spray of conductive material. (*NYT*, 8/10/69, 55)

- William Hines in Washington *Sunday Star* said time was "ripe" for NASA reorganization, "not merely the firing, promoting and transferring of officials, but the functional restructuring of the agency as well." There was some suspicion that no place existed for Dr. Thomas O. Paine, NASA Administrator, who was "not only a Johnson-administration holdover, but a card-carrying Democrat as well." Some observers believed *Apollo 8* Astronaut Frank Borman, "the President's current darling and space confidant," might be next NASA Administrator. "The idea of putting astronauts in charge of the space program may seem incongruous, but it is clearly not beyond the realm of possibility in the image-conscious Nixon regime."

  Paine's fate would determine that of Associate Administrator for Manned Space Flight, Dr. George E. Mueller. With lunar landing, Apollo had changed from developmental to operational effort. Christopher C. Kraft, Director of Flight Operations at MSC, would likely transfer to Washington as Apollo Program Manager—possibly "controlling all space missions for NASA." Hines saw possibility of Apollo Program Director, George H. Hage's moving into position being vacated by L/G Samuel C. Phillips, Apollo Program Director, who was rejoining USAF. "George M. Low, Apollo chief at Houston, may replace the Houston Center director, Robert R. Gilruth, if Gilruth can be prevailed upon to retire." There was talk about moving MSFC Director Dr. Wernher von Braun to Washington "to do what he does best: charm money out of Congress." Dr. Kurt H. Debus, KSC Director, "may retire to make way for Rocco Petrone. . . ." (W *Star*, 8/10/69, G4)

- History of Jet Propulsion Laboratory from 1936, "when the 'crazy scientists' under . . . Dr. Theodore von Kármán lit off one of their rockets in the dry gulch called the Arroyo Seco," to current time, when "JPL's sights are set a little higher—and farther out," was traced by John Lannan in Washington *Sunday Star*. JPL controlled NASA Deep Space Network with jurisdiction over space efforts 10,000 mi from earth and beyond, though its Goldstone facility also had participated with GSFC in support of *Apollo 11*. Its space flight operation facility was "actual directorate" for handling cosmic penetration flights. It was currently gearing for Martian orbiter series in 1971 and for 1973 Viking landers. Future held possibilities of developing and directing missions leading to Grand Tour of planets. JPL was owned by Federal Government and staffed and operated by Cal Tech. (W *Star*, 8/10/69, G3)

- George Gallup released results of first poll of President Nixon's popularity since *Apollo 11* success. Poll showed 65% of U.S. public approved his performance in office. Poll July 11–14, before *Apollo 11* mission, had shown 58% approval. (*W Post*, 8/11/69, A2)
- *Apollo 11* had opened vast market for medals, tokens, and pins to collectors who specialized in commemorative pieces, *New York Times* said. Medals issued by several countries after *Apollo 8* were "few compared to the meteoric shower of commemoratives for the moon landing." Medal by Ralph J. Menconi portrayed *Apollo 11* astronauts on face side; reverse showed Astronauts Neil A. Armstrong and Edwin E. Aldrin, Jr., with LM on lunar surface and earth in background. Medal designed in U.K. by Paul Vincze depicted astronaut on lunar surface with names of *Apollo 11* astronauts. Reverse showed figure of Hermes holding winged staff of flight and olive branch with Saturn V in background. (Haney, *NYT*, 8/10/69)

August 11: Lunar scientist Dr. Harold C. Urey discussed moon findings at San Diego, Calif., press conference after return from study of *Apollo 11* lunar samples at LRL. There was "no evidence whatever" of life on moon and, "at present, no age of the rocks on the surface of the moon is known." Fine dust on lunar surface consisted of 50% glass—10% in form of small beads, 40% as broken-up beads—and "something else" containing minerals similar to those in lunar rocks. Dust and rocks evidenced shock as though produced by "rather large collisions somewhere that dug up material considerably deeper in the moon than the few centimeters they have been investigating." There was no evidence of water. Chemical analyses had shown high concentration of titanium in lunar samples characteristic of area where *Surveyor V* and *Apollo 11* had landed. "No such . . . titanium concentration occurs anywhere that we know of on the surface of Earth."

No one had "anticipated what we are finding. And I believe that no one has a good explanation of how it got that way." It would take "much more" than nine more Apollo manned landings to solve lunar mysteries. He saw no economic value in future lunar colonization but "an observatory on the moon would be a wonderful way to investigate the heavens, because the difficulties of the atmosphere would disappear." (*Space World*, 12/69, 35–42)

- Rep. George P. Miller (D-Calif.), Chairman of House Committee on Science and Astronautics, told House: ". . . I do not at this time wish to commit ourselves to a specific time period for setting sail for Mars. I believe that there are many tasks that can be accomplished that will ultimately provide that capability, but will be less costly and will be necessary in meeting short term objectives." He urged priority attention to intermediate steps and balanced program "that fully exploits the great potential of unmanned spacecraft, while at the same time maintaining a vigorous manned flight program."

He advocated continuation of lunar exploration to obtain "experience of operating a base for science and exploration on another heavenly body"; manned earth-orbital operations leading to long-term space station supported by low-cost shuttle rocket; greater emphasis on applications satellites that "have the greatest potential for economic return in the near term"; larger funding for unmanned planetary exploration, "an area in which the U.S. may soon be overshadowed by the Soviet Union"; continued NERVA development because "improved propulsion is a key to space leadership"; and "special emphasis on ERS satellites, "which promise to yield so much . . . to agriculture and industry." (Text; *CR*, 8/11/69, H7251–4)

- On first day out of quarantine, *Apollo 11* crew visited MSC offices, then enjoyed official day off. NASA spokesman said astronauts had requested their activities be kept secret. (UPI, *W Post*, 8/12/69, A3)
- Approximately nine hours of satellite time had been booked for TV coverage abroad of *Apollo 11* astronaut events during coming week, ComSatCorp said. They included more than two hours live coverage of Aug. 12 MSC news conference, to be relayed via *Intelsat-III F–2* to Western and Eastern Europe and, in part, to Tokyo and Sydney via Pacific *Intelsat-III F–4*. Nearly two hours coverage of New York ticker-tape parade and visit to U.N. would be transmitted to Europe Aug. 13. Ceremonial dinner in Los Angeles, later in day, would be taped for transmission to Europe Aug. 14. (ComSatCorp Release 69–50)
- Sen. J. William Fulbright (D-Ark.) told Senate: "It would be a major step forward if we could now negotiate a new space treaty which would go beyond the disavowal of national claim of sovereignty in the existing treaty and explicitly recognize the United Nations as the 'owner' or sovereign of extraterrestrial bodies and also define the functions and responsibilities of a United Nations space authority, particularly the ways in which it would regulate and coordinate national space exploration programs. The overall objective of such a treaty would be to regulate but not eliminate the competition in space. One benefit of such an arrangement is that it would allow the space powers to reduce their expenditures and so reallocate funds to more pressing domestic and international requirements." (*CR*, 8/11/69, S9633)
- *New York Times* editorial: "On that eventful day when the first men walk on the surface of Mars, they will find much 'magnificent desolation' akin to that seen by Neil Armstrong and Edwin Aldrin when they strolled on the moon last month. That virtual certainty emerges from the brilliantly successful exploration of the red planet just completed by Mariners 6 and 7. Their expedition lacked the human drama of Apollo 11, but the scientific information they returned may well qualify the two Mariners as the most scientifically productive enterprise men have yet carried out in space." (*NYT*, 8/11/69)
- Subcommittee on NASA Oversight submitted to House Committee on Science and Astronautics report *Engineering Management of Design and Construction of Facilities of the National Aeronautics and Space Administration*. Report concluded that closer economic scrutiny of design and construction management at NASA executive level could "yield dividends in more efficient management and lower costs at the field centers." Organization of Office of Facilities was "progressive step." Cost accounting of administrative expenditures needed to be

improved and engineering management costs should be clearly identified for all projects at all centers. (Text)

August 12–18: NASA's 951-lb *Ats V* (ATS-E) Applications Technology Satellite was launched from ETR at 7:01 am EDT by Atlas (SLV-3C)-Centaur booster on mission to conduct carefully instrumented gravity-gradient-orientation experiment for basic design information on stabilization and control of long-lived spacecraft in synchronous orbit and to obtain useful data from onboard experiments during first 30 days in orbit. Spacecraft successfully entered elliptical transfer orbit with 26,737.2-mi (43,020.2-km) apogee, 5,297.0-mi (8,522.9-km) perigee, 686.5-min period, and 17.9° inclination.

Because of anomaly which required excessive fuel to maintain stable spin condition, apogee-kick motor was fired on first apogee instead of second and spacecraft had to be biased so it would drift from position over India to intended station over area west of Ecuador. Maneuver successfully placed *Ats V* into near-synchronous orbit with 22,927-mi (36,889.5-km) apogee, 22,221-mi (35,753.6-km) perigee, 1,464.0-min period, 2.7° inclination, and 6.9° per day westward drift. Active nutation control was overpowered by unidentified force that caused spacecraft to go into flat spin, preventing ejection of motor case without possibility of damage to spacecraft. Controllers were investigating alternatives—stopping spacecraft spin, restoring spacecraft to normal spin mode, or minimizing ejection hazard—which could be executed Aug. 25 when spacecraft became visible to Rosman, N.C., ground station. Spacecraft was not in danger thermally or electronically and was expected to become stable and operational after successful ejection of kick motor.

Ats V was fifth in series of seven ATS satellites designed to investigate and flight-test technological developments common to number of satellite applications and useful to satellites operating in stationary orbits, conduct carefully instrumented gravity-gradient experiments for basic design information, and flight-test experiments peculiar to orbits of various missions. *Ats I* (launched Dec. 6, 1966) had exceeded test objectives and was still operating satisfactorily. *Ats II* (launched April 5, 1967), though judged a failure because of eccentric orbit, had transmitted some useful data before being turned off Oct. 23, 1967. *Ats III* (launched Nov. 5, 1967) had operated successfully and transmitted color photos of earth. *Ats IV* (launched Aug. 10, 1968) had remained in parking orbit when Centaur failed to complete second burn and had reentered Oct. 17, 1968. ATS program was managed by GSFC under OSSA direction. (NASA Proj Off)

August 12: Apollo 11 Astronauts Neil A. Armstrong, Edwin E. Aldrin, Jr., and Michael Collins held first postflight press conference at MSC, narrating 45-min film of mission and answering questions. On meaning of lunar landing, Collins said it was "technical triumph for this country to have said what it was going to do a number of years ago, and then by golly do it. Just like we said we were going to do. Not just . . . purely technical, but also a triumph of the nation's overall determination, will, economy, attention to detail, and a thousand and one other factors that went into it."

To Aldrin mission meant "that many other problems perhaps can be solved in the same way by taking a commitment to solve them in long time fashion. I think that we were timely in accepting this mission of going to the moon. It might be timely at this point to think in many other areas of other missions that could be accomplished."

Armstrong said moon landing heralded "beginning of a new age." He said moon was "stark and strangely different place, but it looked friendly . . . and proved to be friendly." Astronauts had much less trouble than expected on lunar surface. Primary difficulty was that "there was just far too little time to do the variety of things that we would have liked to have done. . . . We had the problem of the 5 year old boy in a candy store. There are just too many interesting things to do."

Armstrong said that during landing they "were concerned about running low on fuel on range extension we did to avoid the boulder field and craters. We used a significant percentage of our fuel margins and we were quite close to our legal limit." On possibility of abort during period they were receiving alarm signals, Aldrin said procedure in preparation simulations had been always to "keep going as long as we could. . . . The computer was continuing to issue guidance . . . and it was continuing to fly the vehicle down in the same way that it was programmed to do. The only thing that was missing . . . is that we did not have some of the displays . . . and we had to make several entries . . . to clear up that area." Armstrong added, "We would have continued the landing so long as the trajectory seemed safe. And landing is possible under these conditions, although with considerably less confidence than you have when you have the information from the ground and the computer in its normal manner available to you." (Transcript)

- Leningrad astronomer Nikolay A. Kozyrev called for lunar laboratories over, under, and on moon's surface. Soviet and American space exploration had made scientists "more confident that this is not a dead accumulation of rocks but a space body with a very interesting history whose life also continues today." Lunar research goals were establishment of astronomical instruments on stable platforms in lunar orbit, permanent scientific laboratory on moon, spacecraft launching centers on moon for planetary exploration, and laboratory stations under lunar surface or in natural caves, "to give reliable protection from dangerous radiation and meteorite hits." (UPI, *NYT*, 8/13/69, 11)

- MSFC announced award of $15,455,800 contract modification to Boeing Co. for continued Saturn V systems engineering and integration. Contract covered work from June 1967 through June 1970 and continued effort through 10 Saturn V boosters. (MSFC Release 69–177)

- New Jersey State Div. of Clean Air and Water requested order from Superior Court, Newark, asking seven airlines to stop polluting air with jet engine exhaust at Newark Airport. Suit called for modification of existing jet engines with air-pollution-control devices or for switching to new smokeless engines and asked imposition of $2,500 fine. In Washington, Air Transport Assn. spokesman said that "it would be hard to make a case for massive retrofit with the absence of a major health hazard." He said studies had shown that jet engine pollution was only one percent of total problem and was case of "visibility" and "esthetics" rather than health danger. United Airlines spokesman said

November 1968 engine modifications to three of airline's Boeing 727s had sharply decreased pollution. (Sullivan, *NYT*, 8/13/69, 1)

- Philadelphia *Evening Bulletin* editorial: "The public ceremonies honoring the astronauts underscore identity in a larger and much more responsible sense—a feeling of community, rooted in a family and expanding to embrace the nation, perhaps ultimately the world. There are other words for it—awareness of a common purpose, a sense of decency both public and private, a common standard of behavior and a common sense of service and loyalty to country. This is what made Apollo succeed, and this is what the nation is recognizing as the celebration begins today." (P *Bull*, 8/12/69)

August 13: NERVA experimental engine (XE) was successfully run through two bootstrap startups in open-loop control and three autostart experiments in Jackass Flats, Nev. Objective was to obtain additional data about engine in startup phase. Engine and test facility operated normally and all test objectives were achieved. (NASA Proj Off; *SBD*, 9/2/69, 4)

- Nike-Apache sounding rocket launched by NASA from WSMR carried GSFC-Dudley Observatory payload to collect micrometeoroids during Perseid meteor shower and to study electric field. Rocket and instruments functioned satisfactorily and data were expected from all experiments. Data would be compared with data from Nike-Apache mission to be launched Aug. 22. (NASA Rpt SRL)

- *Apollo 11* Astronauts Neil A. Armstrong, Edwin E. Aldrin, Jr., Michael Collins, their families, and NASA Administrator, Dr. Thomas O. Paine, flew in presidential jet from Houston to New York, Chicago, and Los Angeles during day of cross-country celebrations.

    Three-hour New York visit included greeting at City Hall by Mayor John V. Lindsay, motorcade to U.N. for greeting by U.N. Secretary General U Thant, and ticker-tape procession to John F. Kennedy International Airport for departure to Chicago. New York Public Events Commissioner John S. Palmer estimated crowds at 4 million; other observers said there were fewer and blamed ahead-of-schedule appearance and TV coverage.

    In Chicago, welcoming crowd was estimated at 3.5 million. Mayor Richard J. Daley greeted party at Civic Center and presented medals symbolic of honorary citizenship to astronauts and Dr. Paine. Illinois Gov. Richard B. Ogilvie said, "To these first citizens of the new epoch, the people of Chicago and Illinois offer their profound admiration and respect." Astronauts spoke to 15,000 young people in Grant Park before returning by helicopter to O'Hare International Airport for flight to Los Angeles.

    Mayor Samuel W. Yorty met party at Los Angeles International Airport. After brief ceremony, party sped to Century Plaza Hotel for reception preceding state dinner. (Lelyveld, *NYT*, 8/14/69, 1; Oberdorfer, *W Post*, 8/14/69, A1; NASA PAO)

- Climaxing day of cross-country celebrations, President and Mrs. Nixon hosted formal state dinner at Century Plaza Hotel in Los Angeles to honor *Apollo 11* astronauts, their wives, and "historic achievement of the first manned landing on the moon." Guests included other astronauts and wives; widows of Astronauts Virgil I. Grissom and Edward

H. White II; Mrs. Esther Goddard, widow of rocket pioneer Dr. Robert H. Goddard; NASA and other space program officials; U.S. and international aviation pioneers; Cabinet members; Chief Justice and Mrs. Warren E. Burger; governors of 44 states; members of Joint Chiefs of Staff; Diplomatic Corps members representing 83 nations; Mrs. Dwight D. Eisenhower, widow of former President; former Vice President and Mrs. Hubert H. Humphrey; and Congressional leaders.

President asked NASA Administrator, Dr. Thomas O. Paine, to read citation of posthumous awards: "The National Aeronautics and Space Administration awards posthumously to Virgil I. Grissom, Edward H. White, and Roger B. Chaffee the NASA Distinguished Service Medals for professional skill, courage, and dedication to duty in Project Apollo. They gave their lives in their country's historic undertaking to realize the goal of landing men on the moon and returning them safely to earth."

President also asked Dr. Paine to read citation of NASA Group Achievement Award to *Apollo 11* Mission Operations Team "for exceptional service in planning and exemplary execution of mission operational responsibilities for . . . first manned lunar landing mission." Award was presented to Apollo Flight Control Engineer Stephen G. Bales, who had made decision to proceed with lunar landing when computers failed just before *Eagle*'s landing on Sea of Tranquility, on behalf of 400,000 persons who had contributed to Apollo program success.

Vice President Spiro T. Agnew, as NASC chairman, presented Medal of Freedom, nation's highest civilian honor, to *Apollo 11* astronauts for participation in "a unique and profoundly important adventure. The accumulated scientific knowledge and technological ability of mankind made man's first step on the moon practicable; the courage and skill of men like these made it possible. Their contributions to this undertaking will be remembered so long as men wonder and dream and search for truth on this planet and among the stars."

Replying to honors, Astronaut Edwin E. Aldrin, Jr., said: "What Apollo has begun we hope will spread out in many directions, not just in space, but underneath the seas and in the cities, to tell us unforgettably that we can do what we will and must and want to do."

During evening orderly crowd of peace and antipoverty protesters gathered outside hotel. (*PD*, 8/18/69, 1141–2, 1148–51; Roberts, *NYT*, 8/15/69, 14; B *Sun*, 8/14/69, A1)

- MSFC announced award of three 10-mo contracts totaling $1,370,000 to McDonnell Douglas Corp., North American Rockwell Corp., and Lockheed Aircraft Corp. to study design concepts and development requirements for nuclear rocket stage that could replace Saturn V 3rd stage (S–IVB) for advanced missions beginning in late 1970s and serve as workhorse for earth orbital and planetary applications.

  McDonnell Douglas received $570,828 to develop and evaluate two alternative stage concepts—one with modified Saturn V hardware, other with new stage design and advanced design techniques. NAR received $511,734 to study modified Saturn V hardware concept only and Lockheed received $287,000 to study advanced design concept only. (MSFC Release 69–180)

- *New York Times* editorial on Aug. 12 *Apollo 11* news conference in

Houston: "What came through most clearly in yesterday's enthralling first-hand report by the Apollo astronauts was the infinitesimal margin by which Eagle escaped either catastrophe or a decision to abort the moon landing. Either of the two major problems that emerged in those nerve-wracking moments before touchdown—the overburdened computer and the near-exhaustion of their fuel supply before Neil Armstrong and Edwin Aldrin found a suitable landing spot—might have forced a very different ending to the historic mission. That all turned out perfectly is a tribute to the astronauts' skill, courage and poise as well as to the ability of the back-up personnel at Mission Control in Houston." (*NYT*, 8/13/69, 40)

- *Apollo 11* commander Neil A. Armstrong stood to inherit 100,000-franc fortune of Mme. Anna E. Guzman, widow of French industrialist, which had been held in trust by Academy of Science of Institute of France since her 1891 death, according to article Rep. James G. Fulton (R-Pa.) inserted in *Congressional Record*. Legacy—once worth $20,000 but currently decreased in value to $290 exclusive of interest—was to be awarded to first scientist to make personal contact with heavenly body other than Mars. (*CR*, 8/13/69, E7023)
- In Senate, Sen. William Proxmire (D-Wis.) called for at least temporary halt in USAF plans to purchase Lockheed C–5A cargo aircraft while U.S. Comptroller General studied aircraft's costs and value to be gained from further purchases. (*CR*, 8/13/69, S9972–8)
- Rep. J. Herbert Burke (R-Fla.) introduced joint resolution calling for redesignation of Cape Kennedy as Cape Canaveral. (*CR*, 8/13/69, H7387)

August 14: U.S.S.R. launched *Cosmos CCXCII* from Plesetsk into orbit with 765-km (475.4-mi) apogee, 745-km (462.9-mi) perigee, 99.9-min period, and 74.0° inclination. (GSFC *SSR*, 8/15/69; *SBD*, 8/20/69, 169; UN Public Registry)

- NASA announced that 8 of 14 aerospace research pilots trained for USAF's Manned Orbiting Laboratory program terminated June 10 would join NASA. One, L/C Albert H. Crews (USAF), would be assigned to Flight Crew Operations Directorate at MSC. Seven would be astronauts, bringing total number of active NASA astronauts to 54: Maj. Karol H. Bobko (USAF), L/CDR Robert L. Crippen (USN), Maj. Charles G. Fullerton (USAF), Maj. Henry W. Hartsfield, Jr. (USAF), Maj. Robert F. Overmyer (USMC), Maj. Donald H. Peterson (USAF), and L/CDR Richard H. Truly (USN). Effective date for new assignments had not been set. Maj. Bobko, Maj. Hartsfield, and Maj. Peterson would complete studies for graduate degrees before assuming astronaut duty. (NASA Release 69–120)
- NASA announced appointment of eight-man failure review committee to determine why *Intelsat-III F–5* comsat did not achieve planned orbit after launch from KSC July 25. (NASA Release 69–119)
- Discovery of x-ray "star" between constellations Centaurus and Lupus from data relayed during July by two Vela nuclear detection satellites launched May 23 had been announced by Los Alamos Scientific Laboratory astronomers, *New York Times* reported. Dr. J. P. Conner, Dr. W. D. Evans, and R. D. Helian said object had twice the intensity of most brilliant x-ray sources previously known—in constellation Scorpius—and had not yet been identified in wavelengths observable by

August 14

human eye. No obvious source of x-ray emissions had been identified, such as stars, stellar explosions, or pulsars. (Sullivan, *NYT*, 8/14/69, 7)

- Cross-country ceremonies for *Apollo 11* astronauts constituted "probably the single greatest peacetime celebration in the nation's history," *New York Times* editorial said. "It was more than a tribute to three courageous and able men; it was also an act of homage to the hundreds of thousands of workers, engineers, technicians and scientists whose hard work over almost a decade made the moon landing possible. At the most fundamental level, perhaps, the outpouring of national rejoicing stemmed from the renewed sense of purpose the Apollo's incredible feat had brought to a nation long torn and depressed by military travail abroad and racial and generational antagonisms at home. The essence of that sentiment was well stated by Mr. Armstrong when he declared at the United Nations that 'we citizens of earth who can solve the problem of leaving earth can also solve the problems of staying on earth.'" (*NYT*, 8/14/69)
- Research submarine *Ben Franklin* surfaced 300 mi south of Nova Scotia, ending 1,200-mi, month-long Gulf Stream Drift by Swiss oceanographer Jacques Piccard and team which included MSFC researcher Chester B. May [see July 14]. During journey team had noted Gulf Stream contained fewer fish, stronger current, and more turbulence than expected. (UPI, W *Star*, 8/14/69, A1; Blakeslee, *NYT*, 8/8/69, 38)

August 15: Results of preliminary qualitative study of *Mariner VI* photos were summarized in *Science* by Dr. Robert B. Leighton, Dr. Norman H. Horowitz, Dr. Bruce C. Murray, and Dr. Robert P. Sharp of Cal Tech; Alan G. Herriman and Dr. Andrew T. Young of JPL; Bradford A. Smith of New Mexico State Univ.; Merton E. Davies of RAND Corp.; and Conway B. Leovy of Univ. of Washington: Surface of Mars "appears similar to that of the Moon, but there are significant differences; some features seen from Earth are characterized; the 'blue haze' hypothesis is disproved; and new phenomena associated with the polar cap are discovered." Mars resembled moon in abundance, form, arrangement, and size of craters, but there appeared to be break in size-distribution curve of craters in some parts of Mars not characteristic of moon—apparently because Mars had more effective weathering and transportation process than moon. Similarities between Martian and lunar surfaces included craters with slump blocks, terrace, and radial dry-debris avalanche chutes on steep inner surfaces; central peaks, polygonal outlines, blocky ejecta rims, and irregular ejecta; and irregularly sinuous ridges. Differences included more subdued relief of many Martian craters, flatter floors, fewer central peaks, more subdued debris blankets, absence of obvious secondary craters and rays, and greater abundance of "ghost" craters. Photos showed no sinuous rilles and no distinctive earth-like phenomena such as mountain ranges, tectonic basins, stream-cut topographs, dune fields, playa flats, or other arid-region features. (*Science*, 8/15/69, 685–90)

- Classical astronomical data on figures of moon and terrestrial planets were being supplemented by new information from Lunar Orbiter program. Comparable future planetary probes would provide fundamental data from simple experiments, Cornell Univ. radiophysicists

Dr. Brian T. O'Leary, Dr. Malcolm J. Campbell, and Dr. Carl Sagan said in *Science*. Lunar Orbiter results had revealed lunar mascons' nonuniform surface distribution that could explain lunar dynamical asymmetries "and perhaps similar asymmetries for Mars and Mercury." (*Science*, 8/15/69, 651-7)

- Astronaut Joseph P. Kerwin was uninjured when faulty landing gear on T-33 jet trainer forced belly landing at Ellington AFB, Tex. (AP, *W Star*, 8/16/69, A2)
- Soviet newspaper said Tu-144, Soviet supersonic transport, had been flying beyond sound barrier "for extended periods of time" with no difficulty, Associated Press reported. (*W Post*, 8/16/69, A2)
- C-5 Galaxy aircraft would demonstrate its cargo and troop delivery capability in joint USAF–USA–Lockheed-Georgia Co. Transport Air Drop and Jettison Test (TADJET) program to begin in early October, DOD announced. Approximately 150 flights from Pope AFB, N.C., would airdrop equipment and men. During transport phase, C-5 would be loaded and unloaded some 50 times and perform mating maneuvers with air-transportable dock that could handle cargo capacity of three C-5s. (DOD Release 683-69)

August 15-17: Second National Air Exposition at Dulles International Airport, Va., featured large static display including first public appearance of Lockheed C-5A, world's largest aircraft, and flying exhibitions by F-111 and other aircraft. (Program; *NYT*, 8/16/69, 46)

August 16: U.S.S.R. launched *Cosmos CCXCIII* from Plesetsk into orbit with 244-km (151.6-mi) apogee, 202-km (125.5-mi) perigee, 88.9-min period, and 51.7° inclination. Satellite reentered Aug. 28. (GSFC *SSR*, 8/31/69; UN Public Registry; *SBD*, 8/20/69, 166)

- Estimated 250,000 persons watched *Apollo 11* astronauts parade in Houston, Tex. Crowd threw confetti, ticker tape, and "moon certificates"—fake $100 and $1,000 paper money—until streets were two to three feet deep in litter. Later, 55,000 persons attended gala in Houston's Astrodome coliseum, which was filled to capacity. Total of 31 astronauts and families rode through cheering throngs. (UPI, *W Post*, 8/17/69, A12)
- Associated Press said Austin, Tex., Judge John R. Brown had granted request of atheist Madalyn Murray O'Hair for three-judge Federal court to hear her suit against NASA seeking to prevent astronauts on duty from practicing religion [see Aug. 6]. (AP, *W Post*, 8/16/69, A3)
- Agnew E. Larsen, space research consultant with Frankford Arsenal, Philadelphia, Pa., died at age 73. He had received 1930 Robert J. Collier Trophy as member of Harold F. Pitcairn's staff for developing and applying autogiro and demonstrating its possibilities for safe aerial transport. (*NYT*, 8/18/69, 31; Neely, *Pegasus*, 12/50, 10)

August 17: *Apollo 11* astronauts discussed possible manned Mars landing by 1982 on CBS TV program "Face the Nation." Astronaut Neil A. Armstrong said, "I am quite certain that goals of the Mars variety are within our range, should we choose . . . that investment of our national resources." First exploratory flights could be combined with earth-orbiting spacecraft to develop long-term capability with same kind of spacecraft. It was "well within our capability" to be prepared for Mars launch in 1981.

Astronaut Edwin E. Aldrin, Jr., said he was "not so sure . . . this is the time that we can accurately set a date like 1981." Setting goal was worthwhile but as intermediate goals were reached "I believe we will be able to better define exactly what our longer term goals are in terms of ten years from now."

Astronaut Michael Collins said, "I don't think 1981 is too soon. I think it is well within our capability to do so." Long-duration trip "requires careful design and testing of the equipment, which could easily be done in Earth orbit with a number of ancillary benefits." He defended Bible reading in space and announced he would not fly in space again, because he found it increasingly difficult "to keep up year after year" with rigorous training required. (*SBD*, 8/19/69, 159; *W Post*, 8/18/69, A2; *NYT*, 8/18/69, 33)

- Japan successfully launched her largest rocket to date—four-stage, 75-ft-long, 4.5-ft-dia, 43.8-ton MU3D—Kyodo News Service reported. Rocket reached 100-mi (160.9-km) altitude in 4½ min, with last stage reaching 1.8 mps—about half speed thought needed to orbit satellite—and splashing down in Western Pacific after 7 min 35 secs of flight. (B *Sun*, 8/18/69, A4)

- President Nixon's post-*Apollo 11* tour of Asia and Romania July 25–Aug. 3, plus his remarks and reactions aboard U.S.S. *Hornet* at splashdown and during welcoming ceremony for astronauts, were recorded in *New York Times Magazine* article by Max Frankel and Robert B. Semple, Jr. Authors were among press accompanying President and Mrs. Nixon on tour. President and party had basked "in reflected moonglow." When President walked toward reviewing stand in Guam, spectator had remarked, "that's his moon walk." *Apollo 11* had given President "new exuberance." (*NYT Magazine*, 8/17/69, 26–9, 76–80)

- Rep. George P. Miller (D-Calif.), as Chairman of House Committee on Science and Astronautics, had forced NASA to pay $5,522 for USAF jet to transport 32 committee members and wives to Aug. 13 *Apollo 11* state dinner in Los Angeles, Rowland Evans and Robert Novak said in *Washington Post*. NASA also had to pay $19,342 for chartered commercial jet for space officials and $2,800 for Aug. 12 Houston luncheon, and White House was charging agency with most of estimated $75,000 cost of state dinner. (*W Post*, 8/17/69, B7)

- Controversy was building up over astronauts' future, Apollo program, and manned space flight generally, Harry Schwartz said in *New York Times*. Three major debates were over whether engineer-astronauts or scientist-astronauts should be sent on future Apollo missions; who should control mission schedules and astronaut activities, "NASA hierarchy" or ground-based scientists in NASA; and whether U.S. should emphasize unmanned probes or crash program to put men on Mars in early 1980s. "The fact that it is the scientists who have been resigning while astronauts with test pilot backgrounds have been receiving unprecedented public acclaim makes it evident where the balance of power lies for the moment within NASA. But the issue is far from settled, since NASA itself must and does use the prospect of scientific advances as a key argument in seeking appropriations for space activities. Hence the dissident scientists could have substantial leverage if they teamed up with Congressmen and others who oppose the space appropriations for other reasons. It would not be surprising . . . if NASA sought to ease the scientists' irritation by satisfying some of their demands." (*NYT*, 8/17/69, D2)

August 18: Swiss physicist Dr. Johannes Geiss, originator of *Apollo 11* experiment to trap atomic particles from solar wind on lunar surface, would use "deliberate speed" in assaying results, *New York Times* reported. NASA courier had delivered square foot of aluminum foil exposed on moon for an hour to catch particles emitted by sun. Dr. Geiss and associates in Berne Univ.'s Physics Institute had devised plan for dual study of foil in Berne and at Federal Polytechnic at Zurich. Analyses, determining components by spectrometer, would require several weeks. NASA would not release remaining three square feet of foil to him until 1970. (*NYT*, 8/18/69, 34)

- British Aircraft Corp. and Sud-Aviation announced completion of second phase of Anglo-French Concorde supersonic transport flight development program. Two prototypes were being readied for transonic phase to push aircraft's speed beyond mach 1 in early September. Two prototypes had logged 104 flying hrs in 39 and 24 flights and had achieved speeds to mach 0.95 and altitudes to 40,000 ft. Concorde 002 was being prepared for supersonic flights to mach 2, or 1,400-mph cruising speed, in tests expected to begin at year's end. (BAC/Sud-Aviation Release 10C/69)

August 19: McDonald Observatory in Texas successfully recorded its first hits on laser reflector left on moon by *Apollo 11* astronauts at 9:30 pm CDT. Scientists said distance at that moment was 232,271.406 mi and moon was 131.2 ft farther from earth than previously believed. Lick Observatory in California had recorded first hits Aug. 1 and had estimated earth-moon distance to be 226,970.9 mi at that time. (AP, W *Star*, 8/21/69, A3)

- U.S.S.R. launched *Cosmos CCXCIV* from Plesetsk into orbit with 343-km (213.1-mi) apogee, 205-km (127.4-mi) perigee, 89.7-mi period, and 65.4° inclination. Satellite reentered Aug. 27. (GSFC *SSR*, 8/31/69; *SBD*, 8/20/69, 166; UN Public Registry)

- NASA announced selection of Chester M. Lee as Apollo Mission Director, succeeding George H. Hage, who had been elected vice president for product development with Boeing Co. Lee, retired USN captain who had served in Polaris missile program and in Directorate of Research and Engineering in Office of Secretary of Defense, had been Assistant Apollo Mission Director since August 1966. (NASA Release 69-122)

August 20: Study of possible Space Technology Applications and Research Laboratory (STARLAB), sponsored by NASA and American Society for Engineering Education, was completed at MSFC. Eleven-week design project focused space-developed technology on earth resources use, crop-maturity prediction, soil analysis, vegetation vigor, sea farming, and other earth problems. Final presentation in project, which had participation of 21 faculty members from 18 colleges and universities, was report on orbiting space laboratory illustrating systems approach that could be valuable in solving major earth problems. (MSFC Release 69-179)

- *Washington Post* published letter from former Secretary of State Dean Rusk. He recommended U.S. abandon idea of space race with U.S.S.R.; "throw wide open the doors on international cooperation"; proceed with development of near-earth space capabilities and activities contributing to understanding of earth; and "take advantage of NASA's extraordinary ability to mobilize scientific, technical, industrial and other talents" for other tasks, like air travel and air pollution problem-solving. "Manned flights to the planets might better be a decision for the next generation." (*W Post*, 8/20/69, A28)

August 21: NASA's X–24A lifting-body vehicle, piloted by Maj. Jerauld R. Gentry, successfully completed third flight after air-launch from B–52 aircraft over South Rogers Lake Bed, Calif. Objectives of unpowered flight were to obtain handling qualities, stability and control derivatives, flow visualization over aft portion of vehicle, and longitudinal trim curves and lift-to-drag ratio at 15° upper-flap setting. Procedural error caused X–24A to be launched 35 secs early and some planned data were not obtained. (NASA Proj Off)

- *Intelsat I (Early Bird)* had been put back into orbital retirement and full communications service via *Intelsat-III F–2* had been restored, ComSatCorp announced. *Intelsat I*, reactivated June 30 after six-month retirement to compensate for failure of *Intelsat-III F–2* until service was restored Aug. 1, would remain in orbit and would be capable of operational service if needed. Restored *Intelsat-III F–2* was handling 620 full-time commercial circuits serving countries in Atlantic area and transatlantic TV programming when ordered. (INTELSAT Release 69–53)

- NASA launched series of three sounding rockets from Wallops Station. Nike-Tomahawk carried GSFC payload to 141.1-mi (227.0-km) altitude to test neutral-mass spectrometer system with unique sample-flow and test-leak subsystem necessary for making high-pressure neutral-constituent measurements on planned Mars and Venus missions and to demonstrate capability of sterilized-mass-spectrometer electronics in flight environment for first time. Rocket and instruments functioned satisfactorily and complete data were obtained.

    Nike-Tomahawk carried GSFC and Univ. of Michigan payload to 197.6-mi (318.0-km) altitude to measure nitrogen density and temperature. Secondary objectives were to evaluate Omegatron system designed for San Marco-C satellite, measure density and temperature of nitrogen simultaneously, compare Pitot-static-probe and thermosphere-probe density in 74.6-mi (120.0-km) region, and validate mass spectrometer nitrogen measurement and electrostatic-probe electron temperature data. Rocket and instruments functioned satisfactorily and complete data were obtained.

    Nike-Apache carried Univ. of Michigan payload to 124.3-mi (200.0-km) altitude to measure neutral atmospheric density by Pitot technique in 18.4- to 74.6-mi (30.0- to 120.0-km) region. Rocket and instruments functioned satisfactorily and all atmospheric measurements made by Pitot probe were excellent. (NASA Rpts SRL)

- Every scientist-astronaut except one—geologist Harrison Schmitt—had been removed from NASA's lunar landing training list, Victor Cohn reported in *Washington Post*. Report was later denied by NASA. Cohn said remaining scientists had been assigned to train for long-duration, earth-orbiting Apollo Applications missions beginning in 1972. (*W Post*, 8/21/69, A1; 8/22/69, A18)

- *Washington Post* published letter from Irene S. Rubin in Lampang, Thailand. Real impact of *Apollo 11* success in Thailand had been "on the group of educated men who have some effect on government. Their primary reaction was not one of shared accomplishment but of shame in the gap thus dramatized between themselves and the developed countries." Though U.S. could not hide technological capacity, "I think we should be more aware of the context into which news of the Apollo mission is received. Far from bringing the world closer together with such performances, we may be arousing bitterness and obstinacy in the misallocation of development funds." (*W Post*, 8/21/69, A18)

August 22: U.S.S.R. launched *Cosmos CCXCV* into orbit with 473-km (293.9-mi) apogee, 270-km (167.8-mi) perigee, 91.9-min period, and 71.0° inclination. Satellite reentered Dec. 1 (GSFC *SSR*, 8/31/69; 12/15/69)

- Nike-Apache sounding rocket launched by NASA from WSMR carried GSFC and Dudley Observatory payload to 98.2-mi (158.0-km) altitude to provide background particle collection for comparison with data from Nike-Apache launched Aug. 13 during Perseid meteor shower. Rocket and instruments functioned satisfactorily. Data were expected from all experiments. (NASA Rpt SRL)
- NASA named Rocco A. Petrone, Director of Launch Operations at KSC since 1966, to succeed L/G Samuel C. Phillips (USAF) as Director of Apollo Program, effective Sept. 1. He would be succeeded by Deputy Director of Launch Operations Walter J. Kapryan. Petrone had been Saturn Project Officer and Apollo Program Manager. His awards included NASA Exceptional Service Award for direction of *Apollo 7* checkout and launch and NASA Distinguished Service Medal, NASA's highest award, for direction of *Apollo 8* checkout and launch. (NASA Release 69–124)
- IAA announced selection of Dr. Charles A. Berry, Director of Medical Research and Operations at MSC, to receive Daniel and Florence Guggenheim International Astronautics Award for 1969. Award and $1,000 prize would be presented during 20th International Astronautical Congress in Argentina in October. (UPI, *W Post*, 8/23/69, B3)
- NASA's alleged neglect of pure science research goals in favor of engineering pursuits and "glamor" had caused undercurrent of dissatisfaction among scientists, *Science* noted. When interviewed by *Science* Dr. F. Curtis Michel, Dr. Donald U. Wise, and Dr. Elbert A. King, who had resigned from NASA recently, declined to attribute their resignations directly to major dissatisfactions with NASA and denied that they had resigned to protest emphasis on engineering rather than scientific research. They did, however, express some dissatisfaction with role of basic science in space exploration and impatience with NASA's management of scientific projects and admitted they were lured from NASA by prospects of new positions that offered more time for scientific research. (*Science*, 8/22/69, 776–8)
- AIAA announced election of Honorary Fellows: Secretary of the Air Force, Dr. Robert C. Seamans, Jr.; German rocket pioneer Hermann Oberth; and Northrop Corp. founder John K. Northrop. Dr. Seamans, former NASA Deputy Administrator, was honored for "organizing the research, development and operational base which produced the Apollo program." Honors would be presented at Oct. 23 banquet in Anaheim, Calif. (AIAA Release)
- In letter advocating postponement of decision on manned Mars landing [see Aug. 20] former Secretary of State Dean Rusk had "gone to the heart of what is bound to become a critical national decision," Richard

August 22

Wilson said in Washington *Evening Star*. "Now that we know we can and will do this thing does it make any difference in the eons of time yet to come when we do it? Fifty years from now might we not have developed far cheaper and more efficient ways to do it? Mars will still be there. . . . The space men have shown us not only the moon, but what a beautiful planet we have in what may otherwise be a wholly desolate solar system—a beautiful planet that needs loving care to preserve it." (W *Star*, 8/22/69, A15)

- In telephone interview, UCLA astronomer Dr. Samuel Herrick, Jr., said planetoid Geographos, due to pass earth at 5.6-million-mi distance Aug. 27, would be best site of all asteroids for eventual space station beyond moon and good spot for manned or unmanned spacecraft landing. Its farthest point from sun in given orbit was least distant from sun and from earth of all minor planets. But astronauts landing on it would have to "dig in and tie themselves down" since its estimated g was so slight "even a sneeze directed at the surface would propel a man off into space." (AP, B *Sun*, 8/23/69, A3)

- In *Science*, MIT Lincoln Laboratory scientists Alan E. E. Rogers and Richard P. Ingalls reported mapping Venus surface reflectivity by radar interferometry at 3.8-cm wavelength for region from $-80°$ to $0°$ longitude and from $-50°$ to $+40°$ latitude. Map was free from twofold range-Doppler ambiguity, presented new features, and clearly delineated features previously observed. It showed large circular regions of significantly lower reflectivity than their surroundings, with size and appearance of lunar maria. (*Science*, 8/22/69, 797–9)

- At National Amateur Astronomers convention in Denver, Colo., six-member panel including Northwestern Univ. astronomer Dr. J. Allen Hynek and Univ. of Arizona physicist Dr. James E. McDonald suggested UFO investigation be taken from USAF and placed with scientific body. Panelists said since UFOs apparently presented no danger to national defense, they were unimportant to USAF. Panel disagreed with 1968 Condon Report on UFOs [see Jan. 9]. Hynek said UFO research should continue. (AP, W *Star*, 8/24/69, A17)

August 23: USAF launched unidentified satellite from Vandenberg AFB by Titan IIIB-Agena booster. Satellite entered orbit with 234.3-mi (377.0-km) apogee, 85.8-mi (138.1-km) perigee, 89.6-min period, and 108.1° inclination and reentered Sept. 7. (GSFC *SSR*, 8/31/69; 9/15/69; *SBD*, 8/26/69, 190)

- Chemical analysis of moon rocks at Lunar Receiving Laboratory had disclosed their age might range from 2 billion to 4.5 billion yrs—far greater than most scientists expected—lunar scientists in touch with LRL colleagues said. It was "almost conclusive evidence that it has been billions of years since these rocks crystallized." Finding might settle difference between geologists who had viewed lunar surface as having had continuous history and those like Dr. Harold C. Urey who believed moon was ancient, undisturbed place made of material which would help unfold history of early planets. Later, MSC Director of Science and Applications, Dr. Wilmot N. Hess, said Dr. Oliver A. Schaeffer and Dr. John Funkhouser of State Univ. of New York, Dr. Joseph Zahringer of Max Planck Institute in Heidelberg, and Dr. Donald Bogard of MSC had measured solar particles trapped in lunar rocks to determine lunar material's age. (Cohn, *W Post*, 8/24/69, A1; UPI, *W Star*, 8/25/69, A4)

- Ten space pioneers were named to first National Space Hall of Fame. Honorees, chosen by Houston City committee, would be feted at first annual awards dinner in Houston, Tex., Sept. 27. They included Astronaut Alan B. Shepard, Jr., first American to journey in space; former Astronaut John H. Glenn, Jr., first American to orbit in space; late Astronaut Edward H. White II, first man to walk in space; Dr. Wernher von Braun, MSFC Director; late Rep. Albert Thomas (D-Tex.), staunch supporter of space program; Dr. Kurt H. Debus, KSC Director; late Dr. Hugh L. Dryden, former NASA Deputy Administrator; Dr. Maxime A. Faget, Director of Engineering and Development at MSC; Dr. Robert R. Gilruth, MSC Director; and late Dr. Robert H. Goddard, father of rocketry. Hall was in Albert Thomas Center in Houston. (UPI, *NYT*, 8/25/69, 8)

August 24: *This Week* published interview with science fiction author Arthur C. Clarke: Most important recent outer space discovery was pulsars—"It's possible that they might be signals from some higher civilization." Scientific surprise in Apollo program was "its immaculate perfection. You don't expect that, no matter how carefully you prepare." Clarke was writing space exploration documentary which would show "whole span of human interest in space, back to the Babylonian astronomers and on up through the colonization of the solar system." It would include Stonehenge which was "as big a burden for the primitive economy that built it—in fact, probably a much bigger burden than the Apollo program is for us." (Bradford, *This Week*, 8/24/69, 7)

- Transfer of USAF MOL officers to NASA astronaut corps [see Aug. 14] was criticized in Washington *Sunday Star* by William Hines: "With the initial moon landing now an accomplished fact, the pace of manned space operations has slowed down to three flights per year. This means that no more than nine men can fly annually, and with 54 astronauts now on board, this, in turn, means an average of six years between flights." Though pace might accelerate in time and future space stations would increase annual number of crew assignments, "the glamor and glory of being an astronaut—particularly a nonflying one—no longer compensates for the enforced idleness imposed by the modified flight schedule." (*W Star*, 8/24/69, D4)

August 25: Postmaster General Winton M. Blount announced that "First Man on the Moon" postage stamp would be issued Sept. 9 in Washington, D.C., in conjunction with National Postal Forum. Printed from master die carried to moon on *Apollo 11* mission (July 16–24), 10-cent airmail stamp would be 50% larger than conventional commemorative stamps and would be dedicated in special ceremony attended by *Apollo 11* Astronauts Neil A. Armstrong, Edwin E. Aldrin, Jr., and Michael Collins. Post Office had received 500,000 first-day cover requests within three weeks after stamp was announced July 9 and was still receiving 60,000–80,000 requests daily—one-fifth from foreign countries. (PO Dept Release 130)

- Dr. Harry H. Hess, Chairman of NAS–NRC Space Science Board and member of NASA's Science and Technology Advisory Committee for

Manned Space Flight since 1963, died of heart attack suffered while attending Board meeting at Woods Hole, Mass. Dr. Hess, Blair professor of geology at Princeton Univ., was one of scientists who had analyzed *Apollo 11* lunar samples. In 1960 he had advanced theory that volcanic activity on ocean floor caused continental drift. He had been past president of Mineralogical Society of America and of Geological Society of America, chairman of site-selection committee for NSF's Project Mohole, and adviser to numerous other Federal agencies. (UPI, *W Post*, 8/27/69, A10; *Science*, 8/29/69, 882)

- Robert E. Bernier, former ComSatCorp systems engineer for Intelsat III program, became NASA European Representative in Office of International Affairs. He replaced Clotaire Wood, who would return to Office of Advanced Research and Technology at NASA Hq. Bernier would begin his duties at American Embassy in Paris in early October. (NASA Release 69-125)
- Scientist-astronaut Dr. William E. Thornton, who had been grounded in spring, received USAF clearance to continue jet pilot training. Thornton had had difficulty landing because of distortion of vision called aniseikonia, which reduced his depth perception. Vision had been corrected with special glasses. (UPI, *W Star*, 8/26/69, A3; *W Post*, 8/26/69, A9)
- *Washington Post* published results of July 30–Aug. 4 Harris survey of 1,577 U.S. households to determine attitude toward spending $4 billion annually for decade to explore moon and other planets. While 53% of those polled approved funding for lunar landing, narrow plurality of 47% was opposed to further $4 billion annually; 44% favored. Persons under 30 favored extension of space program 60% to 34% but those over 50 opposed it 59% to 30%. Black citizens were opposed 68% to 19%. (*W Post*, 8/25/69)

August 25–30: Eighth International Symposium on Space Technology, first major international space meeting since *Apollo 11* launch, was held in Tokyo. In opening speech, general chairman Tsuyoshi Hayashi expressed world's appreciation to U.S. for making "a great leap for mankind" but said many other nations had contributed to scientific knowledge that made lunar landing possible. He asked recognition of moon as international territory.

Among 400 scientists from 19 countries attending meeting were NASA Apollo Applications Program Director William C. Schneider; Dr. Christopher C. Kraft, Jr., MSC Director of Flight Operations, and M. P. Frank from MSC; Herbert A. Wilson, Jr., Chief of Applied Materials Div., from LaRC; OGO Project Manager Wilfred E. Scull from GSFC; Dr. Thomas Vrebalovich from JPL; and Leon C. Hamiter, Jr., MSFC engineer. Hamiter presented paper on increased computer capacity and lighter weight flight hardware. Prof. Masahiko Kido of Japan's Ehime Univ. said legal status should be developed for moon before disputes arose over lunar real estate. Other participants urged steps to outlaw military use of moon.

Dr. Werner J. Kleen, Director of European Space Research and Technology Center, said ESRO had been given permission to put comsat into orbit and would start work in autumn. Japan announced its space development corporation would begin operations Oct. 1 and would launch MS-4 three-stage rocket in early 1970, followed by comsat launch. (Shabecoff, *NYT*, 8/26/69, 11; MSFC Release 69–181; NASA Int Aff)

August 26: Moon landing would change human lives, British novelist and scientist C. P. Snow said in *Look*. "I am afraid that in the long run, perhaps a generation, perhaps longer, it will have a bad effect. It will give us the feeling, and the perfectly justified feeling, that our world has finally closed in. This is forever the end of the mortal frontier." Space enthusiasts thought lunar landing would liberate human imagination but "I believe . . . that human imagination is going to be restricted—as to an extent it was when the last spots on the globe had been visited, the South Pole and the summit of Everest. Nowhere on earth for adventurous man to go. Very soon, there will be no place in the universe for adventurous man to go." (*Look*, 8/26/69, 68–72)

- NASA announced award by LaRC of $2.5-million contract to Ling-Temco-Vought Aerospace Corp. to design, develop, and flight-qualify larger 1st-stage solid rocket motor for Scout booster. New Algol III motor would have 44- or 45-in dia, 4 or 5 in wider than Algol IIB, and would enable Scout to place 400-lb payload, 100 lbs more than IIB capacity, into orbit with 300-mi altitude. (NASA Release 69/126)
- Bright red lights, believed by observers to be meteors, flashed across California, Nevada, and Arizona at 8:50 pm PDT. North American Air Defense Command (NORAD) later identified lights as parts of Soviet booster burning during reentry. Booster had launched *Cosmos CCXCIV* Aug. 19. (AP, W *Star*, 8/27/69, A5; later ed, A13)

August 27: NASA's 148-lb drum-shaped Pioneer E failed on 5:29 pm EDT launch from ETR by Thrust-Augmented Improved Thor-Delta (DSV-3L) booster. Satellite had been intended for solar orbit to collect scientific data on electromagnetic and plasma properties of interplanetary medium near earth's orbital path during six or more passages of solar activity centers.

Jettison of three strap-on solid-propellant rockets, 1st-stage Thor engine cutoff, and 2nd-stage ignition occurred as planned but vehicle began gyrating, veered off course, and was destroyed by Range Safety Officer at 8 min 2 secs GET. Pioneer E and TETR C test and training satellite, carried as secondary payload to test Apollo communications network, splashed into Atlantic about 300 mi southeast of Barbados. Preliminary analysis of data indicated loss of hydraulic pressure during 1st-stage burn had permitted engine nozzle to develop uncontrolled gimbaling and vehicle gyrations. Investigation would be conducted to determine exact cause and action to prevent recurrence.

Pioneer E was last in series of five spacecraft designed to provide continuing measurements over the solar cycle at widely separated points in interplanetary space. *Pioneer VI* (launched Dec. 16, 1965), *Pioneer VII* (launched Aug. 17, 1966), *Pioneer VIII* (launched Dec. 16, 1967), and *Pioneer IX* (launched Nov. 8, 1968) had received 25,000 commands from ground and were still producing useful data from widely scattered positions in heliocentric orbits. Most recent Pioneer missions had provided new information on functions of magnetosphere, additional data on finding that diffuse solar plasma regions appeared to have attraction of their own, measurements of cosmic dust populations, data on changes in electrical and magnetic characteristics

August 27

of solar corona, and targets for precision radar tracking which led to establishment of reliable value for earth-moon-mass ratio and sun-earth-mass ratio. Pioneer program was managed by ARC under OSSA direction. (NASA Proj Off; NASA Release 69-116; *SBD*, 8/29/69, 213)

- Moon was twin planet of earth, formed from same whirling gas cloud, in early view of two LRL scientists studying *Apollo 11* samples. Dr. S. Ross Taylor of Astri National Univ., Canberra, Australia, said, "Moon's composition is unlike the earth's. But it is not outside our experience. It is like the material you would expect if the earth and moon were formed as a double planet." He thought moon was younger twin, while Dr. Oliver A. Schaeffer of State Univ. of New York thought it might be equally old.

  Age of two lunar rocks had been estimated at 3.1 billion yrs, "give or take . . . 200 million years," by measuring proportion of argon 40 to potassium in rocks, Dr. Schaeffer said. Lunar highlands might be 4.5 billion yrs old. Moon, he thought, never grew big enough to melt internally and produce geologic activity to change lunar surface and leave younger rocks. Dr. Taylor's studies had shown unusually high amounts of refractory material and absence or low concentration of volatile materials, implying volatile material had boiled away in melting process. He inferred rock chemistry was different from deep mantle of earth and from cosmic abundances—distribution of elements that would be expected in distant, more primitive planet captured by earth. (Cohn, *W Post*, 8/28/69, A1)

- MSC Deputy Director George S. Trimble announced his resignation, effective Sept. 30, after 2½ yrs with NASA. He had been Director of Advanced Manned Missions Program in NASA Office of Manned Space Flight before appointment to MSC post Oct. 13, 1967. (MSC Release 69-70; *W Post*, 8/28/69, A8; NASA Ann, 10/13/67)

- NASA announced selection of RCA Service Co. to receive two-year, cost-plus-award-fee contract with one-year option for logistic support to Space Tracking and Data Acquisition Network (STADAN), Manned Space Flight Network (MSFN), and NASA Communications Network (NASCOM). Contract was expected to exceed $17 million. (NASA Release 69-127)

- American Airlines began showing NASA color film of *Apollo 11* and distributing free copies of CBS News recording "Man on the Moon" and free cut-out lunar modules for children on "Americana" flights between East Coast and California through Sept. 23. (*NYT*, 8/18/69, 23)

August 28: Leading lunar scientist Dr. Harold C. Urey told conference on nuclear energy at Argonne National Laboratory near Chicago he was "pleased" at discovery that age of lunar rocks might range between 3 billion and 4.5 billion yrs [see Aug. 23] and had "expected this for a long time. . . . But I'm not making any more bets on the moon's origin." He was "puzzled" by once-molten lunar sea material; it might have been formed by huge meteor or asteroid impacts rather than volcanism and moon might have originated out of cluster of such asteroidal debris. (Cohn, *W Post*, 8/29/69, A3)

- Fiftieth anniversary of International Air Transport Assn. (IATA), founded in Amsterdam Aug. 28, 1919. International flying under IATA auspices in 1919 amounted to 3,500 passengers; in 1969 it was expected to total 300 million. Organization was still devoted to original principles: promotion of safe, regular, and economical air transport; collaboration among international carriers; processing of technical matters and common fares; and functioning as clearinghouse for settlement of member airline accounts. From original membership of six airlines, IATA had 103 participating members. (Bamberger, *NYT*, 8/24/69, 86)

August 29: U.S.S.R. launched *Cosmos CCXCVI* from Baikonur into orbit with 299-km (185.8-mi) apogee, 227-km (141.1-mi) perigee, 89.6-min period, and 64.9° inclination. Satellite reentered Sept. 6. (GSFC *SSR*, 8/31/69; 9/15/69; *SBD*, 9/3/69, 9; UN Public Registry)

- Some plants treated with lunar dust in early August were showing unexpected responses. Treated plants—including seedlings of several common food plants like wheat, tomatoes, cucumbers, and limes—were generally huskier and slightly greener than untreated plants. NASA statement said: "The seedlings challenged with lunar materials uniformly look better than the controls (untreated plants). Germination in the presence of lunar soil indicates that it is behaving like a source of nutrients." Plant cells in tissue culture showed "some evidence of subtle change as a result of lunar inoculation." Dr. J. A. Vozzo, plant pathologist at Lunar Receiving Laboratory, emphasized that changes were minor and could not yet be positively attributed to lunar dust. (Cohn, *W Post*, 8/30/69, A1)

- NASA selected General Electric Co. to receive three-year, $4-million, cost-plus-award-fee contract with two-year option to provide engineering and mission-related support to LaRC for Viking Project—series of planetary probes which would begin softlanding on Mars in 1973. (NASA Release 69-128)

- New determination of abundance of water in Mars atmosphere was reported in *Science* by Illinois Institute of Technology astronomers Tobias Owen and Harold P. Mason. New spectrograms of planet had been obtained in region of water-vapor band at 8,200 Å during February and March 1969. Amount of precipitable water was found to be about 15 μ. Abundance reaffirmed that some water was present at current epoch but otherwise had little bearing on evolution of Martian atmosphere. Water vapor did not imply liquid water existed on Martian surface. (*Science*, 8/29/69, 893–5)

August 31: Washington Post Sunday supplement *Potomac* published profile of Dr. Richard T. Whitcomb, head of 8-Foot Tunnels Branch at LaRC. He had won 1954 Robert J. Collier Trophy for design of "coke bottle" aircraft fuselage configuration that enabled aircraft to pass through mach 1 with increased power. More recently he had devised supercritical wing, which would permit subsonic jet aircraft to approach mach 1. If adopted by commercial aircraft manufacturers, wing would cut nearly one hour from current five-hour transcontinental flights. (*Potomac*, 8/31/69, 1, 5–7)

- DOD internal, classified memoranda suggested Government would waste money buying additional Lockheed C–5A aircraft, *Washington Post* article said. Central conclusion was that most efficient and least costly transportation network to support two major and one "brushfire" war "for which military wants to be prepared consists of the existing three squadrons (58) [of] C–5As plus smaller carriers like the C–141 and modern freighters." (Nossiter, *W Post*, 8/31/69, A1)

During August: Pace magazine published articles by Vice President Spiro T. Agnew, also NASC Chairman, and by NASA Administrator, Dr. Thomas O. Paine.

Dr. Paine said, "To improve conditions in our society we need to create more wealth through greater productivity based on new technology. We should be restless and dissatisfied with our slowness in overcoming social ills, and I hope that the space program will continue to spur us onward here. If we can go to the moon, why can't we build great and shining cities? Why can't we eliminate ignorance, crime and poverty? If our space program highlights such questions and helps form a national commitment to find new solutions, it will have served the nation well. Our space advances should embolden the nation to proceed forward with increased confidence in these other areas. Our Apollo program has demonstrated anew what Americans can accomplish given a national commitment, capable leadership and adequate resources.

"Man's future in space is limitless. We have embarked on a new stage of evolution that will engage all future generations of men. We face the unknown in countless areas: What are the effects of sustained zero and artificial gravity? Of time-extending flight at nearly the velocity of light? Of societies genetically selected for extraterrestrial living?

"We must find the answers. We must move vigorously forward in space. The practical benefits alone justify this venture, but there are many other compelling human reasons. Progress in space should continue to spur us onward to find new solutions to our age-old problems here on Spaceship Earth. We must make the blue planet Earth a home base, worthy of men who will set forth one day on journeys to the stars."

Vice President Agnew said: "With the remarkably successful Apollo moon-landing program on the verge of culmination, we are now faced with a need to define just what we should proceed to do to make use most effectively of the results of our past and continuing space-exploration investment. Wealthy as our economy is, rich as our technology has become, we must plan carefully in order to meet a wide range of urgent national requirements. . . . It is our hope that, with a carefully reasoned set of goals adequately funded by the people through their Congress, the nation and the world will reap the maximum possible benefit from mankind's most ambitious undertaking. We must keep our horizons wide and our sights high. Despite its many internal domestic priorities, this nation should never turn inward, away from the opportunities and challenges of its most promising frontier." (*Pace*, 8/69, 2-4)

- Four hundredth anniversary of Mercator's map of the world, published in Rhenish city of Duisberg in 1569 by Gerhard Kremer (known by his Latin name Gerardus Mercator). Map translated earth's sphere into plane on chart on which straight line drawn by navigator cut across all meridians at same angle. Mercator projection was still standard for worldwide sea navigation and for aeronautical charts despite its distortion of northern latitudes. (*NYT*, 8/17/69; EH)

- "Technologically and managerially, Apollo was difficult," Englebert Kirchner said in *Space/Aeronautics* editorial. "Politically and socially, it was simple. Just the reverse is true about the great problems of our society. What is making these so hard to solve is not technology but serious disagreement about goals and priorities, about what is good for whom, who is to get what and who should pay for it. The space program does not hold the answer to these questions. Trying to find them in Apollo will only distort and therefore belittle an incomparable achievement. Apollo took us to the moon, to that shining disk in the sky that looks so unbelievably distant. Isn't that enough?" (*S/A*, 8/69, 27)

- AFSC *Newsreview* editorial commented on *Apollo 11*: "If, like the early Vikings or Columbus at the shores of the New World, Amundsen at Antarctica, Hillary at the peak of Mt. Everest—our astronauts stood alone with their thoughts on unknown soil, they were not alone. With them was the invisible presence of the most extensive, highly trained, professionally competent, and thoroughly dedicated task force we have known. We in the Air Force Systems Command salute the astronauts on their accomplishment. We are proud that we have been able to contribute to their magnificent achievement." (AFSC *Newsreview*, 8/69, 2)

September 1969

September 1: L/G Samuel C. Phillips, NASA Apollo Program Director, became Commander of Air Force Space and Missile Systems Organization (SAMSO). Gen. Phillips had been Director of Minuteman program before assignment to NASA in January 1964 as Deputy Director of Apollo program. He was succeeded by Rocco A. Petrone [see Aug. 22]. (NASA Ann, 7/31/69)

- S. Paul Johnston retired as Director of Smithsonian Institution's National Air and Space Museum. He would represent AIAA on NRC. (*A&A*, 9/69, 15)
- *Scientific Research* article commented on attitude toward science of President Nixon and Dr. Lee A. DuBridge, Presidential Science Adviser: "If there's a Nixon-DuBridge science policy it is this: to revitalize federal support of basic research . . . and to point government-financed applied research toward the solution of the country's many social ills." (*Scientific Research*, 9/1/69, 11–12)

September 2: U.S.S.R. launched *Cosmos CCXCVII* from Plesetsk into orbit with 309-km (192.0-mi) apogee, 204-km (126.8-mi) perigee, 89.6-min period, and 72.8° inclination. Satellite reentered Sept. 10. (GSFC SSR, 9/15/69; SBD, 9/3/69, 8)

- Qatron Corp. announced it had received $275,000 contract from GSFC to build several recorder-receiver switching and preprogrammable patch systems for Apollo program. (*W Star*, 9/2/69, A16)

September 3: NASA's HL-10 lifting-body vehicle, piloted by NASA test pilot William H. Dana, reached 81,000-ft altitude and mach 1.42 after airlaunch from B-52 aircraft west of Rosamond, Calif. Primary objective of flight, 24th in series and first with new engine, was to obtain stability, control, and engine data. (NASA Proj Off)

- Tokyo Univ. scientists successfully launched four-stage Lambda rocket in preparation for launch of Japan's first satellite in late September. (UPI, *W News*, 8/4/69; Harrison, *W Post*, 9/24/69, A9)
- MSFC announced contract awards: McDonnell Douglas Astronautics Co. was awarded $97,340,000 cost-plus-fixed-fee/award-fee contract to provide for two Saturn V Workshops—one for launch in 1972 and second for backup. McDonnell Douglas also received $87,450,000 cost-plus-fixed-fee/award-fee contract modification for continued work on two airlock modules for Apollo Applications (AA) program cluster, including tests, checkout, documentation, and logistics support.

    Boeing Co. received $25,130,376 contract modification extending period for completion of Saturn V 1st stage (S-IC-15) from June 30, 1970, to June 30, 1971. (MSFC Releases 69-199, 69-200, 69-201)

- Swedish aircraft constructor Has Fancher had said that in 1944 Adolf Hitler took delivery of first Junkers 390 aircraft with 14,400-hp engine constructed specially to bomb New York, *Washington Daily News* reported. Fancher, pilot on aircraft's nonstop test flight between Germany and South America, said plane weighed 93 tons with bombs and had planned range for nonstop flights from Bordeaux in occupied France to New York and return. Aircraft, delivered too late for use in war, had been burned by Germans. Comparable aircraft was not built until 1956, Fancher said. (*W News*, 9/3/69)

September 4: Some NASA scientists were helping their communities and hiding their aid projects "as tho they were sinful," Ray Cromley said in *Washington Daily News*. Scientists were using space-acquired skills "to help their fellow men in ways they were uniquely qualified." Projects included applying systems analysis to air pollution problem, planning school expansion to meet population expansion, applying systems concept to town management and to city police force problems, developing new concepts for airport planning and new technique for vandalism prevention, developing improved communications systems for city emergency departments, and helping an agency develop ways of evaluating proposals for study and development contracts with private industry. (*W News*, 9/4/69, 23)

September 5: First measurement of Mars UV dayglow, made during *Mariner VI* Mars flyby July 31, was reported in *Science* by Univ. of Colorado astrogeophysicists C. A. Barth, C. W. Hord, J. B. Pearce, K. K. Kelly, A. I. Stewart, G. E. Thomas, and G. P. Anderson; Johns Hopkins physicist W. G. Fastie; and JPL's O. F. Raper. Emission features from ionized carbon dioxide and carbon monoxide were measured in 1,900 Å to 4,300 Å spectral region. Lyman alpha 1,216 Å line of atomic hydrogen and 1,304 Å, 1,356 Å, and 2,972 Å lines of atomic oxygen were observed. Prime objective of experiment was to search for nitrogen in Martian atmosphere. First analysis had shown no evidence of nitrogen emissions in UV spectrum of upper atmosphere (*Science*, 9/5/69, 1004–5)

- Aerobee 170 sounding rocket was launched by NASA from WSMR carrying Naval Research Laboratory payload to conduct solar physics studies. Mission was unsuccessful. (NASA Proj Off)

- *September 6:* Astronauts Frank Borman, James A. Lovell, Jr., and William A. Anders were named winners of 1969 Harmon International Astronaut's Trophy for December 1968 *Apollo 8* mission. Maj. Jerauld R. Gentry (USAF) was awarded Aviator's Trophy for testing NASA's HL–10 lifting-body vehicle. Harmon trophies were awarded annually to world's outstanding pilots for feats of individual piloting skill. (UPI, *W Star*, 9/7/69, A7)

- *Apollo 11* astronauts attended celebrations in their hometowns. In Wapakoneta, Ohio, Neil A. Armstrong was cheered by crowd estimated at 10 times normal 7,000 population, addressed teen-age rally, and led parade including Gov. James A. Rhodes, Dr. Albert D. Sabin (developer of oral polio vaccine), and comedian Bob Hope—all Ohioans. Edwin E. Aldrin, Jr., on second visit to hometown as astronaut, presented Montclair, N.J., Library with autographed photo of plaque left on moon; Library named its science collection in his honor. Astronaut Michael Collins, who was born in Rome, Italy, visited New Orleans, La., as his adopted hometown. He attended luncheon in his honor and visited NASA's Michoud Assembly Facility. (*W Post*, 9/7/69, A3)

September 7: Self-testing-and-repairing (STAR) computer to direct unmanned spacecraft of multiyear missions to outer planets and intergalactic space had passed preliminary tests and would begin full-scale ground operation at JPL during week, JPL announced. Believed first computer capable of detecting its own failures and repairing itself, STAR had been developed by Dr. Algirdas A. Avizienis, JPL computer expert, who was trying for 90% probability that it would last 15 yrs, to control operations to Neptune or Pluto in solar system Grand Tours scheduled for late 1970s. During 9-to-11-yr minimum lifetime, STAR would automatically switch on up to three backup units to replace defective parts. By 1974, more modest model might replace defective parts twice for use on shorter missions like one to Jupiter. STAR could also aid in hospital and supersonic-aircraft automation. (JPL Release 532)

September 8: NASA's 363-ft-tall Saturn V launch vehicle, tipped with Apollo 12 spacecraft scheduled to carry astronauts toward moon Nov. 14, was placed on launch pad at KSC. (AP, *W Post*, 9/9/69, A2)

- MSFC announced selection of McDonnell Douglas Corp. to receive 11-mo, $2,899,986 contract for preliminary design and planning for 12-man earth-orbital space station for possible mid-1970 launch. Station—initial element of large space base and means of investigating effects of long-duration space flight on man—would have 10-yr lifetime, subject to expendables resupply and crew rotation. Parallel effort was being conducted by MSC and North American Rockwell Corp. (MSFC Release 69–204)

- Mexican President Gustavo Diaz Ordaz announced in Coahuila, Mexico, that President Nixon had accepted invitation for *Apollo 11* astronauts to start round-the-world tour in Mexico. He repeated congratulations to Government and U.S. people on *Apollo 11* success: "The United States gave proof of its greatness when it achieved this triumph, but it became even greater when they understood it and accepted it as a triumph of all humanity." President Nixon was in Mexico to attend dedication of Amistad Dam on Rio Grande. (*PD*, 9/15/69, 1241)

September 9: NASA's X-24A lifting-body vehicle, piloted by Maj. Jerauld R. Gentry (USAF), reached mach 0.6 after air-launch from B-52 aircraft at 40,000-ft altitude over South Rogers Lake Bed, Calif. Purposes of unpowered flight, fourth in series, were to evaluate stability and control derivatives at upper flap positions, determine handling qualities, and obtain flow visualization motion pictures of tufts on vehicle's aft portion. (NASA Proj Off)

- Aerobee 150 MI sounding rocket, launched by NASA from WSMR with VAM-20 booster, carried Cornell Univ. payload to 97.8-mi (157.4-km) altitude to examine sky in far infrared (5 μ–1,600 μ), using copper-doped-germanium, two gallium-doped-germanium, and indium-antimonide detectors. Loss of residual helium at 162 secs disabled attitude-control system. Timing failed in experimental payload and no useful scientific data were obtained. Some useful engineering data were collected. (NASA Rpt SRL)

- FRC announced award of $1.8-million NASA contract to North American Rockwell Corp. for construction of new supercritical aircraft wing. Wing, which utilized airfoil shape with flat top and rear edge curved downward, had been developed by Dr. Richard T. Whitcomb and tested at LaRC. Wind tunnel tests indicated new shape could allow highly efficient cruise flight at nearly 600 mph at 45,000-ft altitude. By increasing cruise speeds without increasing power, wing might significantly reduce operational cost of subsonic jet transport flights and allow faster travel, lower fuel consumption and costs, increased operational range, or increased payload. Wing would be mounted on modified Navy F–J fighter aircraft at FRC for flight testing. (FRC Releases 4–69, 15–69)

- Former NASA Apollo Program Director, L/G Samuel C. Phillips (USAF), received Distinguished Service Medal from Secretary of the Air Force, Dr. Robert C. Seamans, Jr., in Pentagon ceremonies. Award was for achievements with NASA from December 1964 to August 1969. Gen. Phillips had left NASA to become commander of USAF Space and Missile Systems Organization (SAMSO) in Los Angeles. (*AFJ*, 9/27/69, 8)
- At *Apollo 11* splashdown party at Shoreham Hotel in Washington, D.C. —attended by *Apollo 11* astronauts and wives—NASA Administrator, Dr. Thomas O. Paine, announced new Apollo Achievement Award of lapel button and certificate. He presented awards to NASA Associate Administrator for Manned Space Flight, Dr. George E. Mueller; former Apollo Program Director, L/G Samuel C. Phillips (USAF); and former Deputy Director of Apollo Program George H. Hage. (Beale, W *Star*, 9/10/69, F1)
- At first day ceremonies for commemorative moon landing stamp in Washington, D.C., Postmaster General Winton M. Blount presented *Apollo 11* Astronauts Neil A. Armstrong, Edwin E. Aldrin, Jr., and Michael Collins and NASA Administrator, Dr. Thomas O. Paine, with albums containing 32 stamps each. He said: "In the largest sense we pay tribute today to the spirit of man. We cannot separate the accomplishments of Apollo 11 from those of Vostok 1; we cannot separate the contributions of Michael Collins, or Edwin Aldrin or Neil Armstrong from those of Goddard and Einstein, Kepler and Newton, Copernicus and Galileo. We know this. And in the knowing again we find hope. For if men of all nations, together, can achieve dominion over the heavens, men of all nations, together, can achieve peace on earth for men for all time."

Armstrong said astronauts had deferred cancellation of stamps until they were reunited in CM, July 22. They had then grasped canceler simultaneously and pressed it upon die-proof version of commemorative stamp affixed to unaddressed envelope. Cancellation date remained July 20, day of lunar landing. (PO Dept Release 135; Shandler, W *Star*, 9/10/69, A3)

September 10: Nike-Apache sounding rocket launched by NASA from Wallops Station carried Univ. of Illinois and GCA Corp. payload to 127.4-mi (205-km) altitude to measure electron density, collision frequency, and temperature in lower ionosphere on quarterly world day. Payload included dual-frequency propagation experiment. Rocket altitude was nominal but range was only one-fourth that predicted. Instrument performance was excellent and good data were expected from all experiments. (NASA Rpt SRL)

- Paul G. Dembling, NASA General Counsel since January 1967, became NASA Deputy Associate Administrator. Dembling, who had joined NACA in 1945, had been principal drafter of bill which became National Aeronautics and Space Act of 1958 and had received NASA Distinguished Service Medal in 1968 for contributions to development of legal framework of U.S. aeronautical and space activities. (NASA Release 69-131)

Study of lunar samples was "bringing to light as many mysteries as it unravels," *New York Times* editorial said. Theorists were cautious, with evidence from one small area, Tranquility Base. "It is likely that the picture will become still more complex when a representative collection of samples becomes available from ten, twenty or thirty areas spread over the entire lunar surface. But even the limitations of the present data suggest strongly that the moon is very different from earth, and therefore has much to teach human science about the origin and evolution of the solar system. The case for intensive scientific study of the moon—conducted in part by geologists and other scientists sent there for on-the-spot investigation—is strong." (*NYT*, 9/10/69, 40)

September 11: Press conference on results of *Mariner VI* (launched Feb. 24) and *Mariner VII* (launched March 27) was held at NASA Hq. Some 200 TV pictures of Mars were taken by two Mariners, including 57 high- and medium-resolution views of selected Martian surface areas from altitude of only few hundred miles. Spacecraft measured Martian atmospheric temperature, pressure, and chemical constituency and measured surface temperatures in effort to correlate thermal characteristics with features observed in TV pictures. Data indicated Mars was heavily cratered, bleak, cold, dry, nearly airless, and generally hostile to any earth-style life forms.

Dr. Robert B. Leighton of Cal Tech said: "We got nine times the number of far encounter pictures that were originally proposed [few years ago], 20 per cent more near encounter pictures than were proposed, and 1,100 digital pictures which were entirely impossible according to schemes at the time of the proposal. . . . After Mariner 4 Mars seemed to be like the moon. At last Mariners 6 and 7 have shown Mars to be like Mars and have brought out Mars' own characteristic features, some of them unknown and unrecognized elsewhere in the solar system."

Dr. Robert P. Sharp of Cal Tech said Martian terrain could be divided into three types—crater, featureless, and chaotic. Cratered terrain was widespread and common on Mars and resembled moon. Featureless terrain was represented by Hellas area, which appeared to be upland area, 150-mi-wide zone that gently sloped into flat featureless floor. Chaotic terrain had series of "short ridges, little valleys, and irregular, jumbled topography." Chaotic and featureless terrain appeared to be distinctly Martian, suggesting "that on Mars we have either a difference in processes that are operating on the surface or within the crust or we have a difference of material from one place to another on Mars and different than on the moon, or, more likely, a combination of both. . . . We also have good reason for believing that the evolutionary history has been somewhat different. Again, there are scars on the face of Mars that we do not see on the face of the moon. And there have perhaps been episodic events in Martian history that are unique to the planet Mars. We end up with the conclusion that Mars is its own planet."

Dr. George C. Pimentel of Univ. of California at Berkeley said reevaluation of initial data from infrared spectrometer had shown infrared spectral features earlier ascribed to methane and ammonia were actually due to previously undiscovered absorptions of solid carbon dioxide. Reflection peak recorded three times in atmosphere off Mars' bright limb showed presence of solid carbon dioxide at high altitudes and at latitudes north of polar cap. Broad absorption near 9 μ recorded on bright limb was ascribed to solid silica or silicate material and broad absorptions near 12 μ recorded near dark limb were tentatively ascribed in part to solid carbon dioxide above ground. Further experimental work was in progress to refine thermal map.

Initial results of UV spectrometer experiment were detection of ionized carbon dioxide, carbon monoxide, atomic hydrogen, and oxygen. Nitrogen and nitric oxide were not detected and no evidence was found of clouds, blue haze, or any appreciable atmospheric absorption of UV radiation. Dr. Charles A. Barth of Univ. of Colorado said important point "is that the atmosphere of Mars is different than the atmosphere of the earth. If I showed you a spectrum taken the same way from the upper atmosphere of the earth, we would see a plentiful number of nitrogen bands. We could see emissions from nitric oxide. We could see emissions from atomic nitrogen. None of those features is present in the atmosphere of Mars. . . ."

Dr. Norman H. Horowitz of JPL presented biological implications of Mariner 1969 results. "There is nothing in the new data that encourages the belief that Mars is a body of life. But the results don't exclude this possibility. . . . The Mariner 6 and 7 data strengthen the previous conclusion that the scarcity of water on Mars is the most serious limiting factor for life. . . . Mars is a cold desert by terrestrial standards. If there is life on Mars, it must be a form of life that can utilize water in the form of water vapor or ice." (Transcript; NASA News Release)

- Aerobee 150 sounding rocket launched by NASA from WSMR carried Harvard Univ. payload to conduct solar studies. Rocket and instruments functioned satisfactorily. (NASA Proj Off)
- President Nixon announced intention to nominate Secor D. Browne to be member of Civil Aeronautics Board for remainder of term expiring Dec. 31, 1974. He would replace John H. Crooker, who had resigned effective Sept. 30. Browne would also be designated CAB Chairman. (PD, 9/15/69, 1249)

September 12: NASA began distribution, at MSC, of about 18 lbs (8.2 kgs) of lunar material to 106 U.S. scientific investigators and 36 in eight other countries for university, industrial, and governmental laboratory analyses. Lot comprised one-third of lunar samples returned by *Apollo 11*. Another 15% would be kept as examples of Tranquility Base material. Remainder would be held for later scientific experiments, with small amount possibly available on loan for public display. Material had been quarantined in LRL since its July 25 return; tests on animal and plant life had shown no ill effects. Interagency Committee on Back Contamination had approved release of samples to principal investigators or their representatives whose plans for safeguarding material had been approved by MSC officials.

Preliminary LRL examinations had disclosed two basic rock types, compacted lunar soil and igneous rocks. Rocks had been on lunar surface from 10 to 150 million yrs; igneous rocks had crystallized from 3 to 4 billion yrs ago. Approximately 3 kgs of samples would be de-

September 12: NASA *began distribution of 18 pounds of lunar material to scientific investigators in the United States and eight other countries for anaysis. The rock above, one of the samples collected by Astronauts Neil A. Armstrong and Edwin E. Aldrin, Jr., on the moon July 20, was studied at MSC's Lunar Receiving Laboratory.*

stroyed during experiments; residues and remaining 5.1 kgs would be returned to NASA. Results of analyses were to be reported early in 1970.

Among measurements to be made were those of physical properties of rocks or soil to help in understanding optical observations of moon from earth and future seismic experiments; mineralogy and petrology to show mineral content, amount of water present when rocks crystallized, and how surfaces were eroded by particles; chemical composition of rocks and fines to determine concentration of 92 elements occurring on earth and in meteorites, times of crystallization of igneous rocks, and periods rocks had lain on lunar surface. Studies of rare gases in soil would furnish first data on isotopic compositions of solar materials. Biologists and organic chemists would determine structures and abundances of carbon compounds in and on lunar surface and their origin, catalog microstructures in terms of organized elements and microfossils, and define presence or absence of viable lunar organisms. (NASA Release 69-130)

- Nike-Apache sounding rocket launched by NASA from Wallops Station

September 12

carried Univ. of Illinois and GCA Corp. payload to 117.4-mi (189-km) altitude to measure electron density, collision frequency, and temperature in lower ionosphere at midnight. Secondary objective was to test mechanical delay igniter and monitor its performance. Simultaneous launch from Chamical, Argentina, studied particle precipitation and transport effects across equator. Rocket and instruments functioned satisfactorily. (NASA Rpt SRL)

- Spencer M. Beresford, former special counsel of House Committee on Science and Astronautics, was appointed NASA General Counsel succeeding Paul G. Dembling, new NASA Deputy Associate Administrator [see Sept. 10]. (NASA Release 69–173)
- White House announced President's Science Adviser, Dr. Lee A. DuBridge, would visit four Western and two Eastern European countries in September and October to discuss arrangements for international scientific and technological cooperation and explore specific possibilities for strengthening existing arrangements. (PD, 8/15/69, 1251)
- Reuters said NASA had accepted offer of French sculptor Marcel Recher to build 140-ft "Platform for the Conquest of the Cosmos" at KSC as memorial to first lunar landing. Recher was looking for sponsor to contribute $145,000 for project. (W Post, 9/11/69, A3)
- Dec. 15 debut of Boeing 747 would be delayed six to eight weeks, Boeing Co. said. Pratt & Whitney Div. of United Aircraft Corp. had encountered problems in meeting performance goals in 362-passenger aircraft's engines. (NYT, 9/13/69, 46)

September 13: Aerospace Corp. announced election of Dr. T. Keith Glennan, President Emeritus of Case Institute of Technology and first NASA Administrator (1958–1961), as Chairman of Board of Trustees. Sherrod E. Skinner retired as Chairman and L/G James H. Doolittle (USAF, Ret.), Vice Chairman, also retired during annual meeting of Board of Trustees. Skinner and Gen. Doolittle were awarded USAF Exceptional Service Award by Under Secretary of the Air Force John L. McLucas in El Segundo, Calif., ceremony Sept. 12 (Aerospace Release; CR, 9/25/69, E7813)

- Smithsonian Institution Curator of Meteorites, Dr. Kurt Fredriksson, arrived in Washington, D.C., carrying 10 gr of lunar material from LRL in nitrogen-filled plastic bag inside steel briefcase. One of six men in U.S. who had studied lunar samples, he later said Smithsonian scientist Dr. Bryan H. Mason would receive another 10-gr set. (Conroy, W News, 9/16/69, 5)

September 14: NASA announced availability of *Earth Photographs from Gemini VI Through XII* (NASA SP–171), collection of best 250 pictures taken between 1965 and 1967 from altitudes between 99 and 850 mi as Gemini spacecraft orbited earth. First and last views were of Cape Kennedy, with views of principal areas within 30° latitude of equator between. (NASA Release 69–129)

September 15: Space Task Group presented report *The Post-Apollo Space Program: Directions for the Future* to President Nixon at White House. It recommended basic goal of balanced manned and unmanned space program conducted for all mankind, with emphasis on increased utilization of space capabilities for services to man through expanded space applications program; enhancement of U.S. defense posture for world peace and security through exploitation of space techniques for military missions; continuing strong program of lunar and planetary exploration, astronomy, physics, and earth and life sciences; development of new systems and technology for space operations, emphasizing commonality, reusability, and economy through development of new space transportation capability and space station modules; and promotion of world community through program of broad international participation and cooperation.

As focus for development of new capability, Task Group recommended U.S. accept long-range goal of manned planetary exploration with manned Mars mission before end of century. Activities leading to goal should include initial concentration on exploiting existing capability and developing new one while maintaining program balance within available resources; operational phase using new systems and capabilities in earth-moon space, with men living and working in that environment for extended periods; and manned exploration missions out of earth-moon space, using experience of earlier two phases. Schedule and budgetary implications of phases were subject to Presidential choice, with detailed program to be determined in normal annual budget and program review.

Report outlined three possible NASA programs for manned Mars landing before century's end. Option I would launch manned mission in mid-1980s and would establish orbiting lunar station, 50-man earth-orbiting space base, and lunar surface base. Funding would rise from current $4-billion level to $8- to $10-billion level in 1980. Decision to proceed with development of space station, earth-to-orbit shuttle, and space tug would be required in FY 1971. Option II would include Mars mission launch in 1986, allowing for evaluation of unmanned Mars mission results before final designation of landing date and require about $8-billion maximum annual expenditure in early 1980s. Option III would include initial development of space station and reusable shuttles, as in Options I and II, but would defer decision on manned Mars landing date while maintaining goal of after 1980 but before close of century. Concurrent development of space transportation system and modular space stations would require rise in 1976 annual expenditures to $5.7 billion, while their development in series would entail $4- to $5-billion funding level.

Recommended DOD options were: (A) program of full military space capability in case of overt threat to national security, (B) development of efforts to counter known and accepted projections of security threat and increase in development activities if threat increased, and (C) program of lower level system deployment with technology and support effort necessary for contingency planning on assumption that lessening of world tensions would reduce emphasis on national defense.

At White House briefing following presentation, press secretary Ronald L. Ziegler said President Nixon had concurred in Task Group's rejection of two other, extreme space programs, one to land men on Mars as soon as possible, regardless of cost, and one to eliminate manned flight program after completion of Apollo. He did not know when President would make decision on course to follow, but budgetary considerations would be major factor. (Text; *PD*, 9/22/69, 1291; *NYT*, 9/16/69, 1)

- U.S.S.R. launched *Cosmos CCXCVIII* from Baikonur into orbit with

September 15

162-km (100.7-mi) apogee, 127-km (78.9-mi) perigee, 87.3-min period, and 49.6° inclination. Satellite reentered same day. (GSFC *SSR*, 9/15/69; *SBD*, 9/19/69, 81)
- Lunar Rock Conference was held at Smithsonian Institution, with participation of Dr. Thomas O. Paine, NASA Administrator; Dr. Henry J. Smith, NASA Deputy Associate Administrator (Science); and Lunar Receiving Laboratory scientists. During conference NASA released *PET Summary of Apollo 11 Lunar Samples*, report of 60-day preliminary examination of 48 lbs of *Apollo 11* lunar samples in LRL by university and Government scientists on NASA Preliminary Examination Team (PET).

  Report confirmed existence of unexplained erosion process on lunar surface indicated in Ranger, Lunar Orbiter, and Surveyor photos, "unlike any process so far observed on earth"; said unique chemical composition (that of silicate liquid) of Tranquility Base fines and igneous rocks "implies either the composition of the rock from which the liquid was derived differs significantly from that of the mantle of the earth, or that the mechanism by which the liquid was formed differs from analogous terrestrial processes"; and concluded there was "very good chance that the time of crystallization of some of the Apollo 11 rocks may date back to times earlier than the oldest rocks on earth."

  Samples could be divided into fine- and medium-grained crystalline of igneous origin, breccias of complex origin, and fines. Crystalline rocks differed from any terrestrial rock and from meteorites in modal mineralogy and bulk chemistry. Erosion had occurred on lunar surface but there was no evidence it was caused by surface water. Probable presence of assemblage iron-troilite-ilmenite and absence of any hydrated phase indicated crystalline rocks were formed under extremely low partial pressures of oxygen, water, and sulfur. Absence of hydrated minerals suggested absence of any surface water at Tranquility Base since rocks were exposed. Rocks and fines showed evidence of shock or impact metamorphism; all rocks displayed glass-lined surface pits possibly caused by impact of small particles; and fine material and breccia contained gases that indicated they were derived from solar wind. Measurements on igneous rock indicated crystallization 3 billion to 4 billion yrs ago. Rocks had been within one meter of surface for 20 million to 160 million yrs. Level of indigenous volatilizable and/or pyrolyzable organic material was extremely low. All rocks and fines were generally similar chemically. Major and minor constituents were same as in terrestrial igneous rocks and meteorites, but differences in composition were significant. Elements that were enriched in iron meteorites were not observed or were very low in occurrence. No evidence of biological material had been found. Tranquility Base soil was fine grained, granular, cohesive, and incompressible, with hardness increasing at six-inch depth. It was similar in appearance and behavior to soil at Surveyor landing sites. (Program; Text; *Science*, 9/19/69)
- NASA announced withdrawal of three Apollo range instrumentation ships—USNS *Redstone*, *Mercury*, and *Huntsville*—from tracking network supporting Apollo flights. Remaining tracking ship, USNS *Vanguard*, would be continued on station in Atlantic about 1,000 mi southeast of

Bermuda. NASA said reduction of Apollo ship support was based on high success of Apollo missions, particularly their excellent "launch on time" record. (NASA Release 69-133)

- House passed H.J.R. 775, to authorize President "to award appropriate medals honoring those astronauts whose particular efforts and contributions to the welfare of the Nation and of mankind have been exceptionally meritorious." (CR, 9/15/69, H7870-2)

September 16: Astronauts Neil A. Armstrong, Edwin E. Aldrin, Jr., and Michael Collins reported on *Apollo 11* mission to joint session of Congress called in their honor. Astronaut Armstrong said: "Several weeks ago I enjoyed the warmth of reflection on the true meanings of the spirit of Apollo. I stood in the highlands of this Nation, near the Continental Divide, introducing to my sons the wonders of nature and pleasures of looking for deer and for elk. In their enthusiasm for the view they frequently stumbled on the rocky trails, but when they looked only to their footing, they did not see the elk. To those of you who have advocated looking high we owe our sincere gratitude, for you have granted us the opportunity to see some of the grandest views of the Creator. To those of you who have been our honest critics, we also thank, for you have reminded us that we dare not forget to watch the trail."

Astronaut Aldrin said: "Our steps in space have been a symbol of this country's way of life as we open our doors and windows to the world to view our successes and failures and as we share with all nations our discovery. The Saturn, Columbia, and Eagle, and the extravehicular mobility unit have proved . . . that this Nation can produce equipment of the highest quality and dependability. This should give all of us hope and inspiration to overcome some of the more difficult problems here on earth. The Apollo lesson is that national goals can be met where there is a strong enough will to do so."

Astronaut Collins said: "We have taken to the moon the wealth of this Nation, the vision of its political leaders, the intelligence of its scientists, the dedication of its engineers, the careful craftsmanship of its workers, and the enthusiastic support of its people. We have brought back rocks. And I think it is a fair trade. For just as the Rosetta stone revealed the language of ancient Egypt, so may these rocks unlock the mystery of the origin of the moon, of our earth, and even of our solar system."

Astronauts presented Congress with two U.S. flags which previously had flown over Senate and House of Capitol and had been carried to moon aboard *Apollo 11* spacecraft. (CR, 9/16/69, H7937-9)

- At Smithsonian Institution ceremony attended by *Apollo 11* astronauts, Dr. Thomas O. Paine, NASA Administrator, presented two-pound, gray, lunar rock of igneous, breccia type to Smithsonian Secretary, Dr. S. Dillon Riply, for Smithsonian collection. It would be sealed in nitrogen-filled container covered by three-foot glass bubble and displayed to public beginning Sept. 17 for indefinite period in Arts and Industries Building. At presentation, Astronaut Edwin E. Aldrin, Jr., said: "Every human being, every animal who has looked up into the heavens has seen that rock. It is a fortunate time for mankind to look up and be able to say, 'here is the moon.'" (Smithsonian Release SI-150-69; Shelton, *W Post*, 9/17/69, B1)

- Senate passed H.J.R. 775, "to authorize the President to award, in the name of Congress, Congressional Space Medals of Honor to those astronauts whose particular efforts and contributions to the welfare of the Nation and of mankind have been exceptionally meritorious." (CR, 9/16/69, S10630)
- *New York Times* editorial commented on *Apollo 11* and *Mariner VI* and *VII:* "The unprecedented advances in the study both of the moon and of Mars during the past few weeks have produced a stunning crop of surprises about both celestial bodies. On the closest examination yet, these neighbors in space have proved far more complex and strange than previous theories have led men to believe. And the magnificent, lifeless desolation of the lunar and Martian surfaces emphasizes more than ever how wonderful it is and how little science understands why it is that this third planet from the sun is so uniquely green, vibrant and overrunning with life." (*NYT,* 9/16/69, 40)

September 17: Space Task Group report to President on post-Apollo space program [see Sept. 15] was released at White House press conference by Vice President Spiro T. Agnew and Space Task Force Group members Dr. Thomas O. Paine, NASA Administrator; Dr. Robert C. Seamans, Jr., Secretary of the Air Force; Dr. Lee A. DuBridge, Presidential Science Adviser; and William A. Anders, NASC Executive Secretary. Vice President Agnew said Task Group had rejected "crash program of the magnitude that would turn loose every bit of our technological ability" to achieve quickest possible manned Mars landing because "there are competing priorities in a difficult time of inflation." Task Group had also rejected "foregoing the substantial benefits that have come out of the *Apollo* program, the benefits of National prestige."

Dr. Paine said all three options recommended to President in report would enable NASA to "hold together the team" and provide "major challenge."

Dr. DuBridge said all three options held "heavy emphasis on earth applications, satellites, for studying the geology, the geography, the atmosphere of the oceans of the earth and bringing space technology directly and immediately to the benefit of the people on earth. All three programs also . . . include heavy emphasis on scientific programs, to extend our scientific knowledge of the earth itself, of the moon, through additional lunar expeditions, interplanetary space and additional scientific information about the moon and the planets." He also cited emphasis on international collaboration. (Transcript)

NASA released *America's Next Decade in Space: A Report for the Space Task Group.* Major points had been incorporated in Task Group report. (Text)

- Aerobee 150 MI sounding rocket was launched by NASA from WSMR with VAM–20 booster. Rocket carried AFCRL payload to 135.5-mi (218-km) altitude to calibrate Harvard College Observatory spectrometer on board orbiting *Oso VI* by telemetering, grazing incidence, scanning EUV monochromator to study active regions of sun simultaneously at 300 to 1,400 Å. Pointing was marginal but data were 100% satisfactory. (NASA Rpt SRL)
- AH–56A helicopter, under development by Lockheed California Co. for

USA, was destroyed when it broke loose and moved downwind inside wind tunnel at ARC. Flying debris punctured steel wall and injured two men in control room. (NASA Release 69–154)

- First day of public display of lunar rock at Smithsonian Institution attracted 8,200 visitors, including former NASA Administrator James E. Webb. Webb said: "The rock represents all the work and all the submergence of personal ambitions that thousands put into the space effort. It proves we have the scientific, technical and managerial capability of expanding our space values for use under the sea, on the land and in the air." (Schaden, W Star, 9/18/69, B4)
- Senate adopted by 85–0 vote amendment offered by Sen. William Proxmire (D-Wis.) to S. 2546, FY 1970 military procurement authorization, which would require study and review by Comptroller General of profits made by Government agencies, including NASA, on contracts for which there had been no formally advertised competitive bidding. (CR, 9/17/69, S10743–52)
- Rep. George A. Goodling (R-Pa.) introduced H.R. 13838 "to provide for the distribution to the several States, for display to the public . . . samples of the lunar rocks and other lunar materials brought back by the Apollo 11 mission." (CR, 9/17/69, H8098)
- *New York Times* editorial: "The space age is here to stay, but the precise contours of how far and how fast this nation will go in the decades ahead will have to be determined on a pragmatic basis, almost year by year and Administration by Administration." (*NYT*, 9/17/69, 40)

September 18: U.S.S.R. launched *Cosmos CCXCIX* from Baikonur into orbit with 219-km (136.1-mi) apogee, 207-km (128.6-mi) perigee, 89.2-min period, and 64.9° inclination. Satellite reentered Sept. 22. (GSFC SSR, 9/30/69; SBD, 9/19/69, 81)

- NASA's HL–10 lifting-body vehicle, piloted by NASA test pilot John A. Manke, reached 79,000-ft altitude and mach 1.39 after air-launch from B–52 aircraft west of Rosamond, Calif. Purpose of flight, 25th in series and 12th using engine, was to obtain stability and control data at various angles of attack in speed range around mach 1.2. (NASA Proj Off)
- NASA and AEC announced successful completion of NERVA nuclear experimental rocket engine (XE) testing in Jackass Flats, Nev. Tests, from March through August, had included 28 successful engine startups and 3 hrs 48 min cumulative operating time, with 3.5 min at full power (55,000-lb thrust). XE program had explored wide variety of operating modes and pressure and temperature conditions, demonstrated automatic startups using bootstrap techniques, demonstrated stability of nuclear rocket engine performance, and validated design and operation of engine test stand No. 1. XE engine runs concluded series of successful technology tests over several years. Design and development of flight-rated 75,000-lb-thrust NERVA rocket was being initiated on basis of information produced. Nuclear rocket program was managed by AEC–NASA Space Nuclear Propulsion Office. (NASA Release 69–134; AEC–NASA Release M–216)
- President Nixon addressed 24th session of U.N. General Assembly: "Of all man's great enterprises, none lends itself more logically or more compellingly to international cooperation than the venture into space.

Here, truly, mankind is one: as fellow creatures from the planet Earth, exploring the heavens that all of us enjoy. The journey of Apollo 11 to the moon and back was not an end, but the beginning.

"There will be new journeys of discovery. Beyond this, we are just beginning to comprehend the benefits that space technology can yield here on earth. And the potential is enormous. For example, we are now developing earth resource survey satellites, with the first experimental satellite to be launched sometime early in the decade of the seventies. Present indications are that these satellites should be capable of yielding data which could assist in as widely varied tasks as these: the location of schools of fish in the oceans, the location of mineral deposits on land, the health of agricultural crops.

"I feel it is only right that we should share both the adventures and the benefits of space. As an example of our plans, we have determined to take actions with regard to earth resources satellites. . . . The purpose . . . is that this program will be dedicated to produce information not only for the United States, but also for the world community. We shall be putting several proposals in this respect before the United Nations. These are among the positive, concrete steps we intend to take toward internationalizing man's epic venture into space—an adventure that belongs not to one nation but to all mankind." (*PD*, 9/22/69, 1275–81)

September 18: *Dr. Vikram A. Sarabhai (left), Chairman of the Indian Space Research Organization, and Dr. Thomas O. Paine,* NASA *Administrator, signed an agreement at* NASA *Headquarters for a cooperative experiment to broadcast educational TV programs from* NASA's *planned* ATS-F *satellite direct to 5,000 small Indian villages.*

- Dr. Thomas O. Paine, NASA Administrator, and Dr. Vikram A. Sarabhai, Chairman of Indian Space Research Organization, on behalf of India and U.S. signed agreement at NASA Hq. to provide direct TV broadcasts from satellite to some 5,000 small Indian villages. Broadcasts would be first from satellite to small receivers without ground relay. Experiment would utilize ATS-F, sixth in NASA series of Applications Technology Satellites, scheduled for mid-1972 launch. India would use experimental ground station at Ahmedabad and others to transmit TV programs to satellite, which would relay them to village receivers. Increased onboard power and deployable satellite antenna with high pointing accuracy made direct broadcast possible. (NASA Release 69-135)
- Senate began consideration of H.R. 11271, FY 1970 NASA authorization bill passed by House June 10 and reported with amendment in form of substitute bill by Senate Committee on Aeronautical and Space Sciences June 24. Sen. William Proxmire (D-Wis.) introduced new Section 7: "Of the funds authorized . . . $300,000,000 . . . earmarked for operation of the Apollo missions shall not be obligated or expended until the Administrator, in consultation with the State Department, has fully explored the possibilities of international cooperation and cost-sharing in space exploration, and has reported to Congress on the results of these efforts." Efforts should include possibility of establishing international consortium with NASA as manager of operations or possibility of bringing space exclusively within U.N. jurisdiction and control, establishing "United Nations Space Council modeled after the World Health Organization." (CR, 9/18/69, S10895-907)
- List of U.S. attempts during 1969 to effect cooperative space agreement with U.S.S.R. was entered in *Congressional Record:*

    April 30, NASA Administrator, Dr. Thomas O. Paine, forwarded copy of *Opportunities for Participation in Space Flight Investigations* to Academician Dr. Anatoly A. Blagonravov and assured him that proposals by Soviet scientists of experiments to fly on NASA spacecraft would be welcomed. Supplements to NASA document were to be sent routinely to Soviet Academy.

    May 29, Dr. Paine invited Academician Blagonravov to attend *Apollo 11* launch and to discuss, informally, mutual interests in cooperative space projects. Dr. Blagonravov had declined.

    August 21, Dr. Paine invited Academician Prof. Mstislav V. Keldysh to send Soviet scientists to Sept. 11-21 briefing at NASA Hq. for investigators who might wish to propose experiments for 1973 Viking missions to Mars. Dr. Paine suggested meeting serve as opportunity for discussion of planetary exploration plans contributing to coordinated efforts beneficial to both countries. Prof. Keldysh had declined, but asked for copies of meeting materials so Soviet scientists might develop proposals. He had suggested possibility of later discussions. (CR, 9/18/69, S1095-6)
- Post Office Dept. announced delay in delivery of moon landing stamp first day covers because of "unprecedented number of requests." Processing crew of 100—more than twice number normally employed—were working longer shifts with more special canceling equipment than ever before to handle "response from people all over the world." (PO Dept. Philatelic Release 50)

September 18

- Senate swore in Sen. Ralph T. Smith (R-Ill.) to serve unexpired term of late Sen. Everett M. Dirksen (R-Ill.) and adopted resolution assigning him to Senate Committee on Aeronautical and Space Sciences to replace Sen. Charles McC. Mathias (R-Md.), reassigned to Government Operations Committee. (CR, 9/18/69, S10763)
- Senate passed by voice vote S. 1857, FY 1970 NSF authorization of $487,150,000. (CR, 9/18/69, S10764–70)
- Senate passed by record vote of 81 to 5, S. 2546, FY 1970 military procurement authorization which included amendment requiring study and review by Comptroller General of profits on Government contracts for which there had been no advertised competitive bidding [see Sept. 17]. (CR, 9/18/69, S10888–91)

September 19: Canadian Black Brant IV sounding rocket was launched by NASA from Barreira do Inferno, Natal, Brazil, carrying MSC and Univ. of California payload to provide detailed scientific measurements of charged particle environment in South Atlantic Anomaly region. Secondary objectives were to measure magnetic field strength and flight-evaluate payload telemetry-system performance. Rocket reached 532-mi (856-km) altitude, with performance higher than expected. All experiments performed satisfactorily and data were obtained on all channels. (NASA Rpt SRL)

- Senate passed by voice vote H.R. 11271, FY 1970 NASA authorization of $3.716 billion, allocating $3.020 billion for R&D, $58.2 million for construction of facilities, and $637.4 million for research and program management. Total was $250.85 million less than had been passed by House June 10 [see also June 24]. Senate insisted on its amendments and requested conference with House. (CR, S10977–99, 11002; Text)
- White House announced *Apollo 11* astronauts would make 22-nation tour starting Sept. 29, to stress U.S. willingness to share space knowledge. Itinerary would include Mexico City; Bogota, Colombia; Buenos Aires, Argentina; Rio de Janeiro, Brazil; Las Palmas, Canary Islands; Madrid; Paris; Amsterdam and Brussels; Oslo; Cologne, Germany; Berlin; London; Rome; Belgrade, Yugoslavia; Ankara, Turkey; Kinshasa, Congo; Teheran, Iran; Bombay, India; Dacca, Pakistan; Bangkok, Thailand; Darwin and Sydney, Australia; Guam; Seoul; Tokyo; Honolulu; and return to Houston, Tex., Nov. 5. Additional trip to Ottawa and Montreal, Canada, was planned for December. (UPI, *NYT*, 9/20/69, 5)
- U.K.'s first lunar samples—3 oz of moon dust in 16 contamination-proof boxes—arrived in London and were shown to scientists and press at Science Research Council. Dr. S. O. Agrell of Cambridge Univ. and Dr. P. E. Clegg of London Univ. had flown to MSC to collect them. They would be examined by 14 British research teams. (AP, *Kansas City Times*, 9/20/69)
- French scientists, using "world's most powerful laser" at Limeill Weapons Research Center of French Atomic Energy Commission near Paris, had generated succession of tiny thermonuclear explosions, Walter Sullivan said in *New York Times*. It was important step toward taming hydrogen bomb energy. It also underlined concern of some scientists that lasers might simplify design of devastating nuclear weapons. (*NYT*, 9/19/69, 1)
- Report of President Nixon's Task Group on Space [Sept. 15] and Apollo

astronauts' speeches to Congress [Sept. 16] had brought some "rationality back to the discussion of whither the space program," *Washington Post* editorial said. Acceptance by President of recommendation "would eliminate talk of abandoning manned space flight, which would be a foolish course of action, or of proceeding toward Mars in a crash effort to get there as quickly as possible." It was important "for the nation to push ahead on the immediate recommendations of the Task Group—exploring the moon, developing the tools that are needed for systematic exploration of our space travel capability, and extracting from the space program more benefits for those of us who are earthbound." (*W Post*, 9/19/69)

September 20: Economist commented on lunar investigation: "Scientists, unlike engineers, are not at this stage interested in whether the moon can be made habitable; but this, rather than its age, or peculiarities of its composition, is what the astronauts went out to the moon to find. The results are more encouraging than anyone but the confirmed star gazers could have hoped. Space is not unfriendly; nor is the moon, superficially barren though it looks. What we need are more assurances about sources of water which space planners continue to be convinced is trapped in, and can be extracted from, the rocks. Also more data on those surprising experiments where plants have thrived on moon soil. . . . And some idea about whether it will be possible to protect man from ultraviolet radiation up there, without having to put him in a protective pressure suit." (*Economist*, 9/20/69, 17)

September 21: Washington *Sunday Star* commented on display of moon rocks at Smithsonian Institution: "The lunar chunk does indeed look like something that, if it turned up in a Bethesda [Md.] backyard, would not draw a second glance. And yet it is something that, until two months ago, no man had seen before. . . . It is a promise of unimagined things to come." Judging by crowds queued up, "Smithsonian has booked its best act since the Mona Lisa came to town six years ago." (W *Star*, 9/21/69, C1)

- Bert Greenglass, former head of Apollo Program Control Office at KSC and later Deputy Director of Management Systems Div. in NASA Office of Technology Utilization, joined HUD as Director of Management Information and Program Control Systems. (W *Star*, 9/24/69; HUD PIO)

- *Parade* magazine called for establishment of July 4, 1976—200th anniversary of U.S.—as national deadline for conquering some of earth's social problems. "Having harnessed our special strengths—money, men, materials and the organizational genius to control them—we conquered space before 1970. Why can we not conquer some of our social problems on earth by 1976?" (*Parade*, 9/21/69, 1)

- "The notion has occurred to more than one person that NASA, having reached the moon and now fearing its way to the planets possibly blocked by budgetary obstacles, might find the requisite new worlds to conquer right here, at home," William Hines said in Washington *Sunday Star*. It was "fundamental precept of modern technology that anything which can be imagined can be accomplished. A cure for cancer, an end to poverty, a cleanup of the environment, termination of the Vietnam war, even effective nuclear disarmament? If it is conceivable it is achievable." While NASA could be depended on to give good account of itself in scientific and engineering situations, "its ability to handle problems with a big 'people' component is largely untested." Since "people-problems" were predominant these days, maybe NASA wouldn't work out so well after all. "But on the second thought, no other government agency is showing much flair for coping with the human element, either." (W *Star*, 9/21/69, C4)

September 22: USAF launched two unidentified satellites from Vandenberg AFB by Thorad-Agena booster. First entered orbit with 157.2-mi (252.9-km) apogee, 110.0-mi (177-km) perigee 88.7-min, period, and 85.0° inclination and reentered Oct. 12. Second entered orbit with 308.2-mi (495.9-km) apogee, 305.1-mi (490.9-km) perigee, 94.4-min period, and 85.1° inclination. (GSFC *SSR*, 9/30/69; 10/15/69; UPI, *W Post*, 9/23/69, A20; *Pres Rpt* 70 [69])

- Japan failed in fifth attempt to launch satellite when four-stage, unguided Lambda booster malfunctioned. (Harrison, *W Post*, 9/24/69, A9)
- President Nixon announced appointment of NAS President, Dr. Philip Handler, to President's Science Advisory Committee. He would replace Dr. Frederick S. Seitz, President of Rockefeller Univ. in New York. (*PD*, 9/29/69, 1335; *W News*, 9/23/69, 44)
- President Nixon announced establishment of series of Presidential task forces, including Task Force on Oceanography, to review public and private efforts in oceanography and suggest actions to accelerate development of "increasingly important area of exploration"; and Task Force on Science Policy, to review present policy and make recommendations for future scope and direction. (*PD*, 9/29/69, 1304)

September 23: U.S.S.R. launched *Cosmos CCC* into orbit with 189-km (117.4-mi) apogee, 183-km (113.7-mi) perigee, 88.2-min period, and 51.5° inclination. Satellite reentered Sept. 27. (GSFC *SSR*, 9/30/69; *SBD*, 9/25/69, 106)

- Aerobee 150 MI sounding rocket, launched by NASA from WSMR with VAM-20 booster, carried Univ. of Hawaii payload to 108.4-mi (174.4-km) altitude. Objectives were to obtain high-resolution spectra of solar disc from 1,800 to 2,000 Å, using high-resolution echelle-grating spectrograph pointed by Univ. of Colorado biaxial pointing control. Rocket and instruments functioned satisfactorily and photographic spectra were obtained on both camera cycles. (NASA Rpt SRL)
- President Nixon announced decision to continue development of SST. "The supersonic transport is going to be built. The question is whether in the years ahead the people of the world will be flying in American supersonic transports or in the transports of other nations . . . whether the United States, after starting and stopping this program . . . finally decides to go ahead. . . . I have made the decision that we should go ahead . . . because I want the United States to continue to lead the world in air transport. And it is essential to build this plane if we are to maintain that leadership. . . . I have made the decision, also, because . . . through this plane we are going to be able to bring the world closer together in a true physical and time sense. . . . This is a massive stride forward in the field of transport." President said prototype would be flown in 1972. (*PD*, 9/29/69, 1309)
- President Nixon would ask Congress to appropriate $662 million over five years to assist in SST development, Secretary of Transportation John A. Volpe announced. Federal Government would spend estimated

$761 million through FY 1974, including $99 million in funds already appropriated, to construct and flight-test two prototype Boeing SST models. Total development cost was estimated $1.5 billion, with $1.3-billion Government participation to be repaid from sale of approximately 300 aircraft capable of carrying 300 passengers each at maximum 1,800-mph speeds. (DOT Release 21069)

- Modified test-pilot pressure suit delivered by ARC's Dr. Alan Chambers, Hubert Vykukal, and Richard Gallant to Stanford Univ. Hospital saved life of Mrs. Mary Phillips, who was hemorrhaging uncontrollably after minor surgery. G-suit, worn by pilots to avoid blacking out during high-speed maneuvers, applied pressure to counter draining of blood from brain and upper body. Fitted to Mrs. Phillips, suit arrested abdominal bleeding during 10-hr application. (NASA Release 69–168)

- USA Atmospheric Sciences Laboratory helium-filled balloon was successfully launched from WSMR, carrying 70-lb scientific payload to measure ozone concentration, cosmic radiation, and atmospheric pressure, temperature, and density at 160,000-ft altitude. The 600-ft-tall, 1,700-lb balloon drifted to New Mexico where it released payload for recovery on ground. Data would be used for number of WSMR projects. (USAF PIO; UPI, *W News*, 9/24/69, 9)

- Associated Press quoted Col. Edwin E. Aldrin (USAF, Ret.) as saying NASA had rejected his proposal to postpone Apollo 12 and run it in tandem with Apollo 13 so crews could protect or rescue each other in emergency. Aldrin was father of *Apollo 11* Astronaut Edwin E. Aldrin, Jr., and a NASA safety consultant. NASA Manned Space Flight Safety Director Jerome F. Lederer had called proposal impractical, "tremendously expensive, and I don't know if it could be done." Lederer had said there was no question that astronaut rescue capability from lunar surface or orbital emergency must be provided, but it was "out of the picture for Apollo." (Haughland, AP, *W Star*, 9/22/69, A4)

- Fédération Aéronautique Internationale posthumously awarded its highest honor—Gold Medal—to NASA test pilot Joseph A. Walker for "his many enduring contributions to the advancement of aviation made during a 21-year flight research career marked by extraordinary perfection and valor." Award was received by his widow at Edwards AFB ceremony. As FRC chief research pilot, Walker had flown X–15 to its highest altitude, 354,200 ft (67 mi); was first man to fly LLRV astronaut training craft; was author of 20 technical papers and articles; and had taught *Apollo 11* commander Neil A. Armstrong at FRC. (FRC Release 17–69)

- MSFC announced award of $19,073,032 modification to IBM contract for fabrication, checkout, and delivery of 27 instrument units for Saturn IB and Saturn V boosters. Modification revised delivery schedule, extended performance period 15 mos, and provided for assessment of certain MSFC engineering change requests. (MSFC Release 69–214)

- FAA, Air Transport Assn., and manufacturers McDonnell Douglas Corp., Bendix Corp., and Wilcox-Sierra Div. of American Standard, Inc., successfully flight-tested three separate but compatible devices composing aircraft collision avoidance system (CAS) capable of issuing microsecond warning. Tests were held at Martin-Marietta Airport, Baltimore. CAS included cesium atomic clock so precise that watch of similar construction would lose only one second in 67 yrs. System operated like balloon around aircraft which, when penetrated by similarly equipped plane, provided pilots with command to make evasive maneuver. All aircraft would need system for it to be effective. Product of $12-million, 13-yr R&D, system could be operational by 1971. (Yarborough, W *Star*, 9/24/69, A7)

- In Paris press conference Presidential Science Adviser, Dr. Lee A. DuBridge, and French Minister for Industrial and Scientific Development Francois X. Ortoli announced plans to increase flow of scientists and specialists between France and U.S. to broaden scientific and technical cooperation in wide areas, including nuclear research for peaceful purposes. (*W Post*, 9/24/69, A22)

September 24: Two photometers on board NASA's *Ogo V* orbiting geophysical observatory (launched March 4, 1968) had successfully scanned Lyman-alpha radiation, NASA announced. Data were expected to provide new information on Lyman-alpha emission from Milky Way and to help determine what portion of observed radiation was from geocorona and what portion was from outer space. On Sept. 12 *Ogo V* had pointed at sun and spun slowly while scanning mirror in Univ. of Paris experiment rotated, covering 30° of celestial sphere. On Sept. 14 spacecraft returned to normal three-axis-stabilized operation, where it would remain until December when second series of maneuvers would be conducted to cover remaining portion of sky and provide first complete mapping of extraterrestrial Lyman-alpha radiation. Univ. of Colorado photometer, which provided broader coverage of Lyman-alpha radiation at 180° to Univ. of Paris experiment, would be used to confirm measurements and verify calibration levels.

Ogo V had 18 of 24 onboard experiments still operating. It had provided first measurements of electric fields in earth's bow shock and comprehensive data on particles and fields in earth's magnetosphere. (NASA Proj Off; NASA Release 69–137)

- U.S.S.R. launched *Cosmos CCCI* from Baikonur into orbit with 279-km (173.4-mi) apogee, 192-km (119.3-mi) perigee, 89.2-min period, and 65.4° inclination. Satellite reentered Oct. 2. (GSFC *SSR*, 9/30/69; 10/15/69; *SBD*, 9/25/69, 106)

- NASA's X–24A lifting-body vehicle, piloted by Maj. Jerauld R. Gentry (USAF), reached mach 0.62 after air-launch from B–52 aircraft at 40,000-ft altitude over South Rogers Lake Bed, Calif. Purpose of unpowered flight, fifth in series, was to obtain data on upper-flap control effectiveness, handling qualities during change from lower-flap to upper-flap control, and effect of rudder position on air flow around tail. (NASA Proj Off)

- Aerobee 150 MI sounding rocket, launched by NASA from WSMR with VAM–20 booster, carried Univ. of Colorado Laboratory for Atmospheric and Space Physics payload to 124-mi (199.5-km) altitude. Objective was to obtain high-resolution spectra of Carbon IV resonance doublet at 1,548 and 1,550 Å using high-resolution, narrow-band spectrograph with echelle as principle dispersing element and SPARCS solar pointing control. Rocket and instruments functioned satisfactorily. (NASA Rpt SRL)

- Board of Investigation which probed Feb. 17 death of Sealab III Aquanaut Berry L. Cannon had concluded probable cause was carbon dioxide poisoning due to faulty diving gear, USN announced. Gear had lacked substance for filtering carbon dioxide from exhaled breath. Associated Press said Sealab III project had been "shelved" because of insufficient funds for FY 1970 and quoted USN spokesman as saying it would be continued later. Project had been suspended since Cannon death. (DOD Release 794-69; W Star, 9/24/69, A9)

- Shawbury, England, innkeeper Jack Warner had asked U.S. Government for license to open first pub on moon, Associated Press said. He would call it "The Space Inn" or "The Lunatic Tavern." (W Star, 9/24/69, A16)

September 25: Apollo 9 commander James A. McDivitt was appointed Manager of Apollo Spacecraft Program at MSC, replacing George M. Low, who was temporarily on special assignment to MSC Director to plan future MSC programs and work on organizational matters. (MSC Release 69-66)

- East Germany's People's Chamber unanimously ratified nuclear nonproliferation treaty. West Germany had not yet signed. (P Inq, 9/25/69)
- House Committee on Science and Astronautics reported favorably S. 1287, which authorized appropriations for FYs 1970, 1971, and 1972 for metric system study. (CR, 9/25/69, H8488)

September 25-26: National Seminar for Manned Flight Awareness at MSC attracted some 400 representatives of NASA, DOD, and aerospace industry. MSC Director, Dr. Robert R. Gilruth, said: "I think we are all concerned about the period of letdown which tends to occur after a great milestone such as has just been completed," but NASA "must continue to demonstrate . . . that success can follow success."

Lee B. James, Director of Program Management at MSFC, said next moon flights could suffer from lack of proper employee motivation. "We are completing [rocket] stages with welders who know they are going to be laid off." Sheet metal workers in plants with termination papers were working on vital space hardware. Twenty defects attributed to human error had been uncovered in single rocket.

Apollo Program Director Rocco Petrone said future moon landings would be even more demanding than first, with astronauts spending 54 hrs on moon during some. To make missions successful, workers must be motivated to pay greatest attention to detail.

Associate Administrator for Manned Space Flight, Dr. George E. Mueller, said NASA hoped to cut payload launch costs to $200 per pound and reduce number of workers on Saturn V launches from 20,000 to about number required to get Boeing 747 off ground. (MSC Release 69-65; Maloney, H Post, 9/26/69)

September 26: Glazing discovered on lunar surface by *Apollo 11* astronauts was analyzed in *Science* by Thomas Gold of Cornell Univ., senior investigator for close-up photography. Glossy surfaces similar to glass found clumped in centers of small lunar craters appeared to have been swept in after craters had been formed. Glazed areas were also concentrated toward tops of protuberances and, in some cases, droplets appeared to have run down inclined surface and congealed on sides. Glazing might have originated from effect of exhaust of LM descent stage, splashing of liquid drops from larger impact elsewhere, shock heating or volcanism on moon, same impact that created craters in which glazing was found, or intense radiation heating. Intense radiation was most probable cause, Gold said. Source could have been impact fireball on moon, impact fireball on earth, or most likely, solar outburst in geologically recent times. (*Science*, 9/26/69, 1345–9)

- In address at Clarkson College of Technology in Potsdam, N.Y., Dr. Thomas O. Paine, NASA Administrator, said: "We know that hydrogen bombs work (unfortunately)—the Lord made the Universe that way. And, of course, we know that (fortunately) fusion energy keeps the sun shining every day. Our great visionary dream is to find out how to unlock this energy for spaceship propulsion. This is a great challenge but consider the new tools that we have available: giant magnetic fields with superconducting magnets, tremendous power densities from lasers, the great energy of nuclear power, and new high temperature materials. These, with new plasma dynamic developments, may usher in fusion power in the eighties, the nineties, or in the next century. Harnessing fusion power for propulsion is an even farther-out challenge, but it could prove to be one of the more direct applications. We won't have to convert the fusion energy to electricity; just fuse a couple of deuterium atoms and then let them blast out the back of the vehicle!

    "So fusion remains a tantalizing promise for the future. . . . If we ever do achieve such propulsion, we'll be able to move with some ease out from our little 8,000-mile-diameter Solar System. All of the 9 planets, 32 moons and 1600 known asteroids will come within reach of our vehicles. And, indeed, if we could achieve high efficiencies in a fusion propulsion process, we could talk of eventual relativistic velocities, of time compression, and of travel to the nearest stars." (Text)

- Venus, "least understood of the inner planets," should be U.S. space program priority target, Kitt Peak National Observatory physicist Dr. Donald M. Hunten and Harvard Univ. physicist Dr. Richard M. Goody said in *Science*. Some fundamental data were available; quantitative theories had been stated; questions about atmosphere could be answered by feasible missions; and geophysicists' interest had been aroused and offered specialized knowledge needed to understand complex processes. But NASA had no present plans for investigation of Venus' lower atmosphere. Uncertainty as to Soviet intentions had been cited as reason for giving Venus low priority. But until collaboration with U.S.S.R. and other European countries could begin, "we have no choice but to base our judgment upon our own scientific and technical abilities and desires." (*Science*, 9/26/69, 1317–23)

- *Wall Street Journal* editorial: "While there will be debate on the Administration's approval of a go-ahead on the supersonic transport, the President plainly picked the proper method for financing further development of the controversial plane. Earlier there had been talk of setting up a special SST authority that would raise money by selling Government-guaranteed bonds to the public. The idea never had much to recommend it. At the moment the SST faces an uncertain economic future. . . . If the plane is a flop, the Government would be stuck one way or the other. . . . the bond plan would ease the current pressure on the Federal budget. But it also would fool at least part of the public about the financial risk that the Government actually is assuming. . . . Whether one especially relishes the notion or not, supersonic travel is sure to come sooner or later. In heading toward that development, the Administration is wise to avoid financial subterfuge." (*WSJ*, 9/26/69)

September 27: Evidence for detection of high-energy cosmic gamma radiation (above 50 mev) from point source in constellation Sagittarius was reported in *Nature.* Case Western Reserve Univ. and Univ. of Melbourne (Australia) physicists presented preliminary results from collaborative program in which two high-altitude balloon flights were made from Parkes, Australia, Feb. 5–6 and Feb. 26–27. Object was first such point source of gamma rays detected in heavens. Research was supported by NASA, NSF, and Australian Research Grants Committee. (Frye et al., *Nature*, 9/27/69, 1320–1; Sullivan, *NYT*, 10/2/69, 33)

- Satellite system which combined navigational and air traffic control and collision prevention had been proposed to FAA and DOD by TRW Inc. Systems Group, Washington *Evening Star* reported. Proposal called for four satellites, one in permanent orbit with others revolving around it at lower altitudes. Aircraft would radio distinct signal to satellites and its position relative to two or more satellites would be determined by ground computers that could figure latitude and longitude of aircraft within 50 ft. System, based on delicate measurement of time for aircraft signals to reach satellite, would cost estimated $100 million, could be in operation by mid-1970s, and was also being proposed to aviation industry. (Lannan, W *Star*, 9/27/69, A11)

September 28: In telephone call to *Apollo 11* commander Neil A. Armstrong on eve of *Apollo 11* astronauts' round-the-world tour, President Nixon asked astronauts to invite foreign countries to become "partners in space" with U.S. He also invited astronauts and wives to White House dinner Nov. 5. (UPI, W *Star*, 9/29/69, A3)

- *Washington Post* columnist Franklin R. Bruns, Jr., said 10 days after issuance of moon landing airmail stamp in Washington, D.C., "an already tired city post office crew had just passed the two-million first day cover mark." Post Office had gone "all out" to cooperate with those of other countries in returning covers and with Voice of America, NASA, and regular servicers. There was little doubt that "new first day cover record is in the making." (*W Post*, 9/28/69, F9)

September 29: *Apollo 11* astronauts and wives arrived in Mexico City for start of 39-day tour of 22 countries [see Sept. 19]. (AP, B *Sun*, 9/30/69, A1)

- President Nixon approved H.J.R. 775, to authorize President to award Congressional Space Medals of Honor to astronauts [see Sept. 15]. (*CR*, 10/6/69, 1362)

- NASA announced appointment of Daniel J. Harnett as Assistant Administrator for Industry Affairs, effective Oct. 1. He would be responsible for all NASA relationships with industry. Before his appointment he had held executive positions with Northrop Corp. (NASA Release 69–139)

September 30: USAF launched two unidentified satellites from Vandenberg AFB by Thorad-Agena D booster. First entered orbit with 303.2-mi (487.9-km) apogee, 299.5-mi (481.9-km) perigee, 93.8-min period, and 69.6° inclination. Second entered orbit with 586.0-mi (942.9-km) apogee, 574.8-mi (924.9-km) perigee, 103.7-min period, and 70.7° inclination. (GSFC *SSR*, 9/30/69; *SBD*, 10/7/69, 162; *Pres Rpt 70* [69])

- NASA's HL–10 lifting-body vehicle, piloted by Maj. Peter Hoag (USAF), reached mach 0.9 after air-launch from B–52 aircraft at 45,000-ft altitude over FRC. Purposes of flight, 26th in series and first powered flight

for Maj. Hoag, were to provide pilot training and obtain stability and control data. Winds rose from 5 knots at launch to 20 knots at touchdown, but did not interfere with flight. (NASA Proj Off)
- In speech before Chicago Executive Club in Chicago, Ill., Vice President Spiro T. Agnew said operation of military systems in space "to enhance the national defense" must be one objective of U.S. future space program to ensure "there will be no blind reliance on good faith." Vice President Agnew was also Chairman of NASC and of President's Space Task Group.

  Two questions dominated speculation over national space policy, he said: Why space? And Why Mars? "Mars holds the greatest promise of a capability to sustain human life. It is a potential resource and reserve. More important for the present is the fact that the mind of America functions better when it focuses upon a clear target. Manned exploration of the Solar System is too nebulous to capture the public's attention. A manned landing on Mars is as understandable a challenge to the citizen as it is to the scientist. It is a test that can be put in a time frame and its anticipation can be appreciated by all." (Text)
- MSFC announced selection of Bendix Corp. and Boeing Co. for further competitive negotiations on cost-plus-incentive-fee contract for design, development, test, and delivery of four manned lunar roving vehicles for flight to lunar surface aboard descent stage of Apollo LM [see July 11]. First operational vehicle would be delivered in early 1971 for launch late that year.

  MSFC also had awarded $238,400 contract to Bryson Construction and $224,888 contract to Miller and Berry for construction of two clean rooms for Apollo Telescope Mount (ATM) assembly and test and had called for bids for checkout station construction. Bids were due Oct. 21. (MSFC Releases 69–220; 69–221)
- Federal Electric Corp., IT&T Corp. subsidiary, announced it had received $21,321,680 NASA contract for continued work as KSC prime contractor. (UPI, W Star, 9/30/69, B7)
- Washington Airlines had terminated first and only STOL service between Washington and Baltimore after one year and would liquidate its three Dornier aircraft, *Washington Post* said. Company had lost nearly $5,000 weekly and carried 25,000 passengers instead of targeted 108,000 since inauguration of service Sept. 23–25, 1968. (Samuelson, W Post, 9/30/69)

During September: Dr. Robert C. Seamans, Jr., Secretary of the Air Force, defended value of Safeguard ABM defense and also discussed use of space for strategic deterrence, writing in *Air Force/Space Digest:* "In terms of security, the space age presents dangers—but it also affords opportunities for increasing strategic stability." Dangers stemmed from weapons placed in orbit: "It might be possible to trigger such weapons with very little warning, thus increasing the risk of surprise attack." Outer Space Treaty of 1967 might help avoid this danger, "while providing us opportunities for other sorts of military systems that could strengthen deterrence rather than weaken it." Each generation of space vehicles would provide additional improvements in monitoring enemy activities. "We are now working on a satellite early-warning system that would detect missiles as they are launched from land or sea." Dispersed bomber force "would be able to take off from its bases before the impact of enemy weapons, even if the time of flight of the latter were greatly reduced." (*AF/SD*, 9/69, 61–4)

- USAF magazine *Airman* published interview with Secretary of the Air Force, Dr. Robert C. Seamans, Jr., former NASA Deputy Administrator. NASA–USAF X–15 program had been highly successful in providing data for many disciplines and "very good test bed" for atmospheric probe instrumentation. XB–70 program, initiated as manned supersonic bombing system, was "very bold step" in speeds over mach 3. When technological difficulties and rising costs resulted in decision against XB–70 production, NASA with USAF had initiated experimental program. Dr. Seamans felt USAF was not using NASA expertise to fullest extent. USAF needed new manned bomber, new fighter, and modernization of air defense. He was not convinced USAF had yet established "best relationships with industry to get these things done." (*Airman*, 9/69, 7–9)

- "There appears to be much more to be squeezed from Apollo than just the incalculable value of national prestige or scientific discovery," Michael Getler wrote in *Space/Aeronautics*. "The ability to mine these supporting talents, to judge their value outside manned space flight and disseminate and apply them may well prove the most telling and measurable argument in the debate which is bound to continue over Apollo's real value." Apollo had combined "much of what we have with what we do best. Though complex, it was manageable. It had clear goals, was well funded, enjoyed fairly widespread public support, and dangled the element of competition in front of our involuntary reflex. Most importantly, it tapped an industrial base and an enthusiasm for gadgetry that are unmatched anywhere. . . . Because many of today's challenges confront human nature and not technology, Apollo can be made to seem irrelevant. In fact, however, we are not left alone with our behavioral troubles. There is still an economy to keep sound, industry and commerce to be kept competitive, and a government to be made more efficient. Failure to take this extraordinary project apart, piece by piece, and examine its usefulness in these areas would indeed be wasteful." (*S/A*, 9/69, 42–53)

- Dr. Mose L. Harvey in *Science and Public Affairs*, bulletin of atomic scientists, discussed lunar landing and U.S.-Soviet equation: "The capability of the United States to continue in space, and otherwise keep pace with the scientific-technological revolution, depends entirely on continued public faith in the 'military-industrial complex' and the 'scientific and technological elite,' if one wishes to keep using these unfortunate terms. It was only because we were able effectively to organize and use a genuine and mutually rewarding partnership between industry, universities and government that we were able to effect the moon landing and to do the other near-miraculous things we have done in space and in other fields involving advanced science and advanced technology. It is precisely on this partnership that the superiority of the American way over the Soviet way has so far rested." (*Science and Public Affairs*, 9/69, 28–35)

- *Atlas* published translation of article in *L'Espresso*, Rome, by Italian novelist Alberto Moravia on implications of *Apollo 11*. "In Columbus's days, men were offered finite goals, like the discovery of America; or they were offered spiritual aims, like the search for goodness, truth and beauty. At that time, no one could have guessed that Columbus's discovery was only a beginning. That other discoveries would follow, a second America, a third, a fourth, and so on through millions of years and billions of kilometers. But today that is happening. Compared to our new set of goals, the aims of Marx and communism are pure imagination. For the first time the real and the rational are about to become one. We are now at the end of history—and post-history is just beginning." (*Atlas*, 9/69, 40–3)

- *Atlas* said its "Talk of the World" section had "got a little hoarse" on subject of *Apollo 11*. It quoted "a few of the more unusual moonthoughts" from international press. *El Tiempo* in Bogota had reported Colombian wool fabrics were used in *Apollo 11* spacecraft upholstery. Canada's *Kitchener Waterloo Record* had said LM had touched down on moon with legs made in Canada by Montreal firm. *Oiga*, in Lima, Peru, had said when Sputnik went into orbit "it was noted that Pedro Paulet Mostajo had invented a jet-propelled rocket back in 1895." *Atlas* commented that "this could go on forever." (*Atlas*, 9/69, 10)

- In *Astrophysical Journal*, Princeton Univ. astronomers Jeremiah P. Ostriker and Dr. James E. Gunn predicted few pulsars should be found with periods more than 1.5 secs, from results of their quarantine exploration of pulsar model. (*Astro Journ*, 9/69, 1395–1417)

- USAF communications and navigation satellite programs were outlined by L/G John W. O'Neill, Vice Commander of AFSC, in *TRW Space Log*. AFSC was testing new UHF communication terminal in conjunction with *Tacsat I* tactical comsat and had tested UHF shipboard, jeep, van-mounted, and team-pack terminals as receivers for satellite communications. Second-generation defense comsats would have earth coverage antennas and also steerable narrow-beam antennas to direct energy to two "spotlighted" areas on earth's surface, permitting use of small terminals instead of large ground stations. Proposed navigation satellite system would consist of high-altitude satellites transmitting navigation signals with worldwide coverage, ground stations to track and command satellites, and user receiving equipment. Navsat would provide all four armed services with common grid for mobile operations and could be used by aircraft, ships, submarines, and foot soldiers. Potential for aircraft carrier operations was being studied.

    Computer revolution would provide new data processing equipment "with infinite potential for influencing satellite design and function" in future. "We are pressing hard for progress in laser technology, which appears to have excellent potential for communication applications. We are also pushing the development of new sources of power in space in which a breakaway from our heavy reliance upon the solar cell could make future satellite development a whole new game." (*TRW Space Log*, Summer/Fall 69, 3–17)

October 1969

October 1: *Boreas* (*Esro IB*) satellite—designed, developed, and constructed by European Space Research Organization—was successfully launched by NASA from WTR by Scout booster. Orbital parameters: apogee, 237.4 mi (382 km); perigee, 180.8 mi (291 km); period, 91.3 min; and inclination, 85.1°. Primary NASA mission objectives were to place *Boreas* into planned orbit and provide tracking and telemetry support. *Boreas* was backup for and identical to *Aurorae* (*Esro 1A*) successfully launched by NASA Oct. 3, 1968. It carried eight experiments to study aurora borealis (Northern Lights) and related phenomena of polar ionosphere, representing six organizations from U.K., Denmark, Sweden, and Norway.

Boreas was third successful ESRO satellite launched by NASA. First success, *Iris I* (*Esro IIB*), had been launched by NASA May 16, 1968, to replace ESRO IIA, which had failed to enter orbit May 29, 1967. ESRO was responsible for experiment instrumentation, delivery of spacecraft to launch site, equipment and personnel necessary to mate spacecraft to launch vehicle, and spacecraft testing. NASA provided Scout launch vehicle and launch services in second launching on cost-reimbursable basis under Dec. 30, 1966, agreement with ESRO. (NASA Proj Off; GSFC SSR, 10/15/69; SBD, 10/3/69, 147)

- Solid-fuel U.K. Falstaff rocket, carrying equipment to measure vibrations and temperatures, reached mach 5 after launch from Australian monorail launcher at Woomera, Australia, in joint U.K.-Australia research program. (*Interavia*, 11/69, 1751)
- Eleventh anniversary of NASA, established by National Aeronautics and Space Act of 1958. (Space Act)
- Portrait of James E. Webb, second NASA Administrator (1961–1968), was unveiled in anniversary ceremony at Smithsonian Arts and Industries Building. Painted by Gardner Cox, portrait would eventually hang in NASA Hq. (Program; NASA Release 69–140)
- Lockheed C–5A Galaxy, world's largest aircraft, took off from Edwards AFB, Calif., with 410,000-lb load—heaviest ever carried by any aircraft, 21,000 lbs heavier than C–5A was expected to lift even under wartime conditions, and 28,100 lbs heavier than record it established June 16. Aircraft, C–5A No. 3, reached 18,800-ft altitude burning 21,000 lbs of fuel during climb. (*P Inq*, 10/2/69, 3; UPI Service, 10/2/69)
- Sud-Aviation chief test pilot André Turcat flew Anglo-French Concorde 001 supersonic airliner for about nine minutes at mach 1.05 (693 mph), passing sound barrier for first time. Two outer engines were at full force and two inner engines at less than capacity force during 36,000-ft-altitude flight. Concorde flew from Toulouse-Blagnac Airport. (*NYT*, 10/1/69, 1)
- Vice President Spiro T. Agnew said at press conference following tour of

October 1: NASA *launched* Boreas, *European Space Research Organization satellite, to study the aurora borealis and polar atmosphere. In the photo the spacecraft was prepared for launch on a four-stage Scout booster from the Western Test Range.*

JPL: ". . . the cities may benefit more from what's happening right here at the Jet Propulsion Laboratory than they do from what's happening in some community action agency. Now, I'm not downgrading the need to work closely with people . . . but I think it would be a desperate mistake . . . to adopt an attitude that we do not challenge the unknown, we do not move forward simply because we can't predict what we're going to find." Columbus "didn't find what he went for but he found something even better. Maybe that applies to the situation with regard to the ultimate goal . . . about landing a man on Mars. . . . we may do something a lot better."

In reply to question, he said: "We will never reach a point where we'll have enough money for . . . the problems of the cities and of the population. . . . we could spend every resource we have and forego any scientific exploration and forget the need to research and develop new techniques. . . . I suppose you come down to the final determination that you've got to put some of your effort in the future and not all of it in the present. And I don't think the lack of a focus in the space program would be a benefit at all to the future of the cities and the problems you mention." (Transcript)

- Daniel J. Harnett was sworn in as NASA Assistant Administrator for Industry Affairs. He had held executive positions with Northrop Corp. since 1964. (NASA Release 69–139)
- Soviet space scientist Dr. Oleg G. Gazenko said at news conference during meeting of International Academy of Astronautics at Cloudcroft, N.Mex., it would be desirable and technically feasible for cosmonaut to be member of future Apollo moon-landing crew. (AP, B *Sun*, 10/2/69, A15)
- Twenty-third anniversary of Naval Missile Center at Point Mugu, Calif., USN's principal facility for testing and evaluating air-launched missiles and other airborne weapons systems. (PMR *Missile*, 10/3/69, 1)
- William Teir, Saturn IB program manager at MSFC, became deputy director for management of Program Management directorate. (MSFC Release 69–225)
- Japan inaugurated semiofficial Space Development Corp. to coordinate space activities, including orbiting of two satellites—one in 1972 for ionospheric observation and one in 1974 for communications. Corporation would replace system under which projects were undertaken separately by different ministries. Its 539-million yen ($1.5-million) capital had been raised by government fund of 500 million yen and 39 million yen from private industry. (Reuters, *NYT*, 10/5/69, L27)

October 2: Astronaut Alan L. Bean described plans for Apollo 12 extravehicular activities to press at KSC. Mission, to begin Nov. 14, would include two 3½-hr EVA periods on moon. For first EVA main objective was to deploy ALSEP, deploy TV camera and take photos, deploy lunar equipment conveyor, take contingency sample, deploy S-band antenna and solar wind experiment, and collect lunar material.

After rest in LM astronauts would return to lunar surface for second EVA period, to collect good documented sample and document geologically interesting features with photographs, samples, and description. "There's not going to be a lot of time to pick up a rock and think about it . . . but there's going to be time to look at the craters and try to determine what kind they are and where they came from, if this one's different from that one, what part of the crater you want to sample. . . . We're going to take photographs as we see all of it. And when this is finished we're hopefully going to be over near Surveyor." Examination of *Surveyor III* (launched April 17, 1967) was second objective. Crew would retrieve parts for evaluation of how materials withstood long-term exposure in space.

Final surface experiment would be conducted after liftoff from moon when crew crashed LM ascent stage onto surface. Crash would be recorded by seismometer and was expected to provide data from which scientists could make inferences about moon's internal structure. (Transcript)

- Aerobee 150 MI sounding rocket launched by NASA from WSMR with VAM-20 booster carried MIT payload to 97.9-mi (157.5-km) altitude to determine precise position of two or more x-ray sources and evaluate small photoelectric detector. Rocket and instruments—including several bands of proportional counters, slot collimators, modulation collimators, aspect cameras, and attitude control system—functioned satisfactorily. All collimators gave expected rates and modulation on star X-1 in constellation Scorpius (calibration) and on Sagittarius. (NASA Rpt SRL)
- At MSFC ceremony, NASA Administrator, Dr. Thomas O. Paine, presented awards to 117 center employees and industry representatives, mostly in recognition of exceptional service to Apollo program. Employee awards included NASA Medal for Distinguished Service to MSFC Director, Dr. Wernher von Braun; Deputy Director, Technical, Eberhard F. M. Rees; Deputy Director, Management, Harry H. Gorman; Director of Science and Engineering Hermann K. Weidner; Vice Commander, Aeronautical Systems Div., M/G Edmund F. O'Connor (USAF); Director of Program Management Lee B. James; and Deputy Director of Science and Engineering Ludie G. Richard.

NASA Medal for Exceptional Scientific Achievement was presented to James A. Downey, III; Erwin Fehlberg; Gerhard B. Heller; Robert J. Nauman; and Joseph L. Randall. Other awards received were NASA Medal for Exceptional Service by NASA employees and NASA Certificate for Distinguished Public Service by industry personnel. (MSFC Release 69-222)

October 3: *Ogo VI* Orbiting Geophysical Observatory, launched into low-altitude polar orbit June 5, was adjudged successful by NASA. Spacecraft had completed first diurnal cycle and had provided data on global characteristics of neutral atmosphere; association of electric fields with ionospheric irregularities; airglow emissions associated with oxygen, sodium, and molecular nitrogen; and propagation of proton whistlers. Performance of *Ogo VI* subsystems had been excellent and instrumentation for 23 of 25 experiments was operational. Active three-axis stabilization had been maintained since initial acquisition and gas usage was consistent with one-year operation. To extend attitude-stabilized lifetime beyond one year, operations plan had been modified to include manual control of gas jet firing. (NASA Proj Off)

- MSC announced appointment of Astronaut L. Gordon Cooper as Assistant for Space Shuttle Program in MSC's Flight Crew Operations Directorate. Cooper would be responsible for flight crew training program, astronaut inputs into design and engineering, and directorate's part in hardware development and testing for Space Shuttle. He would remain on flight status and eligible for space flight. (MSC Release 69/67)
- MSFC announced that it had issued $10,751,000 contract to General Electric Co.'s Apollo Systems Div. for electrical support equipment for

Apollo Telescope Mount and launch systems for Saturn V Workshop multiple docking adapter and airlock. Work was to be completed June 30, 1975.

MSFC had also issued six-month $56,727 contract to Bionic Instruments, Inc., to develop lunar roving vehicle hazard locator. Locator, which would be installed on lunar rover or dual-model vehicle, would use laser beam to spot rocks, holes, and other obstacles on lunar surface and display on screen warning of obstacles hidden from astronauts' view. (MSFC Releases 69–223, 69–224)

- Spain awarded its Grand Cross of Aeronautic Merit to *Apollo 11* Astronauts Neil A. Armstrong, Edwin E. Aldrin, Jr., and Michael Collins. (UPI, *NYT*, 10/4/69, 23)

October 4: Nike-Tomahawk sounding rocket launched by NASA from Wallops Station carried 223-lb Cal Tech payload to 146-mi (234.9-km) altitude to study intensity, spectrum, and degree of isotropy of diffuse x-ray background in 0.1- to 20-kev range and to study energy spectrum of star X–1 in constellation Scorpius in 0.1- to 15-kev range. Rocket and instruments functioned satisfactorily and good data were obtained in all 12 prime data channels. (NASA Rpt SRL; WS Release)

- NASC Executive Secretary William A. Anders, *Apollo 8* astronaut, told Western Conference of Young Presidents' Organization in Phoenix, Ariz., that U.S. lost $13 billion each year in agricultural production from insects, disease, and fire. Use of satellite sensors could improve surveys of agricultural and forest resources, aid mineral and petroleum prospecting, obtain better inventory of earth's water sources, detect natural and man-made geography changes, sense ocean currents and temperatures to aid fishing industry and improve routing of commercial shipping, and study effects of environmental and water pollution. While NASA space expenditures were averaging about $4.5 billion annually—about 2/3 of 1% of GNP—nearly $4 billion was spent in U.S. for nondurable toys and sport supplies, $4.7 billion for foreign travel, and over $6.5 billion in amusements.

  Anders urged international harmony through space. "We travel through space on a small planet. The Earth looked so tiny in the heavens that there were times during the Apollo 8 mission when I had trouble finding it. If you can imagine yourself in a darkened room with only one clearly visible object, a small blue-green sphere about the size of a Christmas tree ornament, then you can begin to grasp what the Earth looks like from space. I think all of us subconsciously think the Earth is flat or at least almost infinite. Let me assure you that, rather than a massive giant, it should be thought of more as the fragile Christmas tree ball which we should handle with considerable care. . . . From space, the earth is indivisible. There are no flags, no national boundaries. Let us on Earth then use the Communications Revolution to break down the barriers which separate us, so that all of Earth's people will be truly brothers." (*CR*, 10/6/69, E8190–2; NASC PIO)

- *Apollo 11* Astronauts Neil A. Armstrong, Edwin E. Aldrin, Jr., and Michael Collins and wives arrived at Las Palmas, Canary Islands, for two-day rest before proceeding to Madrid and meeting with Gen. Francisco Franco during 38-day goodwill tour. (AP, *NYT*, 10/6/69)

- USAF had awarded three-year, $20-million contract to North American

Rockwell Corp. Rocketdyne Div. to design new rocket engine to power payloads in space, *Business Week* said. United Aircraft's Pratt & Whitney Div. was also working on concept, magazine believed. (*Bus Wk*, 10/4/69)

October 5: GSFC scientists and Smithsonian Institution ecologist Dr. Helmut K. Buechner planned to use female elk named Moe for first experiment in tracking animals by satellite, *New York Times* said. Wearing 23-lb instrumentation around neck, elk was expected to migrate from point in Wyoming to national elk refuge south of Yellowstone Park at Jackson Hole, Wyo.—100-mi distance. (Teltsch, *NYT*, 10/5/69, 16)

October 5–11: International Astronautical Federation (IAF) held 20th Congress in Mar del Plata, Argentina.

Dr. George E. Mueller, NASA Associate Administrator for Manned Space Flight, gave Invited Lecture Oct. 6 on Apollo program, space benefits, Apollo Applications program, and lunar exploration. Apollo Applications program would "study the earth through the use of a six lens multispectral camera installation. On Apollo 9 some of the rudimentary work was done using four lenses. 50% more kinds of information will be brought back by the crews of Apollo Applications. Because almost everything on earth has a different reflective quality, as distinctive as a signature or a fingerprint, results from this photographic exploration can be expected to yield rich rewards. Many of the earth sciences look to this kind of information for answers to previously unanswerable questions. Hydrologists are interested in discovering not only the depth, but also the temperature of the waters of the world, and these conditions can be ascertained from space. Warm water attracts certain species of fish, so fishermen are also interested in these experiments.

"Agronomists believe that soil chemistry will reveal itself on a large scale in certain kinds of filtered photography. Agriculturists know that different crops, at different stages of their development and in different states of health, reflect distinctive amounts of light. They hope, using the photographs we will take, to get some measure of the world's crop, as well as some knowledge of its condition. Geologists, who have found clues to mineral deposits in the Gemini and Apollo pictures, look to this more sophisticated photography for additional information, and for distinctions which will eventually permit certain kinds of prospecting from space. Cartographers and Geodesists are interested in everything from urban sprawl to continental drift."

Sun would be prime target for exploration in Apollo Applications program, with experiments conducted by astronaut-astronomers using Apollo Telescope Mount (ATM) in orbit. "For the first time we will have a massive set of instruments with high resolution for looking in some depth and detail at the activities of the sun from beyond the veil of the earth's atmosphere." Program would consist of three missions into near-earth space, one of 28 days and two of 56 days each. First step would be establishment of Saturn V workshop in orbit, planned for March 1972 launch, followed by launches of solar panels, ATM, and crew of three for 28-day stay. Second manned mission was scheduled for launch about three months after first, with third to follow one month after return of second crew. (Text)

- At Third International History of Astronautics Symposium Oct. 10, organized by IAF in connection with its 20th Congress, A. Ingemar Skoog of Swedish Interplanetary Society traced rocket development in Sweden from 1807, when military first understood possibilities of rockets in warfare, to mid-1860s, when rockets were taken out of service. "An examination of the rockets preserved at the Army Museum in Stockholm, has shown that all 14 rockets are still fitted with their original propellant. This will be tested by propellant experts later this year in order to find out the properties of a 140 years old propellant." (Text)

  W. Geisler of Polish Astronautical Society had submitted paper on history and development of rocket technology and astronautics in Poland before 1949: use of rockets on Polish territory by Tartars in 13th century probably had marked introduction of use of rockets in Europe. (Résumé)

  Frederick I. Ordway, III, of Univ. of Alabama delivered paper "The Alleged Contributions of Pedro E. Paulet to Liquid Propellant Rocketry." "Paulet, a Peruvian chemical engineer-turned-diplomat, spent much of his professional foreign service career in Europe. . . . his claim to being a precursor of liquid propellant rocketry rests in a letter he wrote from Rome on the 23rd of August 1927 that was published in the 7 October 1927 issue of the Lima, Peru, newspaper *El Comercio*. Therein, he describes liquid propellant rocket engine experiments he had conducted . . . in Paris thirty years earlier. . . . Relying on this source and derivatives, many subsequent writers have accorded Paulet a perhaps undeserved place in the history of rocketry." (Summary)

October 6: U.S.S.R. launched *Meteor II* weather satellite from Plesetsk into orbit with 676-km (420.1-mi) apogee, 619-km (384.6-mi) perigee, 97.6-min period, and 81.2° inclination. Powered by solar cells, satellite was designed to trace cloud cover, detect presence of snow on dark and daylight sides of earth, and record radiated and reflected heat from atmosphere. (GSFC *SSR*, 10/15/69; *SBD*, 10/7/69, 166; *Av Wk*, 10/20/69, 191)

- President Nixon announced Ruben F. Mettler, Executive Vice President of TRW Inc., would be Chairman of new Presidential Task Force on Science Policy [see Sept. 22]. (*PD*, 10/13/69, 1376)
- *Apollo 8* Astronaut James A. Lovell, Jr., and wife inspected container vessel *American Astronaut* in London and signed autographs for longshoremen. Mrs. Lovell had christened vessel in April. (Reuters, *C Trib*, 10/7/69)

October 7: Anglo-French supersonic airliner Concorde 001 reached 730 mph during 110-min flight from Toulouse, France. It was second time aircraft had broken sound barrier [see Oct. 1]. (UPI, *W Star*, 10/8/69, A13)

- Space Power Facility, with world's largest high-vacuum chamber (100-ft dia by 120-ft height, containing 800,000 cu ft), and Spacecraft Propulsion Research Facility were opened officially at LeRC. Space Power Facility—to test large, space electric power generating systems and spacecraft—would first test Brayton cycle power system operated with nuclear isotope or solar heat. Facility could produce equivalent of

October 7

vacuum in space 100 to 300 mi above earth. Propulsion Research Facility would be used first for development of Centaur, 2nd-stage rocket vehicle. (LeRC Release 69–54)

- House passed 384 to 5 amended FY 1970 NSF authorization of $477.3 million. Amendments had reduced authorization by $3.3 million. (CR, 10/7/69, H9151–62)
- U.S. and U.S.S.R. in Geneva announced agreement on draft treaty to ban placing of nuclear weapons on world's seabeds. Proposed treaty would be discussed by Geneva Disarmament Conference and U.N. General Assembly. It would come into effect when ratified by U.S., U.S.S.R., and 20 other nations. (Roberts, *W Post*, 10/8/69, A1)
- *Apollo 11* astronauts placed wreath at Christopher Columbus monument in Madrid and later met Gen. Francisco Franco and Prince Juan Carlos during world tour. (UPI, *W Star*, 10/7/69, A6)
- Boeing Co. had responded to Federal budget trimming by eliminating 11,230 jobs between Jan. 1 and Sept. 26 and would continue cutting back 1,000 jobs monthly well into 1970, *Washington Post* said. Reduction in work force from high of 101,554 on Jan. 19, 1968, to 83,765 on Sept. 26, 1969, had been blow to Seattle. Boeing President T. A. Wilson saw problem for next five years as national desire to retrench, with military spending under especially tight rein. (Wilson, *W Post*, 10/7/69, A1)
- DOD announced General Electric Co. was being awarded $1,300,000 supplemental agreement to previously awarded USAF contract for R&D of Mark 12 reentry vehicle. Contract was managed by SAMSO. (DOD Release 842–69)

October 8: France successfully tested Diamant-B carrier rocket 1st stage (L–17) for 112 secs at Vernon, France. Test was first in series of four planned in initial phase of qualification testing. (*Interavia*, 11/69, 1751)

- In Paris during world tour *Apollo 11* astronauts were made chevaliers of French Legion of Honor by Premier Jacques Chaban-Delmas and met with French President Georges Pompidou. Paris Mayor Etienne Royer de Vericourt presented them with city's Gold Medal. (UPI, *C Trib*, 10/9/69)
- Two-man, 15-ft submarine *Nekton* with seven-inch knife taped to its mechanical claw, sliced line to free research submarine *Deep Quest*, and four-man crew trapped on ocean floor off San Diego, Calif., for 12 hrs. *Deep Quest*, developed by Lockheed Aircraft Corp. for undersea research and salvage, had been demonstrating its ability to raise and lower objects when its propeller became snarled. (AP, *W Post*, 10/9/69, A3)
- Sen. William Proxmire (D-Wis.) introduced S. 3003, "to provide for more effective control over the expenditure of funds by the Department of Defense and the National Aeronautics and Space Administration for independent research and development." Measure was referred to Senate Committee on Armed Services. (CR, 10/8/69, S12100)
- Astronaut Anthony W. England suffered symptoms of bends while testing moon-walking equipment in vacuum chamber at MSC. NASA spokesman said cause of symptoms, which disappeared as soon as chamber was returned to full atmospheric pressure, was not known. England was placed under 24-hr observation. (UPI, *W Post*, 10/9/69, A28)

- First FB–111A strategic bomber was delivered to USAF Strategic Air Command (SAC) by AFSC in Carswell AFB, Tex., ceremonies. Aircraft were scheduled to become operational during spring and summer 1970. General Dynamics Corp. was prime contractor. (General Dynamics Release 1475)

October 9: Joseph P. Loftus, Jr., Manager of MSC Program Engineering Office, described plans for future Apollo lunar exploration at MSC press conference. Technological objectives were to increase scientific payload to lunar orbit and to lunar surface, permit high flexibility in landing site selection, increase lunar orbit and lunar surface stay time, increase lunar surface mobility with self-propelled lunar roving vehicle, develop and demonstrate advanced techniques and hardware for expanded manned space mission capabilities, develop techniques for achieving point landings, and demonstrate closed-loop onboard navigation capability as applicable to advanced missions.

Scientific objectives were to investigate major classes of lunar surface features, surface processes, and regional problems; collect samples at each site for analyses on earth; establish network of surface instrumentation to measure seismic activity, heat flow, and disturbance in moon's axis of rotation; survey and measure lunar surface from lunar orbit with high-resolution photography and remote sensing; investigate near-moon environment and interaction of moon with solar wind; map lunar gravitational field and internally produced magnetic field; and detect atmospheric components resulting from neutralized solar wind and micrometeoroid impacts.

Achievement of scientific objectives would be facilitated by addition of scientific instrument module (SIM) under service module sector door. SIM would consist of scientific instruments mounted on shelves behind door, which would be deployed pyrotechnically after crew left lunar surface. Spacecraft would also use new modular equipment storage assembly (MESA). "Unlike the existing MESA it is modular and people, instead of having to take many things out of compartments and stick them into a bag . . . here you simply take hold of a handle . . . and lift out an entire shelf. On that shelf are all the things required for the next EVA period. So, it is one movement instead of a dozen."

One tank of hydrogen and one tank of oxygen would be added to extend mission capability to 16½ days. Since each EVA period would increase up to 5 hrs and total lunar stay time would increase up to 200 hrs, lithium hydroxide would be increased for portable life-support system and for spacecraft, and insulation would be increased on top of spacecraft around docking tunnel. Minor changes would be made to interior garments so crew would have more suitable environment and crew would remove spacesuits and sleep in hammocks during rest periods on lunar surface. (Transcript)

- Boosted Arcas II sounding rocket launched by NASA from Resolute Bay, Canada, carried GSFC payload to 68.4-mi (110-km) altitude to obtain electron-density and collision-frequency profiles of high-latitude quiet D region and positive ion-density measurements. Rocket and instruments performed satisfactorily. (NASA Rpt SRL)

- Lunar scientist Dr. Harold C. Urey said in lecture at Univ. of California at San Diego that new evidence had been uncovered during *Apollo 11* mission that moon had been formed by collision process begun about

4.5 billion yrs ago. Water might have been present on moon temporarily and might still be beneath lunar surface. Information was to be made public by NASA in January 1970. (UPI, *W Post*, 10/10/69, A5)

- *Apollo 11* astronauts, welcomed by crowds in Amsterdam during world tour, presented Queen Juliana with plaque similar to one they left on lunar surface. In Brussels later in day they were decorated by King Baudouin with insignia of the Order of Leopold, nation's highest honor. (AP, *Huntsville Times*, 10/9/69; UPI, *W Star*, 10/10/69, A9)
- ARC Director, Dr. Hans Mark, had announced appointment of Executive Assistant Director Loren G. Bright as Director of newly established Directorate of Research Support, ARC *Astrogram* reported. Divisions in new directorate included computation, research facilities and equipment, and technical services. (ARC *Astrogram*, 10/9/69, 1)
- Sen. Edward W. Brooke (R-Mass.) delivered Landon Lecture at Kansas State Univ.: "With this new generation of weapons [MIRVs] about to sprout from the arsenals of the Soviet Union and the United States, I have been joined by almost half the Senate and a sizable number of House members in calling for a joint moratorium on flight tests of the so-called MIRV systems." (Text)

October 10: Nuclear energy in space was discussed in address opening U.S. Technical Forum at Nuclex 69 in Basel, Switzerland, by Milton Klein, Manager of AEC–NASA Space Nuclear Propulsion Office and Director of AEC Space Nuclear Systems Div.: "A forward looking space program such as that envisioned by the Task Group [see Sept. 15, 17] will mean growing reliance on nuclear systems. . . . As an integral part of a new capability for space transportation, we look to the nuclear rocket to provide the propulsion to move large payloads from low earth orbit to geosynchronous orbit or to lunar orbit to support extended lunar exploration. It also offers major advantages for transporting heavy payloads into deep space." NERVA development was being initiated, with nuclear engine expected to be in operation in late 1970s. "It is also possible to envision ultimately the use of a controlled thermonuclear, or fusion, reaction for producing propulsion energy. Obviously development of such a propulsion system for space would come only after the development of a useful controlled thermonuclear reaction in a ground based plant, a goal that has not yet been achieved." (Text)

- *Apollo 11* astronauts lunched with King Olav of Norway in Oslo before spending weekend at cottage of Norwegian Defense Minister Otto Greig Tidemand during world tour. (UPI, *W Star*, 10/10/69, A9)
- President Nixon announced James H. Wakelin, Jr., former Assistant Secretary of the Navy for R&D, would be chairman of new Presidential Task Force on Oceanography [see Sept. 22]. (*PD*, 10/13/69, 1393-4)

October 11–18: U.S.S.R.'s *Soyuz VI*, carrying Cosmonauts Georgy S. Shonin and Valery N. Kubasov, was successfully launched from Baikonur at 4:10 pm Baikonur time into orbit with 229-km (142.3-mi) apogee, 194-km (120.6-mi) perigee, 88.8-min period, and 51.7° inclination. Tass said spacecraft carried equipment to test methods of welding materials in weightlessness and did not carry docking equipment used on *Soyuz IV* and *Soyuz V* Jan. 14–18. Western speculation, later confirmed, was that *Soyuz VI* would rendezvous with other spacecraft.

Soyuz VII, carrying Cosmonauts Anatoly V. Filipchenko, Vladislav N. Volkov, and Viktor V. Gorbatko, was launched from Baikonur at 3:45 pm Oct. 12 into orbit with 217-km (134.8-mi) apogee, 200-km (124.3-mi) perigee, 88.4-min period, and 51.6° inclination. Tass said spacecraft's mission was "maneuvering in the orbit, staging joint navigation observations of the spaceships ... in group flight, observation of celestial bodies and the horizon of the Earth, determination of the real luminosity of stars, observation of changes in illumination created by the Sun and other scientific experiments."

Third spacecraft, *Soyuz VIII*, carrying Cosmonauts Vladimir A. Shatalov and Dr. Aleksey S. Yeliseyev, was launched from Baikonur at 3:29 pm Oct. 13 into orbit with 278-km (172.7-mi) apogee, 215-km (133.6-mi) perigee, 89.4-min period, and 51.6° inclination. Its mission, Tass said, was "comprehensive simultaneous scientific studies in near-terrestrial space in accordance with an extensive program; testing of the complex system of controlling a simultaneous group flight of three space ships; mutual maneuvering of ships in orbit with the aim of solving a number of problems of developing the piloted space system." TV viewers received announcement of flights about 30 min after launches and saw video-tape recordings of launches about 1½ hrs later. Ultimate goal of three-spacecraft mission was not given. In prelaunch interview shown on TV several hours after launch, Dr. Yeliseyev said group mission would conduct experiments leading to "creation of still more powerful orbiting stations." Tass reported crews had checked out equipment and communicated with each other. *Soyuz VI* was conducting medical-biological tests, *Soyuz VII* was observing and photographing earth and celestial bodies, and *Soyuz VIII* was conducting research on polarization of solar light reflected by atmosphere.

On Oct. 14 Tass reported *Soyuz VII* and *Soyuz VIII* had moved close together, carried out "mutual observation-photography and movie filming," studied "possibility of exchanging information with the aid of light signals and visceral optical methods," conducted series of medical experiments, and observed "effect of erosion by micrometeorites on the condition of illuminators and optic systems of the craft." On Oct. 15 spacecraft approached to within 500 yds during rendezvous, while *Soyuz VI* hovered nearby.

On Oct. 16 Radio Moscow announced *Soyuz VI* had successfully completed mission and had softlanded in preset area 100 mi northwest of Karaganda at 12:52 pm Moscow time. No information on other two Soyuz spacecraft was given until Tass announced softlanding of *Soyuz VII* at 12:36 pm Oct. 17 and of *Soyuz VIII* at 12:10 pm Oct. 18. Western officials speculated missions had not achieved all objectives. Tass said all major tasks were carried out with "high efficiency." (Gwertzman, *NYT*, 10/12–14, 18/69; *SBD*, 10/14/69, 195; 10/21/69, 230; Reuters, *W Post*, 10/15–16/69; AP, *W Star*, 10/16/69, A1; Reston, *LA Times*, 10/19/69, A2; GSFC *SSR*, 10/15/69; 10/31/69)

October 11: Apollo 12 Astronauts Charles Conrad, Jr., Richard F. Gordon, Jr., and Alan L. Bean held press conference at MSC. Code names for Apollo 12 LM and CM had been selected, they said, from entries submitted by North American Rockwell Corp. and Grumman Aerospace Corp. employees. CM would be called "Yankee Clipper" and LM, "Intrepid."

Major differences from *Apollo 11* mission would be increase in geology and photography. Detailed documented sample collection would be team effort. Description would tell geologist about rock's location, including how it was related to other rocks, whether it was partially covered, and why it attracted attention. Astronauts would carry bags and tongs at all times during EVA to collect rocks when sighted instead of being limited to collecting during specific period of EVA.

Photographic activities, described by Conrad as "the world's greatest zero-G juggling act," would include multispectral photography of lunar surface, using four Hasselblad cameras in hatch window. Three cameras would have black-and-white film with filters; fourth would have infrared film. Purpose of camera setup, Gordon explained, was to "photograph the illuminated side of the moon from one minute after sunrise to one minute prior to sunset in a stiff manner." Every 20 secs "cameras would be activated, to give complete strip photography across the surface of the moon. . . . The big step photography I think is probably one of the most important things we're doing on this flight. . . ." (Transcript; MSC Release 69–68)

- At Fourth Annual AVLABS Awards Banquet of USA Aviation Material Laboratories in Fort Eustis, Va., NASA Deputy Associate Administrator for Aeronautics Charles W. Harper discussed NASA–USA cooperation in aeronautics. "Aviation, both military and civil, has suffered in past years from a scarcity of new young blood bringing new ideas and new training to research. For several years NASA has been constrained by personnel limits and so unable to bring this kind of new talent into its aeronautics problem. A joint Army-NASA program may provide a unique opportunity to increase the interest of our universities in the challenges of aviation technology and to bring the new scientists they train into the job of finding solutions." (Text)

October 12: Walter Sullivan described mystery of cosmic rays in *New York Times*. "No natural phenomena with which we have intimate contact are more awesome." Scores of these high-energy particles flashed through human body every second with energy capable of piercing six feet of lead. Some believed them accountable for aging process. Recent Australian observations had shown some particles (mostly protons) hit atmosphere with 100-billion-bev energy. Princeton Univ. physicists Dr. Jeremiah P. Ostriker and Dr. James E. Gunn had suggested pulsars might be superaccelerators of cosmic ray particles. (*NYT*, 10/12/69, 8)

October 12–13: Apollo 11 astronauts visited West Germany during world tour. They were escorted to Berlin Wall by Mayor Klaus Schuetz. (NASA EH; B *Sun*, 10/14/69)

October 13: MSC announced award of $4.1-million modification to IBM Corp. contract for design, development, implementation, maintenance, and operation of real-time computer complex (RTTC) which supported Apollo lunar landing missions. Modification definitized requirements to support Apollo lunar surface experiments package

- (ALSEP) and continued work under multi-inventive arrangement covering cost, performance, schedule, and equipment management. (MSC Release 69–69)
- NASA and DOT award of $165,908 contract to Booz-Allen Applied Research Inc. for six-month study of total U.S. investment in aeronautical R&D since 1945 was announced by Secretary of Transportation John A. Volpe. Joint effort was to develop methods of assessing national benefits which accrued from R&D development expenditures. (DOT Release 22269)
- Flaws in F-111s produced to date were detailed by L/G John W. O'Neill, Vice Commander of AFSC, and L/G George S. Boylan, Jr., Deputy Chief of Staff, Programs and Resources, USAF Hq., in testimony before Senate Committee on Appropriations. Aircraft were seriously short in engine power for nonnuclear missions. Their weight had grown without commensurate engine power increase and thus acceleration time was now four minutes. Outlay of $80 million was necessary to overhaul 450 F-111s to correct serious flaw in wing boxes. (Testimony)
- *Izvestia* published article by Soviet space scientist Alexander Koval, Vice President of International Astronomical Commission. It was "high time" that space knowledge was used for advancement of communications, meteorology, geodesy, and navigation. Since space exploration was not cheap, simpler and more reliable experiments must be tried before complex and expensive ones. Three practical applications of space research were industrial use of high vacuum and absolute cold, industrial use of rocketry innovations, and putting "production-technical complexes" in space. (*W Post*, 10/15/69, A21)

October 14: U.S.S.R. launched *Intercosmos I* into orbit with 626-km (389.0-mi) apogee, 253-km (157.2-mi) perigee, 93.3-min period, and 48.3° inclination. Spacecraft carried experiments from U.S.S.R., East Germany, and Czechoslovakia, including x-ray spectrograph to determine which areas of sun were chief producers of x-rays and whether these emissions were polarized. It reentered Jan. 2, 1970. (GSFC *SSR*, 10/15/69; 1/15/70; *SBD*, 10/15/69, 203; *Interavia*, 11/69, 1751)
- *Apollo 11* astronauts were received at Buckingham Palace by Queen Elizabeth during their visit to U.K. on 22-nation tour. (Reuters, B *Sun*, 10/15/69)

October 15: Astrobee 1500 sounding rocket launched by NASA from Wallops Station carried 102-lb GSFC payload to 1,600.0-mi (2,575-km) altitude and transmitted 35 min of data. Primary objectives were to measure cosmic radio noise intensities over 600-khz to 3-mhz frequency band, continue evaluation of Astrobee performance, and verify new Alcor 1B 2nd stage. Secondary objective was to test receiver system for *Explorer XXXVIII* Radio Astronomy Explorer (launched July 4, 1968). Solid-fueled Astrobee 1500 weighed 11,600 lbs and was being developed to provide research rocket to carry heavy scientific payloads to high altitudes with relatively easy handling. (WS Release 69–17; NASA Rpt SRL)
- NASA acknowledged it was reviewing proposed orbital flight of monkey with transplanted heart. Project was one of two based on work of Dr. Christiaan Barnard, South African surgeon and heart transplant pioneer. Second Barnard-based study would deal with effects of space

October 15

flight on mechanism of rejection of foreign tissues in living organisms. Both proposals had been submitted by General Electric Co. Reentry Systems Div. (*W Post*, 10/16/69)

- *Washington Post* reported interview with L/G Samuel C. Phillips, Commander of USAF Space and Missile Systems Organization (SAMSO) and former NASA Apollo Program Director. Space shuttle was SAMSO's top priority program. New family of spaceships was being roughed out on drawing boards. "From a military standpoint, we need to be aggressive advocates of the capability to take a look at and deal as necessary with space vehicles of another country." USAF and NASA were dividing labor on space shuttle program, which would cost estimated $1 billion. Pace of program would depend on progress toward development of reusable rocket engine which would not burn itself out as it hurtled into space, LeRoy E. Day, Chief of NASA Space Shuttle Task Group, said. (Wilson, *W Post*, 10/15/69, A3)
- USAF Space and Missile Systems Organization (SAMSO) had awarded $14,303,150 contract to General Dynamics' Convair Div. for manufacture of six Atlas (SLV–3C) launch vehicles for NASA and for conversion of one USAF launch vehicle (SLV–3) to later configuration, Convair announced. (Convair Div Release 1481)
- USAF dedicated 365-ft-long, $3.3-million telescope on 9,200-ft elevation in Sacramento Mountains in southeast New Mexico. Described as most important new solar observing facility constructed in U.S. in decade, telescope would expand solar research capability of USAF's Sacramento Peak Observatory and provide unequaled image stability. Telescope would be used for research on solar centers of activity—sunspots, magnetic fields, solar flares, and plage areas. Sacramento Peak solar research was endeavoring to identify causal relationships to predict solar energy variations and environmental disturbances affecting USAF operations. Observatory would participate in Global Flare Patrol Network support of NASA's Apollo 12 mission, scheduled for Nov. 14 launch. Operated by ESSA's Space Disturbance Center in Boulder, Colo., network would give warning of any solar flares dangerous to mission. (USAF Release 10–69–42; AP, *NYT*, 10/19/69, 62; ESSA PIO)
- Eastern Airlines became first U.S. carrier to employ antihijacking system developed by FAA. System combined knowledge of certain behavioral traits of hijackers with weapon screening device. Program was joint effort of FAA, Air Transport Assn., and Eastern, with assistance from U.S. Marshal's Office. (FAA Release 69–119)
- *Washington Post* editorial said: "It is a pity that the Soviet Union continues to wrap its space program in secrecy. The flights of Soyuz 6, 7 and 8 were launched in secret and on a mission also kept secret from the people whose labor supports the government that is paying for it. The contrast with the openness of the American space program could not be sharper and it is our lack of knowledge of what the Russians are doing that makes difficult any assessment of the meaning of these flights." As long as U.S.S.R. preferred to operate in secret, "there is little that Americans can do other than congratulate them on their successes." (*W Post*, 10/15/69, A18)

October 16: NASA estimated total cost of Viking project at $750 million, Dr. John E. Naugle, NASA Associate Administrator for Space Science and Applications, said in testimony before House Committee on Science and Astronautics' Subcommittee on Space Science and Applications. "We have . . . made a substantial effort to accurately determine funding requirements before beginning hardware development. We believe our estimates are sound and that the Viking mission will make major scientific advance in our knowledge about Mars."

While earliest estimates for Earth Resources Technology Satellite (ERTS) program were about $50 million, current preliminary estimates varied from $100 to $200 million, depending on selection of spacecraft, sensors, telecommunications network, and ground data-handling system. "The study and design effort underway will provide for formal cost definition by the beginning of hardware development in mid-1970."

In Applications Technology Satellite (ATS) program, 18 experiments had been chosen in air traffic control and communications, orbiting spacecraft communications and tracking, interference measurements, and meteorology. Instructional TV experiment would be conducted after completion of originally planned ATS-F program [see Sept. 18]. "We are well into the definition phase for ATS F & G, and plan to select a contractor for hardware development later this year with the launch of ATS F in 1972." (Text)

- Aerobee 150 MI sounding rocket launched by NASA from WSMR carried GSFC payload to 127.2-mi (204.7-km) altitude to obtain high-resolution spectrograms of belt and sword stars of Orion. Excellent spectrum of star zeta in Orion was obtained and good spectrum of star epsilon in Orion was recorded. (NASA Rpt SRL)

- *Apollo 11* astronauts and wives on world tour had private audience with Pope Paul VI at Vatican and later met with bishops from around the world. During Papal audience, 200 dissident bishops attending "shadow synod" near Vatican issued denunciation of meeting. Pope had previously refused to see them. Priests said poor people of world would interpret astronauts' audience with Pope to mean "that the Church is ready to link itself with power, of which the astronauts are the symbol, and that she refused to accept direct contact with weakness, of which our modest assembly is a symbol." (AP, P *Bull*, 10/16/69)

- U.S.S.R. Cosmonauts Georgy T. Beregovoy and Konstantin P. Feoktistov would arrive in New York Oct. 20 to begin two-week visit to U.S. as guests of astronauts, *Apollo 8* Astronaut Frank Borman announced. He had extended invitation during his July tour of U.S.S.R. Beregovoy had flown *Soyuz III* mission, Oct. 26–30, 1968; Feoktistov was scientist aboard Oct. 12, 1964, *Voskhod I* flight. Itinerary was expected to include MSC, Grand Canyon, California, Detroit, and Washington, D.C. (NASA Release 69–141; AP, W *Star*, 10/21/69)

- At LeRC technical conference on plasmas and magnetic fields, LeRC Electromagnetic Propulsion Div. Chief Wolfgang E. Moeckel said, "Man has reached the Moon and there is now some talk about going to the stars, but we are not yet ready with propulsion systems that are suitable for exploring our own solar system." Manned flights to distant planets would become reasonable only with advanced nuclear propulsion systems. Thermonuclear fusion rocket, if feasible, could reduce manned Mars mission to four to five months and journey to Jupiter or Saturn to under three years. Solar electric rockets looked promising

for unmanned probes to planets as remote as Jupiter. (LeRC Release 69–58; LeRC PAO)
- AIAA announced former ARC director H. Julian Allen had been named to receive 1969 Daniel Guggenheim Medal. Medal—presented annually by AIAA, American Society of Mechanical Engineers, and Society of Automotive Engineers for achievement in advancement of aeronautics —would be awarded Oct. 21 during AIAA Sixth Annual Meeting and Technical Display in Anaheim, Calif. (AIAA Release)
- Dr. Charles Stark Draper, head of MIT's Instrumentation Laboratory since 1940, said he had been replaced because "they are going to take the lab" out of defense work and "'convert it to civilian purposes.'" MIT had set up committee with veto power over research projects. Prof. Charles L. Miller had been named as Dr. Draper's successor. (AP, *W Post*, 10/17/69, A3)
- Nobel Prize in Physiology or Medicine was awarded in Stockholm, Sweden, to U.S. scientists Dr. Max Delbruck, Cal Tech biologist; Dr. Alfred D. Hershey, director of Carnegie Institution's genetics research unit; and Dr. Salvador E. Luria, MIT microbiologist. Award recognized discoveries concerning viruses and viral diseases including reproductive processes of bacteriophage virus which infected bacteria. Scientists would share $73,000 cash prize for what selection committee called setting "the solid foundation on which molecular biology rests." (Lee, *NYT*, 10/17/69, 1)
- In age of lunar landings there was boom in superstition in U.K., London Express Service reported in *El Paso Herald-Post*. Psychological investigation of 140 Manchester Univ. students had revealed 12% believed in old superstitions and London street survey had shown "pattern of belief incredible a few years ago." One U.K. astrologer received more than 100,000 letters yearly; three-fourths of population studied newspaper horoscopes regularly; and 3 out of 10 winners in premium bond stakes reckoned they owed winnings to "lucky numbers under a particular star." (*El Paso Herald-Post*, 10/16/69, C12)
- Meteor Crater, Arizona, 640-acre site containing crater three miles in circumference and 570 ft deep, was attracting 200,000 visitors annually as "finest example on earth" of what *Apollo 11* astronauts found on moon. (Arline, W *Star*, 10/16/69, A21)

October 17: Mariner Mars 1969 missions were adjudged successful by NASA. Both spacecraft had performed satisfactorily with only minor anomalies in transit to Mars. *Mariner VI* (launched Feb. 24) had encountered Mars July 31 and had returned valuable data on Mars equatorial region. All scientific instruments except one of two channels of infrared spectrometer operated successfully. *Mariner VII* (launched March 27) had encountered Mars Aug. 5 with all scientific instruments operating successfully and had returned good data on Mars southern hemisphere.

Scientific data provided by two spacecraft included more than 2,000 UV spectra and more than 400 infrared spectra of atmosphere and surface. Infrared radiometer returned more than 800 near-encounter and 100,000 far-encounter surface and atmospheric measurements. TV cameras produced 198 high-quality analog pictures of Martian surface. Tracking data provided measurements of mass and ephemeris of Mars, but degree of success of celestial mechanics experiment was yet to be determined. S-band occultation experiment determined electron and temperature profile of ionosphere and temperature and pressure profile of lower atmosphere. (NASA Proj Off)

- U.S.S.R. launched *Cosmos CCCII* from Plesetsk into orbit with 320-km (198.8-mi) apogee, 207-km (128.6-mi) perigee, 89.6-min period, and 65.4° inclination. Satellite reentered Oct. 25. (GSFC *SSR*, 10/31/69; *SBD*, 10/20/69, 226)

- NASA announced appointment of L/G Frank A. Bogart (USAF, Ret.) as MSC Associate Director, succeeding Wesley L. Hjornevik, who had been nominated Deputy Director of Office of Economic Opportunity. Bogart would be succeeded as Deputy Associate Administrator for Manned Space Flight (Management) at NASA Hq. by Harry H. Gorman, MSFC Deputy Director (Management). (NASA Release 69-144)

- At Moscow meeting of Franco-Soviet Grand Commission, formed in 1967 to boost cooperation, France and U.S.S.R. signed protocol to join in attempt to put laser reflector on moon. French laser reflector would be launched by Soviet rocket. (Reuters, B *Sun*, 10/18/69)

- Cambridge Univ. astronomer Fred Hoyle said in *Science* indications that igneous lunar rocks were perhaps as old as meteorites might "lead to the suggestion that the moon experienced a period of intense volcanic activity early in its history." There was no need for volcanoes to have occurred *in situ* on moon. "Considerations of angular momentum show that planetary material probably separated from the sun when the radius of the latter was considerably greater than its present value. Current work on stellar structure requires that the effective surface temperature of the solar condensation be substantially constant at 3500° to 4000°K during this phase, independent of radius. Hence, for comparatively large radii the luminosity would have been very much greater than the present-day value, so that primitive planetary material could well have been considerably hotter than would be estimated for material at corresponding distances from the present-day sun. . . . It will be of great interest to see if the recently acquired samples of lunar material establish the existence of such a hot phase, and, if so, to discover if any features of terrestrial geochemistry, which have hitherto been attributed to igneous activities on the earth itself, really belong to the initial primitive phase of the solar system." (*Science*, 10/17/69, 401)

- Gravity at *Apollo 11* lunar landing site had been determined to be 162,821,680 milligals from data telemetered to earth by LM on lunar surface, MSC scientist Richard L. Nance reported in *Science*. Gravity was measured with pulsed integrating pendulous accelerometer. Measurement could suggest order of magnitude of other anomalies, provide guide for future surveys, and indicate degree of homogeneity of moon. Radius of moon at an observation point could be determined independently of other methods of measurement. (*Science*, 10/17/69, 384-5)

- *Science* published letter from Cornell Univ. astronomer Dr. Brian T. O'Leary, former NASA scientist-astronaut: Comment from NASA officials that he had resigned from NASA program because he did not want to become pilot was oversimplification. "The budgetary delays in plans for scientific space flights and the inability to carry on a reasonable amount of scientific research in the meantime were equally important reasons." (*Science*, 10/17/69, 313)

- Univ. of Iowa astronomers Dr. James A. Van Allen and Dr. Richard S. Yeh reported in *Science* that abstract measurements made by lunar orbiting *Explorer XXXV* during 1967–68 showed it unlikely that alpha-particle emissivity of moon was greater than 0.064 per square cm per sec per steradian. And it was extremely unlikely it was greater than 0.128. (Values were 0.1 and 0.2 of 1966 provisional estimates by H. W. Kraner and others.) Result implied abundance of uranium-238 in outer crust of moon was much less than typical of earth's lithosphere, though it was consistent with abundance of uranium-238 in terrestrial basalt or in chondritic meteorites. (*Science*, 10/17/69, 370–2)
- NASA selected TRW Inc. and General Electric Co. for contract negotiations to conduct competitive studies for Earth Resources Technology Satellite (ERTS) program. Studies would determine how designs of existing spacecraft could be adapted with minimum modifications to ERTS requirements and to evaluate orbital and ground-based processing requirements. Final negotiations were expected to lead to two $500,000, firm-fixed-price contracts. First ERTS flights would be launched in early 1972 to evaluate sensors for monitoring earth resources and to assess scope and requirements of eventual Earth Resources Satellite (ERS) program. (NASA Release 69-142)
- Dr. Charles A. Berry, MSC Director of Medical Research and Operations, received Mission of the Doctor award, presented annually by Italy's Carlo Erba Foundation to a doctor for his "human qualities and talent." (AP, *NYT*, 10/17/69)
- Washington *Evening Star* editorial commented on U.S.-India agreement to provide first direct TV broadcasts from satellite to small receiving stations [see Sept. 18]: "All parties involved in this unusual enterprise seem confident it will succeed. If they are proved right, the whole idea can be broadened to enrich . . . every land on every continent where there is need to spread knowledge and forge unifying links between cities and isolated hamlets in the hinterland." (*W Star*, 10/17/69, A12)
- In interview published by *New York Times*, Boeing Co. Vice President H. W. Withington said SST cost, quoted at $40 million, would climb to $50 million or $60 million if inflation continued at current rate. He did not expect increase to cut heavily into sales. Under current forecast, Boeing would have to raise "about $2-billion to go into production, and that's assuming we get 50 per cent progress payments [half of the purchase price] in the 1973–74 period. I'm not so sure we'll be able to do this." (Lindsey, *NYT*, 10/17/69)
- Sir George Edwards, Chairman and Managing Director of British Aircraft Corp., told American Newcomen Society meeting in New York that Anglo-French Concorde supersonic transport had completed 75 flights and approximately 128 flying hrs. (BAC Release 42/69)
- ComSatCorp reported third-quarter net income of $1,446,000 (14 cents per share), decline from 20 cents per share for second quarter and from $11,760,000 (17 cents per share) for third quarter of 1968. Earnings for first nine months of 1969 totaled $4,947,000 (49 cents per share), down from $5,054,000 (50 cents per share) for first nine months of 1968. ComSatCorp said revenues failed to reach expected level because of interruption of service on *Intelsat-III F-2* June 29

and amortization costs had depreciated because of expansion of comsat system. As of Sept. 30, ComSatCorp was leasing 1,364 circuits full-time, increase of 522 over Sept. 30, 1968. (ComSatCorp Release 59)
- Dr. Caryl P. Haskins, Carnegie Institution President, announced 40-in telescope of advanced design would be erected atop 8,000-ft Las Campanas Mountain in north central Chile. It would be first telescope of Carnegie Southern Observatory, which eventually would house 200-in reflector similar to that at Mt. Palomar Observatory in California. New telescope, expected to be operational in one year, would be equipped with ultrasensitive photoelectric instruments to measure light from remote sources and with digital data system. (Carnegie Institution Release)

October 18: Cosmos CCCIII was launched by U.S.S.R. from Plesetsk into orbit with 457-km (284.0-mi) apogee, 268-km (166.5-mi) perigee, 91.7-min period, and 70.9° inclination. Satellite reentered Jan. 23, 1970. (GSFC *SSR*, 10/31/69; 1/31/70; *SBD*, 10/21/69, 233)
- *Apollo 11* astronauts arrived at Belgrade from Rome during world tour. They were cheered by crowd estimated at 400,000 to 500,000 persons between airport and tomb of Yugoslavia's Unknown Soldier, where they placed wreath. Later, at luncheon in their honor, President Tito said, "I do not like invaders on earth, but I hold in high esteem the conquering of celestial bodies, and I express my wishes for biggest success." (*NYT*, 10/19/69, L69; UPI, W *Star*, 10/19/69, A4)
- Proposed Nixon Administration program of low-budget oceanic and marine science activities was announced in Washington, D.C., by Vice President Spiro T. Agnew. Program sought establishment of coastal zone management program, modest grants for state planning and regulatory mechanisms, and National Oceanic and Atmospheric Agency, currently under consideration by Congress and Ash Commission on Federal Reorganization. Program also proposed coastal laboratories for regional problems, pilot project of lake restoration, International Decade of Ocean Exploration, and accelerated program of Arctic research. (Lannan, W *Star*, 10/19/69, A12)

October 19: Boeing Co. President T. A. Wilson was quoted in *Washington Post* interview as saying 747 airliner was biggest financial gamble his firm had ever taken. Huge aircraft would compete with Lockheed 1011 and McDonnell Douglas DC-10, which were smaller but less expensive to purchase and operate, with three engines instead of four. While airlines were seeking ways to retrench, Boeing Co. was building 747 in plant bigger than vertical assembly building for Saturn V at KSC. To date, Boeing Co. had announced 183 orders for 747—not enough to recoup its expenses with aircraft price at $20 million each. McDonnell Douglas had orders for 97 DC-10s and options on 104 orders. Lockheed had 181 orders for 1011s. (Wilson, *W Post*, 10/19/69, B5)
- John N. Wilford said in *New York Times:* "Even though the Soyuz flights accomplished less than American observers had expected, they served notice that the Soviet Union may have forged ahead in space-station development while the United States was concentrating on the Apollo moon flight preparations." But if U.S.S.R. had meant missions as beginning of first space station, "they fell far short." There were "no link-ups of vehicles, no transfers of crews between ships and no test of the effects of long-duration weightlessness. None of the three ships was

placed in an orbit high enough for the prolonged flights being considered for space stations."

Mission accomplishments could prove useful for construction of future space stations. "They proved that they were able to launch three manned spacecraft in three consecutive days, which the United States has not done. Such precision launchings are necessary for space station deployment." Missions also had demonstrated first welding in space. "It will probably be necessary to weld together sections of a space station, rather than rely solely on the mechanical clamps used for present docking operations, and to make repairs by welding during flight." (*NYT*, 10/19/69, E9)

October 20: Capt. Chester M. Lee (USN, Ret.), Apollo 12 Mission Director, described Apollo 12 plans to press at NASA Hq. [see Oct. 2]. Mission would be launched toward moon at 11:22 am EST Nov. 14, carrying Astronauts Charles Conrad, Jr. (commander), Alan L. Bean (LM pilot), and Richard F. Gordon, Jr. (CM pilot). Primary objectives would be to perform selenological inspection, survey and sample mare area, deploy Apollo lunar surface experiment package (ALSEP), develop techniques for point landing capability, and photograph candidate landing sites.

Basic flight plan would be similar to one for *Apollo 11* (July 16–24) with modifications to improve landing accuracy and increase data return. Launch azimuth had been narrowed from 72°–108° for *Apollo 11* to 72°–96° for Apollo 12 and trajectory had been changed from free-return to hybrid. Lunar orbit stay time for Apollo 12 would be increased from 59.6 hrs for *Apollo 11* to 89 hrs; lunar surface stay time, from 21.6 hrs to 31.5 hrs; EVA from one 2-hr 32-min period to two 3-hr 30-min periods; and total mission time, from 195.3 hrs to 244.7 hrs. LM sleeping arrangements would be improved with hammocks, color TV if repaired by launch date would be used on surface, and LM ascent stage would be crashed onto lunar surface near seismometer after ascent from moon and redocking. To increase navigational accuracy crew would avoid waste and water dumps 10 hrs before landing to prevent thrust that might carry them off course, would approach moon upside-down so landing radar would be operational during entire landing phase, and would speed up descent to conserve fuel at landing. (Transcript)

- Tass reported Soviet scientists had tested "electric jet plasma engine" producing jet flow of gas that reached 75 mps. Engine "needs neither fuel nor oxidizer from the ground. Once the aircraft is taken to the ionosphere it is able to continue flying using only atmospheric nitrogen." Tass said engine might be used to power "superfast" jetliners in upper atmosphere.

  Washington Post quoted U.S. scientists as saying this was misinterpretation. No aircraft could fly high enough to take advantage of such an engine. Soviet development sounded like "very interesting" breakthrough, which U.S. was not even pursuing. It appeared to be new version of ion engine, which NASA would use aboard SERT spacecraft later in year. But SERT engine would carry its own fuel. Electrical power supply to convert nitrogen to electrified gas would be too heavy. (O'Toole, *W Post*, 10/21/69, A23)

- NAS released *Scientific Uses of the Large Space Telescope*, report of ad hoc committee of NAS–NRC Space Science Board appointed to consider scientific functions and practical feasibility of placing in earth orbit or on lunar surface diffraction-limited, optical telescope with 120-in aperture. Committee concluded LST would make dominant contribution to understanding content, structure, scale, and evolution of universe and provide important, decisive information in other astronomical fields; efficient space astronomy program would also require continuing series of smaller telescopes; and most effective utilization of powerful space telescope would require substantial increase in ground-based instruments. (Text)

- Washington *Evening Star* editorial commented on *Soyuz VI, VII,* and *VIII:* "The latest Soviet space spectacular, it would seem, adds up to a small step for mankind and a giant step for the welding trade. There is nothing overwhelmingly impressive about having three spacecraft in orbit at once. All that is required these days is the decision to get them up and the money to build three rockets and launch facilities. Even the highly touted welding experiment does not constitute a major technical breakthrough. It did raise the honorable trade to new heights, but the conditions under which the tests were made could have been—and probably have been—largely duplicated in the laboratory." (W *Star,* 10/20/69)

- Aquanauts Dr. Larry Hallanger, Dr. David Youngblood, Wally Jenkins, and Richard A. Waller had successfully completed two-day experiment in first self-contained undersea laboratory 50 ft below Atlantic near Riviera Beach, Fla., Associated Press reported. Hydro-Lab, 16 ft long and 8 ft in diameter, supplied its own electricity from fuel cell similar to cell used for power in Apollo spacecraft. It was built by Perry Oceanographics, Inc. (W *Star,* 10/20/69, A3)

- San Francisco Board of Supervisors had voted to negotiate with Texas millionaire Lamar Hunt on lease of 22-acre Alcatraz Island, United Press International reported. Hunt planned underground museum in tribute to *Apollo 11,* while restoring island's historic buildings. (W *Star,* 10/20/69, B4)

October 20–23: AIAA's Sixth Annual Meeting and Technical Display was held in Anaheim, Calif. Among 6,559 persons attending were Cosmonauts Georgy T. Beregovoy and Konstantin P. Feoktistov with *Apollo 8* Astronaut Frank Borman, host for their U.S. visit. At press conference Beregovoy said military use of space was not practical: "The aim of our program is the exploration of space. In our minds, space is the scene of peaceful work and investigation." Feoktistov said U.S.–U.S.S.R. space race was "first phase of space flight." Space research was now in second phase. "I think we can say that in that phase Soviet and American scientists are intensely helping each other." U.S.S.R. wanted to send unmanned spacecraft to moon's vicinity, outer planets, and solar orbit and to develop weather and communications satellites. Soviet plans called for manned space stations and manned flights to vicinities of Mars, Venus, and Mercury. Asked when astronauts and cosmonauts might fly side by side, Beregovoy said, "We are going parallel but different ways now, but in principle such a possibility exists. Maybe a year, maybe three years. Maybe as soon as we learn English." (AIAA Release 11/3/69; O'Toole, *W Post,* 10/24/69, A3)

ESSA Administrator Robert M. White described new-generation

weather satellite ITOS that would be launched in prototype by NASA later in 1969 and in operational form by ESSA in early 1970. "It combines in one spacecraft both the stored-data and direct readout cameras of the present two-satellite [TIROS] arrangement. Fewer launches will be required to keep the system in operation. The design allows for growth and the capacity to accommodate a variety of instruments to meet new and developing needs." System would contain high-resolution infrared radiometer for "nighttime observations approaching daytime pictures in resolution and quality" thus providing 24-hr weather satellite coverage. "This step will enable us to meet fully the first national program objective—providing twice-daily observations of the entire earth, both by stored and direct readout systems, day and night." (Text)

Fifty years of transatlantic flight—from USN NC-4 flying boat's Lisbon arrival May 27, 1919, to June 3, 1969, maiden Atlantic crossing of Boeing 747—was traced in paper by Dr. Richard K. Smith, Ramsey Fellow at Smithsonian Institution's National Air and Space Museum. During "the Heroic Years on the North Atlantic between 27 May 1919 and 27 May 1939, there were 175 flights across the Atlantic; 142 were by airplanes, 33 by airships. . . . It is confidently believed that all the persons who flew the Atlantic by airplane during that period could be accommodated aboard a 747, perhaps with some seats to spare. However, at least three more 747s would be required to lift those persons who flew by airship during 1919–1937!" (Text)

MSC Flight Operations Director Christopher C. Kraft, Jr., said Apollo program might be extended to 1973 to allow NASA to fly three space station training missions in 1972. "It's going to be some time before we settle it. But it's going to be difficult to handle both Apollo and Apollo Applications from an operational point of view as well as a people point of view in 1972." Revised launch schedule set Apollo 19 landing back to November 1972 and Apollo 20 to May 1973. Between Apollo 18 launch Feb. 11, 1972, and Apollo 19 were three orbital flights preparatory for establishment of first U.S. permanent space station in 1975. These flights could come just after Apollo 18, in May 1972 and in September 1972. Further reason for Apollo landings delay was possibility they would be made in remote lunar areas like crater Tycho, almost 1,000 mi below moon's equator and rated among top-priority landing sites. (O'Toole, *W Post*, 10/23/69, A21)

Dr. Arthur D. Code, Univ. of Wisconsin astronomer, reported discovery that relatively young stars in constellations Orion and Scorpius were generating heats of 45,000°F—5,000° hotter than had been thought. Finding, from data obtained by NASA's *Oao II* satellite (launched Dec. 7, 1968), suggested young stars might be more massive and maturing faster than originally believed and supported theories of thermonuclear processes in stars' interiors that generated such heats. *Oao II* data also showed younger stars had more complex chemical composition than stars formed near universe's beginning. They contained fairly large amounts of carbon, silicon, magnesium, and heavier elements, as well as predominant hydrogen and helium. Other *Oao II* data indicated particles in interstellar space were variety of complex matter, much of which might be debris ejected from stars. (Wilford, *NYT*, 10/21/69, C19)

MSC Director of Engineering Development Maxime A. Faget outlined concept of space vehicle with two reusable stages—booster rocket and orbiter craft—both with fixed wings, long fuselages, and rear rocket engines. Vehicle would be 225 ft high and weigh 2.5 million lbs at launch. Orbiter vehicle, capable of carrying 25,000 lbs of cargo and passengers, would be attached pickaback to booster stage's upper half. Booster, burning liquid hydrogen and liquid oxygen, would elevate orbiter to fringe of space, separate, drop away, and cruise to landing like jet airliner. Booster probably would be piloted by two astronauts. Orbiter would ignite rocket engine to continue upward to space station orbiting at 300-mi altitude. After link-up with space station, transfer of cargo or 12 passengers, and pickup of new load, orbiter would be steered by pilot into atmosphere and controlled glide to landing on 10,000-ft runways used by aircraft. Key to reentry would be orbiter's "angle of attack" when hitting atmosphere. Faget proposed 60° angle with nose up, so flattened bottom would catch airflow to slow descent. Small jet engines would be fired to help control final descent phases of both orbiter and booster. Outer surface of vehicles would be made of metal alloys that could sustain temperatures to 2,700°F. (Wilford, *NYT*, 10/24/69, 43)

Dr. Robert C. Seamans, Jr., Secretary of the Air Force, discussed DOD space activities: "In communications, meteorology and surveillance . . . we are actively pursuing increased capability. . . . Tactical communications using satellite relays are presently being intensely tested and show great promise in improving control of field units and increasing . . . cooperation between land, sea and air forces. In surveillance, our efforts have produced results. Of first importance is a new satellite early warning system that will do a great deal to further insure our ability to deter nuclear attack." Savings in space travel costs "must be linked to the recovery and reuse of space vehicles. . . . Since major improvement in our ability to use space environment would be of direct assistance to both NASA and the military, we are jointly planning our research and development towards this objective." (Text)

Major AIAA awards presented included Louis W. Hill Space Transportation Award to George M. Low, former Manager of Apollo Spacecraft Program at MSC, currently on special assignment to MSC Director, for "his leadership role in bringing the Apollo Program to fulfillment, and to the thousands of engineers who dedicated their careers, without public recognition, to the conquest of space." Daniel Guggenheim Medal for 1969 was presented to retired ARC Director H. Julian Allen for "personal contributions to outstanding research and development leading to vastly improved re-entry bodies, missiles, satellites, and spacecraft, and for leadership in directing and inspiring a large group of research men at Ames Laboratory." Lawrence Sperry Award was given Edgar C. Lineberry, Jr., Chief of Orbital Mission Analysis Branch, MSC, for "significant advancement in the field of rendezvous mechanics through his development of the space maneuver logic and associated control techniques, and his formulation of the mission plans which contributed decisively to the success of all rendezvous operations conducted during United States manned space flights." De Florez Training Award was given to Gifford Bull, Principal Engineer and Engineering Pilot at Cornell Aeronautical Laboratory, for "establishing

October 20–23

the use of variable stability airplanes as flight dynamics training vehicles for engineering test pilots."

New AIAA Award for Spacecraft Design was awarded Otto E. Bartoe, Jr., Vice President, Aerospace Div., Ball Brothers Research Corp., for "concept and preliminary design of the Orbiting Solar Observatory spacecraft configuration and control systems, the first embodiment of the spin-despin design." Aerospace Communications Award was awarded Dr. Eberhardt Rechtin, Director of DOD Advanced Research Projects Agency and former JPL Assistant Director for Tracking and Data Acquisition, for "development of phase lock systems for space communications, guidance and control, and for contributions to the design, development and operation of NASA's Deep Space Network." (MSC Release; AIAA Releases; AIAA *Booster*, 10/20/69)

Honored as new Honorary Fellows at Honors and Award Banquet Oct. 23 were Dr. Robert C. Seamans, Jr., Secretary of the Air Force and former NASA Deputy Administrator; German rocket pioneer Hermann Oberth; and Northrop Corp. founder John K. Northrop.

Among 29 new Fellows were: *Apollo 8* Astronaut Frank Borman, Deputy Director, Flight Crew Operations Directorate, MSC; former GSFC Director, Dr. Harry J. Goett, Chief Engineer, Space and Re-Entry Systems Div., Philco-Ford Corp.; Najeeb E. Halaby, President, Pan American World Airways, Inc.; Samuel L. Higginbottom, Vice President, Operations Group, Eastern Airlines, Inc.; L/G Samuel C. Phillips, Commander of USAF Space and Missile Systems Organization (SAMSO) and former NASA Apollo Program Director; John G. Borger, Chief Engineer, Pan American World Airways, Inc.; Walter Haeussermann, Director, Astrionics Laboratory, MSFC; Dr. John C. Houbolt, Vice President and Senior Consultant, Aeronautical Research Associates of Princeton. (AIAA Release)

During third AIAA President's Forum, Under Secretary of Interior Russell E. Train said, "Photographs of Earth, taken by our astronauts from the vicinity of the Moon, bring home more forcefully than the words of all scientists of all times the fact that we live on a finite planet with finite but poorly known resources." (*A&A*, 2/70)

October 21: U.S.S.R. launched *Cosmos CCCIV* into orbit with 760-km (472.2-mi) apogee, 742-km (461.1-mi) perigee, 99.8-min period, and 74.0° inclination. (GSFC *SSR*, 10/31/69)

• NASA's Annual Honor Awards Ceremony was held in Washington, D.C., with keynote speaker Dr. Charles H. Townes, Univ. of California at Berkeley physicist and member of President's Task Force on Science Policy.

Recipients of NASA Distinguished Service Medal were: L/G Frank A. Bogart (USAF, Ret.), MSC Associate Director; Robert E. Bordeau, Assistant Director for Projects, GSFC; Dr. John F. Clark, GSFC Director; Charles W. Mathews, Deputy Associate Administrator, OMSF; Ozro M. Covington, Assistant Director for Manned Flight Support, GSFC; George H. Hage, Boeing Co. Vice President for Product Development and former NASA Apollo Program Deputy Director; Dr. George E. Mueller, NASA Associate Administrator for Manned Space Flight; Dr. John E. Naugle, NASA Associate Administrator for Space Science and Applications; Rocco Petrone, Apollo Program Director; L/G Samuel C. Phillips, Commander of USAF Space and Missile Systems Organization (SAMSO) and former Apollo Program Director; Julian W. Scheer, Assistant Administrator for Public Affairs; Dr. Robert C. Seamans, Jr., Secretary of the Air Force and former NASA Deputy Administrator; Willis H. Shapley, Associate Deputy Administrator; and Gerald M. Truszynski, Associate Administrator for Tracking and Data Acquisition.

Distinguished Public Service Medal was awarded to Dr. Harry H. Hess (posthumously), former Princeton Univ. geologist and Chairman of NAS–NRC Space Science Board, and to Dr. Townes.

Exceptional Bravery Medal was awarded to Charles J. Beverlin and Billy B. McClure of General Dynamics Corp. at KSC, each for prompt action during "accidental depressurization of an Atlas launch vehicle" which "prevented destruction of a Mariner Mars space vehicle and potential injury to his comrades."

Special recognition given to employees of NASA and leading aerospace industries for their role in successful Apollo lunar landing resulted in largest number of annual awards to date. Exceptional Scientific Achievement Medal was awarded to 25 persons, Exceptional Service Medal to 97, and Public Service Group Achievement Award to 29. Other awards included Group Achievement Award, Public Service Award, and Certificate of Appreciation. (Program; NASA Release 69–143; NASA Personnel Div)

- MSC Director, Dr. Robert R. Gilruth, announced appointment of Dr. Gene Simmons, MIT professor of geophysics, to new position of MSC Chief Scientist and of Anthony J. Calio as Director of Science and Applications. Dr. Simmons would divide his time between MIT and NASA, reporting directly to MSC Director. At MSC Dr. Simmons would attempt to emphasize strong role of science in future manned flights and lunar exploration and to effect close ties between NASA programs and scientific community. (MSC Release 69–72)
- Visiting Cosmonauts Georgy T. Beregovoy and Konstantin P. Feoktistov received Presidential pens and bronze inaugural medals from President Nixon during brief stop at White House. Later they flew to MSC, where they were honored at dinner by 30 astronauts. *Apollo 8* Astronaut Frank Borman showed cosmonauts NASA T–38 jet aircraft. (*PD*, 10/27/69, 1485; AP, W *Star*, 10/22/69, A3)
- Richard W. Cook, Deputy Director for Operations in Science and Engineering Directorate, MSFC, would become Deputy Director, Management, to MSFC Director, Dr. Wernher von Braun, MSFC announced. He would succeed Harry H. Gorman, new NASA Deputy Associate Administrator for Manned Space Flight (Management), for one or two years, or until permanent replacement became available. (MSFC Release 69–232)

October 22: *Cosmos CCCV* was launched by U.S.S.R. from Baikonur. Satellite entered orbit with 340-km (211.3-mi) apogee, 203-km (126.1-mi) perigee, 88.4-min period, and 51.4° inclination and reentered Oct. 24. (GSFC *SSR*, 10/31/69; *SBD*, 10/27/69, 254)

- Press conference on preliminary science results of *Biosatellite III* (launched June 28) was held at NASA Hq. Spacecraft, carrying male macaque monkey Bonny, had been launched on 30-day mission to investigate physiological problems during space flight but had been deorbited after 8½ days when monkey's condition deteriorated.

John W. Dyer, Manager of Biosatellite Project Operations at ARC, said all of automatically controlled functions performed "beautifully" after launch and spacecraft and instrumentation performed well throughout mission. "The animal was reported to be enthusiastically eating and drinking, and after four or five days, a routine operation capable of going the duration of on-board consumables was projected." Clear data on Bonny and entire flight system were recorded at 16 stations in 180 intervals of 5 to 10 min each and good quality data were returned from onboard tape and film records recovered with capsule.

Bonny remained alert until eighth day, providing much new information on cycles of sleep and wakefulness at zero g. Dr. W. Ross Adey of UCLA, principal experimenter, said one important discovery was alterations in circadian rhythms, daily rhythms in physiological functions. "At least it is clear that if there is not an actual prolongation of the rhythms there is a very significant phase shift so that the animal woke later and later each day." Most interesting discovery about sleep patterns was that dream sleep, which constituted 20% of normal night's sleep on earth and was characterized by rapid eye movements (REM), occurred in space. "It had never been clear in space whether man or animals have this REM sleep, and there is evidence that decreased gravitational inputs, or decreased inputs from the body will seriously disrupt it. So we were very interested to see that REM sleep in the monkey . . . that there was indeed a great deal of sleep of this REM type, and that it occurred in . . . an intermediate stage of sleep."

One of first abnormalities noted in Bonny was pendular eye movement, swinging movement that occurred when vestibular mechanism in inner ear was disturbed. Fluid was moving in some fashion that might indicate pressure or cooling if it occurred on earth. "None of these conditions applied here. But the weightlessness would allow movements of the fluid which were probably unusual. . . . This is interesting because the Apollo astronauts who have now much more room to move around than in . . . the Mercury and Gemini spacecraft have . . . almost uniformly reported some degree of vestibular disturbances."

By eighth day, brain and body temperature and central venous pressure had dropped dangerously. Fluid loss by sweating and diuresis was high, apparently because of redistribution of blood in visceral pools from weightlessness. On recovery Bonny was semicomatose and his temperature was below 35°C. Immediate resuscitation measures with intravenous fluid were begun and monkey's condition improved substantially. "The temperature came up to about 35.8, the heart rate stabilized, and the blood pressure came up. . . . And the animal was lifting his head . . . and making coordinated movements with all four limbs. And then quite suddenly, about twelve hours after recovery, a condition of ventricular fibrillation ensued. It came on very suddenly, and the heartbeat became totally disordered. And there was no recovery. This is a common occurrence in monkeys of this species when recovering from hypothermia . . . and death is almost a certainty."

Autopsy revealed 20% loss in body weight, much higher than 3%–8% reported for astronuats in early days of flights. Restraint, weightlessness, and decreased feedback from peripheral structures had decreased shivering response, and monkey had failed to respond to falling temperature with normal thermogenic response. Restraint and, particularly, weightlessness had led to pooling of blood in thorax and abdomen, raising central venous pressures and leading to loss of fluid through kidney and sweating. Dr. Adey said mission was "highly successful in revealing physiological effects of weightlessness in spite of the reduced duration of the experiment" and indicated "great value of carefully designed animal experiments in collection of important biomedical data relevant to manned flight."

Virtually every piece of information that could be drawn about manned space flight indicated similar changes in astronauts, though to smaller degree, he said. Scientists still did not know whether gravity was necessary for long-duration space flights. "Therefore, I think it is premature to consider the design of space platforms or the larger space stations until we know more from the biomedical point of view about what is absolutely necessary." President's Science Advisory Committee report to President Nixon, being printed, would say necessary biomedical basis for elaborate space platforms and space stations did not exist in NASA or in scientific community. (Transcript)

- Retirement of Dr. Abe Silverstein, LeRC Director, was announced by Dr. Thomas O. Paine, NASA Administrator. He would be succeeded Nov. 1 by OART Acting Administrator Bruce T. Lundin. In requesting retirement, Dr. Silverstein wrote: "As NASA engages in its second ten-year program, it may be important that the men whose decisions initiate the new long-range projects be available to complete them. Since I do not think I can stretch my 40 years of service into 50, it is perhaps best . . . if I bow out now." Few have made so great a contribution to our national space effort," Dr. Paine said.

  Dr. Silverstein had been first Director of Space Flight Programs in NASA Hq. in 1958. Under his leadership first U.S. man-in-space program, Project Mercury, had been planned and groundwork laid for Gemini and Apollo programs. He had joined NASA at Langley Aeronautical Laboratory in 1929 and helped design and later was in charge of Full-Scale Wind Tunnel. Transferring to Lewis Laboratory in 1943, he was responsible for conception, design, and construction of first U.S. supersonic-propulsion wind tunnels. After three years with NASA Hq., 1958–1961, he returned to LeRC as Director.

  Lundin had served NASA since 1943, when he joined staff at Lewis. In 1961 he was appointed Associate Director for Development. In May 1968 he went to NASA Hq. as Deputy Associate Administrator for OART. (NASA Release 69–145; LeRC Release 69–61; LeRC Biog 4/67)

- NASA's X–24A lifting-body vehicle, piloted by NASA test pilot John A. Manke, reached mach 0.6 after air-launch from B–52 aircraft at 40,000-ft altitude over South Rogers Lake Bed, Calif. Objectives of unpowered flight, sixth in series, were to check out new pilot, evaluate handling characteristics with 30° upper flap, and evaluate handling qualities at various roll gain settings of stability augmentation systems (SAS). (NASA Proj Off)

- At Moscow ceremony honoring *Soyuz VI, VII,* and *VIII* cosmonauts, Soviet Communist Party General Secretary Leonid L. Brezhnev said U.S.S.R. had "an extensive space program drawn up for many years." Main road lay in orbital space stations. "Our road of space conquest is the road of solving vital fundamental problems, the problems of science and technology." While U.S.S.R. favored international space cooperation, "a major advance in the development of space techniques has been achieved in the flight . . . just completed. Our science approached the setting up of long-term orbital stations and laboratories —the decisive means of extensive exploration of outer space." (Clarity, *NYT*, 10/23/69, 78)

- Discovery of possible planet, in orbit around NPO-532 pulsar about 6,000 light years from solar system, was reported by Cornell Univ. astronomer Dr. Thomas Gold. Discovery, by astronomers at Arecibo Ionospheric Observatory in Puerto Rico, was indicated by wobble in pulsation rate from NPO-532, pulsar or neutron star in Crab Nebula. Planet was same distance from neutron star as Mercury was from sun. (*W Post*, 10/22/69, A9)

October 23: Apollo 12 countdown demonstration test (CDDT) began at KSC in preparation for launch to moon Nov. 14. (UPI, *NYT*, 10/25/69)

- In Kinshasa, Republic of Congo, *Apollo 11* astronauts were awarded National Order of the Leopard, Congo's highest decoration, by President Joseph D. Mobutu for "setting an example to all mankind." (*C Trib*, 10/24/69)
- Cosmonauts Georgy T. Beregovoy and Konstantin P. Feoktistov told press conference at MSC they had declined invitation to visit Cape Kennedy because they could not reciprocate and invite U.S. astronauts to Soviet launch facilities. (*C Trib*, 10/24/69)
- Dr. George E. Mueller, NASA Associate Administrator for Manned Space Flight, addressed 25th Annual General Meeting of IATA in Amsterdam, Netherlands. "Within the next ten years we can expect the full effect of the space age to be felt. New insights into our complex environment will spur the development of new industries, new products and the new jobs which we need to gainfully employ our expanding population." Air transportation would be one of first areas "dramatically affected" by space experience. "An inertial guidance system fundamentally the same as that which carried Apollo 11 to the moon is now being installed in the Boeing 747. Data concerning the flammability of nonmetallic materials, accumulated for use in the Apollo Modules, is being utilized for interior fittings. . . ." Other space-to-aviation transfers included composite materials to withstand stresses and temperatures not previously encountered, optical tooling for large structures, welding with new materials and by new processes in new aircraft, and general-purpose onboard digital computers. "The use of solid-state large-scale integrated circuits for multiplexing and self-checking circuits will eliminate most cabling and much maintenance."

Desired characteristics of space shuttle, which could be operational by 1976, would anticipate those of next generation of air transports, possibly, global transport "so that no place on earth would be more than an hour from any other." (Text)

October 24: USAF launched unidentified satellite from Vandenberg AFB by Titan IIIB-Agena booster into orbit with 395.8-mi (636.8-km) apogee, 78.3-mi (125.9-km) perigee, 92.1-min period, and 108.1° inclination. Satellite reentered Nov. 8. (GSFC *SSR*, 10/31/69; *Pres Rpt 70* [69])

- U.S.S.R. launched two Cosmos satellites. *Cosmos CCCVI*, launched from Baikonur, entered orbit with 307-km (190.8-mi) apogee, 203-km (126.1-mi) perigee, 89.5-min period, and 64.9° inclination and reentered Nov. 5. *Cosmos CCCVII* entered orbit with 2,157-km (1,340.3-mi) apogee, 213-km (132.4-mi) perigee, 109.1-min period, and 48.4° inclination. (GSFC *SSR*, 10/31/69; 11/15/69; *SBD*, 10/27/69, 255)

- Apollo 12 spacecraft was lashed by 40-mph winds during second day of countdown demonstration test (CDDT) at KSC. Officials said exercise proceeded without difficulty. (*W Post*, 10/25/69, A9; UPI, *NYT*, 10/25/69)

- Dr. Gene Simmons, MSC Chief Scientist, told press at MSC reported conflict between scientists and engineers in U.S. space program was "more apparent than it is real." MSC officials were making great efforts to increase scientific return from nine remaining Apollo missions. Dr. Simmons was "not sure there is this rift." (*W Post*, 10/25/69, A9)

- LaRC announced award of $313,620,000 cost-plus-incentive-fee/award-fee contract to Martin Marietta Corp. for construction of lander portions of two Viking spacecraft scheduled to softland on Mars in summer 1973 and for project integration services. (LaRC Release)

- Soviet Academy of Sciences President, Prof. Mstislav V. Keldysh, said in Stockholm, "We no longer have any time plan for manned moon trips. Right now we are concentrating on constructing big satellite space stations." He predicted expanded scientific cooperation between U.S. and U.S.S.R. with possibility of cosmonauts and astronauts aboard same spacecraft in future. In Sweden for Swedish Academy of Engineering's 50th anniversary, Prof. Keldysh implied U.S.S.R. planned to assemble spacecraft in orbiting stations and launch them from there. He indicated existence of conflicting views in U.S.S.R. as to space research funding. "My personal view is that, when man has taken his first step into space, you cannot stop further development." (UPI, *W Post*, 10/25/69, A3)

- Visiting U.S.S.R. Cosmonauts Georgy T. Beregovoy and Konstantin P. Feoktistov went on "Flight to the Moon" ride at Disneyland, Calif., during U.S. tour. They received Mickey Mouse watches and caps from employees dressed as cartoon characters in spacesuits. Beregovoy said, "Now we are admitted to this particular society." (UPI, *NYT*, 10/25/69, 3)

October 25: *Economist* said of Soviet Soyuz "fiasco" U.S.S.R. had tried "highly ambitious link-up of the components of a permanent, orbiting space station, something the Americans will be in no shape to do for perhaps five years. The fact that the Russians apparently failed must have been humiliating to them, and gratifying to the Americans, but it is unlikely to be more than a temporary check to a programme that has rolled with considerable success since the Russians put the world's first satellite in orbit more than a decade ago. The slow progress reflects the state of Soviet industry, but is steady progress with few loose ends. The American programme, trimmed and squeezed by political pressure and public opinion, is trailing appreciably more loose ends; despite the victory on the moon Nasa will have to set about tidying them up. And it is then that the Russians will have the last laugh." (*Economist*, 10/25/69, 20)

October 26: Estimated 1.5 million persons lined streets of Bombay, India, to welcome *Apollo 11* astronauts. It was largest reception given to astronauts and wives during 22-nation tour. (AP, *B Sun*, 10/27/69, A3)

October 26

- U.S.S.R. Cosmonauts Georgy T. Beregovoy and Konstantin P. Feoktistov attended football game in San Diego, Calif., during two-week U.S. tour. When interpreter failed to explain game clearly, Beregovoy described action as "All fall down, all get up, all fall down." (AP, W *Star*, 10/27/69, A5)
- Eureka, Calif., high school senior Ronald Titus had identified sketch of future space station released by Tass Oct. 11 as identical to drawing in Sperry Gyroscope Co. advertisement in February 1962 *Scientific American*, *New York Times* reported. Original Soviet caption had said it was "Soviet sketch of a future space station." Sketch had appeared in Eureka *Times-Standard* and *New York Times* as concept of Soviet space station. (*NYT*, 10/26/69, 11)
- In $100,000 experiment to evaluate claims that jet noise had caused female minks to panic and eat their young, USAF planned to send jet aircraft through sound barrier over Univ. of Alaska's experimental mink farm near Petersburg. On ground, observers from Dept. of Agriculture and Cornell Univ. would study mink reaction to sonic boom. Since 1962 USAF had paid 25 claims totaling $67,000 for mink problems alleged to have been caused by supersonic aircraft. (Pinto, W *Star*, 10/26/69, A22)
- Comsat issue had emerged as "classic case of governmental indecision—the result of divided federal responsibilities, complicated technical and legal issues, and powerful, competing industrial interests," Robert J. Samuelson said in *Washington Post* article. Nearly year had passed since task force appointed Aug. 14, 1967, by President Johnson to formulate "national communications policy" had submitted report. Nixon Administration had begun separate investigation. Meanwhile, "supporters of a domestic system, frustrated by repeated delay, are pushing the White House and the FCC for a quick decision." ComSatCorp had always been "ardent advocate." In 1965 and 1966, NBC and ABC had "enthusiastically embraced" comsats as means of transmitting TV signals. Recently, CBS President Dr. Frank M. Stanton had become convert, "apparently provoked" by AT&T's new, higher TV signal rates.

   Within month, White House might propose approval of domestic comsat system but FCC would still have difficult task of approving specific system. Only limited number of comsats could be orbited 24,000 mi above equator without interfering with each others' transmission, even if Government were to allow more than one comsat system. If comsats assumed major role in communications, any decision allocating ownership rights could involve multimillions in eventual annual revenue. At first networks had asked to be allowed to create separate system, while ComSatCorp and AT&T advocated single system. Unresolved issue helped explain "extended years of study." However, recent changes made compromise more likely. (*W Post*, 10/26/69, F1)

October 27: NASA's HL-10 lifting-body vehicle, piloted by NASA test pilot William H. Dana, reached 60,500-ft altitude and mach 1.6—vehicle's fastest speed to date—after air-launch from B-52 aircraft west of Rosamond, Calif. Objective of flight, 27th in HL-10 series and 14th with power, was to obtain stability and control data at transonic speeds. (NASA Proj Off; AP, *NYT*, 10/29/69, 12)

- Dr. Persa R. Bell, Chief of MSC Lunar and Earth Sciences Div. and Manager of Lunar Receiving Laboratory, announced his resignation, effective in January 1970. He said he would return to scientific research at Oak Ridge National Laboratory. Until his departure Dr. Bell would act as special assistant to Anthony J. Calio, MSC Director of Science and Applications, evaluating and making recommendations on LRL operations and performing scientific research within LRL. Bryan R. Erb, Deputy Chief of Lunar and Earth Sciences Div., would be acting LRL manager for Apollo 12 mission. (MSC Release 69-74)

- Thomas W. Ray, historian at Defense Communications Agency, became Assistant NASA Historian for Manned Space Flight, replacing William D. Putnam, who had resigned in August to join RAND Corp. (NASA Note)

October 28: Temperature measurements from satellite soundings had proved so valuable they were being used to prepare basic maps of Northern Hemisphere weather, ESSA announced. Soundings, made by 91-lb satellite infrared spectrometer (SIRS) on board NASA's *Nimbus III* (launched April 14), had provided unprecedented coverage of conditions over oceans and other areas where few upper-air measurements were made. Use of SIRS data had proved so successful that special computer-to-computer link had been installed between GSFC and National Meteorological Center in Suitland, Md., to permit regular use of data in analyses of weather conditions over eastern Atlantic and most of Pacific from surface to 53,000-ft altitude. Improved version of SIRS was being readied for flight on next Nimbus launch in spring 1970. (ESSA Release ES 69-67)

- NASA announced selection of Boeing Co. Aerospace Group to receive $19-million, cost-plus-incentive-fee contract to design, develop, test, and deliver four flight-qualified lunar roving vehicles and related test and training equipment. Four-wheeled vehicles would be carried on board Apollo LM descent stages in 1971-72 to transport astronauts on lunar surface and store equipment and lunar samples. (NASA Release 69-147)

- *Apollo 11* astronauts were welcomed by estimated 100,000 persons in Bangkok, Thailand, after flight from Dacca, Pakistan, during world tour. (*C Trib*, 10/29/69)

- Dr. William D. McElroy, NSF Director, announced 36-mo extension of Deep Sea Drilling Project, which was nearing end of 18-mo contract period. To date, holes up to 3,231 ft deep had been drilled at more than 66 sites in Atlantic and Pacific Oceans, in water depths up to 20,140 ft, to retrieve sediment samples that had rained onto ocean floor for millions of years. Project sought data on inner space—history and origins of ocean basins and continents. Contract extension—from June 30, 1970, to June 30, 1973—covered additional drilling during 15 two-month cruises of drilling ship *Glomar Challenger*. (NSF Release 69-144)

October 29: Apollo 12 Astronauts Charles Conrad, Jr., Richard F. Gordon, Jr., and Alan L. Bean achieved mock liftoff at 11:22 am EDT—exact minute planned—completing final phase of initial countdown demonstration test (CDDT) for Nov. 14 launch to moon. No major problems arose in spite of gusty winds and rain that would have delayed actual launch. (*W Post*, 10/30/69, A8)

- Second official public display of lunar rock fragment retrieved from moon by *Apollo 11* astronauts was unveiled at MSC by Dr. Robert R. Gilruth, MSC Director. Exhibit included 12 photograph and text panels

October 29

and four-foot-diameter opaque sphere with 20-in viewing port. (MSC Release 69–73)
- NASA launched Arcas sounding rocket from Wallops Station carrying Pennsylvania State Univ. payload to study ion composition. Mission did not meet minimum scientific requirements. (NASA Proj Off)
- Space shuttle was "first step or keystone to the success and growth of future space flight developments for the exploration and exploitation of space," LeRoy E. Day and B. G. Noblitt of Space Shuttle Task Group, NASA OMSF, said in paper presented at IEEE EASCON Session on Earth Orbiting Manned Space Station in Washington, D.C. "Large experiment modules and unmanned satellites can be placed into low earth orbit and retrieved as desired. Propulsive stages and payloads . . . destined for higher energy orbits can be placed into low earth orbits." On-orbit reusable shuttles like space tugs and nuclear stages "become economically advantageous once propellants can be inexpensively delivered to earth orbit. . . . On-orbit maintenance services can be provided for malfunctioning or inoperative satellites. . . . short duration special purpose orbital missions can be conducted by the space shuttle itself to augment or complement space station activities. The design and operational characteristics of the space shuttle will also provide a potential capability to conduct space rescue missions—a capability that is not practical with conventional expendable launch systems." (Text)
- DOD announced it would scrap B–58 supersonic bomber and cut back continental air defenses in economy drive necessitated by congressional reductions in defense budget. (DOD Release 927–69; Homan, *W Post*, 10/30/69, A9)
- Dr. Glenn T. Seaborg, AEC Chairman, testified before Senate Joint Committee on Atomic Energy during hearings on environmental effects of producing electric power: "In the years ahead today's outcries about the environment will be nothing compared to cries of angry citizens who find that power failures due to a lack of sufficient generating capacity to meet peak loads have plunged them into prolonged blackouts . . . when their health and well being . . . may be seriously endangered. The environment of a city whose life's energy has been cut —whose transportation and communications are dead, in which medical and police help cannot be had, and where food spoils and people stifle or shiver while imprisoned in stalled subways or darkened skyscrapers—all this also represents a dangerous environment that we must anticipate and work to avoid. . . . I believe that the judicial development of nuclear power as a major source of energy for the future is in the public interest and that five, ten and a hundred years hence, men will look back with favor on the course we are taking today." (Text)
- Former astronaut and aquanaut M. Scott Carpenter announced he would become President and Chief Executive of Sea Sciences Corp., private enterprise formed to develop underwater projects. Prince Bernhard of The Netherlands would be board chairman. (UPI, *W Star*, 10/30/69, A16)
- Cosmonauts Georgy T. Beregovoy and Konstantin Feoktistov spent half hour riding into Grand Canyon on mules after arriving in Arizona from San Francisco during two-week U.S. tour. (UPI, *W Star*, 10/30/69, A3)

October 30: Dr. Gary V. Latham of Lamont-Doherty Geological Observatory said in Washington, D.C., that evidence of vibrations recorded by seismic equipment left on moon by *Apollo 11* astronauts "now looks more like volcanoes we see on earth than anything else." Concept of volcanoes caused by isolated pockets of radioactive heating did not conflict with "cold moon" theory but was "new working hypothesis." Among things affecting data received on earth before *Apollo 11* seismic equipment failed were gasping, creakings, and groanings of LM lower stage. (Lannan, W *Star*, 10/31/69, A3)

- LeRC announced award of $25,518,000 cost-plus-incentive-fee contract to General Dynamics Corp. Convair Div. to design and build first of new series of improved Centaur upper stage rockets. General Dynamics would update and simplify Centaur system and make it compatible with Titan III booster. Centaur, first liquid hydrogen, liquid oxygen rocket developed by U.S., had been used with Atlas booster to launch successfully Surveyor, Mariner, OAO, and ATS satellites. (LeRC Release 69–65)

- Dr. Murray Gell-Mann of Cal Tech was named winner of 1969 Nobel Prize in physics for "fundamental work in nearly all domains of his field," especially for "discoveries concerning the classification of elementary particles and their interactions." Dr. Gell-Mann, who would receive $72,800 award, had introduced concept of quark—hypothetical particle from which other particles of nucleus of atom were made—and had proposed eightfold-way theory that many particles might change places and their differences from each other lay in their energy quotient. Dr. Gell-Mann was member of NASA Physics Advisory Committee from 1964 to 1969.

    Chemistry prize was awarded Prof. Derek H. R. Barton of Imperial College of Science and Technology in London and Prof. Odd Hassell of Univ. of Oslo for work to develop and apply concept of conformation in chemistry. (Getze, *LA Times*, 10/31/69; AP, W *Star*, 10/30/69, A1; NASA OSSA)

- Dr. Harold Brown, former Secretary of the Air Force, was inaugurated as president of Cal Tech during day-long ceremonies on campus. He succeeded Dr. Lee A. DuBridge, who had become Presidential Science Adviser. (*LA Times*, 9/4/69)

October 31: NAS–NRC Space Science Board published *Lunar Exploration: Strategy for Research 1969–1975*. Best use of manned lunar-landing capability in lunar exploration would entail shift of emphasis from technological development to exploitation of existing Apollo technology for scientific objectives. Board recommended: immediately focusing lunar program on optimizing returns on investment already made in Apollo through maximum use of existing Apollo technology; giving priority to current Apollo lunar-exploration phase should choice be necessary in early 1970s between funding major new manned exploration technology and funding extensions of Apollo lunar technology; periodic reevaluation of NASA's proposed lunar landing sites; earliest possible extensions of Apollo technology to increase scientific returns and increasing interval up to six months between missions to allow incorporation of improvements; early provision of roving vehicle to transport astronauts at least 3.1–6.2 mi (5–10 km) from LM; inclusion in Apollo 16 to 20 time frame, in high-inclination orbit, of experiments in gravimetry by satellite-to-satellite tracking technique,

October 31

altimetry, magnetometry, x-ray and gamma-ray spectrometry, and medium- and high-resolution photography; and increase in supporting research in orbital sensing techniques, both active and passive.

Board also recommended high-priority consideration for long-distance, remotely controlled, traversing vehicle for later missions; careful review of Apollo program management structure to increase role of scientists in influencing basic policy and mission hardware; specific support to lunar experiment definition as integral part of Apollo program and more substantial investment in developing experiment packages easily modified and interchanged even shortly before flight; and high-priority attention to management, handling, distribution, and analysis of lunar samples brought to earth. (Text; NRC Release)

- NASA announced new quarantine procedures for Apollo 12 lunar landing mission scheduled for launch Nov. 14. Procedures had been recommended by Interagency Committee on Back Contamination. If astronauts' condition was normal at splashdown they would don fresh flight suits and oral-nasal masks instead of biological isolation garments (BIG) used on *Apollo 11*. BIG would be available in case of unexplained crew illness. Swimmer would swab hatch and adjacent areas with liquid decontamination agent and crew would be carried by helicopter to mobile quarantine facility (MQF) on board recovery ship. Subsequent crew quarantine procedures would be same as for *Apollo 11*.

  Spacecraft would be returned to Hawaii by recovery ship, where team would deactivate pyrotechnics and flush and drain fluid systems (except water). It would then be flown to Lunar Receiving Laboratory for storage in special room until release from quarantine Jan. 7, 1970. (NASA Release 69–148A)

- Measurement of transient Faraday rotation of *Pioneer VI* (launched into solar orbit Dec. 16, 1965) as it was occulted by sun Nov. 21 through Nov. 24, 1968, was reported in *Science* by members of JPL and UCLA staff. JPL's 210-ft antenna at Barstow, Calif., equipped with automatic polarization tracking system, was used to observe three large-scale transient phenomena, measurements of which indicated Faraday rotation of 40° had occurred. Duration of each was approximately two hours. Phenomena appeared to be correlated with observations of solar radio bursts with wavelengths in dekametric region. (*Science*, 10/31/69, 596–8)

- Tektite II project, in which more than 50 scientists would spend varying periods in ocean over seven-month span starting in spring 1970, was announced by Secretary of the Interior Walter J. Hickel. Program would include major marine scientific mission and extensive human behavioral studies, with emphasis on problems of small crews in isolation for extended periods under stress conditions. (Interior Dept Release)

- Aerospace industry employment would continue to decline from record achieved in 1968, according to semiannual survey results released by Aerospace Industries Assn. Total employment was expected to drop from 1,388,000 to 1,305,000 (6%) between March 1969 and March 1970 because of reduced expenditures for civilian space program and declining sales of civilian transport aircraft during phase-out of current models and commencement of production on new aircraft. (AIA Release 69–45)

- Walter Rundell, Jr., of Iowa State Univ. Dept. of History described U.S. Government historical programs at National Convention of Southern Historical Assn. in Washington, D.C.: "NASA's historical program began in 1959 under the leadership of Eugene M. Emme. Two years later Emme launched the official publications with his *Aeronautics and Astronautics: An American Chronology of Science and Technology in the Exploration of Space, 1915–1960.* This has been followed by similar annual chronologies and chronologies devoted to the Mercury, Gemini, and Apollo projects. Since 1966 the office has published several monographs, some written under contract by private scholars, rather than being undertaken by the permanent staff. A good example is *This New Ocean: A History of Project Mercury,* by Swenson, Grimwood, and Alexander." (Text)

During October: The *Atlantic* commented on *Apollo 11:* "We enthusiastically join the rest of human kind in applauding the feat of the Apollo 11 astronauts and their NASA sponsors. It is an achievement far greater, even, than the construction and successful launchings of *The Brick Moon,* which happened in these pages exactly one hundred years ago." Story, by Edward Everett Hale, had told of group of New Englanders who had constructed artificial moon of bricks and launched it into orbit to provide perpetual navigation aid for seamen. "So far as we can determine, the launching of *The Brick Moon* provoked considerably less public interest than the latest American lunar expedition." (*Atlantic,* 10/69, 3)

- German space program was described in *International Science Notes* published by Dept. of State. For 1969, German space budget was $88 million, with $51 million for national program and remainder for international programs. Amount was approximately 16.5% of total Science Ministry budget. During 1969 to 1973, program would be directed to extraterrestrial research, development of scientific and applications satellite technology (communications, TV, navigation, and meteorology), and development of launcher technology in cooperation with other European countries. Plans included two German Azur research satellites to measure earth radiation belt and density composition of higher atmosphere, both to be launched by United States rockets; development of two solar probes (Helios) in cooperation with NASA; research comsat (Symphonie) to be developed with France; and participation in ESRO research programs and in development of ELDO launcher. (*Science Policy Bull,* 10/69, 51)

November 1969

November 1: In Sydney, Australia, during world tour of *Apollo 11* astronauts, Astronauts Michael Collins said NASA might someday send women on space flights. "NASA has never barred women," he told news conference. "Flying background is usually what has prevented the young ladies from participating." (AP, *W Post*, 11/2/69, 18)

- Visiting Soviet Cosmonauts Konstantin P. Feoktistov and Georgy T. Beregovoy viewed *Apollo 11* lunar rock sample at Smithsonian Institution during tour of Washington, D.C. (W *Star*, 11/2/69)
- Dr. Lee A. DuBridge, Presidential Science Adviser, addressed Sigma Xi Convention in Palm Springs, Calif.: "Science is now a part of society, is a part of politics, is a part of the social and economic system. Scientists must carefully ponder the relevance of their work to the problems of human beings.... They must ask the question of whether the scientific work in which they are engaged is of sufficient importance to the progress of knowledge and its application to be worthy of public support. They must face the fact that budgetary problems will be difficult in the years ahead and that we must reevaluate priorities and reevaluate the mechanisms that we have adopted for allocating our limited funds." (Text)
- In first large-scale attempt at radar mapping of ice-buried continent surrounding South Pole, scientists of Scott Polar Research Institute in Cambridge, England, would make at least 20 flights with airborne-radar-equipped USN C–130 Hercules aircraft from McMurdo Sound in Antarctic, *New York Times* said. Radar equipment already had penetrated more than 14,800 ft of ice, recording profile of mountains and valleys underneath. (Sullivan, *NYT*, 11/1/69, 30)
- Article in *Komsomolskaya Pravda* acknowledged publicly, for first time, that U.S.S.R. had conducted underground nuclear tests "several years ago" to blast oil storage areas in deep salt bed. In same issue Andronik M. Petrosyants, Chairman of State Committee on Atomic Energy, said he had discussed Soviet program with U.S. in Vienna in April. (UPI, *C Trib*, 11/2/69, 16)

November 2: In *Washington Post* Thomas O'Toole traced NASA plans for planetary exploration from May 1971 launch of two Mariner spacecraft for Mars to Grand Tours of planets beginning in 1977 and 1979. "What comes after the two Grand Tours? Nobody really knows, but the space agency's dreamers about unmanned missions have a plot already written and it reads a little better than most science fiction does." Spacecraft would circulate through asteroid belt for year, making asteroid counts and photographing larger asteroids. Beyond that, scientists would like to rendezvous with a comet and investigate Saturn's rings. (*W Post*, 11/2/69, B3)

November 2–3: AFRCL conducted program to measure polar cap absorption (PCA) event and to determine effects on earth's upper atmosphere.

Thirty-six sounding rockets—Nike-Iroquois, Black Brant, Nike-Javelin, Sidewinder-Arcas, and Arcas—were launched from Churchill Research Range to 37.3- to 124.3-mi (60- to 200-km) altitudes to measure energies and densities of charged particles, atmospheric temperatures and composition, and geomagnetic fluctuations. Background data had been provided from eight sounding rockets launched before program and support was provided by measurements from Vela, Pioneer, ATS, Explorer, OV, and ESRO satellites; from instruments on board KC–135 aircraft; and from ground-based instruments. (*SBD*, 11/19/69, 80)

November 3: NASA's HL–10 lifting-body vehicle, piloted by Maj. Peter C. Hoag (USAF), reached 66,000-ft altitude and mach 1.4 after air-launch from B–52 aircraft at 45,000-ft altitude west of Rosamond, Calif. Objective of powered flight, 28th in series, was to obtain stability and control data at various angles of attack and stability augmentation settings. (NASA Proj Off)

- Analysis of lunar soil samples supported theory that moon once was part of earth and became separated in cataclysmic tidal wave, Dr. John A. O'Keefe, Assistant Chief of GSFC's Laboratory for Theoretical Studies, said in lecture at Northern Illinois Univ. Theory would explain why both earth's crust and moon were deficient in precious metals, nickel, and cobalt and would account for indications of intense heat in moon's formation. Heavy metals would have sunk to earth's core when it was molten; moon had no heavy core, so it must be part of earth's crust. (UPI, LA *Her-Exam*, 11/4/69; *W News*, 11/4/69)

- Saturn V 1st stage (S–IC–12) was successfully fired for 125 secs at Mississippi Test Facility in first ground test since failure of S–IC–11 June 26, which caused test-stand fire. NASA board of investigation had concluded fire had been caused by polyethylene disc dust cover that should have been removed before test and recommended 12 corrections in assembly, preparation, and test procedures. (MSFC Release 69–237)

- NASA published *Mariner-Mars 1969: A Preliminary Report* (NASA SP–225), summarizing results of *Mariner VI* (launched Feb. 24) and *Mariner VII* (launched March 27) and describing Mariner program, spacecraft, flights, equipment, experiments, and observations. (Text)

- *Apollo 11* astronauts on world tour arrived in Seoul, Korea, for overnight stay. They were treated as state guests and decorated in ceremony at presidential mansion. (AP, W *Star*, 11/3/69, A21)

- Richard W. Cook, former MSFC Deputy Director for Operations, Science and Engineering, became Deputy Director for Management, replacing Harry H. Gorman, who had been reassigned to NASA Hq. Ludie G. Richard, former Deputy Director (Technical), Science and Engineering, became Deputy Director of directorate. (*Marshall Star*, 11/12/69, 1)

- USAF issued RFPs to airframe and engine manufacturers for development of B–1 strategic aircraft, formerly designated Advanced Manned Strategic Aircraft (AMSA). Proposals were due by spring 1970. (DOD Release 946–69)

November 4: U.S.S.R. launched *Cosmos CCCVIII* into orbit with 408-km (253.5-mi) apogee, 270-km (167.8-mi) perigee, 91.2-min period, and 71.0° inclination. Satellite reentered Jan. 5, 1970. (GSFC *SSR*, 11/15/69; 1/15/70)

- NASA launched two sounding rockets from WSMR with VAM–20 boosters.

Aerobee 150 carried Naval Research Laboratory payload to 124.0-mi (199.6-km) altitude to photograph images of sun and part of corona in 150–400 Å and 400–650 Å region. Rocket and instruments functioned satisfactorily and excellent spectroheliograms were obtained.

Aerobee 150 MI carried American Science and Engineering, Inc., payload to 112.5-mi (181-km) altitude to obtain high-resolution pictures of time development of solar flare in x-ray region of spectrum. Rocket and instruments functioned satisfactorily, photographing 1B limb flare with appreciable x-ray emission that lasted throughout flight. (NASA Rpts SRL)

- Plans for launch of first of two Jupiter probes from KSC between Feb. 26 and March 15, 1972, were outlined by Charles F. Hall, manager of Pioneer F and G projects at ARC, during meeting of scientists, engineers, and contractors at Mountain View, Calif. Launched by Atlas-Centaur booster, Pioneer F would take 610–770 days to reach Jupiter on voyage through planet's heavy radiation belt to within 90,000 mi of Jupiter surface. It would carry 60 lbs of instruments and photographic equipment to conduct 13 experiments to measure solar wind and space dust and analyze Jupiter's radiation and atmosphere. Probe would be first to travel to asteroid belt between Mars and Jupiter and eventually would leave solar system. Pioneer G Jupiter probe would be launched in early 1973. (UPI, P Bull, 11/5/69)

- House and Senate conferees, in executive session, agreed to file conference report on differences between Senate-passed [see Sept. 19] and House-passed [see June 10] versions of H.R. 11271, FY 1970 NASA authorization. Report would retain Senate-passed authorization of $3.716 billion. (CR, 11/4/69, D1023)

- At televised news conference in Moscow on *Soyuz VI, VII,* and *VIII* mission [see Oct. 12–18], Mstislav V. Keldysh, President of Soviet Academy of Sciences, said U.S.S.R. hoped to have permanent space station in earth orbit "certainly" within 10 yrs "and I think less than five years." Station plans did not rule out manned lunar exploration but, he said, "I can only say such operations are not planned for the coming months. Shall we study the universe with the help of automatic (unmanned) spacecraft? We shall. Shall we fly to the moon and the other planets of the solar system in the future? I think we shall because such travel is one of the further aims of mankind in conquering outer space." (Gwertzman, NYT, 11/5/69, 16)

- AEC–NASA Nuclear Rocket Development Station in Jackass Flats, Nev., was "scaling down operations, laying off workers, husbanding the facilities it already has—and waiting," *New York Times* reported. Like many U.S. space research and test centers, "their futures uncertain amid an increasing emphasis on defense and domestic matters at the expense of space," station's scientists and engineers were "waiting for President Nixon to decide on the direction and pace of the national space effort in the next decade." (Wilford, NYT, 11/4/69, 20)

- *Apollo 11* astronauts on world tour arrived in Tokyo. Tight security measures prompted by fears of possible attack by radical students protesting Japan's defense treaty with U.S. permitted public exposure for only half hour. Police estimated 120,000 people watched brief parade down Ginza, Tokyo's main street. Later astronauts and wives were received by Emperor Hirohito and Empress Nagako. Astronauts were presented cultural medal by Prime Minister Eisaku Sato—first time medal had been presented to foreigners. (Pepper, B *Sun*, 11/5/69, A2)

- Visiting Soviet Cosmonauts Konstantin P. Feoktistov and Georgy T. Beregovoy left New York's John F. Kennedy International Airport for Moscow after two-week U.S. visit. During airport interview Feoktistov said he liked U.S. astronauts' "sincerity and hopeful approach to life." Beregovoy said, "Friendship is a force which will help the world to conquer space." *Apollo 8* Astronaut Frank Borman was at airport to see them off. (*NYT*, 11/5/69, 52)

- National Transportation Safety Board opened hearings in Washington, D.C., on midair collisions. In last 31 yrs of recorded aviation history there had been only 12 midair collisions with passenger fatalities, Clifton F. von Kann, Vice President for Operations and Engineering of Air Transport Assn., testified. "Even with the explosive growth of airline service during the last ten years, there have been just four collisions with airline passenger fatalities." (Sehlstedt, B *Sun*, 11/5/69, A9)

November 5: NASA announced decision to install color TV camera on Apollo 12 LM for operation on lunar surface. Decision had been made after three-hour test using unified S-band transmitter. Engineers had simulated varying return signal strengths to Manned Space Flight Network tracking antennas. Camera had been refurbished after flying in *Apollo 10* CM. (MSC Release 69–75)

- *Apollo 11* Astronauts Neil A. Armstrong, Edwin E. Aldrin, Jr., and Michael Collins, with wives, were welcomed by President Nixon at White House ceremony on their return from 44,650-mi around-the-world goodwill journey. President called 38-day, 29-stop tour of 22 countries "most successful goodwill trip in the history of the United States." First men on the moon had demonstrated they were "the best possible ambassadors America could have on this earth." Armstrong said astronauts had been received everywhere, "not just as individuals, but as representatives of the United States" and of "scientific and technological accomplishment . . . that serves as a symbol . . . of a willing and talented nation assembling its resources and firm in its will . . . to share for the benefits of all mankind."

  During journey, astronauts had given 22 news conferences, been received by 20 heads of state, and received decorations on nine occasions. (*PD*, 11/10/69, 1563; Robertson, *NYT*, 11/6/69, 22)

- NASA had put into operation $2.6-million, six-story-high flight simulator for advanced aircraft (FSAA) at ARC, ARC announced. Largest piloted-aircraft simulator ever built would provide research data on handling characteristics of faster and larger jets to ensure safe control by their pilots. Designed by ARC engineers John C. Dusterberry, Maurice D. White, and Shizuo Doiguchi, simulator had motion-generator with 100 ft of lateral motion, 10 ft of vertical motion, and 8 ft of forward and backward motion, combined with pitch, roll, and yaw motions. It was operated by general-purpose, digital-analog computer, which could be programmed with flight characteristics of most aircraft. FSAA could simulate aerodynamics of hypersonic and supersonic flight by reentry vehicle such as space shuttle. (ARC Release 69–13)

- Research to improve materials and lubricants for aerospace might aid in replacement of human hip joint with prosthesis better than any yet developed, LeRC scientist Robert L. Johnson announced. Bearing properties of materials—like cobalt alloy—used in commercial hip joints could be improved by changing crystal structure of metal to hexagonal form, Johnson had found in NASA studies. Interest in these alloys had led to cooperative program with Cornell Medical Center in New York. (LeRC Release 69–66)

- Dr. Raymond L. Bisplinghoff, Dean of MIT School of Engineering and former NASA Associate Administrator for Advanced Research and Technology, was among persons inside MIT Instrumentation Laboratory when students from MIT and other Boston, Mass., campuses demonstrated against laboratory's defense work and were confronted by police. Later Dr. Bisplinghoff said: "The university has set as policy a deliberate movement from Department of Defense and space research to a different mix: more and more research relevant to societal problems. It's going to take a considerable time to make the transition, however, and if we are going to continue on the same scale of operations . . . we're going to need lots of time and new money." It would be up to Congress to change emphasis, not MIT. Instrumentation Laboratory, meanwhile, would honor commitments to DOD, including work on Poseidon missile. (Nordheimer, *NYT*, 11/9/69, 61)

- Giant $6-million solar furnace had been built by French National Center for Scientific Research at Odeillo in Pyrenees for tasks ranging from testing effects of atomic bomb flashes to production of exotic materials for electronic industries, *New York Times* reported. More than 20,000 mirrors focused sunlight on ½-in steel plate to melt hole through plate in seconds. Furnace also had 63 flat, mirrored panels holding 180 small mirrors to follow motion of sun and parabolic array of 9,000 mirrors, mounted to form north-facing wall of nine-story building. Movable panels on hillside reflected sunlight into paraboloid; tower held material to be exposed to extreme heat at focal point of paraboloid. (Sullivan, *NYT*, 11/5/69, 49)

November 6: Air Force Cambridge Research Laboratories launched 997-ft-tall, 34-million-cu-ft helium balloon—largest ever built—from Holloman AFB, N. Mex. Balloon, part of NASA's cosmic ray ionization program (CRISP), was to have provided high-altitude platform for 13,800-lb payload to measure cosmic radiation in upper atmosphere. It had been scheduled to drift eastward at 108,000-ft altitude for 24 hrs, but descended 3 hrs after launch because of leak. (MSC *Roundup*, 11/14/69, 2; MSC Release 69–70)

- House adopted conference report on H.R. 11271, FY 1970 NASA authorization, and sent it to Senate for further action. Report retained Senate's $3.716-billion authorization—$280 million below FY 1969 NASA authorization and $285 million less than $3.966 billion passed by House June 10. It accepted House amendment requiring astronauts to plant U.S. flag on any planet visited for first time on U.S.-financed mission, Senate amendment requiring public disclosure of names and salaries of employees who exchanged positions between NASA and aerospace industry, and House provision withholding funds to any college which barred military recruiters. (House Rpt 91–609; *CR*, 11/6/69, H10679–84)

- Senate Committee on Appropriations favorably reported with amendments H.R. 12307 FY Independent Offices and HUD bill, which in-

November 6

cluded $3.715 billion NASA appropriation. (*CR*, 11/6/69, D1034; Senate Rpt 91-521)

- Ten NASA employees and four support contractors received special awards at White Sands Test Facility for participation in extensive Apollo propulsion system test programs. (MSC Release 69-76)

November 7: West Germany's 157-lb *Azur* (GRS-A) research satellite was successfully launched by NASA from WTR by four-stage Scout booster. Orbital parameters: apogee, 3,145.4 km (1,954.5 mi); perigee, 383.8 km (238.5 mi); period, 121.9 min; and inclination, 103°. Primary NASA objective was to insert spacecraft into quasi-polar orbit in which it could study Van Allen belt, Northern Hemisphere auroral zones, and spectral variation of solar particles versus time during solar flares.

The 48-in-high, 30-in-dia cylindrical satellite carried seven experiments from five German research institutes to continue studies conducted by NASA Explorer and OGO satellites. Data would be available exclusively to German principal investigators for one year. Launch was first in series of cooperative missions between NASA and German Ministry for Scientific Research (BMWF) under July 1965 agreement. BMWF was responsible for design, development, fabrication, and testing of spacecraft and experiments and for ground-support equipment

November 7: *West German research satellite Azur—first satellite in a cooperative Federal Republic of Germany and NASA program—was orbited from WTR by NASA to study the earth's radiation belts, auroral zone, and effects of solar proton events. The spacecraft was photographed during checkout, before mating to its Scout booster.*

and data-reduction facilities. NASA was responsible for Scout booster, launch, and tracking and data acquisition support. (NASA Release 69-146; NASA Proj Off; GSFC SSR, 11/15/69; AP, C Trib, 11/10/69)
- Apollo 12 countdown began at KSC at 12:00 pm EST for launch toward moon at 11:22 am EST Nov. 14. Countdown, originally scheduled to begin at 7:00 pm EST Nov. 8, had been started early to reduce costs by eliminating overtime that would have been paid to launch crews. (UPI, NYT, 11/8/69, 20)
- Senate adopted conference report on H.R. 11271, FY 1970 NASA authorization bill [see Nov. 6], thus clearing bill for White House. (CR, 11/7/69, S13918-9)
- Number of eminent selenologists had confirmed dating of *Apollo 11* moon rocks at 4.5 billion yrs or more, to presumed formation of solar system, indicating lunar surface was far older than any material originating on earth's surface, John Lannan said in Washington *Evening Star*. Same lunar scientists had "lashed out" at what they claimed were NASA's "restrictions on the free flow of scientific information." Lunar investigators would present findings at January symposium but many felt "information of such significance should be published immediately through normal scientific channels." Counterargument was "that all the investigators get an even break this way." (W Star, 11/7/69, A1)
- Analysis of organic combustion products generated by LM descent engine, deemed major and least controllable source of organic and inorganic contamination in lunar sample collection procedure, was reported in *Science* by B. R. Simoneit and A. L. Burlingame of Univ. of California at Berkeley, D. A. Flory of MSC, and I. D. Smith of MSC's White Sands Test Facility. Major gaseous combustion products found in model engine's exhaust were ammonia, water, carbon monoxide, nitrous oxide, oxygen, carbon dioxide, and nitric oxide. Minor products were acetylene, hydrogen cyanide, ethylene, formaldehyde, propadiene, ketene, cyanous acid, hydrazoic acid, various methylamines, acetaldehyde, methylnitrite, formic acid, nitrous acid, butadiene, nitrilohydrazines, nitromethane, and nitrosohydrazines with other oxidized derivatives of unsymmetrical dimethylhydrazine and hydrazine. Ion intensities of species in all mass spectras were estimated as: gases, 87.7%; compounds of carbon, hydrogen, and oxygen, 6.0%; and compounds of carbon, hydrogen, and nitrogen (with traces of oxygen), 5.8%. (Science, 11/7/69, 733-8)

November 8: Apollo 11 Astronauts Neil A. Armstrong, Edwin E. Aldrin, Jr., and Michael Collins received first Pere Marquette Discovery Award of silver medal from Father Marquette Tercentenary Commission of Marquette Univ. (CR, 12/4/69, E10223)
- *Economist* editorial commented: "If the Russians are ever going to get another prestige victory over the United States, and not the sort of worldwide humiliation they have endured with Luna 15 during the American moon landing, and then with the triple Soyuz anti-climax, they will have to pull something off soon. For if not the Soviet planners may see increasingly less reason to continue spending even 1 per cent of the country's gross national product on its cosmonauts. . . . What is ironic is that a Soviet slowdown could be seen as a mirror image of the running battle in the United States between the scientists and the space engineers, but with the Russian scientists coming out on

November 8

top. The Academy of Sciences is widely regarded as the repository of all that is most stuffily conservative in Soviet science. That may have been the central cause of the Russian flop." It was also blow to Americans who relied for new space funds on Nixon Administration's "continuing fear of what the Russians may yet do." (*Economist*, 11/8/69, 12–3)

- Dr. Vesto M. Slipher, astronomer who headed team that discovered planet Pluto in 1930 and discoverer of aurora-like radiations of night sky, died at age 93. He had been director of Lowell Observatory in Flagstaff, Ariz., 1916–1954. (UPI, *LA Times*, 11/10/69)

November 9: Canadian Black Brant VB sounding rocket was launched by NASA from Wallops Station carrying 450-lb Naval Research Laboratory payload containing telescope instrumented for infrared astronomy. Telescope, completely cooled with liquid helium to keep telescope radiation from interfering with measurements, contained photoelectric detectors sensitive to optical radiation in region between visible light and microwaves. Rocket and instruments functioned satisfactorily and payload was recovered. (WS Release 69–18)

- Apollo 12 Astronauts Charles Conrad, Jr., Richard F. Gordon, Jr., and Alan L. Bean underwent last major preflight medical examination at KSC and were cleared for Nov. 14 launch. Dr. Charles A. Berry, Director of Medical Research and Operations at MSC, said astronauts were rested and in good spirits and showed no evidence of any infectious disease that might interfere with launch. (UPI, *NYT*, 11/10/69, 10)

- NASA published *A Long-Range Program in Space Astronomy: Position Paper of the Astronomy Missions Board* (NASA SP–213), which Board had submitted to NASA in July. Board of 19 leading U.S. astronomers—chaired by Dr. Leo Goldberg, Harvard College Observatory Director—had been appointed by NASA in autumn 1967 to consult scientific community and propose program for 1970s.

  Astronomy had "far greater potential for advancement by the space program than any other branch of science." Astronomical instruments in space could reach regions of electromagnetic spectrum unable to penetrate earth's atmosphere; thus, "most fundamental problems of astronomy may be brought within range of solution."

  Board provided guidelines for minimum balanced program costing $250 million annually in 1970s and optimum program "proceeding at fastest possible rate consistent with available scientific and technical manpower" and costing $500 million annually during same period. Proposals included "careful assignment of priorities and balanced allocation of resources in order to optimize scientific progress on such problems as the origin of the universe; the course of stellar evolution, including the ultimate destiny of the Sun and solar system; the existence of other planetary systems, some of which may support other forms of intelligent life; and other problems with deep philosophical significance which are of great interest to everyone and are therefore properly supported by public expenditure."

  Board recommended increased effort in x-ray and gamma-ray astronomy using Explorer spacecraft with large payload capability; optical UV astronomy program leading to large space telescope in 1980s; R&D of detectors and small cooling systems for infrared astronomy from spacecraft to complement ground and aircraft observations;

observation of astrophysical objects in longwave radio portion of spectrum; more sophisticated solar spacecraft to extend observations of solar surface and study effects of solar activity on earth; continued observations of planets from earth orbit using OAOs and Small Astronomy Satellite; more advanced observation of interplanetary medium, cosmic rays, and magnetic fields, with astronomy instrumentation on planetary and Explorer spacecraft; and augmented experiments using aircraft, balloons, and suborbital rocket launches. (Text; NASA Release 69-149)

- Cal Tech astronomer Dr. J. B. Oke reported discovery of new kind of galaxy with tiny, brilliant nuclei like quasars at visual edge of universe, billions of light years away. Like quasars, galaxies emitted powerful radio signals but, unlike quasars, they seemed to have spiral arms like Milky Way galaxy. Thus, they might be "missing link in the evolution of the universe." First evidence of new galaxies had been found by $250,000, 32-channel photoelectric spectrometer atop 200-in telescope at Mt. Palomar, Calif. (Dighton, AP, *W Post*, 11/10/69, A1)
- *New York Times* editorial: "Apollo 12 must not be taken for granted. It is another extremely important step in man's exploration of earth's natural satellite, a venture accompanied by great risk yet promising large rewards in additional knowledge. The mere fact that the Apollo 12 astronauts are scheduled to spend almost three times as long walking on the moon as their predecessors and to bring back to earth part of Surveyor 3—which landed in 1967—emphasizes how much more complex and ambitious this mission is." Second moon landing might seem less glamorous than first, "but it requires great bravery and skill from all involved, and, if successful, could move lunar science ahead substantially." (*NYT*, 11/9/69, 14)
- Joint U.S.–U.S.S.R. moon flight was, again, advocated by Jess Gorkin in *Parade*. Magazine had proposed cooperative space venture Jan. 9, 1966. "The idea had international appeal. Residents of Moscow spoke warmly of the concept. Letters of approval poured in from all over the United States as well as Europe and Asia. The proposal was supported by newspapers around the world." NASA had "been sharing the country's space knowledge with other nations for years." U.S. scientists had worked closely with scientists from nations whose scientific payloads were launched aboard U.S. rockets. (*Parade*, 11/9/69)

November 10: The Biomedical Foundations of Manned Space Flight: A Report of the Space Science and Technology Panel of the President's Science Advisory Committee was submitted to President's Science Adviser, Dr. Lee A. DuBridge: "In order to define an appropriate mix of manned and unmanned operations, NASA will need to 'qualify man for space flight' in the broadest sense. That is, NASA should pursue a biomedical program which explores the optimization of man's role in space, the limitations on his effectiveness and means to circumvent those limitations . . . a program to determine the best use of man as a space subsystem in interaction with automated subsystems. An effective program directed to this objective exceeds the present capabilities of NASA and involves resources not yet developed in the biomedical community."

Report criticized NASA for not implementing previous recommendations toward strong biomedical research capability. "Pressures for early and reasonably certain success led to placing primary emphasis initially on validation of the capability to construct, orbit, and recover manned spacecraft, with long-range, scientific goals in the biomedical field set aside as potentially interfering with flight objectives. Scientific requirements were thus deemphasized at the very time when development of basic scientific knowledge and a cadre of scientific talent should have been begun." NASA was now faced "with the question of qualifying man for more complex tasks in space, for longer duration flights, and of evaluating man as an integral part of the spacecraft man-machine system, without sound biomedical foundation." In particular, "decisions concerning development directions for the next generation of manned spacecraft systems for use in earth orbit must be made without an adequate basis of understanding."

NASA should plan overall biomedical program using resources, personnel, and experience of other Government areas, especially DOD, and support multidisciplinary environmental medical laboratories within universities. Report favored international cooperation in space exploration for peaceful purposes, close communications with biomedical community, and broadened participation of biomedical scientists in space flight programs. NASA should affiliate biologically and medically trained astronauts with research components of NASA or qualified institutions. They should be assigned to MSC flight-crew organization only when participating in specific missions. Report recommended that NASA consider separating space biomedical research from organization conducting space missions and from flight medical services. (Text)

- NASA announced Apollo 12 LM guidance-system targeting for Site 7 landing point on moon's Ocean of Storms had been changed to improve crew's ability to observe landing point during late portion of descent. Crew would attempt to land close to *Surveyor III* spacecraft at 2.990 south latitude by 23.204 west longitude. Original aiming point was 2.982 south latitude by 23.392 west longitude—1,118 ft northeast of *Surveyor III*. (NASA Release 69–148B)
- NASA announced resignation of Dr. George E. Mueller, NASA Associate Administrator for Manned Space Flight, effective Dec. 10. No successor was named and Dr. Mueller did not reveal his plans for future. Dr. Thomas O. Paine, NASA Administrator, said: "It is due to Dr. Mueller's creative leadership of the magnificent manned space flight organization that the flight of Apollo 11 . . . achieved the national goal set in May, 1961. . . . We regret that Dr. Mueller has made the decision to return to private life, but recognize that decision comes at a time when the task he accepted is complete and a sound foundation for our future national space program has been established."

Dr. Mueller, who had directed manned space flight program for six years, had received NASA Distinguished Service Medal, achieved early operational flight schedule in Gemini program, formulated concept for thorough and comprehensive ground testing permitting all-up flight testing and early operational availability of Saturn-Apollo hardware, introduced improved techniques that established efficient management of nationwide industrial complex for first lunar landing and return on schedule and within cost, and formulated low-cost space transportation and operations which were basis for space shuttle and space station programs. (NASA Release 69-151)

- Senate considered H.R. 12307, FY 1970 Independent Offices and HUD bill which contained $3.715 billion NASA appropriation. It rejected by 22 to 46 amendment by Sen. William Proxmire (D-Wis.) that would have reduced FY 1970 NASA funds by $100 million. Conference report accompanying bill noted considerable concern about "future of funding for manned lunar programs because of budget constraints . . . and a question of National priorities." Conferees suggested Congress, NASA, and Administration review manned lunar program policy for future "and decide and determine policy" and that Committees on Appropriations be "advised at earliest possible date." (CR, 10/11/69, D1046, S13989, S14010-6, S14025-50, S14053; House Conf Rpt 91-649)

- Aircraft collision was "one of the most important specific difficulties facing the continuing expansion of aeronautical services in the United States," NASA OART said in prepared statement for National Transportation Safety Board [see Nov. 14]. Continuing increase in air passenger and cargo transportation and number of general-aviation aircraft would further increase collision risk. Advent of jumbo jet and SST would increase number of fatalities and value of equipment lost in individual collisions. Best long-term solution was "development of an adequate traffic control system, adequate airports, and adequate airspace utilization techniques." Until achieved, other means for preventing collisions must be developed for commercial and general-aviation aircraft. NASA was developing two approaches for needs of general-aviation. (Text)

- Biggest engineering problem facing 108-yr-old MIT at close of 1960s was one of identity, Eric Wentworth said in *Washington Post*, "namely the increasingly disputed place of war-related research within its academic confines." MIT ranked first among U.S. universities in receipt of DOD funding, which in 1968 accounted for almost half of Institute's $17-million budget. Since March 4 day-long work stoppage by students and young faculty members protesting MIT's heavy military involvement, President Howard W. Johnson and other MIT leaders had moved to reassess MIT's role. They had declared temporary halt on accepting new contracts for classified research at Instrumentation Laboratory and Lincoln Laboratory and established commission to review laboratories' role. Commission had recommended continuing basic DOD research but avoiding advanced development of weaponry, reducing secrecy, and shifting efforts toward meeting domestic and social needs.

  Meanwhile, Instrumentation Laboratory had already "come to the end of an era." Its founder and head, Dr. Charles S. Draper, had been "eased out of his job as director although he may continue in other capacities." His accomplishments with laboratory had included on-board guidance navigation systems "that took Apollo 11 spacemen to the moon last summer." (*W Post*, 11/10/69, 2)

November 11: Senate passed 68 to 1, H.R. 12307, FY 1970 Independent Offices and HUD bill containing $3.715 billion NASA appropriation with $3.019 billion for R&D, $58.2 million for construction of facilities, and $637.4 million for research and program management. (CR, 11/11/69, S14099-108)

- Hughes Aircraft Co. executive team held press conference aboard private yacht near Cape Kennedy, Fla., to announce company President Howard R. Hughes' intention to rename *Surveyor III* "Hughes Automated Lunar Observer" (HALO). Hughes' assistant, Peter Maheuson, said Hughes was understood to have asked NASA to use acronym in all air-to-ground communications. Hughes Aircraft Co. had manufactured *Surveyor III*, which landed on moon April 19, 1967. Pieces of unmanned spacecraft were to be returned by Apollo 12 for examination at LRL. (Lannan, W *Star*, 11/12/69, A1)

- West German ambassadors in Moscow, Washington, D.C., and London signed nuclear nonproliferation treaty sponsored by U.S. and U.S.S.R. (Morgan, *W Post*, 11/29/69, A1)
- Secretary of the Interior Walter J. Hickel announced DOI–NASA agreement to enlarge Merritt Island National Wildlife Refuge within KSC area by 25,830 acres, to total 83,796 acres. Agreement put all but KSC's most intensely used land within wildlife sanctuary established in 1963. Merritt Island's eagle population had been depleted by land development until NASA began acquiring land in area. (DOI Release 20242–69)
- Behind "computerized smoothness" of Apollo 12 countdown, there was "smoldering and sometimes explosive struggle" at KSC among scientists, engineers, and Government officials over how to run space program, John Noble Wilford said in *New York Times*. General uncertainty over future of NASA and of space priorities was fueling "internal feud" which began to be exposed after July *Apollo 11* lunar landing. Successful Apollo 12 flight could intensify controversy by raising "stakes" in struggle. Scientists wanted greater emphasis on scientific experiments; scientist-astronauts were disenchanted at slim prospects of space flights; engineers, especially at MSC, resented scientists' attitude. (*NYT*, 11/11/69, 1)

November 12: U.S.S.R. launched *Cosmos CCCIX* from Plesetsk into orbit with 353-km (219.3-mi) apogee, 191-km (118.1-mi) perigee, 89.9-min period, and 65.4° inclination. Satellite reentered Nov. 20. (GSFC SSR, 11/15/69; 11/30/69; SBD, 11/14/69, 59)

- *Oso VI* Orbiting Solar Observatory (launched Aug. 9) was adjudged successful by NASA. All spacecraft systems had operated satisfactorily and all primary objectives had been met. Spacecraft had been operating at slightly lower than expected temperatures and elevation motor current was higher than on previous flights. As result, elevation servo jitter was less than 2 arc secs, zero to peak, best achieved on any OSO flight. (NASA Proj Off)
- Countdown for Apollo 12 lunar landing mission continued at KSC despite discovery of leak in fuel tank. Officials said launch crews would have ample time to replace tank and check out replacement without delaying liftoff Nov. 14. (O'Toole, *W Post*, 11/13/69, A1)
- Observers on earth would have two opportunities to see Apollo 12 after launch Nov. 14, NASA announced. During first opportunity, 57 min after liftoff, oxygen release from Saturn V 3rd stage (S–IVB) would be visible on east coasts of North and South America and western Europe and Africa. During second opportunity, about 15 hrs after liftoff, spacecraft would be visible during translunar coast through telescopes in U.S., Mexico, Central America, South America, Europe, and western Africa. (NASA Release 69–148C)

- Dr. Kurt H. Debus, KSC Director, announced resignation of Albert F. Siepert, KSC Deputy Director (Management), effective Dec. 1. He would become Program Associate at Univ. of Michigan's Institute for Social Research and project manager for large-scale organizational research study sponsored by General Motors Corp. Siepert had been executive officer at National Institutes of Health before joining NASA and had received HEW's Distinguished Service Award (1955), Arthur S. Flemming Award (1950), NASA Exceptional Service Medal (1968), and NASA Distinguished Service Medal (1969). (KSC Release 464–69)
- Rep. Joseph E. Karth (D-Minn.), Chairman of Subcommittee on Space Science and Applications of House Committee on Science and Astronautics, addressed National Space Club in Washington, D.C.: "During the entire spectrum of our second and third decades in space, manned space flight has not merely been emphasized, but has literally cannibalized any hope for major scientific planetary, interplanetary, communications, meteorological, oceanographic (in cooperation with satellites), earth resources satellite programs, etc.! I must conclude . . . that the only valid justification for early development of the [space] shuttle is to supply the newborn 50–100 crew manned space stations and the manned exploration of Mars."

  From 1958 to 1969 U.S. had launched approximately 750 spacecraft, Karth said: USAF, 380; NASA, 241; USN, 91; and joint USAF–USN, 37. (Text)
- Secretary of Transportation John A. Volpe announced new FAA regulation establishing maximum noise level for new subsonic transport aircraft at 93–108 effective perceived noise decibels—up to 10 epndbs less than those for noisiest aircraft already in service. Regulation would be effective Dec. 1, but aircraft like Boeing 747 with high-bypass-ratio engines for which application for type certificate was made before Jan. 7, 1967, would be granted additional time to comply. (DOT Release 69–124)

November 13: NASA's X–24A lifting-body vehicle, piloted by Maj. Jerauld R. Gentry (USAF), reached mach 0.65 after air-launch from B–52 aircraft at 45,000-ft altitude over South Rogers Lake Bed, Calif. Purpose of unpowered flight, seventh in series, was to obtain stability and control data at 30° upper-flap setting and various rudder settings. (NASA Proj Off)
- President Nixon telephoned Apollo 12 Commander Charles Conrad, Jr., to wish crew good luck on eve of launch. (PD, 11/17/69, 1604)
- President Nixon sent to Senate nomination of Dr. George M. Low as NASA Deputy Administrator. Manager of Apollo spacecraft at MSC since 1964, Dr. Low had served space program since 1949, when he joined NACA. He was recipient of NASA's Outstanding Leadership and Distinguished Service Awards and had received Arthur S. Flemming Award in 1963 as one of the 10 outstanding young men in Government. (PD, 11/17/69, 1597)
- Apollo 12 prelaunch press conferences were held at KSC and MSC. Operations were proceeding as planned toward launch from KSC at 11:22 am EST Nov. 14. Astronauts were in good health and weather was expected to be within minimums for launch. Liquid-hydrogen tank that had leaked Nov. 12 had been replaced, liquid hydrogen and liquid oxygen

had been loaded, and conditions were all go. Launch azimuth had been expanded to 160° to provide 120 min additional time for liftoff and greater opportunity for launch on Nov. 14. (Transcript)

- In message transmitting to Congress NASA's Twentieth Semiannual Report, President Nixon said: "During this decade, we have successfully met many challenges and have achieved significant progress in our ability to utilize space for practical applications, scientific exploration, and expansion of man's frontiers. We have subsequently landed astronauts upon the Moon, explored its surface, and returned these men to Earth. This historic event was made possible because of the solid foundation of a broad range of earlier activities, and through the skill and dedication of the many contributors to our space program." (PD, 11/17/69, 1603; CR, 11/13/69, S14274)

- U.S. and U.S.S.R. space programs had common problem—what to do next in space exploration, Associated Press quoted Cosmonaut Konstantin P. Feoktistov as saying in Moscow. Both countries understood need for new space exploration technology, "and we, as well as the Americans, are facing the choice of a new direction." (AP, W Post, 11/14/69)

- House Appropriations Committee approved full $95.9 million requested by President Nixon for continued SST development. (CR, 11/13/69, D1065; AP, B Sun, 11/14/69)

  At 42nd scientific session of American Heart Assn. in Dallas, Tex., Dutch scientist Dr. A. C. Arntzenius described application of jet propulsion principle—for every movement in one direction, there is an equal opposite movement—in experiments to increase blood pumped by animal heart without increasing heart action. With research team from medical faculty of Rotterdam he had strapped piglets and dogs to "frictionless" table that slid in synchronization with animal heartbeat. Movement made it unnecessary for ventricle to expend energy to accelerate blood as it left heart. When blood started toward animal's head, table was moved in direction of its feet, forcing blood in opposite direction without adding to demands on heart. (UPI, W Star, 11/23/69, A11; Am Heart Assn PIO)

- DOD announced General Dynamics Corp. was awarded $66,850,000 supplement to previously awarded USAF contract for production of F-111 aircraft, bringing total obligations to $2,608,785,766. (DOD Release 982-69)

November 14-24: Apollo 12 (AS-507), second manned lunar landing mission, was successfully flown by NASA. Spacecraft's LM-6, *Intrepid*, made pinpoint landing Nov. 19 on lunar surface near *Surveyor III*, which had landed April 19, 1967. Two astronauts deployed experiments, took photos, and collected samples in two EVA periods on moon before lifting off to dock with orbiting CSM-108, *Yankee Clipper*, and return safely to earth.

November 14-18: Spacecraft carrying Astronauts Charles (Pete) Conrad, Jr. (commander), Richard F. Gordon, Jr. (CM pilot), and Alan L. Bean (LM pilot) was launched from KSC Launch Complex 39, Pad A, by Saturn V booster at 11:22 am EST Nov. 14. Launch was watched by 3,000 invited guests, including President and Mrs. Nixon. Weather conditions at launch were minimal: peak ground winds of 14 knots, light rain showers, broken clouds at 800 ft, and overcast at

November 14: *President Nixon and his daughter Tricia (foreground) watched preparations for the launch of* Apollo 12 *from* KSC, *while Dr. Thomas O. Paine,* NASA *Administrator, shielded Mrs. Nixon from rain. Seconds into flight an electrical discharge shut down the spacecraft fuel cells, but the astronauts corrected the condition.*

10,000 ft with tops at 21,000 ft. During ascent observers on ground saw two parallel streaks of lightning flash between clouds and launch pad. NASA reported electrical transients, later attributed to electrical potential discharges from clouds through spacecraft to ground, had suddenly shut off spacecraft's electrical power at 00:36 GET and turned on numerous alarms in CM. Spacecraft automatically switched to backup battery power while crew restored primary power system.

Commander Conrad radioed, "We had everything in the world drop out." Control Center commented, "We've had a couple of cardiac arrests down here, too." "There wasn't time up here," Conrad answered.

Power system remained normal throughout rest of mission. Spacecraft entered planned parking orbit with 118-mi (189.9-km) apogee and 115.0-mi (185.0-km) perigee.

After postlaunch checkout CSM separated from Saturn V 3rd stage (S–IVB) and LM, code-named *Intrepid*. Onboard TV initiated after separation clearly showed CSM transposing and docking with LM at 3:27 GET and ice on windows from rain frozen during liftoff. Crew ejected CSM/LM from S–IVB and conducted first SPS burn for translunar insertion. Slingshot maneuver placed S–IVB into earth orbit with

515,549.4- to 560,429.9-mi (829,519.0- to 901,731.7-km) apogee and 93,213.2- to 109,324.1-mi (149,980.0- to 175,902.5-km) perigee and period of 39 to 45 days, instead of heliocentric orbit planned, because of error in instrument unit. To ensure that electrical transients during launch had not affected LM systems, Conrad and Bean entered LM at 7:20 GET for housekeeping and systems checks. All checks indicated systems were satisfactory. Translunar insertion maneuver was so accurate that midcourse maneuver No. 1 was not necessary. Midcourse correction No. 2, at 30:53 GET, placed spacecraft on desired hybrid circumlunar trajectory with closest approach of 69.1 mi; third and fourth maneuvers were not necessary. Good-quality TV coverage of preparations for and performance of midcourse maneuver was transmitted for 47 min.

Conrad and Bean began transfer to LM during translunar coast ½ hr earlier than planned to obtain full TV coverage through Goldstone tracking station. The 56-min transmission, beginning at 62:52 GET, showed excellent color pictures of CSM, intravehicular transfer, LM interior, earth, and moon. Gordon reported crew was in good condition. TV broadcast scheduled for 81:30 GET before lunar orbit insertion was canceled because of sun angle and glare on spacecraft windows.

November 18–21: Apollo 12 CSM *and* LM *separated in lunar orbit and the* LM Intrepid *made a pinpoint landing on the moon Nov. 19, 600 feet away from* Surveyor III, *which had landed April 19, 1967. Astronauts Charles Conrad, Jr., and Alan L. Bean deployed experiments, took photos, and collected samples in two* EVA *periods. At left Conrad, photographed by Bean, examined* Surveyor III *in the Ocean of Storms, with* Intrepid *on the horizon. Above, a crewman used hand tools from a tool carrier.*

Spacecraft entered lunar orbit with 194.3-mi (312.6-km) apolune and 72.0-mi (115.9-km) perilune at 83:25 GET (10:47 pm EST Nov. 17) after first SPS burn. During first lunar orbit good quality TV coverage of lunar surface was transmitted for 33 min. Crew provided excellent descriptions of lunar features. Second SPS burn circularized orbit with 76.1-mi (122.5-km) apolune and 62.5-mi (100.6-km) perilune at 87:47 GET. Conrad and Bean transferred to LM for 1½-hr housekeeping, voice and telemetry test, and oxygen-purge-system check and then returned to CM.

November 18–21: Conrad and Bean reentered LM and checked out all systems before firing reaction-control-system thrusters at 107:54 GET to separate CSM and LM. Descent-propulsion-system engine propelled LM to position 4.6–5.8 mi north of expected ground track and error was corrected during powered descent maneuver. LM guidance computer was updated during powered descent to compensate for indications that trajectory was 4,200 ft short of target point. At entry into approach phase trajectory was close to nominal. Crew took over manual control at 370 ft, passed over right side of target crater, and flew to left for landing on moon's Ocean of Storms about 600 ft from *Surveyor III* spacecraft at 111:32 GET (1:55 am EST Nov. 19). Landing coordinates were approximately 3.036° S and 23.418° W. Conrad reported extensive dust obscuring view during final descent. After landing he reported sighting CMS orbiting overhead. Gordon, orbiting moon in

CM *Yankee Clipper*, reported sighting *Surveyor III* and *Intrepid* on moon.

Conrad opened LM hatch at 115:11 GET, descended LM ladder, and deployed modularized equipment stowage assembly (MESA) containing camera which recorded his descent to lunar surface. Conrad reported seeing *Surveyor III* spacecraft and said LM had landed 25 ft downrange from lip of crater. Conrad, inches shorter than Neil A. Armstrong, who had stepped onto moon July 21, had difficulty taking last step from ladder. When he touched lunar surface at 115.22 GET (6:44 am EST Nov. 19) Conrad said, "Whoopee! Man, that may have been a small step for Neil, but that's a long one for me." He said lunar surface was soft and loosely packed, causing his boots to dig in as he walked, and sun was bright like a spotlight. LM had landed so gently its shock-absorbing legs were barely telescoped by gentle impact.

Conrad recorded Bean's descent to lunar surface at 115:52 GET (7:14 am EST Nov. 19). Shortly after color TV camera was removed from MESA bracket, transmission was lost and was not regained for remainder of EVA. Crew collected 40- to 50-lb contingency sample and reported mounds resembling volcanoes. Mounds were 4 ft high with flattened tops 5 ft wide and sides sloping out to 15- to 20-ft dia. Conrad said he saw neither breccia nor vesicular rocks, only basalts. Lithium hydroxide canisters and contingency sample were transferred to LM; S-band antenna, solar wind composition experiment, and American flag were deployed as planned. ALSEP with SNAP-27 atomic generator was deployed 600-700 ft from LM. Shortly after deployment passive seismometer transmitted to earth signals from astronauts' footsteps as they returned to LM. During deployment astronauts kicked up dust and some adhered to instruments. Overall effect would be determined through long-term measurements of system's engineering parameters. Conrad and Bean dusted each other off and entered *Intrepid* after 3 hrs 56 min walking on lunar surface.

After resting inside LM and checking plans for second EVA period, astronauts left LM at 131:33 GET (10:55 pm EST Nov. 19), 1 hr 40 min ahead of schedule. Astronauts stored LM TV camera in equipment transfer bag for return to earth for failure analysis. Conrad walked to ALSEP site to check leveling of lunar atmosphere detector. Astronaut movement on surface was recorded on passive seismometer and lunar surface magnetometer. EVA traverse took crew to ALSEP deployment site, Head Crater, Bench Crater, Sharp Crater, Halo Crater, *Surveyor III* site, Block Crater, and back to LM. Astronauts walked 1,500-2,000 ft from LM, covering about 6,000-ft distance, their confidence and speed increasing with experience. After walk Conrad reported he had fallen once but Bean had picked him up without difficulty. Conrad rolled grapefruit-sized rock down wall of Head Crater about 300-400 ft from passive seismometer, but no significant response was detected.

Crew obtained desired photographic panoramas, core samples, trench sample, lunar environment sample, and assorted rock, dirt, bedrock, and molten samples. They reported fine dust buildup on all sides of larger rocks and said soil color lightened as depth increased. Crew reported Surveyor footpad marks were still visible and entire spacecraft looked brown, as if something had rained on it. Glass parts were not broken. Crew retrieved parts of *Surveyor III,* including TV camera and soil scoop. They then retrieved solar wind composition experiment and stowed it in equipment bag, took stereo pictures near LM, and transferred all collected samples, parts, and equipment to LM. They dusted off, reentered LM, and jettisoned equipment according to plan, closing hatch by 134:82 GET (2:44 am EST Nov. 20), after 3 hrs 49 min walking on lunar surface in second EVA period.

While LM was on moon Gordon, orbiting moon in CSM, completed lunar multispectral photography experiment and photographed Wall of Theophilus and future landing sites, Fra Mauro and Descartes. Film would be analyzed to aid scientists in planning for future sample collection and in extrapolating known compositions from returned samples to parts of moon that would not be visited.

At 142:04 GET (9:26 am EST Nov. 20) LM successfully lifted off moon after 31 hrs 31 min on lunar surface, leaving LM descent stage. Astronauts brought back 95 lbs of lunar surface samples, parts of *Surveyor III,* films, and miniature flags of 136 nations, 50 U.S. states, and 4 U.S. possessions. A 1.2-sec overburn of LM ascent propulsion system caused by incorrect manual switching sequence prevented automatic shutdown of engine and resulted in insertion velocity 32 fps greater than planned, placing LM in orbit with 71.4-mi (114.9-km) apolune and 10.6-mi (17.1-km) perilune. Crew quickly recognized discrepancy, manually shut down engine, and used RSC trim maneuver to enter planned orbit with 53.3-mi (85.8-km) apolune and 10.1-mi (16.3-km) perilune.

Rendezvous maneuvers occurred as planned and LM docked with CSM at 145:36 GET (12:58 pm EST Nov. 20). TV was transmitted from CSM for last 24 min of rendezvous sequence. Crew transferred with samples, equipment, and film to CSM; LM ascent stage was jettisoned and intentionally crashed onto lunar surface at 5:17 pm EST Nov. 20 about 44.9 mi (72.2 km) southeast of *Surveyor III.* Crash was detected by seismometer left on moon and produced reverberations lasting for more than 30 min. Crew made CSM plane-change maneuver at 159:05 GET and took high-resolution and stereo-strip photos and tracked landmarks, to conclude planned photography of Fra Mauro, Descartes, and Lalande.

November 21–24: Crew fired SPS engine at 172:27 GET (3:49 pm EST Nov. 21), injecting CSM into transearth trajectory after 89 hrs 2 min (44 revolutions) in lunar orbit. Midcourse correction No. 5, at 188:28 GET, was so accurate that sixth correction was not necessary. TV transmission of receding moon and spacecraft interior was transmitted for 38 min, beginning at 192:27 GET. Good-quality transmission of question-and-answer period with scientists and press was conducted for 37 min, beginning at 224:07 GET. Final midcourse correction— No. 7 at 241:24 GET—resulted in predicted entry velocity of 36,116 fps and flight path angle of $-6.47°$. CM *Yankee Clipper* separated from SM at 244:07 GET. Parachute deployment and other reentry events occurred as planned. *Yankee Clipper* splashed down in mid-Pacific at 244:36 GET (3:58 pm EST Nov. 24), 4.03 mi from recovery ship U.S.S. *Hornet.* Astronauts, wearing flight suits and masks, were carried by helicopter from CM to recovery ship, where they entered mobile quarantine facility (MQF) with recovery physician and technician. Crew, physician, and technician remained inside MQF until it was delivered to Lunar Receiving Laboratory in Houston.

CM was retrieved and mated to MQF transfer tunnel on board recovery ship. From inside MQF/CM containment envelope, MQF engineer removed lunar samples and equipment through decontamination lock and CM was sealed until delivery to LRL. Sample return containers, film, and other data were flown to Pago Pago and to Houston for transport to LRL.

Primary *Apollo 12* mission objectives—to perform selenological inspection, survey, and sampling of mare area, deploy and activate ALSEP, develop techniques for point landing capability, develop man's capability to work in lunar environment, and obtain photos of candidate exploration sites—were achieved. All launch vehicle and spacecraft systems performed according to plan, with only minor and corrected discrepancies. Flight crew performance was outstanding. All three crew members remained in excellent health and good spirits. Accomplishments included first use of hybrid trajectory, largest U.S. payload placed into lunar orbit (72,335 lbs after lunar orbit insertion), first demonstration of point landing capability, first use of two EVA periods, first recharge of portable life-support system, first double-core-tube sample, first return of samples from vehicle previously landed on moon, longest distance traversed on lunar surface, first multispectral photography from lunar orbit, and longest lunar mission to date.

Apollo 12 was ninth Apollo mission to date, sixth manned Apollo mission, and second manned lunar landing mission. *Apollo 11* (July 16–24, 1969) had proved capability to perform manned landing on moon and return and to retrieve lunar samples for study on earth. Apollo program was directed by NASA Office of Manned Space Flight; MSC was responsible for Apollo spacecraft development, MSFC for Saturn V launch vehicle, and KSC for launch operations. Tracking and data acquisition was managed by GSFC under overall direction of NASA Office of Tracking and Data Acquisition. (NASA Proj Off; Mission Commentary; NASA Release 69–148; *FonF*; NYT, 11/15–26/69; *W Post*, 11/15–26/69; *W Star*, 11/15–26/69; *B Sun*, 11/15–26/69)

November 14: President Nixon became first President to witness space launch while in office when he, Mrs. Nixon, and daughter Tricia viewed *Apollo 12* launch at KSC. President called liftoff "spectacular." Later he addressed NASA personnel in Launch Control: ". . . I know there has been a lot of discussion as to what the future of the space program is. . . . I do think you can be assured that in Dr. Paine and his colleagues you have men who are dedicated to this program, who are making the case for it . . . as against other national priorities and making it very effectively. I leaned in the direction of the program before. After hearing what they have had to say with regard to our future plans, I must say that I lean even more in that direction. . . . I realize that in those within the program . . . there are different attitudes as to what the emphasis should be, whether we should emphasize more exploration or more in taking the knowledge we have already acquired and making practical applications of it. . . . We want to have a balanced program, but most important, we are going forward. America . . . is first in space." (Carroll, *B Sun*, 11/15/69, A1; *PD*, 11/17/69, 1601–2)

- Tass carried *Apollo 12* report shortly before launch. Report described mission objectives and ended: "We wish a successful flight and a safe return to the courageous team of the American spaceship Apollo 12." (Reuters, *W Post*, 11/15/69, A6)
- While there was some "feeling of dejavu" at Cocoa Beach, near *Apollo 12* launch site, some residents, especially local business men, believed interest in *Apollo 12* was "just as keen as in any other launching," *New York Times* said. Manager Hal Saunders of Cape Kennedy Hilton Hotel had said, while pace seemed calmer and more organized, people were no less excited than before. He felt there were more tourists at Cape Kennedy this week than in July bcause "fewer were scared by press reports . . . that all facilities would be jammed." Many were seeing launch for first time "and it is still a thoroughly thrilling experience." (Blakeslee, *NYT*, 11/14/69, 33)
- Liverworts grown in lunar soil at LRL had reached three-inch height and were "green and happy," *Washington Post* said, while controls, grown in earth soil, were "puny—not even an inch high—and pale." Difference in plants, commonly found on rocks and in forests and among first to grow in volcanic areas, had been considered small when first noted in August. However, LRL Chief of Preventive Medicine, Dr. William Kemmerer, now called difference "one of the really unexpected results of the lunar mission." Plant tests had been conducted to see if lunar material caused harm to earth plants. Control plants had been grown in desert soil. Tobacco cells and soy bean cells in culture with lunar soil had done better than controls in earth soil; ferns were growing "more spectacularly" than some grown in earth soil. Man's knowledge of trace mineral requirements in plants and animals was "very, very limited," Dr. Kemmerer had said. "I don't think we'll find something going on in lunar nutrition that we don't find on the earth. I do think important questions have been raised." (Cohn, *W Post*, 11/14/69)
- Control of *Azur*, West German research satellite launched by NASA Nov. 7, was phased over to German control center. Turn-on and checkout of all subsystems and instruments had been completed within six days after launch. All housekeeping and scientific data values were well within expected ranges, but spurious commands had been occurring intermittently; program of protective commands had been developed to keep spacecraft in proper mode of operation. (NASA Proj Off)
- NASA and DOD announced signing of agreement between NASA and U.S. Army Materiel Command to expand national capability in low-speed aviation technology through joint use of NASA test chambers, wind tunnels, and other facilities. Agreement expanded effort at ARC and established similar arrangement at LaRC and LeRC. (NASA Release 69–150)
- LeRC announced award of $2.5-million letter contract to United Aircraft Corp.'s Pratt & Whitney Div. to build and test 18 RL–10 A3–3 rocket engines. Engine, developed in 1958 and first liquid-hydrogen and liquid-oxygen engine to be flown in space, would be used on Centaur high-energy upper-stage rockets in early 1970s. (LeRC Release 69–68)
- U.S. astronauts and Soviet cosmonauts, in joint space efforts, could succeed in improving U.S.-Soviet relations where "tired old politicians" had failed, Sen. Barry M. Goldwater (R-Ariz.) said during visit to

November 14

MSC. "Maybe this would be a way to break through to the rulers of Russia. When you get professional people together—scientists, aviators . . . you find you have many common problems." Sen. Goldwater said he hoped *Apollo 8* Astronaut Frank Borman would not seek Senate seat in 1970 but would wait awhile "and take mine." (UPI, *NYT*, 11/16/69, 79)

- USAF announced decision not to fund C–5A procurement beyond FY 1970 program of 23 aircraft. Budget constraints had reduced program requirements from 120 to 81 aircraft. (DOD Release 998–69)
- National Aeronautic Assn. conferred title "Elder Statesman of Aviation" on Adm. Joseph J. Clark (USN, Ret.), Katherine Stinson Oteroo, and William A. Patterson, retired President of United Air Lines. (*W Star*, 11/15/69, A3)

November 15: U.S.S.R. launched *Cosmos CCCX* into orbit with 332-km (206.3-mi) apogee, 203-km (126.1-mi) perigee, 89.8-min period, and 64.9° inclination. Satellite reentered Nov. 23. (GSFC *SSR*, 11/15/69; 11/30/69)

- *Washington Post* editorial on *Apollo 12* launch: "Through the peculiarities of our time, we are all about to become participants, albeit vicariously, in the kind of research and exploration that has been carried out previously by lonely scientists almost in isolation. Radio and television make it possible for us to watch and to hear two explorers as they go about the somewhat mundane business of making and recording observations, picking up rocks, and setting up equipment. As these flights roll on through the next few years, we will watch the slow process of exploration proceed. In a way, it will be like being there when Darwin traveled around the world and made the observations that resulted in his theory of evolution or when Lewis and Clark explored the Northwest. Not too much should be expected from any one trip or any one set of experiments. It is the totality of knowledge, not its fragments, that counts." (*W Post*, 11/15/69, A20)
- Americans had generated only "scant enthusiasm" for *Apollo 12* launch, *New York Times* said. In contrast to *Apollo 11*, July 16, "television sets in bars and business offices drew only small, languid crowds. . . ." Collective sense of anticlimax was "perhaps predictable considering the intense national emotion spent on the first moon landing four months ago." Contributing to indifference might have been "the mounting preoccupation with the war in Vietnam, the peace rallies, the controversy surrounding Vice President Agnew and earthly social problems." However, "there were no reports of the kind of demonstrations that protested the Apollo 11 flight." Marchers in antiwar protests had appeared "to show little interest" in *Apollo 12*. (Reinhold, *NYT*, 11/15/69)
- FAA announced proposal to extend hourly flight quotas in effect at five high-density airports serving New York, Chicago, and Washington, D.C., for 9–12 mos beyond Dec. 31 expiration date. Experience with quota system since June 1 inception indicated it had relieved congestion and reduced delays. (FAA Release 69–125)

November 16: During "Meet the Press" interview shown on NBC TV Dr. Thomas O. Paine, NASA Administrator, said he thought *Apollo 11* spacecraft would have crashed if Astronaut Neil A. Armstrong had not piloted it on final lunar approach. Quick reactions by *Apollo 12* crew had helped save mission from power failure on its launch. "I think that the [*Apollo 12*] launch would have been a success. But we felt a good deal better with the astronauts there." Dr. Paine thought U.S.S.R. would attempt moon trips "in the next few years." (*W Post*, 11/17/69, A3)

- First public display of 21.1-gm *Apollo 11* moon rock at American Museum of Natural History in New York attracted largest crowd in museum's history—42,195 persons. Display would run for 2½ mos as highlight of museum's centennial celebration. (*NYT*, 11/16/69, 66; AP, *W Star*, 11/17/69, A1)
- Merger of U.S. and U.S.S.R. satellite systems into worldwide comsat link for all nations was proposed in report of Sept. 21–25 international conference at Talloires, France, released by sponsors Carnegie Endowment for International Peace and Twentieth Century Fund. In *Communicating by Satellite: An International Discussion*, task force appointed to consider new rules and regulations for international satellite communications said: "Obstacles undoubtedly exist to achieving the goal of an integrated, global system, but technical compatibility between the two major satellite communications systems—Intelsat and Intersputnik (through the Soviet Orbita)—is not difficult to obtain. Their orbital systems are complementary; their frequency plans can be coordinated; a single ground station can operate in either system, and their transmitting and receiving equipment can be adapted for operation in both systems."

  All nations should have access to global, integrated comsat system with "willingness and ability to accept certain technical and administrative requirements" only relevant consideration. No political conditions should be applied to membership. Voting rules should "take account of the special contributions" of countries like U.S. and U.S.S.R. "while recognizing the interests of virtually all nations in the basic rules of operation of the system." No nation should be allowed to broadcast sound and TV into territory of other nations without their consent. (Text; AP, *W Star*, 11/17/69, A12)
- Space race was underway in summer of 1945, as U.S., U.K., and U.S.S.R. searched for key men who had effected Germany's "staggering lead in rocketry," David Owen of *London Daily Telegraph* said in *Washington Post*. Article traced escape of Wernher von Braun and Peenemünde rocket scientists from defeated German Gen. Hans Kammler who had ordered them shot rather than have them captured by Allies; hiding of rocket drawings and records in remote Harz Mountain cave; voluntary surrender of scientists to U.S.; and withdrawal of scientists, documents, and rockets by U.S. before U.K. and U.S.S.R. could capture any but small part of German rocketry effort. (*W Post*, 11/16/69, B3)

November 17: NASA's HL–10 lifting-body vehicle, piloted by NASA test pilot William H. Dana, reached 66,000-ft altitude and mach 1.6 in powered flight after air-launch from B–52 aircraft at 45,000-ft altitude west of Rosamond, Calif. Objectives of flight, 29th in series, were to obtain stability and control data and airspeed calibration. (NASA Proj Off)

- Senate and House conferees on H.R. 12307, FY 1970 Independent Offices and HUD appropriations bill, agreed to $3.006-billion NASA R&D appropriation instead of $3 billion proposed by House and $3.019 billion proposed by Senate. Appropriation for construction of facilities was $53.2 million as proposed by House rather than $58.2 million as proposed by Senate, and research and program management appropriation was accepted at Senate's $637.4 million rather than $643.7 million proposed by House. Conferees also agreed on $440-million FY 1970 appropriation for NSF. (Conference Rpt 91–649; *CR*, 11/17/69, D1078)

- Space program could provide tools and knowledge to help eliminate air pollution, NASC Executive Secretary William A. Anders, former *Apollo 8* astronaut, said in speech before Governors' Conference on California's Changing Environment in Los Angeles. With remote sensors in aircraft "pilot can quickly and accurately map pollution levels over a wide area and range of altitudes." Pollution map of entire U.S. could be generated each day. "Thus, aircraft and satellites with these remote sensors offer the unique advantage of being able to view large areas through new eyeballs very quickly and selectively. Further, by computer, the data can be reduced to formats that can be quickly and easily understood." Aircraft and satellite surveys would provide "data which supplement surface techniques and which, in some cases, can't be gathered in any other way." Rather than fantastically expensive, satellite measurements could "be competitive with surface systems." (Text)

- Soviet space program had been severely set back by "catastrophic explosion of 10-million-lb-thrust prototype booster during preparatory launch operations at Tyuratam last summer," *Aviation Week & Space Technology* reported. U.S.S.R.'s manned orbiting platform (MOP) program, "already at least a year behind schedule and proceeding slowly, has been further retarded because of the . . . failure. Last month's triple Soyuz launch has emerged as a prime example of rescheduling necessitated by the booster's absence and by the aftermath of short-term and conflicting political decisions. . . .

  "Failure of the booster prototype was only one event in a largely chaotic year for the Soviet space program. During 1969 the Russians: shifted much of their large booster inventory to military purposes in connection with the Chinese border crisis"; postponed scheduled late spring launch of *Soyuz VI* "because all remaining non-military facilities were in turn preempted by a highly compressed lunar effort"; and hastily launched *Luna XV* "in an effort to prove, in event of an Apollo failure, that at least a one-way soft landing from lunar orbit could be performed unmanned and, alternately, that failure would not result in loss of human life." (*Av Wk*, 11/17/69, 26–7)

- SST economics were discussed by FAA Administrator John H. Shaffer before Long Island Assn. of Commerce and Industry in Manhasset, N.Y. Study of simulated SST operation in 1980s, with comparison of SST and subsonic jets based on total operating cost rather than direct operating cost and using 1969 values, had shown "SST beats the 707 substantially and comes much nearer to equaling the impressively low 747 costs." Elements of ground support and overhead costs gained advantage of SST's greater productivity in seat-miles per hour. "The 2707–300 [SST] is two thirds as big as the 747 and it flies three times as fast, so it will do twice as much as the 747 (and 4½ times the 707 or DC–8) in the same time period." By 1978, SST introduction date, aircraft's total operating cost "comes within one-tenth of a cent per

seat-mile of matching the 440-seat economy version of the 747." (Text)
- Univ. of California at Berkeley astronomer Dr. David Cudaback reported that observations indicated dust clouds in sky might contain great quantity of diamond grains, each few thousandths of inch in diameter. (AP, W *Star*, 11/18/69, A10)

November 18: President Nixon signed H.R. 11271 (P.L. 91–119), $3.715-billion NASA FY 1970 Authorization Act. It provided for R&D: Apollo, $1.691 billion; space flight operations, $225.6 million; advanced missions, $2.5 million; physics and astronomy, $117.6 million; lunar and planetary exploration, $138.8 million; and bioscience, $20.4 million.

Space applications R&D authorization was $128.4 million. Also authorized were $112.6 million for launch vehicle procurement, $9 million for sustaining university program, $27.5 for space vehicle systems, $33.5 for electronic systems, and $22.1 million for human factor systems.

Act allocated $20.2 million for basic research, $36.9 million for space power and electric propulsion systems, $50 million for nuclear rocket, $22.8 million for chemical propulsion systems, $278 million for tracking and data acquisition, and $5 million for technology utilization.

For construction of facilities, law authorized $8 million for ERC, $670,000 for GSFC, $12.5 million for KSC, $4.7 million for LaRC, $1.7 million for MSC, and $500,000 for Wallops Station. Act authorized $637.4 million for research and program management. (*PD*, 11/24/69, 1643; PL 91–119)

- Senate and House adopted conference report on H.R. 12307, Independent Offices and HUD appropriations bill including FY 1970 NASA funding of $3.697 billion and $440 million for NSF. Bill was forwarded to President Nixon for signature. (*CR*, 11/18/69, S14574–9, H10981–7)
- House by vote of 362 to 25, passed and sent to Senate H.R. 14794, DOT FY appropriations bill containing $96 million for SST development program. (*CR*, 11/18/69, H10990–1034)
- President Nixon approved S. 1857, $480-million National Science Foundation Authorization Act, 1970 (P.L. 91–121). (*PD*, 11/24/69, 1643)
- Sen. William Proxmire (D-Wis.) introduced, for himself and cosponsors, S.J.R. 285, authorizing Senate Foreign Relations Committee to undertake comprehensive study of all possibilities for international cooperation in space exploration. (*CR*, 11/18/69, S14593–4)
- President Nixon announced membership of Task Force on Air Pollution, with Arie Jan Haagen-Smit, Chairman of California Air Pollution Board, as chairman. It would evaluate effectiveness of efforts to curtail air pollution and recommend further actions. (*PD*, 11/24/69, 1624)
- *New York Times* editorial asked: "Why should lunar research be limited to what this country can afford? Even if the Soviet Union is unwilling to cooperate, there are very substantial human and material resources in Western Europe, Japan and other areas that could be mobilized for the task of lunar exploration and settlement that lies ahead. President Nixon could demonstrate high statesmanship by offering to turn the National Aeronautics and Space Administration into the International Aeronautics and Space Administration if others will join to help carry the burdens of the effort and provide additional talents for the

job. And if NASA became IASA, even Moscow—after its recent space disappointments—might see advantages in joining the common effort to make the moon a lever for uniting mankind." (*NYT*, 11/18/69)

- AEC released report of investigation into May 11 fire at AEC plutonium-handling facility at Rocky Flats, Colo. Estimated damage to buildings and equipment was $45 million excluding cost of plutonium recovery. Fire had originated in plutonium storage cabinet; cause was unknown. (AEC Release M-257)

November 18–19: LeRC's work in development of longer lasting, high-temperature-enduring, corrosion-resistant materials was described at conference on Research in Aerospace Materials at Center. Conference was attended by 400 scientists and engineers from universities, industry, and Government. New alloying concept had yielded "WAZ–20," alloy of nickel, tungsten, aluminum, zirconium, and carbon with melting point 150° higher than conventional cast-nickel-base alloys and higher strength at 2,200°F, for possible use in 1st-stage turbine stator vanes. Also reported were use of prealloyed powders to improve properties of nickel-base alloys; dispersion-strengthened nickel-base alloys and solid-state welding of these materials; developments in chromium, molybdenum, and tungsten alloys and fiber composites; and experimental techniques developed to predict properties of materials after 10–20 yrs of use from tests lasting less than year. (LeRC Release 69–70; LeRC PIO)

November 19: In international reaction to *Apollo 12* moon walk Moscow Radio announced landing eight minutes after touchdown, followed with brief progress bulletins, and broadcast seven-minute film on event seven hours after LM landed on moon. Foreigners in Moscow said BBC broadcast was jammed by two stronger stations just before lunar touchdown.

Pope Paul VI at Vatican watched astronauts on TV, then knelt for short prayer. He sent message to President Nixon: "Many, many congratulations to you and the American nation on yet another magnificent step for the human race."

Sir Bernard Lovell, Director of U.K.'s Jodrell Bank Experimental Station, saluted "precision and accuracy" of *Apollo 12* lunar landing. He said, "Until the Russians achieve a rocket with the thrust of the order of Saturn 5, or greater, it is unlikely they will be able to stage any manned flight comparable with Apollo." London's morning newspapers put *Apollo 12* story well down on front pages in contrast to *Apollo 11* banner headlines.

In Helsinki, Finland, Soviet diplomats attending strategic arms limitation talks toasted *Apollo 12* crew.

Tokyo TV dealers expressed unhappiness because repeat of run on color sets after *Apollo 11* landing failed to materialize.

West German Foreign Minister Walter Scheel said *Apollo 12* was an invitation to Europe to organize its own space potential.

Egyptian newspapers gave *Apollo 12* second billing to continuing conflict with Israel. In Poland millions watched lunar landing by direct TV transmission. In Sweden TV reception was poor and viewer interest lagged.

Former Argentine President Arturo U. Illia, in Buenos Aires, called lunar landing "a victory for democracy."

Australians watched landing 10 secs before U.S. because of time lag. In France many viewers were unable to watch landing because of electric utility strikes. Moon walk was one of few programs shown on Italian TV during day of general strike.

Shah of Iran sent congratulations to President Nixon and *Apollo 12* crew for "untiring endeavor." Landing was shown live on Iranian TV.

West Germany's Wickert Public Opinion Institute estimated 82% of German adults watched TV shots of lunar landing, 92% watched *Apollo 11*. It concluded *Apollo 12* lacked suspense that attended first moon shot. (AP, B *Sun*, 11/20/69, A1; Robinson, *NYT*, 11/20/69, 30; *W Post*, 11/20/69, A16; Spencer, W *Star*, 11/21/69, A6)

- Reaction to *Apollo 12* lunar landing was "almost a ho-hum," Associated Press reported. "It seemed to indicate they are now taking moon voyages for granted."

  At White House, President and Mrs. Nixon arose before dawn to watch moon walk on TV. President had commented it was "first time anybody has sung from the moon," as Astronaut Charles Conrad, Jr., hummed while conducting lunar tasks.

  In Philadelphia's Franklin Institute, where thousands had gathered in July to watch *Apollo 11* lunar landing, only 45 persons showed up to watch *Apollo 12*.

  Of 12 New Yorkers interviewed by *New York Times*, only Bernard Granite of the Bronx had watched lunar landing. He said, "The first one was more exciting. This one was anticlimactic, but I still think they are justified in spending the money."

  Portland, Ore., secretary Jeanne Paulson said, "I'm at the point where I think the money should go to the poverty program. There are too many starving people in the slum ghettos."

  Des Moines, Iowa, secretary Eileen Brown said, "The biggest concern seemed to be that they couldn't make the TV work rather than whether they could perform their assignments. What do we think they are up there for, anyway, to put a show on for us?"

  Police in Los Angeles, Calif., attributed sharp drop in reported crime to *Apollo 12* moon landing and walk.

  In Washington, D.C., McKinley High School teacher said, "This doesn't concern us at all. There are far more important things going on here." Question for his pupils was "whether we will survive to the year 2000." (AP, B *Sun*, 11/20/69, A1; Robinson, *NYT*, 11/20/69, 31; *W Post*, 11/20/69, A16; Spencer, W *Star*, 11/21/69, A6)

- NASA Launch Vehicle Review Board recommended resumption of launch operations for spacecraft using Delta booster. In interim report board said recommendation was based on findings of Failure Review Board which had investigated failures of Delta 71 (July 25) and Delta 73 (Aug. 27). Delta 71, carrying *Intelsat-III F-5*, had failed because of motor case rupture or nozzle failure; Delta 73, carrying Pioneer E, had failed because of vibrating relief valve which caused hydraulic oil leak. Failure Review Board suggested additional internal insulation, pressure tests, equipment x-rays, installation of specially tested and selected valves, and new acceptance tests of hydraulic system. Next mission scheduled with Delta was U.K.'s Skynet comsat Nov. 21. (NASA Release 69-152)

- NASA announced selection of Aerojet-General Corp. to receive $5-million,

November 19

cost-plus-award-fee contract for development of advanced optical communications experiment—first laser communications system to be used on satellite. Lasers provided extremely wideband communications function and would be able to transmit hundreds of TV channels around the world, greatly increasing microwave capability. Experiment would be placed on board ATS-F, scheduled for launch in 1972, and could be expanded to include spacecraft-to-spacecraft communications between ATS-F and ATS-G, to be launched in 1974. (NASA Release 69-153)

- Delegation of 10 Soviet scientists visited Brookhaven National Laboratory at Upton, N.Y. During luncheon they peered through microscope at 12-gm piece of lunar matter brought back by *Apollo 11* for study at laboratory's nuclear research facilities. Scientists were on two-week tour arranged through memorandum of cooperation on peaceful uses of atomic energy. Group of U.S. scientists would pay reciprocal visit to U.S.S.R. early in 1970. (Kaufman, *NYT*, 11/20/69, 45)
- DOD announced Martin Marietta Corp. would receive $2,313,740 supplemental agreement to previously awarded USAF contract for design, development, fabrication, and delivery of Titan III booster. Contract would be managed by Space and Missile Systems Organization. (DOD Release 1003-69)

November 20: U.S. newspaper editorials commented on *Apollo 12* landing on moon:

Washington Post: "The sheer joy of these two astronauts . . . shines out over everything else. They are obviously moon-struck. Conrad sounded like the most eager and happiest young geology student in the universe as he hummed and laughed his way from rock to rock, grabbing one after the other and complaining that he couldn't get enough. And who would have thought that one of these highly skilled men would resort, as Bean did, to bonking the television camera with a hammer in an effort to make it work? Unfortunately, the failure of the camera did deprive us of seeing this frolic. . . . There was much to be learned from it about the moon and about the men who combine such cool technical competence with such zest for a serious and dangerous job." (*W Post*, 11/20/69)

Detroit News: "To travel 230,000 miles and hit the target on the button is almost unbelievable except that NASA, its crews and its machines are making believers of us all. But there's a lesson to be taken to heart in the camera's freak failure. Conrad and Bean tried to remedy the defect. . . . But they had to drop that to get on with more essential scientific exploratory experiments. Conrad's oxygen limit was four hours. He hadn't time to spare. . . . that should be warning to those euphoric advocates of setting a target date for a trip to Mars. An astronaut limited to four hours work on the moon is a reminder of the vast project ahead in confounding the elements in this untapped lunar mystery. Prophecies of colonizing the moon should be set against that four-hour limit, even though we no doubt will improve on it." (*D News*, 11/20/69)

Atlanta Constitution: "This second landing, more than 900 miles from where the pioneers of Apollo 11 put down last July, notably extends our knowledge. New landing techniques, a longer lunar stay, nuclear-powered experiments—they all serve to push back a fantastic frontier which only a few short years ago was regarded as a most implausible province of man." (*Atlanta Constitution*, 11/20/69)

Houstin Chronicle: "Our admiration for the coolness and the skill of astronauts Conrad, Bean and Richard F. Gordon, Jr.—the latter still in lonely orbit around the Moon—is boundless. The astronauts continue to perform, one mission after the other, in story-book perfection. We marvel, too, at the ability of the technicians on the ground to quickly respond to unexpected eventualities, like the electrical difficulty just after the launch, and to enable the mission to proceed unimpaired. These achievements inspire all men. They lift our sights for the future." (*H Chron*, 11/20/69, 2 Sec, 5)

Birmingham News: "The flight has been so predictably on schedule that its very success is certain to help push moon travel back into the ho-hum recesses of blasé mankind's mind, as each success in the Mercury and Gemini and early Apollo series made earth orbital spaceflight seem routine." (*B News*, 11/20/69)

- *Apollo 8* Astronaut Frank Borman received gold medal for "distinguished service to humanity" from National Institute of Social Sciences in New York. Institute President Frank Pace, Jr., said, "The great scientific adventure in which you extend the reach of man cannot equal for us the moments in which you expanded the human spirit." Borman replied, "The awesome power of today's technology undoubtedly terrifies many social scientists. But far from subjugating man, I think this technology, if properly applied, is the only chance to preserve the dignity of human life." (AP, *W Star*, 11/21/69, A2)

- Second largest lunar sample shown to public to date—67.7-gm moon rock—went on display at USIA "Education—U.S.A." exhibit at Sekoniki Park in Moscow. One of 14 moon rocks thus far released by NASA for publicity tours, rock later would travel with USIA exhibit to Tashkent, Baku, and Novosibirsk, where it would become first U.S. exhibit ever shown in Siberia. Other moon rocks had been scheduled for showing at 30 U.S. museums. Largest lunar sample—Smithsonian Institution's 478.8-gm rock—was expected to be only permanent display. NASA also had approved display of samples entrusted to U.S. scientists who wished to show them in their home towns. (Lardner, *W Post*, 11/24/69, A1; USIA PIO)

- In NASA-funded project U.S. Bureau of Mines was seeking way to get water and air from lunar rocks, build underground lunar shelters like mines, weld and melt lunar materials, and mine planets like Mars and Venus which might bear substances more valuable than those on moon, Associated Press said. First lunar mining would occur during Apollo 13 mission in March. Astronaut would bore 10-ft hole in lunar crust to determine what lay beneath and to test drill designed for NASA by Martin Corp. Scientists at U.S. Bureau of Mines research center in Bruceton, Pa., and at six other locations had been studying since 1965 possible use of lunar materials to build and support manned lunar station. Project Director Thomas C. H. Hutchinson had said, "Even if we found pure platinum on the moon, it would cost too much to bring it back." Goal was to establish manufacturing processes on moon to speed exploration of planets. (AP, *W Post*, 11/20/69, A78)

- Cosmonaut Konstantin P. Feoktistov said in *Pravda:* "The Americans

have not been entirely rational in adapting aviation control panels to the needs for space flight. The control system of the Soyuz ship, for example, seems to me to be simpler, more logical and therefore more perfect." Feoktistov, who toured U.S. Oct. 20 to Nov. 4, congratulated his new acquaintances in U.S. on *Apollo 12* lunar landing and wished them "complete success." (AP, W *Star*, 11/20/69, A6)
- Moon "loomed large in Chinese tradition," but 700 million people of Communist China had not been told of *Apollo 12* lunar landing, *Christian Science Monitor* said. Though "elite hierarchy in Peking . . . learned of *Intrepid*'s touchdown as fast as the Soviets," story of *Apollo 12* had not passed beyond select group of leaders. "Indeed the Chinese people have yet to be told by their government-controlled press and radio of the flight of Apollo 11. . . ." Some might hear of flight from foreign radio broadcast, but many were jammed by Peking. VOA report to Hong Kong on *Intrepid*'s lunar landing had been "curiously overlaid with martial music from a Chinese Communist radio station on the mainland." (Hughes, *CSM*, 11/20/69)
- *Washington Daily News* reported song called "The Wondrous Telephone," by Thomas P. Westendorf, had been published in 1877 with cover showing group of men talking by telephone to man in the moon. (*W News*, 11/20/69, 52)
- *Apollo 11* CM was being readied at North American Rockwell plant in Downey, Calif., for tour of 50 state capitals and final resting place at Smithsonian Institution in Washington, D.C., Associated Press reported. NASA was sponsoring tour. (*CSM*, 11/20/69)
- North American Rockwell Corp. announced plans to develop for NASA flying lunar excursion experimental platform (FLEEP)—one-man, jet-propelled craft with adjustable hand controls—to transport astronauts across lunar surface. (AP, *NYT*, 11/22/69, 38)
- Use of space technology in "complete and systematic rehabilitation" of Washington, D.C., was proposed by RCA president Robert W. Sarnoff in speech before Fourth Annual Computer Age Conference of National Industrial Conference Board in New York City. "A comprehensive systems effort to revitalize the city and its environs should invoke a nationwide response . . . as broad and enthusiastic as that inspired by the Apollo moon landing." (Text)
- MSFC announced award of $8-million letter contract to McDonnell Douglas Corp. for two sets of structural components for Saturn V 3rd stage (S-IVB) for Saturn V manned orbital workshop. Work would be completed by January 1972. (MSFC Release 69-249)
- Secretary of Transportation John A. Volpe announced award of $279,032 contract to Control Data Corp. Melville Space and Defense Systems Div. for 22-mo study of pilot warning instrument (PWI) systems that would improve pilots' ability to detect other aircraft in flight. (FAA Release 69-126)

November 21-23: U.K.'s *Skynet A* (IDCSP-A) military comsat was successfully launched from ETR at 7:37 pm EST by NASA for USAF and U.K. by Long-Tank Thrust-Augmented Thor-Delta (DSV-3M) booster. Spacecraft entered transfer orbit with 23,045.4-mi (37,080.1-km) apogee, 160.7-mi (258.6-km) perigee, 655.3-min period, and 27.6° inclination. Primary NASA objective was to place spacecraft into synchronous transfer orbit accurate enough for apogee motor to place spacecraft into synchronous equatorial orbit.

On Nov. 23 apogee motor was fired and spacecraft entered circular orbit with 22,216.5-mi (35,746.4-km) apogee, 21,558.3-mi (34,687.3-km) perigee, 1,431-min period, 2.44° inclination, and 1.3° per day eastward orbital drift. Drift rate was increased to 8° per day to ensure arrival on station over Indian Ocean by Dec. 30. All systems except primary communications system had been turned on and were operating satisfactorily.

Skynet A was first of two U.K. military comsats scheduled to be launched over Indian Ocean under DOD–U.K. agreement. USAF managed project for DOD and would reimburse NASA for launch services. U.K. would reimburse USAF. Skynet B would be launched in May 1970. Spacecraft consisted of two concentric cylinders containing apogee motor, solar cells, despun antenna, high-pressure hydrazine stabilization system, and redundant x-band communications system. It was 32 in high, 54 in in dia, and had five-year lifetime.

Initial operation of spacecraft telemetry and command functions were performed from USAF satellite control facility. Control of orbital operations would be transferred to U.K. telemetry command facility after spacecraft reached station. (NASA Proj Off; *SBD*, 11/26/69, 114)

November 21: Apollo 12 Astronaut Richard F. Gordon, Jr., orbiting moon in CSM, set new world record for solo space flight. Gordon piloted CSM alone for 37 hrs 41 min, breaking 30-hr 20-min record set by Astronaut L. Gordon Cooper, Jr., in *Faith 7* May 15–16, 1963. (UPI, *W Star*, 11/21/69, A7)

- NASA's HL–10 lifting-body vehicle piloted by Maj. Peter C. Hoag (USAF) reached 78,000-ft altitude and mach 1.4 in powered flight after air-launch from B–52 aircraft at 45,000-ft altitude west of Rosamond, Calif. Purpose of flight, 30th in series, was to obtain stability and control data at varying stability augmentation gain settings. (NASA Proj Off)

- President Nixon and Prime Minister Eisaku Sato of Japan issued joint statement following White House discussions on international situation. On space, statement said: "The Prime Minister congratulated the President on the successful moon landing of Apollo XII, and expressed the hope for a safe journey back to earth for the astronauts. The President and the Prime Minister agreed that the exploration of space offers great opportunities for expanding cooperation in peaceful scientific projects among all nations. In this connection, the Prime Minister noted with pleasure that the United States and Japan last summer had concluded an agreement on space cooperation. The President and the Prime Minister agreed that the implementation of this unique program is of importance to both countries." (*PD*, 11/24/69, 1633–7)

- Inaccurate signals from controlling gyro mechanism had caused Sept. 17 destruction of AH–56A helicopter inside wind tunnel at ARC, NASA announced. Investigation board, appointed by ARC Director, Dr. Hans Mark, had determined accident did not result from malfunction of rotor control system or of wind-tunnel systems and equipment. Board recommended greater use of computers to analyze rotor systems before test, study of safety systems and hazards and procedures for wind-

November 21

tunnel crews in emergencies, and addition of metal barriers to protect control room and sections near tunnel. It said visual close observation of tests should be minimized. AH-56A was under study for stability and control at USA request. (NASA Release 69-154)

- Total $21.35-billion investment "in the development and demonstration of a national manned lunar landing capability" was detailed by Dr. Thomas O. Paine, NASA Administrator, in letter to Sen. Clinton P. Anderson (D-N. Mex.), Chairman of Senate Committee on Aeronautical and Space Sciences. "At the time of its establishment, the national goal of a manned lunar landing and return in this decade represented the most difficult technological endeavor ever to challenge the American nation. During this period, senior NASA officials in testimony before the Congress estimated the cost . . . at between $20 and $40 billion." Estimate was refined to approximately $19.5 billion in March 1964, with assumption "there would be a timely initiation of a follow-on program which would bear a portion of the relatively fixed cost required to develop and sustain this national capability."

  Estimates then and annual reassessments had reflected total program cost, including cost of initial lunar landing. Lunar objective "was not simply an end in itself but, rather, provided the focus for the effort to attain space supremacy for whatever the national interest required."

  In March 1966 NASA had furnished estimate of $22.718 billion based on assumption that there would not be timely initiation of follow-on program. In April 1969 NASA furnished estimate of $23.877 billion, which would still be valid if negative assumption had materialized. "But with the success of Apollo 11, we have the opportunity to utilize this demonstrated capability in a more meaningful way. By improving payloads and modifying spacecraft to increase lunar surface systems, we can enhance significantly the return of scientific data from both lunar orbit and the lunar surface."

  NASA had met national commitment "at a cost nearer the lower end of the range of estimates" despite "unpredictable substantial inflationary conditions." Actual cost accrued through July 31, 1969, was $21.35 billion, of which $2 billion was value of flight hardware available for future flights. Capital assets included were approximately $2.8 billion and were "of continuing national value."

  Apollo was "triumph in management as well as in technology and engineering which united government, industry and universities in a common peaceful undertaking. At least one-half million people worked on the manned lunar landing program . . . during the eight years from its announcement to its initial success." (Text)

- U.S. radioastronomers had been unable to obtain funds for facilities recommended five years ago and were beginning to fear "that the momentum that has attracted talented researchers from engineering and physics into radio astronomy may soon be lost," Robert W. Holcomb said in *Science*. "Pulsars and the interstellar clouds provide astronomers with the opportunity to extend their ideas about stellar evolution into very early and very late stages, and most of the pertinent research must be done with radio telescopes. These instruments are also required for some of the most important cosmological problems currently being considered." (*Science*, 11/21/69, 984-6)

- At opening day ceremonies of American Bible Week in New York, sponsors, Laymen's Committee of American Bible Society and Catholic Biblical Assn. of America, awarded citation to crew of *Apollo 8* for reading from Genesis as they orbited moon during Dec. 21–27, 1968, mission. (*NYT*, 11/22/69, 40)

- Astronaut promotion policy seemed "certain to discomfit someone after the end of *Apollo 12* no matter whether it's followed or ignored," *Washington Post* said. President Johnson had decreed in 1965 that each military astronaut would get one spot promotion after his first space flight. *Apollo 12* commander Charles (Pete) Conrad, Jr., and CM pilot Richard F. Gordon, Jr., had both received promotions to commander, USN. *Apollo 12* LM Pilot Alan L. Bean, also commander, USN, now rated promotion to captain, but was junior crewman in age, service in grade, and space experience. Bean promotion might "irritate" others but would be fair to Bean. Promotion for all three astronauts would be "unfair to other astronauts who have flown twice or three times and only promoted once." President Nixon might have to make "sticky" decision. (*W Post*, 11/21/69, A17)
- Secretary of Transportation John A. Volpe announced award of $200,000 FAA contract to Univ. of Tennessee at Knoxville for two-year study of methods of reducing sonic boom, to help FAA establish certification standards for new aircraft. (FAA Release 69–127)
- DOD announced General Electric Co. would receive $5,192,730 supplemental agreement to previously awarded USAF contract for R&D of Mark 15 reentry vehicle. Contract would be managed by Space and Missile Systems Organization. (DOD Release 1015–69)

November 22: Weightlessness during extended space trips "reduces the flexibility of men's bones," Soviet scientist V. V. Parin said in interview published in *Komsomolskaya Pravda* in Moscow. "The organism of a cosmonaut dehydrates in weightlessness and calcium leaves the bones." Probable solution would be to create artificial gravity in space, Parin said. (UPI, *W Post*, 11/23/69, A6)

- *Economist* commented on perils of prolonged space voyages: "When astronauts begin to do tours of duty that could run from three months to a year at a time on the moon, or on orbiting earth platforms—still more when they embark on 2½ year trips to Mars—the problems will be much greater. Their bodies may adapt themselves to conditions of weightlessness or low gravity, and to living in atmospheres of rather less density than on earth, and then find it hard to re-adapt back to earth conditions again. If, over time, the changes become irreversible, we could witness the evolution of a new race of space hominoids with whom we could communicate but not cohabit." (*Economist*, 11/22/69)
- *Apollo 12* congratulatory telegrams at MSC suggested "crew members of Apollo 12 will not be lacking their own fan club," *Washington Post* article said. Meanwhile, *Apollo 11* Astronauts Neil A. Armstrong and Edwin E. Aldrin, Jr., and *Apollo 8* crew continued to receive fan mail. Biggest fans of space program appeared to be "little children and supporters of prayers in outer space." (Lardner, *W Post*, 11/22/69, A6)

November 23: ESRO's *Boreas* (*Esro IB*), launched Oct. 1 into lower than planned orbit, reentered atmosphere after 52 days in orbit. During this period spacecraft and all experiments functioned satisfactorily. ESRO accumulated large quantity of scientific data and adjudged mission successful. (GSFC *SSR*, 11/30/69; NASA Proj Off)

- *New York Times* editorial commented on *Apollo 12:* "It is almost incredible that such giant strides could have been made in the few months that separated Apollo 12 from Apollo 11." Navigational capabilities available to Neil Armstrong last July were "so inexact that for days or weeks after his landing there was no certainty even as to just where Eagle had touched down. Intrepid, on the contrary, landed within a few hundred feet of Surveyor 3, the prime target of its crew's planned collection activities. It will take months, perhaps years, to harvest the full scientific gains from Apollo 12. Already, however, it is evident that mankind is still at the stage where the more it learns about the moon, the more mysterious and puzzling that natural satellite appears to be." (*NYT*, 11/23/69, 12)
- Nonspace nations were expressing growing resentment at U.N. over lack of progress on treaty covering damages for space accidents, *New York Times* said. Members of 28-nation Outer Space Committee had complained that they cooperated with space powers on space rescue treaty in 1967 on understanding that treaty on damages would be pushed to completion. U.S. had reassured nonspace nations that it wanted immediate action on damages treaty but U.S.S.R. had been "balking" over provision for binding arbitration when damage claims were not settled by direct negotiations or through commission. Other disagreements included financial ceiling on liability for single accident—U.S. had suggested $500 million—and objections to U.S. and U.S.S.R. tendency to negotiate directly and consult other nations later. (Teltsch, *NYT*, 11/23/69, 73)

November 24: Following successful completion of *Apollo 12* President Nixon issued statement: "This mission has shown conclusively that the system we have developed has enormous scientific potential and we can now look forward to utilizing that capability.... The triumph of Apollo 12 is not only an American triumph. This second voyage to the surface of the moon represents another great victory of the human mind and spirit, one which will lift the sights and raise the spirits of men everywhere." (*PD*, 12/1/69, 1659)
- During telephone call from White House to U.S.S. *Hornet*, shortly after *Apollo 12* splashdown and recovery, President Nixon told *Apollo 12* astronauts Charles Conrad, Jr., Richard F. Gordon, Jr., and Alan F. Bean of their promotions to captain, USN. (*PD*, 12/1/69, 1659)
- Christopher C. Kraft, Jr., MSC Director of Flight Operations, told press at MSC he wanted full report on piloting difficulties in landing on moon before committing Apollo 13 to landing attempt on moon's rugged central highlands. Full discussion of lunar landing—described by Astronaut Charles Conrad, Jr., as "no easy task"—was important aspect of what must be learned during crew debriefing, Kraft said. (Wilford, *NYT*, 11/25/69, 33)
- U.S.S.R. launched two Cosmos satellites from Plesetsk. *Cosmos CCCXI* entered orbit with 467-km (290.2-mi) apogee, 272-km (169.0-mi) perigee, 91.9-min period, and 71.0° inclination and reentered March 10, 1970. *Cosmos CCCXII* entered orbit with 1,180-km (773.2-mi) apogee, 1,141-km (709.0-mi) perigee, 108.5-min period, and 74.0° inclination. (GSFC *SSR*, 11/30/69; 3/31/70; *SBD*, 11/26/69, 112)
- President Nixon signed nuclear nonproliferation treaty in Washington, D.C. Presidium of the Supreme Soviet ratified treaty simultaneously in

Moscow, with President Nikolay V. Podgorny signing document. U.S. and U.S.S.R. were 23rd and 24th nations to ratify treaty, which would become effective after ratification by 43 countries. (*PD*, 12/1/69, 1658; *NYT*, 11/25/69, 1)

- Aurora Expedition—during which ARC's Convair 990 aircraft would make about 12 flights from Fort Churchill, Canada, to study aurora in polar regions—began with first data flight. Twenty-five university, industry, NASA and other U.S. Government, French, and Canadian scientists would operate at altitudes up to 40,000 ft across and parallel to auroral oval in flights until Dec. 18. (NASA Note to Editors, 11/12/69; NASA News Release 69-165)

- U.S. Geological Survey Director William T. Pecora said in Washington, D.C., that rocks brought back by *Apollo 11* were geologic "hors d'oeuvre" but *Apollo 12* rocks "will be a veritable feast." Geologists were delighted "by the onsite descriptions provided by astronauts Conrad and Bean" who were "eager rock hounds." (UPI, *NYT*, 11/25/69, 32)

- Washington *Evening Star* editorial commented on *Apollo 12*'s man-made lunar quake: "It seems that when the lunar lander was sent rocketing down to crash on the Ocean of Storms, it left the moon 'ringing like a gong' for some 30 minutes. This, in the words of one ecstatic scientist, constitutes 'a major discovery . . . quite beyond the range of our experience.'" Scientific curiosity "is a good thing—provided it's kept within reasonable limits. . . . Just suppose that their first reading was right, and that the moon really is some sort of celestial gong. Remember what happened when people got too enthusiastic with the Liberty Bell." (W *Star*, 11/24/69, A12)

- USAF announced successful completion of tests at Arnold Engineering and Development Center at Tullahoma, Tenn., to qualify 41,000-lb-thrust TF-39 turbofan engine—largest U.S. military jet engine—for USAF's C-5 Galaxy transport, world's largest aircraft. (AFSC Release 183.69)

- Senate Interior Committee held hearing on July 10 S.J.R. 133 by Sens. Spessard L. Holland (D-Fla.) and Edward J. Gurney (R-Fla.) to return original name, "Cape Canaveral," to Cape Kennedy. Sen. Gurney said name "Canaveral" "may well be the oldest geographical point in the United States, certainly on the east coast, recorded even before the ancient names of Cape Cod and Jamestown." KSC would retain late President John F. Kennedy's name. Witnesses suggested name change had been made "with high emotions" and without proper legal proceedings. Cape had been renamed for late President by executive order from President Lyndon B. Johnson Nov. 29, 1963. (Greider, *W Post*, 11/25/69, A3)

- Jack C. Swearingen, former Chief of Program Control in MSFC Apollo Applications Program Office, became Assistant Director in Science and Engineering. (*Marshall Star*, 11/12/69, 1)

- Raymond Einhorn, former NASA Director of Audits, became Special Assistant to NASA Acting Associate Administrator for Organization and Management. He would be succeeded by Martin Sacks, Special Assistant to Assistant Administrator for Special Contracts Negotiation and Review. (NASA Ann, 11/20/69)

- Washington *Evening Star* editorial commented on proposal by Carnegie Endowment for International Peace and Twentieth Century Fund conference that U.S. and U.S.S.R. merge comsat systems [see Nov. 16]. "Intelsat . . . is already a going concern—an increasingly successful and expanding one—operating on a global scale. Open to every country on a nonpolitical, nondiscriminatory basis, it has no reason whatever to consolidate itself and its resources with Russia's Intersputnik setup—a system that exists only on paper, and not very clearly at that. . . . In the circumstances, the Talloires panelists would have been on sounder ground had they recommended that the Kremlin join Intelsat. Such a move would better serve not only Russia's interests, but the cause of East-West cooperation as well." (W Star, 11/24/69, A12)

November 25: NASA's X-24A lifting-body vehicle, piloted by Maj. Jerauld R. Gentry (USAF), completed eighth glide flight at FRC. Objective of flight, last glide flight in series, was to obtain stability and control data at 30° upper flap setting and 0° rudder bias setting. (NASA Proj Off)

- Two boxes of lunar samples from *Apollo 12* arrived at Lunar Receiving Laboratory in Houston, where they would be examined and used in experiments. (AP, W Star, 11/26/69, A2)
- Pakistan President Yahya Khan sent "hearty felicitations" to President Nixon and American people on *Apollo 12* success. "May your endeavor lead to increasing expansion of the bounds of human knowledge and bring forth newer possibilities of lasting peace and progress on earth." (NYT, 11/26/69, 24)
- *Washington Post Apollo 12* editorial: "Although it will be weeks or months before we know what the harvest of knowledge from this trip has been, the first impression is that the harvest has been a rich one. The astronauts accomplished everything they had been asked to do and more. The pinpoint landing indicates that the navigational problems are not nearly as great as once thought, clearing the way for future astronauts to go precisely where the scientists want them to. The ease with which Mr. Conrad and Mr. Bean functioned on the moon's surface indicates that their successors will be able to range farther afield and undertake more complicated assignments. Even the accidental fall of Pete Conrad is a plus. It cuts down the fear that a fall would be extremely serious and points out . . . that man may be more adaptable to alien conditions than he dreams." (W Post, 11/25/69, A20)
- At hearing on suit of atheist Madalyn Murray O'Hair to ban broadcast of prayers by U.S. astronauts in space [see Aug. 16], U.S. Attorney Seagel Wheatley and NASA counsel asked dismissal of suit on grounds Mrs. O'Hair and her Society of Separationists lacked necessary legal standing to sue Government. Justice Dept. Attorney James Barnes said, "NASA has no plans to instruct astronauts what to say. The statements that the astronauts made are their own. . . . NASA has no intention of circumscribing . . . the astronauts' rights in the free exercise of religion." (UPI, W Post, 11/25/69, A8)
- NASA was using July 26 edict against unauthorized use of Apollo flight insignia to make "unauthorized" possession of souvenir Apollo flight patches Federal offense, *Washington Post* said. While astronauts felt they should be only ones entitled to distribute patches, businessmen contended they ought to be in public domain. NASA contractors like Grumman Aircraft Engineering Corp. had been granted exceptions to blanket rule. (Lardner, W Post, 11/25/69, A8)

November 26: Explorer XLI Interplanetary Monitoring Platform (launched June 21) was adjudged successful by NASA. Spacecraft was functioning satisfactorily and mission objectives had been exceeded. Nine operational experiments were providing detailed information on galactic and solar cosmic rays, interplanetary medium, and distant magnetosphere. All systems and experiments were functioning satisfactorily and returning useful data except GSFC and Univ. of Maryland plasma experiment and Univ. of Iowa and Univ. of California energetic particle experiment, which malfunctioned during orbits 10 and 15.

Preliminary data indicated that although spacecraft had been launched close to solar maximum, sun had been unusually quiet. However, low-energy galactic cosmic rays appeared to be more strongly modulated than observed previously and interplanetary conditions appeared to be more disturbed. Observations would be continued throughout solar maximum. (NASA Proj Off)

- NASA's *Pioneer VI* (launched Dec. 16, 1965) and *Pioneer VII* (launched Aug. 17, 1966) were performing three new experiments on solar system scale, possible only because spacecraft's extremely long lifetimes had allowed them to reach necessary positions in space, ARC announced. On Nov. 6, when *Pioneer VI* and *Pioneer VII* were 175 million mi apart on common line with sun, scientists had observed changes in behavior of solar wind particles due to passage through space. On Nov. 29, when spacecraft reached far side of sun on common line with earth, engineers would conduct 150-million-mi interplanetary communications experiment. On Dec. 2, when spacecraft reached points on common spiral line leading out from sun, scientists would measure different kinds of solar particles coming from same events on sun. (ARC Release 69-15)

- President Nixon signed H.R. 12307 into P.L. 91-126, Independent Offices and HUD FY 1970 appropriations act, which included $3.697-billion NASA appropriation—down $299 million from $3.995-billion FY 1969 NASA appropriation and $181 million from President Johnson's budget request of $3.878 billion.

  Act allocated $3.006 billion for R&D, down $364 million from FY 1969 allocation and $162 million below Johnson request. Construction of facilities allocation for FY 1970 was $53.2 million, up $31.4 million from FY 1969 but $5 million below Johnson request. Allocation for research and program management of $637.4 million was $34.3-million increase over FY 1969 and $13.5 million below Johnson request.

  Act also contained $440-million NSF FY 1970 appropriation. (*PD*, 12/1/69, 1669; Texts)

- Senate confirmed nomination of George M. Low as NASA Deputy Administrator. (*CR*, 11/26/69, S15140, D1126)

November 27: Geologists at Lunar Receiving Laboratory examined several large, dust-covered, crystalline rocks from first box of *Apollo 12* lunar samples. Largest rock weighed 3–4 lbs and was 5 in long and 4½ in thick. (Rossiter, W *Star*, 11/27/69, A4)

- NASA released first photos taken by *Apollo 12* astronauts on and near lunar surface, including color stills and 16-mm film showing solar eclipse, LM descent to moon, astronauts walking on moon, and LM separating from CSM.

Still photos showed closeup of Ocean of Storms, bleak, dull-gray area strewn with rocks and sprinkled with footprints and an astronaut inspecting *Surveyor III*, with LM in background at top of crater rim. (AP, *NYT*, 11/28/69, 32)

November 28: Geologists at Lunar Receiving Laboratory held press conference on *Apollo 12* lunar samples and expressed surprise at samples' differences from rocks retrieved by *Apollo 11*. Preliminary examination of samples showed they were crystalline and larger than anticipated. Dr. Jeffrey L. Warner, MSC geologist, said rocks returned by *Apollo 11* contained up to 12% titanium oxide, but those from *Apollo 12* contained only about 2%—amount consistent with terrestrial rocks. He said geologists were very puzzled by absence of breccia rocks in *Apollo 12* samples because 75% of rocks from *Apollo 11* were breccia. Crystalline rocks were similar to volcanic rocks found by *Apollo 11*, but some of *Apollo 12* rocks were coated with glass and had protruding crystals up to 2½ in long. (Rossiter, *W Post*, 11/29/69, A3; *W Star*, 11/29/69, A1)

- *Apollo 12* astronauts, enclosed in mobile quarantine facility, arrived at Pearl Harbor, Hawaii, where they were greeted by huge crowd, Marine band, civic and military officials, and hula troupe. (Rossiter, *W Post*, 11/29/69, A3)

- President Nixon announced intention to nominate *Apollo 11* Astronaut Michael Collins as Assistant Secretary of State for Public Affairs to succeed Dixon Donnelly who had resigned in January. Nomination was submitted to Senate Dec. 2. (*PD*, 12/1/69, 1667; 12/8/69, 1702)

- Mars' reddish color might be attributed to carbon suboxide, Univ. of Massachusetts physicists William T. Plummer and Robert K. Carson reported in *Science*. They had found reflection spectrum of Mars could be well matched from 0.2 μ through 1.6 μ and farther by polymers of carbon suboxide. (*Science*, 11/28/69, 1141–2)

- In *Science* article advocating large-scale mobilization of scientists to solve world's "crisis problems," John Platt, Associate Director of Univ. of Michigan Mental Health Research Institute, said human race was on steeply rising "S-curve" of change. "We are undergoing a great historical transition to new levels of technological power. . . . In the last century, we have increased our speeds of communication by a factor of 10^7; our speeds of travel by 10^2; our speeds of data handling by 10^6; our energy resources by 10^3; our power of weapons by 10^6; our ability to control diseases by something like 10^2; and our rate of population growth to 10^3 times what it was a few thousand years ago." Within last 25 years "the Western world has moved into an age of jet planes, missiles and satellites, nuclear power and nuclear terror." But S-curve was beginning to level off. "This means that if we could learn how to manage these new powers and problems in the next few years without killing ourselves by our obsolete structures and behavior, we might be able to create new and more effective social structures that would last for many generations." (*Science*, 11/28/69, 1115–21)

- *Science* editorial said faltering U.S. public support of science pointed to conclusion: "Science has established no secure claim in its own right upon the priorities of our national treasury." At World War II's end, U.S. taxpayer had become leading patron of science. "For the next

20 years, public money flowed in increasing volume to the support of science." Public support "began to level out in 1965, and the support of university science is now down about $250 million from the 1965 peak of $1.3 billion." In foremost U.S. universities "federal funding of science has exerted pressures tending to divide and dissolve that frail community. At best, it has installed and expanded scientific departments . . . without regard to the needs and priorities of the university as a whole. . . . At worst, it has established in the universities entirely inappropriate activities, motivated by the interests of the mission-oriented granting agencies and often inimical to free inquiry and to the humanity of science." (*Science*, 11/28/69, 1101)

November 29: *Apollo 12* astronauts, enclosed in mobile quarantine facility, arrived at Ellington AFB, Tex., where they were greeted by 500 cheering persons, including their families and NASA officials. MQF was moved in motorcade to Lunar Receiving Laboratory, where astronauts would remain until Dec. 10. (DeLone, *W Post*, 11/30/69, A3)

- Conclusion that rings of Saturn were water ice of extraordinarily low temperature was announced by Dr. Gerard P. Kuiper, Director of Univ. of Arizona Lunar and Planetary Laboratory. Researchers had ruled out earlier theory that composition could be ammonia ice by comparing spectra of rings and ices of number of compounds. Team had studied infrared reflection of spectrum of rings with interferometer attached to 61-in NASA telescope at LPL observatory near Tucson. (Univ. of Arizona Release 6911.29 dh)

November 30: Following publication Nov. 29 by *Pravda* of article describing U.S. moon rock exhibit in Moscow, crowd of "thousands," eager to see display, smashed glass door and nearly demolished room in which Apollo films were shown. (Clarity, *NYT*, 12/1/69, 14)

- Worldwide program to provide widest display possible of *Apollo 11* lunar samples was announced by NASA. Of 15 moon rocks weighing between ⅔ and 2½ oz, 8 were for U.S. display and 7 had been released to USIA for overseas showing. In addition to Smithsonian Institution display, exhibits would include rock traveling with *Apollo 11* CM on tour of 50 state capitals and two-month showings in 30 museums and planetariums during year. USIA would show six stones on world tour, and seventh at Expo '70 in Osaka, Japan. (NASA Release 69-155)

- Completion of multimillion-dollar electron microscope, twice as powerful as any previously existing, at Laboratory of Electronic Optics in Toulouse, France, opened question "Will we ever be able to look into the very heart of things?" Walter Sullivan said in *New York Times*. Instrument, without its accelerator, weighed 22 tons and stood 10 ft tall. Interviews with U.S., French, and U.K. scientists had revealed wide feeling that "before long it may be possible . . . to see individual atoms and 'read' the genetic message in a strand of DNA (deoxyribonucleic acid)." Univ. of Chicago biophysicist Dr. Humberto Fernandez-Moran had said when that stage was reached, it might be possible not only to predict, but also to design life at the molecular level. (*NYT*, 11/30/69, E9)

During November: It was "no government secret" that NASA, "preoccupied with putting men on the moon," had neglected aeronautics in favor of space, C. V. Glines said in *Armed Forces Management*. But pressures generated by "airways crisis of 1st year, an ever-growing divergence

During November

between civilian and military aeronautical R&D requirements and applications and concern for loss of U.S. preeminence in aeronautics are demanding that attention be given to national aeronautical research and development policies." Effects of Nixon Administration "study timetable" for long-range NASA–DOT civil R&D program would affect DOT–FAA 1972 budgets. Meanwhile, "quiet gains are being made in the aeronautics side of the NASA house." NASA aircraft technology budget would increase from $94.9 million in FY 1969 to over $100 million in FY 1970, with further increase in FY 1971 and new high in FY 1972, "when the long-range study gives new directions for aeronautical research." (*Armed Forces Management*, 11/69, 34–8)

- *Astronautics & Aeronautics* magazine published special issue containing "1973 Viking Voyage to Mars" by NASA Viking project management. Two orbiters would release two softlanders to search for life, map Martian surface from orbit, monitor Martian weather, study atmosphere, and take three-dimensional color photos from ground. Flight directors on earth would probably choose landing site from orbiter reconnaissance photos taken at arrival. A science satellite in its own right, the orbiter would also relay a record volume of interplanetary data from the lander to earth. Nuclear-powered instrument platform based on Surveyor would push evolution of automation step further during mission. (*A&A*, 11/69, 30–59)

- *Army Digest* published interview on space spinoffs with Harry N. Lowe, Jr., Chief of Extraterrestrial Research Agency of USA Office of Chief of Engineers: Space program should return profit to Nation. Tangible spinoff results were being observed in medicine, communications, engineering, and architecture. Savings in building industry and related engineering fields alone would eventually exceed present cost of entire space program. "We have seen the world's largest building put up at the Cape, and other scientific and engineering marvels at Huntsville and Houston. We have had to work with scientists and others to an extent never before necessary. We have had to develop and apply new concepts and standards of performance."

  Extraterrestrial Research Agency work included design and construction concept for semipermanent lunar base, definition of effort to develop lunar construction capability, and studies with NASA of lunar vehicles and mobility. Analyses of lunar samples indicated no need for change in concepts of lunar construction. With water source and cheaper transportation, "man can colonize the moon." (*Army Digest*, 11/69, 30–1)

- NSF *Reviews of Data on Science and Resources* (NSF 69–36) reported nearly 13,000 scientists and engineers became immigrants into U.S. in FY 1968, less than 4% growth over 1967 level. U.K. and India were largest sources of immigrant scientists and engineers, with 2,400 coming from U.K. and 1,400 from India. (NSF 69–36, No. 18, 1)

December 1969

December 1: NASA announced Australis Oscar-A, 39-lb spacecraft designed and constructed by amateur radio operators at Melbourne Univ. in Australia, had been accepted for launch Jan. 9, 1970, as secondary payload on Tiros-M. Radio Amateur Satellite Corp., group of U.S. amateurs, was preparing satellite for launch, testing and qualifying it to comply with NASA requirements. Australis Oscar-A would transmit low-power signals on two amateur bands—29.45 mhz in 10-meter band and 144.05 mhz in 2-meter band—that would be used by radio amateurs throughout world for training in satellite tracking and for radio propagation experiments.

Australis Oscar-A would be fifth satellite launched under Project Oscar and first accepted by NASA as secondary payload. Four previous satellites had been launched by group of U.S. radio operators on Pacific Coast in conjunction with DOD spacecraft. (NASA Release 69–157)

- In interview released by United Press International Dr. Wernher von Braun, MSFC Director, said he knew of nothing which could feasibly send man to the stars in the predictable future. NERVA prototype, only working model of new-generation engine in U.S., was too bulky for flight. Refined version, scheduled for test flight in mid-1970s, was expected to land man on Mars. Beyond that all was conjecture. "There is no design concept and even the basic idea of a controlled thermonuclear reaction has not been demonstrated." It was not known what thrust could be generated by fusion engine. U.S.S.R. had announced development of plasma-jet or ion-thrust engine which, as designed, would not be useful in deep space or close to earth. U.S. scientists had been working on plasma-jet theory and electric propulsion machine. Science fiction concept of space distortion or warp through which spacecraft could travel in "null space" where distances were shorter and speeds faster had some basis according to Einstein's theory of relativity. In Univ. of Maryland experiments "presence of gravitational radiation has been tentatively discovered . . . [and] can be interpreted as the interaction of such a warp with the sensor used in the experiment." Despite this, "there does not appear to be any possibility of utilizing such interactions as a space propulsion system." (W *Star*, 12/1/69, A5)

- Roger Lewis, President of General Dynamics Corp., announced appointment of Dr. George E. Mueller, NASA Associate Administrator for Manned Space Flight, as General Dynamics Vice President, effective immediately. Dr. Mueller's resignation from NASA would be effective Dec. 10. (General Dynamics Release 1491)

- *Apollo 11* Astronauts Neil A. Armstrong, Edwin E. Aldrin, Jr., and Michael Collins visited Canadian Parliament in Ottawa. Canadian Prime Minister Pierre E. Trudeau said astronauts' contributions to

December 1

science and humanity "will live as long as mankind lives." Astronauts would conclude two-day Canadian visit with Dec. 2 visit to Montreal and tour of nearby factory where LM landing pads were made. (UPI, *W Post*, 12/4/69, B3)

- Eleven scientists and technicians joined *Apollo 12* astronauts in quarantine at Lunar Receiving Laboratory after accidental exposure to lunar samples. (UPI, *NYT*, 12/2/69, 49)
- Role of ERC in electronics-related aeronautical R&D was described by ERC Director James C. Elms in testimony before House Committee on Science and Astronautics' Subcommittee on Advanced Research and Technology. Center had continued research on electronic components and devices, especially application of microelectronics and large-scale integrated circuits to advanced avionics systems; on application of advanced technology to instrumentation for sensing and controlling aircraft motion and attitude; on optics and microwaves; on psychological instrumentations; on data processing; and on electrical power systems. ERC had 14 programs in aeronautical R&D, including air traffic control, V/STOL guidance, navigation and flight control, collision avoidance, use of navigational and communications satellites in aviation, support technology, remote detection of clear air turbulence, physiological monitoring, and NASA intercenter research in high-speed flight using YF-12 aircraft. (Testimony)

December 2: Boeing 747 slated for passenger service with Pan American World Airways beginning mid-February 1970 flew from Seattle, Wash., to John F. Kennedy International Airport with 176 passengers in first public preview of jumbo jet flight. Passengers—PAA officials, flight crew, and press—were startled by shudder as plane's wheels were buffeted by rough spots on runway. Phenomenon was repeated when plane touched down in New York after 4-hr 5-min flight. Passengers also were critical of "interior noise level in the rear half of the plane." (Witkin, *NYT*, 12/3/69, 1)

- Scientist Dr. Edwin C. T. Chao, quarantined with *Apollo 12* astronauts at LRL, said mound on moon photographed by Astronaut Charles Conrad, Jr., during *Apollo 12* moonwalk could be crushed rock ejected from crater. MSC geologist Dr. Robin P. Brett said 1.5-lb rock among *Apollo 12* samples had been formed far beneath surface and could be deepest piece of lunar material man had yet studied. (UPI, *NYT*, 12/4/69, 17)
- Gayle Planetarium in Montgomery, Ala., opened formally with exhibit of lunar sample through Dec. 7. MSFC announced it would display lunar rock during February and March 1970. (MSFC Release 69-258)
- Austin, Tex., District Judge Jack Roberts dismissed suit by atheist Mrs. Madalyn Murray O'Hair and Society of Separationists, Inc., to stop astronauts from saying prayers in space [see Nov. 25]. He ruled prayers were not NASA policy but individual decision by astronauts. If NASA had forbidden astronauts to pray in space it would have been unconstitutional abridgment of their rights. Mrs. O'Hair planned to appeal decision. (*NYT*, 12/3/69, 44)
- *Washington Post* editorial praised Nov. 20 speech of RCA President Robert W. Sarnoff which called for use of space technology in "full-scale pilot program" to rehabilitate Washington, D.C.: Nation "needs to know whether the computer-age technology that sent us to the moon

twice can be used to solve the problems of American cities. That question in a slightly different form has been plaguing the space community. Are vital national resources needed to meet the crisis at home being diverted to the moon and beyond?"

Sarnoff speech "picks up from a series of addresses by Dr. Thomas O. Paine, administrator of the National Aeronautics and Space Administration, suggesting that modern technology can assist in solving the problems of cities, provided there is a national commitment to do so." (*W Post*, 12/2/69, A18)

December 3: Dr. George M. Low was sworn in as NASA Deputy Administrator by Dr. Thomas O. Paine, NASA Administrator. (NASA Release 69–159)

December 3: *Dr. George M. Low—veteran of Mercury, Gemini, and Apollo programs—was sworn in as Deputy Administrator of NASA by Dr. Thomas O. Paine, Administrator, after Nov. 26 Senate confirmation.*

- U.S.S.R. launched *Cosmos CCCXIII* from Plesetsk into orbit with 247-km (153.5-mi) apogee, 197-km (122.4-mi) perigee, 88.9-min period, and 65.4° inclination. Satellite reentered Dec. 15. (GSFC *SSR*, 12/15/69; *SBD*, 12/4/69, 148)
- NASA selected Hughes Aircraft Co. Space Systems Div. and RCA Corp. Astro-Electronics Div. to receive parallel four-month, $250,000 fixed-price study contracts for definition and design of Atmosphere Explorer spacecraft AE–C and AE–D. Spacecraft would carry experiments to study atmospheric composition and characteristics in lower thermosphere. (NASA Release 69–158)
- At opening session of Governors' Conference in Washington, D.C., President Nixon presented to governors mementos from *Apollo 11* mission for people of 50 states, Commonwealth of Puerto Rico, District of

December 3

Columbia, and Trust Territories of Virgin Islands, Guam, and American Samoa. Each presentation, consisting of state flag and chip from moon's surface, bore inscription: "This flag of your state was carried to the Moon and back by Apollo 11, and this fragment of the Moon's surface was brought to Earth by the crew of that first manned lunar landing." (*PD*, 12/8/69, 1696)

- President Nixon announced appointment of *Apollo 11* Astronaut Neil A. Armstrong as Chairman and member of Peace Corps National Advisory Council to succeed W. Thomas Johnson, Jr., who resigned May 27. (*PD*, 12/8/69, 1696)
- DOD announced USAF contract awards. Martin Marietta Corp. received $1,081,000 supplemental agreement to previously awarded contract to design, develop, and fabricate Titan IIIC boosters and associated aerospace equipment. Contract would be managed by Space and Missile Systems Organization (SAMSO).

  Philco-Ford Corp. received $44,375,975 contract for engineering, furnishing, installing, and testing aircraft control and warning system. Contract would be managed by Oklahoma City Air Materiel Area. (DOD Release 1046–69)
- Secretary of Transportation John A. Volpe announced FAA plans to issue proposal to control emission of aircraft smoke in flight in effort to combat air pollution. (FAA Release 69–129)

December 4: USAF launched unidentified satellite on Thor-Agena booster from Vandenberg AFB into orbit with 155.3-mi (249.9-km) apogee, 105.6-mi (169.9-km) perigee, 88.4-min period, and 81.4° inclination. Satellite reentered Jan. 10, 1970. (GSFC *SSR*, 12/15/69; 1/15/70; *Pres Rpt 70* [69])

December 5: NASA launched two sounding rockets from WSMR to conduct stellar x-ray studies. Nike-Apache carried Dudley Observatory payload and Aerobee 150 carried University of Wisconsin payload. (NASA Proj Off)

- NASA and public TV would participate in first full-scale experiments in use of satellites to transmit TV programs domestically, John W. Macy, Jr., President of Corp. for Public Broadcasting announced. Scheduled for near future, tests would transmit public broadcasting programs using *Ats III* spacecraft in orbit and NASA ground stations at Rosman, N.C., and Mojave, Calif., to indicate technical problems and operating costs of satellite transmission. NASA had authorized first regular use of its satellites domestically for other than purely scientific purposes in letter to Macy, Chairman of Satellite Task Force organized by CPB to represent public broadcasting interests in satellite field. (CPB Release)
- Successful results of seeding hurricane Debbie on Aug. 18 and 20 were announced by Secretary of Commerce Maurice H. Stans and Secretary of the Navy John H. Chafee at Washington, D. C., press conference. Analysis of silver iodide seeding had suggested storm was weakened by intervention. While scientists could not state absolutely that hurricanes could be modified, Secretaries pledged their departments to intensified effort in Project Stormfury, joint venture of ESSA and USN. (Schmeck, *NYT*, 12/5/69, 90)
- DOD announced Lockheed Aircraft Corp. was receiving $100,000 initial increment to $2,532,250 fixed-price USAF contract for experimental studies of airflow characteristics of advanced aircraft engines. Contract

would be managed by AFSC Aeronautical Systems Div. (DOD Release 1052-69)
- Analysis of Pesyanoe enstatite achondrite samples confirmed new isotopic composition of xenon in Pesyanoe meteorite was due to presence of component like that in solar gas, Kurt Marti of Univ. of California at San Diego reported in *Science*. Xenon in Pesyanoe meteorite was mixture of several components. Solar-type xenon was new component deficient in neutron-rich isotopes as compared to both trapped chondritic and terrestrial atmospheric xenon. (*Science*, 12/5/69, 1263-5)

December 7: Astronomers were contemplating possibility that universe might be several times larger than previously believed as result of observations from *Oao II*, NASA announced. Spacecraft, launched Dec. 7, 1968, had discovered that many galaxies were much brighter in UV radiation than expected, confirmed that hot stars lost as much as sun's total mass in 100,000 yrs and equivalent of earth's mass in 1 yr, and indicated that if extra mass assumed to exist as unobservable matter in universe was present it did not radiate in UV, suggesting that universe was not closed system.

Oao II's performance during first year in orbit had been exceptional. By Nov. 9, Smithsonian Astrophysical Observatory's sky-mapping instrument package, in 169 days of operation, had taken 5,884 pictures during observations of 2,265 individual square areas of sky. Univ. of Wisconsin experiment had studied 568 specific objects during 1,995 observations in 165 days of operation. Achievements had prompted some astronomers to rank *Oao II* with invention of telescope in its importance to astronomy. (NASA Release 69-156)
- Dr. John M. DeNoyer, former Assistant Director for Research at U.S. Geological Survey, became Director of Earth Observations Programs in NASA Office of Space Science and Applications. (NASA Hq *WB*, 1/5/70)
- "Eyewitness to Space," exhibit of works of over 70 U.S. artists commissioned by NASA to document its activities, opened at National Gallery of Art in Washington, D.C. Show included paintings, drawings, and sculpture by James B. Wyeth, Mitchell Jamison, Norman Rockwell, Lamar Dodd, William Thon, and Robert Rauschenberg. Eight-foot montage of disparate space themes by Rauschenberg was largest lithograph ever made, according to Gallery's Curator of Art H. Lester Cooke. (Constantine, W *Star*, 12/7/69, J8)

December 8: Recording by seismometer on moon of seven external impacts near *Apollo 12* landing site since astronauts left moon Nov. 20 was reported by Columbia Univ. scientist Dr. Gary V. Latham, principal investigator for Apollo program's seismic research. Each impact had produced tremors. Dr. Latham said meteors might have struck Ocean of Storms, whose surface resonated when hit. Lack of internal quakes indicated moon had not been heated substantially for about 4.6 billion yrs. (AP, W *Star*, 12/9/69, A6)
- Edgar L. Piret, U.S. Embassy Scientific Attaché in Paris, accepted Prix Pierre Guzman gold medals on behalf of *Apollo 11* Astronauts Neil A. Armstrong, Edwin E. Aldrin, Jr., and Michael Collins. Award had been established by Mme. Anna Emile Guzman in 1889 for first persons "to find the means of communicating with a heavenly body—Mars excluded" [see Aug. 13]. (AP, *W Post*, 12/9/69, A23)

- *Apollo 11* lunar landing had been selected number one foreign news story of year by Japan's Kyodo news service, Associated Press said. Second was President Nixon's decision to reduce U.S. troops in Vietnam; third was Communist China's ninth party congress in Peking. (*St. Louis G-D*, 12/8/69)
- LeRC's 60-mw test reactor at Plum Brook Station, Ohio, completed 100th cycle of operation. It had begun full-power operation in April 1963. Since then it had participated in 1,100 irradiations of experiments in nuclear propulsion, energy conversion, basic radiation effects, and nuclear physics programs. About 30 active irradiation experiments in progress ranged from nuclear fuel material tests to studies of atomic and molecular structure of matter. (LeRC Release 69-72)
- Rep. Edgar F. Foreman (R-N. Mex.) introduced H.C.R. 464 for Congressional recognition of Goddard Rocket and Space Museum, Roswell, N. Mex., as memorial to Dr. Robert H. Goddard, "who pioneered in rocket experimentation and contributed to America's success in landing men on the moon." (*CR*, 12/8/69, H11865)
- By vote of 330 to 33 House passed H.R. 15090, $69.9-billion DOD FY 1970 appropriations bill. (*CR*, 12/8/69, H11865–909)
- Princeton Univ. scientists had observed sudden speedup in fastest and youngest pulsar, in heart of Crab Nebula, Walter Sullivan reported in *New York Times*. With second discovery of periodic phenomenon, astronomers had taken word "glitch" from electronic engineers and astronauts to describe sudden departure from normal pulsar behavior. (*NYT*, 12/8/69)

December 9: LeRC Director Bruce T. Lundin discussed future interest in transonic speed range—mach 0.7 to mach 1.4—before House Committee on Science and Astronautics' Subcommittee on Advanced Research and Technology during hearings on U.S. aeronautical activities: "Cruise speeds of commercial and military transports have already reached about Mach 0.85, and further increases to about Mach 1.15 are of interest because this speed can be attained without sonic boom effects. The transonic speed range is also critical for a supersonic transport because performance in this range determines subsonic cruise efficiency. Transonic speed characteristics are important to fighter aircraft because they are required to maneuver at these speeds, and they are important to bomber aircraft because they affect the engine size which is necessary to accelerate to higher supersonic speeds. . . . It is in this speed range that many interactions, shock waves, and flow separations occur that become important to the thrust, drag, and stall margin of the engine and on the lift, drag and buffet characteristics of the aircraft. Unfortunately, mathematical prediction techniques do not work well in this speed range and experimental testing becomes of major importance." Only very small models could be used in transonic tunnels "and scaling up results from very small models to full size airplanes is at best difficult and usually impossible. We will be paced here for some time to come by the capabilities of our experimental facilities." (Testimony)

- Sen. Robert J. Dole (R-Kans.) introduced S.R. 167 "providing for the display in the Capitol Building of a portion of the moon." Resolution was referred to Senate Committee on Rules and Administration. (*CR*, 12/9/69, S16142)

December 10: NASA announced it was proceeding with plans and preparations for launch of Apollo 13 manned lunar landing mission to Fra Mauro on March 12, 1970. Decision was based on review of photos taken of Fra Mauro area and successful demonstration of pinpoint landing techniques by *Apollo 12*. Fra Mauro was flat, vast highland area about 110 mi east of *Apollo 12* landing point on Ocean of Storms. (NASA Release 69–162)

- *Apollo 12* Astronauts Charles Conrad, Jr., Richard F. Gordon, Jr., and Alan L. Bean and 25 other persons quarantined in Lunar Receiving Laboratory were released one day ahead of schedule. Dr. Charles A. Berry, Director of Medical Research and Operations at MSC, said astronauts were in good physical condition. (B *Sun*, 12/11/69, A1)
- Electrical fire damaged GSFC offices, laboratories, and equipment including 150-lb Small Scientific Satellite scheduled for 1970 launch in Explorer series. Damage to satellite and laboratory facility was estimated at $400,000. There were no injuries. Origin of fire was being investigated. (W *Star*, 12/10/69, B1; GSFC Historian)
- End of era in satellite watching would come with discontinuation in June 1970 of Smithsonian Institution's optical tracking program, John Lannan said in Washington *Evening Star*. NASA already was terminating 10-yr support of Smithsonian Astrophysical Observatory's efforts at Cambridge, Mass. Optical tracking had cost $4 million annually since 1958, first through NSF, then NASA. Program had led to nationwide "Dial-A-Satellite" telephone net by which tape-recorded messages told callers where satellites could be seen in sky. Service had become "Dial-A-Phenomenon" to note ecological, geophysical, and astronomical events, since decay of only visible satellites, *Echo I* and *Echo II*. Smithsonian Observatory had watched satellites with worldwide battery of Baker-Nunn cameras, which had been phased out except for observing maneuvers in major space flights. Observatory's geodetic program was being expanded, with two lasers operational, one on Mount Hopkins, Ariz., the other in Athens, Greece, but financial support had been halved for current fiscal year and would be cut to $1.3 million in 1971. (W *Star*, 12/10/69, A26; Smithsonian PAO)

December 11: U.S.S.R. launched *Cosmos CCCXIV* from Plesetsk into orbit with 465-km (288.9-mi) apogee, 296-km (183.9-mi) perigee, 91.6-min period, and 71.0° inclination and reentered March 22, 1970. (GSFC *SSR*, 12/15/69; 3/31/70; *SBD*, 12/15/69, 193)

- USAF YF–12A supersonic aircraft made first flight from Edwards AFB, Calif., under joint NASA–USAF sponsorship, opening program to advance U.S. knowledge of aerial defense tactics and future of commercial aviation [see July 11]. (AFSC *Newsreview*, 2/70, 1)
- NASA announced appointment of Charles W. Mathews, Deputy Associate Administrator for Manned Space Flight, as Acting Associate Administrator for Manned Space Flight, replacing Dr. George E. Mueller. (NASA Ann)
- "Into the New Realm," exhibit of documentary history of U.S. Government in space from early balloon ascensions to origination of Apollo program, opened at National Archives in Washington, D.C. Exhibit included original of National Space Act of 1958, which created NASA, and copies of correspondence on establishment of U.S. space program from President Eisenhower; President Kennedy; first NASA Adminis-

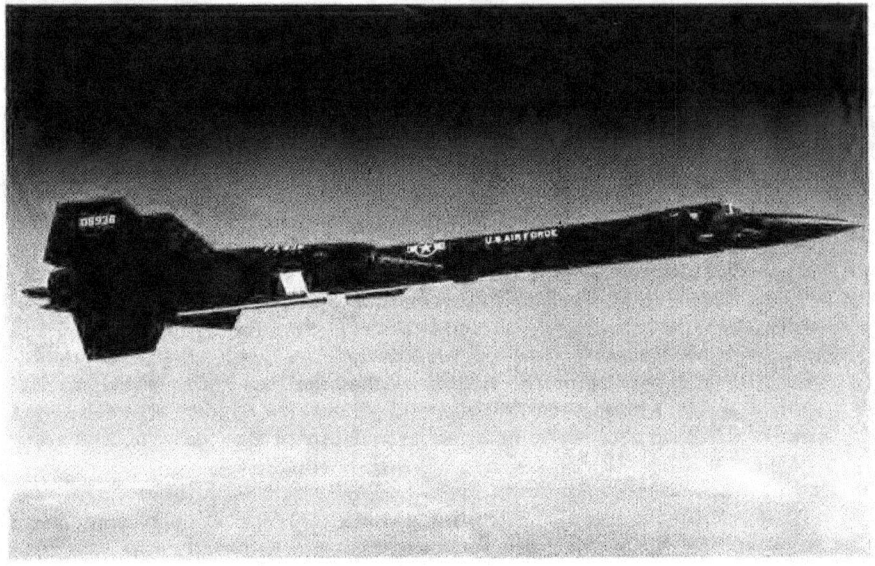

December 11: USAF *YF–12A* supersonic aircraft made its first research flight from Edwards AFB, to open a joint NASA–USAF program to advance knowledge of high-performance flight. Special instrumentation installed by NASA increased the value of the flying test bed for accumulating data. Pilots were from NASA and the Air Force.

trator, Dr. T. Keith Glennan; former NASA Administrator James E. Webb; and others. (Natl Archives PIO)

- President Nixon issued proclamation commemorating Dec. 17 as Wright Brothers Day: "On December 17, 1903, Orville and Wilbur Wright . . . stepped from a homemade contraption onto an ocean beach in the State of North Carolina, after completing the first successful airplane flight.

  "Almost sixty-six years later, another man stepped from another craft onto another plain . . . the waterless Sea of Tranquility on the Moon. Man had not only removed his bondage to the earth, but had expanded his horizons to outer space." (*PD*, 12/15/69, 1732)

- Former astronaut John H. Glenn, Jr., formally announced candidacy for Democratic nomination for U.S. Senate seat being vacated by Sen. Stephen M. Young (D-Ohio). (*NYT*, 12/12/69, 33)

December 12: NASA's HL–10 lifting-body vehicle, piloted by NASA test pilot William H. Dana, reached 80,000-ft altitude and mach 1.4 after air-launch from B–52 aircraft at 45,000-ft altitude west of Rosamond, Calif. Objective of powered flight, 31st in series, was to obtain stability and control data. (NASA Proj Off)

- *Apollo 12* Astronauts Charles Conrad, Jr., Richard F. Gordon, Jr., and Alan L. Bean held press conference at MSC and showed movies and slides of mission. Explaining failure of color TV camera, Conrad said: "What apparently happened . . . is that I took the thing off MESA and set it down. The MESA was in the sun at the time and . . . I apparently either pointed it at the sun or pointed it at the MESA, which was in the sun and a pretty bright reflector . . . and just that few seconds . . . was enough to sunburn the camera. . . . I didn't realize it was as sensitive to light as it is."

Dust on surface was a major problem, Conrad said. Ground was soft and astronauts sank in significantly. It was difficult to keep experiment package clean. "It was almost impossible for us to walk around those packages without our little dust cloud and at any time the dust hit the surface of the package it stuck and the worst thing you could do was try to rub it off because you just smeared it over them even worse. . . ."

To question about difficulty in landing LM on moon, Conrad replied he had misinterpreted correct gage reading and underestimated accuracy of LM's descent. ". . . I felt that that gage was not . . . giving me the proper information. I found it quite hard to believe that I could have gotten the velocity killed quite as well as I did . . . and so I was continually going out the window to get roll information and a general idea of my lateral and horizontal velocities." Although Conrad felt landing on moon was not "an amateur's game," he felt satisfactory landings under completely instrumented flight rule conditions could be made. Crew had suggested adding instrument that would provide automatic nulling of horizontal and lateral velocities but would leave pilot with rate of descent command. Conrad said he did not think dust during descent would be problem on future missions "as long as you get a look at your landing site before you get into the dust, and ascertain that it's all right—it's perfectly all right to go in and land, IFR. . . . I don't recommend any change from the procedures that we're using right now."

Bean said photos of moon were realistic, but photos of earth were not. "As we view the Moon from the Earth, it's white and flat; it's quite pretty. When you get up at the Moon . . . it's much the same; it's either white or concrete colored or grey. . . . But when you look at the Earth . . . it kind of sparkles, and . . . you can't capture it on the film. It's sort of like taking a picture of an emerald . . . and hoping to get it."

Gordon said he had mixture of emotions, "a mixture of thoughts, as to why is it all here? And . . . what the function that each person, each part, each molecule, each body performs, in the overall scheme of why is it, what is it, and from where did it come." (Transcript)

- Lunar Receiving Laboratory scientists told press at MSC rocks retrieved from moon by *Apollo 12* crew were younger, lower in titanium, and higher in nickel than those retrieved by *Apollo 11* and showed no signs of bacteria or toxicity. Dr. Oliver A. Schaeffer of New York State Univ. at Stony Brook said *Apollo 12* samples seemed to be 2.2 billion to 2.5 billion yrs old, compared to 3.8 billion to 4.6 billion yrs old for *Apollo 11* samples. "The younger age means the . . . activity, whether volcanic or meteorite impact, took place over an extended period of time, not on a short time scale."

Dr. S. Ross Taylor of Australian National Univ. said chemistry of *Apollo 12* and *Apollo 11* samples was basically similar but had many interesting differences. "The surface is not completely homogeneous across the moon. There are differences in chemistry probably reflecting in the underlying rock. . . . it looks as though we could have two extremes of differences." *Apollo 12* samples from Ocean of Storms had about half titanium content found in *Apollo 11* samples from Sea of

Tranquility. They had few refractory elements and fewer volatile elements. (Wilford, *NYT*, 12/13/69, 20; Cohn, *W Post*, 12/13/69, A10)
- Senate confirmed nomination of former *Apollo 11* Astronaut Michael Collins to be Assistant Secretary of State for Public Affairs. (*CR*, 12/12/69, D1201)
- Moon was gray and colorless, rather than orange from oxidation of ferrous oxide in rocks by photolytically liberated oxygen, UCLA geophysicist Dr. Willard F. Libby noted in *Science*. He suggested reason "lies in the solar wind's bringing in atomic hydrogen to replace that lost by the photolytic decomposition of water vapor." (*Science*, 12/12/69, 1437–8)
- Results of four radio occultation measurements of Mars' atmosphere, ionosphere, and surface configuration by *Mariner VI* and *VII* were reported in *Science* by JPL's Dr. Arvydas J. Kliore, Dr. Gunnar Fjeldbo, and Boris L. Seidel and Goddard Institute for Space Studies' Dr. S. Ichtiaque Rasool. Measurements had provided refractivity data in Mars atmosphere at four points above its surface. For atmosphere consisting predominantly of carbon dioxide, surface pressures between 6 and 7 millibars at three points of measurement and 3.8 at fourth, indicated elevation of 3.1–3.7 mi (5–6 km). Temperature profile measured by *Mariner VI* near equator in daytime indicated temperatures in stratosphere about 100°K warmer than those predicted by theory. Measurements by *Mariner VI* taken at 79°N at beginning of polar night indicated conditions were favorable for condensation of carbon dioxide at almost all altitudes. *Mariner VII* measurements taken at 58°S in daytime and 38°N at night also showed carbon dioxide condensation was possible at altitudes above about 15.5 mi (25 km). Measurements of electron density in ionosphere showed upper atmosphere was substantially warmer than in 1965, possibly because of increased solar activity and proximity to sun. (*Science*, 12/12/69, 1393–7)
- TOR–SHOK energy-absorbing system designed for Apollo program was being installed by State Highway Commission of Kansas along major arteries in and about Topeka, Republic Steel Corp. announced. Developed by Ara, Inc., system used multiple-telescoping, lightweight, high-strength steel tubing to attach guardrail to its mounting posts, concrete abutments, or other roadside objects. When car struck guardrail, most of impact force was transmitted from rail to TOR–SHOKs, rings that absorbed shock by turning inside out. (Republic Steel Release R–1053)
- NASA was allowing *Apollo 12* crew and colleagues to take part in Jewish National Fund dinner Dec. 14 in Houston "after years of guarding against the use of its astronauts as fund-raisers," Associated Press said. Proceeds of $100-a-plate affair honoring *Apollo 12* and other astronauts would be used to plant trees in Israel. (AP, *W Post*, 12/13/69, A12)
- DOD announced General Electric Co. would receive $5,333,250 supplemental agreement to previously awarded USAF contract for R&D of Mark 12 reentry vehicle. Contract would be managed by Space and Missile Systems Organization (SAMSO). (DOD Release 1068–69)

December 13: NASA successfully launched first two in series of Nike-Apache sounding rockets carrying chemical cloud experiments from NASA Wallops Station. Rocket launched at 5:22 pm EST carried sodium experiment which created reddish-orange cloud visible for hundreds of miles. Rocket launched at 7:30 pm EST carried trimethylaluminum (TMA) experiment which created bluish-white cloud. Primary objective was to measure wind directions and speeds in 25- to 135-mi (40.2- to 217.2-km) region and temporal variations that occurred throughout night. Four TMA experiments and two acoustic grenade experiments, postponed because of cloud cover, would be rescheduled. (WS Releases 69-19, 69-20)

- Aerobee 150 sounding rocket carrying GSFC payload was launched by NASA from WSMR to conduct stellar UV studies. (NASA Proj Off)
- Pan American World Airways took delivery of first Boeing 747 to be turned over to commercial line in ceremony at Boeing Field in Seattle, Wash. Later in day, 362-passenger transport left on six-hour flight to Nassau, Bahamas. It was scheduled to fly to New York same day. (AP, W Star, 12/14/69, A25)
- Smithsonian Astrophysical Observatory in Cambridge, Mass., said comet Tago-Sato-Kosaka, discovered Oct. 10 by Japanese astronomers, was being observed by astronomers in Southern Hemisphere and should be visible to naked eye throughout U.S. in mid-January 1970. It was first comet visible without telescope in more than two years. (UPI, W Star, 12/14/69, A33)

December 15: At meeting of American Geophysical Union in San Francisco, Columbia Univ. scientist Dr. Gary V. Latham, principal investigator for Apollo program's seismic research, proposed detonation of nuclear device on moon to assist analysis of lunar interior. He announced tentative plans for placement of one- to five-kiloton bomb on lunar far side by unmanned Atlas-Agena rocket between Apollo 15 and 16 missions in November 1970 and envisioned possible cooperative effort with U.S.S.R. "I expect to run into a large number of political snags. But these problems are not insurmountable if we include the Russians . . . and agree to share the data with them. . . . I would like to ask them to put the bomb on the moon." Proposal was to be submitted to NAS for approval. Nuclear blast would send seismic waves through moon's core. These would be measured to yield information on nature and origin of lunar interior. (Reuters, *W Post,* 12/16/69, A2)

- NASA established 14-member Apollo Orbital Science Photographic Team to provide scientific guidance in design, operation, and data utilization of photographic systems for Apollo lunar orbital science program. Chairman was Frederick J. Doyle of U.S. Geological Survey. (Apollo Prog Off)
- NRC Panel on Remote Atmospheric Probing issued *Atmospheric Exploration by Remote Probes,* Volume 1, *Summary and Recommendations,* of final report to NAS-NRC Committee on Atmospheric Sciences. Report recommended that National Center for Atmospheric Research organize scientific committee to formulate with NASA "an integrated remote atmospheric probing program at the unique Wallops Island facility; and that NASA establish procedures for the utilization of the facility and make widely known its availability for atmospheric research." Report also recommended inclusion of atmospheric probing among scientific missions of Haystack radar at MIT and Millstone radar and urged development of new facilities for remote atmosphere probing by Doppler radar. (Text)

- Senate, by vote of 85 to 4, passed H.R. 15090, $69.3-billion DOD FY 1970 appropriations bill. (*CR*, 12/15/69, S16743–4, S16750–1, S16782, S16784–96)
- Gen. James Ferguson (USAF), AFSC Commander, said in keynote address before Air Force Fatigue and Fracture Conference at Miami Beach, Fla.: "Our potential adversaries are pushing on all frontiers of technology. We cannot safely do less, and yet we must achieve our technological goals *with* less in the way of money, manpower, and facilities." Each aircraft "will have to be better, more capable, stronger, more durable, and preferably less costly." (Text)

December 16: NASA announced appointment of Robert N. Lindley, Vice President Program General Manager of McDonnell Douglas Corp., as Special Assistant to NASA Associate Administrator for Manned Space Flight. (NASA Ann)

- Senate Committee on Appropriations voted in closed session to cut approximately $16 million from House-recommended allocation of $96 million for SST development under H.R. 14794, DOD FY 1970 appropriations bill. (Reuters, B *Sun*, 12/17/69, A9)
- Man Will Never Fly Memorial Society held 10th annual meeting at Kitty Hawk, N.C. on eve of 66th anniversary of Wright brothers' first airplane flight. Society's highest honor, National Anti-aviation Citation Presentation, was awarded in absentia to President Nixon for his decision to spend $1.3 billion on two prototype SSTs. "Since Britain, France and Russia have already built SST prototypes," citation said, "an all-American SST will . . . give the United States a smashing aviation fourth."

  Julian Scheer, NASA Assistant Administrator for Public Affairs and founding member of society, narrated films showing simulations of lunar surface used by NASA for training. They indicated "that you can really fake things on the ground—almost to the point of deception." Society's 600 members included aviation pioneers, military pilots, airline executives, and aerospace newswriters. (Wilford, *NYT*, 12/18/69, 30)
- Secretary of Transportation John A. Volpe announced selection of Robert P. Selfridge, Director of Greater Rockford (Ill.) Airport, to receive FAA Award for Distinguished Service for his "contribution in developing and promoting the cause of General Aviation" and for his "constructive suggestions, imagination, dedication and service to the aviation users, above and beyond the role of an Airport Director." (FAA Release 69–67)

December 16–17: USNS *Vanguard*, which provided only sea-going link in NASA tracking and communications network during Apollo flights, was on display at Port of Baltimore, Md. (*W Post*, 12/15/69, B8; NASA PAO)

December 17: SNAP–27 nuclear generator placed on moon by *Apollo 12* astronauts had continued to operate during first 28-day-and-night lunar cycle despite temperatures from $-291°F$ to $+283°F$, AEC announced. It was producing more than 70 w of electricity to power five instruments deployed Nov. 19 to transmit lunar data. (AEC Release M–274)

- NASA Aerobee 150 MI sounding rocket launched from WSMR with VAM–20 booster carried AFCRL payload to 127-mi (204.3-km) altitude. Objective was to calibrate Harvard College Observatory spectrometer on board orbiting *Oso VI* by studying active regions of sun simultaneously at 300–400 Å with telemetering, grazing incidence, scanning EUV monochromator. Rocket and instruments functioned satisfactorily. (NASA Rpt SRL)

- USAF took delivery of first operational C–5 Galaxy aircraft in ceremony at Lockheed-Georgia Co. plant in Marietta, Ga. Later, aircraft was flown to Altus AFB, Okla., for use by Military Airlift Command. (DOD Release 1078–69; USAF PIO)
- Termination of Project Blue Book—USAF investigation of UFOs—was announced by Dr. Robert C. Seamans, Jr., Secretary of the Air Force. Project's continuation could not be justified "on the ground of national security or in the interest of science." Decision was based on evaluation of Jan. 9 report by Univ. of Colorado, NAS review of Univ. of Colorado report, and USAF investigations of UFO reports since 1948. USAF had concluded: no UFO reported, investigated, and evaluated by USAF had threatened U.S. security; there had been no evidence that "unidentified" sightings represented technological developments or principles beyond range of scientific knowledge; and there had been no evidence that "unidentified" sightings were extraterrestrial vehicles. Project Blue Book records would be retired to USAF archives at Maxwell AFB, Ala. (DOD Release 1077–69)
- National Investigations Committee for Aerial Phenomena (NICAP), 10,000-member private UFO-investigating group, released statement on termination of USAF Project Blue Book: "Congress can now discuss the possibility of turning over UFO investigations to a civilian federal agency or a private scientific organization." USAF decision "increases the need for an active program of research and education." (NICAP Release)
- London-to-Sydney Air Race commemorating first England-to-Australia flight in 1919 by Ross and Keith Smith started at Gatwick Airport near London, with 89 entries competing for $112,000 in prizes. (*NYT*, 1/5/70, 61)
- Dr. Thomas O. Paine, NASA Administrator, in letter to Sen. Clinton P. Anderson, Chairman of Senate Committee on Aeronautical and Space Sciences, summarized recent NASA efforts to improve international cooperation in space. He had visited major European capitals and Canada and hoped to visit Australia and Japan, to explain "planning for U.S. space activities in the next decades." Foreign space authorities had been invited to "sponsor their own industrial participation in the NASA conference on space shuttle concepts" held in October. "Additional mechanisms are being developed to permit foreign space interests to keep in touch with and even contribute to our studies over the next year, especially in the space shuttle and station programs."

  In letters to President Mstislav V. Keldysh and Academician Anatoly A. Blagonravov of Soviet Academy of Sciences Dr. Paine had invited Soviet proposals for experiments on U.S. spacecraft, offered use of laser reflector left on moon by *Apollo 11* astronauts, invited Soviet proposals for lunar sample analysis, invited Soviet scientists to Viking Mars 1973 conference, offered to discuss coordination of planetary programs, and reiterated U.S. readiness "to meet anytime, any place, to consider any possibilities for cooperation or coordination between

us." There had been "no substantial response." (CR, 1/21/70, S259)
- Senate, by vote of 58 to 22, rejected amendment submitted by Sen. William Proxmire (D-Wis.) to strike from H.R. 14794, DOT FY 1970 appropriations bill, $80 million for SST development. (CR, 12/17/69, S17003, S17005–48)
- Maj. Alexander P. de Seversky, long-time advocate of air power, received USAF Exceptional Service Award at Pentagon ceremony for "exceptionally meritorious service" to U.S. aviation from 1918 to 1969. (W Star, 12/18/69, A2)
- Surprise had been expressed over enthusiastic reception in Moscow of U.S. lunar rock exhibit, Washington *Evening Star* editorial noted. There was really nothing surprising, however. "From a narrowly nationalistic point of view, the Russian people and the Soviet government can claim that, were it not for the early Russian successes in space, that piece of rock would still be resting on the moon. But it is more likely that the Russians, and the other peoples of the Earth, see the first moon landing for what it truly was: A triumph for the collective ingenuity of Man." (W Star, 12/17/69)

December 18: Selection of investigators and investigation areas for 1973 Viking mission to Mars was announced by NASA. LaRC would manage overall project and be responsible for lander portion of spacecraft. JPL would manage orbiter portion and be responsible for tracking and data acquisition. Viking, follow-on to 1964–1965, 1969, and 1971 Mariner Mars flights, would consist of two instrumented spacecraft in Mars orbit, each of which would detach landing capsule for softlanding on Mars. Mission objectives included detection of life if it existed. (NASA Release 69–166)
- Hot firing of twin RL–10 Centaur engines marked first use of LeRC's new Spacecraft Propulsion Research Facility, B–2 stand, at Plum Brook Station, Ohio [see Oct. 7], and first Centaur tests at Plum Brook since structural tests in mid-1960s. Since then, Centaur, configured with 1st-stage Atlas, had launched seven Surveyor spacecraft to moon, two Mariner spacecraft to Mars, one OAO, and one ATS. (LeRC Release 69–76)
- NASA announced appointment of John A. Whitney as Assistant General Counsel for Procurement Matters. (NASA Release 69–167)
- Planned U.S. participation in 1971 Paris Air Show was announced in Washington, D.C., by Secretary of Transportation John A. Volpe and Secretary of Commerce Maurice H. Stans. DOT would coordinate exhibition of U.S. aircraft on flight line. (DOT Release 26669)
- Senate and House cleared for President's signature H.R. 15090, $69.6-billion DOD FY 1970 appropriations bill. (CR, 12/18/69, S17181–6; H12706–8)
- Tom Wicker commented in *New York Times* on suggested detonation of nuclear device on moon [see Dec. 15]. "Aside from the obvious questions about the effects of nuclear fallout in the moon's atmosphere, what might be the total environmental consequences of such an explosion—for the moon itself, for those who will be visiting it from earth, for other objects in the solar system?" Past performances had indicated "some unexpected and probably unwelcome result." Indus and Ganges River irrigation systems had contaminated soil in India with salt that rose from earth with water and Aswan Dam was spreading disease

with irrigation waters in Egypt and damaging fertile Nile delta by interfering with ancient silting process. In U.S., oil leaks from ocean floor set off by scientific drilling techniques had ruined Santa Barbara, Calif., coast. "Monstrous engines" of Boeing 747, Concorde supersonic airliner, C–5A Galaxy jet, and SST would "spew their poisons on mankind." Could man master anything that really mattered? "Certainly not nature, and least of all himself; rather it is altogether likely that if the Biblical flood someday engulfs the earth, it will flow from seeded clouds. That might even be a fitting end." (*NYT*, 12/18/69, 46)

December 19: Results of *Apollo 11* experiment to trap atomic particles from solar wind on lunar surface [see Aug. 18] were reported in *Science* by originator Dr. Johannes Geiss and Univ. of Berne, Switzerland, team. Helium-4 solar wind flux during *Apollo 11* excursion was 5.1 million to 7.5 million atoms per sq cm per sec. Solar wind direction and energy were essentially not perturbed by moon. Evidence for solar wind albedo had been found. (*Science*, 12/19/69, 1502–3)

December 20: U.S.S.R. launched *Cosmos CCCXV* into orbit with 541-km (336.2-mi) apogee, 518-km (321.9-mi) perigee, 95.2-min period, and 74.0° inclination. (GSFC SSR, 12/31/69)

- *Apollo 11* Astronaut Neil A. Armstrong arrived in Bangkok, Thailand, to meet with Peace Corps officials on Presidential mission. (AP, *W Post*, 12/21/69, A17; W *Star*, 12/21/69, D7)
- Lunar rock sample weighing 34.1 gm left Lisbon, Portugal, for Washington, D.C., after eight-day exhibit that attracted some 60,000 persons, including 400 leaders of Portuguese scientific, military, and educational world. Rock, shown on Portuguese TV to estimated 9 million viewers, would be sent to Prague. (*NYT*, 12/21/69, 22)
- Bell Aerosystems Co. announced it would become Bell Aerospace Co. Div. of Textron, Inc., effective Jan. 3, 1970, when Bell Aerospace Corp. would merge into Textron. (Bell Aerosystems Release 52)

December 21: *Apollo 12* astronauts and wives attended Sunday services at White House and later were guests at informal dinner given by President and Mrs. Nixon. They stayed overnight at White House. (*PD*, 12/22/69, 1763; AP, B *Sun*, 12/22/69, A3)

- *Philadelphia Inquirer* editorial criticized Dec. 15 proposal of Columbia Univ. scientist Dr. Gary V. Latham, principal investigator of seismic experiments for NASA Apollo missions, to set off nuclear blast on moon: "... the whole idea seems fundamentally repugnant and whether scientifically sound or not likely to lead into ramifications outweighing any possible return in scientific knowledge.... Reaction against such an experiment would be even more intense, if it were undertaken in cooperation, as proposed, with the Soviet Union. Nuclear devices are not toys to be used for scientific game playing anywhere on earth or in space." (*P Inq*, 12/21/69)

December 22: USAF ordered temporary halt to F–111 operations following 15th F–111 crash since aircraft's inception in 1965. Accident occurred 45 mi northwest of Nellis AFB, Nev., when aircraft, in shallow dive to drop smoke bombs and fire rockets, struck ground past target and exploded, killing USAF pilots Maj. James L. Anthony and L/C Thomas J. Mack. (Witkin, *NYT*, 12/23/69, 13)

- Washington *Evening Star* editorial commented on Dec. 17 USAF termination of Project Blue Book: "Loyal saucer people, undeterred by bureaucratic huffiness, will now look forward to that impending twilight when a strange galaxy of glowing, humming, reds, greens and lavenders will settle over the [USAF] Academy at Colorado Springs, spirit it away to Arcturus or Andromeda and put the specimens in a zoo." (W *Star*, 12/22/69, A14)

December 23: U.S.S.R. launched two Cosmos satellites. *Cosmos CCCXVI* entered orbit with 1,638-km (1,017.8-mi) apogee, 147-km (91.3-mi) perigee, 102.8-min period, and 49.4° inclination. *Cosmos CCCXVII* entered orbit with 296-km (183.9-mi) apogee, 191-km (118.7-mi) perigee, 89.3-min period, and 65.4° inclination; it reentered Jan. 5, 1970. (GSFC *SSR*, 12/31/69; 1/15/70)

• *Apollo 11* Astronaut Neil A. Armstrong joined comedian Bob Hope in entertaining 15,000 U.S. troops at field headquarters of 25th Infantry Div. near Saigon, South Vietnam. (UPI, W *Star*, 12/24/69, A5)

• Astronauts Alan L. Bean, R. Walter Cunningham, and Joseph P. Kerwin were grounded by NASA for 30 days for minor infractions of aircraft flight rules. Groundings—apparent attempt by NASA to emphasize importance of safety during jet training flights—were first for astronauts; in past, infractions had been discussed at weekly pilot meetings. (O'Toole, *W Post*, 1/7/70, A3)

• House adopted H.R. 765, authorizing Smithsonian Institution to display U.S. flag presented to House by *Apollo 11* astronauts. (*CR*, 12/23/69, H13090)

• Dr. Robert C. Seamans, Jr., Secretary of the Air Force, announced selection of McDonnell Douglas Corp. as prime contractor for development and production of F–15 advanced tactical fighter aircraft. Initial award would be $1,146,385,000 fixed-price contract for engineering, design, and fabrication of 20 aircraft. First wing of 107 aircraft would be produced under later contract, not to exceed $936,591,000. (DOD Release 1095–69)

December 24: Tiros Operational Satellite (TOS) System—joint effort of NASA and ESSA to provide systematic, daily, worldwide cloud-cover observations—was adjudged successful by NASA. TOS system was operated in support of National Operational Meteorological Satellite System (NOMSS). It included four advanced vidicon-camera-system stored-data spacecraft for remote global cloud-cover observations and five automatic-picture-transmission (APT) direct-readout spacecraft for local cloud-cover observations. Nine spacecraft had been successfully launched and operated in orbit; tenth had been qualified but had not been launched because of extended life of earlier spacecraft. (NASA Proj Off)

• *Nimbus III*, launched by NASA April 14, was adjudged successful. Spacecraft had achieved primary and secondary objectives July 14 when it completed three months in orbit with all experiments operating successfully. Spacecraft had acquired representative number of global samples from which vertical temperature profiles of atmosphere were derived, mapped global radiative energy balance of earth atmosphere and cloud cover, demonstrated feasibility of surface tracking and obtaining remote data from platforms on earth's surface and in atmosphere via IRLS system and temperature profile determination by infrared spectrometry, and demonstrated SNAP–19 as auxiliary power system. (NASA Proj Off)

• *Ats V* Applications Technology Satellite, launched Aug. 12, was adjudged unsuccessful by NASA. Primary objective—deployment and exercise of gravity-gradient stabilization system—had not been achieved because spacecraft failed to despin properly. Secondary objectives, except for operation of ion engine, were being achieved. (NASA Proj Off)

- NASA's *Biosatellite III*, launched June 28, was adjudged unsuccessful by NASA. Although data obtained from mission had been significant, primary objective—to operate spacecraft for more than 15 days to provide suitable physiological environment for instrumented monkey—had not been achieved. Spacecraft had been deorbited after 8½ days because of monkey's rapid physiological deterioration. Monkey had died shortly after recovery July 7. (NASA Proj Off)
- House Committee on Science and Astronautics' Subcommittee on Space Science and Applications published *Future of the Bioscience Program of the National Aeronautics and Space Administration*. Report concluded bioscientific experimentation on man and animals in first decade of space flight had been "relatively meager." Scientific research on manned flight had been "minimal" and had not included biomedical experimentation of kind done in Biosatellite program. More research was urgently needed on biological influence of zero gravity and effects of change or elimination of rhythmic daily periodicity of animal and plant life on earth. Space technology now permitted advanced scientific research on these and related biological factors.

  While professors exchanged scientific theories freely, NASA representatives inclined toward "protecting jurisdiction authority and extending organizational power within the framework of their agency." Complete information exchange should prevail, especially in bioscience, "which eventually may affect human life." "Rivalry" between science and engineering was "real." In NASA, science was "used" but "rarely accepted in its own right by the engineering management."

  While knowledge from 14-day missions was adequate to proceed with planning of 28-day AAP mission, it was inadequate to proceed safely with proposed 56-day flight or longer planetary flights without monitoring astronauts on 28-day mission in experiments "beyond any yet undertaken in manned flight." Further biosatellite exploration with primates and instruments was "merited if not indeed prescribed." Final evaluation of *Biosatellite III* would determine new mechanisms and improved monitoring of test mammals to be programmed.

  Subcommittee endorsed proposal of Dr. W. Ross Adey, NASA's principal investigator for *Biosatellite III*, that basic measurements be required on all flight crews and recommended reinstitution of biosatellite experiments with animals in earth orbit, extension of bioscience research to planets when technologically feasible, uprating of science role as mission objective, and NASA and Congress delegation of "highest relative priority" to "bioscience programs which offer the most immediate return to the taxpayers through ... medicine and physiology." Subcommittee further recommended NASA examine and implement "to fullest practicable extent" recommendations of President's Scientific Advisory Panel for "new level of biomedical research capability," that OMSF conduct biomedical experiments on astronauts during 28-day AAP missions, and that NASA's scientific investigations be planned "with close and continued participation of the academic research institutions." (Text)
- USAF changed Dec. 22 temporary suspension of F–111A aircraft to com-

plete ban on all F-111A flights as result of investigation to date of Dec. 22 crash at Nellis AFB, Nev., which killed two pilots. Preliminary study indicated cause of crash as "failure of its left wing prior to its impact with the ground." (*CSM*, 1/5/70; W *Star*, 12/25/69, A3)

December 25: U.S.S.R. launched *Intercosmos II* into orbit with 1,169-km (726.4-mi) apogee, 200-km (124.3-mi) perigee, 98.4-min period, and 48.4° inclination to study ionosphere. Tass said equipment on board was designed in Bulgaria, Czechoslovakia, East Germany, and U.S.S.R. (B *Sun*, 12/27/69; GSFC *SSR*, 12/31/69)

December 26: Studies of shock and thermal metamorphism of olivine trachybasalt by nuclear explosion at Nevada Test Site for comparison studies of *Apollo 11* lunar samples were described in *Science*. Preliminary examination of *Apollo 11* lunar surface material suggested many samples would be mafic igneous rocks showing shock metamorphism produced by meteorite impact. (James, *Science*, 12/26/69, 1615–9)

December 26–31: American Assn. for the Advancement of Science held 136th meeting in Boston, Mass. Panel discussion of future of U.S. space program was disrupted by some 50 student protesters who converged on panel chairman Dr. Charles Stark Draper, former director of MIT's Instrumentation Laboratory, and participants. Protesters said they objected to space program because funds could be better spent to solve social ills on earth. They littered podium with signs and papier-mache moon rock. Dr. Draper justified continuation of space program because of its technological and prestige contributions. "Space is a bargain."

Outgoing AAAS president Dr. Walter Orr Roberts met continued objections in his attempt to justify space program on basis of applications and was almost shouted down when he suggested U.S.-U.S.S.R. cooperation in internationalized space stations. In presidential address Dr. Roberts, President of University Corp. for Atmospheric Research, said space technology had "vastly enlarged the arsenals of the United States and the U.S.S.R., bringing to reality hitherto fanciful modes of military surveillance, communications, and weaponry." Time had come "for us to take a bold new step in space. I propose that this nation call upon the Soviet Union to join hands in space, with a jointly conducted, earth-oriented space program that will put the new-found Soviet and American skills in space to work for the direct benefit of man."

Dr. J. Allen Hynek, Northwestern Univ. astronomer and consultant to USAF Project Blue Book, expressed fear that USAF would destroy classified UFO records because of Dec. 17 termination of project. "I do not believe we are being visited by little green men. . . . I do believe a phenomenon exists, and that it is worthy of scientific attention."

Dr. Lewis M. Branscomb, Director of National Bureau of Standards, and Dr. Gordon J. F. MacDonald, Vice Chancellor of Univ. of California at Santa Barbara, called for hard, new look at manned space flight spending and urged moratorium on pressure for early manned missions to Mars.

Dr. S. Fred Singer, Deputy Assistant Secretary of Interior, said, "If we downgrade the manned space program we may find we don't have any space program."

Panel of young scientists from Harvard Univ. and MIT, invited by AAAS, attacked "misuse" of science in formal papers presented before meeting. Allen S. Weinraub, graduate student at Harvard, said science was controlled by Federal Government and large corporations that held pursestrings. Panel member Larry W. Beeferman of MIT called NASA "the national aerospace subsidy administration."

At press conference following meeting, Presidential adviser on hunger Dr. Jean Mayer said, "We can feed everybody—and feed them very well—and go to the moon as well."

Dr. Frank J. Low, Univ. of Arizona scientist, reported discovery of "Irtrons"—cells of creation—at centers of 12 galaxies so far, including Milky Way. He believed both matter and antimatter were created in each and then annihilated each other, continuously spraying out debris to fill universe, making stars, planets, and new worlds. Theory had originally been proposed by British astronomer Sir James Jeans early in century. (Lannan, W *Star*, 12/28/69, A7; *W Post*, 12/27/69, B6; 12/29/69, A1, A7; 12/30/69, A2; *Science*, 1/2/70, 11–6)

December 27: Scientists from U.S., Canada, and France, participating in NASA-sponsored 25-day probe of aurora borealis Nov. 24 to Dec. 18, had discovered very high ratios of red oxygen emissions to blue nitrogen emissions in earth's upper atmosphere, NASA announced. Higher measured ratios indicated many solar electrons that bombarded atmosphere might have been low energy. There might be more oxygen in polar atmosphere than at other latitudes, or high temperatures in upper polar atmosphere. Studies in airborne laboratory—NASA Convair 990 jet aircraft *Galileo*—also detected bombardment of earth's atmosphere by protons. Comparison of light emission caused by protons and electrons would help explain particles' origin. Measurement from six photometers and three spectrometers would provide explanation of newly discovered phenomena. (NASA Releases 69–165, 69–169)

December 28: Sigurd A. Sjoberg became MSC Director of Flight Operations, succeeding Dr. Christopher C. Kraft, Jr. Sjoberg had been Deputy Director of Flight Operations since 1963. (MSC Release 70–1)

December 29: NASA announced decision to close Electronics Research Center at Cambridge, Mass. Dr. Thomas O. Paine, NASA Administrator, told staff: ". . . we must effect reductions and consolidations across the board if we are to reshape our programs to meet the nation's future needs in aeronautics and space. . . . We are simply faced with the hard fact that NASA cannot afford to continue to invest broadly in electronics research as we have in the past." Phasing down of ERC work would begin immediately. Final plans were in preparation for placement of personnel and disposition of real property in Cambridge. Center, which opened Sept. 1, 1964, had 850 employees. Six buildings, representing $30-million investment, were in final phases of construction. (NASA Release 69–171)

- NASA exhibit at Expo 70 in Osaka, Japan, March 15 through September would feature F–1 rocket engine, models of *Explorer VII, VIII,* and *XI,* tools being developed for Saturn V Workshop, and Saturn V launch vehicle digital computer, MSFC announced. Exhibit also would display weather and communications satellites, Apollo spacecraft, spacesuits, flight cameras, and moon rocks. (MSFC Release 69–273)

- Five-year program to improve detection of clear air turbulence (CAT)— with cooperation of NASA, DOT, DOD, and ESSA—was announced in Washington, D.C., by ESSA Administrator, Dr. Robert M. White. Proj-

ect would seek improvement of long-range weather forecasting and development of onboard portable detection devices for aircraft to locate CAT in time to permit pilot to take evasive action or prepare for penetration. (UPI, *NYT*, 12/29/69, 57)
- President Nixon announced intention to appoint Abbott M. Washburn as U.S. Representative to Plenipotentiary Conference on Definitive Arrangements for the International Telecommunications Satellite Consortium, with rank of Ambassador, succeeding William W. Scranton, whose resignation would be effective Dec. 31. (*PD*, 1/5/70, 5)

December 30: Cambridge, Mass., city officials and industrialists set up study group to protect their interests in $60-million ERC, which NASA was closing [see Dec. 29]. Municipal officials called NASA move "a clear breach of faith on the part of the Federal Government" and hinted possible court action to stop closing, on breach of contract grounds. They said they had not been consulted about NASA's decision. (Fenton, *NYT*, 12/31/69, 10)
- In *Pravda*, Soviet Academician and automation expert Dr. Boris N. Petrov said U.S.S.R., in effort to conserve resources, had decided not to compete with U.S. in manned lunar exploration, but to concentrate on launching unmanned vehicles into outer space and on creation of permanent space stations around earth. "Our program by no means excludes manned flights to the moon, but at the present time, we attach prime importance to lunar exploration by unmanned vehicles. The economic side is of no small importance. Unmanned vehicles are many times less expensive than manned." (Gwertzman, *NYT*, 12/31/69, 11)
- Soviet scientist Nikolay A. Kozyrev told Soviet Committee for Inventions and Discoveries he had discovered active volcano on moon. He had detected glowing rocks in Aristarchus crater in 1955 and obtained spectrogram of crater's gas in 1958. Comparing spectrogram with others obtained from active volcanos in Soviet Far East, he had found them similar. (UPI, *W News*, 12/31/69, 9)
- Secretary of Transportation John A. Volpe announced 10-mo extension, to Oct. 25, 1970, of FAA rule setting hourly flight quotas at five high-density airports serving New York, Chicago, and Washington, D.C. Decision was based "on operational experience . . . which clearly establishes that the traveling public has benefited substantially from this measure." (FAA Release 69-137)

December 31: President Nixon in San Clemente, Calif., announced 1969 recipients of National Medal of Science, Federal Government's highest award for distinguished achievement in science, mathematics, and engineering, including: Purdue Univ. chemist Herbert C. Brown, for "discovery and exploration of hydroboration reaction and for developing it into a major and powerful tool in chemical synthesis"; Princeton Univ. mathematician William Feller, "for original and definitive contributions to pure and applied mathematics, for making probability available to users, and for pioneering work in establishing *Mathematical Reviews*"; Jack S. C. Kilby, Texas Instruments Inc., for "original conceptions and valuable contributions in the production and application of integrated circuits"; and Wolfgang K. H. Panofsky, Director of Stanford Univ. Linear Accelerator Center, for "classic experiments probing the elementary particles of matter and for contributions to advancing the means of experimentation in this challenging field." (*PD*, 1/5/70, 10–1)

During December: NASA issued *Apollo 11: Preliminary Science Report* (NASA SP–214), summarizing scientific findings of first manned lunar landing mission July 16–24. In foreword Dr. Thomas O. Paine, NASA Administrator, said: "The concept of traveling across the vastness of space to new worlds has stirred the imagination of men everywhere. . . . The success of this mission has opened new fields of exploration and research . . . which will lead to a greater understanding of our planet and provide a new insight into the origin and history of the solar system."

Report included photographic review of *Apollo 11* mission with observations by crew—Astronauts Neil A. Armstrong, Edwin E. Aldrin, Jr., and Michael Collins. Descriptions of geologic setting of lunar material, soil mechanics investigation, passive seismic experiment, laser ranging retroreflector, and solar wind composition experiment were provided by principal investigators.

In addition to major findings reported Sept. 15 in summary by NASA Preliminary Examination Team (PET), SP–214 reported passive seismic experiment package deployed on moon had operated satisfactorily for 21 days and had detected seismic signals from astronaut activity or LM motions. Whether actual lunar seismic events had been detected was uncertain. Laser reflector deployed on moon had been used as target for earth-based lasers and distance to moon had been measured to within four-mile accuracy. Future studies would be made on distance variation to study motion of moon and earth. Preliminary analysis had been made on part of aluminum foil and showed helium, neon, and argon; isotopic composition of each element had been measured. (Text; NASA Release 69–160)

- Dr. Robert R. Gilruth, MSC Director, was named to receive first "at large" award in 17-yr history of $10,000 Rockefeller Public Service Awards for "distinguished service to the government of the United States and to the American people." Other 1969 recipients included Dr. John W. Evans, Director of USAF's Sacramento Peak Observatory, in science, technology, or engineering category. (*AF/SD*, 1/70, 24–5)
- Vice President Spiro T. Agnew, NASC Chairman, discussed future of U.S. space program in *Space Age News:* "It is my personal belief that a vigorous space effort is essential to the welfare of this country, particularly since it has contributed so much to our international prestige, our national security and our economy. I can assure you that the President shares this view." Through 11 yrs of space activity U.S. had "gained one of its most priceless resources—trained, experienced professionals capable of creating, managing, and operating a complete range of space systems dedicated to bringing the benefits of space within reach of all of us here on Earth." U.S. had "just begun its space program. The success of the first decade indicates an enlarged effort in the decade to come." (*Space Age News*, 12/69)
- Coming age of economy space flight was discussed by J. S. Butz, Jr., in *Air Force and Space Digest*. Space shuttle was key to opening of space "much as the railroads opened a stream of travel into the American West." It was "complex and expensive system that is within our grasp

if the nation's technology and management are willing to meet challenges of herculean proportions." In prospect was day when astronauts would be as numerous as present day airline pilots, space flights would be scheduled almost daily, and "virtually any young man who yearns to voyage into space will be able to do so at some point in his life." (*AF/SD*, 12/69, 37–44)

- *Space/Aeronautics* described 1969 as U.S.S.R.'s "most active and most frustrating" year in space. "In the fall of 1968 it was possible to believe . . . that the Soviet Union might fly men around the moon before the United States. A year later it was possible to ask: 'Where are the Russians?'" Most failures were in attempts to prepare man's way to moon. "The inglorious performance of Luna 15, just as America was landing men on the lunar surface, could almost be called the least of Russia's lunar problems. By far the worst . . . must be repeated technical troubles with the long-awaited 'super booster,' the rough equivalent of Saturn 5. If semi-official rumors are to be accepted as accurate, the Russian vehicle is in a great deal of difficulty.

  "The . . . launchings of Soyuz 6, 7, and 8 were impressive in many ways—but Western officials still counted the group flight as a disappointment to the Russians. . . . The most skeptical Western estimates now are that the Russians may have abandoned the big booster, that they will not land men on the moon before 1972 . . . and that no large Soviet station will be assembled in orbit before the last half of the '70s." (*Space/Aeronautics*, 12/69, 24–6)

- "Dethronement" of Dr. Charles Stark Draper as head of MIT Instrumentation Lab had resulted from campaign waged by "coterie of self-appointed zealots who insist that defense research is 'war research' and cannot be tolerated," William Leavitt said in *Air Force and Space Digest* article. On campus these days, "genius is not enough to keep you on the job. Your work has to be politically, morally, and socially acceptable" to these zealots. (*AF/SD*, 12/69, 46–50)

- NSF published *American Science Manpower, 1968*, report of National Register of Scientific and Technical Personnel: Almost 298,000 scientists—nine percent of whom were women—reported to National Register in 1968. Of these, three-fifths were in physical and mathematical sciences, one-fifth in life sciences, remainder in behavioral and social sciences. Registration was up 23% over 242,800 in 1966. Industry employed 32% of 1968 registrants, down from 34% in 1966. Scientists employed in educational institutions increased from 36% to 40%. Those in Federal Government remained at 10% both years. Federal Government provided funds for at least some of work of 127,400 scientists, or 43% of total registrants. (Text)

- In *American Scholar* architect R. Buckminster Fuller said: "It seems eminently clear that we not only must put our space programs on highest priority of attention and resource investment but that all humanity must be accredited and financed to enter into a new re-educational system that is geared to develop our most prominent awareness, that we indeed *are in space* and that all of our concern is with the fact that our space-vehicle Earth and its life-energy-giving Sun, and the tide-pumping Moon can provide ample sustenance and power for all humanity's needs to be derived from our direct energy income without further robbing our fossil fuels energy savings account. In reality, the

Sun, the Earth and the Moon are nothing else than a most fantastically well-designed and space-programmed team of vehicles. All of us *are,* always *have been,* and so long as we exist, *always will be—nothing else but—astronauts.* Let's pull our heads out of the brain benumbing, mind frustrating, misinformedly conditioned reflexes. If it is going to be 'All ashore who's going ashore,' once more intent to return to nonspace DOWN HERE ON EARTH, humanity is doomed." (*American Scholar,* Winter 1969–70, 27–47)

During 1969: In 1969—the year man first set foot on another celestial body—U.S. orbited 54 spacecraft and U.S.S.R., 70. U.S. total included 31 orbited by DOD. NASA's 23 included 2 satellites orbited for DOD as secondary payloads.

Highlight of 1969 was achievement of major goal in NASA's Apollo program—to land man on the moon and return him safely to earth. Lunar landing system completed final phases of test and demonstration when *Apollo 9* demonstrated operation of LM in earth orbit and *Apollo 10* demonstrated operation of complete Apollo spacecraft in lunar orbit, with LM making two passes within 47,000 ft of lunar surface. Apollo program climaxed with *Apollo 11,* July 16–24, when LM *Eagle* landed on moon and Astronauts Neil A. Armstrong and Edwin E. Aldrin, Jr.—first men on the moon—performed tasks on lunar surface, redocked with CSM in lunar orbit, and returned safely to earth with lunar samples. *Apollo 12* successfully accomplished second manned lunar landing with pinpoint accuracy, touching down close to *Surveyor III,* which had landed on moon April 19, 1967. Crew set up first ALSEP on lunar surface and returned safely to earth with lunar samples and parts of *Surveyor III.*

Unmanned *Mariner VI* and *Mariner VII* passed within 2,000 mi of Mars, transmitting more than 200 times more data about Mars than were obtained by *Mariner IV* in 1964. Spacecraft measured chemical composition and temperature of atmosphere and surface and took 198 high-quality pictures, providing new insight into Mars surface characteristics. Other scientific achievements included orbiting of *Oso V, Oso VI, Ogo VI,* and *Explorer XLI.* Applications satellites included *Intelsat-III F–3, Intelsat-III F–4,* and *Intelsat-III F–5* for ComSatCorp; *Essa IX* meteorological satellite for ESSA; *Nimbus III;* and *Ats V. Biosatellite III* provided significant biological data on instrumented monkey on board, but mission was terminated early after monkey's physiological condition deteriorated, and monkey died shortly after recovery. Pioneer E, scheduled to enter solar orbit, and its secondary payload TETR–C were destroyed shortly after launch when Delta booster malfunctioned.

Some 189 meteorological sounding rockets, 85 scientific sounding rockets, and 60 high-altitude balloons were launched.

NASA–USAF XB–70 and X–15 programs were concluded and new joint program was initiated to continue aeronautical R&D, using YF–12 supersonic aircraft. First operational C–5A, military cargo transport and world's largest aircraft, was delivered to USAF and by November seven C–5As had accumulated 1,320 hrs flying time. Aeronautical R&D continued to improve quality and quantity of options available in aircraft technology, and aircraft industry made first flights of several wide-bodied jumbo jet transports for civil use, including Boeing 747.

New SST design with delta wing was approved by FAA and prototype construction was recommended by President Nixon.

NASA's HL-10 lifting-body vehicle completed 3 successful glide flights and 15 powered flights included first supersonic flight at mach 1.1 and 54,000-ft altitude. USAF's X-24A lifting-body vehicle completed 7 flights.

DOD space program included orbiting of 2 Vela nuclear-detection satellites, 7 Orbiting Vehicle research satellites, *Tacsat I* tactical comsat, *Egrs XIII* (*Secor XIII*) geodetic satellite (launched by NASA), and U.K.'s *Skynet A* comsat (launched by NASA). Titan IIIM development was terminated and USAF's Manned Orbiting Laboratory (MOL) program was canceled because of budget cutbacks.

NASA and AEC completed ground tests of XE experimental rocket engine with 28 successful startups and 2.8 hrs of operation, including 3.5 min at full thrust, and validated test stand No. 1 in Jackass Flats, Nev. SNAP-27 generator was used as electrical power source for *Apollo 12* experiments left on moon and SNAP-19 was auxiliary power system for *Nimbus III*. SNAP-29 program was terminated. New liquid-propellant rocket engine was developed and new tripropellant system produced record performance for chemically powered engines.

U.S.S.R.'s 70 payloads included 55 Cosmos satellites and 2 Intercosmos, 2 Meteor, 1 Luna, 1 Zond, 5 Soyuz, and 2 Venus spacecraft, and 2 Molniya I comsats. Manned *Soyuz IV* docked with manned *Soyuz V* and two cosmonauts transferred from *Soyuz V* to *Soyuz IV* before returning to earth. Manned *Soyuz VI, VII,* and *VIII,* launched on three successive days, rendezvoused in orbit and performed experiments including photography, communications, and welding. *Luna XV,* apparent unmanned challenger to NASA's *Apollo 11* for first lunar landing and return, crashed onto lunar surface in July. U.S.S.R.'s Tu-144 delta-wing supersonic transport successfully reached speeds up to 900 mph and was demonstrated publicly in 90-min test flight. (*Pres Rpt 70* [69]; NASA Release 69-161; GSFC *SSR*, 12/31/69)

- Year 1969 was one of critical decision and realignment of programs for NASA as primary goal of first decade was achieved through Apollo program. Post-Apollo planning had preceded the coming of the Nixon Administration and *Apollo 11*. Budget retrenchment accelerated throughout Government as cost of Vietnam war continued high and White House sought to lessen inflation by fiscal economy. Nationwide reevaluation of societal priorities emphasized problems of pollution, poverty, and crime.

  Successful lunar landings of *Apollo 11* and *12*—seen and acclaimed worldwide—mushroomed U.S. enthusiasm for manned space achievement. NASA announced development of plans for further lunar missions and post-Apollo planetary exploration. But year was turning point, with completion of a major goal and examination of possible new goals and priorities for the next decade. Phasing down of Apollo program after climax brought shifts in program emphasis and in personnel. Most of initial Saturn V launch vehicles had been completed. Hardware requirements would decrease as missions slowed. Peak costs during 1960s had built NASA base of people, technology, and facilities as national capability to conduct space missions.

  Future space objectives and budget options were examined, within

Government and outside, for balance between more manned missions and more unmanned scientific exploration and applications of knowledge already gained. Criticism of past emphasis on engineering rather than on science was partly answered by decision to provide more time for scientific study and planning between the next Apollo lunar missions. Program plans included Apollo Applications, missions to Mars and Jupiter, long-range studies of Venus-Mercury mission and three-planet grand tours, development of permanent space station, joint NASA–DOD studies of space transportation system with reusable shuttle, and accelerated aeronautics activities.

NASA FY 1970 appropriation of $3.697 billion was $181 million below budget proposed by outgoing President Johnson and $19 million below President Nixon's amended request for $3.716 billion. Cutbacks in Government spending forced NASA to announce 1970 closing of Electronics Research Center and DOD to cancel Manned Orbiting Laboratory (MOL) program. Dr. Thomas O. Paine, NASA Administrator, told ERC employees in December that agency had to "effect reductions and consolidation across the board if we are to shape our programs to meet the nation's future needs in aeronautics and space." Reductions in work force—largely in contractor personnel—were brought by planned phasing out of Apollo program as well as by budget cuts. Year of man's first landing on moon, despite its global impact, ended with space goals and levels of effort for the next decade an open question. (*Pres Rpt 70* [69]; *A&A 69*)

- In its international cooperation program, NASA successfully launched three foreign satellites—Canadian *Isis I*, third of series to conduct ionospheric measurement; *Boreas* (*Esro 1B*) for European Space Research Organization on reimbursable basis; and West German *Azur* under cooperative project to conduct radiation belt measurements. Agreements for new projects were reached with U.K., Italy, Netherlands, and Canada. U.K. agreement called for 1971 launch of satellite on NASA Scout vehicle to measure ionospheric particles and radio noise and experiment to fly on board NASA's Nimbus-E in 1972. Italian agreement provided for 1970 launch of San Marco-C satellite on NASA Scout vehicle from Italian San Marco range in Indian Ocean and use of range for launching NASA spacecraft into equatorial orbit.

  Under Netherlands agreement two Dutch scientists would participate as observers in OAO program. Canadian agreement called for reimbursement launching of Canadian comsat. Project Helios, cooperative endeavor with West German Ministry for Scientific Research to place two solar probes (1974–1975) closer to sun than any other spacecraft, would obtain data on solar processes and earth-sun relationship. Under another agreement with West Germany, Project Aeros, NASA would launch German-developed satellite in 1972 for aeronomy measurements. In agreement with Indian space agency, ATS–F satellite would be made available for experimental educational TV broadcasting directly and by relay to receivers in 5,000 Indian villages.

  Additional cooperative projects during 1969 included launch of one Italian, two U.K., and two French experiments; continuation of sounding rocket projects with Australia, Brazil, Canada, India, Japan, Norway, Pakistan, Spain, and Sweden; inclusion of Canadian and French experiments on Convair 990 flights to study auroras in northern latitudes; and extension of aeronautical research projects with Canada, France, West Germany, and U.K. Agreement with Spain for NASA Madrid tracking station was extended in June until 1984 and negotiations were begun with Australia for extension of tracking arrangements until same year.

Lunar surface material returned by *Apollo 11* astronauts was distributed to 39 principal investigators from nine countries—Australia, Belgium, Canada, Finland, West Germany, Japan, South Africa, Switzerland, and U.K.

International highlights in 1969 included around-the-world tour of *Apollo 11* astronauts to dramatize conviction that successful lunar landing was triumph for all mankind; travels of NASA Administrator, Dr. Thomas O. Paine, to inform ministerial and space agency officials abroad of U.S. space plans and invite their participation; and participation of foreign experts in NASA conference on space shuttle concepts.

Efforts by NASA to develop cooperative programs with U.S.S.R. continued during 1969, but met with limited success. (*Pres Rpt 70* [69])

- Drone aircraft was reportedly impaled with spear of light from experimental laser and destroyed in three milliseconds at Air Force Special Weapons Laboratory, Kirtland AFB, N. Mex. Account in *Aviation Week & Space Technology* was later denied by USAF officials.

  Tri-service laser weapon research programs, principally sponsored by ARPA, were studying number of military applications of lasers, according to magazine. USAF had increased planned funding of high-energy gas laser development in FY 1969 from $0.5 million to $2.5 million to make use of power output increase achieved by United Aircraft Corp. in 1968. And $1.5 million had been given to exploratory development of high-power solid-state lasers. For FY 1970 USAF had asked $8 million to exploit United Aircraft development. (*Av Wk*, 1/12/70, 16–7; USAF PIO; *A&A*, 4/70, 16–21)

- Aerospace industry sales of $28.3 billion in 1969 were second highest in industry's history despite anticipated decline of 4.1% from 1968 record of $29.5 billion. All areas of aerospace endeavor leveled off except nonaerospace sales by aerospace companies, which increased. Commercial aerospace sales, principally jet transports, were $5,800 million, down from $6,429 million in 1968. Trend resulted from phasing out of current transport models while jumbo jet production was beginning. Sales by aerospace industry to DOD were $16.2 million in 1969 and $16.6 billion in 1968. Military aircraft sales totaled $10 billion in 1969 and $10.7 billion in 1968. Space sales dropped to $4,499 million from $5,108 million in 1968, because of approaching completion of Apollo program hardware phase and cancellation of MOL. (*Aerospace*, Winter, 1970)

- U.S. scheduled airlines carried 159 million passengers; in 1949 they had carried less than 17 million. In 1969 airlines accounted for 75% of common carrier passenger-miles between U.S. cities; in 1949 they had accounted for 14%. In 1969 18.3 million passengers flew between U.S. and other countries; in 1949 there were 2.2 million. U.S. scheduled airlines produced 3.2 billion ton-miles of air freight service in 1969, increase of 15.7% over 1968. (*1970 Air Transport Facts & Figures*)

- Air carrier accidents decreased from 73 in 1968 to 68 in 1969. Accident rate was down from 1.12 per 100,000 hrs flown to 0.98. Fatal accidents

had dropped from 16 to 10 and total fatalities from 352 to 160, lowest since 1957. Fatality rate was 2.3 per 100,000 hrs flown; it had been 5.4 in 1968. Total air carrier accidents had declined steadily from 102 in 1959 to 1969's 68. (FAA Release 70-2)
- Number of active pilots in U.S. reached 720,028, rise of 4% over 1968 total. Student pilots dropped 3% from end of 1968, to 203,520 at end of 1969. (FAA Release 70-37)
- Best seller of books published by *New York Times* during year was *We Reach the Moon* by John Noble Wilford. Book sold 1 million copies in U.S. and abroad. (*NYT*, 2/22/70)

Appendix A

SATELLITES, SPACE PROBES, AND MANNED SPACE FLIGHTS

A CHRONICLE FOR 1969

The following tabulation was compiled from open sources by Leonard C. Bruno of the Science and Technology Division of the Library of Congress. Sources included the United Nations Public Registry; the *Satellite Situation Report* issued by the Operations Control Center at Goddard Space Flight Center; public information releases of the Department of Defense, NASA, ESSA, and other agencies, as well as those of the Communications Satellite Corporation. Russian data are from the U.N. Public Registry, the *Satellite Situation Report*, translations from the Tass News Agency, statements in the Soviet press, and international news services' reports. Data on satellites of other foreign nations are from the U.N. Public Registry, the *Satellite Situation Report*, government announcements, and international news services' reports.

This tabulation lists payloads that have (*a*) orbited; (*b*) as probes, ascended to at least the 4,000-mile altitude that traditionally has distinguished probes from sounding rockets, etc.; or (*c*) conveyed one or more humans into space, whether orbit was attained or not. Furthermore, only flights that have succeeded—or at least can be shown by tracking data to have fulfilled our definition of satellite or probe or manned flight—are listed. Date of launch is referenced to local time at the launch site. An asterisk by the date marks those dates that are one day earlier in this tabulation than in listings which are referenced to Greenwich Mean Time. A double asterisk by the date marks dates of Soviet launches which are a day later in this compilation than in listings which are referenced to Greenwich Mean Time.

World space activity declined for the second straight year. There was a decline in the total successful launches—110 against 119 in 1968—and a decline in total payloads orbited—124 against 138 in 1968. The difference between launches and payloads is of course accounted for by the multiple-payload launches (DOD is the principal user of this system, with 7 multiple-payload launches orbiting a total of 19 payloads and as many as 5 payloads on one Titan IIIC; NASA had 2 multiple launches totaling 4 payloads in 1969 and the same in 1968; the U.S.S.R., whose last multiple-payload launch was in 1965, had none).

Of the 1969 world total, the United States launched 40 boosters carrying 54 payloads (compared to 45 and 64 in 1968). Of these totals, DOD was responsible for 19 launches and 31 payloads. The 40 launches made the lowest U.S. total since 1963, and the 54 payloads are the fewest since 1961. Eight of NASA's total were non-NASA missions—*Intelsat-III F-3, Intelsat-III F-4, Intelsat-III F-5, Boreas, Azur, Skynet A, Essa IX,* and *Egrs XIII.* The U.S.S.R. launched 70 boosters in 1969; it had launched 74 in 1968.

In 1969 manned space flight peaked both in quantity and in achievement. Four U.S. Apollo flights put 12 astronauts in space and five U.S.S.R. Soyuz flights orbited 11 cosmonauts, for a record 23 men in space in one year. *Apollo 11* was naturally the most dramatic, with the United States achieving its manned lunar landing goal. In steady progression, the successes of *Apollo 9* in earth orbit and *Apollo 10* in lunar orbit prepared the way for the *Apollo 11* lunar touchdown on July 20, 1969. The *Apollo 12* follow-up lunar landing four months later further demonstrated man's ability to function in the lunar environment.

The Soviet Union performed two lunar missions in 1969, both unmanned. *Luna XV* impacted the moon and *Zond VII* was the third successful Soviet circumlunar flight, reentry, and recovery. The U.S.S.R. also launched two Venus probes, both of which landed on the planet. The five Soviet manned flights in earth orbit performed rendezvous and docking and accomplished the first crew exchange in orbit.

As we have cautioned in previous years, the "Remarks" column of these appendixes is never complete because of the inescapable lag behind each flight of the analysis and interpretation of the results.

Launch Date	Name, Country, International Designation, Vehicle	Payload Data	Apogee (st mi)	Perigee (st mi)	Period (min)	Inclination (degrees)	Remarks
Jan. 5	Venus V (U.S.S.R.) 1969-1A Not available	Total weight: 2,491 lbs. Objective: Softland spacecraft on planet Venus. Payload: Not available.	Impacted on Venus				Venus V (Venera V) jettisoned instrument capsule 5/16/69 before entry into Venusian atmosphere. Capsule decelerated by parachute and transmitted data during 53-min descent to planet's night side. Measured chemical composition, pressure, density, and temperature of planet's atmosphere. Results reinforce findings of Venus IV: 93%-97% carbon dioxide and pressure 20 times heavier than earth. Capsule apparently failed to reach planet's surface with transmitters still operating.
Jan. 10	Venus VI (U.S.S.R.) 1969-2A Not available	Total weight: 2,491 lbs. Objective: Softland spacecraft on planet Venus. Payload: Not available.	Impacted on Venus				Venus VI (Venera VI), launched 5 days after Venus V, reached Venusian atmosphere 5/17/69. Instrumented capsule ejected, decelerated, and transmitted 51-min of data which, like that received from Venus V, indicated planet to be inhospitable to man. Capsule also apparently failed to reach surface of Venus with transmitters still operating.
Jan. 12	Cosmos CCLXIII (U.S.S.R.) 1969-3A Not available	Total weight: Not available. Objective: Continuation of Cosmos scientific satellite series. Payload: Not available.	225	129	89.6	65.4	Reentered 1/20/69.
Jan. 14	Soyuz IV (U.S.S.R.) 1969-4A Not available	Total weight: Not available. Objective: Perform wide range of tests in interaction of two orbiting spaceships. Payload: 3 unit spacecraft with	139	132	88.8	51.7	Launch of Soyuz IV, carrying Cosmonaut Vladimir Shatalov, shown on Moscow TV video recording. During 34th orbit, Shatalov switched to

ASTRONAUTICS AND AERONAUTICS, 1969

Launch Date	Name, Country, International Designation, Vehicle	Payload Data	Apogee (st mi)	Perigee (st mi)	Period (min)	Inclination (degrees)	Remarks
		2 crew cabins, TV cameras, 2 large wing-like solar panels.					manual control and rendezvoused and docked with *Soyuz V* (launched 1/15/69), carrying Cosmonauts Yevegeny Khrunov, Boris Volynov, and Alexey Yeliseyev. Link-up covered on live TV. Two cosmonauts (Khrunov and Yeliseyev) after 1 hr EVA, entered service compartment of *Soyuz IV* accomplishing first crew exchange in orbit. After 4 hrs 35 min of docked flight, spacecraft undocked and went into separate orbits. *Soyuz IV* landed safely 1/17/69 with 3-man crew.
Jan. 15	*Soyuz V* (U.S.S.R.) 1969-5A Not available	Total weight: Not available. Objective: Perform wide range of tests in interaction of two orbiting spaceships. Payload: 3 unit spacecraft with 2 crew cabins, TV cameras, 2 large wing-like solar panels.	132	122	88.6	51.7	*Soyuz V* launched one day after *Soyuz IV*, carrying Cosmonauts Yevegeny Khrunov, Boris Volynov, and Alexey Yeliseyev. Rendezvoused with *Soyuz IV* and docked during 18th orbit. Cosmonauts Khrunov and Yeliseyev completed first manned transfer (to *Soyuz IV*) after 1 hr EVA. Spacecraft uncoupled after 4 hrs 35 min of docked flight and went into separate orbits. *Soyuz V* with Cosmonaut Volynov aboard, landed safely 1/18/69.
Jan. 22	*Oso V* (United States) 1969-6A Thor-Delta	Total weight: 636 lbs. Objective: Obtain high-resolution spectral data from pointed experiments within range 1 A to 1,250 A during a solar rotation, including raster scans of solar disc in selected wavelengths.	353	338	95.8	32.9	*Oso V* launched into nominal circular orbit; all spacecraft systems operated satisfactorily; all experiments functioned and returned good data. By 2/25/69 primary objective achieved. Still in orbit, transmitting on command.

Jan. 22	DOD spacecraft (United States) 1969-7A Titan IIIB-Agena D	Payload: Top part of spacecraft a 22"-radius semicircular sail continuously pointed at sun with 3 experiments, 4 sq ft of n-on-p solar cells; lower part a 44" x 9" 9-sided revolving wheel containing 5 nonpointed experiments that scan sun every 2 secs; controls, telemetry, recorder, batteries.	673	92	96.9	106.1	Reentered 2/3/69.
Jan. 23	Cosmos CCLXIV (U.S.S.R.) 1969-8A Not available	Total weight: Not available. Objective: Continuation of Cosmos scientific satellite series. Payload: Not available.	183	129	89.5	69.9	Reentered 2/5/69.
Jan. 30	Isis I (United States-Canada) 1969-9A TAID	Total weight: 520 lbs. Objective: Study topside of ionosphere above electron peak of F region. Continue U.S.-Canadian program of ionospheric studies by combining sounder data with correlative direct measurements for time sufficient to cover latitudinal and diurnal variations during period of high solar activity. Payload: 50" (dia) x 42" 8-sided oblate spheroid spacecraft with frame consisting of central 16" (dia) thrust tube which supports 8 radial ribs; 16 solar cell panels (11,136 n-on-p solar cells) and 8 equatorial panels mounted on ribs. Contains 10 experiments: 2 crossed dipoles 62' and 240' tip-to-tip for ionospheric sounding, 4-pole turnstile array for telemetry and command, quadraloop antenna, 3 nickel-cadmium batteries, spin maintenance and spin attitude control by magnetic torquing.	2,188	357	128.3	88.5	Isis I, 3rd in series of 5 cooperative U.S.-Canadian space efforts to develop better understanding of physics of ionosphere, launched by Thrust-Augmented Improved Thor-Delta. All spacecraft systems performing satisfactorily; 9 of 10 experiments operational. Still in orbit, transmitting on command.

Launch Date	Name, Country, International Designation, Vehicle	Payload Data	Apogee (st mi)	Perigee (st mi)	Period (min)	Inclination (degrees)	Remarks
Feb. 5	DOD spacecraft (United States) 1969-10A LTTAT-Agena D and	Total weight: Not available. Objective: Develop space-flight techniques and technology. Payload: Not available.	171	92	88.7	81.6	Launched by Long-Tank, Thrust-Augmented Thor-Agena D. Reentered 2/24/69.
	DOD spacecraft 1969-10B	Total weight: Not available. Objective: Develop space-flight techniques and technology. Payload: Not available.	895	867	114.1	80.4	Still in orbit.
Feb. 5*	Intelsat-III F-3 (United States) 1969-11A LTTAT-Delta	Total weight: 332 lbs (in synchronous orbit; 632 lbs at lift-off, including 310 lbs of apogee motor fuel. Objective: Place satellite and apogee motor into proper transfer orbit, provide tracking and telemetry and backup calculations through transfer orbit so satellite can be injected into synchronous orbit to provide commercial voice and TV communications. Payload: 78" x 56" (dia) cylindrical satellite, including mechanically despun receive/transmit and omnidirectional command and telemetry antennas on top of spacecraft; outer cylindrical sleeve covered with 10,720 solar cells; 2 frequency-translation-mode repeaters with design capability for 1,200 high-quality, 2-way voice or 4 TV channels, or a mix; attitude and control system; 1 battery.	23,497 After apogee motor firing, 22,200	157 22,190	671.9 1,436.4	29.8 1.3	Intelsat III F-3 put into good transfer orbit by Long-Tank, Thrust-Augmented Thor-Delta. Apogee motor fired 2/7/69, putting satellite into synchronous orbit over Pacific Ocean Commercial operations began 2/16/69. During March, satellite suffered degradation of performance making it unsuitable for heavy traffic load over Pacific. After replacement launch of Intelsat-III F-4 5/21/69, Intelsat-III F-3 shifted 7,000 mi westward to less busy station over Indian Ocean. Still in orbit.
Feb. 7	Cosmos CCLXV (U.S.S.R.) 1969-12A Not available	Total weight: Not available. Objective: Continuation of Cosmos scientific satellite series. Payload: Not available.	284	169	91.8	71.0	Reentered 5/1/69.

432

Feb. 9	*Tacsat I* (United States) 1969-13A Titan IIIC	22,387	22,332	1,446.6	0.6

Total weight: 1,600 lbs.
Objective: Test feasibility of using synchronous satellites for tactical communications with small ground stations, aircraft, and ships; determine best frequency bands to be used for tactical purposes.
Payload: 25' x 9' (dia) cylinder; 3 antenna systems consist of 5 helical UHF antennas, 2 microwave horns for X-band communication, and biconical horn for telemetry and command; all mechanically despun; 60,000 solar cells.

Tacsat I, largest and most powerful experimental communications satellite to date, put into synchronous orbit over Pacific. Communications capacity of spacecraft comparable to 10,000 2-way telephone channels. Still in orbit.

Feb. 24*	*Mariner VI* (United States) 1969-14A Atlas-Centaur		Heliocentric orbit		

Total weight: 840 lbs.
Objective: Conduct flyby mission to make exploratory investigation of Mars that will set basis for future experiments, particularly those relevant to search for extraterrestrial life; provide topographic information about Martian surface, provide atmospheric profile information, and measure principal atmospheric constituents.
Payload: 4½'-high x 19' spacecraft (cruise position with solar panels deployed), basically octagonal structure with 8 bays; top of spacecraft supports omnidirectional low-grain pole antenna and high-gain dish antenna, attitude-control system, Canopus star tracker, sun sensors, and 4 solar panels (17,742 photovoltaic solar cells); scientific experiments (2 TV cameras, infrared spectrometer, ultraviolet spectrometer, infrared radiometer, and 2 planet sensors) mounted on 2° of freedom scan platform on bottom side of spacecraft; telemetry and command; 31-lb silver-zinc battery.

Mariner VI, first of 2 spacecraft in Mariner-Mars 1969 mission, flew past Mars equator at 2,131-mi altitude on 7/31/69 after 2/28/69 midcourse correction. All scientific experiments except 1 of 2 channels of infrared spectrometer operated successfully. Excellent quality live TV pictures taken. Revealed 3 types of Martian terrain: heavily cratered, resembling moon; featureless, desert-like, flat floor covering 1,200 mi; chaotic, irregular terrain similar to landslide. Instruments determined radius of Mars at 2,125 mi and diameter 4,250 mi. Thin atmosphere discovered with CO_2 its major constituent and less than 1% nitrogen. Solid CO_2 found at high altitudes. Polar cap either solid CO_2 or shrouded by thin cloud of solid CO_2. Martian pressure equivalent to atmospheric pressure 30,500-35,000 mi above earth. Daytime temperature about 62°F at equator and −45°F at polar cap edge; nightside of equator −100°F. Lowest temperature recorded −193°F at polar cap. "Canals" considered to

Launch Date	Name, Country, International Designation, Vehicle	Payload Data	Apogee (st mi)	Perigee (st mi)	Period (min)	Inclination (degrees)	Remarks
							be alignments of impact craters with dark floors. Life on Mars, if existing, considered almost certainly microbial. Still in heliocentric orbit.
Feb. 25	Cosmos CCLXVI (U.S.S.R.) 1969–15A Not available	Total weight: Not available. Objective: Continuation of Cosmos scientific satellite series. Payload: Not available.	209	126	89.8	72.0	Reentered 3/6/69.
Feb. 26	Essa IX (United States) 1969–16A TAID	Total weight: 320 lbs. Objective: Place and operate advanced vidicon camera system in sun-synchronous orbit with local equator crossing time between 2:15 and 2:35 pm, so that daily AVCS pictures of entire globe can be obtained regularly and dependably. Payload: 22" x 42" (dia) 18-sided hatbox-shaped polygon, with 18" receiving antenna and 4 22" transmitting whip antennas; containing 2 AVCS cameras, FM transmitters, 2 spin-control systems (magnetic coil; solid-propellant rockets); 4 solar radiation sensors; 63 nickel-cadmium batteries; 10,-020 n-on-p solar cells.	935	884	115.2	101.8	Essa IX was placed into nearly polar sun-synchronous orbit by Thrust-Augmented Improved Thor-Delta as back-up for Essa VII, which, although still functioning, had lost its spare camera. Essa IX spin-down anomaly resulted in spin rate of 20 rpm in lieu of expected 10 rpm. Correct spin-down achieved and satellite turned over to ESSA after complete checkout. Still in orbit, still transmitting.
Feb. 26	Cosmos CCLXVII (U.S.S.R.) 1969–17A Not available	Total weight: Not available. Objective: Continuation of Cosmos scientific satellite series. Payload: Not available.	206	126	89.8	65	Reentered 3/6/69.
Mar. 3	Apollo 9 (United States) 1969–18A Saturn V	Total weight: 289,970 lbs (weight in earth orbit, including S-IVB stage, instrument unit, spacecraft LM adapter, LM, and CSM). Objective: Demonstrate crew/	120	118			Apollo 9 launched successfully carrying Astronauts James A. McDivitt, David R. Scott, and Russell L. Schweickart. First manned flight of complete Apollo spacecraft. After inser-

Date	Name	Weight (lbs)	Perigee (mi)	Apogee (mi)	Incl. (deg)	Remarks
Mar. 4	DOD spacecraft (United States) 1969-19A Titan IIIB-Agena D	280				space vehicle/mission support facilities performance during manned Saturn V mission with CSM and LM; demonstrate LM/crew performance; demonstrate nominal and selected backup Lunar Orbit Rendezvous mission activities; assess CSM/LM consumables. Payload: 114'-long S-IVB/IU spacecraft-LM adapter/Block II command and service modules/lunar module; cameras; telemetry.
						tion into earth orbit, CSM separated from S-IVB, and LM, transposed, and docked with LM; spacecraft separated from 3rd stage. McDivitt and Schweickart entered LM on 3rd day, conducted 1st firing of LM descent propulsion system (DPS), and returned to CSM. Both entered LM on 4th day; Schweickart performed 37 min EVA; both returned to CSM. On 5th day, CSM separated from LM containing McDivitt and Schweickart. LM descent stage jettisoned. LM ascent stage docked with CSM after 6½ hrs separation and max distance of 114 mi. LM ascent stage jettisoned. Crew performed 3 more days of experiments, tracking, and photographs and reentered during 152nd revolution. Splashed down in Atlantic 3/13/69, 241 hrs 53 secs after launch. Reentered 3/18/69.
Mar. 5	Cosmos CCLXVIII (U.S.S.R.) 1969-20A Not available	1,343	131	109.1	48.4	Total weight: Not available. Objective: Develop space-flight techniques and technology. Payload: Not available. Reentered 5/9/70.
Mar. 5	Cosmos CCLXIX (U.S.S.R.) 1969-21A Not available	337	326	95.2	74.0	Total weight: Not available. Objective: Continuation of Cosmos scientific satellite series. Payload: Not available. Still in orbit.
Mar. 6	Cosmos CCLXX (U.S.S.R.) 1969-22A Not available	205	124	89.8	65.4	Total weight: Not available. Objective: Continuation of Cosmos scientific satellite series. Payload: Not available. Reentered 3/14/69.
Mar. 15	Cosmos CCLXXI (U.S.S.R.) 1969-23A Not available	193	116	89.7	65.4	Total weight: Not available. Objective: Continuation of Cosmos scientific satellite series. Payload: Not available. Reentered 3/23/69.
Mar. 17	Cosmos CCLXXII (U.S.S.R.) 1969-24A Not available	752	733	109.3	73.9	Total weight: Not available. Objective: Continuation of Cosmos scientific satellite series. Payload: Not available. Still in orbit.

Launch Date	Name, Country, International Designation, Vehicle	Payload Data	Apogee (st mi)	Perigee (st mi)	Period (min)	Inclination (degrees)	Remarks
Mar. 18	OV 1-17 (United States) 1969-25A Atlas F	Total weight: 312 lbs. Objective: Measure incoming solar electromagnetic radiation and its reaction with earth's upper atmosphere. Payload: Spacecraft carries 12 experiments.	288	247	93.1	99.1	Four spacecraft carrying total of 41 experiments launched with single booster. OV 1-17 reentered 3/5/70.
	and OV 1-18 1969-25B	Total weight: 275 lbs. Objective: Gather information on ionosphere, radio interference, electrical fields, and radiation. Payload: Spacecraft carries 16 experiments.	362	288	95.0	98.8	Still in orbit.
	and OV 1-19 1969-25C	Total weight: 273 lbs. Objective: Study events resulting in and sustaining trapped radiation in Van Allen belts and hazards of incoming and trapped radiation to man. Payload: Spacecraft carries 12 experiments.	3,593	288	153.5	104.7	Still in orbit.
	and OV 1-17A 1969-25D	Total weight: 487 lbs. Objective: Study unusual transmission of radio waves through ionosphere. Payload: Spacecraft consists of OV 1-17 propulsion module and NRL 2-beacon ORBIS-CAL II experiment.	233	107	89.8	99.0	Reentered 3/24/69.
Mar. 19	DOD spacecraft (United States) 1969-26A LTTAT–Agena D	Total weight: Not available. Objective: Develop space-flight techniques and technology. Payload: Not available.	157	103	86.6	82.9	Launched by Long-Tank, Thrust-Augmented Thor-Agena D. Reentered 3/24/69.
	and DOD spacecraft 1969-26B	Total weight: Not available. Objective: Develop space-flight techniques and technology. Payload: Not available.	319	313	94.7	83.0	Still in orbit.

Date	Name	Payload				Remarks
Mar. 22	Cosmos CCLXXIII (U.S.S.R.) 1969-27A Not available	Total weight: Not available. Objective: Continuation of Cosmos scientific satellite series. Payload: Not available.	208	89.8	65.4	Reentered 3/30/69.
Mar. 24	Cosmos CCLXXIV (U.S.S.R.) 1969-28A Not available	Total weight: Not available. Objective: Continuation of Cosmos scientific satellite series. Payload: Not available.	186	89.5	64.9	Reentered 4/1/69.
Mar. 26	Meteor I (U.S.S.R.) 1969-29A Not available	Total weight: Not available. Objective: Trace cloud cover, detect presence of snow both on dark and daylight sides of earth, and record amount of radiated and reflected heat energy from atmosphere. Payload: Not available.	426	97.9	81.1	Meteor I meteorological satellite still in orbit.
Mar. 27	Mariner VII (United States) 1969-30A Atlas-Centaur	Total weight: 840 lbs. Objective: Conduct flyby mission to make exploratory investigation of Mars that will set basis for future experiments, particularly those relevant to search for extraterrestrial life; provide topographic information about the Martian surface, provide atmospheric profile information, and measure principal atmospheric constituents. Payload: 4½'-high x 19' spacecraft (cruise position with solar panels deployed); basically octagonal structure with 8 bays; top of spacecraft supports omnidirectional low-gain pole antenna and high-gain dish antenna, attitude-control system, Canopus star tracker, sun sensors, and 4 solar panels (17,742 photovoltaic solar cells); scientific experiments (2 TV cameras, infrared spectrometer, ultraviolet spectrometer, infrared radiometer, and 2 planet sensors) mounted on 2° of freedom scan platform on bottom of spacecraft; telemetry and command; 31-lb silver-zinc battery.	Heliocentric orbit			Mariner VII, second of 2 spacecraft in Mariner-Mars 1969 mission, flew by Mars polar cap at 2,130-mi altitude on 8/4/69 after 4/8/69 midcourse correction. All scientific experiments operated successfully despite apparent meteorite hit. Excellent quality live TV pictures transmitted. Revealed 3 types of Martian terrain: heavily cratered resembling moon; featureless, desert-like, flat floor covering 1,200 mi; chaotic, irregular terrain similar to landslide. Instruments determined radius of Mars at 2,125 mi and diameter 4,250 mi. Thin atmosphere discovered with CO_2 its major constituent and less than 1% nitrogen. Solid CO_2 found at high altitudes. Polar cap either solid CO_2 or shrouded by thin cloud of solid CO_2. Martian pressure equivalent to atmospheric pressure 30,500–35,000 mi above earth. Daytime temperature about 62°F at equator and −45°F at polar cap edge; nightside of equator −100°F. Lowest temperature

Launch Date	Name, Country, International Designation, Vehicle	Payload Data	Apogee (st mi)	Perigee (st mi)	Period (min)	Inclination (degrees)	Remarks
							recorded −198°F at polar cap. "Canals" considered to be alignments of impact craters with dark floors. Life on Mars, if exists, considered almost certainly microbial. Still in heliocentric orbit.
Mar. 28	Cosmos CCLXXV (U.S.S.R.) 1969-31A Not available	Total weight: Not available. Objective: Continuation of Cosmos scientific satellite series. Payload: Not available.	473	171	95.1	70.9	Reentered 2/7/70.
Apr. 4	Cosmos CCLXXVI (U.S.S.R.) 1969-32A Not available	Total weight: Not available. Objective: Continuation of Cosmos scientific satellite series. Payload: Not available.	231	124	90.1	81.3	Reentered 4/11/69.
Apr. 4	Cosmos CCLXXVII (U.S.S.R.) 1969-33A Not available	Total weight: Not available. Objective: Continuation of Cosmos scientific satellite series. Payload: Not available.	290	166	91.8	70.9	Reentered 7/6/69.
Apr. 9	Cosmos CCLXXVIII (U.S.S.R.) 1969-34A Not available	Total weight: Not available. Objective: Continuation of Cosmos scientific satellite series. Payload: Not available.	198	126	89.6	65.4	Reentered 4/17/69.
Apr. 11	Molniya I-11 (U.S.S.R.) 1969-35A Not available	Total weight: Not available. Objective: Develop and improve satellite and TV communications system. Payload: Satellite with transmitter, command system, orientation system, orbit correction device, power supply.	24,608	300	712.1	64.9	Satellite to relay telephone and telegraph communications and TV broadcasts to far north, Siberia, Central Asia, and Far East. Still in orbit.
Apr. 12*	DOD spacecraft (United States) 1969-36A Atlas-Agena D	Total weight: Not available. Objective: Develop space-flight techniques and technology. Payload: Not available.	24,391	20,302	1,486.0	10.2	Still in orbit.
Apr. 14	Nimbus III (United States) 1969-37A LTTAT-Agena D	Total weight: 1,269 lbs. Objective: Demonstrate satisfactory operation of active, 3-axis, earth-oriented spacecraft for at	703	662	107.3	80.1	Nimbus III launched into near-polar sun-synchronous orbit by Long-Tank, Thrust-Augmented Thor (Thorad)

438

ASTRONAUTICS AND AERONAUTICS, 1969

	and	least 3 mos; acquire representative number of global samples of infrared spectra from which vertical temperature profiles of atmosphere may be derived. Payload: 10' x 11' (dia) butterfly-shaped spacecraft consisting of 56" (dia) sensory ring forming base, connected by truss structure to smaller hexagonal-shaped package and flanked by 2 3' x 8' solar paddles covered with 10,962 n-on-p solar cells; containing image-dissector camera system, IRIS instrumentation, SIRS instrumentation, MRIR instrumentation, HRIR instrumentation, MUSE instrumentation, SNAP-19 generator; active 3-axis control system, 8 nickel-cadmium batteries, transmitters, receivers, tape recorders, thermal control system.	704	667	107.3	Agena D. Also launched U.S. Army geodetic satellite *Egrs XIII* as secondary payload. *Nimbus III* heaviest U.S. experimental weather satellite to date and first civilian use of atomic energy in space. Replaced unsuccessful *Nimbus B*. Excellent data returned and all objectives achieved. Demonstrated that satellites can obtain global sounding data and cover inaccessible regions of globe. Obtained atmospheric temperatures at varying levels for first time and provided worldwide day-night weather coverage. Still in orbit, transmitting on command.
Apr. 15	*Egrs XIII* (Secor XIII) (United States) 1969-37B	Total weight: 46 lbs. Objective: Continue geodetic measurements program. Payload: 9" x 11" x 13" rectangular box-shaped structure; solar cell power supply.			99.9	*Egrs XIII*, U.S. Army's SECOR (Sequential Collation of Range) satellite, carried as secondary payload on launch vehicle's Agena D second stage and injected into orbit 48 min after *Nimbus III* separation. Still in orbit.
Apr. 15	*Cosmos CCLXXIX* (U.S.S.R.) 1969-38A Not available	Total weight: Not available. Objective: Continuation of Cosmos scientific satellite series. Payload: Not available.	217	127	89.8	51.8	Reentered 4/24/69.
Apr. 15	DOD spacecraft (United States) 1969-39A Titan IIIB-Agena D	Total weight: Not available. Objective: Develop space-flight techniques and technology. Payload: Not available.	293	79	89.9	108.7	Reentered 4/30/69.
Apr. 28	*Cosmos CCLXXX* (U.S.S.R.) 1969-40A Not available	Total weight: Not available. Objective: Continuation of Cosmos scientific satellite series. Payload: Not available.	156	123	88.1	51.0	Reentered 5/6/69.

Launch Date	Name, Country, International Designation, Vehicle	Payload Data	Apogee (st mi)	Perigee (st mi)	Period (min)	Inclination (degrees)	Remarks
May 2	DOD spacecraft (United States) 1969-41A LTTAT–Agena D and	Total weight: Not available. Objective: Develop space-flight techniques and technology. Payload: Not available.	202	105	89.5	64.9	Launched by Long-Tank, Thrust-Augmented Thor-Agena D. Reentered 5/23/69.
	DOD spacecraft 1969-41B	Total weight: Not available. Objective: Develop space-flight techniques and technology. Payload: Not available.	283	256	93.2	65.7	Reentered 2/16/70.
May 13	Cosmos CCLXXXI (U.S.S.R.) 1969-42A Not available	Total weight: Not available. Objective: Continuation of Cosmos scientific satellite series. Payload: Not available.	187	117	89.3	65.4	Reentered 5/21/69.
May 18	Apollo 10 (United States) 1969-43A Saturn V	Total weight: 295,150 lbs (weight in earth orbit, including S-IVB stage, instrument unit, spacecraft LM adapter, LM, and CSM). Objective: Demonstrate crew/space vehicle/mission support facilities performance during manned lunar mission with CSM and LM; evaluate LM performance in cislunar and lunar environment. Payload: 114'-long S-IVB/IU/spacecraft-LM adapter/Block II command and service modules/lunar module; cameras; telemetry.	190 Lunar orbit, 196	184 69			*Apollo 10* launched successfully carrying Astronauts Thomas P. Stafford, John W. Young, and Eugene A. Cernan. First lunar orbital mission with complete Apollo spacecraft. After insertion into orbit, S-IVB burn initiated translunar injection. Crew successfully transposed CSM and docked with LM. S-IVB slingshot to earth-escape velocity. 5 live color TV transmissions during translunar coast. Lunar orbit insertion accomplished 5/21, orbit circularized, and Cernan transferred to LM for 2 hrs of housekeeping and communications tests. On 5th day, Stafford and Cernan entered LM, undocked from CSM, and descended within 47,000 ft of lunar surface. Photographed potential landing sites. LM descent stage jettisoned. After 8-hr separation, LM docked

Date	Name/Designation	Payload/Objective	Orbit params			Remarks
May 20	*Cosmos CCLXXXII* (U.S.S.R.) 1969-44A Not available	Total weight: Not available. Objective: Continuation of Cosmos scientific satellite series. Payload: Not available.	199	125	89.7	65.4	Reentered 5/28/69.
May 21	*Intelsat-III F-4* (United States) 1969-45A LTTAT–Delta	Total weight: 382 lbs (in synchronous orbit, 632 lbs at liftoff, including 310 lbs of apogee motor fuel). Objective: Place satellite and apogee motor into proper transfer orbit, provide tracking and telemetry and backup calculations through transfer orbit so satellite can be injected into synchronous orbit for commercial voice and TV communications. Payload: 78" x 56" (dia) cylindrical satellite, including mechanically despun receive/transmit and omnidirectional command and telemetry antennas on top of spacecraft; outer cylindrical sleeve covered with 10,720 solar cells; 2 frequency-translation-mode repeaters with design capability for 1,200 high-quality, 2-way voice or 4 TV channels, or a mix; attitude and control system; battery.	22,803 After apogee motor firing, 22,164	183		29.1 5.0	*Intelsat-III F-4* put into good transfer orbit by Long-Tank, Thrust-Augmented Thor-Delta. Apogee motor fired 5/23/69, putting satellite into synchronous orbit over Pacific Ocean. Replaced *Intelsat-III F-3* which was shifted to less busy Indian Ocean station after performance degradation. Still in orbit.
May 28	*OV V-5* (United States) 1969-46A Titan IIIC and	Total weight: 25 lbs. Objective: Measure power spectrum of magnetic and electrical field fluctuations at magnetospheric boundary. Payload: 11"-side octahedral.	69,427	10,480	3,119.6	32.9	Five spacecraft launched with single booster. Still in orbit, transmitting on command.

Launch Date	Name, Country, International Designation, Vehicle	Payload Data	Apogee (st mi)	Perigee (st mi)	Period (min)	Inclination (degrees)	Remarks
	OV V-6 1969-46B and	Total weight: 25 lbs. Objective: Provide research data on solar processes that influence near-earth environment. Payload: 11"-side octahedral.	69,427	10,480	3,119.6	32.9	Still in orbit, transmitting on command.
	OV V-9 1969-46C and	Total weight: 25 lbs. Objective: Measure energetic particles emitted by sun and study trapped electrons and protons at near-synchronous altitude. Payload: 11"-side octahedral.	69,427	10,480	3,119.6	32.9	Still in orbit, transmitting on command.
	Vela IX 1969-46D and	Total weight: 571 lbs. Objective: Orbit nuclear detection sensors capable of monitoring x-ray, gamma-ray, neutron, optical, electromagnetic-pulse, and air-fluorescence emissions. Payload: Cylindrical 26-sided polyhedron (50" dia) consisting of reaction wheel plus gas jets for stabilization, solar cell array, 28 detectors.	69,387	68,653	6,718.5	32.7	Still in orbit.
	Vela X 1969-46E	Total weight: 571 lbs. Objective: Orbit nuclear detection sensors capable of monitoring x-ray, gamma-ray, neutron, optical, electromagnetic-pulse, and air-fluorescence emissions. Payload: Cylindrical 26-sided polyhedron (50" dia) consisting of reaction wheel plus gas jets for stabilization, solar cell array, 28 detectors.	69,614	68,774	6,707.6	32.8	Still in orbit.
May 27	Cosmos CCLXXXIII (U.S.S.R.) 1969-47A Not available	Total weight: Not available. Objective: Continuation of Cosmos scientific satellite series. Payload: Not available.	938	122	102.0	81.9	Reentered 12/10/69.

ASTRONAUTICS AND AERONAUTICS, 1969

May 29	Cosmos CCLXXXIV (U.S.S.R.) 1969-48A Not available	Total weight: Not available. Objective: Continuation of Cosmos scientific satellite series. Payload: Not available.	185	127	89.5	51.7	Reentered 6/6/69.
June 3	Cosmos CCLXXXV (U.S.S.R.) 1969-49A Not available	Total weight: Not available. Objective: Continuation of Cosmos scientific satellite series. Payload: Not available.	306	165	92.1	71.0	Reentered 10/7/69.
June 3	DOD spacecraft (United States) 1969-50A Titan IIIB-Agena D	Total weight: Not available. Objective: Develop space-flight techniques and technology. Payload: Not available.	265	86	89.8	110.0	Reentered 6/14/69.
June 5	Ogo VI (United States) 1969-51A LTTAT-Agena D	Total weight: 1,393 lbs. Objective: Conduct correlative studies of latitude-dependent atmospheric phenomena during period of maximum solar activity; acquire data from attitude-stabilized platform as function of altitude and local time (rotation of orbital plane of about 180°). Payload: 67″ x 32″ x 31″ rectangular spacecraft (49′ x 60′ x 30′ fully deployed); extending from ends are 2 22′ booms and 4 4′ booms, all of which mount experiments; deployed from sides are 2 large rotatable solar paddles with 30,000 n-on-p solar cells; 2 box-like solar oriented experimental packages mounted on solar panels and 2 box-like orbital plane experimental packages supported by trusses at each end of spacecraft; 2 omnidirectional antennas and 2 30° dipole antennas; 2 nickel-cadmium batteries; thermal control system; active 3-axis stabilization and spin stabilization; total of 25 experiments for atmospheric and ionospheric studies, solar radiation measurements, air-glow and auroral studies, energetic particle measurements, and magnetic and electrical fields studies.	682	246	99.8	82.0	Ogo VI, sixth and last in NASA's OGO series, launched in low-altitude, near-polar orbit by Long-Tank Thrust-Augmented Thor (Thorad)-Agena D. Heaviest of OGO series, spacecraft completed first diurnal cycle (180° rotation of orbital plane) on 9/6/69 with all observatory subsystems performing satisfactorily, 23 experiments operational; 2 failed. Provided new and significant data in practically all principal areas of investigation, especially on global characteristics of neutral atmosphere; association of electric fields with ionospheric irregularities; airglow emissions associated with oxygen, sodium, and molecular nitrogen; and propagation of proton whistlers. Still in orbit, transmitting on command.

Launch Date	Name, Country, International Designation, Vehicle	Payload Data	Apogee (st mi)	Perigee (st mi)	Period (min)	Inclination (degrees)	Remarks
June 15	Cosmos CCLXXXVI (U.S.S.R.) 1969-52A Not available	Total weight: Not available. Objective: Continuation of Cosmos scientific satellite series. Payload: Not available.	200	122	89.7	65.4	Reentered 6/23/69.
June 21	Explorer XLI (United States) 1969-53A TAID	Total weight: 174 lbs. Objective: Place spacecraft into orbit with apogee of 92,000 mi or greater; obtain for approximately 90 days adequate measurements from plasma and energetic particle experiments to allow continuation and extension of studies of environment within and beyond earth's magnetosphere during period of high solar activity. Payload: 10" x 28" (dia) octagonal aluminum-honeycomb equipment platform with 8 radial support struts around hollow center tube; 11 of 12 experiments carried inside platform; 2 6' double-hinged fiber glass booms extend from base of spacecraft, one carrying remaining experiment, other used for balance; 4 16" telemetry antennas affixed to top of spacecraft; 4 solar paddles with 6,144 n-on-p solar cells; silver-cadmium battery; experiment categories are magnetic fields (1), solar and galactic cosmic radiation (8), and solar plasma (3).	110,723	213	4,996	87.0	Explorer XLI, launched successfully by Thrust-Augmented Improved Thor-Delta during power blackout at WTR, carried 12 experiments, greatest number ever on IMP (Interplanetary Monitoring Platform) spacecraft. By 11/26/69, adjudged a success. 9 of 12 experiments operational. Although launched during peak of solar activity cycle, sun unusually quiet. However, data received indicated that low-energy galactic cosmic rays appeared more strongly modulated than previously observed, with consequent increase in disturbed interplanetary conditions. Still in orbit, transmitting on command.
June 24	Cosmos CCLXXXVII (U.S.S.R.) 1969-54A Not available	Total weight: Not available. Objective: Continuation of Cosmos scientific satellite series. Payload: Not available.	158	116	88.9	51.7	Reentered 7/2/69.

ASTRONAUTICS AND AERONAUTICS, 1969

Date	Name	Payload/Objective	Col4	Col5	Col6	Col7	Remarks
June 27	Cosmos CCLXXXVIII (U.S.S.R.) 1969-55A Not available	Total weight: Not available. Objective: Continuation of Cosmos scientific satellite series. Payload: Not available.	168	124	89.2	51.7	Reentered 7/5/69.
June 28*	Biosatellite III (United States) 1969-56A LTTAT-Delta	Total weight: 1,535 lbs. Objective: Provide suitable environment for instrumented monkey and measure either central nervous system function or cardiovascular and metabolic function. Payload: 7'-long spacecraft with 2 major sections; 48" x 57" (dia) cylindrical cone adapter section containing electrical power system (1 large silver-zinc battery, 2 small silver-zinc batteries), 4 small silver-zinc batteries), attitude control system, gaseous storage tanks, telemetry receivers and transmitters, and thermal control system; 49" x 40" (dia) blunt cone reentry vehicle containing experimental capsule with primate and necessary life support, heat shield, thrust cone with solid retromotor, recovery parachutes, and recovery telemetry and beacon.	245	224	92.1	33.6	Biosatellite III carrying Bonny, male pigtailed monkey, launched into low-altitude orbit by 2-stage Long-Tank, Thrust-Augmented Thor-Delta. First U.S. spacecraft to provide 2-gas atmosphere (20% oxygen, 80% nitrogen). Spacecraft in orbit 8½ days with all life support parameters within specification before deteriorating physiological conditions of monkey required recovery of capsule on 7/7/69. Semicomatose on recovery, primate responded initially to intensive care, but died suddenly 8 hrs later. Autopsy indicated heart failure due to effects of weightlessness and low body temperature; 20% loss of body weight and pooling of blood in thorax and abdomen, contributing factors.
July 10	Cosmos CCLXXXIX (U.S.S.R.) 1969-57A Not available	Total weight: Not available. Objective: Continuation of Cosmos scientific satellite series. Payload: Not available.	201	129	89.6	65.4	Reentered 7/15/69.
July 13	Luna XV (U.S.S.R.) 1969-58A Not available	Total weight: Not available. Objective: Conduct further scientific exploration of moon and near lunar space. Payload: Not available.	Impacted on moon				Luna XV entered lunar orbit 7/16/69. Conducted 4 orbital changes. Speculation that spacecraft would softland, obtain lunar soil sample, and return to earth. After 52 revolutions and 86 communications sessions, spacecraft crashed into Sea of Crisis 7/21/69.

445

Launch Date	Name, Country, International Designation, Vehicle	Payload Data	Apogee (st mi)	Perigee (st mi)	Period (min)	Inclination (degrees)	Remarks
July 16	Apollo 11 (United States) 1969-59A Saturn V	Total weight: 297,865 lbs (weight in earth orbit, including S-IVB stage, instrument unit, spacecraft, LM adapter, LM, and CSM). Objective: Perform manned lunar landing and return. Payload: 114'-long S-IVB/IU/ spacecraft LM adapter/Block II command and service modules/ lunar module; cameras; telemetry.	119 Lunar orbit, 194	119 71			Apollo 11, first manned lunar landing mission, launched successfully carrying Astronauts Neil A. Armstrong, Michael Collins, and Edwin E. Aldrin, Jr. Entered lunar orbit 7/19. LM undocked, initiated descent, and touched down in southwestern part of Mare Tranquillitatis 4:17:43 pm EDT 7/20. Armstrong took man's first step on moon 10:56:15 pm EDT 7/20. Crew performed limited selenological inspection and photography. All scientific objectives achieved as crew deployed solar wind composition experiment, passive seismic experiment package, laser ranging retroreflector, and obtained contingency sample, core sample, and bulk sample of lunar surface. Live TV transmitted during entire EVA. Total astronaut EVA time, 4 hrs 22 min 1 sec. LM liftoff from moon at 1:54:00 pm EDT 7/21. Total lunar stay time, 21 hrs 36 min 17 secs. CSM/LM docking, crew transfer, and transearth injection (TEI) accomplished. CM splashed down in Pacific 12:50:35 EDT 7/24. Total flight time, 8 days 3 hrs 18 min 35 secs.
July 22	Cosmos CCXC (U.S.S.R.) 1969-60A Not available	Total weight: Not available. Objective: Continuation of Cosmos scientific satellite series. Payload: Not available.	201	119	89.6	65.4	Reentered 7/30/69.

Date	Spacecraft	Description	Weight (lbs)				Remarks
July 22	Molniya I-12 (U.S.S.R.) 1969-61A Not available	Total weight: Not available. Objective: Develop and improve satellite and TV communications system. Payload: Not available.	24,560	308	711	64.9	Still in orbit.
July 23	DOD spacecraft (United States) 1969-62A Thor-Burner II	Total weight: Not available. Objective: Develop space-flight techniques and technology. Payload: Not available.	532	488	101.3	98.8	Still in orbit.
July 24	DOD spacecraft (United States) 1969-63A LTTAT-Agena D	Total weight: Not available. Objective: Develop space-flight techniques and technology. Payload: Not available.	136	111	88.4	74.9	Launched by Long-Tank, Thrust-Augmented Thor-Agena D. Reentered 8/23/69.
July 25*	Intelsat-III F-5 (United States) 1969-64A LTTAT-Delta	Total weight: 332 lbs (in synchronous orbit; 632 lbs at lift-off, including 310 lbs of apogee motor fuel). Objective: Place satellite and apogee motor into proper transfer orbit, provide tracking and telemetry and backup calculations through transfer orbit so satellite can be injected into synchronous orbit for commercial communications. Payload: 78" x 56" (dia) cylindrical satellite including mechanically despun receive/transmit and omnidirectional command and telemetry antennas located on top of spacecraft; outer cylindrical sleeve covered with 10,720 solar cells; 2 frequency translation mode repeaters with design capability for 1,200 high quality, 2-way voice or 4 TV channels, or a mix; attitude and control system; battery.	3,355	167	146.7	30.3	Intelsat-III F-5 was unable to reach planned synchronous orbit when 3rd-stage motor (TE-364) of Long-Tank, Thrust-Augmented Thor-Delta malfunctioned. Spacecraft unable to operate satisfactorily. Originally scheduled for October 1969, mission had been rescheduled for July to replace Intelsat-III F-2 over Atlantic, which had stopped operating. Still in orbit.
July 31	DOD spacecraft (United States) 1969-65A LTTAT-Agena D	Total weight: Not available. Objective: Develop space-flight techniques and technology. Payload: Not available.	333	289	94.6	75.0	Launched by Long-Tank, Thrust-Augmented Thor-Agena D. Still in orbit.
Aug. 6	Cosmos CCXCI (U.S.S.R.) 1969-66A Not available	Total weight: Not available. Objective: Continuation of Cosmos scientific satellite series. Payload: Not available.	327	89	91.2	62.2	Reentered 9/8/69.

Launch Date	Name, Country, International Designation, Vehicle	Payload Data	Apogee (st mi)	Perigee (st mi)	Period (min)	Inclination (degrees)	Remarks
Aug. 8**	Zond VII (U.S.S.R.) 1969-67A Not available	Total weight: Not available. Objective: Continue study of moon and near-lunar space, photograph lunar surface, and test improved onboard systems and design of rocket-space complex. Payload: Not available.		Circumlunar flight			Zond VII third successful Soviet circumlunar flight, reentry, and recovery. After circling moon and photographing lunar surface, spacecraft reentered atmosphere by skipping across outer layers, reducing its entry speed. Softlanded in predetermined area in northern Kazakhstan 8/14/69.
Aug. 9	Oso VI (United States) 1969-68A LTTAT–Delta	Total weight: 640 lbs. Objective: Obtain high-resolution spectral data from pointed experiments within range from 10 to 20 kev and 1 Å to 1,300 Å during solar rotation, including raster scans of solar disc in selected wavelengths; spacecraft is to permit acquisition of useful data from pointed and nonpointed experiments beyond one solar rotation for extended observations of single lines and solar flares. Payload: Top part of spacecraft 22″-radius semi-circular sail continuously pointed at sun with 2 experiments, 4 sq ft of n-on-p solar cells; lower part 44″ x 9″ 9-sided revolving wheel containing 5 nonpointed experiments that scan sun every 2 secs; controls; telemetry, recorder; batteries.	348	308	95.2	32.9	Oso VI launched into nearly circular orbit by 2-stage Long-Tank, Thrust-Augmented Thor-Delta; also carried Pac stabilization experiment on second stage. Oso VI first to incorporate point and raster scan of solar disc; able to study more than 16,000 points of sun's surface. All systems operated satisfactorily and all experiments returned good data. Adjudged success 11/12/69. Still in orbit, transmitting on command.
	Pac 1969-68B	Total weight: 55 lbs. Objective: Flight-test long-life, low-power, 3-axis earth-stabilized control system for Delta second stage to demonstrate	340	300	94.2	32.9	Pac (Package Attitude Control) system launched as secondary payload on Oso VI mission. Rigidly attached to Delta 2nd stage, system be-

ASTRONAUTICS AND AERONAUTICS, 1969

Date	Name	Payload/Objective	Weight	Apogee	Perigee	Period	Remarks
		feasibility of using stage as experiment platform. Payload: Pencil-shaped semiactive gravity-gradient system (SAGS) attached to Delta second stage; takes up 1 cubic ft of space; power system; thermal control system; communications and data-handling system; attitude-sensing and control system.					haves like pendulum, tending to point long axis toward earth's center. Operating within design specifications. Still in orbit.
Aug. 12	Ats V (United States) 1969–69A Atlas-Centaur	Total weight: 951 lbs. Objective: Conduct carefully instrumented gravity-gradient orientation experiment to provide basic design information for stabilization and control of long-lived spacecraft in synchronous orbit; obtain useful data from onboard experiments during first 30 days in orbit. Payload: 72" x 56" (dia) cylinder which consists of 2 360° circular cylindrical solar panel sections separated by 25" cylindrical midsection compartment; 22,344 n-on-p solar cells; 4 124' booms which deploy in X configuration and stabilize spacecraft; 2 45' damper booms; telemetry and command antennas; payload and control sensors; TV camera; gravity-gradient, communications, and scientific experiments; nickel-cadmium batteries.	26,737 After apogee motor firing, 22,927	5,297	686.5	17.9	Ats V launched successfully into elliptical transfer orbit by Atlas-Centaur. After maneuver into near-synchronous orbit, high spacecraft spin rate could not be dampened and spin ultimately deteriorated into opposite direction from that planned, making despin impossible. Unable to perform primary gravity-gradient objective, although most secondary experiments returned good data. Still in orbit, transmitting on command.
						2.7	
Aug. 14**	Cosmos CCXCII (U.S.S.R.) 1969–70A Not available	Total weight: Not available. Objective: Continuation of Cosmos scientific satellite series. Payload: Not available.	475	463	99.9	74.0	Still in orbit.
Aug. 16	Cosmos CCXCIII (U.S.S.R.) 1969–71A Not available	Total weight: Not available. Objective: Continuation of Cosmos scientific satellite series. Payload: Not available.	152	127	88.9	51.7	Reentered 8/28/69.
Aug. 19	Cosmos CCXCIV (U.S.S.R.) 1969–72A Not available	Total weight: Not available. Objective: Continuation of Cosmos scientific satellite series. Payload: Not available.	218	127	89.7	65.4	Reentered 8/27/69.

Launch Date	Name, Country, International Designation, Vehicle	Payload Data	Apogee (st mi)	Perigee (st mi)	Period (min)	Inclination (degrees)	Remarks
Aug. 22	Cosmos CCXCV (U.S.S.R.) 1969-73A Not available	Total weight: Not available. Objective: Continuation of Cosmos scientific satellite series. Payload: Not available.	294	168	91.9	71.0	Reentered 12/1/69.
Aug. 23	DOD spacecraft (United States) 1969-74A Titan IIIB-Agena D	Total weight: Not available. Objective: Develop space-flight techniques and technology. Payload: Not available.	234	86	89.6	108.1	Reentered 9/7/69.
Aug. 29	Cosmos CCXCVI (U.S.S.R.) 1969-75A Not available	Total weight: Not available. Objective: Continuation of Cosmos scientific satellite series. Payload: Not available.	186	141	89.6	64.9	Reentered 9/6/69.
Sept. 2	Cosmos CCXCVII (U.S.S.R.) 1969-76A Not available	Total weight: Not available. Objective: Continuation of Cosmos scientific satellite series. Payload: Not available.	192	127	89.6	72.8	Reentered 9/10/69.
Sept. 15	Cosmos CCXCVIII (U.S.S.R.) 1969-77A Not available	Total weight: Not available. Objective: Continuation of Cosmos scientific satellite series. Payload: Not available.	101	79	87.3	49.6	Reentered 9/15/69.
Sept. 18	Cosmos CCXCIX (U.S.S.R.) 1969-78A Not available	Total weight: Not available. Objective: Continuation of Cosmos scientific satellite series. Payload: Not available.	136	129	89.2	64.9	Reentered 9/22/69.
Sept. 22	DOD spacecraft (United States) 1969-79A LTTAT-Agena D and	Total weight: Not available. Objective: Develop space-flight techniques and technology. Payload: Not available.	157	110	88.7	85.0	Launched by Long-Tank, Thrust-Augmented Thor-Agena D. Reentered 10/12/69.
	DOD spacecraft 1969-79B	Total weight: Not available. Objective: Develop space-flight techniques and technology. Payload: Not available.	308	305	94.4	85.1	Still in orbit.

ASTRONAUTICS AND AERONAUTICS, 1969

Date	Name	Description				Notes
Sept. 23	*Cosmos CCC* (U.S.S.R.) 1969-80A Not available	Total weight: Not available. Objective: Continuation of Cosmos scientific satellite series. Payload: Not available.	117	114	88.2	51.5	Reentered 9/27/69.
Sept. 24	*Cosmos CCCI* (U.S.S.R.) 1969-81A Not available	Total weight: Not available. Objective: Continuation of Cosmos scientific satellite series. Payload: Not available.	173	119	89.2	65.4	Reentered 10/2/69.
Sept. 30	DOD spacecraft (United States) 1969-82A LTTAT–Agena D and	Total weight: Not available. Objective: Develop space-flight techniques and technology. Payload: Not available.	303	300	93.8	69.6	Launched by Long-Tank, Thrust-Augmented Thor-Agena D. Still in orbit.
	DOD spacecraft 1969-82B	Total weight: Not available. Objective: Develop space-flight techniques and technology. Payload: Not available.	586	575	103.7	70.7	Still in orbit.
Oct. 1	*Boreas* (U.S.–ESRO) 1969-83A Scout	Total weight: 189 lbs. Objective: Place satellite in earth orbit to perform integrated study of high-latitude ionosphere, including particle experiments, auroral photometry, and ionospheric experiments. Payload: 39" x 30" (dia) cylinder with sensor booms extending tip-to-tip 95.6" and 60" in height; carries 8 experiments; 6,990 n-on-p solar cells; battery; high- and low-speed telemetry and command system; passive magnetic stabilization.	287	181	91.2	85.1	*Boreas* (also called *Esro IB*) achieved lower than planned circular orbit when Scout 4th stage malfunctioned. Despite shortened lifetime, all experiments returned good data and objective was achieved. ESRO responsible for experiment instrumentation, equipment and personnel necessary for mating, and spacecraft testing. NASA provided launch vehicle and launch services. Reentered 11/23/69.
Oct. 6	*Meteor II* (U.S.S.R.) 1969-84A Not available	Total weight: Not available. Objective: Trace cloud cover, detect presence of snow on both dark and daylight sides of earth, and record amount of radiated and reflected heat energy from atmosphere. Payload: Not available.	420	385	97.6	81.2	*Meteor II* meteorological satellite still in orbit.
Oct. 11	*Soyuz VI* (U.S.S.R.) 1969-85A Not available	Total weight: Not available. Objective: Conduct further testing of systems of manual control.	142	121	88.8	51.7	*Soyuz VI* launched successfully carrying Cosmonauts Georgy S. Shonin and Valery N. Kubasov. Tested welding

ASTRONAUTICS AND AERONAUTICS, 1969

Launch Date	Name, Country, International Designation, Vehicle	Payload Data	Apogee (st mi)	Perigee (st mi)	Period (min)	Inclination (degrees)	Remarks
		Payload: 3-unit spacecraft with 2 crew cabins; TV cameras; 2 large wing-like solar panels; not equipped for docking; carried welding test equipment.					techniques in depressurized spacecraft chamber. Conducted manual control tests and medico-biological research. Hovered nearby as *Soyuz VII* and *Soyuz VIII* performed rendezvous maneuvers. Soft-landed in preset area, northwest of Karaganda 10/16/69.
Oct. 12	*Soyuz VII* (U.S.S.R.) 1969-86A Not available	Total weight: Not available. Objective: Maneuver in orbit, stage joint navigation observations of spaceships in group flight, and observe celestial bodies and earth. Payload: 3-unit spacecraft with 2 crew cabins; TV cameras; 2 large wing-like solar panels; docking equipment.	135	124	88.4	51.6	*Soyuz VII* launched successfully carrying Cosmonauts Vladislav N. Volkov, Anatoly V. Filipchenko, and Viktor V. Garbatko. Performed rendezvous maneuvers with *Soyuz VIII*, approaching within 500 yds of spacecraft. Carried out mutual observation-photography, conducted series of medical experiments, and observed effect of micrometeoroid erosion on craft. Observed and photographed earth and celestial bodies. Speculation was that unexpected manual control problems may have prevented docking with *Soyuz VIII*. Softlanded safely 10/17/69.
Oct. 13	*Soyuz VIII* (U.S.S.R.) 1969-87A Not available	Total weight: Not available. Objective: Test complex system of controlling simultaneous group flight of 3 ships; mutual maneuvering of ships in orbit to solve number of problems of developing piloted space system. Payload: 3-unit spacecraft with 2 crew cabins; TV cameras; 2 large wing-like solar panels; docking equipment.	173	134	89.4	51.6	*Soyuz VIII*, third manned craft in 3 days, launched successfully carrying Cosmonauts Vladimir A. Shatalov and Alexey S. Yeliseyev. Performed rendezvous maneuver within 500 yds of *Soyuz VII*, but speculation was that unexpected control problems prevented actual docking. Carried out mutual observation-photography, series of medical experiments, and observed

ASTRONAUTICS AND AERONAUTICS, 1969

Date	Name	Weight (lb)	Perigee/Apogee	Period (min)	Inclination	Remarks
Oct. 14	*Intercosmos I* (U.S.S.R.) 1969-88A Not available	Total weight: Not available. Objective: Study effect of solar radiation on structure of upper atmosphere. Payload: Spacecraft equipment includes alpha-particle photometer, x-ray and optical photometer, and spectroheliograph and polarimeter.	389	157	93.3	48.3	effects of micrometeoroid erosion on craft. Conducted research on polarization of solar light reflected by atmosphere. Softlanded safely 10/18/69. *Intercosmos I* carried instrumentation from Soviet Union, East Germany, and Czechoslovakia. Reentered 1/2/70.
Oct. 17	*Cosmos CCCII* (U.S.S.R.) 1969-89A Not available	Total weight: Not available. Objective: Continuation of Cosmos scientific satellite series. Payload: Not available.	199	129	89.6	65.4	Reentered 10/25/69.
Oct. 18	*Cosmos CCCIII* (U.S.S.R.) 1969-90A Not available	Total weight: Not available. Objective: Continuation of Cosmos scientific satellite series. Payload: Not available.	284	167	91.7	70.9	Reentered 1/23/70.
Oct. 21	*Cosmos CCCIV* (U.S.S.R.) 1969-91A Not available	Total weight: Not available. Objective: Continuation of Cosmos scientific satellite series. Payload: Not available.	472	461	99.8	74.0	Still in orbit.
Oct. 22	*Cosmos CCCV* (U.S.S.R.) 1969-92A Not available	Total weight: Not available. Objective: Continuation of Cosmos scientific satellite series. Payload: Not available.	211	126	88.4	51.4	Reentered 10/24/69.
Oct. 24	*Cosmos CCCVI* (U.S.S.R.) 1969-93A Not available	Total weight: Not available. Objective: Continuation of Cosmos scientific satellite series. Payload: Not available.	191	126	89.5	64.9	Reentered 11/5/69.
Oct. 24	*Cosmos CCCVII* (U.S.S.R.) 1969-94A Not available	Total weight: Not available. Objective: Continuation of Cosmos scientific satellite series. Payload: Not available.	1,340	132	109.1	48.4	Still in orbit.
Oct. 24	DOD spacecraft (United States) 1969-95A Titan IIIB-Agena D	Total weight: Not available. Objective: Develop space-flight techniques and technology. Payload: Not available.	396	78	92.1	108.1	Reentered 11/8/69.

453

Launch Date	Name, Country, International Designation, Vehicle	Payload Data	Apogee (st mi)	Perigee (st mi)	Period (min)	Inclination (degrees)	Remarks
Nov. 4	*Cosmos CCCVIII* (U.S.S.R.) 1969–96A Not available	Total weight: Not available. Objective: Continuation of Cosmos scientific satellite series. Payload: Not available.	254	168	91.2	71.0	Reentered 1/5/70.
Nov. 7*	*Azur* (U.S.–West Germany) 1969–97A Scout	Total weight: 157 lbs. Objective: Inject satellite into elliptical, quasi-polar orbit which will permit studies of inner Van Allen belt, auroral zones of Northern Hemisphere, and spectral variations of solar particles versus time during solar flares. Payload: 48" x 30" (dia) cylinder with conical top; circular instrument platform mounted to central tube forms basic internal structure; 4 whip turnstile antennas on spacecraft's flat bottom, from which protrudes 33"-long magnetometer boom; contains 7 experiments; practically entire exterior envelope covered with solar cells; 1 silver-cadmium battery; command and telemetry system; attitude and control system.	1,955	239	121.9	103.0	*Azur*, first satellite in cooperative U.S.–West German space program, functioning properly in near-polar orbit. Spacecraft control was passed to German control center Nov. 14. Still in orbit, transmitting on command.
Nov. 12	*Cosmos CCCIX* (U.S.S.R.) 1969–98A Not available	Total weight: Not available. Objective: Continuation of Cosmos scientific satellite series. Payload: Not available.	219	119	89.9	65.4	Reentered 11/20/69.
Nov. 14	*Apollo 12* (United States) 1969–99A Saturn V	Total weight: 300,269 lbs (weight in earth orbit, including S-IVB stage, instrument unit, spacecraft LM adapter, LM, and CSM). Objective: Make selenological inspection, survey, and sampling in mare area; deploy and activate ALSEP; develop techniques for point landing capability;	118 Lunar orbit, 194	115 72			*Apollo 12*, second manned lunar landing mission, launched successfully carrying Astronauts Charles Conrad, Jr., Richard F. Gordon, Jr., and Alan L. Bean. Momentary power loss 36 secs GET, after electrical potential discharge from clouds passed through space vehicle to ground.

develop man's capability to work in lunar environment; obtain photographs of candidate exploration sites.
Payload: 114'-long S-IVB/IU/ spacecraft LM adapter/Block II command and service modules/ lunar module; cameras; telemetry.

After translunar injection, attempted S-IVB slingshot maneuver achieved geocentric orbit rather than desired heliocentric orbit. TLI maneuver placed spacecraft on first hybrid lunar approach trajectory (nonfree return). Spacecraft entered lunar orbit 11/17. LM undocked, initiated descent, and touched down in Ocean of Storms, 600' from *Surveyor III* spacecraft 1:54:35 am EST 11/19. Crew performed 2 EVAs. During first EVA (3 hrs 56 min), TV transmission was lost for remainder of mission, contingency sample obtained, and solar wind composition experiment and ALSEP with SNAP-27 atomic generator deployed. During second EVA (3 hrs 49 min), Astronauts Conrad and Bean obtained photographic panoramas, stereo photographs, and assorted lunar samples; made geological traverse totaling 6,000' and retrieved *Surveyor III* soil scoop. Total astronaut EVA time, 7 hrs 45 min. LM liftoff from moon at 9:26:00 am EST 11/20. Total lunar stay time, 31 hrs 31 min. After CSM/LM docking and crew transfer, LM ascent stage deorbited and impacted near landing site. Crash was registered by seismometer left on moon and produced reverberations for more than 30 min. Transearth injection initiated and 2 midcourse corrections required during transearth coast. CM splashed down in Pacific 3:58:25 pm EST 11/24. Total flight time 10 days 4 hrs 36 min 24 secs.

Launch Date	Name, Country, International Designation, Vehicle	Payload Data	Apogee (st mi)	Perigee (st mi)	Period (min)	Inclination (degrees)	Remarks
Nov. 15	Cosmos CCCX (U.S.S.R.) 1969-100A Not available	Total weight: Not available. Objective: Continuation of Cosmos scientific satellite series. Payload: Not available.	206	126	89.8	64.9	Reentered 11/28/69.
Nov. 21*	Skynet A (U.S.-U.K.) 1969-101A LTTAT-Delta	Total weight: 535 lbs. Objective: Place into synchronous orbit over Indian Ocean a United Kingdom communications spacecraft as part of initial defense communications satellite program (augmented) in response to U.S.-U.K. agreement. Payload: 32" x 54" (dia) spacecraft constructed of 2 concentric cylinders with apogee motor within cylinder; solar cells mounted on outside surface of outer cylinder; despun antenna system, partially mounted within inner cylinder, on spacecraft bottom; high-pressure hydrazine system used for spin-stabilization and positioning; command and telemetry system.	23,045 After apogee motor firing, 22,217	161 21,558	655.3 1,431	27.6 2.4	Skynet A launched successfully by Long-Tank, Thrust-Augmented Thor-Delta. First of 2 U.K. military comsats to be launched over Indian Ocean under DOD-UK agreement. USAF managed project for DOD and reimbursed NASA for launch services. U.K. would reimburse USAF. On 5th apogee of transfer orbit, spacecraft apogee motor circularized orbit at near-synchronous altitude. All systems operating as designed and adjudged success on 1/16/70. Operational control given to U.K. 1/29/70. Still in orbit.
Nov. 24	Cosmos CCCXI (U.S.S.R.) 1969-102A Not available	Total weight: Not available. Objective: Continuation of Cosmos scientific satellite series. Payload: Not available.	290	169	91.9	71.0	Reentered 3/10/70.
Nov. 24**	Cosmos CCCXII (U.S.S.R.) 1969-103A Not available	Total weight: Not available. Objective: Continuation of Cosmos scientific satellite series. Payload: Not available.	733	709	108.5	74.0	Still in orbit.
Dec. 3	Cosmos CCCXIII (U.S.S.R.) 1969-104A Not available	Total weight: Not available. Objective: Continuation of Cosmos scientific satellite series. Payload: Not available.	153	122	88.9	65.4	Reentered 12/15/69.

ASTRONAUTICS AND AERONAUTICS, 1969

Date	Spacecraft	Description				Notes
Dec. 4	DOD spacecraft (United States) 1969-105A LTTAT-Agena D	Total weight: Not available. Objective: Develop space-flight techniques and technology. Payload: Not available.	155	106	88.4	81.4	Launched by Long-Tank, Thrust-Augmented Thor-Agena D. Reentered 1/10/70.
Dec. 11	Cosmos CCCXIV (U.S.S.R.) 1969-106A Not available	Total weight: Not available. Objective: Continuation of Cosmos scientific satellite series. Payload: Not available.	289	184	91.6	71.0	Reentered 3/22/70.
Dec. 20	Cosmos CCCXV (U.S.S.R.) 1969-107A Not available	Total weight: Not available. Objective: Continuation of Cosmos scientific satellite series. Payload: Not available.	336	322	95.2	74.0	Still in orbit.
Dec. 23	Cosmos CCCXVI (U.S.S.R.) 1969-108A Not available	Total weight: Not available. Objective: Continuation of Cosmos scientific satellite series. Payload: Not available.	1,018	91	102.8	49.4	Still in orbit.
Dec. 23	Cosmos CCCXVII (U.S.S.R.) 1969-109A Not available	Total weight: Not available. Objective: Continuation of Cosmos scientific satellite series. Payload: Not available.	184	119	89.3	65.4	Reentered 1/5/70.
Dec. 25	Intercosmos II (U.S.S.R.) 1969-110A Not available	Total weight: Not available. Objective: Study effect of solar radiation on structure of upper atmosphere. Payload: Spacecraft equipment includes alpha-particle photometer, x-ray and optical photometer, and spectroheliograph and polarimeter.	726	124	98.4	48.4	Intercosmos II carried instrumentation from Soviet Union, East Germany, Czechoslovakia, and Bulgaria.

* Local time at site; 1 day later by Greenwich time.
** Local time at site; 1 day earlier by Greenwich time.

Appendix B

CHRONOLOGY OF MAJOR NASA LAUNCHES, 1969

This chronology of major NASA launches in 1969 is intended to provide an accurate and ready historical reference, compiling and verifying information previously scattered in several sources. It includes launches of all rocket vehicles larger than sounding rockets launched either by NASA or under "NASA direction" (e.g., in 1969 NASA provided vehicles and launch facilities and launched ComSatCorp's three Intelsat III satellites, ESSA's *Essa IX*, USA's *Egrs XIII* as a secondary payload, ESRO's *Boreas*, West Germany's *Azur*, and U.K.'s *Skynet A*, as well as *Isis I* in joint U.S.-Canadian program). NASA sounding rocket launches are published annually by the Goddard Space Flight Center Historian in *Goddard Projects Summary: Satellites and Sounding Rockets*.

An attempt has been made to classify performance of both the launch vehicle and the payload and to summarize total results in terms of primary mission. Three categories have been used for evaluating vehicle performance and mission results—successful (S), partially successful (P), and unsuccessful (U). A fourth category, unknown (Unk), has been added for payloads when vehicle malfunctions did not give the payload a chance to exercise its main experiments. These divisions are necessarily arbitrary; many of the results cannot be neatly categorized. Also they ignore the fact that a great deal is learned from missions that may have been classified as unsuccessful.

Date of launch is referenced to local time at the launch site. Open sources were used, verified when in doubt with the project offices in NASA Headquarters and with NASA Centers. For further information on each item, see Appendix A of this volume and the entries in the main chronology as referenced in the index. Prepared April 1970 by William A. Lockyer, Jr., Historical and Library Services Branch, Kennedy Space Center.

Date	Name (NASA Code)	General Mission	Launch Vehicle (Site)	Performance Vehicle	Performance Payload	Performance Mission	Remarks
Jan. 22	Oso V (OSO-F)	Scientific satellite, solar physics	Delta DSV-3C (ETR)	S	S	S	Oso V was launched into circular earth orbit (353/338) to study solar radiation in x-ray gamma ray, and ultraviolet regions of solar spectrum. All 8 experiments operational.
Jan. 30	Isis I (ISIS-A)	Scientific satellite, ionospheric physics	Delta DSV-3E (WTR)	S	S	S	Isis I was launched into polar orbit (2,188/357) in continuing joint U.S.-Canadian program toward better understanding upper atmosphere. Canadian-built satellite.
Feb. 5*	Intelsat-III F-3	Commercial communications satellite	Delta DSV-3M (ETR)	S	S	S	Launched by NASA into elliptical orbit. Transferred by ComSatCorp into synchronous orbit (22,200/22,190) over Pacific; maneuvered to station over Indian Ocean June 24 (62.5° east long.). Partially crippled by equipment failure, but capable of handling traffic load at present location.
Feb. 24*	Mariner VI (Mariner F)	Scientific interplanetary probe	Atlas-Centaur (ETR)	S	S	S	First of 2 Mars 69 spacecraft flew by planet at 2,131 sm July 31. Transmitted 74 TV pictures to earth and investigated Martian atmosphere.
Feb. 26	Essa IX (TOS-G)	Meteorological satellite	Delta DSV-3E (ETR)	S	S	S	Launched into near-polar, sun-synchronous orbit (935/884) to obtain daily global cloud photos and measure atmospheric heat balance. NASA launch, non-NASA mission.
Mar. 3	Apollo 9 (AS-504, CSM-104, LM-3)	Earth-orbital manned Apollo flight	Saturn V (KSC)	S	S	S	First manned flight test of complete Apollo spacecraft; first manned testing of LM including firing of ascent and descent engines; active LM rendezvous with CSM. James A. McDivitt, David R. Scott, and Russell L. Schweickart. 37-min EVA by Schweickart; 10 days 1 hr 1 min flight duration.
Mar. 27	Mariner VII (Mariner G)	Scientific interplanetary probe	Atlas-Centaur (ETR)	S	S	S	Second of 2 spacecraft in Mars 69 program flew by planet at 2,130 sm Aug. 4. Transmitted 91 TV pictures to earth and investigated Mars atmosphere.
Apr. 14	Nimbus III (Nimbus B-2)	Experimental meteorological satellite	Thor-Agena (Long-Tank, Thrust-Augmented) (WTR)	S	S	S	Fourth of 7 spacecraft in planned Nimbus program was injected into near-polar, sun-synchronous orbit (703/662). Carried 7 experiments to improve satellite meteorological observations. Also carried experimental SNAP-19 power supply to assess operational capability of radioisotope power for space applications.

and

	Egrs XIII (*Secor XIII*)	Geodetic satellite		S	Unk	U.S. Army SECOR (Sequential Collation of Range) satellite was carried as a secondary payload on launch vehicle's Agena 2nd stage and was injected into orbit 48 min after separation of *Nimbus III* spacecraft. Non-NASA mission.
May 18	*Apollo 10* (AS-505, CSM-106, LM-4)	Lunar-orbital manned Apollo flight	Saturn V (KSC)	S	S	Manned flight test of complete Apollo spacecraft and systems in cislunar and lunar environment. LM descended to within 50,000 ft of lunar surface. 19 color TV transmissions made during mission, Eugene A. Cernan, John W. Young, Thomas P. Stafford. 8 days 3 min duration.
May 21*	*Intelsat-III F-4*	Commercial communications satellite	Delta DSV-3M (ETR)	S	S	Launched by NASA into elliptical transfer orbit, then maneuvered by ComSatCorp into synchronous orbit (22,164/21,887) over Pacific at 170° east long. Replaced *Intelsat-III F-3* on this station.
June 5	*Ogo VI* (OGO-F)	Geophysical investigations satellite	Thor-Agena (Long-Tank, Thrust-Augmented) (WTR)	S	S	Launched into low-altitude (682/246) polar orbit. Provided data on global characteristics of neutral atmosphere, association of electrical fields with ionospheric irregularities, airglow emissions, and propagation of proton whistlers.
June 21	*Explorer XLI* (IMP-G)	Interplanetary Monitoring Platform	Delta DSV-3E (WTR)	S	S	Seventh of 10 missions in current IMP program. Injected into highly elliptical (110,723/213) orbit. Carried 12 experiments, cosmic rays, solar plasmas, and magnetic fields in interplanetary space. First launch from new NASA WTR facility (SLC-2W).
June 28*	*Bios III* (BIOS-D)	Biological experimentation in space	Delta DSV-3N (ETR)	S	P	Launched into low-altitude (245/224) orbit, satellite included reentry module containing pigtailed monkey. Planned for maximum orbital stay time of 30 days, but reentry commanded on July 7, when instrumentation showed physical problems. Primate died after recovery. Autopsy indicated heart failure due to weightlessness and low body temperature.
July 16	*Apollo 11* (AS-506, CSM-107, LM-5)	Apollo manned lunar landing and return	Saturn V (KSC)	S	S	Man's first landing on the moon. Neil A. Armstrong, commander; Michael Collins, CM pilot; Edwin E. Aldrin, Jr., LM pilot. Liftoff from Pad A, LC-39, KSC, 9:32 am EDT July 16. LM touchdown on moon 4:17:43 pm July 20. Neil Armstrong took man's first step on moon 10:56:15 pm EDT July 20. Edwin Aldrin stepped onto lunar surface 11:16:15 pm EDT July 20. Aldrin returned to LM 1:01:39 am EDT July 21; Armstrong returned 1:11:12 am July 21. Astronauts outside LM total of 4 hrs 22 min 1 sec. LM liftoff from moon, 1:54:00 pm EDT July 21. Total lunar stay time, 21 hrs 36 min 17 secs. CM landed in Pacific Ocean 12:50:35 pm EDT July 24. Recovery by U.S.S. *Hornet*. Total flight time 8 days 3 hrs 18 min 35 secs.

Date	Name (NASA Code)	General Mission	Launch Vehicle (Site)	Performance Vehicle	Performance Payload	Performance Mission	Remarks
July 25*	Intelsat-III F-5	Commercial communications satellite	Delta DSV-3M (ETR)	U	U	U	Intended to replace Intelsat-III F-4, launched Dec. 18, 1968. Launch delayed 8 days by launch vehicle problems. Injected into an abnormally low (8,355/167) orbit. Spacecraft separated, but not transmitting.
Aug. 9	Oso VI (OSO-G) and	Scientific satellite, solar physics	Delta DSV-3N (ETR)	S	S	S	First OSO with offset pointing and rastering capability, permitting studies of ultraviolet and x-ray spectra at any point on solar disc and within a few arc minutes above limb. Near-circular (348/308) orbit.
	Pac	Experimental spacecraft control system			S	S	Secondary payload, Packaged Attitude Control (Pac) spacecraft rigidly attached to Delta 2nd stage, injected into orbit (340/300). Flight test of system to convert 2nd stage into 3-axis earth-stabilized platform for various future piggy-back payloads.
Aug. 12	Ats V (ATS-E)	Applications technology satellite	Atlas-Centaur (ETR)	S	U	U	Successfully launched into transfer orbit; maneuvered into near-synchronous (22,927/22,221) orbit, but high spacecraft spin rate about yaw axis could not be dampened. Usefulness of primary experiments in doubt.
Aug. 27	PIONEER E and	Scientific, interplanetary investigations	Delta DSV-3L (ETR)	U	Unk	U	Fifth and last spacecraft of current Pioneer program. Liftoff normal, but launch vehicle veered off course after 8 min of powered flight; destroyed by Range Safety Officer. Spacecraft fell into Atlantic off Barbados.
	TETR C	Test and training satellite			Unk	U	Intended for test and calibration of equipment and crew training for Manned Space Flight Network (MSFN). Fell into Atlantic when launch vehicle destroyed by Range Safety Officer.
Oct. 1	Boreas (Esro IB)	Non-NASA mission; ionospheric physics	Scout (WTR)	S	S	S	Launched into lower-than-planned, near-polar orbit (287/181). Third successful ESRO spacecraft launched by NASA. Boreas carried 8 experiments from 6 European research organizations to study aurora borealis and related ionospheric phenomena. All experiments functioned successfully during 52-day lifetime of satellite, which reentered atmosphere Nov. 23, 1969.

ASTRONAUTICS AND AERONAUTICS, 1969

Nov. 7*	*Azur* (GRS-A) German Research Satellite	Non-NASA mission: particles and fields	Scout (WTR)	S	S	S	Launched into quasi-polar orbit (1,955/239) to study inner Van Allen belt, northern auroral zones, and solar flares. All 7 experiments functioned satisfactorily. Satellite was turned over to West German Ministry for Science and Education Nov. 14.
Nov. 14	*Apollo 12* (AS-507, CSM-108, LM-6)	Manned lunar landing	Saturn V (KSC)	S	S	S	Second manned lunar landing mission. Astronauts Charles Conrad, Jr., CDR; Richard F. Gordon, Jr., CMP; Alan L. Bean, LMP. Launched from KSC at 11:22 EST Nov. 14. Demonstrated LM point landing ability by landing within 600 ft of *Surveyor III* spacecraft (3.036° S, 2.418° W) at 1:55 am EST Nov. 19. Astronauts Conrad and Bean each performed two EVAs totaling 15 hrs 30 min. Deployed ALSEP, obtained lunar samples, made 6,000-ft lunar traverse to investigate *Surveyor III* (landed April 19, 1967) and remove samples. Astronaut Gordon photographed future exploration sites from CSM in lunar orbit. LM lifted off from lunar surface at 9:26 am EST Nov. 20. CM landed in Pacific Ocean near U.S.S. *Hornet* at 3:58 pm EST Nov. 24. Total mission time: 10 days 4 hrs 36 min 36 secs.
Nov. 21*	*Skynet A* (IDCSP-A)	Non-NASA mission; communications satellite	Delta DSV-3M (ETR)	S	S	S	Launched by NASA for USAF and U.K. into elliptical transfer orbit. On Nov. 23, spacecraft's apogee motor was fired to achieve synchronous transfer orbit (22,-217/21,553) over equator. Satellite was adjusted to final orbit over Indian Ocean Jan. 7, 1970. Operational control was given to United Kingdom Jan. 29, 1970.

* Time at launch site; one day later by Greenwich time.

Appendix C

CHRONOLOGY OF MANNED SPACE FLIGHT, 1969

This chronology contains basic information on all manned space flights during 1969 and, taken with Appendix C to the 1965, 1966, and 1968 editions of this publication, provides a summary record of manned exploration of the space environment through 1969. The information was compiled by Leonard C. Bruno of the Science and Technology Division of the Library of Congress.

The year 1969 was manned space flight's most prolific year to date. Four Apollo flights and five Soyuz flights put a record 23 men into space. The Soviets accomplished the first crew transfer in orbit. The United States achieved its goal of landing a man on the moon and returning him safely to earth—and did it again four months later.

None of the five Soviet flights left earth orbit. Rather, all performed earth-related experiments. *Soyuz IV*, carrying one cosmonaut, docked with *Soyuz V*, carrying a 3-man crew. Two members from *Soyuz V* transferred to and remained in *Soyuz IV* until reentry. Ten months later, *Soyuz VI* hovered near *Soyuz VII* and *Soyuz VIII* as they performed close rendezvous maneuvers but did not dock. *Soyuz VI* tested welding techniques in its depressurized spacecraft chamber.

The early March earth orbital flight of *Apollo 9* was the first successful manned flight of the Apollo LM, and *Apollo 10* successfully demonstrated the complete system during its circumlunar flight. *Apollo 11*'s Neil Armstrong and Edwin E. Aldrin, Jr., returned safely to earth with lunar samples after becoming the first men to walk on the moon. And the *Apollo 12* crew took an extensive lunar walk and retrieved a soil scoop from *Surveyor III*, which had been on the moon for 2½ years.

By the end of 1969, the United States had conducted a total of 22 manned space flights—16 in earth orbit, 2 in lunar orbit and 2 lunar landings—with a total of 24 different crewmen. Of the 24 American astronauts, 10 had participated in 2 flights each, and 5 had flown three times. The Soviet Union had conducted a total of 15 manned flights, all in earth orbit, with 21 cosmonauts. Three cosmonauts had flown twice each. Cumulative totals for manned spacecraft hours on flight had reached 2,303 hours 56 minutes for the United States and 1,054 hours 8 minutes for the Soviet Union. Cumulative total man-hours in space were 5,833 hours 57 minutes for the United States and 1,698 hours 47 minutes for the U.S.S.R.

Data on U.S. flights are the latest available to date within NASA. Although minor details are subject to modification as data are refined, major aspects of all U.S. manned flights remain subject to direct observation by interested citizens of the world, with a significant portion of recent missions seen live on worldwide television.

Date Launched	Date Recovered	Designation (NASA Code)	Crew	Weight (lbs)	Revolutions	Maximum Distance from Earth (st mi)	Duration	Remarks
Jan. 14	Jan. 17	*Soyuz IV*	Vladimir Shatalov	Not available	48	157	71 hrs 14 min	*Soyuz IV*, code-named "Amur," launched from Tyuratam 1/14/69 carrying Cosmonaut Vladimir Shatalov. *Soyuz V*, code-named "Baikal," launched 1/15/69 carrying Cosmonauts Boris Volynov, Alexey Yeliseyev, and Yevegeny Khrunov. After both craft reached their initial orbital positions, both were repositioned into prerendezvous orbits. On 34th orbit of *Soyuz IV* and 18th orbit of *Soyuz V*, the ships rendezvoused. Shatalov then achieved docking by manual control. Live TV of docking transmitted. Khrunov and Yeliseyev exited from *Soyuz V* for 1-hr EVA to perform experiments and observations. They then joined Shatalov in *Soyuz IV*. After 4 hrs 35 min docked flight, *Soyuz IV* separated and returned to earth 1/17/69 with 3 crewmen. *Soyuz V*, carrying Volynov, re-entered safely 1/18/69. Mission accomplished first docking in space of 2 independent, manned space vehicles; first men to switch spacecraft while in orbit. Entire mission lasted 96 hrs 30 min.
Jan. 15	Jan. 18	*Soyuz V*	Yevegeny Khrunov Boris Volynov Alexey Yeliseyev	Not available	50	157	72 hrs 48 min	
Mar. 3	Mar. 13	*Apollo 9* (AS-504, CSM-104, LM-3)	James A. McDivitt David R. Scott Russell L. Schweickart	289,970	152	313	241 hrs 1 min	First manned flight of complete Apollo spacecraft; launched by Saturn V booster. After insertion into initial orbit, CSM separated from S-IVB and LM, transposed, and docked with LM, and spacecraft separated from 3rd stage. 4 SPS burns tested structural dynamics of docked CSM-LM and changed orbit. Live TV transmission. McDivitt and Schweickart entered LM on 3rd day, conducted 1st firing of LM descent propulsion system (DPS) (1st manned throttling of engine in space), and returned to CSM. Live TV transmission from LM. On 4th day, Schweickart's scheduled 2-hr EVA

shortened to 36 min because of nausea. On 5th day, CSM separated from LM containing McDivitt and Schweickart. LM descent engine fired twice, placing LM 114 mi from CSM. LM descent stage jettisoned for first-time firing in space of ascent-stage engine. After 6½-hr separation, CSM and LM redocked; ascent stage jettisoned to fuel depletion. Crew performed 3 more days of experiments, landmark tracking, and photographs. On 10th day, reentry was extended 1 revolution because of heavy seas in prime recovery area. Splashed down safely in Atlantic and recovered by U.S.S. *Guadalcanal* on 3/13/69. Successfully simulated in earth orbit LM landing and takeoff from lunar surface and rejoining CM, qualifying last major spacecraft component for lunar landing mission.

| May 18 | May 26 | *Apollo 10* (AS-505, CSM-106, LM-4) | Thomas P. Stafford Eugene A. Cernan John W. Young | 295,150 | 31 (of moon) | 248,490 | 192 hrs 3 min | First lunar orbital mission with complete Apollo spacecraft; launched by Saturn V booster. After insertion into orbit, S-IVB burn initiated translunar injection. Crew successfully transposed CSM and docked with LM. S-IVB slingshot to earth-escape velocity. During translunar coast, only 1 of 4 midcourse corrections necessary. About 76 hrs into mission, lunar orbit insertion occurred with firing of SPS; second firing circularized orbit. Cernan transferred to LM for 2 hrs of housekeeping and communications tests. At about 100 hrs Stafford and Cernan entered LM, undocked from CSM, and briefly flew stationkeeping lunar orbit. LM flew over landing site 2 of Apollo 11 mission and simulated lunar landing by descending within 47,000 ft of lunar surface. With jettison of LM descent stage, gyration of ascent stage was caused by incorrect switch position. LM proper attitude restored quickly by manual control. After 8-hr separation, LM docked with CSM, crew transferred, and LM ascent stage was jettisoned to fuel depletion. After 61.5 hrs in lunar orbit, spacecraft was injected into |

Date Launched / Recovered	Designation (NASA Code)	Crew	Weight (lbs)	Revolutions	Maximum Distance from Earth (st mi)	Duration	Remarks
July 16 / July 24	*Apollo 11* (AS-506, CSM-107, LM-5)	Neil A. Armstrong, Edwin E. Aldrin, Jr., Michael Collins	297,865	30 (of moon)	242,540	195 hrs 19 min	First manned lunar landing mission. Spacecraft launched by Saturn V booster. After insertion into initial orbit, CSM separated from S-IVB and LM, transposed, docked with LM, and spacecraft separated from 3rd stage. S-IVB slingshot to earth-escape velocity. During translunar coast, only 1 of 4 midcourse corrections necessary. Live TV transmissions during coast. Lunar orbit insertion initiated about 76 hrs into the mission. After thorough LM checkout, LM undocked from CSM, initiated descent, and touched down in southwestern part of Mare Tranquillitatis 4:17:43 pm EDT 7/20/69. Neil Armstrong took man's first step on moon 10:56:15 pm EDT. Obtained contingency sample and photographs. TV cameras deployed and transmitted for entire duration of EVA. Aldrin on lunar surface at 11:15:15 pm EDT. Solar wind composition experiment deployed. Bulk sample collected. Passive seismic experiment package and laser ranging retroreflector deployed south of LM. Core sample obtained. Aldrin returned to LM 1:01:39 am EDT 7/21/69. Armstrong entered at 1:11:12 am EDT. Total astronaut EVA time, 4 hrs 22 min 1 sec. LM liftoff from moon at 1:54:00 pm EDT 7/21/69. Total lunar stay time, 21 hrs 36 min 17 sec. CSM/LM docking, crew transfer, LM ascent-stage jettison, and transearth injection accomplished. Total time in lunar orbit, 59 hrs 28 min. Only 1 of 3 scheduled midcourse corrections necessary. Live TV transmissions during coast. CM separated transearth trajectory. One midcourse correction required. CM separated, splashed down in Pacific 5/26/69, and was recovered by U.S.S. *Princeton*.

Date		Mission	Crew			Comments
Oct. 11		Soyuz VI	Georgy S. Shonin, Valery N. Kubasov	Not available	143	118 hrs 21 min	from SM and landed in Pacific 12:50:35 pm EDT 7/24/69. Recovery by U.S.S. Hornet.

Date	Mission	Crew	Weight (lb)	Alt. (mi)	Duration	Comments
Oct. 11	Soyuz VI	Georgy S. Shonin, Valery N. Kubasov	Not available	143	118 hrs 21 min	Soyuz VI, VII, and VIII operated in group flight without actually docking. 3 craft launched on successive days from Tyuratam. Soyuz VI, first to orbit, hovered nearby as Soyuz VII and Soyuz VIII performed rendezvous techniques and mutual observation-photography. While Soyuz VIII approached within 1,500 ft of Soyuz VII, speculation is that unexpected control problems prevented actual docking. Soyuz VI, which carried no docking equipment, tested welding techniques in depressurized spacecraft chamber; also conducted manual control tests and medico-biological research. Soyuz VII observed and photographed earth and celestial bodies. Soyuz VIII conducted research on polarization of solar light reflected by atmosphere. Both VII and VIII conducted medical experiments and observed effect of micrometeoroid erosion on their craft. Videotape of all launches shown on Soviet TV. While Western officials speculated missions had not achieved all objectives, Tass said all major tasks were carried out with "high efficiency." All spacecraft landed safely in preset areas of Soviet Union on successive days.
Oct. 12	Soyuz VII	Vladislav N. Volkov, Anatoly V. Filipchenko, Viktor V. Gorbatko	Not available	140	118 hrs 43 min	
Oct. 13	Soyuz VIII	Vladimir A. Shatalov, Alexey Yeliseyev	Not available	173	118 hrs 51 min	
Nov. 14	Apollo 12 (AS-507, CSM-108, LM-6)	Charles Conrad, Jr., Richard F. Gordon, Jr., Alan L. Bean	300,269	238,340	244 hrs 36 min	Second manned lunar landing mission. Spacecraft launched by Saturn V booster. During insertion into initial orbit, spacecraft experienced momentary power loss at 36 secs GET, after electrical potential discharge from clouds passed through space vehicle to ground. Power quickly restored. After S-IVB initiated translunar injection, CSM separated from S-IVB and LM, transposed, docked with LM, and spacecraft separated from 3rd stage. Attempted S-IVB slingshot maneuver achieved heliocentric orbit rather than desired geocentric orbit. TLI maneuver placed satellite on 1st hybrid lunar

Date Launched Recovered	Designation (NASA Code)	Crew	Weight (lbs)	Revolutions	Maximum Distance from Earth (st mi)	Duration	Remarks
							approach trajectory (nonfree return). Live TV transmission during translunar coast. Spacecraft entered lunar orbit 11/17/69. LM undocked, initiated descent, and touched down in Ocean of Storms 1:54:35 am EST 11/19/69, 600 ft from *Surveyor III* spacecraft, which had softlanded 4/19/67. Conrad and Bean performed 2 EVAs. During 1st EVA (3 hrs 56 min), TV transmission lost for remainder of lunar stay. Contingency sample obtained, S-band erectable antenna and solar wind composition experiment deployed. ALSEP with SNAP-27 atomic generator deployed 600–700 ft from LM. Crew reentered LM. During 2nd EVA (3 hrs 49 min), crew walked 1,500–2,000 ft from LM, totaling 6,000 ft in distance traversed. Also obtained photographic panoramas, stereo photos, and assorted lunar samples. *Surveyor III* inspected and soil scoop retrieved. Crew reentered LM. Total astronaut EVA time 7 hrs 45 min. LM liftoff from moon at 9:26:00 am EST 11/20/69. Total lunar stay time, 31 hrs 31 min. After CSM/LM docking and crew transfer, LM ascent stage deorbited and impacted near landing site. Crash registered by seismometer left on moon, produced reverberations for more than 30 min. Total time in lunar orbit, approximately 89 hrs 36 min. 2 of 8 scheduled midcourse corrections necessary. Live TV during coast. CM separated from SM and splashed down in Pacific 3:58:25 pm EST 11/24/69. Recovery by U.S.S. *Hornet*.

Appendix D

ABBREVIATIONS OF REFERENCES

Listed here are abbreviations for sources cited in the text. This list does not include all sources provided in the chronology, for some of the references cited are not abbreviated. Only references that appear in abbreviated form are listed below. Abbreviations used in the chronology entries themselves are cross-referenced in the Index.

A&A	AIAA's magazine, *Astronautics & Aeronautics*
A&A 69	NASA's *Astronautics and Aeronautics 1969* [this publication]
ABC	American Broadcasting Company
AEC Release	Atomic Energy Commission News Release
Aero Daily	*Aerospace Daily* newsletter
Aero Tech	*Aerospace Technology* magazine (formerly *Technology Week*)
AF Mgmt	*Armed Forces Management* magazine
AFFTC Release	Air Force Flight Test Center News Release
AFHF Newsletter	*Air Force Historical Foundation Newsletter*
AFJ	*Armed Forces Journal* magazine
AFNS Release	Air Force News Service Release
AFOSR Release	Air Force Office of Scientific Research News Release
AFRPL Release	Air Force Rocket Propulsion Laboratory News Release
AFSC *Newsreview*	Air Force Systems Command's *Newsreview*
AFSC Release	Air Force Systems Command News Release
AF/SD	*Air Force and Space Digest* magazine
AFSSD Release	Air Force Space Systems Division News Release
AIA Release	Aerospace Industries Association News Release
AIAA *Facts*	American Institute of Aeronautics and Astronautics' *Facts*
AIAA *News*	American Institute of Aeronautics and Astronautics' *News*
AIAA Release	American Institute of Aeronautics and Astronautics News Release
AIP News	*American Institute of Physics News*
Amer Av	*American Aviation* magazine (formerly *Aerospace Technology*)
AP	Associated Press news service
ARC *Astrogram*	NASA Ames Research Center's *Astrogram*
ARC Release	NASA Ames Research Center News Release
Atlanta J/C	*Atlanta Journal and Constitution* newspaper
Av Daily	*Aviation Daily* newsletter
Av Wk	*Aviation Week & Space Technology* magazine
B News	*Birmingham News* newspaper
B Sun	Baltimore *Sun* newspaper
Bus Wk	*Business Week* magazine
C Daily News	*Chicago Daily News* newspaper
C Trib	*Chicago Tribune* newspaper
Can Press	Canadian Press news service
CBS	Columbia Broadcasting System
C&E News	*Chemical & Engineering News* magazine

471

ComSatCorp Release	Communications Satellite Corporation News Release
CQ	*Congressional Quarterly*
CR	*Congressional Record*
CSM	*Christian Science Monitor* newspaper
CTNS	Chicago Tribune News Service
D Post	*Denver Post* newspaper
DJ	Dow Jones news service
DOD Release	Department of Defense News Release
DOT Release	Department of Transportation News Release
EH	NASA Historical Division (Code EH)
ERC Release	NASA Electronics Research Center News Release
ESSA Release	Environmental Science Services Administration News Release
FAA Release	Federal Aviation Administration News Release
FonF	*Facts on File*
FRC Release	NASA Flight Research Center News Release
FRC X-Press	NASA Flight Research Center's *FRC X-Press*
GE Forum	*General Electric Forum* magazine
Goddard News	NASA Goddard Space Flight Center's *Goddard News*
GSFC Release	NASA Goddard Space Flight Center News Release
GSFC SSR	NASA Goddard Space Flight Center's *Satellite Situation Report*
H Chron	*Houston Chronicle* newspaper
H Post	*Houston Post* newspaper
JPL Lab-Oratory	Jet Propulsion Laboratory's *Lab-Oratory*
JPL Release	Jet Propulsion Laboratory News Release
JSR	American Institute of Aeronautics and Astronautics' *Journal of Spacecraft and Rockets* magazine
KC Star	*Kansas City Star* newspaper
KC Times	*Kansas City Times* newspaper
KSC Release	NASA John F. Kennedy Space Center News Release
LA Her-Exam	*Los Angeles Herald-Examiner* newspaper
LA Times	*Los Angeles Times* newspaper
Langley Researcher	NASA Langley Research Center's *Langley Researcher*
LaRC Release	NASA Langley Research Center News Release
LATNS	Los Angeles Times News Service
LC Info Bull	Library of Congress *Information Bulletin*
LeRC Release	NASA Lewis Research Center News Release
Lewis News	NASA Lewis Research Center's *Lewis News*
M Her	*Miami Herald* newspaper
M News	*Miami News* newspaper
M Trib	*Minneapolis Tribune* newspaper
Marshall Star	NASA George C. Marshall Space Flight Center's *Marshall Star*
MJ	*Milwaukee Journal* newspaper
MSC Release	NASA Manned Spacecraft Center News Release
MSC Roundup	NASA Manned Spacecraft Center's *Space News Roundup*
MSFC Release	NASA George C. Marshall Space Flight Center News Release
NAA News	National Aeronautic Association *News*
NAC Release	National Aviation Club News Release
NAE Release	National Academy of Engineering News Release
NANA	North American Newspaper Alliance
NAR Release	North American Rockwell Corp. News Release
NAR Skywriter	North American Rockwell Corp. *Skywriter*
NAS Release	National Academy of Sciences News Release
NAS–NRC Release	National Academy of Sciences–National Research Council News Release
NAS–NRC–NAE News Rpt	National Academy of Sciences–National Research Council–National Academy of Engineering *News Report*
NASA Ann	NASA Announcement
NASA Hq PB	NASA Headquarters Personnel Bulletin
NASA Hq WB	NASA Headquarters *Weekly Bulletin*

NASA Int Aff	NASA Office of International Affairs
NASA *LAR* VIII/8	NASA *Legislative Activities Report*, Vol. VIII, No. 8
NASA Proj Off	NASA Project Office
NASA Release	NASA Headquarters News Release
NASA Rpt SRL	NASA Report of Sounding Rocket Launching
NASA SP-4014	NASA Special Publication #4014
NASC Release	National Aeronautics and Space Council News Release
N News	*Newark News* newspaper
Natl Obs	*National Observer* magazine
NBC	National Broadcasting Company
NGS Release	National Geographic Society News Release
NMI-	NASA Management Instruction-
NN	NASA Notice
NSC Release	National Space Club News Release
NSF Release	National Science Foundation News Release
N Va Sun	*Northern Virginia Sun* newspaper
NY News	*New York Daily News* newspaper
NYT	*New York Times* newspaper
NYTNS	New York Times News Service
O Sen	*Orlando Sentinel* newspaper
Oakland Trib	*Oakland Tribune* newspaper
Omaha W-H	*Omaha World-Herald* newspaper
P Bull	Philadelphia *Evening* and *Sunday Bulletin* newspaper
PAO	Public Affairs Office
PD	National Archives and Records Service's *Weekly Compilation of Presidential Documents*
P Inq	*Philadelphia Inquirer* newspaper
PIO	Public Information Office
PMR *Missile*	USN Pacific Missile Range's *Missile*
PMR Release	USN Pacific Missile Range News Release
Pres Rpt 70 [69]	*Aeronautics and Space Report of the President, Transmitted to the Congress January 1970* (report of activities during 1969)
SA	*Space Aeronautics* magazine
SBD	*Space Business Daily* newsletter
SAO Release	Smithsonian Astrophysical Observatory News Release
Sci Amer	*Scientific American* magazine
Sci Serv	Science Service news service
SD	*Space Digest* magazine
SD Union	*San Diego Union* newspaper
SF	*Space Flight* magazine
SF Chron	*San Francisco Chronicle* newspaper
SP	*Space Propulsion* newsletter
SR	*Saturday Review* magazine
SSN	*Soviet Sciences in the News*, publication of Electro-Optical Systems, Inc.
St Louis G-D	*St. Louis Globe-Democrat* newspaper
St Louis P-D	*St. Louis Post-Dispatch* newspaper
Testimony	Congressional testimony, prepared statement
Text	Prepared report or speech text
Transcript	Official transcript of news conference or Congressional hearing
UPI	United Press International news service
USGS Release	U.S. Geological Survey News Release
US News	*U.S. News & World Report* magazine
W News	*Washington Daily News* newspaper
W Post	*Washington Post* newspaper
W Star	Washington *Evening Star/Sunday Star* newspaper
WH Release	White House News Release
WJT	*World Journal Tribune* newspaper
WS Release	NASA Wallops Station News Release
WSJ	*Wall Street Journal* newspaper

INDEX AND LIST OF ABBREVIATIONS AND ACRONYMS

A

AA. See Apollo Applications program.
AAS. See American Astronautical Society.
AAAS. See American Assn. for the Advancement of Science.
ABC. See American Broadcasting Co.
Abelson, Dr. Philip H., 179–180
Aberdeen, Md., 49
Abernathy, Rev. Ralph D., 201, 209, 225
ABM. See Antiballistic missile system and Safeguard.
The ABM and the Changed Strategic Military Balance, U.S.A. vs. U.S.S.R. (American Security Council report), 131
Abbot, Dr. Charles G., 147
ABRES. See Advanced Ballistic Reentry System.
Accident
  aircraft, 19, 40–41, 362, 424–425
    AH–56A (helicopter), 260, 308–309, 389–390
    F–111A, 155, 413, 415–416
    T–33, 283
    Tu–144, 121
    XV–4B, 80
    X–15, 177
    XB–70, 11
  launch vehicle
    Saturn V, 188, 195
    U.S.S.R. booster, 382
  lifting-body vehicle, 29
  Lunar Landing Training Vehicle, 5, 173
  Sealab III, 51, 52, 57, 77, 188–189, 316–317
  spacecraft, 392
    Apollo AS–204, 206
    Biosatellite III, 162
Adams, Harold W., 209
Adams, Maj. Michael J. (USAF), 177
Adey, Dr. W. Ross, 348–349
ADS. See Airport Data System.
Adulyadej, King Bhumibol (Thailand), 251
Advanced Ballistic Reentry System (ABRES), 10
Advanced Manned Strategic Aircraft (AMSA). See B–1.
Advanced Marine Vehicles Meeting, Second, 156

Advanced Research Projects Agency (ARPA), 119, 346, 424
Advanced vidicon camera system (AVCS), 57–58
AEC. See Atomic Energy Commission.
AEC–NASA Nuclear Rocket Development Station, 361
AEC–NASA Space Nuclear Propulsion Office, 76, 87, 94, 309, 332
Aero Club of Washington, 29
Aerobee (sounding rocket)
  150
    airglow experiment, 49
    solar astronomy, 48–49, 302, 360–361
    stellar data, 71, 79, 179, 402, 409
    ultraviolet astronomy, 48–49, 78, 409
    x-ray astronomy, 78, 179, 402
  150 A, 139
  150 MI
    infrared astronomy, 299
    solar astronomy, 104, 108, 111, 113, 140–141, 308, 314, 316, 361, 410–411
    stellar data, 31–32, 118, 184, 326, 337
    ultraviolet astronomy, 6, 31–32, 43, 140–41, 179, 184, 308, 410-411
    upper-atmosphere data, 38, 42, 179
    x-ray astronomy, 65–66, 70, 104, 108, 118, 186, 326, 361
  170, 298
  350, 27–28
Aeroflot, 255
Aerojet-General Corp., 87, 94, 154, 168, 241, 385–386
Aeronautics, 113–114, 369
  anniversary, 129, 133, 292–293, 410
  award, 21, 72–73, 87, 101, 127, 132, 150, 178, 200, 209, 283, 293, 298, 315, 338, 345–346, 410
  cooperation, 204–205, 321, 334, 379, 397–398, 417–418, 421–422, 423–424
  employment, 77, 150, 356
  exposition, 52, 161–162, 173, 283, 412
  funds for, 14–15, 60, 70, 78, 113, 119–120, 193, 205
  general aviation. See General aviation.
  military, 14–15, 52, 66, 111–112, 131, 207, 397–398
  NASA program, 14–16, 38, 44, 66, 70, 84, 119–120, 204–205, 397–398, 400, 421–424
  noise abatement. See Noise, aircraft.

research (see also X-15, X-24, XB-70, XV-4B, YF-12, etc.), 11, 44, 70, 72, 78, 84-85, 92, 103, 120-121, 123, 155, 184, 193, 205, 299-300, 315-316, 321, 335, 350, 362, 369, 397-398, 400, 402-403, 404, 417-418, 421-422, 423-424
   statistics, 34, 40-41, 70, 102, 124, 136, 150, 176, 193, 356, 424-425
   U.S.S.R., 120
Aeronautics and Astronautics: An American Chronology of Science and Technology in the Exploration of Space, 1915-1960, 357
Aeros, Project, 423
Aerospace Corp., 304
Aerospace Facts and Figures, 1969, 193
Aerospace Industries Assn. (AIA), 109, 124, 150, 193, 356
Aerospace industry, 35, 77, 80, 95-96, 97, 124, 150, 164, 193, 228, 317, 330, 347, 356, 361, 363, 424
Aerospace Medical Assn., 130-131
Aerospace Systems Laboratory (Princeton Univ.), 208-209
AFCRL. See Air Force Cambridge Research Laboratories.
Africa, 101
AFSC. See Air Force Systems Command.
Agathadaemon canal (Mars), 262
Agena (booster). See Atlas-Agena; Long-Tank Thrust-Augmented Thor-Agena; Thor-Agena; Thorad-Agena; and Titan IIIB-Agena.
Agnew, Vice President Spiro T., 54, 83, 135, 191, 232, 380
   Apollo 9 mission, 65
   Apollo 11 mission, 210, 224, 225
   awards by, 91-92, 100
   oceanography, 56, 341
   press conference, 65, 308, 323-325
   space program, 65, 152, 224, 225, 231, 271, 294, 308, 320, 323-325, 419
Agreement. See International cooperation; International cooperation, space; and Treaty.
Agrell, Dr. S. O., 312
Agriculture, Dept. of, 14, 76, 106, 141, 352
AH-56A (Cheyenne) (helicopter), 106, 115, 148, 150, 160, 260, 308-309, 389-390
Ahmedabad, India, 311
AIA. See Aerospace Industries Assn.
AIAA. See American Institute of Aeronautics and Astronautics.
Air cargo, 113-114, 369, 424
Air Force Academy, 170, 225
Air Force Armament Development and Test Center, 186
Air Force Assn., 86-87
Air Force Cambridge Research Laboratories (AFCRL), 164, 308, 359-360, 410-411
   Lunar Laser Observatory, 192

Air Force Fatigue and Fracture Conference, 410
Air Force Museum, 38, 119, 177
Air Force Review of the C-5A Program, 251
Air Force Special Weapons Laboratory, 424
Air Force Systems Command (AFCS), 140, 260, 295, 331, 410
   Aeronautical Systems Div., 402-403
Air pollution, 147, 278-279, 285-286, 383, 402, 412-413
Air Traffic Activity Report, 102
Air traffic control, 15, 32, 55, 369
   FAA regulation, 32, 43-44, 188, 253, 380
   satellite use in, 37, 47, 319
   statistics, 34, 102, 109, 369
Air Traffic Controllers Organization, 44
Air Transport Assn. of America (ATA), 80, 278, 315, 336
Air transportation. See Air traffic control; Aircraft; Airlines; Airports; General aviation; and Supersonic transport.
Aircraft (see also individual aircraft, such as C-5A, F-111, X-15, XB-70).
   accident, 11, 19, 40-41, 80, 121, 155, 177, 260, 283, 308-309, 362, 389-390, 413, 415-416, 424-425
   air pollution, 278-279, 402, 413
   air show, 146, 168, 169, 173, 283, 412
   antisubmarine, 15, 131
   award, 72-73, 87, 178, 200, 209, 298, 315, 345-346
   bomber, 4, 15, 86, 99, 111-112, 117, 131, 133, 155, 169, 183, 195, 197, 205, 297-298, 321, 360, 413, 415-416
   cargo, 113-114, 369, 424
   carrier, 206
   collision study, 70, 315-316, 319, 362, 369
   cost, 88, 177, 186, 314-315, 340, 341, 410
   delta-wing, 195, 422
   exhibit, 38, 119, 177
   fighter, 1, 4, 9, 13, 15, 39, 86, 129, 131, 155, 197, 201, 283, 372, 413, 414
   flying boat, 133, 344
   foreign, 1, 5, 9, 13-14, 61, 71, 121, 159-160, 161-162, 168, 173, 197, 255, 283, 297-298, 323, 329, 340
   general-aviation, 14-15, 40-41, 70, 102, 109, 124, 150, 369
   helicopter, 35, 79, 106, 115, 124, 129, 131, 141, 148, 150, 160, 260, 308-309, 389-390
   hovercraft, 156
   hypersonic, 15
   interception, 73, 205, 400, 405, 421
   noise. See Noise, aircraft.
   reconnaissance, 73
   record, 79, 129, 151, 158, 177, 323
   regulations, 3, 43-44, 57, 188, 253, 371, 380, 402

research (see also X-15, X-24, XB-70, XV-4B, YF-12, etc.), 11, 44, 70, 72-73, 78, 85, 92, 123, 155, 183-184, 193, 205, 299-300, 315-316, 321, 335, 350, 362, 369, 397-398, 400, 402-403, 404, 421-422, 423-424
safety, 9, 40-41, 70, 148, 208, 362, 369, 424-425
seaplane, 133
sonic boom. See Sonic boom.
statistics, 34, 40-41, 70, 102, 113-114, 136, 150, 193, 356
STOL, 29, 32, 50, 84, 89, 114, 320
supersonic. See Supersonic transport, Concorde, F-8, F-14A, F-111, Tu-144, X-15, XB-70, YF-12, etc.
tracking, 140
traffic control. See Air traffic control.
training, 283
transport (see also Supersonic transport), 40-41, 57, 70, 75, 113-114, 121, 136, 150, 162, 168-169, 193, 208-209, 356, 362, 369, 424
   jet, 45-46, 47, 121, 168-169, 175, 193, 212, 278-279, 299-300, 304, 341, 350, 369, 371, 400, 409, 421-422, 424
   military, 38, 107, 123-124, 128-129, 130, 138, 151, 207, 212, 281, 283, 293, 323, 359, 380, 411, 421
   STOL, 29, 32, 50, 84, 89, 114, 320
Vietnam war use, 99
V/STOL, 15, 75, 84, 92
VTOL, 75, 80, 114, 129
wind-tunnel testing, 183-184, 299, 379, 389
Aircraft Owners and Pilots Assn., 138-139
Airglow, 49, 171-173
Airlines, 9, 29, 40-41, 50, 70, 84, 176, 424-425
Airlock, 128
Airport Data System (ADS), 117
Airports (see also Air traffic control; Noise, aircraft; and individual airports, such as Washington National Airport), 32, 54, 253
   FAA plans, 156
   facilities, 55, 88-89, 113-114, 117, 176, 369
   fog, 42
   funds for, 32, 123, 156, 182
   NASA role in, 253
   noise control, 3, 263-265, 371
   pollution control, 278-279
   regulation, 3, 32, 43-44, 57, 369, 380, 418
   runway research, 29
   statistics, 34, 70, 102, 109, 117, 176
Ajaccio, Corsica, 244
Alabama, Univ. of, 329
Alaska, 179
Alaska, Univ. of, 83, 352
Alcatraz Island, 343
Alcock, John W., 129

Aldrin, Col. Edwin E., Jr. (USAF), 315, 414
Apollo 11 mission, 307, 402
   celebrations for, 279-280, 282, 283, 298, 300
   commemorative medal, 275
   commemorative stamp, 202, 273-274, 289, 300
   Congress, report to, 307
   extravehicular activity, 198-199, 212, 217-220, 255, 267-268, 273-274, 277-278
   flight, 212-224, 277-278, 421
   lunar landing, 232, 262, 271, 277-278, 281, 421
   medical aspects, 137, 156-157, 245, 273
   Nixon, President Richard M., messages and welcome to, 209, 242
   preparations for, 5, 108, 114, 156-157, 168, 182, 196, 204
   press conference, 198-199, 207, 267-268, 277-278, 280-281, 362
   quarantine, 76, 141, 223, 247, 267, 273, 276
   record, 250-251
   significance of, 277-278, 281
   splashdown, 222-223
   TV interview, 283-284
awards and honors, 147, 233, 237, 251, 255, 279-280, 327, 365, 403
Canadian visit, 312, 399-400
commemorative stamp ceremony, 289, 300
tribute to, 228, 243, 279-280, 391, 399-400
White House visit, 319, 362
world tour, 312, 319, 327, 330, 332, 334, 335, 337, 341, 350, 351, 353, 359, 360, 361-362
Alexander, Charles S., 357
Algae, 159
Algiers, 234
Algol IIB (rocket engine), 291
Algol III, 291
Alioto, Mayor Joseph, 211
Allen, H. Julian, 53, 338, 345
Allen, Richard J., 68-69
Alley, Carroll, 71
Alouette I (Canadian satellite), 30
Alouette II, 30
Alphanumeric system, 156
ALSEP. See Apollo lunar surface experiment package.
Altus AFB, Okla., 411
Alvizienis, Dr. Algirdas A., 299
American Academy of Achievement, 203
American Airlines, 292
American Assn. for the Advancement of Science (AAAS), 2-3, 159, 184, 416-417
American Assn. of School Administrators, 46
American Astronautical Society (AAS), 183
American Bible Society, Laymen's Committee, 390-391

American Bible Week, 390-391
American Broadcasting Co. (ABC), 146, 352
American Cancer Society Seminar, 96
American Chemical Society, 110, 184
American Field Service, 238
American Geophysical Union, 116, 409
American Heart Assn., 372
American Institute of Aeronautical Sciences, 187
American Institute of Aeronautics and Astronautics (AIAA), 21, 60, 79, 297
   award, 21, 72-73, 209, 338, 345-346
   Honorary Fellows, 287, 346
   meetings, 21, 72-73, 78, 134, 343-346
   The Post-Apollo Space Program: An AIAA View, 151-152
   President's Forum, 346
American Management Assn., 56
American Museum of Natural History, 381
American Physical Society, 40, 120
American Rocket Society, 226
American Samoa, 401-402
American Science & Engineering, Inc., 104, 361
American Science Manpower, 1968 (NSF report), 420
American Security Council, 131
American Society for Engineering Education, 285
American Society of Biological Chemists, 184
American Society of Mechanical Engineers, 338
American Standard, Inc., Wilcox-Sierra Div., 315
American Systems, Inc., 105
American Telephone & Telegraph Co. (AT&T), 352
American Univ., 136
America's Next Decade in Space: A Report for the Space Task Group, 308
Ames Research Center (ARC) (NASA), 315, 348, 379
   award, 21, 338
   experiment, 395
   flight simulator, 362
   management, 190, 292
   personnel, 53, 332
   research, aircraft, 123, 362, 393
   Thermo and Gas-Dynamics Div., 200
Amistad Dam, 299
AMSA (advanced manned strategic aircraft). See B-1.
Amsterdam, Netherlands, 312, 332, 350
Amundsen, Capt. Roald, 251
An-22 (U.S.S.R. turboprop transport), 121, 159, 162, 168
Anaheim, Calif., 287, 338, 343
Ancient Order of Hibernians, 72
Anders, William A., 308
   air pollution, 382
   Apollo 8 mission, 6-7, 59, 71, 87, 203, 327

Apollo 11 mission, 5
   appointment to NASC, 141, 184
   awards and honors, 2, 6, 10, 68, 87, 100-101, 132, 203, 298, 390-391
   receptions for, 9, 13
   technology utilization, space, 327, 382
   White House visit, 6, 32
Anderson, Sen. Clinton P., 24, 29, 118, 390, 411
Anderson, G. P., 298
Anderson, Minister for Supply Kenneth McC. (Australia), 34
Andoya, Norway, 18, 20
Andrews AFB, Md., 262
Andromeda galaxy (M31), 99
Andromeda Strain, The, 181
Animal experiments, space, 18-19, 139, 189-190, 200, 201-202, 261, 335-336, 347-349, 415, 421
Ankara, Turkey, 312
Anniversary
   aircraft, 129, 133, 410, 411
   manned space flight, 8, 202
   NASA, 323
   Naval Missile Center, 325
   satellite, 33, 51, 59, 83
Antarctica, 159, 257, 359
Antenna, 29, 44, 148, 179, 187, 189, 356
   spacecraft, 137, 143, 259, 322
Anthony, Maj. James L. (USAF), 413
Antiballistic missile (ABM) system (see also Safeguard), 33, 83, 89, 320
   congressional consideration, 39, 46-47, 52, 53, 81, 103, 111-112, 124, 131
   funds for 48, 53, 131
   Nixon, President Richard M., views on, 43, 80, 81
   opposition to, 40, 46-47, 89, 103, 114, 124, 131, 136
   sites, 39, 43, 80
   U.S.S.R., 43, 48, 50, 53, 62, 80, 89, 129, 131, 140
Antihijacking system, 336
Apollo (program) (see also Apollo Applications program) 6, 70, 77, 195, 207, 256, 294-295, 308, 325, 357
   astronaut. See Astronaut.
   award, 68-69, 117, 280, 300, 326, 345-346, 346-347
   cost, 6, 43, 46, 88, 262, 390, 422
   criticism, 2, 273
   funds for, 15, 134, 138, 249-250, 311, 355-356, 422-423
   landing site, 77, 345
   launch
      Apollo 9 (AS-504), 62-65, 421
      Apollo 10 (AS-505), 142-145, 421
      Apollo 11 (AS-506), 168, 212-224, 421
      Apollo 12 (AS-507), 372-378, 421
   management, 202, 274, 287, 297, 313, 356, 422-423
   plans for, 66-67, 73, 77, 81-82, 102, 344, 351, 422-423

policy, 37, 51, 81–82, 86–87, 111, 136, 271–272, 273, 284, 355–356, 422–423
press comment (see also Apollo missions), 1, 250, 321
progress, 6, 65, 76, 77, 144–145, 223–224, 246–247, 420, 421–423
tracking, 140, 306–307
Apollo (spacecraft), 7, 56–57, 59, 73, 149, 197
 ascent propulsion system (APS), 64, 143, 377
 command and service module. See Command and service module.
 command module. See Command module.
 computer error, 174
 control, 26–27, 232, 375
 crewman optical alignment sight (COAS), 64
 debris, 89
 descent propulsion system (DPS), 64, 143, 213, 375
 escape device, 94
 equipment, 14, 77, 91, 102, 111, 168, 201, 373–377
 exhibit, 105, 161–162, 417
 heat shield, 27
 landing system, 26–27, 378, 405
 launch. See Apollo (program).
 launch vehicle. See Saturn.
 life-support system, 123, 206, 267, 378
 lunar module. See Lunar module.
 reaction control system (RCS), 64, 143, 213, 222, 375
 recovery, 141, 223, 378
 service module, 143
 service propulsion system (SPS), 64, 143, 212–213, 222, 373, 375, 377
 test, 65, 77, 143–144
Apollo 4 mission, 65
Apollo 5 mission, 65
Apollo 6 mission, 65
Apollo 7 (spacecraft), 20–21
Apollo 7 mission, 65, 68, 72–73, 144, 223, 287
Apollo 8 (spacecraft), 105, 161–162, 168
 computer error, 174
Apollo 8 mission, 37, 71, 141, 149, 155, 158–159, 391
 award, 2, 6, 10, 32, 68, 87, 100–101, 132, 136, 203, 287, 298, 390–391
 commemorative medals, 265
 commemorative stamp, 8, 46, 59, 129–130
 Congress, report to, 7
 Johnson, President Lyndon B., 6–7, 13, 232
 moving picture color film, 8, 46
 Nixon, President Richard M., 20, 32, 34, 61–62
 orbit, 73
 physiological aspects, 28
 press comment, 1, 2, 26
 press conference, 7, 25, 54
 reaction to, European, 54

 reception, 13
 record, 208
 religious aspects, 268, 283, 390–391, 394, 400
 results, 111, 116, 180
 significance, 2, 6–7, 16, 20, 39, 65, 73, 92, 129–130, 144–145, 223
 success, 47
 U.S.S.R., comment on, 25, 26
Apollo 9 mission, 161–162, 328
 award, 91–92
 biological aspects, 61
 extravehicular activity, 5, 33, 64, 81, 91
 launch, 62–65, 67, 421
 moving picture color film, 81
 Nixon, President Richard M., 65
 photographs, 64, 81, 116–117
 preparations for, 5, 16, 23, 26–27, 33, 58, 61
 press comment, 67, 72, 80, 89, 90
 press conference, 26–27, 33, 61, 77, 91, 116–117
 significance of, 65, 145, 223, 421
 spacecraft debris, 89
 splashdown, 64
 success, 130
 TV coverage, 50, 67
 U.S.S.R., comment, 80
Apollo 10 mission, 116, 138, 169, 174, 180, 200
 launch, 142–143, 421
 moving pictures, 159
 Nixon, President Richard M., 157
 photographs, 90, 142–144, 149, 159
 preparations for, 16, 23, 33, 73, 77, 82–83, 90, 104, 108, 118, 137, 138, 142
 press comment, 147–148, 149–150, 152–154, 155, 157
 press conference, 33, 108
 significance of, 152–154, 158, 223, 421
 splashdown, 143
 success, 230
 TV coverage, 90, 108, 142–143, 147, 149
 U.S.S.R., comment, 157, 158
Apollo 11 mission, 135, 170–171, 187, 188, 342, 368, 370, 388
 achievement, 223, 233, 235–236, 242–243, 245–249, 250–251, 309–310, 378, 390, 391, 419, 421, 422, 424
 awards and honors, 233, 237, 242, 243–244, 251, 279–280, 282, 283, 284, 298, 327, 343, 365, 403–404
 biological aspects, 141, 156–157, 245, 261, 273, 293, 356, 391
 book, 135–136, 252
 commemorative medals, 275
 commemorative stamps, 202, 233, 273–274, 289, 300, 311, 319
 Congress, report to, 307
 cosmonaut medals, 228, 230
 cost, 186, 271, 390
 criticism, 201, 205–206, 210, 211, 230–231, 235, 273

experiments, 259
  laser, 114, 220, 223, 237, 259, 261, 285, 411, 419
  seismic, 114, 223, 229–230, 237, 240–241, 249, 354–355, 419
  solar wind, 219, 223, 230, 285, 413, 419
extravehicular activity, 74, 108, 114, 182, 198, 208, 212, 217–220, 223, 243, 255–256, 267–268, 273–274, 278
Eyewitness to Space (art program), 226, 241, 403
implications of, 200–201, 238–239, 239–240, 240–241, 242–243, 261–262, 268–269, 277–278, 282, 284, 286–287, 291, 300, 307, 309–310, 321–322
launch, 168, 212–224, 421
lunar landing, 114, 168, 181–182, 215–220, 232–236, 238–239, 277–278, 280–281, 370, 378, 380, 404, 421, 422
lunar rock samples, 75–76, 114, 168, 223, 247, 250, 260, 261–262, 263, 266–267, 267–268, 270, 275, 288–289, 290, 292, 306, 353–354, 359, 365, 378, 386, 387, 397, 402, 407–408, 421
medical aspects, 130–131, 137, 156–157, 204, 206, 223, 245, 261, 266–267, 273
moving picture film, 247, 252, 277, 292
museum memorial, proposed, 343
Nixon, President Richard M., 190, 196–197, 200, 204, 209, 219–220, 223, 224, 228, 230–231, 242, 244, 246, 249–250, 261, 262, 275, 279–280, 284, 402
observance of, 206, 208, 209–211, 225–226, 232–234, 239, 244, 380
photographs, 212–223, 241, 247, 252–253, 255–256, 419
preparations for, 5, 23, 105, 108, 114, 137, 141, 144, 146, 149, 156–157, 168, 169–170, 178, 179–180, 181–182, 195, 196–200, 203, 204, 205–206, 209–210
press comment, 157, 193, 207, 208, 210–211, 225–226, 227–228, 228–229, 231, 235–236, 239–240, 244, 246–247, 249–250, 261–262, 271–272, 276, 279, 280–281, 282, 295, 308, 357
  foreign, 193, 210–211, 225–226, 227–228, 229, 231, 236, 240, 242, 247, 248, 251–252, 321–322
press conference, 108, 182, 198–199, 199–200, 204, 207, 231–232, 242–243, 250, 267–268, 277–278, 280–281
quarantine, 75–76, 141, 223, 247, 266–267, 273, 276
records, 250–251
religious aspects, 206–207, 232, 235, 242, 246–247, 253, 268, 283–284, 391, 394, 400
splashdown, 168, 190, 211, 222–223, 242–244
tracking, 204, 224
TV broadcasts, 182, 201, 210, 212–220, 222–223, 230
TV coverage, 212, 217, 225–226, 232–234, 244–245, 276
U.S.S.R. and, 206, 225, 230, 233, 238, 242, 244, 251, 256, 273, 311, 386
wager, 160–161, 233
Apollo 11: Preliminary Science Report, 419
Apollo 12 mission, 370
  achievement, 378, 392, 394, 421, 422
  biological aspects, 356, 377–379, 391
  experiments, 406–407, 410
    seismic, 325–326, 342, 376–377, 403
    solar wind, 376–377
  extravehicular activity, 243, 325–326, 334, 342, 372, 376–377, 378, 386–387, 394, 395–396
  launch, 372–378, 421
  lunar landing, 368, 375, 395–396, 407, 421, 422
  lunar rock samples, 334, 376–377, 378, 393, 394, 395, 396, 407–408, 421
  moving picture films, 395, 407
  Nixon, President Richard M., 371, 372, 378, 384, 385, 389, 392
  photographs, 334, 376–378, 395–396, 407
  power failure, 373–374, 380–381
  preparations for, 105, 243, 299, 315, 342, 350, 353, 362, 366, 368, 370, 371–372
  press comment, 367, 379, 380, 386–387, 391, 392, 393, 394
  press conference, 243, 325–326, 333–334, 342, 371–372, 406–407
  public reaction to, 379, 380, 384–385, 388, 391, 396
  quarantine, 356, 377–378, 396, 397, 399, 405
  records, 378, 389
  splashdown, 377–378
  TV broadcasts, 342, 362, 373–376, 377, 380, 384–385, 386
  U.S.S.R. and, 379, 384, 387–388
Apollo 13 mission, 182, 267, 315, 387, 392, 405
Apollo 14 mission, 267
Apollo 15 mission, 409
Apollo 16 mission, 409
Apollo 18 mission, 344
Apollo 19 mission, 344
Apollo 20 mission, 344
Apollo Achievement Award (NASA), 300
Apollo Applications program, 55, 87, 200, 258, 393
  contract, 59, 77, 85, 94, 154, 270–271, 297
  funds for, 15, 67, 82, 109, 119, 138, 163
  management, 344
  plans for, 37, 66–67, 110, 122, 128, 151–152, 286, 328, 423
Apollo lunar surface experiment package (ALSEP), 334–335
  Apollo 12 mission, 230, 243, 325, 342, 376, 421

Apollo 13 mission, 267
Apollo 14 mission, 267
Apollo Orbital Science Photographic Team, 409
Apollo Telescope Mount (ATM), 16, 17, 237–238
   contract, 11, 71, 77, 177, 270–271, 320
   experiment, 328
   funds for, 15–16
   test, 48–49
Applications Technology Satellite (ATS) program, 23, 139, 150, 360, 402, 414–415, 421, 423
   contract, 241, 337
   experiment, 337
   funds for, 15
Applied Physics Laboratory (Johns Hopkins Univ.), 37
APS. See Ascent propulsion system.
APT. See Automatic picture transmission.
Aquanaut, 51, 86, 343, 354
   accident, 51, 52, 57, 77, 188–189, 316–317
   record, 110
Ara, Inc., 408
Arcas (sounding rocket), 18, 20, 28, 33–34, 41, 79, 331, 354
Arctic, 101, 341
Arctowski, Henry, Medal, 121
Arcturus (star), 175
Arecibo (Puerto Rico) Ionospheric Observatory, 121, 350
Arenosillo, Spain, 94, 132, 133
Argentina, 79, 287, 304, 384
Argentine Radio Astronomy Institute, 181
Argonne National Laboratory, 24, 182, 292
Ariel IV (U.K. satellite), 59
Aristarchus (lunar crater), 139–140, 213
Arizona, 168, 291
Arizona, Univ. of, 16–17, 288, 417
   Lunar and Planetary Laboratory, 397
   Steward Observatory, 164
Armed Forces Day, 140
Armed Services Board of Contract Appeals, 160
Arms Control and Disarmament Agency, 30
Armstrong, Neil A., 315, 414
   Apollo 11 mission, 178, 402
      celebrations for, 279–280, 282, 283, 284, 298, 300
      commemorative medal, 275
      commemorative stamp, 202, 273–274, 289, 300
      Congress, report to, 307
      extravehicular activity, 198–199, 212, 217–220, 255, 267–268, 273–274, 278, 421
      flight, 212–224, 278, 380, 421
      lunar landing, 232, 239–240, 241, 252–253, 262, 271, 277–278, 280–281, 380, 421
      medical aspects, 137, 156–157, 245, 273
      Nixon, President Richard M., 209, 242, 319
      preparations for, 5, 108, 114, 156–157, 168, 181–182, 196, 204
      press conference, 198–199, 207, 267–268, 277–278, 280–281, 362
      quarantine, 76, 141, 223, 247, 267–268, 273, 276
      record, 250–251
      significance of, 278, 282
      splashdown, 223
      TV interview, 283
   appointment, 402
   awards and honors, 233, 237, 246, 251, 255, 279–280, 281, 327, 365, 403
   Canadian visit, 312, 399–400
   legacy, 281
   Thailand visit, 413
   tribute to, 228, 243–244, 391, 392, 399–400
   White House visit, 362
   world tour, 312, 319, 327, 330, 332, 334, 335, 337, 341, 350, 351, 353, 359, 360, 361–362
Armstrong, Neil A., Aerospace Museum, 246
Arnold Engineering Development Center, 138, 183–184, 393
Arnold, Henry H., Trophy, 87
Arntzenius, Dr. A. C., 372
ARPA. See Advanced Research Projects Agency.
Artificial horizon, 263
ARTS. See Automated radar tracking systems.
Ascent propulsion system (APS) (Apollo), 64, 143, 377
Asia, 101, 140
Aspis-Pronoia insurance company, 226
Associated Industries, Inc., 117, 125
Asteroid, 288, 359
Astrobee 1500 (sounding rocket), 335
Astrobotanist, 103
Astrology, 338
Astronaut (see also Cosmonaut; Extravehicular activity), 20, 71, 83, 105, 168, 173, 178, 185, 186, 329, 337, 342, 343, 347, 394
   accident, 206, 283
   achievements, 17, 50, 420
   Apollo mission. See Apollo missions (8, 9, 10, 11, 12, 13).
   appointment, 83, 141, 184, 317, 326, 396, 402
   Canada, visit to, 399–400
   crew assignment, 5, 105, 187–188
   former, 43, 75, 134, 327, 406
   fund raising by, 408
   goodwill tour, 32, 37, 38, 41, 46, 49, 51, 52, 155, 195–196, 197, 199, 202, 204, 299, 312, 319, 327, 330, 332, 334, 335, 337, 341, 350, 351, 353, 359, 360, 361–362, 424
   hazards, 28, 179–180, 205, 207, 278, 281, 380–381, 391

honors, 2, 6–7, 9, 10, 13, 39, 53, 68, 87, 91, 101–102, 132, 136, 200, 203, 233, 237, 246, 251, 255, 279–280, 282, 283, 289, 298, 307, 319, 327, 330, 332, 346, 347, 350, 360, 361–362, 365, 387, 390–391, 396, 403
lunar landing story, contract, 196
memorial, 53, 233, 245, 246, 304
physiology, 61, 71, 81, 130–131, 137, 142, 204, 206, 245, 273, 284, 317, 330
political aspirations, 168
Presidential mission, 413
press conference, 7, 26–27, 41, 53–54, 61, 75, 158–159, 196, 198–199, 204, 207, 267–268, 277–278, 280–281, 325–326, 333–334, 362, 396, 406–407
promotion, 245, 391, 392
quarantine, 76, 141, 223, 247, 267, 273, 276, 356, 377–378, 396, 397, 400, 405
record, 208, 250–251, 389
religion, 268, 391, 394, 400
resignation, 267
scientist-astronaut, 284, 286, 290, 339, 351, 355–356
Smithsonian Institution ceremony, 307
space rescue. See Space rescue treaty.
training, 114, 142, 155, 156–157, 181–182, 284, 414
tributes to, 228, 242, 246, 276, 279–281, 282, 362, 385, 392
USAF, 281, 289
White House liaison, 230, 232
White House visit, 32, 56, 157, 191, 204, 232, 319, 362, 413
women as, 359
Astronautics Engineer Award, 68
Astronauts Memorial Commission (proposed), 53, 245
Astronomy (see also individual observatories, planets, sounding rockets, and satellites such as *Mariner VI, Mariner VII, Oao II;* Pulsar; Radioastronomy; Star; Telescope), 109
award, 121
gamma ray, 2, 124, 318–319, 366–367
NASA program, 15–16, 66, 69, 76, 94–95, 124, 136, 138, 156, 256–257, 263, 361, 366–367, 403, 422–423
solar, 90–91, 121, 160, 336, 366–367, 395
stellar, 42, 134–135, 160, 322, 344, 350
ultraviolet, 3, 99, 124, 298
U.S.S.R. program, 103, 120, 278
x-ray, 104, 118, 120, 124, 133–134, 160, 281–282, 366–367
Athens, Greece, 405
ATA. See Air Transport Assn. of America.
Atlantic II (communications satellite). See *Intelsat-II F–3.*
Atlantic City, N.J., 46, 89
Atlantic Ocean, 18, 30, 64, 133, 137, 291, 343, 344, 353
Atlantis (undersea laboratory), 123
Atlas (booster), 347, 355

F, 10, 83
Atlas-Agena (booster), 106, 409
Atlas-Centaur (booster), 114, 154, 177, 355, 361, 412
AC–19, 92
SLV–3C, 55, 277
ATM. See Apollo Telescope Mount.
Atmosphere, 9, 18, 19, 20, 21, 23, 24, 25, 27, 33–34, 38, 41–42, 83, 88, 94, 204–205, 329, 357, 359–360, 409, 414
Atmosphere Explorer AE–C (spacecraft), 401
Atmosphere Explorer, AE–D, 401
Atmospheric Exploration by Remote Probes (NRC report), 409
Atomic Energy Commission (AEC) (see also AEC–NASA Space Nuclear Propulsion Office; NERVA; Rover; SNAP; and Vela programs), 106, 110
Argonne National Laboratory, 24, 182, 292
Brookhaven National Laboratory, 386
budget, 14–15, 42, 110, 193, 361
contract, 47–48, 94
cooperation, 27, 43, 88, 105, 177–178, 309, 332, 422
deep-water test, 189
magnet, superconducting, 24
nuclear power, peaceful use of, 25, 43, 130, 189, 354
nuclear-powered deep submergence research vehicle, 27
nuclear reactor, 28, 130, 177–178, 189
nuclear rocket engine, 76, 87, 105, 130, 309, 422
personnel, 30, 125, 184
Rocky Flats, Colo., facility fire, 135, 384
Space Nuclear Systems Div., 332
Atoms for Peace program, 41
ATS. See Applications Technology Satellite program.
Ats I (Applications Technology Satellite), 150, 277
Ats II, 150, 277
Ats III, 88, 277, 402
Ats IV, 277
Ats V (ATS–E), 86, 148, 277, 414–415, 421
ATS–F, 86, 118, 241, 311, 337, 423
ATS–G, 86, 118, 337
Atwood, J. Leland, 147
Aurora, 43, 48, 51–52, 54, 58, 72, 83, 90–91, 171–173, 253, 257, 323, 364, 393, 417, 423–424
Aurora 7 (spacecraft), 134
Aurora borealis, 323, 393, 417
Aurora Expedition, 393, 417
Aurorae (Esro IA) (satellite), 323
Austin, Tex., 268, 283, 400
Australia, 25, 43, 233, 244, 319, 334, 385
communication via satellite, 140, 202
international cooperation, space, 34, 137, 177, 318–319, 323, 399, 411, 423–424

tracking station, 34, 148, 189, 424
Australian National Univ., 292, 407
Australian Research Grants Committee, 319
Australian Weapons Research Establishment, 117
Australis Oscar-A (spacecraft), 399
Austria, 210
Autogiro, 283
Automated radar tracking systems (ARTS), 55
Automatic picture transmission (APT), 414
Avco Corp., 191
AVCS. See Advanced vidicon camera system.
Aviation Material Laboratories (AVLABS), 334
Aviation Progress Committee, 88–89
Aviation/Space Writers Assn., 138
Aviator's Trophy, 298
Awards, 338, 355
  civic, 10, 50, 279, 289, 330, 403, 419
  Government, 1–2, 54, 178, 251, 255, 307, 308, 330, 332, 350, 410, 418–419
    NASA, 10, 34, 38, 91–92, 154, 265, 280, 300, 326, 346–347, 371
  institutions, 168, 287, 338, 365
  military, 101, 300, 304, 334, 412
  society, 345–346
    achievement, 122–123, 127, 138, 203
    aeronautics, 21, 72–73, 87, 101, 127, 132, 150, 178, 200, 209, 283, 293, 298, 315, 338, 345–346, 410
    astronautics, 21, 68–69, 72–73, 87, 100–101, 132, 134, 182–183, 237, 251, 298, 345–346
    exploration, 251
    science, 136
    technology, 54, 117, 345–346
AX (close support aircraft), 131
Ayer, Prof. Alfred J., 239
Azcarrago, Gen. Luis (Spain), 180
Azur (W. German satellite), 357
Azur (GRS–A), 364–365, 379, 423

B

B–1 (advanced manned strategic aircraft, AMSA), 86, 131, 169, 360
B–52 (Stratofortress), 15, 86, 183
  HL–10 flights, 113, 117, 133, 148, 159, 174, 267, 297, 309, 319, 352, 360, 381, 389, 406
  X–24A flights, 101, 113, 286, 299, 316, 349, 371
B–58 (supersonic bomber), 354
Back contamination, 76, 136–137, 141, 156–157, 175, 176, 179–180, 181, 205, 235, 261, 270, 273, 302, 356
Bacteria, 159
Baikonur, U.S.S.R., 195–196

launch
  Cosmos, 58, 105, 186, 237, 267, 293, 305–306, 316, 347, 350–351
  Luna XV, 195, 206, 224
  Molniya 1–12, 237
  Soyuz IV, 11
  Soyuz V, 11
  Soyuz VI, 332
  Soyuz VII, 333
  Soyuz VIII, 333
Baker, Norman, 146
Baker-Nunn camera, 405
Baku, U.S.S.R., 387
Baldeschwieler, Dr. John D., 117
Bales, Stephen G., 280
Ball Brothers Research Corp. Aerospace Div., 346
Balloon, 86, 156, 205, 257, 315, 319, 363, 421
Ballute (balloon-parachute), 239
Baltimore, Md., 320, 410
Bangkok, Thailand, 243, 251, 312, 353, 413
Bantam Books, 252
Barbados, West Indies, 88, 291
Barbados Oceanographic and Meteorological Experiment (BOMEX), 88
Barium Cloud Experiment (BCE), 82, 105
Barker, Tom, 134
Barnard, Dr. Christiaan, 335–336
Barnard's Star, 109
Barnes, James, 394
Barreira do Inferno, Natal, Brazil, 312
Barry, Marion, 230–231
Barstow, Calif., 356
Barth, Dr. Charles A., 261, 298, 302
Barth, Robert A., Jr., 52
Bartoe, Otto E., Jr., 346
Basel, Switzerland, 332
Baudoin I, King of Belgium, 46, 55, 332
BBC. See British Broadcasting Corp.
BCE. See Barium Cloud Experiment.
Beam, Ambassador Jacob D., 197
Bean, Capt. Alan L. (USN)
  Apollo 12 mission
    extravehicular activity, 243, 325–326, 333–334, 342, 372, 376–377, 386–387, 394, 395
    flight, 372–378
    lunar landing, 375–376, 385, 386–387
    medical examination, 366
    Nixon, President Richard M., 371, 392
    plans for, 105, 325–326, 342, 353, 365, 366, 368, 371–372
    press conference, 325–326, 333–334, 406–407
    quarantine, 356, 377–378, 396, 397, 405
    splashdown, 377
  fund-raising dinner, 408
  grounded, 414
  promotion, 391, 392
  White House visit, 413

Beecher, William, 50
Beeferman, Larry W., 417
Beggs, James M., 52, 59, 79, 95, 178
Beil, David A., 260
Belgium, 32, 37, 46, 55, 132, 234, 312, 424
Belgrade, Yugoslavia, 312, 341
Bell Aerosystems Co., 103, 413
Bell, Dr. Persa R., 352–353
Bellcomm, Inc., 38
Ben Franklin (PX–15) (research submarine), 103, 282
Bench Crater (moon), 376
Bendix Corp., 111, 177, 315, 320
Bendix Field Engineering Corp., 196
Benedict, Howard, 104
Benetnasch (star), 79
Beniele, Max, 239
Bennett, Dr. Ivan L., Jr., 174
Benoit, Dr. Robert, 159
Beregovoy, M/G Georgy T., 195, 197
  Apollo 10 mission message, 158
  press conference, 343, 350, 362
  U.S. visit, 337, 343, 347, 350, 351, 352, 354, 359, 362
Beresford, Spencer M., 304
Berkeley, Calif., 383
Berlin, West, 37, 49, 55, 334
Berne, Univ. of, 413
  Physics Institute, 285
Bernhard, Prince (Netherlands), 354
Bernier, Robert E., 290
Berry, Dr. Charles A., 110
  Apollo 9 mission, 61
  Apollo 11 mission, 130–131, 200, 204, 206, 273
  Apollo 12 mission, 366, 405
  award, 287, 340
  interview, 206
  press conference, 61
Betadine (disinfectant), 156–157
Bethpage, N.Y., 26
Beverlin, Charles J., 347
Biehl, Richard E., 239
Biloxi, Miss., 225
Biological isolation garments (BIG), 141, 156–157, 356
The Biomedical Foundations of Manned Space Flight: A Report of the Space Science and Technology Panel of the President's Science Advisory Committee, 367–368
Bionic Instruments, Inc., 327
Biosatellite (program), 23
Biosatellite I, 190
Biosatellite II, 78, 160, 190
Biosatellite III, 162, 189–190, 200, 201–202, 347–349, 415, 421
Biosatellite-F, 138, 163
Bird, John D., 4
Bisplinghoff, Dr. Raymond L., 178, 363
Black Arrow (booster), 190
Black Brant (Canadian sounding rocket), 360
Black Brant IIIB, 59, 127
Black Brant IV, 312
Black Brant VB, 366
Blagonravov, Dr. Anatoly A., 80, 185–186, 311, 411
Block Crater (moon), 376
Blount, Postmaster General Winton M., 129–130, 202, 274, 289, 300
Blue Book, Project, 411, 413–414, 416
BMWF. See Germany, West, Ministry of Scientific Research.
BOB. See Budget, Bureau of.
Bobko, Maj. Karol H. (USAF), 281
Boeing Co., 122, 164
  Aerospace Group, 353
  booster, Saturn V, 47, 278, 297
  contract, 1, 25–26, 47, 278, 297, 320
  employment, 330
  jet passenger transport. See Boeing 707 and Boeing 747.
  lunar roving vehicle, 320, 353
  personnel, 285
  space station, 25–26
  supersonic transport, 17, 314–315
  Vertol Div., 92
Boeing 707 (jet transport), 88, 382
Boeing 747 (jet passenger transport), 203, 304, 317, 344, 350, 371, 382–383
  maiden flight, 45–46, 421
  transatlantic, 169, 344
  orders, 341
  Paris Air Show exhibit, 162, 169, 175
  preview flight, 400
Boffey, Philip M., 189
Bogard, Dr. Donald, 288–289
Bogart, L/G Frank A. (USAF, Ret.), 346
Bogota, Colombia, 226, 312
Bohr, Niels, Library, 164
Bolender, Carroll H., 91
Bolger, Philip H., 21
Bologna, Univ. of, 272
Bombay, India, 312, 351
BOMEX. See Barbados Oceanographic and Meteorological Experiment.
Bonn, W. Germany, 37, 48, 55, 177
Bonny (space monkey), 190, 200, 201–202, 347–349, 415, 421
The Book of Mars, 89
Boosted Arcas II (sounding rocket), 28, 331
Booth, William, 2
Booz-Allen Applied Research, Inc., 335
Bordeau, Robert E., 346
Bordeaux, France, 297–298
Boreas (Esro IB) (satellite), 323, 391, 423
Borger, John G., 346
Borman, Col. Frank (USAF), 71, 229, 274, 337, 343, 347, 362, 380
  Apollo 8 mission, 6–7, 32, 34, 39, 68, 71, 268
  appointment, 131
  awards and honors, 2, 6–7, 9, 10, 39, 68, 87, 100–101, 132, 136, 203, 298, 346, 387, 390–391

goodwill tour, 32, 37, 38, 41, 46, 48, 49, 51, 52, 53–54
moon, exploration of, 102
political aspirations, 168, 380
press conference, 7, 41, 53–54, 158–159, 196, 204
receptions for, 9, 10, 13
space, exploration of, 158–159
visit to Czechoslovakia, 136, 155
visit to U.S.S.R., 195–196, 197, 199, 202, 204, 210, 211, 245, 337
White House activities, 6, 32, 56–57, 230, 232
Boston, Mass., 29, 105, 112, 363, 416
Boulder, Colo., 90, 259, 336
bps: bits per second
Bradley International Airport, 263
Bramley, Eric, 136
Branscomb, Dr. Lewis M., 416
Brantford, Canada, 234
Bray, Rep. William G., 253
Brayton Cycle space power generator, 58
Brazil, 175, 179, 186, 234, 423–424
Brett, Dr. Robin P., 400
Brevard County, Fla., 197, 210, 225
Brewster, Wash., 179
Brezhnev, Leonid I., 23, 127, 129, 251, 349–350
The Brick Moon (novel), 357
Bright, Loren G., 332
Bristol, U.K., 105
British Aircraft Corp., 61, 105, 285, 340
British Broadcasting Corp. (BBC), 225
British External Telecommunication Executive, 32
British Interplanetary Society, 187
Bromine, 128
Bronk, Dr. Detlev W., 1–2
Brooke, Sen. Edward W., 183, 332
Brooks, Harvey, 257
Brookhaven National Laboratory, 386
Brown, Arthur W., 129
Brown, Eileen, 385
Brown, Dr. Harold, 355
Brown, Dr. Herbert C., 418
Brown, Judge John R., 283
Brown Univ., 168
Browne, Secor D., 302
Bruceton, Pa., 387
Bruns, Franklin R., Jr., 319
Brussels, Belgium, 37, 46, 55, 234, 312
Bryson Construction Co., Inc., 320
Bucharest, Romania, 261, 262
Buchwald, Art, 235
Buckhorn, Calif., 174
Buckingham Palace, U.K., 335
Budapest, Hungary, 234
Budget, Bureau of (BOB), 5–6, 32, 108–109
Buechner, Dr. Helmut K., 328
Buenos Aires, Argentina, 312, 384
Buffum, Ronald J., 260
Buffum, William B., 242
Bulgaria, 24, 229
Bull, Gifford, 345–346

Bullpup Cajun (sounding rocket), 178
Burcham, Dr. Donald P., 135
Burcher, Eugene S., 138
Bureau of Fisheries, 9
Bureau of Sport Fisheries and Wildlife, 14
Burke, Rep. J. Herbert, 281
Burlingame, A. L., 365
Bush, Dr. Vannevar, 184
Butz, J. S., Jr., 192, 419–420

C

C–5 (Galaxy) (military cargo transport), 393, 411
C–5A, 57, 212, 293, 411, 413
contract, 17, 123–124, 128–129, 130, 132, 138, 251, 380
cost, 17, 107, 128–129, 138, 251, 281
Paris Air Show, 162
record, 151, 323
static test, 207
test flights, 57, 151, 283, 323, 421
C–130 (Hercules) (transport aircraft), 359
C–141 (military transport), 293
CAB. See Civil Aeronautics Board.
California, 177, 291
California Air Pollution Board, 383
California Institute of Technology (Cal-Tech), 117
award, 338, 355
computer (self-testing-and-repairing), 298–299
galaxies, discovery of, 367
Mariner VI, 269, 282, 301
Mariner VII, 265, 269–270, 301
pulsar signal research, 121
California Museum of Science and Technology, 123
California, Univ. of, 42, 179, 238, 395
Berkeley, 53, 72, 110, 167, 186, 261, 301–302, 346, 365, 383
Lawrence Radiation Laboratory, 53, 110
Lick Observatory, 42, 237, 259, 261, 285
Los Angeles (UCLA), 174, 288, 348–349, 356, 408
San Diego, 8, 331–332, 403
Santa Barbara, 263, 416
Calio, Anthony J., 347, 353
Calle, Paul, 202
Cambridge, Mass., 405, 417, 418
Cambridge, U.K., 359
Cambridge Univ., 42, 312
Camden, N.J., 54
Camera, 77, 91, 116–117, 405
Apollo 11, 201, 217
Apollo 12, 334, 362, 376–377, 386
Essa IX, 57–58
Mariner VI, 55, 252, 253–254, 338
Mariner VII, 252, 259, 338
Cameron, Dr. Roy E., 159

Campbell, Dr. Malcolm J., 180
Canada, 155, 399–400
  Apollo 11 reaction, 236, 240
  cooperation, 30, 132, 393, 411, 417, 423–424
  satellite, 30, 241, 423
  sounding rocket (see also Sounding rocket, international programs), 59, 127, 312, 366, 423–424
Canaveral Council of Technical Societies, 83
Canberra, Australia, 34, 189, 292
Cancer, 96
Candau, Dr. M. G., 254–255
Cannon, Berry L., 52, 57, 77, 188–189, 316–317
Canopus (star), 55, 92, 99, 117
Cap Pistol (astronaut maneuvering unit), 239
Cape Canaveral, Fla., 180–181, 203–204, 229, 242, 281, 393
Cape Kennedy, Fla. (see also Eastern Test Range and Kennedy Space Center), 228, 304, 398
  Apollo 11 launch, 183, 186, 204, 205–206, 224–226, 249
  name controversy, 180–181, 203–204, 229, 242, 281, 393
Cape Kennedy Regional Airport, Fla., 263
Cape Keraudren, Australia, 25, 43
Cape Parry, Canada, 71
Carbon dioxide, 57, 128, 365, 408
Carbon suboxide, 396
Cardiff, Wales, 134
Cardiovascular pressure transducer, 128
Carlos, Prince Juan (Spain), 330
Carnarvon Tracking Station, Australia, 34
Carnegie Endowment for International Peace, 381, 393–394
Carnegie Institution of Washington, 341
Carnegie Southern Observatory, 341
Carpenter, Cdr. M. Scott (USN, Ret.), 21–22, 86, 134, 354
Carpentier, Dr. William R., 245
Carson, Robert K., 396
Carswell AFB, Tex., 331
CAS. See Collision avoidance system.
CAS. See Cooperative Applications Satellite.
Case Institute of Technology, 125, 304
Case Western Reserve University, 319
Castel Gondolfo, Italy, 206, 225, 232, 242
CAT. See Clear air turbulence.
Catholic Biblical Assn. of America, 390–391
Catterson, Dr. A. Duane, 142
Cayey, Puerto Rico, 27
CBS Laboratories, 102, 201
CDDT. See Countdown demonstration test.
Ceausescu, President Nicolae (Romania), 261
Centaur (booster upper stage) (see also Atlas-Centaur), 330, 355, 412
Centaurus (constellation), 256, 281
Central America, 370
Centralization of Federal Science Activities (House report), 161
Cerberus canal (Mars), 262
Cernan, Cdr. Eugene A. (USN), 16, 90, 108, 142–144, 152–153, 191, 267
Cerro Tololo, Chile, 133–134
Certificate of Appreciation (NASA), 347
CF-6 (turbofan engine), 212
CH-54H (Flying Crane) (helicopter), 35
Chaban-Delmas, Premier Jacques (France), 330
Chafee, Secretary of the Navy John H., 4, 402
Chaffee, L/Cdr Roger B. (USN), 228
Chamant, Jean, 162
Chamberlain, Dr. Owen, 238
Chambers, Dr. Alan B., 315
Chamical, Argentina, 304
Chanute, Octave, Award, 73
Chao, Dr. Edwin C. T., 400
Chapman, Dr. Dean R., 200
Chappell, Rep. William, 242
Charles, Robert H., 123–124
Charlie Brown (*Apollo 10* CSM). See Command and service module.
Charyk, Dr. Joseph V., 137, 179
Chayes, Abram, 131
Chicago Executive Club, 320
Chicago, Ill., 24, 28, 34, 44, 263, 279, 320, 380, 418
Chicago, Univ. of, 121, 191, 397
Childs, Marquis, 154
Chile, 234, 341
China, Communist, 234, 236, 388, 404
  missile threat, 43, 80, 181
  nuclear test, 181
China, Nationalist, 234
Chlorine, 128
Chrysler Corp., 117–118, 154
  Space Div., 49, 94, 101, 113, 123, 154
Churchill Research Range, Canada (see also Fort Churchill, Canada), 21, 23, 31, 38, 41–42, 43, 48, 49, 51–52, 54, 58, 72, 107, 113, 360
Circadian rhythms, 180
Civil Aeronautics Board (CAB), 136, 244, 302
Civil Air Patrol, 49
Clark, Evert, 69
Clark, Adm. Joseph J. (USN, Ret.), 380
Clark, Dr. John F., 346
Clark Univ., Robert Hutchings Goddard Library, 147
Clarke, Arthur C., 39, 289
Clarkson College of Technology, 318
Clayton, James, 249
Clear air turbulence (CAT), 417–418
Clegg, Dr. P. E., 312
Cleveland, Ohio, 34
Clifford, Secretary of Defense Clark M., 4

Clifton, Dr. H. Edward, 51, 86, 110
Cloudcroft, N. Mex., 325
CM. See Command module.
Coahuila, Mex., 299
COAS. See Crewman optical alignment sight.
Cochran, John, 105
Code, Dr. Arthur D., 99, 344
Cohn, Victor, 81–82, 136, 286
College, Alaska, 257
Collier, Robert J., Trophy, 101, 132, 283, 293
Collins, Michael, 414
  Apollo 11 mission, 401–402
    celebrations for, 279–280, 282, 283, 298, 300
    commemorative medal, 275
    commemorative stamp, 202, 273–274, 289, 300
    Congress, report to, 307
    flight, 212–224, 277, 421
    lunar landing, 277
    medical aspects, 137, 156–157, 245, 273
    Nixon, President Richard M., 209, 242
    preparations for, 5, 156–157, 168, 196, 204
    press conference, 198, 207, 277, 280–281, 362
    quarantine, 76, 141, 223, 247, 273, 276
    record, 250–251
    significance of, 277
    splashdown, 223
    TV interview, 283–284
  appointment, 396, 408
  awards and honors, 233, 237, 251, 255, 279–280, 327, 365, 403
  Canadian visit, 312, 399–400
  commemorative stamp ceremony, 289, 300
  promotion, 245
  tribute to, 228, 399–400
  White House visit, 362
  world tour, 312, 319, 327, 330, 332, 334, 335, 337, 341, 350, 351, 353, 359, 360, 361–362
Collins Radio Co., 189
Collision avoidance, aircraft, 70, 315–316, 319, 362, 369
Collision avoidance system (CAS), aircraft, 315–316
Cologne, W. Germany, 312
Colombia, 226
Colon de Carvajal y Maroto, Cristobal, 52
Colorado Springs, Colo., 134, 170, 225
Colorado, Univ. of, 5, 22, 34, 41, 72, 127, 179, 189, 261, 298, 302, 314, 411
  Laboratory for Atmospheric and Space Physics, 316
Columbia (*Apollo 11*). See Command module.
Columbia Radiation Laboratories, 71
Columbia Univ., 71, 117, 174, 230, 250, 273, 354–355, 403
Columbus, Christopher, 226, 330
Columbus, Diego, 52
Comet, 359, 409
Comet (aircraft), 162
Command and Data Acquisition (CDA) station (ESSA), 58
Command and service module (CSM)
  Apollo 8, 7
  Apollo 9 (*Gumdrop*) (CSM-104), 5, 62–65, 72
  Apollo 10 (*Charlie Brown*) (CSM-106), 16, 90, 108, 142–144, 159
  Apollo 11 (*Columbia*) (CSM-107), 108, 212–218, 222, 255–256
  Apollo 12 (*Yankee Clipper*) (CSM-108), 372–376, 377, 389, 395
  contract, 70
Command module (CM)
  Apollo 8, 7
  Apollo 9 (*Gumdrop*), 5, 26–27, 62, 64, 65, 81, 91–92
  Apollo 10 (*Charlie Brown*), 362
  Apollo 11 (*Columbia*), 141, 198–199, 212, 222–223, 226, 388, 397
  Apollo 12 (*Yankee Clipper*), 333–334, 372–373, 377–378.
Commerce, Dept. of, 50, 88, 199, 402
Commission on Federal Reorganization, 341
Commission on Human Rights (New York), 2
Commission on Marine Science, Engineering and Resources, 9, 56
Committee on Scientific and Technical Communication (SATCOM), 174–175
Committee on Space Research (COSPAR), 136–137
Commonwealth Club, San Francisco, 259–260
Commonwealth Scientific and Industrial Research Organization, 121
Communications by Satellite: An International Discussion, 381
Communications satellite (see also individual satellites: *Echo II, Intelsat I, Intelsat-III F-2, Molniya I-11*, etc.), 140, 343
  benefits, 33, 37, 47, 137
  conference, 24, 88
  contract, 61, 66, 171, 241
  cooperation, 402
  international, 88, 117, 132, 157, 357, 381, 393–394
  ground station, 27, 61, 135, 137, 171
  launch
    failure, 245, 281, 385
    Intelsat-III F-3, 39–40, 151
    Intelsat-III F-4, 150–151
    Intelsat-III F-5, 245, 281, 385
    Molniya I-11, 106
    Molniya I-12, 237
  plans for, 23, 114, 290–291
  Skynet A (IDCSP-A), 388–389

Taccomsat I, 44
military, 44, 66, 131, 140, 322, 346, 388–389, 422
rates, 30
use of, 18, 47, 61, 75, 111, 137, 140, 202, 253, 257, 286, 340–341, 402
U.S. policy, 253, 352
Communications Satellite Act of 1962, 53
Communications Satellite Corp. (ComSatCorp), 290
Annual Meeting of Shareholders, 137
Apollo 11 TV coverage, 225–226, 244–245, 276
Atlantic II. See *Intelsat-II F–3*.
contract, 171
cooperation, 179
Early Bird. See *Intelsat I*.
FCC regulation, 30, 352
ground station, 61, 137, 171
INTELSAT, 101–102
Intelsat I (Early Bird), 61, 137, 191, 259, 286
Intelsat-II F–3 (Intelsat II–C; Atlantic II), 191, 259
Intelsat-III F–1, 40, 135, 151
Intelsat-III F–2, 23, 30, 40, 61, 151, 191, 245, 259, 276, 286, 340–341
Intelsat-III F–3, 39–40, 137, 140, 151, 421
Intelsat-III F–4, 140, 150–151, 276, 421
Intelsat-III F–5, 245, 281, 385, 421
Intelsat IV, 114
rates, 30
revenues, 54, 61, 231, 340–341
satellite program, 114, 137
services, 18, 50, 137, 179, 244–245, 276, 286, 340–341
Computer, 71, 87, 139, 174, 177, 203, 204–205, 290, 298–299, 322, 334–335, 350, 362
Computer Sciences Corp., 48, 203
ComSatCorp. See Communications Satellite Corp.
Concorde (U.K.-France supersonic transport), 13–14, 26, 146, 197, 255, 413
flights, 61, 71, 105, 162, 173, 285, 323, 329, 340
Condon, Dr. Edward U., 5, 8, 20, 50, 288
Cone, Clarence D., Jr., 96
Congress, 47–48, 68, 210, 232, 311, 314–315, 341, 361, 411
Apollo 8 mission, 7
Apollo 11 mission, 186, 207, 279–280, 307, 312–313
ComSatCorp report to, 61
Defense, Dept. of, 99, 104, 123–124, 178
Federal Aviation Administration, 156
NASA's *Twentieth Semiannual Report* to, 372
President's messages
airports, 182
budget, 14–15, 107
State of the Union, 13
Science, Secretary of (proposed), 3
space program, 51, 80, 81–82, 95, 104, 196, 294
Congress, House of Representatives, 108–109, 114, 124, 169, 186, 224, 238, 270, 307, 332, 361, 381–382
bills introduced, 2, 18, 52, 53, 113, 138, 202, 242, 245, 253, 309, 404
bills passed, 53, 176, 186, 307, 312, 330, 363, 383, 404, 412, 414
Committee on Appropriations, 184, 369, 372, 383
Committee on Armed Services, 39, 99
Committee on House Administration, 245
Committee on Interstate and Foreign Commerce, 40, 253
Committee on Judiciary, 202
Committee on Science and Astronautics, 18, 66–67, 69, 70, 113, 115, 117, 147, 245, 257, 275–276, 284, 317
Subcommittee on Advanced Research and Technology, 76, 84, 92, 400, 404
Subcommittee on Manned Space Flight, 71, 73–74, 91
Subcommittee on NASA Oversight, 202, 276–277
Subcommittee on Science, Research, and Development, 52–53, 161, 255
Subcommittee on Space Science and Applications, 74, 76, 78, 84–86, 118, 162, 336–337, 371, 415
Congress, Senate, 57, 78, 103, 131, 186, 224, 276, 281, 309, 312, 332, 361, 363, 369, 381–382, 383, 390, 406, 412
bills introduced, 22, 26, 29, 118, 150, 183, 330, 383, 404
bills passed, 13, 246, 255, 308, 312, 361, 365, 369, 383, 410, 412
Committee on Aeronautical and Space Sciences, 13, 79–80, 312
NASA budget, 24, 118, 186–187, 270, 311
testimony, 119–120, 123, 127, 130, 131, 134, 160, 266
Paine, Dr. Thomas O., letter to, 390, 411–412
Paine, Dr. Thomas O., nomination, 79–80
Committee on Appropriations, 135, 363–364, 369, 410
Committee on Armed Services, 86, 111–112, 330
Committee on Banking and Currency, 115, 251
Committee on Commerce, 133, 150
Subcommittee on Aviation, 88–89
Committee on Foreign Relations, 22, 53, 57, 383
Committee on Government Operations, 312
Committee on Interior and Insular Affairs, 203–204, 393

Committee on Labor and Public Welfare, 42
Committee on Rules and Administration, 404
Joint Committee on Atomic Energy, 354
nominations approved and confirmed, 43, 79-80, 87, 118, 130, 184, 395, 408
nominations submitted to, 91, 141, 180, 371
resolution, 133, 203-204, 229
Congressional Medal of Honor, 2
Congressional Space Medals of Honor, 307, 308, 319
Conklin, Dr. Edward K., 175
Conner, Dr. J. P., 281-282
Conrad, Capt. Charles (Pete), Jr. (USN)
 Apollo 12 mission
  extravehicular activity, 333-334, 342, 372, 376-377, 386-387, 394, 395
  flight, 372-378
  lunar landing, 375-376, 385, 386-387
  medical examination, 366
  Nixon, President Richard M., 371, 392
  plans for, 105, 342, 353, 365, 366, 368, 371-372
  press conference, 333-334, 406-407
  quarantine, 356, 377-378, 396, 397, 405
  splashdown, 377-378
 fund-raising dinner, 408
 promotion, 391, 392
 White House visit, 413
Constan, Dr. George N., 256, 260
Contact: The Story of the Early Birds, 52
Contract (see also under agencies, such as NASA, USAF), 25
 cost-plus-award-fee, 30, 59, 196, 292, 293, 385-386
 cost-plus-fixed-fee, 1, 38, 85, 87-88, 154
 cost-plus-fixed-fee/award-fee, 297, 351
 cost-plus-incentive-fee, 150, 160, 191, 320, 353, 355
 fixed-price, 14, 32-33, 48, 401, 402-403, 414
 fixed-price-incentive-fee, 71, 212
 study, 25-26, 46, 47, 94, 105, 111, 120, 128, 241, 270, 280, 335, 340, 401
Control Data Corp., Melville Space and Defense Systems Div., 388
Convair (*Galileo*) (jet aircraft), 133, 393, 417, 423-424
Convention of Cooperation for the Security of Air Navigation, 148
Convocation on Ecology and the Human Environment, 132
Cook, Richard W., 347, 360
Cooke, H. Lester, 403
Cooke, W. J., 164
Coons, Roy G., 266-267
Cooper, Col. L. Gordon (USAF), 326
Cooperative Applications Satellite (CAS), 86
Copernicus (moon crater), 253
Copernicus, Nicolaus, 300
Cornell Aeronautical Laboratory, 2, 42, 345-346
Cornell Medical Center, 362-363
Cornell Univ., 114, 121, 180, 299, 317, 339, 350, 352
Coronagraph, 127
Corporation for Public Broadcasting, 402
Cortright, Edgar M., 21
Cosmic ray, 29, 134-135, 177, 187, 334, 395
Cosmic Ray Ionization Program (CRISP), 363
Cosmonaut, 41, 48, 249, 365-366
 Apollo 10 mission message, 158
 Apollo 11 mission message, 244
 astronauts, meeting with, 168, 195-196, 197, 199
 awards and honors, 23-24, 39, 349-350
 film, 104
 interview, 25, 159-160, 372, 387-388
 medals placed on moon, 228, 230
 Soyuz IV mission, 11-12, 23-24, 332, 422
 Soyuz V mission, 11-12, 23-24, 332, 422
 Soyuz VI mission, 332-333, 336, 341-342, 343, 349-350, 361, 365-366, 382, 420, 422
 Soyuz VII mission, 333, 336, 341-342, 343, 349-350, 361, 365-366, 420, 422
 Soyuz VIII mission, 333, 336, 341-342, 343, 349-350, 361, 365-366, 420, 422
 space cooperation, 196, 325, 351
 U.S. visit, 337, 343, 347, 350, 351, 352, 354, 359, 362, 387-388
Cosmos (U.S.S.R. satellite), 422
Cosmos CCLXIII, 9
Cosmos CCLXIV, 24
Cosmos CCLXV, 43
Cosmos CCLXVI, 56
Cosmos CCLXVII, 58
Cosmos CCLXVIII, 67
Cosmos CCLXIX, 67
Cosmos CCLXX, 70
Cosmos CCLXXI, 80-81
Cosmos CCLXXII, 82
Cosmos CCLXXIII, 89
Cosmos CCLXXIV, 90
Cosmos CCLXXV, 94
Cosmos CCLXXVI, 101
Cosmos CCLXXVII, 101
Cosmos CCLXXVIII, 105
Cosmos CCLXXIX, 109
Cosmos CCLXXX, 116
Cosmos CCLXXXI, 137
Cosmos CCLXXXII, 148
Cosmos CCLXXXIII, 158
Cosmos CCLXXXIV, 160
Cosmos CCLXXXV, 169
Cosmos CCLXXXVI, 181
Cosmos CCLXXXVII, 186
Cosmos CCLXXXVIII, 189
Cosmos CCLXXXIX, 203

Cosmos CCXC, 237
Cosmos CCXCI, 267
Cosmos CCXCII, 281
Cosmos CCXCIII, 283
Cosmos CCXCIV, 285, 291
Cosmos CCXCV, 287
Cosmos CCXCVI, 293
Cosmos CCXCVII, 297
Cosmos CCXCVIII, 305–306
Cosmos CCXCIX, 309
Cosmos CCC, 314
Cosmos CCCI, 316
Cosmos CCCII, 339
Cosmos CCCIII, 341
Cosmos CCCIV, 346
Cosmos CCCV, 347
Cosmos CCCVI, 350–351
Cosmos CCCVII, 350–351
Cosmos CCCVIII, 360
Cosmos CCCIX, 370
Cosmos CCCX, 380
Cosmos CCCXI, 392
Cosmos CCCXII, 392
Cosmos CCCXIII, 401
Cosmos CCCXIV, 405
Cosmos CCCXV, 413
Cosmos CCCXVI, 414
Cosmos CCCXVII, 414
COSPAR. See Committee on Space Research.
Council of Economic Advisers, 3
Countdown Apollo (U.S. Paris Air Show theme), 161–162
Countdown demonstration test (CDDT), 108, 118, 182, 195, 350, 351, 353
Covington, Ozro M., 117, 346
Cox, Gardner, 323
Crab Nebula, 16–17, 42, 71, 78, 118, 164, 350, 404
Cracow, Poland, 233
Creason, R. L., 57
Crewman optical alignment sight (COAS), 64
Crews, L/C Albert H. (USAF), 281
Crichton, Michael, 181
Crimea, U.S.S.R., 195
Crippen, L/Cdr Robert L. (USN), 281
CRISP. See Cosmic Ray Ionization Program.
Cromley, Ray (Raymond Avolon), 298
Crossfield, A. Scott, 29
Crooker, John H., 302
Crowley, Mrs. Peggy, 46
CSM. See Command and service module.
Cuba, 51, 234
Cudaback, Dr. David, 383
Cunningham, R. Walter, 21, 72–73, 105
Curtiss-Wright Corp., 239
Cyanogen, 139–140
Czechoslovak Academy of Sciences, 136
Czechoslovakia, 155, 233, 249, 251, 335
Czechoslovakian Communist Party, 242

D

Dacca, East Pakistan, 312, 353
Daddario, Rep. Emilio Q., 114
Dade County, Fla., 9
Dai Chi Chinei (Japanese freighter), 198
Daley, Mayor Richard J., 279
Dallas, Tex., 2–3, 372
Dana, William H., 117, 148, 240, 297, 381, 406
Daniel and Florence Guggenheim International Astronautics Award, 287
Daniel Guggenheim Fund for the Promotion of Aeronautics, 226
Darwin, Australia, 312
Davies, Merton E., 282
Day, LeRoy E., 131, 336, 354
Day, Melvin S., 135
Dayton, Ohio, 138
DC–8 (jet transport), 88, 382
DC–10 (jet transport), 212, 341
Dearborn, Mich., 197
"Debrief: Apollo 8" (color film), 8
Debus, Dr. Kurt H., 10, 122, 274, 289, 371
Deception Island, Antarctica, 159
Deep Quest (research submarine), 330
Deep Space Network (DSN), 34, 180, 189, 274–275, 346
Deep Submergence Rescue Vehicle, 112
Deep submergence research vehicle, 27
Defence Research Board (DRB) (Canada), 41
Defense Communications Agency, 353
Defense, Dept. of (DOD) (see also U.S. Air Force, U.S. Army, and U.S. Navy), 5, 116, 316–317
   Advanced Research Projects Agency, 119, 346, 424
   aircraft, 4, 11, 15, 35, 39, 86, 99, 106, 107, 111–112, 115, 123–124, 128–129, 130, 131, 135, 151, 155, 183, 201, 251, 260, 283, 293, 319, 320–321, 354, 372, 379, 389–390, 417–418, 424
   anniversary, 59
   award, 99–100, 101, 346
   budget, 14–15, 42, 86, 99, 107, 109–110, 131, 178, 181, 193, 354, 410, 412
   communications satellite system, 131, 319, 345, 385, 388–389
   contract, 48, 123–124, 130, 135, 136, 201, 251, 330, 372, 386, 408
   cooperation, 88, 399, 417–418
      NASA, 11, 34, 38, 58–59, 86–87, 88, 119, 151–152, 153, 176–177, 200, 317, 345, 368, 379, 388–389, 417–418, 421, 422, 423
   cooperation, international, 388–389
   facilities, 43
   missile program, 10, 15, 33, 39, 43, 46–47, 53, 80, 81, 89, 103, 112, 124, 131, 136, 183, 190–191, 229, 320–321

MOL, 15, 21, 86–87, 110, 151–152, 176–177, 178, 181, 200, 260, 281, 289, 423
  personnel, 4, 119, 193, 251, 285, 346
  R&D, 11, 14, 15, 40, 67, 88, 106, 124, 136, 164, 257, 330, 345, 363, 379, 420, 424
  space program, 15, 59, 86–87, 131, 151–152, 164, 167, 176–177, 193, 200, 304–305, 319, 345, 388–389, 419–420, 421, 422, 423
Defense Satellite Communications System (DSCS), 66, 131
De Florez Training Award, 345–346
De Gaulle, President Charles (France), 41
Delbruck, Dr. Max, 338
Delphi, Greece, 187
Delta (booster) (see also Thor-Delta), 135, 385, 421
Dembling, Dr. Paul G., 300–301
Denebola (star), 79
Denmark, 323
DeNoyer, Dr. John M., 403
Denver, Colo., 75, 183, 288
de Seversky, Alexander P., 412
Des Moines, Iowa, 243, 385
Descartes (moon), 377
Descent propulsion system (DPS), 64, 143, 213, 375
Detroit, Mich., 337
Dewart, Prof. Leslie, 39
Diaz Ordaz, President Gustavo (Mexico), 299
Dighton, Ralph, 185
Dirksen, Sen. Everett M., 246, 271, 312
Disarmament, 13, 21, 30, 33, 67, 81, 86, 131, 155, 174, 181, 183, 190–191, 197, 332
Discoverer I (satellite), 59
Disney, M. J., 164
Disneyland, Calif., 351
Distinguished Public Service Award (USN), 138
Distinguished Public Service Certificate (NASA), 326
Distinguished Public Service Medal (NASA), 154, 347
Distinguished Service Medal (USAF), 300
Distinguished Service Medal (NASA), 10, 13, 91, 101, 154, 280, 287, 326, 346–347
DLRV. See Dual-mode lunar roving vehicle.
Dobbins AFB, Ga., 80, 151
Dobrynin, Ambassador Anatoly F., 206
Docking, 83
  Apollo 9, 62, 64
  Apollo 11, 212, 222, 421
  Apollo 12, 373, 377
  Soyuz IV and *Soyuz V*, 11–12, 19, 120
Dodd, Lamar, 226, 403
Dodd, Sen. Thomas J., 184
Doiguchi, Shizuo, 362
Dole, Sen. Robert J., 404
Donnelly, Dixon, 396

Doolittle, L/G James H. (USAF, Ret.), 304
DOT. See Transportation, Dept. of.
Douglas Aircraft Co., 209
Downey, Calif., 388
Downey, James A., III, 326
Downs, Dr. George S., 56
Doyle, Frederick J., 409
DPS. See Descent propulsion system.
Drake, Dr. Frank D., 121
Draper, Dr. Charles Stark, 338, 369, 416, 420
DRB. See Defence Research Board (Canada).
Dryden, Dr. Hugh L., 184, 289
Dryden, Hugh L., Memorial Fellowship, 68–69
DSCS. See Defense Satellite Communications System.
DSN. See Deep Space Network.
Dual-mode lunar roving vehicle (DLRV), 111
Dublin, Ireland, 121
Dubridge, Dr. Lee A., 54, 355
  ABM system, 83
  appointment, 42, 43
  basic research, 52–53, 68, 297
  international cooperation, 78, 304, 308, 316
  press conference, 49–50, 316
  science, political aspects of, 42, 122, 297
  space program, national, 6, 17–18, 38, 49–50, 52–53, 68, 78, 82, 134, 197–198, 308
Dudley Observatory (Albany, N.Y.), 159, 279, 287, 402
Duff, Brian M., 115–116
Duke, Capt. Charles M., Jr. (USAF), 267
Duke Univ., 19, 109
Dulles International Airport, Va., 89, 283
Dusterberry, John C., 362
Dyer, John W., 348

E

Eagle (*Apollo 11* LM). See Lunar module.
Eaker, L/G Ira C. (USAF, Ret.), 191
Early Apollo scientific experiment package (EASEP), 230
Early Bird (communications satellite). See *Intelsat I*.
Earth (see also Earth Resources Technology Satellite), 20, 49, 76, 136, 158, 198, 394
  crust, 360
  magnetic field, 171–173, 177, 185
  magnetosphere, 115, 185, 257, 291, 316
  mapping, 124, 173
  motion, 175
  photographs of, 59, 62–63, 64, 81, 90, 116–117, 143, 149, 222, 232, 304, 328, 346, 406–407

resources measurement, 37, 62–63, 75, 85, 100, 116–117, 245, 276, 304, 308, 420–421
shape, 83
Earth Photographs from Gemini VI Through XII (NASA SP–171), 304
Earth Resources Technology Satellite (ERTS) program, 85, 139, 156, 157, 245
  benefits, 33, 37, 119–120, 162, 276
  contract, 151, 340
  cost, 162, 337
  funds for, 15, 66, 69, 138, 164
  international cooperation, 310, 416
EASEP. See Early Apollo scientific experiment package.
Eastern Airlines, Inc., 29, 336, 346
Eastern Test Range (ETR) (see also Cape Kennedy and Kennedy Space Center), launch, 23
  Apollo 9 (AS–504), 33, 62
  Apollo 10 (AS–505), 90, 142
  Apollo 11 (AS–506), 212
  Apollo 12 (AS–507), 372–373
  Atlas-Agena, 106
  Atlas-Centaur, 55, 92, 277
  failure, 245, 291
  Long-Tank Thrust-Augmented Thor-Delta, 40, 150, 189–190, 245, 388
  Thor-Delta, 22
  Thor-Delta N, 272
  Thrust-Augmented Improved Delta, 291
  Thrust-Augmented Thor-Delta, 57
  Titan IIIC, 44, 155
Echo I (communications satellite), 405
Echo II, 175, 405
Eclipse, solar, 395
Edison, Thomas A., Memorial Lecture, 74
EDP Technology, Inc., 2
Edwards AFB, Calif., 38, 177, 205, 315, 323, 405
Edwards, Sir George, 340
Eglin AFB, Fla., 186
Egrs XIII (*Secor XIII*) (Sequential Collation of Range satellite), 107, 422
Egypt. See United Arab Republic.
Eiffel Tower, 41
Einhorn, Raymond, 393
Einstein, Prof. Albert, 182, 300
Eisele, L/C Donn F. (USAF), 21, 72–73
Eisenhower, President Dwight D., 47, 60, 196, 405
Eisenhower, Mrs. Dwight D., 280
El Segundo, Calif., 304
ELDO. See European Launcher Development Organization.
ELDO F–8 (ELDO satellite), 196
Electric propulsion, 337–338, 342
Electron microscope, 397
Electronics Research Center (ERC) (NASA), 128
  aeronautical research, 70, 263, 400
  appropriations, 113, 383

closing, 417, 418, 423
Elizabeth II, Queen of Great Britain, 38, 233, 335
Elk experiment, 328
Ellice Island, 239
Ellington AFB, Tex., 5, 103, 182, 283, 397
Ellington, Duke, 233
Elms, James C., 139, 400
Emme, Dr. Eugene M., 196, 357
Emmerton, Bill, 183
Engine (see also individual engines, such as F–1, H–1)
  aircraft, 19, 34, 35, 342, 402
  jet, 25, 32–33, 176, 212, 371, 393, 413
  pollution by, 278–279, 413
  Quiet Engine Research Program, 25, 32–33, 212
  supersonic transport, 17, 173, 323, 329, 413
  turbofan, 45–46, 212, 304, 393
  electric, 17, 134, 342
  nuclear (see also NERVA), 28, 190
  rocket, 70, 79, 101, 104, 105, 162, 185, 327–328, 379, 417
  fire, 195
  test, 104
Engineering, 1, 97
The Engineering Profession: A New Profile, 97
Engineers, 97, 99, 179, 193, 257, 351, 398
Engineers Joint Council, 97
England, Dr. Anthony W., 330
Engle, Capt. Joseph H. (USAF), 267
Environmental Science Services Administration (ESSA)
  budget, 14, 15
  Command and Data Acquisition station, 58
  cooperation, 9, 402, 414, 417
  personnel, 259
  satellite, 15, 88, 205, 343–344, 353, 414, 421
    launch, 57–58, 107
  Space Disturbance Center, 90, 336
  weather modification, 402
epndb: effective perceived noise in decibels
Equal Employment Opportunity Commission, 77
Erb, Bryan R., 353
Erba, Carlo, Foundation, 340
ERC. See Electronics Research Center.
ERTS. See Earth Resources Technology Satellite program.
ERTS–A (Earth Resources Technology Satellite), 66, 85, 151
ERTS–B, 66, 85
Escape system, 94
ESRO. See European Space Research Organization.
Esro IB. See *Boreas*.
Esro IIA (ESRO satellite), 323
Esro IIB. See *Iris I*.
ESSA. See Environmental Science Services Administration.

Essa VII (meteorological satellite), 58
Essa IX (TOS-G), 57–58, 421
Ethiopia, 205, 251
ETR. See Eastern Test Range.
Eupatoria, U.S.S.R., 199
Eureka, Calif., 352
Eurocontrol, 148
Europa (booster), 196
Europe (see also International cooperation and International cooperation, space), 74, 135, 137, 140, 262, 370
European Launcher Development Organization (ELDO), 114, 196, 357
European Space Research and Technology Center, 290
European Space Research Organization (ESRO), 357, 360
  launch, satellite
    Boreas (*Esro IB*), 323, 391, 423
  satellite, 82, 290
    Heos I, 82
    ESRO IIA, 323
    Iris I (*Esro IIB*), 323
EUV. See Ultraviolet, extreme.
EVA. See Extravehicular activity.
Evans, Albert J., 84
Evans, Dr. John W., 419
Evans, Llewellyn J., 92, 262
Evans, L/Cdr Ronald E. (USN), 267
Evans, Rowland, 284
Evans, Dr. W. D., 281
Ewing, Dr. Maurice, 230
Exceptional Bravery Medal (NASA), 347
Exceptional Civilian Service Award (USA), 256
Exceptional Scientific Achievement Medal (NASA), 10, 13, 326, 347
Exceptional Service Award (USAF), 304, 412
Exceptional Service Medal (NASA), 91, 326, 347
Exhibit, 52, 161–162, 173, 226, 283, 309, 313, 400, 403, 405–406, 410, 412, 417
Expanded Use of Federal Research Facilities by University Investigators (report), 56
Experiment module, 128
Experimentoy Corp., 159
Explorer (program), 50, 138, 360, 364, 366–367, 405
Explorer I (satellite), 38, 77
Explorer VII, 417
Explorer VIII, 417
Explorer XI, 417
Explorer XXXI, 30
Explorer XXXIII (IMP-D), 185
Explorer XXXIV (IMP-F), 129
Explorer XXXV (IMP-E), 185, 340
Explorer XXXVIII (Radio Astronomy Explorer RAE-A), 115, 335
Explorer XLI (IMP-G), 185, 395, 421
Explorers Club, 2
Expo 70, 417
Extraterrestrial life, 152, 200, 289
  Mars, 55, 90, 92, 103, 184–185, 271, 302
  moon, 261, 270
  Venus, 103
Extraterrestrial Research Agency (USA), 398
Extravehicular activity (EVA), 331
  Apollo 9, 5, 33, 62, 64, 81, 91
  Apollo 11, 74, 108, 182, 198, 208, 212, 217–220, 223, 243, 255, 274
  Apollo 12, 243, 325, 334, 342, 376–377, 395
  Apollo 13, 267
  Apollo 14, 267
Eyewitness to Space (NASA art program), 226, 241, 403

F

F-1 (rocket engine), 79, 105, 162, 185, 417
F-4 (Phantom II) (fighter aircraft), 129, 201
F-4E, 9
F-8 (ELDO spacecraft), 196
F-8 (supersonic carrier fighter), 44, 155
F-8B, 91
F-8C, 91
F-14A (supersonic fighter aircraft), 13, 15, 39, 131
F-15 (fighter aircraft), 15, 169
F-15A, 131
F-106 (research jet aircraft), 155
F-111 (supersonic fighter), 135, 164, 283, 372
F-111A, 49, 52, 123, 155, 413, 415–416
F-111B, 15, 39
FAA. See Federal Aviation Administration.
Fabiola, Queen of Belgium, 46
Faget, Dr. Maxime A., 289, 345
Fairbanks, Alaska, 58, 83
Falstaff (U.K. rocket), 323
Fancher, Has, 297–298
Farmer, Dr. C. B., 90
Fasi, Mayor Frank F., 247
Fastie, William G., 359
FB-111 (supersonic bomber), 86
FB-111A, 331
FCC. See Federal Communications Commission.
Federal-aid Airport Program, 156
Federal Air-Sea Interaction Research Program, 88
Federal Airport Act, 156
Federal Aviation Administration (FAA)
  accident investigation, 19
  air pollution, 402
  air traffic control, 15, 19, 34, 43–44, 55, 57, 148, 156, 188, 253, 315–316, 319, 388, 418
  airports, 57, 117, 123, 156, 176
  antihijacking system, 336
  award, 54, 123, 127, 178, 410
  budget, 15, 398

contract, 25, 46, 55, 89, 388, 391
cooperation, 315
forecast, 70
landing system, 89
noise, aircraft, 3, 25, 32, 46, 371, 391
personnel, 58, 123, 141
regulations, 3, 9, 19, 32, 43–44, 57, 138–139, 188, 371, 402, 418
transport, supersonic (see also Supersonic transport), 15, 123
design and development, 17, 422
Federal Communications Commission (FCC), 30, 102, 253, 352
Federal Council for Science and Technology, 56
Federal Electric Corp., 320
Federal Polytechnic, Zurich, 285
Federal Support to Universities and Colleges, Fiscal Year 1967 (NSF report), 106
Fédération Aéronautique Internationale, 315
Fedorov, Dr. Yevgeny K., 112
Fehlberg, Erwin, 326
Feller, Dr. William, 418
Felver, Edward R., 202
Feoktistov, Konstantin P., 195, 197, 372
Apollo 11 and *Apollo 12* mission comment, 233, 387–388
U.S. visit, 337, 343, 347, 350, 351, 352, 354, 359, 362, 387–388
Ferguson, Gen. James (USAF), 260, 410
Fernandez-Moran, Dr. Humberto, 397
Filipchenko, Anatoly V., 333
Filton Airfield, U.K., 105
Finger, Harold B., 20, 69, 87, 91, 115, 118
Finland, 384, 424
Finney, John, 178
Fire, 405
"First Man on the Moon" (commemorative postage stamp), 202, 273–274, 289, 300, 311, 319
Fisher, Adrian S., 30
Fisher, Paul C., 48
Fjeldbo, Dr. Gunnar, 408
Flagstaff, Ariz., 366
FLEEP. See Flying lunar excursion experimental platform.
Flemming, Arthur S., Awards, 50, 371
Flight Research Center (FRC) (NASA), 11, 44, 59, 92, 119, 123, 299–300, 315
Flight simulator for advanced aircraft (FSAA), 362
Flight Test, Simulation, and Support Conference, 3rd, 72
Florida Legislature, 180, 229
Flory, D. A., 365
Flying Baton (artificial horizon device), 263
Flying lunar excursion experimental platform (FLEEP), 388
Flying Tiger Pilot Trophy, 200
Flying Tigers, 200

FOBS. See Fractional Orbiting Bombardment System.
Fog Drops, Project, 42
Foreman, Rep. Edgar F., 404
Formosa, 234
Fort Churchill, Canada, 393
Ft. Davis, Tex., 90
Fort Eustis, Va., 334
Foster, Dr. John S., Jr., 4, 11, 99, 131
Foster, William C., 30
Four Corners, Calif., 113, 133, 148, 159
Fra Mauro (moon), 377, 405
France, 168, 397
aircraft, 29, 162, 197
Apollo missions, reaction to, 16, 210, 244, 385
astronauts visit to, 32, 37
communications satellite system, 56, 114, 132
Concorde (U.K.-France supersonic transport), 13–14, 26, 61, 71, 105, 146, 162, 173, 197, 255, 285, 323, 329, 340, 413
international cooperation, 196, 339, 393, 411, 417, 423–424
laser, 339
satellite, 86
test, 330
Franco Bahamonde, Gen. Francisco (Spain), 327, 330
Franco-Soviet Grand Commission, 339
Frank, M. P., 290
Frankel, Max, 284
Frankford Arsenal, 283
Franklin Institute, 161, 385
FRC. See Flight Research Center.
Fredriksson, Dr. Kurt, 304
French Atomic Energy Commission, 312
French Legion of Honor, 330
French National Center for Scientific Research, 363
Frey, Rep. Louis, Jr., 53, 242, 245
Fricker, John, 164
Frye, William R., 147
FSAA. See Flight simulator for advanced aircraft.
Fubini, Dr. Eugene G., 173
Fuchlow, Capt. William D. (USAF), 49
Fuel, 138, 342, 355, 379
Fuel cell, 343
Fulbright, Sen. J. William, 29, 276
Fuller, R. Buckminster, 420
Fullerton, Maj. Charles G. (USAF), 281
Fulton, Rep. James G., 281
Fungus, 159
Funkhouser, Dr. John, 288
Fusion energy, 318
Future of the Bioscience Program of the National Aeronautics and Space Administration, 415

G

Gagarin, Col. Yuri A. (U.S.S.R.), 41, 51, 104, 199, 228

Galabert International Astronautics Prize, 237
Galilei, Galileo, 300
Galileo (Convair 990 jet aircraft), 133, 393, 417, 423
Gallant, Richard P., 315
Gallup, George, poll, 270, 275
Gamma ray, 2, 45, 124, 318–319, 366
Gandhi, Prime Minister, Mrs. Indira (India), 158, 234
GAO. See General Accounting Office.
Garber, Paul, 133
Gardner, Prof. Richard N., 248
Garner, Howell D., 4
GARP. See Global Atmospheric Research Program.
Garwin, Dr. Richard L., 117
Gast, Dr. Paul, 250
Gates, Dr. Clarence R., 188
Gayle Planetarium, 400
Gazenko, Dr. Oleg G., 325
GCA Corp., 34, 49, 113, 300, 304
Geisler, W., 329
Geiss, Dr. Johannes, 285, 413
Gell-Mann, Dr. Murray, 117, 355
Gemini (program), 110, 160, 257, 304, 357
Gemini VII mission 71
Gemini XII mission, 71
General Accounting Office (GAO), 47–48, 138
General aviation
 aircraft, 70, 102, 109, 124,
 award, 410
 collision avoidance system, 396
 employment, 150
 The Magnitude and Economic Impact of General Aviation (study), 109
 research, 15
 safety, 40–41, 369
 tax, 182
General Dynamics Corp., 399
 Atlas-Centaur, 114, 336, 355
 award, 347
 contract, 47, 270, 336, 355, 372
 Convair Div., 114, 336, 355
 F–111, 164, 372
General Electric Co.
 Aircraft Engine Group, 212
 award, 147
 contract, 32, 191, 293, 330, 340, 391, 408
 laboratory, undersea, 51
 MIRV missile, 191
 neutron radiography research, 168
 quiet jet engine, 32
 Re-Entry Systems Div., 336
 reentry vehicle, 391, 408
 spacecraft, 293, 340
 supersonic transport engine, 17
 turbofan engine, 17
General Motors Corp., 371
General Telephone & Electronics International, 171
Geneva Disarmament Conference, 81, 155
Geneva, Switzerland, 189, 197, 198

Gentry, Maj. Jerauld R. (USAF), 101, 113, 133, 286, 298, 299, 371
Geodetic satellite, 33, 86, 107, 422
Geographos (planetoid), 288
Geological Society of America, 290
Geomagnetism, 96
Georgadze, Mikhail P., 197
GEOS–C (geodetic satellite), 86
Gerathewohl, Dr. Siegfried J., 180
Germany, East, 251, 317, 335
Germany, West
 Apollo flights, reaction to, 16, 193, 225, 244, 384
 astronaut visit to, 32, 37, 48, 49, 334
 cooperation, space, 76, 114, 132, 177, 196, 364, 379, 423–424
 Ministry of Scientific Research (BMWF), 364
 space program, 76, 114, 177, 357, 364, 379, 423
Gesell, Gerhard A., 138–139
GET: ground elapsed time
Getler, Michael, 321
Ghana, 234
Ghiorso, Albert, 110
Gilbert Island, 239
Gill, C. James, 38
Gilruth, Dr. Robert R., 10, 274, 289, 317, 347, 353, 419
Glasstone, Dr. Samuel, 89
Glendale, Ariz., 39
Glenn, Col. John H., Jr. (USMC, Ret.), 43, 289, 406
Glennan, Dr. T. Keith, 125, 184, 406
Glines, Carroll V., 397
Global Atmospheric Research Program (GARP), 88, 204
Global Flare Patrol Network, 336
Glomar Challenger (drilling ship), 273, 353
Goddard Institute for Space Studies, 2, 174, 200, 408
Goddard Memorial Dinner, 68
Goddard, L/C P. M. (RN), 129
Goddard, Dr. Robert H., 147, 226, 280, 289, 300, 404
Goddard, Mrs. Robert H., 147, 280
Goddard, Robert H., Award, 21
Goddard, Robert H., Historical Essay competition, 69
Goddard, Robert H., Memorial Trophy, 68
Goddard, Robert Hutchings, Library, 147
Goddard Space Flight Center (GSFC)
 Apollo 9, 65
 Apollo 10, 145
 Apollo 11, 224, 274–275
 Apollo 12, 358
 award, 117, 154, 346
 buoy-tracking experiment, 199
 contract, 271, 297
 cooperation, 353
 Extraterrestrial Physics Branch, 50
 facilities, 383, 405
 fire, 405

Laboratory for Theoretical Studies, 360
mapping, 50
National Space Science Data Center, 1, 271
patent, 82
personnel, 259, 290
satellite animal-tracking experiment, 328
satellite monitoring
  Essa IX, 58
  Explorer XLI (IMP–G), 185
  Mariner V, 127
  Oao II (OAO–A2), 99
  Ogo VI (OGO–F), 171
  Oso V (OSO–F), 22
  Relay I and *Relay II*, 124
Satellite Tracking Center, 37
Small Scientific Satellite, 405
sounding rocket experiments
  astronomical, 107, 140, 184, 335, 337, 408
  atmospheric data, 9, 18, 19, 21, 25, 31, 33–34, 38, 41, 94, 132, 133, 286
  electric fields, 71, 279, 331
  instrumentation test, 286
  vehicle performance test, 28, 335
  weather data analyses, 353
Goett, Dr. Harry J., 346
Gold, Dr. Thomas, 121, 317, 350
Goldberg, Dr. Leo, 156, 256, 366
Golden Key Award, 46
Goldmark, Dr. Peter C., 102, 201
Goldstein, Dr. Richard M., 252
Goldstone Tracking Station, 56, 137, 179, 252, 274, 374
Goldwater, Sen. Barry M., 13, 379–380
Goldwater, Rep. Barry M., Jr., 169
Golovin, Dr. Nicholas E., 119
Goodell, Sen. Charles E., 13
Goodling, Rep. George A., 309
Goody, Dr. Richard M., 318
Goodyear Aerospace Corp., 239
Gorbatko, Victor V., 333
Gordon, Capt. Richard F., Jr. (USN)
  Apollo 12 mission
    flight, 372, 374, 375–376, 377, 389
    medical examination, 366
    Nixon, President Richard M., 371, 392
    plans for, 105, 342, 353, 365, 366, 368, 371–372
    press conference, 333, 406
    quarantine, 356, 377–378, 396, 397, 405
    splashdown, 377
  fund-raising dinner, 408
  promotion, 391, 392
  record, 389
  White House visit, 413
Gorkin, Jess, 367
Gorman, Harry H., 326, 347, 360
Gorton, Prime Minister John G. (Australia), 117, 244
Gottlieb, Dr. Peter, 116
Governors' Conference, 401
Governor's Conference on California's Changing Environment, 382
Graham, Billy (William F.), 246
Grand Canyon, Ariz., 337, 354
Grand Cross of Aeronautic Merit, 327
Granite, Bernard, 385
Grant, 3, 32, 52, 209, 319
Gravity, 71, 78, 86, 182, 190, 206
  artificial, 117–118, 391
Great Salt Lake, 49
"Great Transatlantic Air Race of 1969," 129
Greece, 226
Green, Dr. Robert L., 46
Greenberg, D. S., 114–115
Greenglass, Bert, 313
"Greenhouse" effect (Venus), 128
Grimwood, James M., 357
Grissom, L/C Virgil I. (USAF), 228
Grissom, Mrs. Virgil I., 279
Gromyko, Foreign Minister Andrey A. (U.S.S.R.), 236
Groton, Conn., 27
Group Achievement Award (NASA), 34, 154, 347
Grumman Aircraft Engineering Corp., 13, 26, 39, 42, 46, 92, 111, 262, 394
GSFC. See Goddard Space Flight Center.
Guam, 273, 284, 312, 402
Guam International Airport, 246
Guggenheim, Daniel, Medal, 338, 345
Guggenheim, Harry F., 226
Gulf of Mexico, 157
Gulf Stream Drift Mission, 103, 209, 282
Gumdrop (*Apollo 9* command and service module). See Command and service module.
Gunn, Charles R., 54
Gunn, Dr. James E., 322, 334
Gurney, Sen. Edward J., 203, 393
Gutenberg, Johann, 161
Guzman, Mme. Anne E., 281, 403
Gwertzman, Bernard, 152, 181

H

Haagen-Smit, Arie Jan, 383
Haeussermann, Dr. Walter, 346
Hage, George H., 33, 108, 182, 274, 285, 300, 346
Haggerty, James J., Jr., 89, 95
Haggerty, Dr. Patrick E., 117
The Hague, Netherlands, 37
Haile Selassie, Emperor of Ethiopia, 205
Haise, Fred W., Jr., 267
Halaby, Najeeb E., 255, 346
Hale, Edward Everett, 357
Haley Astronautics Award, 72
Halifax, Nova Scotia, 103
Hall, Charles F., 361
Hall, Mayor Chuck, 9
Hallanger, Dr. Larry, 343
Hallgren, Dr. Richard E., 50

HALO. See Hughes Automated Lunar Observer.
Halo Crater (moon), 376
Hamiter, Leon C., Jr., 290
Hand, Alfred, 54
Handler, Dr. Philip, 19
Haney, Paul P., 116, 118
Haredi, Sheikh Ahmand, 226
Harlow, Dr. James G., 70
Harmon International Aviator's Trophy, 298
Harnett, Daniel J., 319, 325
Harper, Charles W., 21, 334
Harrier (VTOL aircraft), 129, 162
Harrington, Michael, 4
Harris, Louis, poll, 22, 208, 290
Harris, Dr. Robert, 253
Hartford, Conn., 263
Hartsfield, Maj. Henry W., Jr. (USAF), 281
Harvard College, 77
Harvard College Observatory, 256, 272, 308, 366, 411
Harvard Univ., 89, 131, 238, 257, 302, 318, 416–417
Harvey, Dr. Mose L., 321
Hassell, Prof. Odd, 355
Hawaii, 134, 190
Hawaii, Univ. of, 314
Hawaiian Telephone Co., 32
Hawker Siddeley Aviation Co., 162
Haworth, Leland J., 184
Hayashi, Tsuyoshi, 290
Hayden Planetarium, 243
Haynes, Charles G., 19
Haynos, Joseph G., 82
Head Crater (moon), 376
Health, Education, and Welfare, Dept. of (HEW), 83, 106, 110, 141, 274
Heart transplant, 1
Hearth, Donald P., 76
Heavy ion linear accelerator (HILAC), 110
Heavy lift helicopter (HLH), 131
Hedin, Sven, 50
Heinemann, President Gustav (West Germany), 244
Helian, R. D., 281
Helicopter, 79, 129
 accident, 260, 308–309, 389
 astronaut pickup, 141
 civil, 35, 124
 military, 106, 115, 131, 148, 150, 160, 260, 389
 record, 79, 129
 U.S.S.R., 35, 79
Heliodyne Corp., 271
Helios, Project (sun probe), 76, 177, 357, 423
Heliports, 176
Helium, 156
Hellas (Mars), 269
Heller, Gerhard B., 76, 326
Helsinki, Finland, 384
Heos I (ESRO satellite), 82

Hero of the Soviet Union, 1
Herr, Dr. Kenneth C., 269
Herrick, Dr. Samuel, Jr., 288
Herriman, Alan G., 282
Hershey, Dr. Alfred D., 338
Hess, Dr. Harry H., 167, 289–290, 347
Hess, Dr. Wilmot N., 259, 288
HEW. See Health, Education, and Welfare, Dept. of.
Heyerdahl, Thor, 146
Hibbs, Albert R., 34
Hickam AFB, Hawaii, 201, 245
Hickel, Secretary of the Interior Walter J., 356
Hidayatullah, Mohammad, 255
Higginbottom, Samuel L., 346
HILAC. See Heavy ion linear accelerator.
Hill, Louis W., Space Transportation Award, 345
Hill, Ambassador Robert C., 180
Hill, William, Organization, 161
Hillary, Sir Edmund, 251
Hines, William M., 28, 90, 119, 176, 186, 207, 262, 274, 289, 313
Hirohito, Emperor (Japan), 234, 361
Hirondelle (turboprop aircraft), 162
"Historical Perspectives on Apollo," 196
Hitler, Adolf, 297
Hjornevik, Wesley L., 339
HL-10 (lifting-body vehicle), 29
 test flight, 422
 award, 298
 glide, 117, 174, 422
 powered, 113, 133, 148, 159, 184, 240, 267, 297, 309, 319, 352, 360, 381, 389, 406, 422
HLH. See Heavy lift helicopter.
Hoag, Maj. Peter C. (USAF), 174, 319, 360, 389
Hoagland, Hudson, 50
Hodgson, Alfred S., 19
Holcomb, Robert W., 390
Holland, Sen. Spessard L., 203, 393
Holloman AFB, N. Mex., 363
Honolulu, Hawaii, 99, 273, 312
Hooker, Dr. Stanley, 21
Hoover Dam, 87
Hope, Bob, 414
Hord, Dr. Charles W., 269, 298
Hornig, Dr. Donald F., 3
Horowitz, Dr. Norman H., 282, 302
Hotz, Robert B., 191
Houbolt, Dr. John C., 346
Housing and Urban Development, Dept. of (HUD), 237
 airport study, 263–265
 funding, 42, 184, 369, 383, 395
 Operation Breakthrough, 188
 personnel, 87, 91, 115, 313
Houston, Tex., 137, 148, 183, 249, 255, 261, 398
 Apollo 8 stamp ceremony, 129
 Apollo 11 mission
  celebration, 244, 283

press conference, 280–281
  astronauts at, 75, 267, 279, 281, 283, 312, 408
  employment, 231
  Lunar Receiving Laboratory, 223, 378, 394
  Lunar Science Institute, 8, 227
  National Space Hall of Fame, 289
Houston Welfare Rights Organization, 233
Hovercraft, 156
Hoyle, Fred, 339
Hsieh, Jen-chao, 234
HUD. See Housing and Urban Development, Dept. of.
Hughes Aircraft Co., 61, 241, 369
  Space Systems Div., 401
Hughes Automated Lunar Observer (HALO) (Surveyor III), 370
Hughes, Howard R., 370
Human factor systems program, 154
Humphrey, Vice President Hubert H., 20, 233, 280
Humphrey, Mrs. Hubert H., 280
Hungary, 46, 234, 251
Hunt, Sir John, 251
Hunt, Lamar, 343
Hunten, Dr. Donald M., 318
Huntsville, Ala., 170, 244, 398
Hurd, Peter, 226
Hurricane Debbie, 402
Hutchinson, Thomas C. H., 387
Hyderabad, India, 258
Hydrogen bomb, 312
Hydro-Lab (underwater laboratory), 343
Hydroxyl radical, 189
Hynek, Dr. J. Allen, 28, 288, 416
Hypatia Rille (moon), 159
Hypersonic aircraft, 15, 87, 362

I

IAA. See International Academy of Astronautics.
IAF. See International Aeronautical Federation and International Astronautical Federation.
IATA. See International Air Transport Assn.
IBM. See International Business Machines Corp.
ICBM. See Intercontinental ballistic missile.
IEEE. See Institute of Electrical and Electronics Engineers.
Illia, President Arthuro U. (Argentina), 384
Illinois, 28
Illinois Institute of Technology, 293
Illinois Institute of Technology Research Institute, 11
Illinois, Univ. of, 34, 41, 113, 300, 304

ILRV. See Integral Launch and Reentry Vehicle.
ILS. See Instrument landing system.
Ilyin, Lt. (U.S.S.R.), 33, 39
IME. See Interplanetary Meteoroid Experiment.
IMP. See Interplanetary Monitoring Platform.
IMP-D, 185
IMP-E, 185
IMP-G, 185
Imperial College of Science and Technology, 355
IMS. See Ion mass spectrometer.
Inchon, South Korea, 237
Independence Day, 197–198
India, 190, 234, 398, 412
  cooperation, 118, 120, 258, 311, 340, 423
  Nixon, President Richard M., visit to, 255
Indian Ocean, 25, 140, 263, 389, 423
Indian Space Research Organization, 311
Indiana Univ., 102
Indonesia, 200, 249
Inertial reference integrating gyro (IRIG), 14
Ingalls, Richard P., 288
Institute for Soviet-American Relations, 195
Institute for Strategic Studies, 106
Institute of Electrical and Electronics Engineers (IEEE), 177, 354
Instituto Geofisico del Peru, 267
Instituto Nacional de Técnica Aerospacial, 180
Instrument landing system (ILS), 89
Integral Launch and Reentry Vehicle (ILRV), 47
INTELSAT. See International Telecommunications Satellite Consortium.
Intelsat I (*Early Bird*) (communications satellite), 61, 137, 191, 259, 286
Intelsat-II F–3 (Intelsat II–C; *Atlantic II*), 191, 259
Intelsat-III F–1, 50, 135, 151
Intelsat-III F–2, 30, 40, 61, 151, 191, 226, 245, 259, 276, 286, 340
Intelsat-III F–3, 23, 39–40, 137, 140, 151, 421
Intelsat-III F–4, 23, 140, 150–151, 276, 421
Intelsat-III F–5, 245, 281, 385, 421
Intelsat IV, 114
Interagency Committee on Back Contamination, 141, 273, 302, 356
Intercontinental ballistic missile (ICBM) (see also Multiple independently targetable reentry vehicle), 48, 140
Intercosmos I (U.S.S.R. satellite), 335, 422
Intercosmos II, 416, 422
Interior, Dept. of, 14, 51, 76, 141, 267, 356

International Academy of Astronautics (IAA), 325
International Aeronautical Exposition, 52
International Aeronautical Federation (IAF), 208, 250
International Air Transport Assn. (IATA), 292–293, 350
International Astronautical Congress, 287
International Astronautical Federation (IAF), 328
International Astronomical Commission, 335
International Aviation Service Award (FAA), 54
International Business Machines Corp. (IBM), 25, 79, 91, 315, 334
  Federal Systems Div., 71
  Watson Laboratory, 117
International cooperation (see also Disarmament and Treaty), 51, 261, 262, 300, 392, 393, 417
  air transportation, 121, 148, 292–293
  aircraft (see also Concorde), 161–162, 423–424
  astronomy, 181, 319, 417
  meteorology, 45, 78–79, 204–205
  military, 117, 132, 289
  nuclear power, 25, 43, 316, 386, 409
  oceanography, 133, 197, 330, 341
  science and technology, 78, 102, 304, 316
International cooperation, space (see also European Launcher Development Organization; European Space Research Organization; International Telecommunications Satellite Consortium; Space rescue treaty), 1, 39, 43, 55, 75, 101–102, 103, 117, 158, 167, 175, 238, 248, 255, 285, 290, 308–310, 312, 349–350, 379, 423–424
  law, 189, 198, 229, 392
  military, 117, 132, 389
  satellite, 102, 169
    communications, 61, 171
    DOD–U.K., 388–389
    earth resources, 33, 310
    Europe, 132, 196, 323, 357, 416
    NASA
      -Australia, 177, 399
      -Canada, 30, 241, 423–424
      -ESRO, 323, 423
      -Germany, West, 76, 114, 177, 357, 364, 379, 423–424
      -India, 311, 340, 423
      -Italy, 177, 423
      -Netherlands, 423
      -U.K., 23, 59, 323, 385, 388–389, 422, 423
      -U.S.S.R., 19, 196, 381, 394, 411, 416
    U.S.
      -NATO, 132
      -U.S.S.R., 25, 393–394
  sounding rocket. See Sounding rocket, international programs.
  space research, 4, 43, 44, 102, 168–169, 290
    Europe, 74, 248, 383, 411, 416
    France-U.S.S.R., 339
    U.S.-Australia, 137, 177, 319, 423
      -France, 423–424
      -Germany, West, 177
      -Italy, 177, 423
      -Japan, 383, 389, 411
      -U.K., 423–424
      -U.S.S.R., 25, 26, 41, 78, 95–96, 103, 195, 202, 211, 231, 233, 238, 240, 245, 246, 247, 248, 311, 318, 325, 343, 351, 367, 383, 409, 411–412, 424
  tracking
    U.S.-Australia, 34, 148, 189, 424
      -Peru, 267
      -Spain, 187, 189, 424
      -U.K., 59
International Decade of Ocean Exploration, 341
International Geomagnetic Reference Field, 173
International History of Astronautics Symposium, Third, 329
International Platform Assn., 237
International Rice Research Institute, 163
International Symposium on Space Technology, Eighth, 290
International Telecommunications Satellite Consortium (INTELSAT), 381, 394
  conference, 24, 26, 56, 88, 101
  membership, 61
  satellite, 40, 114, 150
International Union of Radio Science, U.S. National Committee, 115
Interplanetary Meteoroid Experiment (IME), 105
Interplanetary Monitoring Platform (IMP), 30, 185
Interrogation, recording, and location (IRLS) system, 107
Intersputnik (U.S.S.R. communications satellite system), 381, 394
Intrepid (Apollo 12 LM). See Lunar module.
Iodine, 128
Ion mass spectrometer (IMS), 241
Ion propulsion, 48, 342, 399, 415
Ionosphere, 30, 33–34, 41, 84, 113, 135, 160, 188, 257, 300, 304, 323, 339, 423
Iowa, 28
Iowa State Univ., Dept. of History, 357
Iowa, Univ. of, 263, 340, 395
Iran, 385
IRIG. See Inertial reference integrating gyro.
Iris I (Esro IIB) (International Radiation Investigation Satellite), 323
Irwin, Maj. James B. (USAF), 105
Isis 1 (ISIS-A) (International Satellite for Ionospheric Studies), 30, 241, 423

Island Creek Coal Co., 267
Israel, 226, 384, 408
"Issues and Answers" (TV program), 146
Italy, 155, 244, 384–385
  Borman, Col. Frank (USAF), visit to, 32, 37
  cooperation, space, 132, 177, 423
  ELDO F-8 spacecraft, 196
  satellite, 423
IT&T Corp., 320
ITOS (meteorological satellite), 344

J

J-2 (rocket engine), 70, 101, 104–105, 162, 185
Jackass Flats, Nev., 87, 178, 279, 309, 361, 422
Jackson Hole, Wyo., 328
Jackson, Nelson P., Award, 69
Jaffe, Dr. Leonard D., 85, 116, 139, 141, 189
Jakarta, Indonesia, 249
James, Francis, 181
James, Lee B., 188, 251, 317
Jamison, Mitchell, 403
Japan, 82, 140, 168, 175
  Apollo 11 and Apollo 12 missions, reaction to, 226, 234, 244, 249, 361
  booster, 284, 297, 314
  Cabinet, 9
  international cooperation, space, 118, 120, 383, 389, 411, 423–424
  International Symposium on Space Technology, 290
  launch, rocket, 284, 297
  National Defense Council, 9
  space debris damage, 198
  space program, 118, 120, 290, 314, 325
Jastrow, Dr. Robert, 2, 200
Javelin (sounding rocket), 188
Jeans, Sir James, 417
Jenkins, Wally, 343
Jerusalem, Israel, 226
Jet belt, 103
Jet Propulsion Laboratory (JPL) (CalTech), 85, 184, 248
  Agnew, Vice President Spiro T., visit to, 323–325
  award, 203, 346
  computer, 299
  Deep Space Network, 34, 180, 189, 274–275, 346
  Goldstone Tracking Station, 56, 137, 179, 252, 274, 374
  history, 274
  lunar research
    mascons, 116, 120
    surface, 34–35, 189
  Mariner Project, 55–57, 90, 133, 250, 252, 253–254, 256, 259, 261–262, 265–266, 269, 282, 298, 301–302, 408
  personnel, 6, 135, 188, 290
  Pioneer VI, 356
  Viking Project, 57, 111, 274–275, 412
Jewish National Fund, 408
Jobe, Capt. Robert Earle (USAF), 49
Jodrell Bank Experimental Station (U.K.), 158, 224, 232, 233, 237, 238, 244, 384
Johns Hopkins Univ., 1, 37, 43, 46, 49, 298
  Dept. of Biology, 184
Johnson, President Lyndon B., 3, 20
  Apollo 8 mission, 13
  Apollo 11 mission
    Cape Kennedy visit, 209, 224
    message, 196
  astronauts' promotion, 391
  awards by, 1, 6
  budget, 14, 40, 109
    NASA, 14–16, 109–110, 395, 423
  communications policy, 352
  Kennedy Space Center, 393
  memoirs, 136
  space program, 6–7, 10, 20, 60, 231–232
  State of the Union Message, 13
Johnson, Mrs. Lyndon B., 209
Johnson, Robert L., 363
Johnson, W. Thomas, Jr., 402
Johnston Island, 223
Johnston, S. Paul, 297
Joint Chiefs of Staff, 180, 280
Joint Commission on Scientific and Technical Communications (proposed) (NAS–NAE), 175
Joint Computer Conference, 139
Jones, Dr. Norman D., 261
Jones, Robert J., 202
Jordan, Sen. Len B., 13
JPL. See Jet Propulsion Laboratory (CalTech).
Juliana, Queen (Netherlands), 332
Junkers 390 (bomber), 297
Jupiter (planet), 16, 76, 115, 157, 192, 263, 299, 338, 361, 423
Justice, Dept. of, 130, 394

K

Kahn, Richard S., 157
Kamm, Robert W., 20
Kammler, Gen. Hans (Germany), 381
Kansas State Highway Commission, 408
Kansas State Univ., 332
Kapryan, Walter J., 287
Kapustin Yar, U.S.S.R., 67
Karaganda, U.S.S.R., 333
Karth, Rep. Joseph E., 371
Kauai, Pacific Ocean, 201
Kaufman, Harold R., 134
Kazakhstan, U.S.S.R., 196, 271
Kazakov, Vasily, 159
Kedrov, Boniface, 157

Keldysh, Prof. Mstislav V., 25, 202, 229, 244, 311, 351, 361, 411
Kelly, K. K., 298
Kelly, Orr, 4
Kennedy, Sen. Edward M., 131, 147, 148, 175, 224
Kennedy, President John F., 20, 38, 44, 60, 175, 196, 212, 240, 393, 405
Kennedy, John F., International Airport, N.Y., 263, 279, 362, 400
Kennedy, John F., Medal for National Civic Service, 72
Kennedy Space Center (KSC), 393
  accident, 162
  Apollo/Saturn (see also Apollo missions), 118, 195, 299, 350, 351
  astronaut memorial (proposed), 53, 245
  astronauts at, 5, 207, 325, 371
  award, 10, 122, 154, 289
  budget, 122
  contract, 122, 320
  facilities, 341, 383
  launch operations (see also Launch Complex 34, 37, 39; and Apollo missions), 65, 122, 137, 145, 173, 361, 371
  lunar landing memorial, 304
  meeting, 33
  personnel, 122, 154, 287, 313, 371
  press conference, 207, 325, 371
  spacecraft delivery and shipments to, 149, 182
  visits to, 136
    Agnew, Vice President Spiro T., 65, 210, 224
    Johnson, President Lyndon B., 209, 224
    Johnson, Mrs. Lyndon B., 209
    Nixon, President Richard M., 378
    Nixon, Mrs. Richard M., 378
    Nixon, Miss Tricia, 378
Kepler, Johann, 300
Kerwin, Cdr. Joseph P. (USN), 283, 414
Key Biscayne, Fla., 38
Keyhoe, Maj. Donald E. (USMC, Ret.), 8
Khan, President Yahya (Pakistan), 244, 259, 394
Khrunov, Yevegeny, 11–12, 23
Khrushchev, Premier Nikita (U.S.S.R.), 197, 249
Kiesinger, Chancellor Kurt G. (W. Germany), 48, 270
Kilby, Jack S. C., 418
King, Dr. Elbert A., 273, 287
King, Dr. Martin Luther, 231
Kinshasa, Congo, 312, 350
Kirchner, Englebert, 294
Kirkman, Don, 196
Kirtland AFB, N. Mex., 424
Kiruna, Sweden, 18, 20, 24, 27
Kitt Peak National Observatory, 17, 318
Kitty Hawk Flyer (aircraft), 177
Kitty Hawk, N.C., 88, 410
Kleen, Dr. Werner J., 290

Klein, Milton, 130, 332
Kliore, Dr. Arvydas J., 408
Knight, Maj. William J. (USAF), 87
Knoxville, Tenn., 391
Knutson, Don, 169
Komarov, Col. Vladimir M. (U.S.S.R.), 199, 228
Koran, 226
Korea, South, 237, 244
Korolev, Sergey P., 199
Kosygin, Premier Alexey N. (U.S.S.R.), 233
Koval, Alexander, 335
Kozlov, Mikhail V., 1
Kozyrev, Dr. Nikolay A., 139, 418
Kraft, Christopher C., Jr., 10, 274, 290, 344, 392, 417
Kramer, James J., 50
Kraner, H. W., 340
Kranz, Eugene F., 91
Krause, Dr. Helmut G., 183
Kremlin, 33, 39
  Borovitsky Gate, 23
Kryter, Karl D., 26
KSC. See Kennedy Space Center.
Kubasov, Valery N., 332
Kubat, Jerald R., 130
Kuiper, Dr. Gerard P., 397
Kuznetsov, Vasily V., 197

L

L–1011 airbus, 341
Laboratory for Electronics, Inc., 89
Laboratory of Electrical Optics, Toulouse, France, 397
La Canada-Flintridge, Calif., 85
Lagos, Nigeria, 226
Lahore, Pakistan, 259
Lalande (moon), 377
Laird, Secretary of Defense Melvin R., 11, 34
  C–5A, 123, 138
  defense budget, 86, 99
  press conference, 4, 33
  Sentinel ABM, 33, 39, 53
  space program, 146
Lambda (Japanese booster), 297, 314
Lamont-Doherty Geological Observatory, 174, 230, 249, 273, 354–355
Lampang, Thailand, 286
Landon Lecture, 332
Langley Aeronautical Laboratory, 349
Langley Research Center (LaRC) (NASA), 379
  Aircraft Noise Reduction Laboratory, 120
  budget, 383
  buffeting research, 123
  contract, 94
  cooperation, 379
  jet shoes, 4
  Molecular Biophysics Laboratory, 96

patent, 4
personnel, 21, 290, 293
supercritical wing, 44, 293, 299
Viking, Project, 59, 111, 293, 412
Youth Science Congress, 105
Lannan, John, 21, 40, 153, 167, 365, 405
La Plata, Argentina, 181
Lapp, Dr. Ralph E., 37
LaRC. See Langley Research Center (NASA).
Larsen, Agnew E., 283
Las Palmas, Canary Islands, 312, 327
Las Vegas, Nev., 49
Laser
  lunar experiments, 192, 339
    Apollo 11, 114, 220, 223, 237, 259, 261, 285, 411, 419
  use of, 37, 127, 312, 318, 322, 327, 386, 405, 424
Latham, Dr. Gary V., 230, 249, 354, 409, 413
Latin America, 101, 226
Launch Complex 34, 122
Launch Complex 37, 122
Launch Complex 39, 122
  Apollo launches from
    Apollo 9, 62
    Apollo 10, 73, 108, 142
    Apollo 11, 149, 212
    Apollo 12, 372
La Violette, Paul E., 160
Lawrance, Charles L., Award, 138
Lawrence Radiation Laboratory (Univ. of Calif.), 53, 110
Leavitt, William, 420
Le Bourget Airport, Paris, 162, 169
Lecky-Thompson, S/L Tom (RAF), 129
Lederberg, Dr. Joshua, 205
Lederer, Jerome F., 315
Lee, Capt. Chester M. (USN, Ret.), 285, 342
Leicester, Univ. of, 22
Leighton, Dr. Robert B., 269, 282, 301
Lenin, Vladimir I., 199
Leningrad, U.S.S.R., 196, 278
Leonov, L/C Aleksey A. (U.S.S.R.), 48, 158
Leovy, Conway B., 282
LeRC. See Lewis Research Center (NASA).
Lesher, Dr. Richard L., 135
Lewis Research Center (LeRC) (NASA), 75
  award, 50, 134, 154
  booster, 56, 355, 412
  Brayton Cycle space power system, 58
  contract, 212, 355
  cooperation, 379
  Electromagnetic Propulsion Div., 134
  ion propulsion, 48, 342
  noise abatement, 33, 120, 212
  Plum Brook Station, 404
  Propulsion Systems Acoustics Branch, 50
  research, 155, 362–363, 384, 404, 412
  Space Power Facility, 329
  Spacecraft Propulsion Research Facility, 329–330, 412
Lewis, Roger, 399
Ley, Willy, 187
Libby, Dr. Willard F., 408
Library of Congress, 184
  Legislative Reference Service, 95
  Science Policy Research Div., 161
Lick Observatory, Calif., 42, 237, 259, 261, 285
Lifting-body vehicle
  HL-10, 29, 113, 117, 133, 148, 159, 174, 184, 240, 267, 297, 298, 319, 352, 360, 381, 389, 406, 422
  M2-F2, 29
  M2-F3, 29
  X-24, 29
  X-24A, 101, 113, 133, 286, 299, 316, 349, 371, 422
Light Intratheater Transport (LIT) aircraft, 92
Lima, Peru, 267, 322
Limeill Weapons Research Center, 312
Lincoln Center, N.Y., 8
Lincoln Univ., 168
Lindbergh, Charles A., 100, 177
Lindley, Robert N., 410
Lindsay, Mayor John V., 279
Lindsey, Robert, 156
Lineberry, Edgar C., Jr., 345
Lingenfelter, Dr. Richard, 174
Linweaver, Cdr. Paul G. (USN), 188
Lisbon, Portugal, 37, 53, 344, 413
Lisitzin, Dr. Aleksandr P., 273
LIT. See Light Intratheater Transport aircraft.
Little, Brown & Co., 135
Little, Stephen, 90
Lloyd's of London, 244
LLRV. See Lunar Landing Research Vehicle.
LLTV. See Lunar Landing Training Vehicle.
LM. See Lunar module.
Lockheed Aircraft Corp., 79, 157–158
  AH-56A (helicopter), 106, 115, 148, 150, 160, 260, 308, 389–390
  aircraft engine, 402
  C-5A (cargo transport), 17, 57, 107, 123–124, 128, 130, 132, 138, 151, 162, 207, 251, 281, 283, 293, 323, 380, 411, 421
  contract, 128, 130, 132, 150, 160, 251, 280, 402
  L-1011 airbus, 341
  nuclear rocket, 280
  XV-4B (VTOL aircraft), 80
Lockheed, Allan H., 157
Lockheed-California Co., 115, 260, 308
Lockheed-Georgia Co., 57, 151, 283
Lockheed, Malcolm, 157
Lockheed Missiles & Space Co., 47
Loftus, Joseph P., Jr., 331
Logan, Joseph, Jr., 234

London-to-Sydney Air Race, 411
London, U.K., 14, 37, 55, 56, 129, 134, 312, 329, 355, 370, 411
London Univ., 312
Long, Dr. Franklin A., 114, 121
Long, James E., 192
Long Island Assn. of Commerce and Industry, 382
A Long-Range Program in Space Astronomy: Position Paper of the Astronomy Missions Board (NASA SP-213), 366
Long-Tank Thrust-Augmented Thor (Thorad)-Agena D (booster), 107, 171, 173, 319
Long-Tank Thrust-Augmented Thor (LTTAT)-Delta (booster), 40, 150, 189-190, 245, 272, 388
Loprete, Joseph F., 239
Los Alamos (N. Mex.) Scientific Laboratory, 272, 281
Los Angeles, Calif., 123, 209, 300, 382
  aerospace industry, 71
  Apollo flights, reaction to, 235-236, 385
  astronaut's dinner and visit to, 276, 279, 284
Los Angeles International Airport, 102, 279
Lounsberry, Ernest D., 4
Lovelace, W. Randolph, II, Award, 182
Lovell, Sir Bernard, 158, 224, 232, 233, 237, 238, 244, 384
Lovell, Capt. James A., Jr. (USN), 71, 329
  Apollo 8 mission, 32, 68
  Apollo 11 mission, 5
  Apollo 13 mission, 267
  appointment, 83
  awards and honors, 2, 6-8, 10, 68, 87, 100, 101, 132, 298, 391
  press conference, 6
  record, 208
  receptions for, 9, 13
  White House visit, 6, 32
Lovell, Mrs. James A., Jr., 329
Low, Dr. Frank J., 417
Low, Dr. George M., 108, 274, 317, 345, 371, 395, 401
Lowe, Henry N., Jr., 398
Lowell Observatory, Ariz., 209, 366
Lown, Dr. Bernard, 128
LRL. See Lunar Receiving Laboratory.
LST. See Lunar surface telescope.
LTTAT. See Long-Tank Thrust-Augmented Thor-Delta.
LTV Aerospace Corp., 11, 91, 291
Lucian, 232
Luebke, President Heinrich (W. Germany), 158
Luna XV (U.S.S.R. lunar probe), 195, 207-208, 224, 229, 232, 238, 382
  launch, 206
  moon landing, 236, 251, 422
  press comment, 211, 420

Lunar and Planetary Laboratory (Tucson, Ariz.), 397
Lunar Exploration: Strategy for Research, 1969-1975 (Space Science Board report), 355
Lunar Landing Research Vehicle (LLRV), 315
Lunar Landing Training Vehicle (LLTV), 5, 103, 173, 182
Lunar Laser Observatory (Tucson, Ariz.), 192
Lunar module (LM), 111, 169, 182, 185, 205, 320, 353, 365
  Apollo 9 (Spider) (LM-3), 5, 26-27, 33, 62-65, 67, 72, 81, 89-90, 91, 130, 421
  Apollo 10 (Snoopy) (LM-4), 16, 90, 142-144, 159, 421
  Apollo 11 (Eagle) (LM-5), 21, 46, 108, 141, 162, 179, 196, 198-199, 210-213, 215, 217-222, 237, 255-256, 355
    moon landing, 199-200, 212-213, 215, 232, 252, 317, 322, 421
    moon liftoff, 220, 222
    plaque on, 196
  Apollo 12 (Intrepid) (LM-6), 333-334, 342, 362, 368, 372-379, 395-397
    moon landing, 375-376, 407
    moon liftoff, 377
  contract, 42-43
  exhibit, 105, 161-162
Lunar Orbiter (program), 187, 209, 282-283
Lunar Orbiter I Preliminary Results (NASA SP-197), 187
Lunar Receiving Laboratory (LRL) (NASA), 45, 103, 259, 370
  Apollo 11 postage stamp, 202
  astronauts at, 75-76
    Apollo 11, 247, 260, 267, 273
    Apollo 12, 378, 397, 400, 405
  criticism of, 95, 176
  lunar samples, 75-76, 95
    Apollo 11, 222-223, 247, 250, 253, 260, 261, 263, 270, 275, 288, 302-304, 306
    Apollo 12, 356, 378, 394, 395-396, 407
  personnel, 260, 266, 273, 352-353, 400, 405
Lunar Rock Conference, 306
Lunar roving vehicle (LRV), 205, 320, 327, 353, 355-356
  dual-mode (DLRV), 111
Lunar Science Institute, 8, 75, 227
Lunar surface telescope (LST), 123, 343
Lundin, Bruce T., 349, 404
Lupus (constellation), 256, 281
Luria, Dr. Salvador E., 338
Lutine Bell (Lloyd's of London), 244
Lyman-alpha radiation, 316
Lyons, Kenneth T., 253

M

M2–F2 (lifting-body vehicle), 29
M2–F3, 29
McCarthy, Joseph W., 81
McClure, Billy B., 347
McCollum-Pratt Institute, 184
McConnell, Gen. John P. (USAF), 112, 245
McDivitt, Col. James A. (USAF), 5, 26–27, 51, 63–65, 72, 91, 161, 168, 187–188, 317
MacDonald, Dr. Gordon J. F., 263, 416
McDonald, Dr. James E., 288
McDonald Observatory, Tex., 17, 90, 237, 259, 285
McDonnell Douglas Corp., 410
  airlock, 128, 297
  collision avoidance system, 315–316
  contract, 14, 91, 123, 201, 241, 280, 297, 299, 388, 414
  employment, 185
  F–4 Phantom, 9, 201
  Saturn V, 117, 280, 388
  space station, 241, 270, 299
  STOL aircraft, 29
  Titan IIIC, 14
McDonnell Douglas Astronautics Co., 297
McDowell, James, 38
McElroy, Dr. William D., 184, 189, 353
Mack, L/C Thomas J. (USAF), 413
Mackin, Dr. Robert J., Jr., 135
MacLeish, Archibald, 20
McLucas, John L., 304
McMurdo Sound, 359
Macy, John W., Jr., 402
Madrid, Spain, 37, 52, 180, 187, 189, 312, 327, 330, 424
Magnet, superconducting, 24
Magnetic field, 171, 173, 177, 185, 230
Magnetosphere, 115, 185, 257, 291, 316, 395
The Magnitude and Economic Impact of General Aviation (study), 109
Magnuson, Sen. Warren G., 133
Maheuson, Peter, 370
Mahlberg, Dr. Paul, 103
Mahnken, Conrad V. W., 51, 86, 110
Mahon, Joseph B., 118
Mailer, Norman, 135–136
Man of Achievement Award (American Academy of Achievement), 203
"Man on the Moon" (CBS News recording), 292
Man on the Moon (pamphlet), 234
Man Will Never Fly Memorial Society, 410
Management, 148–149, 192–193, 202
Manchester, U.K., 122
Manchester Univ. (U.K.), 338
Mandelbaum, Leonard, 51
Manhasset, N.Y., 382
Manhattan Project, 53, 161, 227

Manila, Philippines, 248
Manke, John A., 113, 133, 159, 184, 267, 309, 349
Manned Orbiting Laboratory (MOL), 110, 151–152, 200, 260
  appropriations, 15, 110
  cancellation, 176, 178, 179, 181, 186, 191, 257–258, 281, 422, 423, 424
  and NASA, 21, 86–87, 104, 146
  pilots, 281, 289
Manned orbiting platform (MOP) program (U.S.S.R.), 382
Manned space flight (see also Apollo program, *Apollo 8, 9, 10, 11,* and *12* missions; Gemini program; Astronaut; Cosmonaut; Manned Orbiting Laboratory; *Soyuz IV, V, VI, VII,* and *VIII* missions; Space biology; and Space station), 71
  achievements, 17, 39, 47, 65, 72–73, 91, 101, 212, 246–247, 259, 262, 372, 378, 384
  advantages, 17–18, 74–75, 242
  cooperation, 248
  criticism, 82, 167, 204, 262, 289, 423
  EVA. See Extravehicular activity.
  funding, 14–15, 28, 29, 109–110, 356
  hazards, 80, 157, 281, 327
  long-duration, 94, 178, 200, 337–338
  lunar landing, manned. See Moon landing, manned.
  military aspects, 86–87, 104, 200
  policy and plans
    U.S., 15–16, 18–19, 37, 47, 51, 66–67, 72, 74, 80, 81–82, 86–87, 94–95, 104, 119–120, 134, 167–168, 214–215, 268–269, 304–305, 313, 320, 367–369, 371, 384, 405, 416, 422–423
    U.S.S.R., 120, 170–171, 241–242, 292
Manned Space Flight Network (MSFN), 10, 117, 196, 362
Manned Spacecraft Center (MSC) (NASA), 27, 74, 185, 231, 233, 245, 249, 288, 289, 290, 312, 337, 339, 365, 366
  Apollo spacecraft, 65, 145, 223–224, 230
  astronauts at, 61, 75–76, 114, 198, 200, 273, 276, 277, 330, 391
  award, 10, 91, 287, 289, 340, 345–346, 419
  facilities, 114, 334, 383
  Lunar Receiving Laboratory. See Lunar Receiving Laboratory.
  lunar rock sample exhibit, 353–354
  management, 47, 65, 223–224, 378
  National Seminar for Manned Flight Awareness, 317
  personnel, 131, 259, 282, 292, 317, 326, 347, 417
  press conference, 61, 77, 108, 198, 242–243, 277–278, 280–281, 333–334, 350, 371–372, 392, 396, 406–408
  quarantine procedures (see also Lunar

Receiving Laboratory), 75-76, 141, 179-180, 273, 356
real-time computer complex, 334
space station, 299
spacecraft test. See Apollo (spacecraft).
visits to
  cosmonauts, 337, 350
  Haile Selassie, Emperor of Ethiopia, 205
Mannheim, W. Germany, 225
Mansfield, Sen. Michael J., 224, 296
Marcos, President Ferdinand E. (Philippines), 248
Mar del Plata, Argentina, 327, 328
Mare Imbrium (moon), 180
Mare Marginis (moon), 180
Mare Occulum (Hidden Sea) (moon), 180
Marietta, Ga., 411
Mariner (program), 15, 23, 56, 138, 256, 359
Mariner (spacecraft), 359
Mariner III (Mars probe), 55
Mariner IV (Mars probe), 54, 89, 210, 253, 421
Mariner V (Venus probe), 127, 160
Mariner VI (Mariner F) (Mars probe), 117, 167, 187, 250
  launch, 55
  photographs, 90, 210, 250, 252, 253, 262, 266, 282
  results, 133, 261, 262, 266, 269, 298, 301-302, 338-339, 408, 421
Mariner VII (Mariner G) (Mars probe), 55, 117, 167, 187
  control, 99, 103, 253, 254, 259
  launch, 92
  photographs, 90, 210, 252, 253-254, 259, 261, 262-263, 265, 266, 269
  results, 133, 262, 265-266, 301-302, 338-339, 407-408, 421
Mark, Dr. Hans M., 53, 332, 389
Mark 12 (reentry vehicle), 330, 408
Mark 15, 391
Marks, Leonard H., 56, 88
Marquette, Father, Tercentenary Commission, 365
Marquette, Pere, Discovery Award, 365
Marquette Univ., 365
Mars (planet) (see also *Mariner III, Mariner IV, Mariner VI, Mariner VII*, and Viking program), 105, 146, 320
  atmosphere, 55, 75, 92-93, 184-185, 250, 261, 266, 269, 293, 298, 301, 338, 398, 408, 421
  The Book of Mars, 89
  canals, 261, 262, 266
  color, 396
  craters, 261, 262, 269-270, 282, 301
  ephemeris, 338
  equatorial region, 338
  exploration of, 74-75, 94, 147, 169, 176, 187, 197, 235-236, 247, 260, 262, 268-269, 271-272, 276, 337
  benefits, 337
  cost, 336-337
  funding, 67, 262, 270, 271
  international cooperation, 248
  manned, 95, 178, 205, 224, 231, 242, 247, 262, 266, 268, 269, 271, 276, 277, 283, 288, 305, 320, 325, 344, 371, 399
  plans for, 15-16, 57, 136, 205, 231, 293, 343, 344, 359, 398, 399, 412, 422-423
  spacecraft, 15-16, 59, 160, 205, 338, 359, 412
  unmanned, 256, 271, 283, 421
  ionosphere, 408
  life on, 55, 90, 94, 185, 261, 269-270, 271, 302, 308, 398
  mass, 339
  photographs of, 55, 89, 90, 210, 250, 252, 259, 261, 262-263, 265, 266, 282, 301, 339, 398, 421
  poles, 252, 262-263, 265, 266
  surface, 55, 92, 250, 252, 261, 265, 269-270, 282, 301, 338, 398, 408, 421
  temperature, 55, 250, 261, 268
  water on, 90, 261, 269, 293, 302
Marshall, Justice Frederick M., 2
Marshall Space Flight Center (MSFC) (NASA), 71, 105, 116, 122, 195, 270, 290, 317, 417
  Apollo Telescope Mount, 12, 71, 177, 320, 327
  Astrionics Laboratory, 347
  award, 10, 69, 147, 288, 326
  contract, 11, 25, 58, 17, 91, 105, 111, 113, 177, 205, 270, 271, 297, 299, 320, 327
  Saturn, 25, 47, 49, 77, 87, 91, 94, 101, 104, 113, 117, 132, 315, 388
  Engineering Laboratory, 83
  launch vehicle. See Saturn.
  lunar rock sample exhibit, 400
  Lunar roving vehicle (LRV), 111, 205, 320, 327
  management, 64, 113, 146, 223-224, 378
  meeting, 122, 156
  Neutral Buoyancy Simulator, 17
  personnel, 13, 77, 103, 185, 187-188, 251, 255, 325, 360, 393
  Saturn I Workshop, 114, 297, 327, 388
  Space Sciences Laboratory, 76-77
  Space Technology Applications and Research Laboratory (STARLAB), 285
  space station, 128, 270, 299
Marsten, Dr. Richard B., 139
Marti, Kurt, 403
Martin, James S., Jr., 111
Martin-Marietta Airport, Baltimore, 315
Martin Marietta Corp., 85, 160, 270, 351, 386, 387, 402
Martin, Minta, Lecture, 75
Maryland, Univ. of, 71, 75, 182, 395, 399
Mascon (mass concentration of gravitational pull), 73, 116, 120, 180, 283

Mason, Dr. Bryan H., 304
Mason, Harold P., 293
Mason-Rust Co., 58
Massachusetts Institute of Technology (MIT), 35, 89, 131, 178, 347, 409, 416, 417
   award, 178
   Div. of Sponsored Research, 14
   experiment, 118
   Instrumentation Laboratory, 338, 363, 369, 416, 420
   Lincoln Laboratory, 288, 369
   Martin, Minta, Lecture, 75
   military research, demonstrations against, 67, 136, 363, 369
   Union of Concerned Scientists, 40
Massachusetts, Univ. of, 396
Materials technology, 58, 85, 203, 350, 362, 363, 384
Mathematics, 1
Mathews, Dr. Charles W., 131, 346, 406
Mathias, Sen. Charles McC., Jr., 13, 312
Mattingly, L/Cdr Thomas K., II (USN), 267
Maurer camera, 217
Maus, Hans H., 256
Maxwell AFB, Ala., 411
Maxwell, M/G Jewell C. (USAF), 127, 186
May, Chester B., 103, 282
Mayer, Dr. Jean, 417
Mead, Dr. Margaret, 201
Medal of Freedom, 280
"Meet the Press" (TV program), 145
Melbourne, Univ. of, 319, 399
Melville Space and Defense Systems Div., Control Data Corp., 388
Memorandum of Understanding, 30, 176
Men of the Year, 2
Menconi, Ralph J., 275
Mercator, Gerardus, 294
Mercator projection, 294
Mercury (planet), 16, 69, 76, 110, 119, 134, 138, 147
Mercury (program), 357
Mercury compounds, 128
Meredith, Scott, 135, 136
MESA. See Modularized equipment stowage assembly.
Meteor, 279, 287
Meteor I (U.S.S.R. meteorological satellite), 91, 422
Meteor II, 329, 422
Meteor Crater, Ariz., 338
Meteorite, 28, 45, 403, 416
Meteorological satellite (see also individual satellites, such as *Ats III, Ats IV, Essa IX, Meteor I, Meteor II, Nimbus I, Nimbus II, Nimbus III*), 33, 43, 47, 111, 157, 357
   Barbados Oceanographic and Meteorological Experiment, 88
   cooperation, 45, 88, 205
   Global Atmospheric Research Program, 88, 204
   ITOS program, 344
   Nimbus program, 18, 23, 86, 88, 107–108, 139, 205, 353, 414, 421, 422, 423
   Tiros program, 23, 57, 85, 344, 399, 414
   U.S.S.R., 91, 329, 422
Meteorology, 42, 44, 78, 86, 88, 132, 139, 204, 353
Metric system, 317
Metroliner (high-speed train), 17
Mettler, Ruben F., 329
Metzenbaum, Howard M., 411
Mexico, 146, 299, 370
Mexico City, Mexico, 312, 319
Meyer, Karl E., 161
Mi-10 (U.S.S.R. helicopter), 35
Miami, Fla., 103, 123, 148, 206
Miami Beach, Fla., 410
Miami, Univ. of, 123
   Institute of Marine Science, 273
Michel, Dr. F. Curtis, 267, 287
Michigan, 28
Michigan, Univ. of, 136, 286
   Mental Health Research Institute, 396
Michoud Assembly Facility (MSFC), 58, 118, 256, 260, 299
Micrometeoroid, 159, 279, 287
Microminiaturization, 196
Midway Island, 112
MIG (U.S.S.R. fighter aircraft), 1
Military Airlift Command, 411
Milky Way (galaxy), 115, 118, 316
Milledgeville, Ga., 88
Miller, Prof. Charles L., 338
Miller, Rep. George P., 18, 113, 138, 275
Miller, Dr. Joseph S., 42
Miller, Dr. Maynard M., 2
Miller and Berry, 320
Mineralogical Society of America, 290
Mink experiment, 352
Minneapolis, Minn., 206
Minnesota, Univ. of, 31
   Institute of Technology, 38, 42
Minuteman (missile), 230, 297
Mirage (French fighter aircraft), 162
Mirage (French supersonic fighter-bomber aircraft), 197
MIRV. See Multiple independently targetable reentry vehicle.
Missile
   antiballistic missile (ABM), 33, 39, 40, 43, 47, 52, 62, 80, 81, 83, 89, 103, 112, 117, 124, 131, 136, 230, 320
   award, 345
   contract, 191, 362
   detection, 183, 321
   foreign
      Communist China, 43, 80, 181
      U.S.S.R., 10, 43, 48, 50, 53, 62, 89, 116, 129, 131, 140, 181, 183, 229, 332
   intercontinental ballistic (ICBM), 48, 140, 230, 297
   limitation of, 13, 21, 67, 81, 86, 131, 155, 173, 181, 183, 190, 196, 332
   multiple independently targetable re-

entry vehicle (MIRV), 10, 116, 183, 190, 191
nuclear, 10, 81, 86, 129, 135, 155, 181
test, 173, 181, 183, 191, 325
underwater-to-surface, 155, 197, 362
Mission Control Center (NASA), 137, 215, 220, 374
Mission of the Doctor award, 341
Mississippi Test Facility (MTF), 24, 104, 182, 185, 188, 360
MIT. See Massachusetts Institute of Technology.
Mitchell, Cdr. Edgar D. (USN), 267
Mitchell, Jesse L., 154
Mobile quarantine facility (MQF), 223, 247, 356, 377–378, 396
Mobutu, President Joseph D. (Congo), 350
Modularized equipment stowage assembly (MESA), 217, 331, 376, 407
Moe (elk), 328
Moeckel, Wolfgang E., 338
Mogadiscio, Somalia, 234
Mohler, Dr. Stanley R., 180
Mohole, Project, 290
Mojave, Calif., 402
MOL. See Manned Orbiting Laboratory.
Molniya I-11 (U.S.S.R. communications satellite), 106
Molniya I-12, 237
Moltke Crater (moon), 159
Monkey experiment, 190, 200, 201–202, 335, 347–348, 415, 421
Monochromator, 309, 411
Monroney, Sen. A. S. Mike, 88–89
Montclair, N.J., 243–244
Montclair, N.J., Library, 298
Montgomery, Ala., 400
Montreal, Canada, 312, 322
Moon (see also Apollo missions, Lunar Orbiter, Lunar Receiving Laboratory, etc.), 131, 136, 142, 420
base, 123, 160, 169, 250, 305, 387, 398
colonization of, 102, 134, 191, 397
color, 217, 408
contamination from, 76, 136–137, 141, 156, 175, 176, 179, 181, 205, 235, 261, 270, 273, 274, 302, 303, 356
crater, 173, 200, 208, 209, 215, 249, 345, 376
distance from earth, 261
exploration of, 2, 14, 47, 48, 50, 59, 73, 74–75, 82, 96, 110, 111, 113, 119, 129, 147, 148, 149, 160, 164, 167, 169, 170, 175, 177, 185, 308, 313, 317, 331, 355
international cooperation, 102
manned, 67, 108
globe, 208
gravity, 157, 340
laboratory, 278
landing, 169, 197
equipment, 80, 111, 388
manned, 33, 37, 43, 46, 53, 55, 67, 72, 74, 75, 77, 82, 90, 91, 92, 103, 106, 108–109, 111, 120, 122, 134, 137, 143, 144, 146, 147, 149, 262, 267, 271, 272, 278, 281, 287, 325, 351, 405, 411, 418, 420, 421, 423
anniversary, 202
commemorative stamp, 202, 273, 289
criticism, 204
implications of, 238, 307, 311, 322
legal aspects, 290
memorial sculpture, 304
plans for, 168, 169, 170, 176, 178, 179, 182, 185, 187, 188, 325, 334, 342, 354, 356, 368
U.S., 199, 208, 212, 252, 362, 368, 370, 372, 384, 385, 389, 390, 391, 392, 398
commemoration of, 196, 254
plans for, 205, 207, 243
U.S.S.R., 246, 361, 365
soft, 141, 382
unmanned, 35, 141, 170, 189, 339, 365, 370, 419, 423
U.S.S.R., 236, 251, 382
landing site, 7, 35, 67, 77, 82, 90, 116, 142, 342, 345, 356, 368, 369, 377, 378
laser experiment, 114, 192, 219, 223, 236, 237, 259, 261, 285, 339, 419
life on, 103, 167, 261, 270, 275, 313
lunar orbit, record, 208, 250
magnetic field, 230
mascon, 93, 116, 120, 180, 283
mining, 191, 387
nuclear explosion on, proposed, 167, 409, 412, 413
observatory, 169
origin of, 198, 241, 292, 332, 360, 409
passenger flight to, 241, 244
photographs of, 90, 92, 116, 159, 179, 187, 201, 208, 218, 219, 220, 241, 252, 253, 255, 306, 325, 334, 343, 376, 395, 405, 407, 409, 419
probe, *Luna XV*, 200
seismic experiment, 114, 223, 230, 237, 240, 248, 325, 342, 354, 376, 377, 419
solar wind experiment, 219, 223, 230, 285, 375, 413, 419
station, 167, 388
surface, 7, 20, 28, 34–35, 103, 141, 146, 152, 159, 160, 174, 187, 189, 240, 308, 338, 340, 419
analysis of, 198, 250, 360
composition, 250, 376
glazing, 317
sample, 195, 200, 208, 219–220, 223, 243, 246, 247, 250, 253, 254, 260, 262, 267, 275, 285, 290, 301, 302, 304, 306, 307, 312, 325, 339, 343, 354, 356, 360, 365, 373, 376, 379, 380, 386, 387, 393, 398, 412
biological tests, 261, 293
exhibit, 309, 314, 401, 412

preliminary study, 270, 288, 292, 306
tavern, 317
telescope on, 123, 332, 342
TV telecasts from, 210, 217, 362, 373, 374
volcanic activity, 139, 187, 212, 339, 418
water on, 167, 275, 313, 332, 398
"Moon In" (Central Park, New York), 233
"Moon Maiden" (song), 279
Moore, David, 204
Moore, George S., 123
Moore, Wendell F., 161
Moorer, Adm. Thomas H. (USN), 180
MOP. See Manned orbiting platform.
Moravia, Alberto, 321
Morea, Saverio F., 185
Moritz, Bernard, 130
Morocco, 146
Moscow, 12, 152, 185, 339, 362
  antimissile defense, 50
  astronaut visit, 195, 202
  cosmonaut ceremony, 23, 33, 349
  lunar rock exhibit, 412
  nuclear nonproliferation treaty signing, 370, 392–393
  press conference, 21, 372
Moser, Dr. Jürgen K., 121
Moslems, 226, 234
Motorcycle, 129
Mt. Everest, 303
Mt. Hopkins, Ariz., 405
Mt. Palomar Observatory, 341, 367
Mountain View, Calif., 361
MQF. See Mobile quarantine facility.
MS-4 (Japanese rocket), 290–291
MSC. See Manned Spacecraft Center.
MSFC. See Marshall Space Flight Center.
MSFN. See Manned Space Flight Network.
MU3D (Japanese rocket), 284
Mueller, Dr. George E., 131, 274, 405
  Apollo Applications, 67
  Apollo program, 73, 74, 77, 88, 262
  award, 10, 300, 346
  earth resources satellite, 33
  lunar exploration, 67, 74, 123, 157, 246
  manned space flight, 73, 74
  press conference, 29, 77, 242–243
  resignation, 368, 399
  reusable launch and space vehicles, 266
  space program, 29, 243, 262
  space shuttle, 350
  space station, 74, 131, 267
  U.S.S.R. space program, 29, 207–208
Muller, Paul M., 116, 120
Multiple docking adapter, 128
Multiple independently targetable reentry vehicle (MIRV), 10, 116, 183, 190, 332
Mumford, Lewis, 201
Murphy, Sen. George L., 150

Murray, Dr. Bruce C., 282
Muscat and Oman (sultanate), 48
Museum, space (proposed), 346

N

NAA. See National Aeronautics Assn.
NAE. See National Academy of Engineering.
NAFEC. See National Aviation Facilities Experimental Center.
Nagako, Empress (Japan), 361
Nance, Richard L., 339
NAS. See National Academy of Sciences.
NASA. See National Aeronautics and Space Administration.
NASA Apollo Lunar Exploration Office, 74
NASA Apollo Program Office, 68–69
NASA Astronomy Missions Board, 156, 256, 366
NASA Communications Network (NASCOM), 292
NASA Historical Advisory Committee, 178
NASA Launch Vehicle Review Board, 385
NASA Office of Advanced Research and Technology (OART), 21, 84, 87, 212, 368
NASA Office of Manned Space Flight (OMSF), 22, 145, 223, 292, 378, 415
NASA Office of Public Affairs, 10
NASA Office of Space Science and Applications (OSSA), 76, 78, 118, 173, 190, 245, 277, 292, 403
  Lunar and Planetary Programs Div., 55
NASA Office of Tracking and Data Acquisition (OTDA), 65, 145, 224, 378
NASA Physics Advisory Committee, 355
NASA Preliminary Examination Team (PET), 306, 419
NASA Science and Technology Advisory Committee for Manned Space Flight, 94
NASA Space Shuttle Task Group, 336, 354
NASC. See National Aeronautics and Space Council.
NASCOM. See NASA Communications Network.
NAS–NRC Space Science Board. See National Academy of Sciences and National Research Council.
Nassau, Bahamas, 409
Natal, Brazil, 179, 186, 188
National Academy of Engineering (NAE), 99, 189
  Committee on Public Engineering Policy, 257
National Academy of Sciences (NAS), 2, 184, 314, 410
  advisory committee to HUD, 189
  annual meeting, 121
  antiballistic missile (ABM) system, 89

applications satellite study, 47
award, 121
Committee on Atmospheric Sciences, 409
cooperation, 141
food management for aerospace vehicles conference, 110
Lunar Science Institute, 8, 75–76, 227
president, election of, 19
report, 19, 136, 257, 263, 343, 355
space program, 19, 136, 226, 263, 355
Space Science Board, 19, 94, 124, 136, 164, 167, 263, 390, 343, 347
Committee on Space Medicine, 164
Universities Organizing Committee for Space Sciences, 8
UFO study review, 5, 411
World Data Center A for Rockets and Satellites, 1
National Aeronautic Assn. (NAA), 101
National Aeronautics and Space Act, 300, 323, 405
National Aeronautics and Space Administration (NASA) (see also NASA centers, programs, satellites, and related headings, such as Ames Research Center, Apollo program, *Essa IX*), 75, 79, 86, 87, 135, 300, 304, 313, 323, 337
accomplishments, 1, 7, 10, 13, 14, 16, 18, 25, 38, 64, 65, 68, 71, 77, 80, 92, 145, 160, 212, 230, 242, 246, 251, 372, 421, 422–423
agreement. See International cooperation, space; and Treaty.
anniversary, 33, 51, 83, 323
Apollo Orbital Science Photographic Team, 410
Apollo 204 Review Board. See Apollo 204 Review Board.
astronaut. See astronaut.
Astronomy Missions Board, 4, 156, 256
awards and honors, 2, 6, 8, 10, 22, 32, 33, 38, 43, 46, 47, 49, 59, 68–69, 72, 87, 100–101, 117, 118, 123, 138–139, 168, 200, 204, 237, 251, 255, 266, 279–280, 288–289, 326, 337, 345–346, 346–347, 350, 360, 362, 364, 371, 387, 388, 403, 419
Apollo Achievement Award, 362
Distinguished Public Service Certificate, 326
Distinguished Public Service Medal, 154, 346–347
Distinguished Service Medal, 10, 13, 91, 154, 280, 287, 326, 346
Exceptional Bravery Medal, 347
Exceptional Scientific Achievement Medal, 10, 13, 326, 347
Exceptional Service Medal, 91, 287, 326, 347
Group Achievement Award, 154, 346
Outstanding Leadership Award, 371
Public Service Award, 92, 347
budget, FY 1970, 3, 5–6, 14–16, 22, 28, 38, 106, 109–110, 115, 116, 122, 147, 163, 167, 204, 208, 305, 390, 423
bills signed, 383, 395
House consideration
appropriations, 184, 186, 381–382
authorization, 18, 66, 67, 69, 70, 71, 74, 78, 84, 85, 91, 108–109, 112–113, 117, 118, 138, 147, 176, 276, 329, 337, 361, 363
press comment, 4, 28, 81–82, 95, 108, 155, 163, 421
Senate consideration
appropriations, 364–365, 369, 382, 383
authorization, 24, 118, 120, 123, 127, 130, 131, 134, 160, 186, 265–266, 270, 307, 311, 312, 330, 361, 365
conference, 17, 106, 111, 122, 169, 317, 337, 375, 384, 411
contract, 25, 35
aeronautics, 335
aircraft, 300
communications system, 241, 386
computer services, 71, 88, 177, 203, 335
engine, 32, 70, 79, 85, 91, 212, 291, 379–380
facilities, 271, 320
guidance and navigation, 13
instrumentation, 11, 24, 76, 91, 253
launch vehicle, 47, 49, 70, 77, 88, 91, 94, 100, 113, 117, 133, 154, 278, 297, 315, 355, 389
life support system, 123, 173
nuclear propulsion, 93, 105
space equipment, 21, 92–94, 111, 297, 320, 327, 353
space shuttle, 26, 48
space station, 11, 26, 240, 270, 299
spacecraft, 30, 43, 46, 59, 70, 154, 160, 297, 340, 351, 401
study, 38, 47, 105, 111, 128, 151, 187, 240, 270, 340
support services, 1, 24, 30, 38, 47, 59, 85, 88, 91, 100, 104, 114, 292, 293
telescope, 71, 177, 320, 327
tracking, 189, 196
cooperation, 103, 111, 170, 402
AEC, 88, 105, 178, 310, 332, 422
Commerce, Dept. of, 88
ComSatCorp, 178
DOD, 11, 34, 38, 87, 88, 146, 152, 153, 176–177, 345, 368, 379, 389, 417
ESSA, 356, 414, 417
Interior, Dept. of, 51, 356
NSF, 88, 356
Transportation, Dept. of, 88, 417
USA, 49, 334, 379
USAF, 21, 29, 86, 119, 146, 176, 192, 205, 258, 321, 336, 363, 389, 405, 421

U.S. Bureau of Mines, 387
U.S. Coast Guard, 51
USN, 44, 51, 119
cooperation, international. See International cooperation, space; and Sounding rocket, international programs.
criticism, 176, 204, 205, 231, 272, 284, 287, 289, 340, 364, 367–368, 398, 413, 415, 418, 423
employment, 66, 69, 197, 231, 361, 363, 390, 422–423
exhibit, 105, 162, 403, 405, 417
facilities, 176, 245, 277, 312, 329, 362, 383, 417, 418, 423
history, 351, 353, 357
inaugural parade float, 21
launch, 23, 420
  Apollo 9 (AS–504), 62
  Apollo 10 (AS–505), 142
  Apollo 11 (AS–506), 211–223
  Apollo 12 (AS–507), 372–378
  failure
    Intelsat-III F–1, 135
    Intelsat-III F–5, 245, 281
    Pioneer E, 291
  postponed, 61, 62, 107, 108
  probe
    Mariner VI (Mariner F), 55
    Mariner VII (Mariner G), 94
  satellite, 22–23, 30, 39–40, 57, 107, 173, 185, 189–190, 245, 323, 364, 388–389
    Ats V (ATS–E), 276
    Azur (GRS–A), 364
    Biosatellite III (Biosatellite-D), 189–190
    Boreas (Esro IB), 323
    Essa IX (TOS–G), 58
    Explorer XLI (IMP–G), 185
    Intelsat-III F–3, 39–40
    Isis I (ISIS–A), 30
    Nimbus III (Nimbus-B2), 107
    Ogo VI (OGO–F), 171
    Oso V (OSO–F), 22
    Oso VI (OSO–G), 272
    Skynet A (IDCSP–A), 389
  sounding rocket
    Aerobee 150, 49, 78, 80, 179, 302, 361, 402, 409
    Aerobee 150A, 139
    Aerobee 150 MI, 9, 38, 42, 43, 65–66, 71, 104, 108, 111, 113, 118, 141, 179, 184, 186, 299, 309, 314, 316, 326, 337, 361, 410
    Aerobee 170, 298
    Aerobee 350, 27
    Arcas, 18, 19, 33, 41, 354
    Astrobee 1500, 335
    Black Brant IIIB, 59, 127
    Black Brant IV, 312
    Black Brant VB, 366
    Boosted Arcas II, 28, 332
    Bullpup Cajun, 178
    Javelin, 188
    Nike-Apache, 33–34, 41, 43, 48, 49, 51–52, 54, 58, 106, 113, 118, 120, 135, 159, 176, 189, 279, 286, 287, 300, 303, 402, 408
    Nike-Cajun, 9, 18, 19, 21, 23, 24, 25, 27, 28, 30, 31, 33–34, 38, 41, 94, 132, 133, 135
    Nike-Tomahawk, 27–28, 29, 49, 71, 72, 83, 112–113, 133, 211–212, 286, 327
    Orion II, 79
    Pacemaker, 203
    Sidewinder-Arcas, 79
legal suit, 270, 283, 394, 401
Lunar Landing Training Vehicle, 5, 173, 182
Lunar Planetary Missions Board, 94
management, 15, 24, 148, 167, 195, 203, 254, 276, 287, 390
Management Advisory Council, 70
manpower. See Employment.
memorial, lunar landing, 304
organization, 139, 167, 274, 332
patents, 4, 82
Performance Evaluation Board, 70
personnel, 9, 42, 46, 67–68, 80, 87, 100, 103, 115, 116, 129, 136, 139, 188, 260, 274, 281, 289, 290, 291, 293, 299, 300, 304, 323, 360, 362, 371, 394, 395, 396, 399, 400, 417
  appointment, 7, 19, 20, 53, 70, 76, 83, 173, 185, 251, 285, 290, 304, 317, 319, 325, 326, 332, 339, 353, 403, 405, 410, 412
  resignation, 135, 251, 259, 267, 292, 297, 340, 347, 352, 368, 371
  retirement, 13, 19, 53, 256, 349
procurement, 25
programs
  aeronautics, 11, 15, 16, 39, 42, 44, 65, 70, 78, 84, 110, 113, 120, 121, 123, 138, 177, 205, 212, 263, 321, 334, 362, 368, 379, 398, 400, 417, 421–422
  art, 226, 403
  astronomy, 3, 15, 16, 22, 23, 28, 35, 37, 55, 57, 66, 73, 76, 85, 89, 92, 94, 111, 119, 123, 136, 138, 160, 192, 257, 263, 266, 276, 283, 305, 318, 328, 337, 359, 361, 367, 412
  communications, 23, 37, 39, 150, 151, 402
  earth resources, 15, 16, 33, 37, 47, 49, 66, 69, 85, 119, 138, 151, 161, 163, 305, 308, 328, 337, 339, 357
  history, 357
  international, 7, 21, 26, 30, 59, 74, 85, 103, 167, 169, 177, 202, 268, 305, 309, 311, 411, 423–424
  manned space flight, 6, 15, 16, 18, 19, 23, 27, 28, 33, 47, 64, 66, 74, 77, 82, 86, 91, 94, 104, 109, 111, 119, 134, 151, 163, 178, 208, 223, 237, 260, 266, 308, 312, 321, 331,

343–344, 355, 367, 371, 378, 390, 421
  meteorology, 23, 47, 57, 69, 85, 107, 151, 204, 344, 353, 409, 414
  nuclear propulsion, 16, 37, 77, 105, 109, 130, 138, 260, 268, 269, 309, 332, 383, 399, 404, 422
  sounding rocket, 124, 421
  space medicine, 18, 19, 23, 51, 72, 76, 78, 123, 128, 130–131, 136, 138, 154, 164, 189, 190, 200, 201–202, 206, 273, 335–336, 348–349, 367–368, 415
  space rescue, 21, 315
  space science, 15, 37, 77, 78, 120, 134, 136, 151, 167, 171, 185, 305, 308, 331, 337, 338, 355
  space station, 15, 16, 17, 66, 73, 83, 86, 94, 111, 120, 138, 146, 167, 169, 237, 241, 260, 266, 268, 269, 305, 328, 343, 344
  technology utilization, 23, 66, 70, 75, 127, 202, 274, 337, 362
  tracking and data acquisition, 37, 149, 187, 210, 267, 307, 423
  Science and Technology Advisory Committee for Manned Space Flight, 94
  Semiannual Report to Congress, 371
  Space Task Group report, 305, 308, 312–313
  test, 17, 403
    aircraft, 263, 405, 421
    ion engine, 47
    launch vehicle, 103, 185, 195, 196, 272, 360, 412
    lifting-body vehicle, 29, 101, 112, 117, 133, 148, 159, 174, 184, 240, 267, 286, 297, 299, 300, 316, 319, 349, 352, 360, 371, 381, 389, 394, 407, 421
    Lunar Landing Training, 103, 182
    nuclear, 87, 178, 190, 294, 310, 410
    parachute, 79, 183
    spacecraft, 117, 195, 196, 351
  universities, 3, 25, 56, 106, 109, 156, 227, 304, 363, 368, 382, 415
  X–15. See X–15.
National Aeronautics and Space Council (NASC)
  award, 87
  Chairman, 20, 54, 66, 91, 294, 320, 419
  Executive Secretary, 49, 87, 147, 184, 257, 281, 327
  President's Space Task Group recommendation, 167, 171
National Air and Space Museum, 297, 344
National Air Exposition, Second, 283
National Amateur Astronomers convention, 288
National Anti-aviation Citation Presentation, 410
National Archives, 405
National Assn. of Government Employees, 253

National Aviation Club, Award for Achievement, 127
National Aviation Facilities Experimental Center (NAFEC), 89
National Aviation Planning Commission (proposed), 26
National Broadcasting Co. (NBC), 146, 357
National Center for Atmospheric Research, 167, 409
National Civil Service League, 122
National Collegiate Athletic Assn., 83
National Conference on Public Administration, 1969, 148
National Council on Marine Resources and Engineering Development, 56, 99
National Day of Participation (*Apollo 11* mission), 224, 230
National Flight Data Center (NFDC), 117
National Gallery of Art, Washington, D.C., 403
National Geodetic Satellite program, 86
National Geographic Society, 100, 101, 204
National Industrial Conference Board, 388
National Institute of Social Sciences, 387
National Institutes of Health (NIH), 2, 110
National Investigations Committee on Aerial Phenomena (NICAP), 8, 411
National Medal of Science, 1, 418
National Meteorological Center, 353
National Oceanic and Atmospheric Agency (NOAA) (proposed), 14, 150, 341
National Oceanographic Center, 9
National Operational Meteorological Satellite System (NOMSS), 86, 422
National Order of the Leopard, 350
National Postal Forum, 289
National Press Club, 268
National Register of Scientific and Technical Personnel, 420
National Research Council (NRC), 205, 297
  Committee on Radio Frequency Requirements for Scientific Research, 125
  Geophysical Research Board, Committee on Solar-Terrestrial Research, 257
  Panel on Remote Atmospheric Probing, 409
  Space Science Board, 18, 94, 124, 136, 164, 167, 263, 289, 343, 347, 355, 356
National Science Board, 121
National Science Foundation (NSF), 32, 35, 53, 182, 290, 405
  American Science Manpower, 1968, 420
  cooperation, 88, 90
  Deep Sea Drilling Project, 353
  Federal Support to Universities and

Colleges, Fiscal Year 1967, 106
  funds for, 15, 40, 110, 113, 184, 312, 329, 383, 395
  grants, 319
  personnel, 114, 121, 189
R&D Activities of Local Governments, 1966 and 1967, 97
Reviews of Data on Science and Resources, 398
Scientific Activities of Nonprofit Institutions, 1966, 96
Scientific and Technical Personnel in the Federal Government, 1967, 198
National Science Teachers Assn., 105
National Sea Grant Program, 9
National security, 35, 258, 306, 411, 419
National Security Council, 2
National Security Industrial Assn., 33, 178
National Seminar for Manned Flight Awareness, 317
National Society of Professional Engineers, 37
National Space Club, 16, 67, 68, 111, 268, 371
National Space Hall of Fame, 289
National Space Science Data Center, 1, 271
National Telemetry Conference, 117
National Transportation Safety Board, 19, 40, 362, 369
National Urban Coalition, 115, 116
NATO. See North Atlantic Treaty Organization.
Naugle, Dr. John E., 69, 84, 127, 139, 179, 336, 346
Nauman, Robert J., 326
Naval Missile Center, 325
Naval Research Laboratory, 22, 74, 78, 84, 111, 124, 171, 272, 298, 361, 366
  Atmosphere and Astrophysics Div., 2
Naval Weapons Center, 34, 41
Navigation satellite, 33, 37, 47, 190, 322, 357
NBC. See National Broadcasting Co.
NC-4 (flying boat), 133, 344
Near East, 140
Need for Improved Guidelines in Contracting for Research with Government-Sponsored Nonprofit Contractors, 47-48
Neilson, Thomas H., 124
Nekton (submarine), 330
Nellis AFB, Nev., 49, 155, 413, 416
Nelson, Bryce, 59-60, 163
Neptune (planet), 263, 299
NERVA. See Nuclear Engine for Rocket Vehicle Application.
Ness, Dr. Norman F., 50
Netherlands 32, 132, 423
Neugebauer, Dr. Gerry, 269
Neutrography, 168-169
Nevada, 291
Nevada Test Site, 416
New Delhi, India, 234, 255, 258

New Jersey, Div. of Clear Air and Water, 278
New Mexico, 315
New Mexico, Univ. of, 272
New Orleans, La., 96, 105, 298
New Orleans Chamber of Commerce, 140
New York, N.Y., 90-91, 129, 190, 232-233, 363
  Apollo 11 moon rock exhibit, 381
  astronauts in, 8, 276, 279, 387
  cosmonaut visit, 337
  air services and traffic, 13-14, 34, 43-44, 418
  meeting, 40, 387, 388
  train services, 17, 105
New York City Medal, 8
New York Society of Security Analysts, 29
New York State Supreme Court, 2
New York State Univ., 288
  Stony Brook, 407
New York Stock Exchange, 150, 238, 243
New York Univ., 121
Newark, N.J., 72
Newark, N.J., Airport, 9, 278
Newcomen Society in North America, 340
Newell, Dr. Homer E., Jr., 110, 111, 178
News conference. See Press conference.
Newton, Sir Isaac, 300
NFDC. See National Flight Data Center.
Niagara Falls International Airport, 103
NICAP. See National Investigations Committee on Aerial Phenomena.
Nigeria, 226
NIH. See National Institutes of Health.
Nike-Apache (sounding rocket)
  auroral data, 43, 48, 51-52, 54, 58
  cosmic radiation, 107
  electron measurement, 33-34, 113, 300, 303-304
  hydroxyl radical measurement, 189
  ionospheric experiments, 33-34, 41, 113, 135, 300
  micrometeoroid sampling, 159, 176, 279, 287
  upper-atmosphere data, 41, 49, 135, 286, 408-409
  x-ray astronomy, 118, 120, 402
Nike-Cajun (sounding rocket), 133
  upper-atmosphere data, 9, 18, 19, 20, 21, 23, 25, 27, 28, 30-31, 33-34, 38, 40-41, 94, 132, 133, 135
Nike-Iroquois (sounding rocket), 360
Nike-Javelin (sounding rocket), 360
Nike-Tomahawk (sounding rocket)
  contract, 253
  electron measurement, 71, 113
  instrumentation test, 27-28, 286
  upper atmosphere data, 29, 49, 72, 83, 286
  x-ray data, 134, 211-212, 327
Nikolayev, Adrian G., 158
Nimbus (program), 23, 139

Nimbus I (meteorological satellite), 108
Nimbus II, 18, 108
Nimbus III (Nimbus-B2), 85, 88, 107–108, 205, 353, 414, 421, 422
Nimbus D, 85
Nimbus E, 85, 423
Nimbus F, 85
Nippon Electric Co., Ltd., 82
Nitric oxide, 34, 41, 179, 365
Nitrogen, 139–140, 179
Nix Olympica (Mars crater), 262
Nixon, President Richard M., 4, 26, 56–57
  Air Force Academy address, 170
  Amistad Dam dedication, 299
  antiballistic missile (ABM), 39, 46–47, 80, 81, 83, 89, 121, 129
  Apollo 8 mission, 20, 32, 34, 61–62
  Apollo 9 mission, 64–65
  Apollo 10 mission, 157
  Apollo 11 mission, 196, 224, 249, 250, 261, 262, 275, 284, 401–402
    astronauts
      communications with, 209, 219–220, 230
      dining with, 200, 204
      state dinner for, 279–280
      tribute to, 228, 242, 246, 362
      U.S.S. *Hornet* greeting, 223
    messages to, 244
    moon plaque, 196
    National Day of Participation, 224, 230–231
  Apollo 12 mission, 371, 372, 378, 384, 385, 389, 392
  appointments and nominations by, 6, 9, 17–18, 30, 46, 68, 117, 121, 141, 184, 189, 302, 314, 329, 332, 392, 402, 418
  arms limitation, 183, 190–191
  Asian tour, 199, 247, 248, 251, 255
  astronaut goodwill tour, 32, 299, 319
  astronauts, visit with, 191, 413
  award to, 410
  awards by, 68, 101, 418–419
  bills signed, 383, 395
  budget, 107, 109–110, 138, 186–187
  ComSatCorp report to, 61
  Congressional Space Medal of Honor approved, 319
  European tour, 50, 55, 56–57, 61–62
  inauguration, 18, 20–21
  international cooperation, space, 20, 32, 43, 238, 309–310
  MIRV missile, 190–191
  nuclear nonproliferation treaty, 41, 392–393
  office performance, 275
  science, 40, 54, 56, 297
  space program, national, 9, 20, 32, 38, 43, 49–50, 78, 80, 82, 95, 104, 108–109, 110, 115, 117, 119–120, 238, 271, 371, 372, 423
  Space Task Force Group report to, 167, 304–305, 308, 312–313, 332

  supersonic transport, 32, 43, 58–59, 82, 95, 123, 137, 314–315, 318, 372, 410, 422
  task forces, 314, 329, 332
  U.N. address, 309–310
  Vietnam War, 404
  White House religious service, 232
  world tour, 247, 251, 259, 261, 262, 284
  Wright Brothers Day, 406
Nixon, Mrs. Richard M., 261, 279, 284, 372, 378
Nixon, Miss Tricia, 378
NOA: new obligational authority (in budget)
Nobel, Alfred B., Prize in Chemistry, 355
Nobel, Alfred B., Prize in Physics, 355
Nobel, Alfred B., Prize in Physiology or Medicine, 338
Noblitt, B. G., 354
Noise, aircraft, 3, 15, 25, 32, 46, 78, 120, 173, 212, 263, 265, 371, 391, 400
NOMSS. See National Operational Satellite System.
NORAD. See North American Air Defense Command.
Noren, Rev. Paul H. A., 206–207
Normyle, William J., 108–109, 115
Norris, Henry W., 111
North American Air Defense Command (NORAD), 89, 291
North American Rockwell Corp. (NAR)
  Aerospace and Systems Group, 195
  aircraft, 11, 85, 169, 177
  Apollo spacecraft, 59, 388
  Atomics International Div., 195
  Autonetics Div., 135
  award, 69, 147
  contract, 47, 85, 94, 132, 135, 270, 280
  escape-to-orbit vehicle, 94
  flying lunar excursion experimental platform (FLEEP), 388
  Power Systems Div., 195
  Rocketdyne Div., 70, 79, 85, 101, 104, 195, 327–328
  Saturn V, 24, 132, 280
  Space Div., 70, 77, 241
  space shuttle, 47
  space station, 270, 299
North Atlantic Treaty Organization (NATO), 132
North Pole, 251
Northern Illinois Univ., 360
Northrop Corp., 287, 319, 325, 346
Northrop, John K., 157–158, 287, 346
Northwestern Univ., 288, 416
  Astronomy Dept., 28
Norton, W. W. & Co., 252
Norway, 18, 20, 132, 323, 423
Nossiter, Bernard D., 107, 135
Nova Scotia, Canada, 282
Novak, Robert D. S., 284
Novikov, Kirill, 21
Novosibirsk, U.S.S.R., 195, 387
Noyes, Crosby S., 154–155, 177

NR-1 (nuclear-powered deep submergence research vehicle), 27
NRC. See National Research Council.
NSF. See National Science Foundation.
Nuclear Engine for Rocket Vehicle Application (NERVA), 105, 266, 332, 399
  contract, 94
  funds for, 16, 66, 109, 120, 163
  NERVA I, 94
  NERVA XE, 177–178, 279, 309, 422
  test, 87, 177–178, 279, 309, 422
Nuclear explosion, 167, 409, 412–413
Nuclear fallout, 412–413
Nuclear generator, 28, 130, 177–178, 189, 190, 329–330, 414, 422
Nuclear nonproliferation treaty, 13, 41, 52, 57, 58, 78, 317, 370, 392–393
Nuclear power, 4, 37, 130, 189, 354
Nuclear propulsion, 37, 66, 76, 87, 94, 130, 156, 200, 260, 280, 332, 337–338, 399, 404, 422
Nuclear test, 155, 181, 359
Nuclear test ban treaty, 13, 81
Nuclear weapons (see also Disarmament; Missile; and Treaty), 81, 86, 131, 135, 136, 155, 181, 330

O

Oak Ridge National Laboratory, 45, 353
OAO. See Orbiting Astronomical Observatory.
Oao II (Orbiting Astronomical Observatory), 134
  awards for contributions to, 154
  disorders, 106, 173
  experimental data from, 3, 99, 344, 403
OART. See NASA Office of Advanced Research and Technology.
Oberth, Prof. Hermann, 188, 287, 346
Oberth, Hermann, Society of Nuremberg, 188
Ocean of Storms (moon), 243, 368, 375, 393, 396, 403, 407–408
Oceanography (see also Aquanaut; Project Tektite I and II; and Sealab)
  award, 50
  contract, 353
  cooperation, 51, 88, 356
  international aspects, 133, 197
  manned flight contributions to, 160
  record, 86, 110
  research, 17, 27, 51, 88, 103, 110, 209, 273, 282, 330, 343
  satellite use in, 88, 199
  U.S. program, 9, 56, 133, 150, 314, 341, 353, 356
O'Connell, Joseph J., Jr., 40
O'Connor, M/G Edmund F. (USAF), 188, 251
Odeillo, France, 363
OECD. See Office for Economic Cooperation and Development.

Office for Economic Cooperation and Development (OECD), 255
  Directorate for Scientific Affairs, 59–60
Office of Economic Opportunity, 339
Office of Science and Technology (President's), 42, 43, 106, 119, 174
OFO. See Orbiting Frog Otolith.
OGO. See Orbiting Geophysical Observatory.
Ogo I, 171
Ogo III, 171
Ogo IV, 171
Ogo V, 171, 316
Ogo VI (OGO–F), 171, 326, 421
O'Hair, Mrs. Madalyn Murray, 268, 283, 394, 400
O'Hair, Richard, 268
O'Hare International Airport, 28, 102, 263–264, 279
Ohio, 406
Ohio Historical Society, 246
Ohio State Univ., 156
Ojai, Calif., 200
Oke, Dr. John B., 367
O'Keefe, Dr. John A., 50, 360
O'Keefe, William J., 263
Olav V. King (Norway), 332
O'Leary, Dr. Brian T., 180, 339
OMSF. See NASA Office of Manned Space Flight.
On the Edge of the Moon (London *Times* supplement), 169–170
O'Neill, L/G John W. (USAF), 322
Operation Breakthrough (HUD), 188
"Operation Paperclip," 13
Operations Research Society of America, 183
Orbiting Astronomical Observatory (OAO), 3, 23, 106, 154, 266, 344, 367, 487
Orbiting Frog Otolith (OFO) (spacecraft), 154
Orbiting Geophysical Observatory (OGO), 171, 173, 364
Orbiting Solar Observatory (OSO), 23
Orbiting Vehicle (OV) (research satellite), 83–84, 155, 164
Order of Lenin Medal, 24
Order of Leopold, 332
Ordway, Frederick I., III, 329
Orion (constellation), 337, 344
Orion II (sounding rocket), 79
Orlando, Fla., 271
Ortoli, Francois X., 316
Osaka, Japan, 417
Oscar, Project, 399
Osgood, John, 52
Oslo, Norway, 312, 355
Osman, Armstrong Abdurahman, 234
OSO. See Orbiting Solar Observatory.
Oso I (Orbiting Solar Observatory), 22–23, 272
Oso II, 22–23, 272
Oso III, 23, 272

Oso IV, 23, 272
Oso V (OSO-F), 22, 23, 28, 56, 108, 272, 421
Oso VI (OSO-G), 272, 308, 370, 411, 421
OSO-C, 272
OSSA. See NASA Office of Space Science and Applications.
Ostriker, Dr. Jeremiah P., 322, 334
Oteroo, Katherine Stinson, 380
O'Toole, Thomas, 46, 62, 92, 95, 359
Otopeni Airport, Romania, 261
Ottawa, Canada, 312, 399
Our Nation and the Sea (report), 9
Outer Solar System: A Program for Exploration (Space Science Board report), 263
Outstanding Leadership Award (NASA), 371
OV 1-16 (orbiting vehicle research satellite), 164
OV 1-17, 83-84
OV 1-17A, 83
OV 1-18, 83
OV 1-19, 83
OV V-5, 155
OV V-6, 155
OV V-9, 155
Overmeyer, Maj. Robert F. (USMC), 281
Owen, David, 381
Owen, Tobias, 293
Owens, Miss Heather A., 266-267
Oxford, Pa., 168
Oxford Univ., 239

P

Pac (Package Attitude Control) system, 272
Pace, Frank, Jr., 387
Pacemaker (sounding rocket), 203
Pacific Ocean
  Apollo 10, 143
  Apollo 11, 168, 223
  Apollo 12, 377
  Biosatellite III, 201
  communications satellite, 18, 137, 140
  Deep Sea Drilling Project, 353
  Sealab III project, 51
  U.S.S.R. rocket test, 112, 116, 156
Packer, Dr. Leo S., 83
Pago Pago, 378
Paine Field, Wash., 45
Paine, Dr. Thomas O., 274, 306, 349, 368
  aeronautics, 66
  Apollo 11 mission, 199-200, 209-210, 223, 231-232, 279, 380-381
  appointment, 46, 68, 87, 100, 401
  appointments by, 70
  awards and honors, 168, 300
  awards by, 6, 32, 132, 154, 280, 326
  budget, 15-16, 66, 109-110, 119-120
  Deep Space Network facility, Madrid, 180
  Electronics Research Center (ERC), 417, 423
  international cooperation, space, 74-75, 169, 268, 311, 411-412, 424
  legal suit, 268, 283, 394, 400
  management, 148-149
  space program, national, 15-16, 66, 68, 74-75, 100, 119-120, 145-146, 169, 175-176, 192, 199-200, 207, 259-260, 266, 268, 294, 417, 424
  space station, 169
  U.S.S.R. space program, 120, 145-146, 169, 199-200, 266
  space station, 75, 120
Pakistan, 190, 244, 259, 394, 423
Pakistan Space and Upper Atmosphere Research Committee, 99
Palm Springs, Calif., 359
Palmer, John S., 279
Palo Alto, Calif., 257
Palomar, Calif., 134
Palos Verdes, Calif., 54
Pan American World Airways, Inc. (Pan Am), 81, 175, 241, 346, 400, 409
Panofsky, Wolfgang K. H., 418-419
Parachute, 143, 183
PARD. See Pilot Airborne Recovery Device.
Parin, Dr. Vassily V., 158, 391
Paris, France, 159, 197, 244, 290, 312, 316
  astronaut visit, 37, 41
  award, 237, 403
  Nixon, President Richard M., visit, 55, 61-62
Paris Air Show, 161-162, 168, 173, 412
  U.S. exhibit, 105, 161-162, 168, 169, 175
  U.S.S.R. exhibit, 146-147, 159-160, 162, 168
Paris, Univ. of, 22, 316
Park, Chauncey C., 266-267
Park, President Chung Hee (South Korea), 244
Park, William C., 73
Parker, Dr. Eugene N., 121
Parker, Jack S., 147
Parkes, Australia, 137, 319
Pasadena, Calif., 158, 261
Pasadena Chamber of Commerce, 158
Patent, 4, 82, 159, 239
Patterson, William A., 380
Paul VI, Pope
  Apollo 10 flight, 152
  Apollo 11 flight, 206, 225, 232, 242, 244
  Apollo 12 flight, 384
  astronauts visit, 51, 337
  Nixon, President Richard M., visit, 61
Paulet, Pedro E., 329
Paulet Mostajo, Pedro, 322
Paulson, Jeanne, 385
Paumalu, Hawaii, 179
PCA. See Polar cap absorption.
Peace Corps, 413

Peace Corps National Advisory Council, 402
Pearce, J. B., 298
Pearl Harbor, Hawaii, 396
Pearson, Drew, 204
Peary, Adm. Robert E. (USN), 101, 251
Pecora, William T., 393
Pegasus III (meteoroid detection satellite), 64, 263
Pell, Sen. Claiborne, 22
Penn Central Co., 17, 105
Pennsylvania, Univ. of, 67
Pentagon, 34, 412
Perez-Marin, Gen. Antonio (Spain), 180
Perl, Dr. Martin L., 124
Perry Oceanographics, Inc., 343
Perseid meteor shower, 279, 287
Peru, 267, 322
PET. See NASA Preliminary Examination Team.
PET Summary of Apollo 11 Lunar Samples, 306
Petersburg, Alaska, 352
Peterson, Maj. Donald H. (USAF), 281
Petrone, Rocco A., 10, 274, 297, 317, 346
Petrosyants, Andronik M., 359
Petrov, Dr. Boris N., 418
Petrov, Prof. Georgy I., 233
Phantom F-4E (jet fighter), 9
Phelps, Robert H., 82
Philadelphia, Pa., 283, 385
Philco-Ford Corp., 402
 Education and Technical Services Div., 87-88
 Space and Re-Entry Systems Div., 346
Philippines, 190, 200, 248
Phillips, Mrs. Mary, 315
Phillips, L/G Samuel C. (USAF), 274, 297
 Apollo 8, 16
 Apollo 11, 178, 182, 243
 awards and honors, 10, 68, 87, 300, 346-347
 space shuttle, 336
Phoenix, Ariz., 327
Physics, 77, 160, 256-257, 355
Physics of the Earth in Space: The Role of Ground-Based Research (NRC report), 256-257
Physiology in the Space Environment (Space Science Board report), 18-19
Piccard, Dr. Jacques, 103, 209, 282
Pickering, Dr. William H., 184-185, 188, 203, 248
Pilot Airborne Recovery Device (PARD), 239
Pilot warning instrument (PWI), systems, 388
Pilots, 425
Pimentel, Dr. George C., 269, 301-302
Pin Main, Canada, 71
Pioneer (interplanetary probe), 89, 360
Pioneer VI, 291, 356, 395
Pioneer VII, 291, 395
Pioneer VIII, 180, 291
Pioneer IX, 159, 291

Pioneer E, 23, 291, 385, 421
Pioneer F, 76, 361
Pioneer G, 76, 361
Piret, Edgar L., 403
Pitcairn, Harold F., 283
Plamondon, Joseph A., 133
Plan for U.S. Participation in the Global Atmospheric Research Program (NRC report), 204-205
Planetary Explorer (program), 163
Planetoid, 288
Plate, Thomas G., 231
Plateau Elysium (Mars), 262
Platt, John, 396
Plesetsk, U.S.S.R.
 Cosmos launch, 67, 70, 80-81, 82, 91, 101, 137, 148, 181, 203, 237, 281, 283, 285, 297, 339, 341, 370, 392, 401, 405
 Meteor I launch, 91
 Meteor II launch, 329
Plum Brook Station, Ohio, 404, 412
Plummer, William T., 396
Pluto (planet), 192, 263, 299, 366
Plutonium, 135, 384
Plymouth, England, 133
pndb: perceived noise in decibels
Podgorny, President Nikolay V. (U.S.S.R.), 24, 210, 211, 244, 251, 392-393
Point Barrow, Alaska, 9, 19, 30, 38
Point Mugu, Calif., 325
Poland, 210, 225, 233, 244, 251, 384
Polar cap absorption (PCA), 359-360
Polish Astronautical Society, 329
Pompidou, President Georges (France), 330
Poor People's Campaign, 205-206, 209-210
Pope AFB, N.C., 283
Porter, Daniel R., 246
Porter, Dr. Richard W., 136-137
Portland, Ore., 385
Portugal, 32, 37, 413
Poseidon (missile), 363
The Post-Apollo Space Program: An AIAA View, 151-152
The Post-Apollo Space Program: Directions for the Future (Space Task Group report), 304-305
Post Office Dept., 8, 129-130, 202, 273-274, 300, 311, 319
Potsdam, N.Y., 318
Prague, Czechoslovakia, 413
Pratt, Perry W., 21
Pratt & Whitney Div., United Aircraft Corp., 25, 32-33, 46, 304, 328
Presidential Task Force on Science Policy, 314, 329, 346
President's Council on Youth Opportunity, 135, 170
President's Science Advisory Committee, 42, 73-74, 117, 151, 184, 314
President's Space Task Group, 78, 131, 134, 167, 173, 192, 320

The Post-Apollo Space Program: Directions for the Future (report), 304–305, 308, 312–313, 332
Press, Dr. Frank, 230
Press comment
  antiballistic missile (ABM) system, 46–47, 129
  Apollo 8 mission, 1, 2, 26
  Apollo 9 mission, 67, 72, 80, 90
  Apollo 10 mission, 147–148, 149–150, 152–154, 155, 157
  Apollo 11 mission
    foreign, 193, 210–211, 225–226, 227–228, 229, 231, 236, 240, 242, 247, 248, 251–252, 321–322
    moon plaque, 196–197
    U.S., 157, 174, 192–193, 207, 208, 210–211, 225–226, 227, 228–229, 231, 235–236, 239–240, 244, 246–247, 249–250, 261–262, 271–272, 276, 279, 280–281, 282, 295, 308, 357
  Apollo 12 mission, 367, 379, 380, 386–387, 391, 392, 393, 394
  astronaut ceremonies, 279, 282
  Blue Book, Project, 413–414
  C–5A (cargo aircraft), 293
  communications satellite, 26, 394–395
  disarmament, 190–191
  docking (*Soyuz IV* and *Soyuz V*), 19
  international cooperation, space, 26, 43, 175, 248, 367, 383–384
  lunar exploration, 152–153, 154, 313, 383–384
  lunar rock samples, 261–262, 313, 412
  Mariner VI, 256, 262, 271, 308
  Mariner VII, 256, 262, 271, 308
  Mars, manned flight to, 231, 262, 271
  MOL, 179, 181, 186, 289
  moon
    contamination from, 175, 176
    international research, 4–5
    nuclear blast on, 412–413
  NASA Administrator, 46
  objects in orbit, 149–150
  pulsars, 134–135
  Science, Secretary of (proposed), 3
  science and technology, 1, 2–3, 154, 162–163, 321, 396–397, 400–401
  space biology, 391
  space program, national, 4–5, 43, 80, 81–82, 95, 119, 154–155, 161, 163, 248, 271–272, 289, 309, 313, 321, 359, 419–421
  space results, 203, 227, 228, 242, 256
  space station, 80, 391
  supersonic transport (SST), 137, 318, 413
  UFO's, 8–9, 20, 413–414
  U.S.S.R. space program, 336, 342, 343, 351, 365–366
Press conference
  aircraft, 128–129
  antiballistic missile (ABM) system, 33, 43
  Apollo 8 mission, 7, 25, 54
  Apollo 9 mission, 26–27, 33, 61, 77, 91, 116–117
  Apollo 10 mission, 33, 108
  Apollo 11 mission, 108, 182, 198–199, 204, 207, 231, 242–243, 267–268, 275, 277–278, 280–281, 407–408
  Apollo 12 mission, 243, 325–326, 333–334, 342, 371–372, 392, 406–408
  Apollo 13 mission, 392
  astronaut. See Astronaut.
  Biosatellite III, 347–349
  cosmonaut, 343, 350, 372
  Defense, Dept. of, 4, 33, 99, 128–129
  disarmament, 21, 174
  earth resources program, 116–117
  Haney, Paul, 116
  international cooperation, space, 25, 41, 102, 158, 169, 196, 238, 245, 325
  manned space flight, 29, 53–54
  Mariner VI, 261, 269–270, 301–302
  Mariner VII, 269–270, 301–302
  Mars, manned flight to, 224, 243, 283–284, 305
  moon
    lunar surface samples, 250, 267–268, 275, 396, 407–408
    nuclear explosion on, 167
  Nixon, President Richard M., 43
  Oao II, 129
  President's Space Task Group report, 308
  space failure, 262
  space program, national, 49–50, 54, 77, 122, 148, 158–159, 170–171, 224, 231–232, 242–243, 248, 305, 323–325, 417
  space station, 169
  supersonic transport (SST), 81, 186, 197
  Surveyor III, 369–370
  UFOs, 8
  U.S.S.R. space program, 29, 170–171, 204, 246, 343, 350, 361
  women as astronauts, 359
Press, Dr. Frank, 230
Preston, Lancashire, U.K., 161
Price, Don K., 136
Pride, Inc., 230–231
Princeton Univ., 127, 184, 290, 322, 334, 404, 418
  Aerospace Systems Laboratory, 208–209
Prix Pierre Guzman, 403
Probe (see also individual probes, such as *Mariner VI*, *Mariner VII*, *Venus V*, and *Venus VI*), 15, 66, 89
  interplanetary, 66, 76, 136–137, 187, 192, 366–367, 421
  Jupiter, 16, 76, 192, 263
  lunar. See *Luna XV*, *Lunar Orbiter I*, *Zond V*, *Zond VI*, and *Zond VII*.
  Mars, 16, 18, 57, 76, 95, 111, 119, 136–137, 224, 231, 266, 343
  Mercury, 16, 69, 76, 343

Neptune, 192, 263
Pluto, 192
Saturn, 192, 263
sun, 37, 177, 357
Uranus, 192, 263
U.S.S.R., 42, 75, 120, 136–137, 140, 141–142, 160, 171
Venus, 16, 69, 76, 136–137, 318, 343
"Progress, Coexistence, and Intellectual Freedom" (essay), 163
Propulsion Joint Specialist Conference, Fifth, 134
Proton, 171, 417
Proxmire, Sen. William, 130, 281, 309, 311, 330, 369, 383, 412
Public Service Award (NASA), 92, 347
Pueblito de Allende, Mexico, 45
Puerto Rico, 121, 401–402
Pulkovo Observatory, 139–140
Pulsar, 16–17, 42, 78, 121–122, 134–135, 289, 322
  acceleration, 56, 121, 334, 404
  optical, 164
Purcell, Joseph, 99, 129, 154
Purdue Univ., 418
Putnam, William D., 353
PWI. See Pilot warning instrument systems.

Q

Qantas Airlines, 239
Qatron Corp., 297
Quamme, Hal J., 80
Quarantine of astronauts
  Apollo 11 mission, 75–76, 141, 223, 273, 276
  Apollo 12 mission, 356, 377–378, 396, 397, 400, 405
Quark (theoretical elementary particle of matter), 355
Quasar (quasi-stellar object), 99, 367
Quiet Engine Research Program, 25, 32–33, 212

R

Rabi, Dr. Isidor I., 39
Radar, 156
  aircraft, 156, 359
  mapping use, 288
  sidelooking, 47
  tracking, 55, 64, 65
  U.S.S.R., 50, 62
Radiation (see also Ultraviolet and X-ray)
  cosmic, 107, 155, 184, 318–319, 403
  effects, 78, 83, 96, 136, 164
  gamma, 318–319
  lunar, 313, 317
  measurement, 83, 84, 175, 187
  solar, 83, 107, 133, 187, 269
  trapped, 84, 173

Radio Amateur Satellite Corp., 399
Radio Corporation of America (RCA), 88, 113, 388, 400, 401
  Astro-Electronics Div., 139, 401
Radio signal, 115, 367
Radioastronomy, 123, 181, 390
Radioisotopes, 132
Radiometer, 175, 254, 344
Radiophysics, 161
Radiotelescope, 42
Radome, 331
R&D. See Research and development.
R&D Activities of Local Governments, Fiscal Years 1966 and 1967 (NSF 69–14), 97
R&D in the Aircraft and Missiles Industry, 1957–68 (NSF 69–15), 164
RAND Corp., 282, 353
Randall, Joseph L., 326
Randall, Judith, 1
Raper, O. F., 298
Rapid eye movements (REM), 348
Raska, Dr. Karel, 254
Rasool, Dr. S. Ichtiaque, 127–128, 408
Raspevin, K., 1
Rat experiment, 139
Rathjens, George W., 131
Rauschenberg, Robert, 226, 403
Ray, Thomas W., 353
Raymond, John M., Jr., 180
RCA. See Radio Corporation of America.
RCA Service Co., 113, 292
RCS. See Reaction control system.
RDT&E (research, development, test, and engineering). See Research and development.
Reaction control system (RCS), 64, 143, 213, 222, 375
Real-time computer complex (RTCC), 334
Ream, Harold E., 103
Recher, Marcel, 304
Rechtin, Dr. Eberhardt, 346
Reconnaissance satellite, 192
Record
  aircraft, 79, 129, 152, 157, 177, 323
  oceanographic, 86, 110
  spacecraft, 208, 250–251, 378, 389
Redstone Arsenal, Ala., 256
Reentry vehicle, 192, 330, 389, 408
Rees, Dr. Eberhard F. M., 10, 188, 326
Regency Corp., 75
Regulus (star), 79
Reichley, Paul, 56
Relativity, theory of, 71
Relay I (satellite), 124
Relay II, 124
Religious Society of Friends, 232
REM. See Rapid eye movements.
Rendezvous
  U.S., 5, 27, 64, 91, 198, 249, 256, 345, 377
  U.S.S.R., 11, 12, 120, 332, 333, 422
Republic Steel Corp., 408
Request for proposals (RFP), 120, 128, 151, 205

Rescue of astronaut. See Space rescue.
Research and development (R&D), 4, 151, 175, 255
  aeronautics, 15–16, 113, 148, 164, 193, 335
  benefits, 40, 75, 128, 335
  computer, 139
  employment, 96, 97, 150, 231
  Federal support, 3, 29, 35, 40, 42, 257, 297
  funds for, 14–16, 96–97, 110, 164, 257
    DOD, 14, 15, 40, 124, 136, 330
    FAA, 15
    NASA, 14–15, 47, 108–109, 113, 138, 176, 177, 184, 193, 257, 312, 330
  U.S.S.R., 60
Research and Development in Industry, 1967: Funds, 1967; Scientists and Engineers, January 1968 (NSF 69–28), 257
Resolute Bay, Canada, 28, 331
Resor, Secretary of the Army Stanley R., 4, 148
Reusable launch and space vehicles, 25, 29, 47, 82, 254, 260, 266, 268, 345, 354, 423
Reviews of Data on Science and Resources (NSF 69–36), 398
Reynolds, Orr E., 78
RFP. See Request for proposals.
Rhodes, Gov. James A., 298
Rice Univ., 58, 76, 253, 267
Richard, Ludie G., 326, 360
Richardson, Robert, 50
Rickover, V/A Hyman G. (USN), 132
Rindner, Dr. Wilhelm, 128
Rio de Janeiro, Brazil, 312
  Museum of Modern Art, 234
Ripley, Dr. S. Dillon, 307
Rivers, Rep. L. Mendel, 39
Riviera Beach, Fla., 343
RL–10 A3–3 (rocket engine), 379
Roberts, Judge Jack, 400
Roberts, Dr. Walter Orr, 44, 167, 416
Rockaway Beach, N.Y., 133
Rockefeller Public Service Award, 419
Rockefeller Univ., 1, 19, 314
Rocket belt, 161
Rockwell, Norman, 403
Rocky Flats, Colo., 135, 384
Rogers, Alan E. E., 289
Rogers, Secretary of State William P., 52, 174
Rolls-Royce, Ltd., Bristol-Siddeley Engine Div., 21
Rom, Frank E., 156
Romania, 46, 234, 249, 251, 261, 262, 284
Rome, Italy, 37, 55, 298, 312, 341
Romney, Secretary of Housing and Urban Development George W., 87, 237
Roosa, Maj. Stuart A. (USAF), 267
Rosamond, Calif., HL–10 (lifting-body vehicle) flight, 184, 240, 267, 297, 309, 352, 360, 381, 406
Rosman, N.C., 402
Rossi, Dr. Bruno, 89
Rossini, Mayor Pascal, 244
Royal Crown Cola International, 43
Royal Geographical Society, 251
Royal Navy (U.K.), 129
Royal Society of Scientists, 38
Royer de Vericourt, Mayor Etienne, 330
RTCC. See Real-time computer complex.
Rubey, William C., 76
Rubin, Irene S., 286
Rudolph, Dr. Arthur, 10, 13, 182
Rundell, Walter, Jr., 357
Rusk, Dean, 285, 287
Rusk, Dr. Howard A., 274
Russell, Lord Bertrand A. W., 210
Rutgers Univ., 272
Ryan, Gen. John G. (USAF), 180
Ryan, Rep. William F., 124
Ryle, Sir Martin, 42

S

S–3A (antisubmarine aircraft), 15, 131, 260
S–IC. See Saturn V (booster), stage, 1st.
S–II. See Saturn V (booster), stage, 2nd.
S–IVB. See Saturn V (booster), stage, 3rd.
Saavedra, Albert, 202
Sabin, Dr. Albert, 298
SAC. See Strategic Air Command.
Sacks, Martin, 393
Sacramento Peak Observatory, 336, 419
Safeguard (formerly Sentinel) (antiballistic missile system), 33, 43, 83, 114, 320
  congressional consideration, 39, 46–47, 52, 53, 103, 131, 229
  name change, 89
  Nixon, President Richard M., views on, 43, 80, 81
  press comment, 46–47, 81, 129
Sagan, Dr. Carl, 180
Sagittarius (constellation), 318–319, 326
Saigon, South Vietnam, 254, 414
St. Alban's School, Washington, D.C., 132
St. John, Virgin Islands, 51, 110
St. Peter's Basilica, 152
Sakharov, Prof. Andrey D., 163
Salon Internationale de l'Aéronautique et de l'Espace. See Paris Air Show.
Salzburg, Austria, 188
SAMSO. See USAF Space and Missile Systems Organization.
Samuelson, Robert J., 101, 352
San Clemente Island, Calif., 51, 189
San Diego, Calif., 57, 225, 273, 275, 330, 331, 352
San Francisco, Calif., 130, 211, 225, 259–260, 354, 409
  Board of Supervisors, 343

San Marco-C (Italian satellite), 286, 423
Sanders Associates, Inc., 91
Santa Monica, Calif., 20
Santa Susana, Calif., 28
Santiago, Chile, 106, 234
Sarabhai, Dr. Vikram A., 310
Saragat, President Giuseppe (Italy), 244
Sarnoff, Robert W., 388, 400–401
SAS. See Stability augmentation system and Small Astronomy Satellite.
SATCOM. See Committee on Scientific and Technical Communication.
Satellite infrared spectrometer (SIRS), 353
Sato, Prime Minister Eisaku (Japan), 244, 361–362, 389
Saturn (planet), 192, 337–338
 rings, 359, 397
Saturn I Workshop (spacecraft), 15, 16, 17, 104, 114
Saturn IB (booster), contract, 25, 49, 94, 101, 113, 154, 315
Saturn V (booster), 228, 275, 317
 capability, 96, 170–171, 384
 contract, 25, 47, 70, 77, 87, 91, 113, 132, 278, 280, 297, 315, 388
 engine
  F-1, 105, 185
  J-2, 105, 185
 exhibit, 105, 417
 launch
  AS-504, 62–65, 67
  AS-505, 16, 142–145
  AS-506, 108, 168, 195, 196, 212–224
  AS-507, 299, 372–378
  AS-508, 182
 program, 13, 67, 69, 76–77, 109–110, 119–120, 134, 138, 163–164, 251, 422
 stage
  1st (S-IC), 182, 188, 297
   test, 118, 188, 195, 360
  2nd (S-II), 70, 77, 132, 182
   test, 24, 104, 185
  3rd (S-IVB), 89, 105, 117, 128, 142, 182, 187, 197, 202–203, 280, 370, 373–374, 388
Saturn V Workshop (spacecraft), 237–238, 270–271, 297, 326–328, 388, 417
Saunders, Hal, 379
Saunders, Stuart T., 17
Savannah, Ga., 233
Savchenko, Boris, 152
Saxbe, Sen. William B., 13
Schaeffer, Dr. Oliver A., 288, 292, 407
Scheel, Walter, 384
Scheer, Julian W., 7, 116, 118, 347, 410
Scherer, Capt. Lee R. (USN, Ret.), 74
Schiller, Karl, 162
Schirra, Walter M., Jr., 21, 46, 72–73, 75, 173
Schmitt, Harrison H., 286
Schneider, William C., 10, 290
Schorn, Dr. Ronald, 90
Schriever, Gen. Bernard A. (USAF, Ret.), 188
Scheutz, Mayor Klaus (W. Germany), 334
Schwartz, Harry, 284
Schweickart, Russell L., 5, 26, 62–65, 81, 91, 161–162, 168
Sciacca, Prof. Michele Federico, 239
Science (see also National Academy of Sciences), 11, 65, 122, 148–149
 award, 1–2
 benefits, 68, 162–163, 321
 Government support of, 28, 29, 42, 359, 396–397
 human needs, 359
 national policy and goals, 2–3, 40, 52–53, 67, 132, 161, 314–315
 Presidential Task Force on Science Policy, 314, 329, 346
 President's Science Advisory Committee, 42, 73, 117, 151–152, 184, 314
 U.S.S.R., 59–60, 250
Science Policy in the USSR (OECD study), 59–60
Science Research Council (SRC) (U.K.), 18, 20, 24, 27, 59, 312
Science, Secretary of (proposed), 2–3
Science, Technology, and Public Policy During the Ninetieth Congress (House report), 255
Scientific Activities on Nonprofit Institutions, 1966 (NSF 69–16), 96–97
Scientific and Technical Advisory Committee (STAC) (University-NASA), 10
Scientific and Technical Personnel in the Federal Government, 1967 (NSF 69–26), 193
Scientific instrument module (SIM), 331
Scientific Study of Unidentified Flying Objects, 5
Scientific Uses of the Large Space Telescope (Space Science Board report), 342–343
Scientist-astronaut, 284, 286, 290, 339, 351, 356, 370
Scientists, 40, 52–53, 96–97, 124, 178–179, 193, 257, 273, 284, 287, 316, 351, 356, 398, 420
Scientists, Engineers, and Physicians from Abroad, Fiscal Years 1966 and 1967 (NSF 69–10), 178–179
Scientists for Social and Political Action, 124
SCLC. See Southern Christian Leadership Conference.
Scorpio X-1 (star), 133–134
Scorpius (constellation), 65–66, 184, 281–282, 326, 327, 344
Scott, Col. David R. (USAF), 5, 26–27, 62–65, 91, 105, 161–162, 168
Scott, Sen. Hugh D., 29
Scott Polar Research Institute, 359
Scout (booster), 59, 86, 291, 323, 364–365, 423
Scranton, William W., 104, 418
Scripps Institution of Oceanography, 273
Scull, Wilfred E., 290

Sea of Crises (moon), 237
Sea of Okhotsk, 175
Sea of Tranquility (moon), 159, 168, 215, 221–222, 237, 249, 271, 280, 301, 306, 407–408
Sea Sciences Corp., 354
Seaborg, Dr. Glenn T., 30, 184, 354
Sealab III (underwater laboratory), 22, 51, 57, 77, 188–189, 316–317
Seamans, Secretary of the Air Force Robert C., Jr., 167, 192, 300, 308, 414
  AIAA presidency, 21, 79
  appointment, 4
  awards and honors, 287, 346, 347
  Blue Book, Project, 411
  C–5A, 128
  military space program, 13, 21, 320, 321, 345
  MOL, 21, 87, 104
  space program, national, 21, 60, 75, 87, 104
  supersonic bomber, 111–112, 183, 321
  U.S.S.R. missile and space program, 21, 140
  X–15, 177, 321
Seaplane bases, 176
Seattle, Wash., 45, 330, 400, 409
Seattle-Tacoma Airport, Wash., 169
Secor XIII (*Egrs XIII*) (Sequential Collation of Range satellite), 107, 422
Securities and Exchange Commission (SEC), 132
Sedov, Prof. Leonid I., 241–242
Seidel, Boris L., 408
Seim, Sandra E., 160
Seismometer experiment, lunar
  Apollo 11 mission, 114, 223, 229–230, 237, 240–241, 249, 354–355, 419
  Apollo 12 mission, 326, 342, 376, 403
Seitz, Dr. Frederick S., 8, 19, 154, 314
Selfridge, Robert P., 410
Self-testing and repairing (STAR) computer, 298–299
Semiconductors, 128
Semple, Robert B., Jr., 284
Sentinel (antiballistic missile system). See Safeguard.
Seoul, S. Korea, 312, 360
Serenitatis (moon), 180
SERT I (Space Electric Rocket Test), 48
SERT II, 48, 342
Serv-Air, Inc., 59
Service module (SM), 7, 142
Service propulsion system (SPS), 64, 142, 143, 212–213, 375
Seversky, Maj. Alexander de, 184, 412
Shabad, Theodore, 39
Shaffer, John H., 58, 156, 382–383
Shannon, Dr. James V., 3
SHAPE (Supersonic High Altitude Parachute Experiment) Project, 183
Shapley, Willis H., 347
Sharp Crater (moon), 376
Sharp, Dr. Robert P., 265, 282, 301

Sharpe, Mitchell R., 69
Shatalov, Vladimir A., 11–12, 23, 158, 159–160, 168, 333
Shawbury, U.K., 317
Sheldon, Dr. Charles S., II, 95–96
Shepard, Capt. Alan B., Jr. (USN), 132, 267, 289
Sheremetyevo Airport, U.S.S.R., 152, 195
Shoes, jet, 4
Shonin, Georgy S., 332
Shriver, Ambassador R. Sargent, 41, 161–162
Shub, Anatole, 33
Sidewinder-Arcas (sounding rocket), 79, 360
Sidey, Hugh, 174
Siegel, Dr. Peter V., 180
Sigma Xi Convention, 359
Significant Achievements in Space Science, 1967 (SP–167), 160
Sikorsky, Igor I., 150
Silverstein, Dr. Abe, 349
SIM. See Scientific instrument module.
Simat, Helliesen, & Eichner, Inc., 113
Simmons, Dr. Gene, 347, 351
Simoneit, B. R., 365
Singapore, 234
Singer, Dr. S. Fred, 416
Singlinger, A., & Co., 177
SIRS. See Satellite infrared spectrometer.
Sjoberg, Sigurd A., 417
Sjogren, William L., 116
Skinner, Sherrod E., 304
Skoog, A. Ingemar, 329
Skynet A (IDCSP-A) (U.K. communications satellite), 385, 388–389, 422
Skynet B, 389
SLA. See Spacecraft-lunar-module-adapter panels.
Slipher, Dr. Vesto M., 366
Sloop, John L., 178
SM. See Service module.
Small Astronomy Satellite (SAS), 367
Smelt, Dr. Ronald, 79
Smith, Bradford A., 282
Smith, Dr. F. Graham, 121–122
Smith, Ambassador Gerard C., 30, 81
Smith, Dr. Harlan J., 237
Smith, Dr. Henry J., 306
Smith, I. D., 365
Smith, Keith, 411
Smith, Sen. Margaret C., 118
Smith, Dr. Paul S., 232
Smith, Sen. Ralph T., 312
Smith, Dr. Richard K., 344
Smith, Sir Ross Macpherson, 411
Smithsonian Astrophysical Observatory (Cambridge, Mass.), 3, 79, 129, 403, 405, 409
Smithsonian Institution, 29, 147, 297, 304, 328, 344
  astronaut dinner, 7
  exhibit
    aircraft, 119, 133
    flag, U.S. (*Apollo 11*), 414

lunar rock sample, 304, 306, 307, 309, 313, 359, 387, 397
spacecraft, 388
TV equipment, 201
Webb, James E., portrait unveiling, 323
Smoke hood, 9
SMS. See Synchronous Meteorological Satellite.
SNAP (Systems for Nuclear Auxiliary Power) program, 28
SNAP-3A (nuclear generator), 190
SNAP-19, 414, 422
SNAP-27, 410, 422
SNAP-29, 422
Snoopy (Apollo 10 LM). See Lunar module.
Snow, Lord Charles (C.P.), 39, 291
Society of Automotive Engineers, 338
Society of Separationists, Inc., 268, 394, 400
Sodium experiment, 408–409
Software Fix (SOFIX), 139
Sokolov, Oleg M., 146
Solar cell, 82, 322, 329
Solar flare, 90–91, 187, 361, 364
Solar furnace, 363
Solar physics, 132–133, 136–137, 160, 366–367
    sounding rocket, 104, 108, 111, 140–141, 176, 298, 302, 308, 314, 360–361, 410–411
Solar wind, 177, 219, 223, 230, 285, 376–377, 408, 413, 419
Solid propellant, 291
Somali Republic, 234, 247
Sonic boom, 11, 26, 58–59, 183–184, 255, 352, 391
Sonnett, Dr. Charles P., 21
Sounding rocket (see also individual sounding rockets: Aerobee, Arcas, Astrobee 1500, Black Brant, Boosted Arcas II, Nike-Apache, Nike-Cajun, Nike-Iroquois, Nike-Tomahawk, Orion II, Pacemaker, Sidewinder-Arcas), 3, 124, 203
    foreign, 139–140, 156, 421–422
        Canada, 59, 127, 312, 366, 423–424
    funds for, 124
    international programs
        NASA-Argentina, 303–304
        -Australia, 423
        -Brazil, 179, 186, 188, 423
        -Canada, 21, 23, 31, 38, 41–42, 43, 48, 49, 51–52, 54, 58, 59, 71, 72, 113, 127, 312, 331, 366, 423–424
        -India, 118, 120, 423
        -Japan, 118, 120, 411, 423
        -Norway, 18, 20, 423
        -Pakistan, 423
        -Spain, 94, 132, 133, 423
        -Sweden, 18, 20, 24, 27, 423
        -U.K., 18, 20, 24, 27
    U.K.-Australia, 323

Sounding Rockets: Their Role in Space Research (Space Science Board report), 124
South Africa, 424
South America, 370
South Atlantic Anomaly region, 312
South Florida, Univ. of, 110
South Pole, 251, 359
South Rogers Lake Bed, Calif., 133, 286, 299, 316, 349, 371
Southern Christian Leadership Conference (SCLC), 201, 205–206
Southern Historical Assn., 357
Southwest Center for Advanced Studies, 43, 48, 51–52, 54, 188
Soviet Academy of Sciences, 25, 202, 229, 244, 351, 361, 366, 411–412
Soviet Committee for Inventions and Discoveries, 418
Soviet Institute for Cosmic Research, 233
Soyuz (U.S.S.R. spacecraft), 109
Soyuz III mission, 337
Soyuz IV mission, 11–12, 23, 332, 422
Soyuz V mission, 11–12, 23, 332, 422
Soyuz VI mission, 332, 336, 341–342, 343, 361, 365–366, 382, 420, 422
    launch, 332
Soyuz VII mission, 333, 336, 341–342, 343, 361, 365–366, 382, 420, 422
    launch, 333
Soyuz VIII mission, 333, 336, 341–342, 343, 361, 365–366, 382, 420, 422
    launch, 333
Space accident liability, 392
Space Age Management, 192–193
"Space-Age Management and City Administration" (conference paper), 148–149
Space biology, 61, 62, 72, 81, 96, 178, 180, 200, 302, 336
    animal experiments, 18, 139, 154, 189–190, 200, 201–202, 261, 335, 347–349, 415, 421
    atmosphere, artificial, 206
    environment, effects, 18–19, 71, 130–131, 138
    life support system, 12, 123, 190, 206, 378
    lunar dust experiment, 261, 293, 379
    medical benefits, 75, 78, 96
    motion sickness, 142
    NASA program, 138, 367–368, 381–382, 415
    nutrition, 110
    Physiology in the Space Environment (NAS–NRC report), 18
    radiation effects, 78, 96, 136–137, 164, 190
    Space Resources for Teachers: Biology, Including Suggestions for Classroom Activities and Laboratory Experiments, 72
    weightlessness, effects, 17, 18–19, 48, 117–118, 136, 201, 206, 348–349, 391
Space debris. See Spacecraft debris.

Space Development Corp. (Japan), 325
Space Disturbance Center, 90, 336
Space Exploration Day (proposed), 202
Space law, 189, 290
Space law treaty, 13, 229, 276, 320
Space manufacturing module (proposed), 83
Space, military use of, 290, 320–321, 343
  communications, 44, 66, 131, 322, 346, 389, 422
  history, 329
  navigation, 322
  reconnaissance, 345
  U.S., 13, 21, 186, 190–191, 200, 336, 343, 345, 416
  U.S.S.R., 21, 50, 190–191, 416
Space, peaceful use of, 6–7, 11, 13, 14, 20, 24, 157–158, 191, 196, 219–220, 224, 244, 249, 343–344, 390
Space program, national (see also individual programs, such as Apollo program and National Aeronautics and Space Administration, budget), 96, 208–209, 252, 390
  achievements, 6–7, 10, 59, 71, 148–149, 372, 421–424
    manned space flight, 2, 13, 14, 17–18, 25, 39, 62–65, 68, 71–72, 77–78, 80, 91, 101, 148–149, 239–240, 242–243, 244, 245, 251–252, 277–278, 280, 282, 298–299, 386–387, 392–393, 394
  Agnew, Vice President Spiro T., 65, 100, 152, 224–225, 231, 271, 294, 308, 320, 323–325, 419
  benefits. See Space results.
  budget, 14–16, 22, 38, 49–50, 67, 81–82, 95, 109, 167, 270, 305, 366, 383, 395, 422–423
  cost, 4, 5–6, 38, 42–43, 47, 87–88, 95–96, 154, 158, 208, 262, 271, 327, 390, 416, 422–423
  criticism, 2–3, 151–152, 176, 201, 204, 273, 287, 367–368, 385, 416–417, 423
  Eisenhower, President Dwight D., 47, 60, 250, 405
  employment, 66, 69, 80, 422–423
  exhibit, 403, 405–406
  international aspects (see also International cooperation) 4, 6–7, 20, 26, 32, 41, 43, 48, 54, 75, 168, 199, 235–236, 238, 244–246, 249, 262, 268, 290, 309–310, 327, 362, 367, 381, 389, 393–394, 411–412, 416, 423–424
  Johnson, President Lyndon B., 1, 6–7, 10, 13, 14–16, 17, 20, 60, 110, 192, 224–225, 231, 232, 395, 423
  Kennedy, President John F., 6, 44, 60, 174, 175, 212, 240, 405
  lunar landing. See Moon, landing.
  management, 24, 57, 84, 111, 148–149, 202, 274, 276–277, 368, 370
  military aspects, 9, 21, 38, 51, 75, 80, 86–87, 104, 146, 167, 174, 176–177, 179, 185, 200, 304–305, 320, 416

Nixon, President Richard M., 9, 20, 32, 38, 43, 49, 61, 65, 68, 78, 80, 81, 95, 104, 108, 109, 110, 115, 117, 119, 224, 238, 246, 250, 262, 271, 304–305, 309–310, 362, 371, 372, 378, 423
  objectives, 6, 37, 44, 49–50, 73–75, 76, 77–78, 81–82, 85–86, 130, 134, 151–152, 273, 275–276, 283–284, 285, 304–305, 308, 318, 331, 421, 422–423
  policy, 10, 24, 49–50, 51, 60, 74–75, 80–81, 86–87, 104, 111, 116, 119–120, 147, 148–149, 167–168, 174, 177, 208, 232, 237, 239–240, 242–243, 248–250, 256, 271–272, 284, 287–288, 290, 294–295, 320–321, 351, 370, 378, 422–423
  post-Apollo, 38, 44, 47, 66–67, 69, 77, 118, 119–120, 151–152, 167, 178, 266, 270, 271, 283–284, 305, 312–313, 328, 337–338, 344, 359, 361, 372, 399, 412, 422–423
  budget, 28, 69, 110, 119–120, 147, 365–366
  cost, 162, 262, 336–337
  suggested programs, 37, 85–86, 94–95, 123, 235–236, 259–260, 263, 266, 268–269, 275–276, 304–305, 308, 312–313, 342–343, 355–356, 366–368, 381, 397–398, 415
  significance, 1, 2, 16, 20, 39, 51, 52–53, 65, 68, 129–130, 174, 175–176, 197–198, 200–201, 224–225, 231–232, 235–236, 242–243, 262, 268–269, 277–278, 294–295, 307, 321–322, 346, 380, 420–421
  U.S.S.R. vs. U.S. See Space race.
Space race, 95–96, 285–286, 321, 343, 365–366, 381
  booster, 95–96, 170
  earth applications, 74–75
  funds, 48
  lunar exploration, 74–75, 96, 170–171, 195, 211, 238, 246, 249, 256, 271–272, 273
  manned space flight, 341–342
  military, 48
  planetary flights, 74–75, 84–85, 96
  space station, 74–75, 96, 341–342, 351
Space rescue, 21, 262, 315, 354
Space rescue treaty, 130, 392
Space Resources for Teachers: Biology, Including Suggestions for Classroom Activities and Laboratory Experiments, 72
Space results (see also Earth; Moon; Mars; Venus; individual probes, satellites, and sounding rockets), 17, 75, 170, 268–269, 298, 312–313, 321, 394, 419
  agriculture, 37, 119, 276, 285, 327
  aircraft, 75, 350, 400
  architecture, 398
  astronomy, 28, 37, 75, 90, 99, 115,

120, 121–122, 124, 160, 256, 266, 308, 408, 419
  communications, 2, 33, 37, 47, 75, 100, 111, 235, 298, 335, 398
  earth sciences, 33, 37, 49, 73, 75, 100, 116–117, 119–120, 124, 203, 235, 245, 276, 285, 298, 308, 327, 382
  economic benefits, 2, 75, 167, 203, 227, 238, 276, 350, 419
  education, 17, 47, 75, 298
  engineering, 203, 238, 267, 335, 350, 398, 400, 408
  geology, 2, 33, 37, 203, 308, 327, 328, 407–408
  international relations, 65, 94, 102, 261–262, 308, 419
  materials technology, 2, 75, 203, 235, 350
  medicine, 47, 203, 235, 274, 315, 262–263, 398
  meteorology, 33, 47, 111, 235, 308, 327, 335, 414
  military, 75, 174
  navigation, 33, 47, 70, 100, 203, 335, 400
  oceanography, 160, 199, 203, 327, 346
  photography, 328
  religion, 39
  science, 68, 230, 240–241, 308, 356, 392
  social science, 147, 148–149, 242, 294, 323–325, 400–401
  technology, 2, 37, 174, 203, 230
Space Science Board (NAS–NRC). See National Academy of Sciences and National Research Council.
Space Science Education Conference, 17
Space shuttle, 25, 29, 47, 73–74, 109, 120, 131, 138, 262, 266, 276, 305, 326, 350, 354, 362, 368–369, 371, 419–420
  cooperation, 29, 173, 336, 411, 424
  nuclear, 268
  reusable, 23, 29, 47, 74, 82, 260, 268–269, 345, 354, 423
Space station (see also Manned Orbiting Laboratory; Saturn I Workshop; Saturn V Workshop), 21, 47, 262, 276, 289, 351, 354, 371
  contract, 25–26, 120, 241, 270, 299
  design, 17, 28, 120, 128, 285, 349
  international cooperation, 102, 169, 248, 411, 416
  military, 86–87, 104
  NASA program, 66–67, 73–74, 83, 91, 94, 104, 108–109, 111, 131, 138, 146, 167, 169, 178, 237–238, 260, 266, 268–269, 305, 343–344, 368–369, 423
  press comment, 80, 351, 391
  press conference, 169
  U.S.S.R., 12, 24, 75, 120, 146, 271, 278, 333, 341, 343, 349–350, 351, 361, 382, 420, 422
Space Task Group. See President's Space Task Group.

Space Technology Applications and Research Laboratory (STARLAB), 285
Space Tracking and Data Acquisition Network (STADAN), 292
Space transportation (see also Space shuttle), 169–170, 177, 270
Space tug, 262, 266, 305
Spacecraft (see also individual spacecraft, such as Apollo, Lunar Orbiter, Luna, Mariner, Surveyor)
  accident, 29, 162, 206
  communications, 26, 44, 117, 159, 210, 241
  control, 11–12, 26–27, 56, 58, 59, 92, 106, 142–143, 173, 259, 278, 291, 333
  debris, 89, 130, 143, 198, 200
  design, 17, 120, 151, 171, 241, 266, 270, 272, 276–277, 343–345, 388, 397–398, 401
  development testing, 44
  electrical systems, 11–12, 27, 37, 114, 135, 173, 373–374
  environment control system, 17, 141–142, 173–174
  equipment, 2, 54–56, 102, 122, 142, 143, 217–218, 298–299, 332
  escape system, 94, 239
  exhibit, 105, 161–162, 417
  extravehicular equipment, 4–5, 11, 239, 270, 331
  hazards, 149–150
  heating, 26–27, 263
  instrumentation, 14, 55, 58, 91, 107, 115, 133, 151–152, 171, 316, 326, 338, 340
  landing system, 26–27, 29, 140, 141–142, 160, 181–182, 183, 378, 405
  life support system, 12, 123, 173–174, 206, 378
  propulsion. See Engine and individual launch vehicles, such as Saturn.
  recovery, 190, 201–202, 222–223
  reentry control system, 47, 271
  reusable (see also Space shuttle), 25–26, 47, 266
Spacecraft debris, 28, 89, 198, 228, 291
Spacecraft-lunar-module-adapter (SLA) panels, 202–203
Spacesuit, 12, 77, 91, 206, 313
Spain, 94, 132, 133, 180, 187, 189, 226, 327, 330, 423–424
Spanish Space Research Council, 180
Spartan (missile), 62
Speas, R. Dixon, Associates, 109
Spectrograph, 90
Spectrometer, 38, 42, 55, 77, 171, 179, 254, 261, 269, 286, 301–302, 367, 410–411, 414, 417
Sperry Gyroscope Co., 352
Sperry, Lawrence, Award, 345
Sperry Rand Corp., 42
  Sperry Gyroscope Div., 42
  UNIVAC Federal Systems Div., 55
Spica (star), 79

Spider (*Apollo 9* LM). See Lunar module.
Spilhaus, Dr. Athelstan F., 159
Spirit of St. Louis (aircraft), 177
Spivak, Lawrence E., 184
Sproul Observatory, 109
SPS. See Service propulsion system.
Sputnik I (U.S.S.R. satellite), 20, 161
SR-71 (strategic reconnaissance aircraft), 73
SRC. See Science Research Council (U.K.).
SS-9 (U.S.S.R. missile), 116, 140, 229
SSRC. See Swedish Space Research Committee.
SST. See Supersonic transport.
SST Authority (proposed), 137
Stability augmentation system (SAS), 349
STAC. See Scientific and Technical Advisory Committee (University-NASA).
STADAN. See Space Tracking and Data Acquisition Network.
Stafford, Col. Thomas P. (USAF), 16, 90, 108, 142–144, 152–153, 191
Stamp, commemorative, 8, 46, 59, 129–130, 202, 233, 273–274, 289, 300, 311, 319
Stamy, James L., 260
Stanford Research Institute, Calif., 26, 257
Stanford Univ., 117, 124, 136, 174, 205
Linear Accelerator Center, 418
Stanford Univ. Hospital, 315
Stanton, Dr. Frank M., 352
Star, 3, 6, 16, 32, 42, 55, 66, 79, 99, 133–134, 134–135, 136, 184, 256, 326, 327, 337, 344, 359
STAR. See Self-testing and repairing computer.
Star City (Vzyozdni Gorodok, U.S.S.R.), 195–196
Star Tracking Rocket Attitude Positioning (STRAP) system, 6, 43
STARLAB. See Space Technology Applications and Research Laboratory.
State, Dept. of, 7, 24, 25, 43, 88, 130, 311, 357
Statler Hilton Hotel, Washington, D.C., 132
Stehling, Kurt R., 192
Sterne, Joseph, 78
Stewart, A. I., 298
Stewart, M/G James T. (USAF), 260
Stockholm, Sweden, 338
STOL (short takeoff and landing) aircraft, 29, 32, 50, 84, 89, 113–114, 320
STOLport, 89
Stoltenberg, Dr. Gerhard, Minister for Scientific Affairs (W. Germany), 48, 177
Stone, Dr. Robert G., 115
Stonehenge, U.K., 289
Storms, Ocean of (moon), 243, 368, 375, 405, 478
Strategic Air Command (SAC), 331

Study of Air Cargo and Air Passenger Terminal Facilitation (DOT report), 113
A Study of Technology Assessment: Report of the Committee on Public Engineering Policy, National Academy of Sciences, 257
Stuhlinger, Dr. Ernst, 77
Submarine, 27, 99–100, 112, 140, 330
missile-carrying, 195
Sudan, 247
Sud-Aviation (France), 61, 71, 173, 197, 285, 323
Suez Canal, 234
Suharto, President (Indonesia), 249
Suitland, Md., 353
Sullivan, Walter S., 28, 90, 181, 312, 334, 397, 404
Summer Space Education Program for the Cities, 170
Sun (see also Eclipse, solar; Radiation, solar; Solar flare; Solar wind; etc.), 37, 49, 121, 288, 299, 329, 335, 336, 420–421
satellite data, 28, 111, 115, 133, 155, 185, 271–273, 364–365, 395
sounding rocket data, 111, 113, 124, 298, 361–362, 410
Sunblazer (program), 163
Supersonic transport (SST) (see also Concorde and Tu-144)
award, 178
benefits, 81, 314–315, 382–383
cost, 186, 314–315, 340, 382–383
criticism, 82, 413
design and development, 1, 11, 404, 421–422
economic aspects, 382–383
flight plans, 146–147, 255, 285
foreign, 1, 13–14, 61, 84, 146–147, 197, 255, 422
funds for, 15, 81, 82, 110, 137, 186, 314–315, 318, 372, 383, 410, 412
hazards, 369
Nixon, President Richard M., 32, 43, 58, 82, 95, 123, 137, 314–315, 318, 372, 383, 410, 421–422
press comment, 137, 318, 413
press conference, 81, 186, 197
sonic boom, 11, 26, 58–59, 183–184, 255, 352
Surveyor (program), 34–35, 141, 187, 191
Surveyor III (spacecraft), 372, 377, 396
examination of, plans for, 243, 325–326
renaming of (proposed), 369–370
retrieval of parts, 342, 367, 376–377, 421
Surveyor V (lunar probe), 250, 275
Sutton, Dr. George, 230
Swarthmore College, 109
Swartz, Reginald W., 239
Swearingen, Jack C., 393

Sweden, 18, 20, 24, 27, 50, 226, 227–228, 236, 240, 252, 323, 384, 423
Swedish Academy of Engineering, 351
Swedish Interplanetary Society, 329
Swedish Space Research Committee (SSRC), 18, 20, 24, 27
Swenson, Loyd S., Jr., 357
Swigert, John L., Jr., 267
Switzerland, 424
Sydney, Australia, 121, 137, 312, 359
Symphonie (W. German comsat), 357
Symposium on Military Oceanography, 160
Synchronous Meteorological Satellite (SMS), 69, 86, 139
Systems engineering, 35, 388
Syvertson, Clarence A., 53

T

T–33 (jet trainer), 283
Tacsat 1 (tactical communications satellite), 44, 322, 422
Tactical Satellite Communications Program, 131, 140
TACV. See Tracked Air Cushion Vehicle.
TADJET. See Transport Air Drop and Jettison Test.
Tago-Sato-Kosaka comet, 409
TAID. See Thrust-Augmented Improved Thor-Delta booster.
Taipei, Formosa, 234
Talkeetna, Alaska, 171
Talloires, France, 381, 394
Tampa, Fla. 110
Tape, Dr. Gerald F., 112, 117
Tappaan, Francis D., 147
Tashkent, U.S.S.R., 387
Task Force on Air Pollution, 383
Task Force on Oceanography, 314
Task Force on Science Policy, 314
Taylor, D. J., 164
Taylor, Henry J., 138
Taylor, Dr. S. Ross, 292, 407
TCFM. See Temperature-control flux monitor.
Teague, Rep. Olin E., 71
Technology, 11, 53, 132, 135, 136, 146, 148–149, 151, 162–163, 257, 321, 396
Technology in Retrospect and Critical Events in Science (TRACES) (report), 11
Technology: Processes of Assessment and Choice (NAS report), 257
Technology utilization, space, 34, 44, 70, 119–120, 157, 298, 309–310, 335, 388, 416
Teheran, Iran, 226, 312
Teir, William, 325
Tektite, 200
Tektite I, Project, 51, 86, 110, 138
Tektite II, Project, 356
Tel Aviv, Israel, 226
Telemetry, 117
Telescope, 29, 133–134, 257, 259, 272, 336, 341, 342–343, 390, 403

gamma-ray, 127
lunar surface, 123, 331, 342–343
space, 81, 122, 342–343, 366–367
spectrographic, 90
Television, 146, 227
Apollo 9, 50, 67
Apollo 10, 90, 108, 142–144, 147–148, 149
Apollo 11, 182, 201, 207, 210, 212–223, 244–245, 267, 276, 279
Apollo 12, 342, 362, 374–376, 377, 380, 384–385, 386
color, 102, 210, 342, 362
educational, 337, 423
Nixon, President Richard M., 18, 50
Soyuz IV, 11–12
Soyuz V, 11–12
space probe use of, 55, 210, 259, 261, 262, 301, 338
tube, 105–106
via satellite, 18, 30, 39–40, 47, 50, 61, 137, 151, 179, 286, 311, 340, 402, 423
Teller, Dr. Edward, 167
Temperature-control flux monitors (TCFM), 133
Tennessee, 225
Tennessee, Univ. of, 391
Space Institute, 20
Tepper, Dr. Morris, 139
Ter Horst, J. F., 9
TERLS. See Thumba Equatorial Rocket Launching Station.
TETR C (test and training satellite), 291, 421
Texas, 397
Texas Instruments, Inc., 117, 418
Texas, Univ. of, 17, 90, 237, 259
McDonald Observatory, 17, 90, 259
Textron, Inc., Bell Aerospace Co., 413
TF–39 (turbofan engine), 212, 393
Thailand, 190, 251
Thant, U, U.N. Secretary General, 158, 244, 279
Thimann, Kenneth V., 162–163
Thiokol Chemical Corp., 177
This New Ocean: A History of Project Mercury, 250, 357
Thomas, Rep. Albert, 289
Thomas, David D., 58
Thomas, David E., Jr., 4
Thompson, Dr. Floyd L., 21
Thon, William, 403
Thor-Agena (booster), 23, 40, 59, 86, 127, 243, 255, 402
Thor-Burner (booster), 240
Thor-Delta (booster), 22, 54
Long-Tank, Trust-Augmented, 40, 150, 189–190, 245, 272, 388
Thrust-Augmented, 57
Thrust-Augmented Improved, 30, 185, 291
Thorad. See Long-Tank Thrust-Augmented Thor-Agena D.
Thorad-Agena (booster), 48, 314

Thorad-Agena D (booster), 107, 171–173, 319
Thoren, Kip S., 121
Thorn, Dr. Oakleigh, II, 225
Thornton, Dr. William E., 290
Threlfall, David, 161, 233
Thrust-Augmented Improved Thor-Delta (TAID) (booster), 30, 185, 291
Thrust-Augmented Thor-Delta (booster), 57
Thumba Equatorial Rocket Launching Station (TERLS), 118, 120
Tibet, 50
Tidemand, Defense Minister Otto Grieg (Norway), 332
Time, Men of the Year award, 2
Time-Life, Inc., 196
Tiros (meteorological satellite), 23, 139
Tiros-M, 85–86, 399
Tiros-N, 85–86
Tiros Operational Satellite (TOS) system, 57, 414
Titan III (booster), 15, 355, 386
Titan IIIB-Agena, 288, 350
Titan IIIB-Agena D, 23, 66, 109, 169
Titan IIIC, 14, 44, 118, 155, 402
Titan IIIM, 185, 422
Titan-Centaur (booster), 118, 192
Titanium, 275
Tito, President Josip Broz (Yugoslavia), 341
Titov, L/C Gherman S. (U.S.S.R.), 158, 195, 197
Titus, Ronald, 352
TMA. See Trimethylaluminum experiment.
Tokyo, Japan, 82, 290, 312, 361–362, 384
Tokyo Univ., 297
Tolson, Clyde A., 239
Toowoomba, Queensland, Australia, 148
Torrey, Volta W., 138
TOR–SHOK (energy-absorbing system), 408
TOS. See Tiros Operational Satellite (TOS) system.
Toulouse, France, 329, 397
Toulouse-Blagnac Airport, France, 13, 61, 323
Townes, Dr. Charles H., 82, 94, 167, 346
Toynbee, Prof. Arnold J., 239
TRACES. See *Technology in Retrospect and Critical Events in Science.*
Tracked Air Cushion Vehicle (TACV), 163
Tracking, 47–48, 59, 199, 204, 322, 399
  aircraft, 140
  animal, 328
  Apollo 11 mission, 204, 224
  deep space (DSN), 34, 180, 189, 274–275, 346
  laser, 37, 405
  MSFN, 10, 33–34, 117, 196, 292, 362
  NASCOM, 292
  radar, 55, 291–292
  ship, 306–307, 410

STADAN, 292
station
  Australia, 34, 148, 424
  Peru, 267
  Spain, 424
  U.S., 142, 274–275, 374, 405
Train, Russell E., 346
Tranquility Base. See Sea of Tranquility.
Trans World Airlines, 81, 241, 244
Transformation of Imagery (TRIM), 139
Transit IV–A (navigational satellite), 190
TransPlan, Inc., 113
Transport Air Drop and Jettison Test (TADJET), 283
Transportation, 17, 105, 113–114, 163, 208, 209
Transportation, Dept. of (DOT), 58, 113–114, 410
  air cushion vehicle, 163
  air traffic control, 32, 88–89, 123, 388, 418
  airports, 32, 123
  budget, 14, 78, 383, 398, 412
  contract, 89, 163, 335, 388
  cooperation, 88, 417–418
  exhibit, 412
  noise abatement, 263–265
  personnel, 52, 79
  R&D, 335, 417–418
  supersonic transport, 32, 43, 95, 123, 146–147, 412
  TurboTrain, 105
Transportation Facilities Committee, 113
Treaty
  damages for space accidents, 392
  missile control, 21, 33, 81, 131, 174, 181, 183, 190–191, 332
  nuclear nonproliferation, 13, 41, 52, 57, 78, 317, 370, 392–393
  nuclear test ban, 13, 81
  outer space, 13, 320
  seabed, weapon ban on, 81, 86, 155, 197, 330
  space rescue, 21, 130, 392
Triesneck Crater (moon), 159
TRIM. See Transformation of Imagery.
Trimble, George S., 292
Trimethylaluminum (TMA) experiment, 49, 408–409
Trubshaw, Brian, 105
Trudeau, Prime Minister Pierre E. (Canada), 399–400
The True History (by Lucian), 232
Truly, L/Cdr Richard H. (USN), 281
Truszynski, Gerald M., 10, 210, 347
TRW Inc., 58, 66, 329, 340
  Systems Group, 113, 319
Tu-22 (U.S.S.R. supersonic bomber), 1
Tu-144 (U.S.S.R. supersonic transport), 1, 5, 121, 146–147, 152, 159, 173, 255, 283, 422
Tu-154 (U.S.S.R. jet transport), 121
Tucson, Ariz., 157–158, 192, 397
Tullahoma, Tenn., 138, 183–184, 393

Tunney, Rep. John V., 202
TurboTrain, 105
Turcat, André, 61, 71, 173, 323
Turkevich, Anthony, 191
"2001: A Space Odyssey" (film), 241
Twentieth Century Fund, 393–394
Tycho (lunar crater), 200
Tyuratam, U.S.S.R., 382

U

U.A.R. See United Arab Republic.
Udall, Rep. Morris K., 168
Uebel, Theodore C., 54
UFO. See Unidentified flying object.
U.K. See United Kingdom.
Ultraviolet (UV), 3, 6, 31–32, 43, 55, 78, 99, 124, 134, 136, 179, 184, 298, 302, 366, 403, 409
  extreme (EUV), 140, 308, 411
U.N. See United Nations.
Underground nuclear test, 359
Unidentified flying object (UFO), 5, 8–9, 20, 50–51, 288, 411, 413–414, 416
Unidentified satellite, 23, 40, 66, 86, 106, 109, 127, 169, 240, 243, 255, 288, 314, 319, 350, 402
Union of Concerned Scientists, 40
Union of Soviet Socialist Republics. See U.S.S.R.
United Air Lines, 278–279, 380
United Aircraft Corp., 21, 105, 173, 424
  Pratt & Whitney Div., 25, 32, 46, 304, 328, 379
United Arab Republic (U.A.R.), 140, 226, 384, 413
United Kingdom (U.K.), 16, 106, 168–169, 381
  aircraft, Concorde, 13–14, 26, 61, 71, 105, 146, 162, 173, 197, 255, 285, 323, 329, 340, 410, 413
  Apollo 11 and *Apollo 12* missions, reaction to, 210, 225, 233, 236, 238, 244, 384
  astronauts
    award to, 251
    visit by, 32, 38, 312, 335
  booster, 190, 196
  cooperation, defense, 132
  cooperation, space, 18, 20, 23, 24, 27, 59, 132, 196, 323, 385, 388–389, 422, 423–424
  Defence Ministry, 195
  House of Commons, 38
  launch, satellite, 388
  lunar sample exhibit and study, 312, 423
  nuclear nonproliferation treaty, 78
  satellite, 23, 59, 385, 388–389, 422, 423
  science and technology, 398
  space program, 59, 132, 323, 385, 388–389, 423–424
United Nations (U.N.), 8, 24, 26, 175, 276, 279, 282, 311
  Committee on the Peaceful Uses of Outer Space, 189, 198, 392
    Scientific and Technical Subcommittee, 157
  Committee on the Peaceful Uses of the Seabed and Ocean Floor, 22
  Disarmament Committee, 197
  General Assembly, 197, 309, 330
  Nixon, President Richard M., address, 309–310
  Secretary General, 244
  Security Council, 242
  Space Council (proposed), 311
  Space Institute (proposed), 248
United Nations Educational, Scientific, and Cultural Organization, 255
United Nations Space Council (proposed), 311
United States and Soviet Rivalry in Space: Who Is Ahead and How Do the Contenders Compare?, 95
United States Space Science Program, 136
United Technology Center, 48
Universe, 403
Universities, 168
  and space effort, 8, 171, 227
  Federal support, 32, 40, 54, 56, 106, 397
  grants to, 3, 32, 52
  military research, 40, 67, 136, 363, 369
  NASA program, 3, 25, 56, 106, 109, 156, 171, 227, 302, 334, 363, 368, 383, 415
Universities Organizing Committee for Space Sciences, 8
Universities Space Research Assn. (USRA), 227
University College (London), 22, 272
University Corp. for Atmospheric Research, 44, 416
Upton, N.Y., 386
Uranus (planet), 192, 263
Urey, Dr. Harold C., 238, 275, 288, 292, 331
United States (U.S.) (see also appropriate agencies and Congress)
  award, 1–2, 50, 54, 100–101, 121, 418
  budget, 3, 4, 14–16, 22, 86, 99, 107, 109–110
  communications, 24, 56, 88, 101–102, 381
  defense, 4, 40, 46–47, 48, 86, 99, 130, 132, 183
  disarmament, 13, 30, 81, 86, 155, 197
  education, 3, 105–106, 136, 147, 168, 174
  health, 147, 266–267
  housing, 188, 237
  international cooperation, 4, 22, 24–25, 43, 56, 78–79, 88, 101–102, 133, 204, 258, 261, 386, 417
  medical research, 96, 128
  meteorology, 78, 88
  nuclear nonproliferation treaty. See Nuclear nonproliferation treaty.

oceanography, 9, 22, 51, 56, 99–100, 110, 133, 197, 209, 314, 341, 353
pollution abatement, 147, 267, 278–279, 285–286, 354, 383, 402, 412–413
research and development, 3, 28, 54–56, 96–97, 136, 164, 257
science and technology, 3, 11, 14–15, 19–20, 28–29, 40, 42, 43, 106, 117–118, 121–122, 125, 174–175, 178–179, 184, 255, 297, 313–315, 321, 346–347, 359, 396–397, 419
space program. See Space program, national.
space rescue treaty. See Space rescue treaty.
transportation (see also Supersonic transport), 17, 32, 52, 79, 105, 163
Vietnam war. See Vietnam war.
U.S. Aeronautics and Space Activities for 1968 (President's report), 17
U.S. Air Force (USAF) (see also individual bases, centers, and commands, such as Air Force Systems Command, Arnold Engineering Development Center, Edwards AFB), 68, 75, 79, 138, 163–164, 167, 177, 245, 284, 355, 410
  aircraft (see also individual aircraft, such as C–5A, C–130, F–111A, X–15, XB–70), 4, 17, 38, 49, 52, 57, 80, 106–107, 111–112, 115, 119, 123–124, 128, 131, 138, 151, 169, 183, 201, 205, 207, 239, 251, 281, 283, 293, 321, 323, 331, 352, 354, 360, 372, 380, 405, 411, 413, 415–416, 421–422
  award, 87, 101, 304, 346–347, 412, 419
  booster, 386
  budget, 15, 424
  communications satellite, 322
  contract, 14, 48, 66, 85, 123–124, 128, 135, 150, 191, 327–328, 330, 336, 372, 386, 391, 402–403, 408, 414
  cooperation, 21, 29, 86–87, 119, 146, 176, 192, 205, 257–258, 283, 321, 336, 388–389, 405, 421–422
  laser, 424
  launch, 371
    balloon, 363
    reentry vehicle, 10
    satellite, 23, 40, 66, 83, 86, 106, 109, 127, 155, 169, 240, 243, 255, 288, 314, 319, 350, 402
  lifting-body vehicle, 421–422
  missile program, 43, 229, 297
  MOL, 15, 87, 104, 110, 146, 151, 176–177, 178–179, 181, 186, 191, 200, 257–258, 260, 281, 289, 422, 423, 424
  navigation satellite, 322
  personnel, 4, 9, 21, 124, 130, 141, 180, 188, 260, 274, 281, 289
  reentry vehicle, 391
  research, 164, 183–184
  satellite, 191
  sonic boom test, 352
  space program, 13, 193
  space shuttle, 336
  telescope, 336
  UFO, 5, 8–9, 288, 411, 413–414, 416
U.S. Army (USA), 235
  aircraft, 131, 148, 150, 160, 308–309
  contract, 148, 150
  cooperation, 49, 283, 334, 379
  personnel, 4, 398
  satellite, 107–108, 422
  Transport Air Drop and Jettison Test (TADJET) program, 283
  universities research, 136
U.S. Army Atmospheric Sciences Laboratory, 315
U.S. Army Aviation Material Laboratories (AVLABS), 334
U.S. Army Ballistics Laboratory, 49
U.S. Army Collateral Investigation Board, 260
U.S. Army Materiel Command, 379
U.S. Bureau of Mines, 387
U.S. Coast and Geodetic Survey, 117
U.S. Coast Guard, 9, 51
U.S. Comptroller General, 281, 309
U.S. Deep Sea Drilling Project, 273
U.S. Geological Survey, 393, 403, 409
U.S. Information Agency (USIA), 234, 387, 397
U.S. Lake Survey, 9
U.S. Marshal's Office, 336
U.S. Naval Academy, 101
U.S. Navy (USN), 123, 206
  aircraft, 13, 15, 39, 44, 91, 119, 133, 155, 201, 260
  anniversary, 133, 325
  aquanaut inquiry, 57, 77, 316–317
  award, 10, 138
  contract, 13, 39, 91, 201, 260
  cooperation, 27, 44, 51, 119, 402
  deep submergence research vehicle, 27, 112
  launch, 371
  missile, 325
  navigation satellite system, 37, 190
  Project Tektite, 51, 138, 356
  personnel, 4, 134
  Sealab III experiment, 22, 51–52, 57, 77, 316–317
  spacecraft recovery, 64, 143, 190, 223, 242, 244, 245–246, 284, 377, 392
  Stormfury Project, 402
  tracking ship, 306
U.S. Patent Office, 239
U.S. Post Office, 273, 289
U.S. Public Health Service, 76
U.S. Supreme Court, 7
U.S. Weather Bureau, 157
USAF. See U.S. Air Force.
USAF Aeronautical Chart and Information Center, 209
USAF Museum, 11
USAF Space and Missile Systems Organization (SAMSO), 297, 300, 336, 346–347, 386, 391, 408

USNS Huntsville, 306
USNS *Mercury*, 306
USNS *Redstone*, 306
USNS *Vanguard*, 306, 410
USRA. See Universities Space Research Assn.
U.S.S. *Elk River*, 22, 52
U.S.S. *Gary*, 206
U.S.S. *Guadalcanal*, 64
U.S.S. *Hornet*, 190, 223, 242, 244, 245, 246, 284, 377, 392
U.S.S. *Princeton*, 143
U.S.S. *Yorktown*, 10
U.S.S.R. (Union of Soviet Socialist Republics) (see also Soviet Academy of Sciences, etc.), 21, 39, 49, 78, 89, 129, 130, 131, 136–137, 145–147, 175–176, 197, 370
  aircraft, 1, 5, 35, 79, 121, 146–147, 152, 159, 162, 168, 173, 225, 283, 422
  antiballistic missile system, 62, 89
  Apollo 8 mission (comment), 25, 26
  Apollo 9 mission (comment), 80
  Apollo 10 mission (comment), 157
  Apollo 11 mission (comment), 206, 225, 230, 233, 242, 244, 251, 256, 273
  Apollo 12 mission (comment), 384
  award, 23–24
  booster, 95–96, 170–171, 382
  Borman, Col. Frank (USAF), visit to, 195–196, 197, 199, 202, 204, 210, 211, 337
  budget, 48
  communications satellite, 106, 381, 422
  communications satellite conference, 24, 26
  cooperation, space, 19, 25, 26, 41, 56, 74, 78, 96, 101–102, 189, 195, 196, 202, 211, 229, 231–232, 233, 238, 240, 245, 246, 247, 248, 311, 318, 325, 339, 343, 351, 367, 379–380, 381, 383–384, 409, 411–412, 416, 424
  cosmonaut. See Cosmonaut.
  disarmament, 21, 33, 81, 86, 174, 181, 197, 330
  electric engine, 342
  Foreign Ministry, 21
  launch, 422
    probe, 351
      Luna XV, 206, 382
      Venus V, 2, 8
      Venus VI, 7–8
      Zond VII, 271
    satellite
      Cosmos, 9, 24, 43, 56, 58, 67, 70, 80–81, 82, 89, 90, 94, 101, 105, 109, 116, 137, 148, 158, 160, 169, 181, 186, 189, 203, 237, 267, 281, 283, 285, 287, 293, 297, 305–306, 309, 314, 316, 339, 341, 346, 347, 350–351, 360, 370, 380, 392, 401, 405, 413, 414
      Intercosmos I, 335
      Intercosmos II, 416
      Meteor I, 91
      Meteor II, 329
      Molniya I–11, 106
      Molniya I–12, 237
      Soyuz IV, 11–12
      Soyuz V, 11–12
      Soyuz VI, 332–333
      Soyuz VII, 332–333
      Soyuz VIII, 332–333
  lunar rock exhibit, reception of, 412
  May Day celebration, 127, 129
  missile and rocket program, 10, 43, 48, 50, 52, 53, 106, 116, 129, 131, 140, 229
  nuclear nonproliferation treaty, 41, 78, 392–393
  nuclear test explosion, 112, 359
  meteorological satellite, 91, 329
  Paris Air Show exhibit, 146–147, 168
  probe, 2, 7–8, 42, 75, 78, 136–137, 140, 141–142, 160, 171, 271, 351, 422
    Luna XV, 195, 206, 207–208, 211, 224, 229, 232, 236–237, 238, 251, 382
  rocket test, 112, 156
  science and technology, 59–60, 103, 273, 278
  space program, 24, 29, 48, 72, 74–75, 86, 95–96, 104, 120, 169, 170–171, 185–186, 195, 200, 241–242, 246, 249, 256, 266, 268, 278, 318, 335, 336, 341–342, 343, 349–350, 351, 361, 365–366, 372, 382, 383–384, 399, 418, 420, 421, 422
  space station, 11–12, 24, 75, 146, 271, 278, 332–333, 341–342, 343, 349–350, 351, 361, 382, 420, 422
  spacecraft. See U.S.S.R., satellite; and individual spacecraft, such as *Luna XV*, *Molniya I–11*, *Soyuz IV*, *Zond VII*.
  spacecraft debris, 130, 198, 291
  State Committee on Atomic Energy, 359
  supersonic transport, 1, 5, 121, 146–147, 152, 159, 173, 255, 283, 422
  weapons, 4, 10, 48, 80, 95, 116, 129, 181, 191, 330, 332, 416
  Weather Bureau, 112
UV. See Ultraviolet.

V

V–12 (U.S.S.R. helicopter), 79
Vacuum chamber, 329–330
VAM–20 (booster), 31, 65–66, 108, 111, 179, 184, 308, 326, 360–361, 410
Van Allen, Dr. James A., 90, 167, 263, 340
Van Allen radiation belt, 84, 90–91, 364
van de Kamp, Dr. Peter, 109
Van Derwalker, John G., 51, 86, 110
Van Nuys, Calif., 102

Van Praagh, David, 199
Vandenberg AFB, Calif. (see also Western Test Range), launch
  Advanced Ballistic Reentry System, 10
  satellite launch vehicle
    Atlas-F, 10, 83
    Thor-Agena, 40, 86, 127, 243, 255, 402
    Thor-Burner, 240
    Thorad-Agena, 314
    Thorad-Agena D, 319
    Titan IIIB-Agena, 288, 350
    Titan IIIB-Agena D, 23, 66, 109, 169
Vanguard I (U.S. satellite), 83
Vanguard II, 51
Varsavsky, Dr. Carlos, 118
Vatican City, 55, 337
Vecchietti, George J., 132
Vega (aircraft), 158
Vega (star), 92
Vegesack (German freighter), 228
Vela (constellation), 56
Vela (nuclear test detection satellite), 281, 360, 422
Vela IX, 155, 256
Vela X, 155, 256
Venus (planet), 260
  atmosphere, 2, 6, 43, 75, 127–128, 140, 141–142, 160, 318
  exploration of, 66, 69, 76, 84, 119, 136–137, 256, 286, 318, 343, 423
  magnetic field, 160
  map, 288
  pressure, 140, 141–142, 160, 171
  probe, 2, 7–8, 16, 89, 119–120, 127–128, 140, 141–142, 160, 171
  surface, 103, 171, 288
  temperature, 140, 141–142, 171
Venus (U.S.S.R. interplanetary probe), 422
Venus III (*Venera III*), 89
Venus IV (*Venera IV*), 2, 89, 140, 160
Venus V (*Venera V*), 2, 8, 23, 42, 141–142, 171
Venus VI (*Venera VI*), 7–8, 23, 42, 140, 141–142, 171
Verne, Jules, 199
Very Low Frequency Propagation Satellite, 131
Vidal, Eugene Luther, 54
Vienna, Austria, 359
Vietnam, North, 245
Vietnam war, 4, 14, 22, 48, 57, 99, 404, 422
Vernon, France, 330
Viking, Project
  contract, 59, 160, 293, 351
  experiments, 311
  funding, 15, 336–337
  landing system, 160, 351, 412
  plans for, 16, 57, 111, 118, 275, 336–337, 398, 411, 412
Villard, Henry Serrano, 52
Vincze, Paul, 275
Virgin Islands, 401–402

Voice of America (VOA), 234, 249
Volcano, 159, 418
Volkov, Vladislav N., 333
Volpe, Secretary of Transportation John A., 58, 371, 402, 410, 412, 418
  contract award, 89, 335, 388, 391
  SST, 43, 81, 95, 314–315
Volynov, Boris, 11, 23
von Braun, Dr. Wernher, 13, 77, 188, 210, 274, 381
  Apollo 11 mission, 187, 247–248
  awards and honors, 10, 147, 289, 326
  European visit, 187
  Mars mission, 266, 270
  press conference, 170–171
  U.S.S.R. space program, 170–171
von Kann, Clifton F., 362
von Kármán Gas Dynamics Facility (Tullahoma, Tenn.), 138
von Kármán, Theodore, 274
Voskhod I mission, 337
Vostok I (U.S.S.R. spacecraft), 199
Vostok I mission, 104
Vozzo, Dr. J. A., 293
Vrebalovich, Dr. Thomas, 290
V/STOL aircraft, 15, 75, 84, 92
VSX (antisubmarine aircraft). See S–3A.
VTOL aircraft, 75, 80, 114, 129, 199
Vykukal, Hubert C., 315

W

Waddell, Jack, 45–46
Wakelin, James H., Jr., 332
Wald, Dr. George, 89, 238–239
Walker, Joseph A., 315
Wall, Dr. Frederick T., 8
Wall of Theophilus (moon), 377
Waller, Richard A., 51, 86, 110, 343
Wallops Station (NASA), 58, 409
  contract, 88, 154, 253
  funding, 383
  launch, sounding rocket
    Aerobee 150 A, 139
    Aerobee 350, 27
    Arcas, 33–34, 41, 354
      Boosted Arcas II, 28
    Astrobee 1500, 335
    Black Brant IIIB, 59, 127
    Black Brant VB, 366
    Bullpup Cajun, 178
    Nike-Apache, 33–34, 41, 49, 113, 135, 159, 176, 189, 286, 300, 303–304, 408–409
    Nike-Cajun, 18, 25, 33–34, 41, 94, 132, 133
    Nike-Tomahawk, 27–28, 29, 49, 211–212, 286, 327
    Pacemaker, 203
    Sidewinder-Arcas, 79
  recovery experiment, 79
Walters, Robert, 115
Wampler, Dr. E. Joseph, 42

Wapakoneta, Ohio, 243, 298
Warner, Jack, 317
Warner, Dr. Jeffrey L., 263, 396
Warren, Chief Justice Earl, 168
Warren, Dr. Shields, 164
Warsaw, Poland, 211, 244, 251
Washburn, Abbott M., 418
Washington Academy of Sciences, 54
Washington Airlines, 50, 320
Washington, D.C., 2, 80, 180, 196, 204, 225, 355, 362, 385, 388, 392, 393, 417
  Apollo 11 postage stamp, 289, 319
  astronauts in, 7, 32, 56–57, 157, 191, 204, 231, 232, 307, 413
  awards presented in, 50, 54, 100–101, 117, 127
  cosmonauts in, 337, 359
  exhibit, 133, 388, 403, 406, 413
  meetings, 24, 32, 54, 56, 68, 72, 74, 78, 111, 115, 116, 120, 121, 133, 148, 170, 178, 237, 268, 357, 371
  nuclear nonproliferation treaty signed, 370
  press conference, 9, 182, 224, 301, 355, 402
Washington National Airport, 43, 380, 418
Washington National Gallery of Art, 226
Washington, Univ. of, 273, 282
Watchers of the Skies, 187
Watson, James Craig, Medal, 121
Watson, Postmaster General W. Marvin, 8
We Reach the Moon, 252, 425
Weapon systems, 41, 42, 136, 155, 191, 257, 330, 332, 424
Weather modification, 42, 205
Webb, James E., 7, 42, 184, 406
  luncheon, for, 209
  portrait of, 323
  Space Age Management, 192
Weber, Dr. Joseph, 182
Weeghman, Richard B., 164
Weidner, Hermann K., 326
Weightlessness, effects of, 19, 48, 117
  animals, 201, 348–349
  human beings, 17, 136, 206, 391
Weinberg, Steven, 131
Weinraub, Allen S., 417
Welsh, Dr. Edward C., 39, 87, 141, 257
Wenk, Dr. Edward, Jr., 99
Wentworth, Eric, 369
West German Air and Space Research Institute, 48
West Palm Beach, Fla., 209
West Virginia Univ., 70
Westendorf, Thomas P., 388
Western Electric Co., 233
Western Test Range (WTR) (see also Vandenberg AFB, Calif.), launch, 23
  Long-Tank Thrust-Augmented Thor (Thorad)-Agena D, 107, 171
  Scout, 59, 323, 365
  Thorad-Agena, 48

Thrust-Augmented Improved Thor-Delta, 30, 185
Westinghouse Astronuclear Laboratory, 87
Westinghouse Electric Corp., 108, 144
Weston Instruments, Inc., Electro Mechanical Research Aerospace Sciences Div., 41
Wethe, Jay D., 195
Wetherill, John Price, Medal, 161
Whalen, Rep. Charles W., Jr., 116
Wheatley, Seagel, 394
Wheaton, Md., 204
Wheeler, Gen. Earle G. (USA), 180
"Where the Legend Starts" (U.S.S.R. film), 104
Whitcomb, Dr. Richard T., 44, 293
White, L/c Edward H., II (USAF), 228, 280, 289
White, Mrs. Edward H., II, 280
White, Maurice D., 362
White House, 34, 83, 109, 178, 304, 364, 422
  Apollo 11 mission, 190, 196, 219, 224, 228, 230, 232, 284
  Apollo 12 mission, 385, 392
  appointments, 49, 68, 91
  astronauts at, 32, 72, 157, 191, 200, 204, 232, 319, 362, 413
  awards presented at, 6, 68
  press conference, 32, 43
  space program, 38, 167, 238, 253, 270, 304–305, 352, 389
White, Dr. Robert M., 344, 417
White Sands Missile Range (WSMR), N. Mex., 119
  launch
    Aerobee 150
      solar astronomy, 48, 302, 361
      stellar data, 71, 79, 402, 409
      ultraviolet astronomy, 48–49, 78, 409
      x-ray astronomy, 78, 402
    Aerobee 150 MI
      infrared data, 299
      solar astronomy, 104, 108, 111, 140, 308, 314, 316, 361, 410
      stellar data, 31, 118, 184, 326, 337, 409
      ultraviolet astronomy, 6, 31, 43, 140, 179, 184, 308, 410
      x-ray astronomy, 65–66, 71, 104, 108, 118, 325, 361
    Aerobee 170, 298
    balloon, 315
    Nike-Apache, micrometeoroid sampling, 279, 287
White, Gen. Thomas D., Space Trophy, 101
White Sands Test Facility, 365
Whitney, John A., 412
Whittaker, Philip N., 104, 123, 130, 132
Whittier College, 232
WHO. See World Health Organization.

Wicker, Tom (Thomas Grey), 412
Wickert Public Opinion Institute, 384
Wiesner, Dr. Jerome B., 35, 131, 254
Wilcox-Sierra Div. of American Standard, Inc., 315
Wild, Dr. J. P., 121
Wilford, John Noble, 252, 273, 370
Wilhelm, Janusz, 164
Williams, George E., 260
Williams, Rev. Hosea, 205
Wilson, Rep. Charles H., 52
Wilson, Gill Robb, Trophy for Arts and Letters, 87
Wilson, Prime Minister Harold (U.K.), 38, 158, 233
Wilson, Herbert A., Jr., 290
Wilson, Richard, 288
Wilson, Riley, 266–267
Wilson, T. A., 330, 341
Wimberley, Robert C., 76
Wing, aircraft, 44, 46, 84, 207, 293, 299
Wings Club, 150
Winte, Ralph F., 19
Winter Study on Uses of Manned Space Flight, 1975–1985, Proceedings, 94
Wisconsin, 28
Wisconsin, Univ. of, 3, 29, 80, 99, 129, 344, 402, 403
Wise, Dr. Donald U., 287
Withington, H. W., 340
Wolf Research and Development Corp., 271
Wolff, Rep. Lester L., 177
Wollenhaupt, Wilbur R., 174
Wollongong, Australia, 233
Women as astronauts, 359
"The Wondrous Telephone" (song), 388
Wood, Clotaire, 290
Woods Hole, Mass., 290
Woods Hole Oceanographic Institution, 273
Woomera Rocket Range, Australia, 99, 190, 196, 323
Worcester Foundation for Experimental Biology, 50
Worden, Capt. Alfred M. (USAF), 105
World Center for Exploration Foundation, 2
World Data Center A for Rockets and Satellites, 1
World Health Organization (WHO), 311
 Communicable Disease Div., 254
World War I, 133
World Weather Program, 78
Wray, James, 28
Wright brothers, 88, 177, 226, 236
Wright Brothers Day, 406, 410
Wright, Orville, 52, 406, 410
Wright, Wilbur, 406, 410
Wright-Patterson AFB, Ohio, 11, 38, 119, 177
WSMR. See White Sands Missile Range.
WTR. See Western Test Range.

Wyeth, James B., 226
Wyld, James H., Propulsion Award, 134

X

X–1 (star), 326, 327
X–15 (rocket research aircraft), 87, 119, 133, 177, 205, 315, 321, 421
X–24 (lifting-body vehicle), 29
X–24A, 101, 113, 133, 286, 299, 316, 349, 371, 421
XB–70 (supersonic aircraft), 11, 38, 205, 321, 421
XE (nuclear rocket engine), 177–178, 279, 309, 422
Xenon, 403
X-ray, 168
 experiment, 22, 71, 78, 179, 326, 335
 mirror, 104
 polarimeter, 71
 radiation, 66, 71, 104, 108, 179, 184, 327, 361
 source, 99, 104, 118, 121, 124, 127, 133–134, 136, 160, 186, 281–282, 326, 335, 366, 402
XV–4B (Hummingbird) (VTOL aircraft), 80

Y

Yak-40 (U.S.S.R. trijet), 168
Yamaguchi, Japan, 135
Yankee Clipper (*Apollo 12* command module). See Command module.
Yarborough, Sen. Ralph W., 270
Yeh, Dr. Richard S., 340
Yeliseyev, Dr. Aleksey S., 11–12, 23, 159, 168, 333
Yellowstone National Park, 107, 328
Yelyan, Eduard V., 1, 152
YF–12 (jet interceptor), 400, 421
YF–12A, 72, 205, 405
Yorty, Mayor Samuel W., 279
Young, Dr. Andrew T., 282
Young, Cdr. John W. (USN), 16, 90, 108, 142, 152, 159, 191, 267
Young Presidents' Organization, 327
Young, Sen. Stephen M., 405
Youngblood, Dr. David, 343
Youth Science Congress, 105
Yugoslavia, 24, 249, 341

Z

Zacharias, Jerrold R., 254
Zähringer, Dr. Joseph, 288
Zakharov, Aleksey V., 242
Zakharov, Matvey V., 129

Zamayatin, Leonid, 21
Zholudev, Gen. Leonid V. (U.S.S.R.), 121
Ziegler, Henri, 197
Ziegler, Ronald L., 305
Ziehl, Dr. Donald, 85

Zoeckler, M/G John L. (USAF), 260
Zond V (U.S.S.R. space probe), 271
Zond VI, 271
Zond VII, 271, 422
Zvezdny Gorodok (Star City) (U.S.S.R.), 195

NASA HISTORICAL PUBLICATIONS

HISTORIES
- Robert L. Rosholt, *An Administrative History of NASA, 1958–1963*, NASA SP–4101, 1966, $4.00.*
- Loyd S. Swenson, James M. Grimwood, and Charles C. Alexander, *This New Ocean: A History of Project Mercury*, NASA SP–4201, 1966, $5.50.
- Constance McL. Green and Milton Lomask, *Vanguard: A History*, NASA SP–4202, 1970; also Washington: Smithsonian Institution Press, 1971, $12.50.
- Alfred Rosenthal, *Venture Into Space: Early Years of Goddard Space Flight Center*, NASA SP–4301, 1968, $2.50.
- Edwin P. Hartman, *Adventures in Research: A History of the Ames Research Center, 1940–1965*, NASA SP–4302, 1970, $4.75.

HISTORICAL STUDIES
- Eugene M. Emme (ed.), *History of Rocket Technology* (Detroit: Wayne State University, 1964).
- Mae Mills Link, *Space Medicine in Project Mercury*. NASA SP–4003, 1965, $1.00.
- *Historical Sketch of NASA*, NASA EP–29, 1965 and 1966.
- Katherine M. Dickson (Library of Congress), *History of Aeronautics and Astronautics: A Preliminary Bibliography*, NASA HHR–29, Clearinghouse for Federal Scientific and Technical Information, Springfield, Va. 22150, $3.00.
- William R. Corliss, *History of NASA Sounding Rockets*, NASA SP–4401 (1970).

CHRONOLOGIES
- *Aeronautics and Astronautics: An American Chronology of Science and Technology in the Exploration of Space, 1915–1960*, compiled by E. M. Emme, Washington: NASA, 1961.
- *Aeronautical and Astronautical Events of 1961*, published by the House Committee on Science and Astronautics, 1962.
- *Aeronautical and Astronautical Events of 1962*, published by the House Committee on Science and Astronautics, 1963.
- *Astronautics and Aeronautics, 1963*, NASA SP–4004, 1964.
- *Astronautics and Aeronautics, 1964*, NASA SP–4005, 1965.
- *Astronautics and Aeronautics, 1965*, NASA SP–4006, 1966.
- *Astronautics and Aeronautics, 1966*, NASA SP–4007, 1967, $1.50.
- *Astronautics and Aeronautics, 1967*, NASA SP–4008, 1968, $2.25.
- *Astronautics and Aeronautics, 1968*, NASA SP–4010, 1969, $2.00.
- James M. Grimwood, *Project Mercury: A Chronology*, NASA SP–4001, 1963.
- James M. Grimwood and Barton C. Hacker, with Peter J. Vorzimmer, *Project Gemini Technology and Operations: A Chronology*, NASA SP–4002, 1969, $2.75.
- Ivan D. Ertel and Mary Louise Morse, *The Apollo Spacecraft: A Chronology*, Vol. I, *Through November 7, 1962*, NASA SP–4009, $2.50.
- Mary Louise Morse and Jean Kernahan Bays, *The Apollo Spacecraft: A Chronology*, Vol. II, *November 8, 1962–September 30, 1964*, NASA SP–4013 (1971)

---

* All titles with prices can be ordered from the Superintendent of Documents, Government Printing Office, Washington, D.C. 20402.

www.ingramcontent.com/pod-product-compliance
Lightning Source LLC
Chambersburg PA
CBHW081713170526
45167CB00009B/3571